SIGNAL PROCESSING NOISE

THE ELECTRICAL ENGINEERING AND APPLIED SIGNAL PROCESSING SERIES
Edited by Alexander Poularikas

The Advanced Signal Processing Handbook: Theory and Implementation for Radar, Sonar, and Medical Imaging Real-Time Systems
Stergios Stergiopoulos

The Transform and Data Compression Handbook
K.R. Rao and P.C. Yip

Handbook of Multisensor Data Fusion
David Hall and James Llinas

Handbook of Neural Network Signal Processing
Yu Hen Hu and Jenq-Neng Hwang

Handbook of Antennas in Wireless Communications
Lal Chand Godara

Noise Reduction in Speech Applications
Gillian M. Davis

Signal Processing Noise
Vyacheslav P. Tuzlukov

Forthcoming Titles

Propagation Data Handbook for Wireless Communications
Robert Crane

The Digital Color Imaging Handbook
Guarav Sharma

Applications in Time Frequency Signal Processing
Antonia Papandreou-Suppappola

Digital Signal Processing with Examples in MATLAB®
Samuel Stearns

Smart Antennas
Lal Chand Godara

Pattern Recognition in Speech and Language Processing
Wu Chou and Bing Huang Juang

Nonlinear Signal and Image Processing: Theory, Methods, and Applications
Kenneth Barner and Gonzalo R. Arce

THE ELECTRICAL ENGINEERING AND APPLIED SIGNAL PROCESSING SERIES

SIGNAL PROCESSING NOISE

Vyacheslav P. Tuzlukov

CRC PRESS

Boca Raton London New York Washington, D.C.

Library of Congress Cataloging-in-Publication Data

Tuzlukov, V. P. (Viacheslav Petrovich)
Signal processing noise / Vyacheslav P. Tuzlukov.
p. cm. — (The electrical engineering and applied signal processing series)
Includes bibliographical references and index.
ISBN 0-8493-1025-3 (alk. paper)
1. Signal processing—Digital techniques. 2. Electronic noise—Prevention. 3. Noise control. I. Title. II. Series.

TK5102.9 .T88 2002
621.382′24—dc21 2002017487

Direct all inquiries to CRC Press LLC, 2000 N.W. Corporate Blvd., Boca Raton, Florida 33431.

International Standard Book Number 0-8493-1025-3
Library of Congress Card Number 2002017487
Printed in the United States of America 1 2 3 4 5 6 7 8 9 0
Printed on acid-free paper

To the undying memory of Dr. Peter G. Tuzlukov,
my dear father and teacher

Preface

The performance of complex signal processing systems is limited by the additive and multiplicative noise present in the communication channel through which the information signal is transmitted.

Multiplicative noise is distortion in the amplitude and phase structure of the information signal. Multiplicative noise can occur in the generation, transmission, and processing of the information signal. The main characteristics of complex signal processing systems that are used, for example, in radar, communications, wireless communications, mobile communications, sonar, acoustics, underwater signal processing, remote sensing, navigation systems, geophysical signal processing, and biomedical signal processing, deteriorate as a result of the effect of multiplicative noise. The impact of multiplicative noise on the main characteristics of complex signal processing systems in various areas of signal processing is great in those cases in which complex signal processing systems use signals with complex phase structure, for example, frequency-modulated signals, phase-modulated signals, and so on, or when complex signal processing systems use signal processing of coherent signals of large duration.

In recent years the problems of signal processing that result from the combined stimulus of additive Gaussian noise and multiplicative noise are of great interest for systems that deploy complex signal processing and coherent signal processing.

In this book we discuss the following problems:

- The main statistical characteristics of multiplicative noise
- The main statistical characteristics of the signals distorted by multiplicative noise
- The main statistical characteristics of the process at the output of linear systems impacted by multiplicative noise
- The main principles of the generalized approach to signal processing in additive Gaussian noise and multiplicative noise
- The main statistical characteristics of the signal at the output of the generalized detector impacted by multiplicative noise
- Impact of multiplicative noise on the detection performances of the signals that are processed by the generalized detector, on the estimation of measurement of the signal parameters, and on the signal resolution

As a starting point, in Chapter 1 we discuss the main concepts of probability and statistics upon which all results and all conclusions in this book are based. The main results and conclusions discussed in this book are based on the generalized approach to signal processing in the presence of additive Gaussian noise and multiplicative noise. This is based on a seemingly abstract idea: the introduction of an additional noise source (that does not carry any information about the signal) in order to improve the qualitative performance of complex signal processing systems. Theoretical and experimental studies carried out by the author lead to the conclusion that the proposed generalized approach to signal processing impacted by additive Gaussian noise and multiplicative noise allows us to formulate a decision-making rule based on the determination of the *jointly sufficient statistics of the mean and variance* of the likelihood function (or functional). Classical and modern signal processing theories allow us to define only the *sufficient statistic of the mean* of the likelihood function (or functional). Additional information about the statistical characteristics of the likelihood function (or functional) leads to better quality signal detection in compared with the optimal signal detection algorithms of classical and modern theories.

The generalized approach to signal processing in the presence of additive Gaussian noise and multiplicative noise allows us to extend the well-known boundaries of the potential noise immunity set by classical and modern signal detection theories. Employing complex signal processing systems constructed on the basis of the generalized approach to signal processing in the presence of additive Gaussian noise and multiplicative noise allows us to obtain better detection of signals with noise components present compared with complex signal processing systems that are constructed on the basis of classical and modern theories. The optimal and asymptotic signal detection algorithms (of classical and modern theories), for signals with amplitude-frequency-phase structure characteristics that can be known and unknown *a priori*, are the components of the signal detection algorithms that are designed on the basis of the generalized approach to signal detection theory.The problems discussed in this book show that it is possible to raise the upper boundary of the potential noise immunity for any complex signal processing system including signal processing systems with associated noise immunity defined by classical and modern signal detection theories.

To better understand the fundamental statements and concepts of the generalized approach to signal processing in the presence of additive Gaussian noise and multiplicative noise the reader should consult my two earlier books: *Signal Processing in Noise: A New Methodology* (IEC, Minsk, 1998) and *Signal Detection Theory* (Springer-Verlag, New York, 2001).

I am extremely grateful to my colleagues in the field of signal processing for very useful discussion about the main results, in particular, Prof. V. Ignatov, Prof. A. Kolyada, Prof. I. Malevich, Prof. G. Manshin, Prof. V. Marakhovsky, Prof. B. Levin, Prof. D. Johnson, Prof. B. Bogner, Prof. Yu. Sedyshev,

Prof. J. Schroeder, Prof. Yu. Shinakov, Prof. V. Varshavsky, Prof. A. Kara, Prof. X. R. Lee, Prof. Y. Bar-Shalom, Dr. V. Kuzkin, Dr. A. Dubey, and Dr. O. Drummond.

I thank my colleagues at the University of Aizu, Japan, for very valuable discussion about the main statements and concepts of the book.

I especially thank my dear mother, Natali Tuzlukova, and my lovely wife, Elena Tuzlukova, for their understanding, endless patience, and tremendous support during the course of my work on this book.

I also wish to express my life-long, heartfelt gratitude to Dr. Peter G. Tuzlukov, my father and teacher, who introduced me to science.

Vyacheslav P. Tuzlukov

The Author

Vyacheslav P. Tuzlukov, Ph.D., is visiting professor at the University of Aizu, Japan and chief research fellow at the Institute of Engineering Cybernetics of the National Academy of Sciences, Belarus. He is also a full professor in the Electrical and Computer Engineering Departments of the Belarussian State University and Institute of Modern Knowledge in Minsk. He is actively engaged in research on radar, communications, signal processing, and image processing and has more than 25 years experience in these areas. Dr. Tuzlukov is the author of more than 120 journal articles and conference papers and four books—one in Russian and three in English—on signal processing.

Introduction

At present, the receivers or detectors of complex signal processing systems in various areas of signal processing are constructed under the following condition: they must be optimal for signal processing in the presence of additive noise when the additive noise is Gaussian. This approach is based on the fact that additive Gaussian noise will always persist at some magnitude. The receivers and detectors of complex signal processing systems are designed in order to solve the problem of signal detection and to estimate the signal parameters.

However, the presence of the multiplicative noise impacts essentially on the main qualitative characteristics of complex signal processing systems. The main characteristics of the functioning of any complex signal processing system are defined by an application area and are often specific for distinctive types of complex signal processing systems. In the majority of cases the main characteristics of complex signal processing systems are defined by some initial characteristics describing a quality of signal processing in the presence of noise: the precision of signal parameter measurement, the definition of resolution intervals of the signal parameters, the probability of detection of the signals, and the probability of false alarm. The main goal of this book is the analysis of the effect of multiplicative noise on signal processing by complex signal processing systems constructed on the basis of the generalized approach to signal processing in the presence of additive Gaussian noise and multiplicative noise.

In line with this statement two problems—analysis and synthesis—arise. The first problem (analysis)—the problem of study of the stimulus of multiplicative noise on the main principles and characteristics under the generalized approach to signal processing—is an analysis of the impact of multiplicative noise on the main characteristics of complex signal processing systems constructed on the basis of the generalized approach to signal processing in the presence of additive Gaussian noise and multiplicative noise, i.e., complex signal processing systems employ the generalized detector or receiver. This problem is very important in practice. Analysis of the stimulus of multiplicative noise under the above-mentioned conditions allows us to define limitations of the use of complex signal processing systems and to quantify the impact of multiplicative noise relative to other noise and interference present in the system.

If we are able to conclude that the presence of multiplicative noise is the main factor or one of the main factors limiting the performance of any complex signal processing system then the second problem—a definition of structure and the main parameters and characteristics of the generalized detector or receiver under a dual stimulus of additive Gaussian noise and multiplicative noise (the problem of synthesis)—arises.

Distortions in amplitude and phase of the signals caused by the multiplicative noise and effects that are caused by these distortions are analogous to the above-mentioned distortions that can occur in many areas of signal processing: radar, communications, mobile communications, wireless communications, sonar, acoustics, underwater signal processing, remote sensing, navigation systems, geophysical signal processing, biomedical signal processing, and so on. For this reason, the main results and conclusions discussed in this book can be applied to the above-mentioned areas of signal processing.

This book is devoted to the generalized approach to signal processing in the presence of additive Gaussian noise and multiplicative noise and, in particular, to signal detection under a dual stimulus of additive Gaussian noise and multiplicative noise that allows us to establish a new viewpoint on the noise immunity of complex signal processing systems designed on the basis of the generalized approach. This book summarizes the investigations carried out by the author over the last 20 years.

The book comprises nine chapters. Chapter 1 discusses the main concepts and definitions in the theory of probability and statistics. The concept of probability is defined and the main properties of the probability are detailed. The main characteristics of the probability distribution function and the probability distribution density for continuous and discrete variables are considered. Special attention is paid to stochastic processes and to estimations of statistical characteristics such as the mean, the variance, and the correlation function of random variables and stationary and non-stationary stochastic processes. The main purpose of Chapter 1 is to summarize briefly for the reader the main concepts and definitions in the theory of probability and statistics because the content of this book is based on the statistical decision-making rules of signal processing in the presence of additive Gaussian noise and multiplicative noise. The information presented in this chapter provides a useful basis for the rest of the book.

Chapter 2 is a concise introduction to classical and modern signal detection theories and the main avenues of investigation in these areas. The features of classical and modern signal detection theories and various approaches to signal detection problems in the presence of additive Gaussian noise and multiplicative noise are briefly discussed.

Chapter 3 focuses on the main characteristics of multiplicative noise and the classification of the noise and interference. Sources of multiplicative noise are defined and discussed. Classification of multiplicative noise and the associated main properties are studied. The correlation function and energy spectrum are defined for the cases of deterministic and quasideterministic multiplicative noise. The notion of stationary fluctuating multiplicative noise is defined. The correlation function of the noise modulation function of stationary fluctuating multiplicative noise is also studied. The notion of pulse-fluctuating multiplicative noise is defined more rigorously in this chapter. The energy spectrum of the noise modulation function of pulse-fluctuating multiplicative noise is investigated. The generalized statistical model of multiplicative noise is introduced for discussion. This model allows us to define the amplitude envelope and phase of the received signal for all types of communication channels used by complex signal processing systems on a unified basis.

Chapter 4 explores the statistical characteristics of the signals under the stimulus of multiplicative noise. Multiplicative noise that can be thought of as the result of deterministic or quasideterministic processes, including periodic multiplicative noise and non-periodic deterministic multiplicative noise, is also investigated. For the case of stationary fluctuating multiplicative noise the statistical characteristics of the undistorted component of the signal and the noise component of the signal are discussed. The statistical characteristics of ensemble (counterpart) and individual realizations of the signal are also analyzed and compared under the stimulus of multiplicative noise. This comparison enables us to establish how the statistical characteristics defined for an ensemble of realizations of the signal distorted by multiplicative noise can provide us with knowledge about distortions of individual realizations of the signal and how we can use a single realization of the signal for the purpose of defining the statistical characteristics of an ensemble of the signal. This is very important in practice. The probability distribution density of the signal in the presence of additive Gaussian noise under the stimulus of multiplicative noise is defined for the cases when distortions in amplitude and phase of the signal are both independent and functionally related. The multivariate probability distribution density of instantaneous values of the signal under the stimulus of fluctuating multiplicative noise is also discussed.

Chapter 5 is devoted to the main theoretical principles of the generalized approach to signal processing in the presence of additive Gaussian noise and multiplicative noise. Basic concepts of the signal detection problem are discussed. The criticism of classical and modern signal detection theories from the viewpoint of the definition of the jointly sufficient statistics of the mean and variance of the likelihood function (or functional) is explored. Modifications and initial premises of the generalized approach to signal processing in the presence of additive Gaussian noise and multiplicative noise are considered. The likelihood function (or functional) possessing the jointly sufficient statistics of the mean and variance under the generalized approach to signal processing in the presence of additive Gaussian noise and multiplicative noise is investigated. Engineering interpretation of the generalized approach to signal processing in the presence of additive Gaussian noise and multiplicative noise is discussed. The model of the generalized detector for the cases of both slow and rapid fluctuating multiplicative noise is studied. The probability distribution density of the process at the output of the generalized detector under the stimulus of multiplicative noise is defined.

Chapter 6 deals with an analysis of the signals distorted by multiplicative noise at the output of linear systems with the constant parameters of complex signal processing systems constructed on the basis of the generalized approach to signal processing in the presence of additive Gaussian noise and multiplicative noise. The main statistical characteristics of the signal at the output of input linear system of the generalized detector are analyzed under combined stimulus of additive Gaussian noise and multiplicative noise. The cases of deterministic and quasideterministic multiplicative noise and stationary fluctuating multiplicative noise are also discussed. The main statistical characteristics of the signal at the output of the generalized detector under the stimulus of multiplicative noise are defined for periodic and fluctuating multiplicative noise. Functional relationships between the parameters of the noise modulation function of multiplicative noise and the main statistical characteristics of the signal noise component at the output of the generalized detector under the stimulus of stationary fluctuating multiplicative noise for some types of signals at the input of the linear system at the generalized detector are investigated, including, for example, the signals with the constant radio frequency carrier and with a square wave-form or bell-shaped amplitude envelope, frequency-modulated signals with the bell-shaped amplitude envelope, and the signals with the phase-manipulated code. Statistical characteristics of the signal noise component at the output of the generalized detector under the stimulus of multiplicative noise are discussed for cases of both slow and rapid fluctuating multiplicative noise. The probability distribution density of the signal at the output of the generalized detector under a combined stimulus of additive Gaussian noise and multiplicative noise is defined.

Chapter 7 focuses on the generalized approach to signal detection in the presence of additive Gaussian noise and multiplicative noise. The first part of Chapter 7 is devoted to the stimulus of multiplicative noise on statistical characteristics of the signal at the output of the generalized detector that is used under detection of the signals in the presence of additive Gaussian noise. The models of the generalized detector for the signals with known and unknown initial phase are investigated. The main statistical characteristics of the signal at the output of the generalized detector when the signal has both known and unknown initial phases for the cases of periodic and fluctuating multiplicative noise are discussed. The second part of Chapter 7 is concerned with methods and techniques intended for the generalized approach to signal processing in the presence of additive Gaussian noise and multiplicative noise along with the definition of the detection performances of signals under the use of these methods and techniques. The detection performances are defined for the signals with known and unknown initial phases for the cases of periodic and fluctuating multiplicative noise. Comparative analysis under definition of the radar range by complex signal processing systems constructed on the basis of the generalized approach to signal processing in the presence of additive Gaussian noise and multiplicative noise, and by complex signal processing systems designed on the basis of the modern signal detection theory, is carried out. The detection performances of the signals under the use of the multi-channel generalized detector for the case of the known correlation function of the noise modulation function of multiplicative noise are studied. The detection performances of the signals under the use of the one-channel generalized detector—the autocorrelation channel of the generalized detector—are defined and compared with the detection performances of the signals under employment for optimal detectors of the modern signal detection theory. The example of the diverse complex multi-channel signal processing system constructed on the basis of the generalized approach to signal processing in the presence of additive Gaussian noise and multiplicative noise is discussed.

Chapter 8 is concerned with an investigation of the stimulus of fluctuating multiplicative noise jointly with additive Gaussian noise on measurement precision of non-energy parameters of the signal under the generalized approach to signal processing in the presence of additive Gaussian noise and multiplicative noise. The problems of a single signal parameter measurement under a combined stimulus of additive Gaussian noise and weak multiplicative noise are discussed. The measurement precision of frequency and appearance time of the signal is analyzed. Estimations of precision of the signal frequency and signal appearance time measurement are defined for the signals with the bell-shaped amplitude envelope, constant radio frequency carrier, and the frequency-modulated signals. The problem of simultaneous measurement of two signal parameters under a combined stimulus of additive Gaussian noise and weak multiplicative noise is discussed. In particular, the problem of simultaneous measurement of frequency and appearance time of the signal is considered for the case of the frequency-modulated signal. The problem of a single signal parameter measurement under a combined stimulus of additive Gaussian noise and strong multiplicative noise is studied.

Chapter 9 is devoted to the problems of signal resolution under the generalized approach to signal processing in the presence of additive Gaussian noise and multiplicative noise. The signal appearance time resolution interval and the signal frequency resolution interval are both defined on the basis of the Woodward criterion and on the basis of the statistical criterion. Comparative analysis of definition of the resolution intervals by the statistical criterion and the Woodward criterion allows us to estimate the use of the Woodward criterion and to explain some equivalent (conditional) statistical sense of the Woodward criterion.

This book demonstrates that it is possible to raise the upper boundary of the potential noise immunity for complex signal processing systems in various areas of signal processing under the use of the generalized approach to signal processing in the presence of additive Gaussian noise and multiplicative noise in comparison with the noise immunity defined by classical and modern signal detection theories.

Contents

1

Probability and Statistics

1.1 Probability: Basic Concepts

The essence of probability approach lies in the fact that the event A is the random event under the given condition S, i.e., the event A may occur or not occur. A characteristic feature of random events is that their regularities can be found only under repeated testing (the condition S). The case in which event A will always occur under the definite condition S is called the certain event; the case in which event A never occurs under the definite condition S is called the impossible event; and the case in which event A may occur under the definite condition S (during testing) is called the random event.

Probability is a numerical characteristic of the extent to which any definite event A is possible under the given condition S repeated an unlimited number of times.

Suppose that some testing is repeated n times. The event A either occurs uniquely or absolutely does not occur as a result of each testing incident. The event A may occur m times during n times testing. Then the value

$$\mathcal{V}(A) = \frac{m}{n} \tag{1.1}$$

is called the frequency of the event A that occurs during testing. If each set of testing is sufficiently high we can think that the frequency $\mathcal{V}(A)$ is approximately equal to the probability $P(A)$, where

$$P(A) = P(A \mid S) \tag{1.2}$$

is the probability that the event A occurs under the condition S. This is the statistical definition of the probability.

The probability of the certain event A is equal to 1. The probability of the impossible event A is equal to zero. For the random event the following inequality is true:

$$0 \leq P(A) \leq 1. \tag{1.3}$$

However, one must bear in mind the following truism. If it is known that the event A is the certain event the probability of the event A is equal to unity: $P(A) = 1$. At the same time the equality $P(B) = 1$ does not mean that the event B is the certain event. This equality means that the random event B will always occur under the given condition. An analogous remark can be made relative to the equality $P(B) = 0$. Because of this, the probability of the random event A is bounded within the limits

$$0 \leq P(A) \leq 1 \tag{1.4}$$

but not within the limits

$$0 < P(A) < 1. \tag{1.5}$$

This fact is very important for understanding the basic concepts of probability.

If under the given testing an appearance of the random event A excludes the possibility of appearance of the random event B then the random events A and B are called incompatible. If under the given testing the random event A can occur in association with the random event B then such random events are called compatible.

The random event $\overline{A}$ implying that the random event A does not occur during testing is called opposite. The probability of the random opposite event $\overline{A}$ is determined in the following form:

$$P(\overline{A}) = 1 - P(A). \tag{1.6}$$

If the probability of two random events does not depend on whether or not another random event occurs or does not occur, then these random events are called independent; otherwise, these random events are called dependent. Notions of dependence and compatibility characterize different features of random events and they must not be confused. Incompatible random events are dependent. Compatible random events may be both dependent and independent.

The probability of the random event A occurring under the condition that the random event B has occurred is denoted as $P(A|\,B)$, and is called the conditional probability of the random event A. If the random events A and B are independent then the following equality

$$P(A \mid B) = P(A) \tag{1.7}$$

is true.

The product of the random events $A_1, A_2, \ldots, A_n$ is called the complex random event lying in the fact that all random events occur: both the random event $A_1, \ldots,$ and the random event A_n. The product of the random events has the following form:

$$A_1 \cdot A_2 \cdot \cdots \cdot A_n = \prod_{i=1}^{n} A_i. \tag{1.8}$$

The sum of the random events $A_1, A_2, \ldots, A_n$ is called the complex random event lying in the fact that if, and only if one random event occurs: either the random event $A_1, \ldots,$ or the random event A_n. The sum of the random events takes the following form:

$$A_1 + A_2 + \cdots + A_n = \sum_{i=1}^{n} A_i. \tag{1.9}$$

The random events $A_1, A_2, \ldots, A_n$ forming a complex random event are called elementary. In many books[1–6] the product of the random events is called the intersection or superposition and denoted in the following manner:

$$A_1 \bigcap A_2 \bigcap \cdots \bigcap A_n = \bigcap_{i=1}^{n} A_i \tag{1.10}$$

and the sum of the random events is called the integration and denoted in the following manner:

$$A_1 \bigcup A_2 \bigcup \cdots \bigcup A_n = \bigcup_{i=1}^{n} A_i. \tag{1.11}$$

The probability of the product of random events $A_1, A_2, \ldots, A_n$ can be determined using one of the following formulae:

$$P\left\{\prod_{i=1}^{n} A_i\right\} = P(A_1)P(A_2 \mid A_1)P(A_3 \mid A_1 \cdot A_2) \times \cdots \times P(A_n \mid A_1 \cdot A_2 \cdot \cdots \cdot A_{n-1}); \tag{1.12}$$

$$P\left\{\prod_{i=1}^{n} A_i\right\} = \sum_{i=1}^{n} P(A_i) - \sum_{i=1}^{n-1}\sum_{j=i+1}^{n} P(A_i + A_j) + \sum_{i=1}^{n-2}\sum_{j=i+1}^{n-1}\sum_{k=j+1}^{n} P(A_i + A_j + A_k) - \cdots + (-1)^{n-1} P\left\{\sum_{i=1}^{n} A_i\right\}; \tag{1.13}$$

$$P\left\{\prod_{i=1}^{n} A_i\right\} = 1 - P\left\{\sum_{i=1}^{n} \overline{A_i}\right\}. \tag{1.14}$$

The random events are called independent in population if the sums composed of any combinations of these random events are independent. For the random events that are independent in population the formula

$$P\left\{\prod_{i=1}^{n} A_i\right\} = \prod_{i=1}^{n} P(A_i) \tag{1.15}$$

is true. The probability of the sum of the limited number of the random events is determined by the following formulae:

$$P\left\{\sum_{i=1}^{n} A_i\right\} = \sum_{i=1}^{n} P(A_i) - \sum_{i=1}^{n-1}\sum_{j=i+1}^{n} P(A_i \cdot A_j) + \sum_{i=1}^{n-2}\sum_{j=i+1}^{n-1}\sum_{k=j+1}^{n} P(A_i \cdot A_j \cdot A_k) - \cdots + (-1)^{n-1} P\left\{\prod_{i=1}^{n} A_i\right\}; \tag{1.16}$$

$$P\left\{\sum_{i=1}^{n} A_i\right\} = 1 - P\left\{\prod_{i=1}^{n} \overline{A_i}\right\}. \tag{1.17}$$

The random events are called incompatible in population if an appearance of one random event excludes an appearance of other random events. For random events that are incompatible in population we can write

$$P\left\{\sum_{i=1}^{n} A_i\right\} = \sum_{i=1}^{n} P(A_i). \tag{1.18}$$

The probability of the sum of the denumerable set of random events, if this probability exists, is determined by analogous formulae as $n \to \infty$.

If under testing only the incompatible in population random events A_1, $A_2, \ldots, A_n$ occur then these random events form an exhausting set or complete group of the random events and the equality

$$P\left\{\sum_{i=1}^{n} A_i\right\} = 1 \tag{1.19}$$

is true.

If the random event H occurs every time the random event A has occurred we can say that the random event H is a result of the random event A or the random event H follows from the random event A (the designation: $A \subset H$ or $H \supset A$). For these random events the inequality

$$P(A) \leq P(H) \tag{1.20}$$

is true.

If the random event H being a result of the random event A is the sum of the exhausting set of the random events $H_1, H_2, \ldots, H_n$ then the probability of the random event A can be presented in the following form:

$$P(A) = \sum_{i=1}^{n} P(H_i) P(A \mid H_i). \tag{1.21}$$

This formula is the formula of the total probability. The probability $P(H_i)$ of the hypothesis H_i $(i = 1, \ldots, n)$ determined irrespective of the random event A is called the *a priori* probability. The conditional probability $P(H_i \mid A)$ of the hypothesis H_i $(i = 1, \ldots, n)$ determined under the assumption that the random event A has occurred is called the *a posteriori* probability. *A priori* and *a posteriori* probabilities are related by the formula

$$P(H \mid A) = \frac{P(H_i)P(A \mid H_i)}{\sum_{j=1}^{n} P(H_j)P(A \mid H_j)} = \frac{P(H_i)P(A \mid H_i)}{P(A)}, \tag{1.22}$$

where $i = 1, \ldots, n$. Equation (1.22) is called the theorem of hypotheses or Bayes formula.

1.2 Random Variables

A variable, the value of which changes randomly from testing to testing, is called the random variable. The random variable is described by the probability distribution function. The probability distribution function of the random variable is accepted as given if:

- A set of possible values of the random variable is defined
- A method of definition of the probability that the random variable is in an arbitrary region of a set of possible random variables is specified

To determine the probability of the random variable there is a need to define the probability distribution function of the random variable.

1.2.1 Probability Distribution Function

The probability distribution function of the random variable X is called the function $F(x)$ defining the probability of the random event $X < x$, i.e., the probability that the random variable X is less than the certain value x:

$$F(x) = P(X < x). \tag{1.23}$$

The probability distribution function has the following features:

- The function $F(x)$ is the non-decreasing function

$$F(x_2) \geq F(x_1) \quad \text{at } x_2 > x_1. \tag{1.24}$$

- The function $F(x)$ is the continuous function

$$F(x) = \lim_{\varepsilon \to 0} F(x - \varepsilon), \quad \varepsilon > 0. \tag{1.25}$$

- The function $F(x)$ approaches zero as $x \to -\infty$

$$\lim_{x \to -\infty} F(x) = 0. \tag{1.26}$$

- The function $F(x)$ tends to 1 as $x \to +\infty$

$$\lim_{x \to +\infty} F(x) = 1. \tag{1.27}$$

One can see from Eqs. (1.23) and (1.25) that under the condition $x > x_0$ the following equality

$$\lim_{x \to x_0} [F(x) - F(x_0)] = P(X = x_0) \tag{1.28}$$

is true and for the condition $x < x_0$ the following equality

$$\lim_{x \to x_0} [F(x) - F(x_0)] = 0 \tag{1.29}$$

is true.

Thus the difference $F(x) - F(x_0)$ tends to the probability that the random variable X takes the value x_0, if the value x tends to the value x_0 from the right, and tends to zero if the value x tends to the value x_0 from the left.

In References 5, 7, and 8, the probability distribution function is defined in the following form:

$$F_1(x) = P(X \leq x). \tag{1.30}$$

The probability distribution function $F_1(x)$ has the same characteristics as the probability distribution function $F(x)$ except for the second feature (see Eq. (1.25)). This statement can be formulated in the following way. The function $F_1(x)$ is the continuous function from the right if the following equality

$$F_1(x) = \lim_{\varepsilon \to 0} F_1(x + \varepsilon), \quad \varepsilon > 0 \tag{1.31}$$

is true.

One can see from Eq. (1.30) that under the condition $x > x_0$ the following equality

$$\lim_{x \to x_0} [F_1(x_0) - F_1(x)] = 0 \tag{1.32}$$

is true and at the condition $x < x_0$ the following equality

$$\lim_{x \to x_0} [F_1(x_0) - F_1(x)] = P(X = x_0) \tag{1.33}$$

is true, i.e., the difference $F_1(x_0) - F_1(x)$ tends to the probability that the random variable X takes the value x_0, if the value x tends to the value x_0 from the left, and tends to zero if the value x tends to the value x_0 from the right.

The probability distribution functions $F(x)$ and $F_1(x)$ are functionally related by the following relationship:

$$F_1(x) = F(x) + P(X = x). \tag{1.34}$$

If the probability distribution function $F(x)$ is the continuous function then the probability distribution function $F_1(x)$ is the continuous function, too, and vice versa. In the process, $P(X = x) = 0$ and the functions coincide with each other.

The random variables are divided into discrete, continuous, and mixed random variables. The discrete random variable can take only a finite or denumerable set of the values $x_1, x_2, \ldots$. The continuous random variable can take arbitrary values within the limits of some closed or opened interval and including an unlimited interval. A range of the continuous random variable can contain some non-crossing intervals.

The probability distribution function of the discrete random variable is the step-function with jumps at the points $x_1, x_2, \ldots$. The probability distribution function of the continuous random variable is the continuous function. The probability distribution function of the mixed random variable is the partially continuous function with a denumerable number of jumps.

The jump of the probability distribution function at the point $x_i, i = 1, 2, \ldots$ (see Eq. (1.28)) is equal to the probability p_i that the random variable X takes the value x_i:

$$F(x_i + 0) - F(x_i) = P(X = x_i) = p_i. \tag{1.35}$$

For the discrete random variables, a sequence of the probabilities $p_1, p_2, \ldots$ defines completely the probability distribution function

$$F(x) = \sum_{x_i < x} p_i, \tag{1.36}$$

where the summation is carried out for all i satisfying the condition $x_i < x$. The fourth characteristic of the probability distribution function (see Eq. (1.27)) takes the following form under the above-mentioned condition:

$$\sum_{i=1}^{n} p_i = 1. \tag{1.37}$$

For the continuous random variables we can write

$$F(x_i + 0) - F(x_i) = 0, \tag{1.38}$$

i.e., the probability that the continuous random variable takes the definite value x_i is equal to zero.

1.2.2 Probability Distribution Density

The probability distribution density is the differential probability distribution function:

$$f(x) = \frac{dF(x)}{dx}. \tag{1.39}$$

The probability distribution density has the following characteristics:

- The function $f(x)$ is the non-negative function

$$f(x) \geq 0. \tag{1.40}$$

- The integral of the function $f(x)$ within the infinite limits is equal to 1

$$\int_{-\infty}^{\infty} f(x)\,dx = 1. \tag{1.41}$$

The probability distribution function of the continuous random variable is functionally related to the probability distribution density by the following formula:

$$F(x) = \int_{-\infty}^{x} f(t)\,dt. \tag{1.42}$$

The probability that the continuous random variable X is within the limits of the closed or opened interval $[x_1, x_2]$ can be determined using the probability distribution density

$$P(x_1 < X < x_2) = \int_{x_1}^{x_2} f(t)\,dt. \tag{1.43}$$

The probability distribution function of the mixed random variable can be determined as the sum

$$F(x) = F^{(1)}(x) + F^{(2)}(x), \tag{1.44}$$

where $F^{(1)}(x)$ is the continuous and differentiable function; $F^{(2)}(x)$ is the sum of jumps from the left of point x.

The identity function determined by the following form

$$\mathcal{I}(x) = \begin{cases} 1 & \text{at } x > 0; \\ 0 & \text{at } x \leq 0 \end{cases} \tag{1.45}$$

can be used for the uniformity of writing form for the probability distribution function of the continuous, discrete, and mixed random variables and formalization of operations with the probability distribution functions.

In terms of Eq. (1.45) the probability distribution function of the discrete random variable (see Eq. (1.36)) has the following form:

$$F(x) = \sum_{i=1}^{n} p_i \mathcal{I}(x - x_i) \tag{1.46}$$

and the probability distribution function of the mixed random variable has the following form:

$$F(x) = F^{(1)}(x) + \sum_{i=1}^{n} p_i \mathcal{I}(x - x_i). \tag{1.47}$$

The delta function $\delta(x)$ (see Appendix I) is used sometimes for uniformity. The delta function has the following features:

$$\delta(x - x_0) = \begin{cases} 0 & \text{at } x \neq x_0; \\ \infty & \text{at } x = x_0, \end{cases} \tag{1.48}$$

$$\int_a^b \delta(x - x_0)\, dx = \begin{cases} 0 & \text{at } x_0 < a; \\ 1 & \text{at } a < x_0 < b; \\ 0 & \text{at } b < x_0. \end{cases} \tag{1.49}$$

Under the conditions $x_0 = a$ and $x_0 = b$, the last integral is equal to 0.5. Using the delta function to describe the probability distribution function we can assume that under the condition $\alpha > 0$ the following equality

$$\int_{-\infty}^{x_0 - 0} \delta(x - x_0)\, dx = \lim_{\alpha \to 0} \int_{-\infty}^{x_0 - \alpha} \delta(x - x_0)\, dx = 0 \tag{1.50}$$

is true.

The delta function can be considered a differential of the identity function in Eq. (1.45)

$$\delta(x) = \frac{d\mathcal{I}(x)}{dx}. \tag{1.51}$$

Using Eqs. (1.48)–(1.51), the probability distribution density of the discrete random variable takes the following form:

$$f(x) = \sum_{i=1}^{n} p_i \delta(x - x_i). \tag{1.52}$$

It should be stressed that Eq. (1.52) has only a symbolic sense.

Consequently, we can use Eqs. (1.39) and (1.42) for the discrete random variables, understanding that the probability distribution function of the discrete random variable is determined by Eq. (1.46) and the probability distribution density of the discrete random variable is determined by Eq. (1.52).

For the mixed random variable we can write

$$f(x) = f^{(1)}(x) + f^{(2)}(x), \tag{1.53}$$

where

$$f^{(1)}(x) = \frac{dF^{(1)}(x)}{dx} \quad \text{and} \quad f^{(2)}(x) = \sum_{i=1}^{N} p_i \delta(x - x_i). \tag{1.54}$$

1.2.3 Numerical Characteristics

The probability distribution function is the complete characteristic of the random variable. Often it is sufficient to indicate some numerical parameters characterizing individual essential features of the random variable distribution law. The mean, mode, and median characterize some values, and all possible values of the random variable are grouped about these characteristics.

The mean of the random variable X is determined by:

- In the case of the discrete random variable

$$M[X] = m_X = \sum_{i=1}^{n} x_i p_i. \tag{1.55}$$

- In the case of the continuous random variable

$$M[X] = m_X = \int_{-\infty}^{\infty} x f(x)\, dx. \tag{1.56}$$

- In the case of the mixed random variable

$$M[X] = m_X = \int_{-\infty}^{\infty} x f^{(1)}(x)\, dx + \sum_{i=1}^{n} x_i p_i. \tag{1.57}$$

The median is called the value x_{med} of the random variable X, which results in

$$P(X < x_{med}) = P(X > x_{med}) = 0.5. \tag{1.58}$$

In the case of the continuous random variable X the median is determined in the following form:

$$F(x_{med}) = 0.5 \tag{1.59}$$

or

$$\int_{-\infty}^{x_{med}} f(x)\,dx = \int_{x_{med}}^{\infty} f(x)\,dx = 0.5. \tag{1.60}$$

In the case when

$$F(x) = 0.5 \quad \text{at } x_1 \leq x \leq x_2 \tag{1.61}$$

the median is not uniquely defined: any value, which is within the limits of the interval $[x_1, x_2]$, may be considered the median. In the case of discrete random variables, the median is not uniquely defined and not used in practice.

The mode is called the value x_{mod} of the random variable X such that the probability $P(X = x_{mod})$ (the case of the discrete random variable) or the probability distribution density $f(x_{mod})$ (the case of the continuous random variable) is maximal. In the case of the single value x_{mod}, the probability distribution density is called unimodal. In the case of several values x_{mod}, the probability distribution density is called multi-modal.

A characteristic grouping of the random variables is described by moments. There are the initial and central moments of the random variable. The initial moment of the k-th order of the random variable X is determined by the following formula:

$$m_k[X] = M[X^k]. \tag{1.62}$$

Using Eqs. (1.55)–(1.57) we can conclude that the mean m_X is the initial moment of the first order of the random variable X. The difference between the random value X and the mean m_X

$$\Delta_X = X - m_1[X] \tag{1.63}$$

is called the deviation of the random variable X. Moments of the probability distribution density of the random variable deviation are called the central moments and are determined by the following formula

$$M_k[X] = M\{(X - m_1)^k\}. \tag{1.64}$$

In accordance with the general definition (see Eq. (1.62)) the initial moment of the second order is equal to:

- For the continuous random variable X

$$m_2[X] = \int_{-\infty}^{\infty} x^2 f(x)\,dx. \tag{1.65}$$

- For the discrete random variable X

$$m_2[X] = \sum_{i=1}^{n} x_i^2 p_i. \tag{1.66}$$

The central moment of the second order is called the variance of the random variable X and is equal in accordance with Eq. (1.64) to:

- For the continuous random variable X

$$M_2[X] = D[X] = \int_{-\infty}^{\infty} (x - m_1)^2 f(x)\, dx. \tag{1.67}$$

- For the discrete random variable X

$$M_2[X] = D[X] = \sum_{i=1}^{n} (x_i - m_1)^2 p_i. \tag{1.68}$$

The value

$$\sigma_X = \sqrt{D[X]} \tag{1.69}$$

is called the root mean square or standard deviation of the random variable X from the mean. The central and initial moments of the second order are functionally related by the following formula:

$$M_2[X] = m_2[X] - m_1^2. \tag{1.70}$$

Using Eq. (1.70) and in terms of Eqs. (1.55), (1.56), and (1.62), we can write

$$D[X] = M[X^2] - (M[X])^2. \tag{1.71}$$

The asymmetry of the probability distribution density is determined by the following formula

$$k = -\frac{M_3[X]}{\sqrt{\{M_2[X]\}^3}}, \tag{1.72}$$

where k is the coefficient of asymmetry. The flatness of the probability distribution density is determined by the following formula

$$\gamma = \frac{M_4[X]}{\{M_2[X]\}^2} - 3, \tag{1.73}$$

where γ is the coefficient of kurtosis.

1.3 Stochastic Processes

The stochastic process is characterized by the changing of some physical random variables in a certain space. In the process, the changing random variable is described by the probability laws. A specific example of the stochastic process during certain experiments is called the realization of the stochastic process.

1.3.1 Main Definitions

The stochastic process is described by the stochastic function. The process $\xi(t)$ is called the stochastic process, which is the random variable for any fixed values of an argument. The scalar stochastic process $\xi(t)$ is the stochastic process, the range of which is a set in real space. The vector stochastic process $\vec{\xi}(t)$ is the stochastic process, the range of which is a set in corresponding coordinate space.

There are five main forms of the stochastic process according to whether a random variable of the stochastic process $\xi(t)$ and its parameter t take a continuous or discrete set of values:

- The discrete random sequence (the discrete stochastic process with a discrete time) is the stochastic process, for which the range and domain are the discrete sets; in this case the parameter t (time) takes the discrete values $t_0, t_1, \ldots, t_i, \ldots, t_M$, and the random variable $x_i = \xi(t_i)$ can only take the discrete set of the values $x_0, x_1, \ldots, x_k, \ldots, x_K$; the sets of the values $\{t_i\}$ and $\{x_k\}$ can be finite or infinite.
- The random sequence (the continuous stochastic process with a discrete time) is the stochastic process, for which the range is the continuous set and the domain is the discrete set; this process differs from the one mentioned above in that the random variable $x_i = \xi(t_i), i = 1, 2, \ldots, M$ can take an infinite number of values.
- The discrete (discontinuous) stochastic process (the discrete stochastic process with the continuous time) is the stochastic process, for which the range is the discrete set and the domain is the continuous set; in this case the stochastic process can take the discrete values $x_k, k = 1, 2, \ldots, K$ and $t \in [0, T]$, where T is the time interval, within the limits of which the stochastic process $\xi(t)$ is defined.
- The continuous stochastic process is the stochastic process, for which the range and domain are the continuous sets; in this case the function $\xi(t)$ takes values from the continuous space and the argument t is continuous.
- The stochastic point process is the random point sequence, for example, on the time axis.

There are more complex mixed forms of the stochastic processes. For example, the stochastic process may be determined in the following form:

$$\xi(t) = F(t, \lambda_1(t), \lambda_2(t)), \tag{1.74}$$

where $F(t, \lambda_1(t), \lambda_2(t))$ is the deterministic function of the first argument t and the parameters $\lambda_1(t)$ and $\lambda_2(t)$ are the stochastic processes involved. If, for example, $\lambda_1(t)$ is the discrete stochastic process and $\lambda_2(t)$ is the continuous stochastic process then the stochastic process $\xi(t)$ may be called the stochastic process in the discrete-continuous or mixed form. In the particular case, when

$$\xi(t) = F(t, \lambda_1, \lambda_2), \tag{1.75}$$

i.e., the parameters λ_1 and λ_2 do not depend on time and are the random variables of the stochastic process $\xi(t)$ (see Eq. (1.75)) and this is called the quasideterministic process. In a general case this is the stochastic process and its realizations are defined by the function $F(t, \lambda_1, \lambda_2, \ldots, \lambda_n)$, which contains one or more random parameters $\lambda = \{\lambda_1, \lambda_2, \ldots, \lambda_n\}$ independent of time.

1.3.2 Probability Distribution Function and Density

Assume that there is an ensemble of the stochastic process $\xi(t)$. Then the probability that a random variable of the stochastic process is less than the value x_1 under the condition $t = t_1$ is determined in the following form:

$$P[\xi(t) < x_1] = F(x_1; t_1). \tag{1.76}$$

The function $F(x_1; t_1)$ is the one-dimensional probability distribution function. The word "one-dimensional" underlines the fact that the random variables of the stochastic process are considered only at the fixed instant. For this reason, we use the writing form $F(x_1; t)$.

The function determined by

$$f(x_1; t_1) = \frac{\partial}{\partial x_1} F(x_1; t_1) \tag{1.77}$$

is called the one-dimensional probability distribution density of the stochastic process (the stochastic function). The pure value $f(x_1; t_1)\, dx_1$ is equal to the probability that the random variable $x_1 = \xi(t_1)$ is within the limits of the interval $x_1 \leq \xi(t_1) < x_1 + dx_1$:

$$f(x_1; t_1) = P[x_1 \leq \xi(t_1) < x_1 + dx_1]. \tag{1.78}$$

The one-dimensional probability distribution function, like the one-dimensional probability distribution density, is a very important but incomplete characteristic of the stochastic process. Using these functions gives us information about the stochastic process only at the fixed instants; however, we do not know how the random variable $x_1 = \xi(t_1)$ impacts on further characteristics of the stochastic process under the condition $t_2 > t_1$. One can say that the one-dimensional probability distribution density, or the one-dimensional probability distribution function, characterizes the stochastic process statically but not dynamically.

The two-dimensional probability distribution function and the two-dimensional probability distribution density are more complete characteristics of the stochastic process. These functions allow us to define the probability distribution function between values of the stochastic process at two instants t_1 and t_2. The function

$$F_2(x_1, x_2; t_1, t_2) = P[\xi(t_1) < x_1; \xi(t_2) < x_2] \tag{1.79}$$

is called the two-dimensional probability distribution function. The function

$$f_2(x_1, x_2; t_1, t_2) = \frac{\partial^2}{\partial x_1 \partial x_2} F_2(x_1, x_2; t_1, t_2) \tag{1.80}$$

is called the two-dimensional probability distribution density and the pure value

$$f_2(x_1, x_2; t_1, t_2)\, dx_1\, dx_2$$

defines the probability that two inequalities are jointly satisfied:

$$\begin{cases} x_1 \le \xi(t_1) < x_1 + dx_1 \\ x_2 \le \xi(t_2) < x_2 + dx_2 \end{cases} \tag{1.81}$$

or

$$f_2(x_1, x_2; t_1, t_2)\, dx_1\, dx_2 = P[x_1 \le \xi(t_1) < x_1 + dx_1; x_2 \le \xi(t_2) < x_2 + dx_2]. \tag{1.82}$$

In a general case the two-dimensional probability distribution function, or the two-dimensional probability distribution density, does not allow us to know complete information regarding the stochastic process. These functions define only a relationship between probable values of the stochastic process at two instants only. A more complete and detailed definition of the stochastic process or the stochastic function can be given by the multivariate probability distribution density or the multivariate probability distribution function.

For a definition of the joint probability of two or more stochastic processes the joint probability distribution functions and the joint probability distribution densities are introduced. For example, two stochastic processes are defined by the following relationships:

$$\begin{aligned} &F_{n+m}(x_1, \ldots, x_n, y_1, \ldots, y_m; t_1, \ldots, t_n, t_1', \ldots, t_m') \\ &\quad = P[\xi(t_1) < x_1, \ldots, \xi(t_n) < x_n; \eta(t_1') < y_1, \ldots, \eta(t_m') < y_m]; \end{aligned} \tag{1.83}$$

$$\begin{aligned} &f_{n+m}(x_1, \ldots, x_n, y_1, \ldots, y_m; t_1, \ldots, t_n, t_1', \ldots, t_m') dx_1 \ldots dx_n\, dy_1 \ldots dy_m \\ &\quad = P[\xi(t_1) < x_1 + dx_1, \ldots, x_n \le \xi(t_n) < x_n + dx_n; \\ &\qquad y_1 \le \eta(t_1') < y_1 + dy_1, \ldots, y_m < \eta(t_m') < y_m + dy_m], \end{aligned} \tag{1.84}$$

where n and m are the non-negative integers.

Two stochastic processes $\xi(t)$ and $\eta(t)$ are called independent if a population of values of the first stochastic process $\xi(t_1), \ldots, \xi(t_n)$ does not depend on the population of values of the second stochastic process $\eta(t_1'), \ldots, \eta(t_m')$ under all $t_1, \ldots, t_n, t'_1, \ldots, t'_m$. A necessary and sufficient condition of independence of the stochastic processes is that the joint probability distribution density in Eq. (1.84) is divided into a product of the probability distribution densities for each stochastic process:

$$f_{n+m}(x_1, \ldots, x_n, y_1, \ldots, y_m; t_1, \ldots, t_n, t'_1, \ldots, t'_m)$$
$$= f_n(x_1, \ldots, x_n; t_1, \ldots, t_n) f_m(y_1, \ldots, y_m; t'_1, \ldots, t'_m). \tag{1.85}$$

The conditional probability distribution density can be introduced for a definition of the stochastic process. For example, the random variable $\xi(t_1) = x_1$ of the stochastic process under the known value of this process at another instant $\xi(t_2) = x_2$ is defined by the conditional probability distribution density

$$f(x_1; t_1 \mid x_2; t_2) = \frac{f_2(x_1, x_2; t_1, t_2)}{f_1(x_2; t_2)}, \tag{1.86}$$

where

$$f_1(x_2; t_2) = \int_{-\infty}^{\infty} f_2(x_1, x_2; t_1, t_2)\, dx_1. \tag{1.87}$$

The conditional probability distribution density $f_2(x_1; t_1 \mid x_2; t_2)$ has more information regarding the stochastic process $\xi(t_1)$ in comparison with the unconditional probability distribution density $f(x_1; t_1)$. The extent, to which the information about the stochastic process $\xi(t_1)$ increases with the result that the random variable $\xi(t_2) = x_2$ is known, depends on specific conditions. In some cases the information regarding the stochastic process $\xi(t_1)$ is not added at all. This means that

$$f_1(x_1; t_1 \mid x_2; t_2) = f(x_1; t_1) \tag{1.88}$$

and

$$f_2(x_1, x_2; t_1, t_2) = f(x_1; t_1) \cdot f(x_2; t_2). \tag{1.89}$$

Equation (1.89) ensures the necessary and sufficient condition of independence of random variables of the stochastic process $\xi(t)$ at two instants t_1 and t_2.

1.3.3 Characteristic Function

The stochastic process can be defined by the characteristic function

$$\Theta_n(j\vartheta_1, \ldots, j\vartheta_n; t_1, \ldots, t_n) = M\{\exp(j\vartheta_1\xi_1 + \cdots + j\vartheta_n\xi_n)\}$$
$$= \int_{-\infty}^{\infty} \cdots \int_{-\infty}^{\infty} \exp[j(\vartheta_1 x_1 + \cdots + \vartheta_n x_n)] p_n(x_1, \ldots, x_n; t_1, \ldots, t_n)\, dx_1 \ldots dx_n \tag{1.90}$$

instead of the probability distribution density, where $\xi_i = \xi(t_i)$. It is well known that the probability distribution function, the probability distribution density, and the characteristic function of the stochastic process $\xi(t)$ are functionally related to one another by the following relationship:

$$f_n(x_1, \ldots, x_n) = \frac{\partial^2 F_n(x_1, \ldots, x_n)}{\partial x_1 \ldots \partial x_n}$$
$$= \frac{1}{(2\pi)^n} \int_{-\infty}^{\infty} \cdots \int_{-\infty}^{\infty} e^{-j(\vartheta_1 x_1 + \cdots + \vartheta_n x_n)} \Theta_n(j\vartheta_1, \ldots, j\vartheta_n)\, d\vartheta_1 \ldots d\vartheta_n. \quad (1.91)$$

1.3.4 Moments of Stochastic Process

Initial moments of the stochastic process $\xi(t)$ defined within the limits of some interval are the functions $m_{\nu_1}(t), m_{\nu_1\nu_2}(t_1, t_2), \ldots, m_{\nu_1\nu_2\ldots\nu_n}(t_1, t_2, \ldots, t_n)$ being symmetric with respect to all arguments that are the mean of corresponding products:

$$m_{\nu_1}(t) = M[\xi^{\nu_1}(t)] = \int_{-\infty}^{\infty} x^{\nu_1} f(x; t)\, dx;$$

$$m_{\nu_1\nu_2}(t_1, t_2) = M[\xi^{\nu_1}(t_1)\xi^{\nu_2}(t_2)] = \int_{-\infty}^{\infty}\int_{-\infty}^{\infty} x_1^{\nu_1} x_2^{\nu_2} f_2(x_1, x_2; t_1, t_2)\, dx_1\, dx_2;$$

$$\cdots\cdots\cdots\cdots\cdots\cdots\cdots\cdots\cdots\cdots$$

$$m_{\nu_1\nu_2\ldots\nu_n}(t_1, t_2, \ldots, t_n) = M[\xi^{\nu_1}(t_1)\xi^{\nu_2}(t_2)\ldots\xi^{\nu_n}(t_n)]$$
$$= \int_{-\infty}^{\infty} \cdots \int_{-\infty}^{\infty} x_1^{\nu_1} x_2^{\nu_2} \ldots x_n^{\nu_n} f_n(x_1, \ldots, x_n; t_1, \ldots, t_n)\, dx_1 \ldots dx_n, \quad (1.92)$$

where ν_i $(1 \le i \le n)$ is the non-negative integer. The moment $m_{\nu_1\nu_2\ldots\nu_n}$ $(t_1, t_2, \ldots, t_n)$ considered as a function of n non-coincided arguments $t_1, t_2, \ldots, t_n$ is called the n-dimensional moment of the $(\nu_1 + \nu_2 + \cdots + \nu_n)$-th order.

We can consider the central moments determined in the following form:

$$M_{\nu_1\nu_2\ldots\nu_n}(t_1, t_2, \ldots, t_n) = M\{[\xi(t_1) - m_1(t_1)]^{\nu_1} \cdots [\xi(t_n) - m_1(t_n)]^{\nu_n}\}$$
$$= \int_{-\infty}^{\infty} \cdots \int_{-\infty}^{\infty} [x_1 - m_1(t_1)]^{\nu_1} \cdots [x_n - m_1(t_n)]^{\nu_n}$$
$$\times f_n(x_1, \ldots, x_n; t_1, \ldots, t_n)\, dx_1 \ldots dx_n \quad (1.93)$$

instead of initial moments of the stochastic process.

The correlation (cumulant) functions $R_1(t_1)$, $R_2(t_1, t_2)$, $R_3(t_1, t_2, t_3), \ldots$, are defined by the Maclaurin series expansion of the characteristic function logarithm. In the n-dimensional case we can write

$$\Theta_n(j\vartheta_1, \ldots, j\vartheta_n; t_1, \ldots, t_n) = \exp\left\{ j \sum_{\mu=1}^{n} R_1(t_\mu)\vartheta_\mu + \frac{j^2}{2!} \sum_{\mu,\nu=1}^{n} R_2(t_\mu, t_\nu)\vartheta_\mu\vartheta_\nu \right.$$
$$\left. + \frac{j^3}{3!} \sum_{\mu,\nu,\lambda=1}^{n} R_3(t_\mu, t_\nu, t_\lambda)\vartheta_\mu\vartheta_\nu\vartheta_\lambda + \cdots \right\}. \quad (1.94)$$

Under the condition

$$t_1 = t_2 = \cdots = t_n = t, \quad n = 1 \quad (1.95)$$

Equation (1.94) takes the form

$$\Theta(j\vartheta) = \exp\left\{ \sum_{\mu=1}^{\infty} \frac{\kappa_\nu}{\nu!}(j\vartheta)^\nu = \exp\left[jm_1\vartheta - D \cdot \frac{\vartheta^2}{2} \right] \exp\left[\sum_{\nu=3}^{\infty} \frac{\kappa_\nu}{\nu!}(j\vartheta)^\nu \right], \right. \quad (1.96)$$

where

$$\kappa_\nu = j^{-\nu} \left[\frac{d^\nu \ln \Theta(j\vartheta)}{d\vartheta^\nu} \right]_{\vartheta=0} \quad (1.97)$$

is the ν-th order semi-invariant or the ν-th order cumulant.

Using Eq. (1.94), we can write

$$jR_1(t) = \frac{\partial}{\partial\vartheta} \ln \Theta_1(j\vartheta_i; t)|_{\vartheta=0}; \quad (1.98)$$

$$j^2 R_2(t, t) = \frac{\partial^2}{\partial\vartheta^2} \ln \Theta_1(j\vartheta; t)|_{\vartheta=0}; \quad (1.99)$$

$$j^2 R_2(t_1, t_2) = \frac{\partial^2}{\partial t_1 \partial t_2} \ln \Theta_2(j\vartheta_1, j\vartheta_2; t_1, t_2)|_{\vartheta_1=\vartheta_2=0}. \quad (1.100)$$

Reference to Eqs. (1.98)–(1.100) shows that

$$\begin{cases} m_1(t) = R_1(t); \\ m_{11}(t_1, t_2) = R_2(t_1, t_2) + R_1(t_1)R_1(t_2); \\ m_{111}(t_1, t_2, t_3) = R_3(t_1, t_2, t_3) + [R_1(t_1)R_2(t_2, t_3) + R_1(t_2)R_2(t_1, t_3) \\ \qquad + R_1(t_3)R_2(t_1, t_2)] + R_1(t_1)R_1(t_2)R_1(t_3). \end{cases} \quad (1.101)$$

Using Eq. (1.101) under the condition

$$t_1 = t_2 = t_3 = t, \quad (1.102)$$

we can write

$$\begin{cases} m_1 = \kappa_1; \\ m_2 = \kappa_2 + \kappa_1^2; \\ m_3 = \kappa_3 + 3\kappa_1\kappa_2 + \kappa_1^2, \end{cases} \tag{1.103}$$

where the first cumulant κ_1 coincides with the mean, the second cumulant coincides with the variance, and so on. Thus, the above-mentioned moments are functionally related to the correlation functions uniquely.

The mean of the stochastic process is determined in the following form:

$$m_\xi(t) = m_1(t) = M[\xi(t)] = \int_{-\infty}^{\infty} x f(x;t)\,dx. \tag{1.104}$$

The initial moment $m_{11}(t_1, t_2)$, called the covariance function of the stochastic process, is determined in the following form:

$$K_\xi(t_1,t_2) = m_{11}(t_1,t_2) = M[\xi(t_1)\xi(t_2)] \\ = \int_{-\infty}^{\infty}\int_{-\infty}^{\infty} x_1 x_2 f(x_1, x_2; t_1, t_2)\,dx_1\,dx_2. \tag{1.105}$$

The central moment $M_{11}(t_1, t_2) = R_2(t_1, t_2)$, called the correlation function of the stochastic process, is determined in the following form:

$$R_\xi(t_1,t_2) = M_{11}(t_1,t_2) = M\{[\xi(t_1) - m_\xi(t_1)][\xi(t_2) - m_\xi(t_2)]\} \\ = \int_{-\infty}^{\infty}\int_{-\infty}^{\infty} [x_1 - m_\xi(t_1)][x_2 - m_\xi(t_2)]\, f_2(x_1, x_2; t_1, t_2)\,dx_1\,dx_2. \tag{1.106}$$

Reference to Eqs. (1.105) and (1.106) shows that

$$K_\xi(t_1,t_2) = R_\xi(t_1,t_2) + m_\xi(t_1)m_\xi(t_2). \tag{1.107}$$

The study of features and properties of stochastic processes defined by the above-mentioned characteristics is called the correlation theory of stochastic processes. The correlation theory of stochastic processes allows us to completely define the Gaussian processes.

1.3.5 Classification of Stochastic Processes

The stochastic process $\xi(t)$ is called stationary in a narrow sense if all finite-dimensional probability distribution functions of any order are invariant with respect to the shift in time, i.e., under all n and t_0 the following equality

$$F_n(x_1, \ldots, x_n; t_1 - t_0, \ldots, t_n - t_0) = F_n(x_1, \ldots, x_n; t_1, \ldots, t_n) \tag{1.108}$$

is true. This means that two stochastic processes $\xi(t)$ and $\xi(t - t_0)$ have the same probability characteristics for all t_0 values. The probability distribution density must satisfy the analogous equality:

$$f_n(x_1, \dots, x_n; t_1 - t_0, \dots, t_n - t_0) = f_n(x_1, \dots, x_n; t_1, \dots, t_n). \tag{1.109}$$

Also this equality must be true for all statistical characteristics, the moments, and the correlation functions of the stochastic process that is stationary in a narrow sense. Stochastic processes that do not satisfy this condition are called non-stationary processes in a narrow sense.

Two stochastic processes $\xi(t)$ and $\eta(t)$ are called jointly stationary in a narrow sense if their joint probability distribution functions of any order are invariant with respect to the shift in time:

$$\begin{aligned} &F_{\xi\eta}(x_1, \dots, x_m; y_1, \dots, y_n; t_1, \dots, t_m; t_1', \dots, t_n') \\ &\quad = F_{\xi\eta}(x_1, \dots, x_m; y_1, \dots, y_n; t_1 - t_0, \dots, t_m - t_0; t_1' - t_0, \dots, t_n' - t_0). \end{aligned} \tag{1.110}$$

But the condition that the processes $\xi(t)$ and $\eta(t)$ are stationary does not mean that these stochastic processes are jointly stationary in a narrow sense.

In particular, from the definition of the stationary state (see Eq. (1.108)) it follows:

$$\begin{cases} f(x; t_1) = f(x; t_1 - t_1) = f(x); \\ f_2(x_1, x_2; t_1, t_2) = f_2(x_1, x_2; t_1 - t_1, t_2 - t_1) = f_2(x_1, x_2; \tau); \\ \tau = t_2 - t_1. \end{cases} \tag{1.111}$$

Thus, the probability distribution density, the moments, and the correlation functions of the n-th order of the stationary in a narrow sense stochastic process depend on $n-1$ instants only, not n instants. One can see from Eq. (1.111) that the probability distribution density of the first order of the stationary in a narrow sense stochastic process is not a function of time. It is obvious that a definition of the stationary in a narrow sense stochastic process, using only the probability distribution density of the first order, is not complete.

The mean of the stationary in a narrow sense stochastic process does not depend on time:

$$m_\xi = M[\xi(t)] = \int_{-\infty}^{\infty} x f_1(x)\, dx. \tag{1.112}$$

The covariance $K_\xi(t_1, t_2)$ and correlation $R_\xi(t_1, t_2)$ functions of the stationary in a narrow sense stochastic process depend only on the difference $\tau = t_2 - t_1$:

$$\begin{aligned} R_\xi(\tau) &= M\{[\xi(t) - m_\xi][\xi(t+\tau) - m_\xi]\} \\ &= \int_{-\infty}^{\infty}\int_{-\infty}^{\infty} (x_1 - m_\xi)(x_2 - m_\xi) f_2(x_1, x_2; \tau)\, dx_1\, dx_2 \\ &= M[\xi(t)\xi(t+\tau)] - m_\xi^2 = R_\xi(\tau) - m_\xi^2. \end{aligned} \tag{1.113}$$

The variance of the stationary in a narrow sense stochastic process is determined in the following form:

$$D_\xi = \sigma_\xi^2 = M\{[\xi(t) - m_\xi]^2\} = R_\xi(0)$$
$$= \int_{-\infty}^{\infty} (x - m_\xi)^2 f_1(x)\,dx = M[\xi^2(t)] - m_\xi^2. \tag{1.114}$$

The variance in Eq. (1.114) does not vary in time and is equal to the correlation function under zero argument $R_\xi(0)$.

In practice the multivariate probability distribution densities of the stochastic process are not considered; only the mean and the covariance (correlation) functions are considered. For this reason, there is the concept of the stationary state in a broad sense.

The stochastic process $\xi(t)$ with the finite variance is called stationary in a broad sense if the mean and the covariance (correlation) function of the stochastic process $\xi(t)$ are invariant with respect to the shift in time, i.e., the mean does not depend on time and the covariance function depends only on the difference $t_2 - t_1$:

$$m_\xi = constant \quad \text{and} \quad K_\xi(t_1, t_2) = K_\xi(t_2 - t_1). \tag{1.115}$$

Notice that the sum of two non-stationary stochastic processes may be the stationary in a broad sense stochastic process. Let $A_1(t)$ and $A_2(t)$ be the independent stationary in a broad sense stochastic processes with zero means and the same correlation functions. Then the stochastic processes

$$\begin{cases} \xi_1(t) = A_1(t)\cos\omega_0 t \\ \xi_2(t) = A_1(t)\sin\omega_0 t \end{cases} \tag{1.116}$$

are non-stationary, where ω_0 is the frequency that does not vary in time. The summary process

$$\xi(t) = \xi_1(t) + \xi_2(t) \tag{1.117}$$

is the stationary in a broad sense stochastic process.

Based on Eqs. (1.112) and (1.113) we can conclude that the stationary in a narrow sense stochastic process is stationary in a broad sense forever. But the reverse statement is not correct in a general sense.

The concepts of the stationary state in narrow and broad senses for the Gaussian processes coincide completely if their probability distribution densities are completely defined by the mean and the correlation function.

Two stochastic processes $\xi(t)$ and $\eta(t)$ are called jointly stationary in a broad sense if their mutual covariance function is invariant with respect to the shift in time:

$$K_{\xi\eta}(t_1, t_2) = M[\xi(t_1)\eta(t_2)] = M[\xi(t_1 - t_1)\eta(t_2 - t_1)] = K_{\xi\eta}(\tau), \tag{1.118}$$

where

$$\tau = t_2 - t_1. \tag{1.119}$$

Notice if the stochastic processes $\xi(t)$ and $\eta(t)$ are stationary in a broad sense it does not mean that the stochastic processes $\xi(t)$ and $\eta(t)$ are jointly stationary in a broad sense.

There are other concepts of the stationary state of stochastic processes:

- The stochastic process $\xi(t)$ is called stationary of the k-th order if Eqs. (1.108) and (1.109) are true only under the condition $n \leq k$.
- The stochastic process $\xi(t)$ is called asymptotic stationary in a narrow sense if there is the following limit:

$$\lim_{t_0 \to \infty} f_n(x_1, \ldots, x_n; t_1 + t_0, \ldots, t_n + t_0). \tag{1.120}$$

- The stochastic process $\xi(t)$ is called stationary in a narrow sense within the limits of the finite interval if Eq. (1.109) is true for all instants within the limits of this interval.
- The stochastic process $\xi(t)$ is called the process with stationary in a narrow sense differentials if the difference $\xi(t+\tau) - \xi(t)$ for each fixed τ is the stationary in a narrow sense stochastic process.
- The stochastic process $\xi(t)$ is called periodically stationary if Eq. (1.109) is true only under the following condition $t_0 = mT, m = 1, 2, \ldots;$ this means that the random variables $\xi(t), \xi(t+T), \ldots, \xi(t+mT)$ obey the same probability distribution density.

1.3.6 Ergodic and Non-Ergodic Stationary Processes

The above-mentioned characteristics and properties of the stationary stochastic process can be defined by averaging only under the use of a large duration realization. For example, the estimation of the mean m_ξ of the stationary stochastic process $\xi(t)$ is determined in the following form:

$$\widehat{m} = \lim_{T \to \infty} \frac{1}{T} \int_0^T \xi(t)\, dt. \tag{1.121}$$

The estimations of the variance D_ξ and the correlation function $R_\xi(\tau)$ of the stationary stochastic process are determined in the following form:

$$\widehat{D} = \lim_{T \to \infty} \frac{1}{T} \int_0^T \left[\xi(t) - m_\xi^2\right]^2 dt, \tag{1.122}$$

$$\widehat{R}_\xi(\tau) = \lim_{T \to \infty} \frac{1}{T} \int_0^T [\xi(t+\tau) - m_\xi][\xi(t) - m_\xi]\, dt. \tag{1.123}$$

In practice the time interval $[0, T]$ is taken as a finite interval with duration as large as possible.

This action is caused by the fact that the stationary stochastic process is uniform in time. Because of this, only the large duration realization of the stochastic process may contain all information about properties and characteristics of the stochastic process. This phenomenon can be explained in another way. Imagine that the large duration realization is divided into equal time intervals. The stationary stochastic process observed within the limits of this interval can be considered as an individual realization of the stationary stochastic process and a component of statistical ensemble of stochastic process realizations. In this case the stationary stochastic processes are called ergodic. In other words, the stationary stochastic processes possess an ergodic character. Consequently, the ergodic character of the stationary stochastic process lies in the fact that the probability characteristics of the stationary stochastic process can be defined using only the large duration realization of this process.

The stationary stochastic process $\xi(t)$ is called ergodic in a rigorous sense if, under the probability equal to unity, all probability characteristics of the stationary stochastic process $\xi(t)$ can be defined using only the realization of the stochastic process $\xi(t)$. In other words, the stationary stochastic process $\xi(t)$ is ergodic if results of averaging over time coincide with results of averaging over ensembles, i.e., with the mean.

In practice there is no need to know all characteristics and properties of the stationary stochastic process. For example, often it is sufficient to know the mean, the correlation function, and the probability distribution density. It is obvious that the stationary stochastic process can be ergodic with respect to one characteristic (parameter) and non-ergodic with respect to other characteristics (parameters). It is possible to introduce the concept of ergodicity of the stationary stochastic process with respect to individual characteristics or parameters:

- The stationary stochastic process $\xi(t)$ has an ergodic property with respect to the mean

$$M[\xi(t)] = m_\xi = \lim_{T\to\infty} \frac{1}{T} \int_0^T \xi(t)\, dt \tag{1.124}$$

if and only if the following equality

$$\lim_{T\to\infty} \frac{2}{T} \int_0^T \left(1 - \frac{\tau}{T}\right) R_\xi(\tau)\, d\tau = 0 \tag{1.125}$$

is true, where $R_\xi(\tau)$ is the correlation function of the stationary stochastic process $\xi(t)$.

- The stationary stochastic process $\xi(t)$ possesses an ergodic property with respect to the correlation function

$$R_\xi(\tau) = \lim_{T\to\infty} \frac{1}{T} \int_0^T [\xi(t) - m_\xi][\xi(t+\tau) - m_\xi]\, dt \tag{1.126}$$

if and only if the following equality

$$\lim_{T\to\infty} \frac{2}{T} \int_0^T \left(1 - \frac{\tau'}{T}\right) R_\eta(\tau')\, d\tau' = 0 \tag{1.127}$$

is true, where $R_\eta(\tau')$ is the correlation function of the stationary stochastic process

$$\eta(t) = \xi(t+\tau)\xi(t). \tag{1.128}$$

- The stationary stochastic process $\xi(t)$ possesses an ergodic property regarding the one-dimensional probability distribution function $F_1(x)$

$$F_1(x) = \lim_{T\to\infty} \frac{1}{T} \int_0^T \eta(t)\, dt, \tag{1.129}$$

$$\eta(t) = \begin{cases} 0, & \text{at } \xi(t) > x; \\ 1, & \text{at } \xi(t) \le x \end{cases} \tag{1.130}$$

if and only if the following equality

$$\lim_{T\to\infty} \frac{2}{T} \int_0^T \left(1 - \frac{\tau}{T}\right) \left[F_2(x, x; \tau) - F_1^2(x)\right] d\tau = 0 \tag{1.131}$$

is true. However, Eq. (1.131) is true if the following condition

$$\lim_{\tau\to\infty} F_2(x, x; \tau) = F_1^2(x) \tag{1.132}$$

is satisfied or the random variables $\xi(t+\tau)$ and $\xi(t)$ are independent for all values of τ.

Notice that the stochastic random process $\xi(t)$ is stationary in a broad sense in Eq. (1.125) and in a narrow sense in Eqs. (1.126) and (1.131).

The stationary stochastic process is not always ergodic. For example, consider the stochastic process

$$\xi(t) = A(t)\cos(\omega_0 t + \varphi_0), \tag{1.133}$$

where the amplitude $A(t)$ and the initial phase φ_0 are the independent random variables that take different sets of values for various realizations. If the initial phase φ_0 is uniformly distributed within the limits of the interval $[-\pi, \pi]$ then the mean and the correlation function determined by averaging over an ensemble are equal to

$$\begin{cases} m_\xi = 0; \\ R_\xi(\tau) = M[A^2(t)] \cos \frac{\omega_0 \tau}{2}. \end{cases} \tag{1.134}$$

Consequently, the stochastic process $\xi(t)$ is stationary in a broad sense. But this process is not ergodic because the mean and the correlation function determined by averaging over time are equal to

$$\begin{cases} m_\xi^* = 0; \\ R_\xi^*(\tau) = \left(\frac{A_i^2(t)}{2}\right) \cos \omega_0 \tau, \end{cases} \tag{1.135}$$

where $A_i(t)$ is one of possible values of the random variable $A(t)$ in the i-th realization, the use of which allows us to carry out averaging over time.

1.4 Correlation Function

As was previously mentioned in Section 1.3.5, the correlation or covariance function is used for solving practical problems. The correlation function of the real stochastic process is defined by Eqs. (1.105)–(1.107). The mean and the correlation and covariance functions of the complex-valued stochastic process $\xi(t)$ have the following form:

$$m_\xi(t) = M[\xi(t)] = \int_{-\infty}^{\infty} x f(x; t)\, dt; \tag{1.136}$$

$$\begin{aligned} R_\xi(t_1, t_2) &= M\{[\xi(t_1) - m_\xi(t_1)][\xi^*(t_2) - m_\xi^*(t_2)]\} \\ &= \int_{-\infty}^{\infty}\int_{-\infty}^{\infty} [x_1 - m_\xi(t_1)][x_2^* - m_\xi^*(t_2)] f_2(x_1, x_2; t_1, t_2)\, dx_1\, dx_2; \end{aligned} \tag{1.137}$$

$$\begin{aligned} K_\xi(t_1, t_2) &= M[\xi(t_1)\xi^*(t_2)] = \int_{-\infty}^{\infty}\int_{-\infty}^{\infty} x_1 x_2^* f_2(x_1, x_2; t_1, t_2)\, dx_1\, dx_2 \\ &= R_\xi(t_1, t_2) + m_\xi(t_1) m_\xi^*(t_2), \end{aligned} \tag{1.138}$$

where the complex conjugate functions are denoted by an asterisk $(*)$. It is obvious that corresponding definitions and results for the real stochastic process are obtained automatically from Eqs. (1.136)–(1.138), reasoning that an imaginary term of the corresponding random variables is equal to zero.

Comparing Eqs. (1.137) and (1.138), one can see that the covariance function is distinguished from the correlation function by the deterministic term $m_\xi(t_1)m_\xi^*(t_2)$. Under the condition $m_\xi(t) \equiv 0$ the covariance and correlation functions coincide. For this reason, the covariance function contains additional information regarding only the mean of the stochastic process in comparison with the correlation function.

The values of the stochastic process $\xi(t)$ at the instants t_1 and t_2 are called uncorrelated if the correlation function $R_\xi(t_1, t_2) = 0$, i.e.,

$$K_\xi(t_1, t_2) = m_\xi(t_1)m_\xi^*(t_2). \tag{1.139}$$

The values are called orthogonal if the following condition

$$K_\xi(t_1, t_2) = 0 \tag{1.140}$$

is satisfied.

Assume that there are two stochastic processes $\xi(t)$ and $\eta(t)$ possessing the correlation functions $R_\xi(t_1, t_2)$ and $R_\eta(t_1, t_2)$ or the covariance functions $K_\xi(t_1, t_2)$ and $K_\eta(t_1, t_2)$. In addition to these functions we can operate with two mutual correlation or covariance functions

$$\begin{cases} R_{\xi\eta}(t_1, t_2) = M\{[\xi(t_1) - m_\xi(t_1)][\eta^*(t_2) - m_\eta^*(t_2)]\}; \\ R_{\eta\xi}(t_1, t_2) = M\{[\eta(t_1) - m_\eta(t_1)][\xi^*(t_2) - m_\xi^*(t_2)]\} \end{cases} \tag{1.141}$$

or

$$\begin{cases} K_{\xi\eta}(t_1, t_2) = M[\xi(t_1)\eta^*(t_2)]; \\ K_{\eta\xi}(t_1, t_2) = M[\eta(t_1)\xi^*(t_2)]. \end{cases} \tag{1.142}$$

Consequently, a correlation between sample values of the stochastic processes $\xi(t)$ and $\eta(t)$ at two arbitrary instants is defined by the correlation or covariance matrix:

$$\mathbf{R} = \begin{Vmatrix} R_\xi(t_1, t_2) & R_{\xi\eta}(t_1, t_2) \\ R_{\eta\xi}(t_1, t_2) & R_\eta(t_1, t_2) \end{Vmatrix}; \tag{1.143}$$

$$\mathbf{K} = \begin{Vmatrix} K_\xi(t_1, t_2) & K_{\xi\eta}(t_1, t_2) \\ K_{\eta\xi}(t_1, t_2) & K_\eta(t_1, t_2) \end{Vmatrix}. \tag{1.144}$$

In a general case for n stochastic processes these matrices are the matrices of the n-th order.

Two stochastic processes $\xi(t)$ and $\eta(t)$ are called uncorrelated when the mutual correlation function at two arbitrary instants is equal to zero:

$$R_{\xi\eta}(t_1, t_2) = 0 \tag{1.145}$$

or

$$K_{\xi\eta}(t_1, t_2) = m_\xi(t_1)m_\eta^*(t_2). \tag{1.146}$$

Two stochastic processes $\xi(t)$ and $\eta(t)$ are called orthogonal when the covariance function is equal to zero:

$$K_{\xi\eta}(t_1, t_2) = 0. \tag{1.147}$$

In many practical instances it is more convenient to consider the normalized correlation function $r_\xi(t_1, t_2)$ determined in the following form:

$$r_\xi(t_1, t_2) = \frac{R_\xi(t_1, t_2)}{\sqrt{D_\xi(t_1)D_\xi(t_2)}} \tag{1.148}$$

or the normalized mutual correlation function determined in the following form:

$$r_{\xi\eta}(t_1, t_2) = \frac{R_{\xi\eta}(t_1, t_2)}{\sqrt{D_\xi(t_1)D_\eta(t_2)}}. \tag{1.149}$$

These functions quantitatively define a degree of linear dependence between corresponding values of one or two stochastic processes.

The reader can find more detailed information regarding the correlation function in References 1–3, 5–8, and 10–12. In the process, it is useful to bear in mind that

$$R_\xi(t_1, t_2) = R_{\xi\eta}(t_1, t_2); \tag{1.150}$$

$$r_\xi(t_1, t_2) = r_{\xi\eta}(t_1, t_2). \tag{1.151}$$

Because of this, all characteristics of the mutual correlation function $R_{\xi\eta}(t_1, t_2)$ are true for the correlation function $R_\xi(t_1, t_2)$:

- The correlation function has the Hermitian property

$$R_{\xi\eta}(t_1, t_2) = R_{\eta\xi}^*(t_1, t_2); \tag{1.152}$$

$$r_{\xi\eta}(t_1, t_2) = r_{\eta\xi}^*(t_1, t_2). \tag{1.153}$$

  Hence it follows that the correlation function of the real stochastic process $\xi(t)$ is symmetric with respect to the arguments

$$R_\xi(t_1, t_2) = R_\xi(t_2, t_1). \tag{1.154}$$

- The Cauchy–Schwarz–Bunyakovsky inequality is true for the correlation function

$$\begin{cases} |R_{\xi\eta}(t_1, t_2)|^2 \leq M[|\xi(t_1)|^2]M[|\eta(t_2)|^2] = D_\xi(t_1)D_\eta(t_2); \\ |r_{\xi\eta}(t_1, t_2)| \leq 1. \end{cases} \tag{1.155}$$

- Equalities

$$\begin{cases} |R_{\xi\eta}(t_1, t_2)| = \sqrt{D_\xi(t_1)D_\xi(t_2)}; \\ |r_{\xi\eta}(t_1, t_2)| = 1 \end{cases} \tag{1.156}$$

are true if and only if there are constant values $a \neq 0$ and b and the considered values of the stochastic processes $\xi(t_1)$ and $\eta(t_2)$ with the probability equal to unity are functionally related by the linear dependence determined by

$$\eta(t_2) = a\xi(t_1) + b. \tag{1.157}$$

- The correlation function $R_\xi(t_1, t_2)$ has the fundamental property of non-negative definiteness in the following sense. Let $t_1, \ldots, t_n$ be any finite number of points and $z_1, \ldots, z_n$ be arbitrary complex-valued numbers. Then the Hermitian form

$$\sum_{i,k=1}^{n} R_\xi(t_i, t_k) z_i z_k^* = M\left\{ \sum_{i,k=1}^{n} \xi(t_i)\xi^*(t_k) z_i z_k^* \right\}$$
$$= M\left\{ \left| \sum_{i=1}^{n} \xi(t_i) z_i \right|^2 \right\} \geq 0 \tag{1.158}$$

is real and positive forever. Here we suppose that the mean of the stochastic process $\xi(t)$ is equal to zero.

- The above-mentioned property of non-negative definiteness is the characteristic of all correlation functions. This means that if any correlation function $R_\xi(t_1, t_2)$ possesses this property then the stochastic process $\xi(t)$ with the correlation function $R_\xi(t_1, t_2)$ exists.

Now consider properties of the correlation function of the real stationary in a broad sense stochastic process $\xi(t)$. Recall that the following relationships are true in the case of the stationary in a broad sense stochastic process $\xi(t)$:

$$M[\xi(t)] = m_\xi = constant; \tag{1.159}$$

$$M\{[\xi(t) - m_\xi]^2\} = D_\xi = \sigma_\xi = constant; \tag{1.160}$$

$$R_\xi(\tau) = D_\xi r_\xi = M\{[\xi(t) - m_\xi][\xi(t+\tau) - m_\xi]\}; \tag{1.161}$$

$$r_\xi(\tau) = M\left[\frac{\xi(t) - m_\xi}{\sqrt{D_\xi}} \cdot \frac{\xi(t+\tau) - m_\xi}{\sqrt{D_\xi}} \right] = M[\tilde{\xi}(t)\tilde{\xi}(t+\tau)], \tag{1.162}$$

where $\tilde{\xi}(t)$ and $\tilde{\xi}(t+\tau)$ are the normalized random variables.

- The absolute value of the correlation function for all τ cannot exceed its value at $\tau = 0$, i.e.,

$$|R_\xi(\tau)| \leq D_\xi \tag{1.163}$$

and

$$|r_\xi(\tau)| \leq 1. \tag{1.164}$$

- The correlation function of the real stationary process $\xi(t)$ is the even function

$$R_\xi(\tau) = R_\xi(-\tau) \tag{1.165}$$

and

$$r_\xi(\tau) = r_\xi(-\tau). \tag{1.166}$$

- If the correlation function is the continuous function under the condition $\tau = 0$ then one is the continuous function for all τ.
- If the equality

$$\lim_{\tau \to \infty} R_\xi(\tau) = 0 \tag{1.167}$$

is true for the stationary stochastic process then this process is ergodic.
- The Fourier transform of the correlation function is the non-negative function

$$\int_{-\infty}^{\infty} R_\xi(\tau) \cdot \exp^{-j\omega\tau} \geq 0. \tag{1.168}$$

Based on the above-mentioned properties and characteristics of the correlation function, we can draw the following conclusions:

- The correlation function of the stationary stochastic process is the even function with respect to the argument τ.
- The correlation function of the stationary stochastic process has the maximum equal to the variance D_ξ under the condition $\tau = 0$.
- The correlation function of the stationary stochastic process is continuous for all τ if and only if it is continuous under the condition $\tau = 0$.
- The correlation function of the stationary stochastic process tends to approach zero as $\tau \to \infty$.

The above-mentioned properties and characteristics of the correlation function of the stationary in a broad sense stochastic process are true for the covariance function owing to the equality

$$K_\xi(\tau) = R_\xi(\tau) + m_\xi^2. \tag{1.169}$$

In practice the correlation interval or the correlation time

$$\tau_c = \frac{1}{2} \int_{-\infty}^{\infty} |r_\xi(\tau)|\, d\tau = \int_{0}^{\infty} |r_\xi(\tau)|\, d\tau \tag{1.170}$$

is often used instead of the exact analytical definition of the normalized correlation function.

Note the simple properties of the correlation and covariance functions:

- The mutual correlation function of two jointly stationary in a broad sense real stochastic processes $\xi(t)$ and $\eta(t)$ is real and asymmetric

$$R_{\xi\eta}(\tau) = R_{\xi\eta}(-\tau) \tag{1.171}$$

and

$$r_{\xi\eta}(\tau) = r_{\xi\eta}(-\tau). \tag{1.172}$$

- The absolute value of the mutual correlation function is limited:

$$R^2_{\xi\eta}(\tau) < R_\xi(0)R_\eta(0) = D_\xi D_\eta \tag{1.173}$$

and

$$2|R_{\xi\eta}(\tau)| \leq D_\xi + D_\eta. \tag{1.174}$$

- The covariance function of the stochastic process

$$\zeta(t) = \xi(t) + \eta(t), \tag{1.175}$$

where the stochastic processes $\xi(t)$ and $\eta(t)$ are stationary in a broad sense, is determined in the following form:

$$\begin{aligned} K_\zeta(\tau) &= M\{[\xi(t) + \eta(t)][\xi^*(t+\tau) + \eta^*(t+\tau)]\} \\ &= K_\xi(\tau) + K_\eta(\tau) + K_{\xi\eta}(\tau) + K_{\eta\xi}(\tau). \end{aligned} \tag{1.176}$$

If the stochastic processes $\xi(t)$ and $\eta(t)$ are orthogonal then the following equality

$$K_\zeta(\tau) = K_\xi(\tau) + K_\eta(\tau) \tag{1.177}$$

is true. Equation (1.177) is true for the correlation functions when the means of the stochastic processes $\xi(t)$ and $\eta(t)$ are equal to zero.

- The covariance function of the stochastic process

$$\zeta(t) = \xi(t)\eta(t) \tag{1.178}$$

cannot be expressed by the two-dimensional moments of these processes. If the stochastic processes $\xi(t)$ and $\eta(t)$ are independent of each other then the random variables $\xi(t)$ and $\xi(t+\tau)$ are also independent of the random variables $\eta(t)$ and $\eta(t+\tau)$. Because of this, in the case of the independent stochastic process we can write

$$\begin{aligned} K_\zeta(\tau) &= M\{[\xi(t)\eta(t)][\xi^*(t+\tau)\eta^*(t+\tau)]\} \\ &= M[\xi(t)\xi^*(t+\tau)]M[\eta(t)\eta^*(t+\tau)] = K_\xi(\tau)K_\eta(\tau). \end{aligned} \tag{1.179}$$

If the condition

$$\zeta(t) = a(t)\xi(t) \tag{1.180}$$

is true, where $a(t)$ is the deterministic function, then the following equality

$$K_\zeta(t_1, t_2) = a(t_1)a^*(t_2)K_\xi(\tau) \tag{1.181}$$

is also true.

Notice that under signal processing and signal detection in the presence of noise we often use the following correlation functions:

- For the pulse signals we use the following form of the correlation function

$$R_S(\tau) = k\int_0^T S(t)S^*(t+\tau)\,dt. \tag{1.182}$$

- For the signals that are real functions of time we use the following form of the correlation function

$$R_S(\tau) = k\int_0^T S(t)S(t+\tau)\,dt. \tag{1.183}$$

Here $[0, T]$ is the time interval, within the limits of which the signal is observed; k is the constant coefficient; and τ is the shift in time of the signal. The function $R_S(\tau)$ defines a degree of correlation between the amplitude of the signal $S(t)$ and the amplitude of the model signal $S(t+\tau)$ shifted in time by the value τ with respect to the amplitude of the signal $S(t)$ and possesses the correlation function properties, i.e., it has the maximum value under the condition $\tau = 0$ and decreases in value with an increase in the value τ. The function $R_S(\tau)$ is called the correlation function of the deterministic signal. The mutual correlation function between two deterministic signals $S_1(t)$ and $S_2(t)$ is determined in the following form:

$$R_{S_1S_2}(\tau) = k\int_0^T S_1(t)S_2^*(t+\tau)\,dt. \tag{1.184}$$

1.5 Spectral Density

The spectral density $S(f)$ of the stationary in a broad sense stochastic process $\xi(t)$ is defined as the Fourier transform of the covariance function in the following form:

$$S(f) = \int_{-\infty}^{\infty} K(\tau)\cdot e^{-j2\pi f\tau}\,d\tau. \tag{1.185}$$

Based on the inverse Fourier transform we can write

$$K(\tau) = \int_{-\infty}^{\infty} S(f) \cdot e^{j2\pi f \tau} \, df. \tag{1.186}$$

Thus, the spectral density and the covariance function of the stationary stochastic process are the components of the mutual Fourier transform.

The spectral density $S_0(f)$ and the correlation function $R(\tau)$ of the stationary in a broad sense centered stochastic process

$$\xi_0(t) = \xi(t) - m_\xi \tag{1.187}$$

are functionally related to each other in an analogous way:

$$S_0(f) = \int_{-\infty}^{\infty} R(\tau) \cdot e^{-j2\pi f \tau} \, d\tau; \tag{1.188}$$

$$R(\tau) = \int_{-\infty}^{\infty} S_0(f) \cdot e^{j2\pi f \tau} \, df, \tag{1.189}$$

where

$$m_\xi = M[\xi(t)] \tag{1.190}$$

is the mean of the above-mentioned process.

Substituting Eq. (1.169) in Eq. (1.185) and taking into account Eq. (I.26) (see Appendix I), we can write

$$S(f) = S_0(f) + m_\xi^2 \delta(f). \tag{1.191}$$

One can see that the spectral density of the stationary stochastic process with the non-zero mean is distinguished from the spectral density of the corresponding centered stationary stochastic process only by a discrete spectral line at the zero frequency. Bearing this fact in mind, we can see that Eqs. (1.185)–(1.189) are equivalent. These formulae are called the Wiener–Heanchen formulae.

Notice that the spectral density $S(f)$ of the stationary in a narrow sense stochastic process $\xi(t)$ can be determined using the two-dimensional probability distribution density $f_2(x_1, x_2; \tau)$. For this purpose we introduce the Fourier transform for this probability distribution density with respect to the variable τ in the following form:

$$G(x_1, x_2; f) = \int_{-\infty}^{\infty} f_2(x_1, x_2; \tau) \cdot e^{-2\pi f \tau} \, d\tau. \tag{1.192}$$

It is well known that by definition the covariance function can be determined in the following form:

$$K(\tau) = \int_{-\infty}^{\infty}\int_{-\infty}^{\infty} x_1 x_2 f_2(x_1, x_2; \tau)\, dx_1\, dx_2. \tag{1.193}$$

Substituting Eq. (1.193) into Eq. (1.185) and changing a sequence of integration in terms of Eq. (1.192), we can write

$$\begin{aligned} S(f) &= \int_{-\infty}^{\infty}\int_{-\infty}^{\infty} x_1 x_2 \int_{-\infty}^{\infty} f_2(x_1, x_2; \tau) \cdot e^{-j2\pi f \tau}\, d\tau\, dx_1\, dx_2 \\ &= \int_{-\infty}^{\infty}\int_{-\infty}^{\infty} x_1 x_2 G(x_1, x_2; f)\, dx_1\, dx_2 \end{aligned} \tag{1.194}$$

to explain a physical sense of the spectral density. If the process $\xi(t)$ is the stochastic (fluctuating) voltage or current, for example, then the values $S(f)$ and $S_0(f)$ in Eqs. (1.185) and (1.188) have the dimensional representation of energy. In many practical cases this is not true. For example, this statement is not true when the stochastic process $\xi(t)$ defines random oscillations of the amplifier coefficient, random delay of the target return signal, frequency and phase fluctuations of the signal, and so on.

If the condition $\tau = 0$ is satisfied in Eq. (1.189) we can write

$$D = R(0) = \int_{-\infty}^{\infty} S_0(f)\, df. \tag{1.195}$$

Thus, the variance (the total energy) of the stationary centered stochastic process is equal to an area limited by a spectral density curve. The value $S_0(f)\, df$ is the part of energy of the signal, which is concentrated within the limits of the frequency interval $[f - \frac{df}{2}, f + \frac{df}{2}]$. For example, if the stationary in a broad sense stochastic process $\xi(t)$ with the spectral density $S_\xi(f)$ comes in at the input of stationary linear system with the pure complex-valued frequency characteristic $K(j\omega)$ then the spectral density $S_\eta(f)$ of the stochastic process $\eta(t)$ forming at the output of the system under stationary conditions is determined in the following form:

$$S_\eta(f) = S_\xi(f)|K(j2\pi f)|^2. \tag{1.196}$$

Equation (1.196) indicates the advisability of introducing the spectral density of the stochastic process.

Assume that the ideal band filter with the pure complex-valued frequency characteristic shown in Fig. 1.1 is the linear system. Then the variance of the

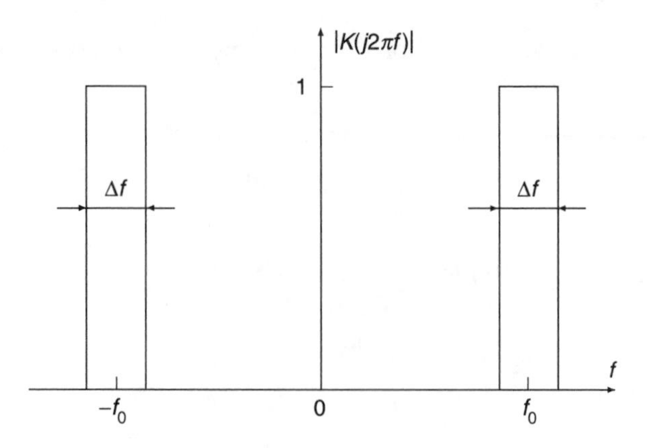

FIGURE 1.1
Amplitude-frequency response of ideal band filter.

stochastic process at the filter output in terms of Eq. (1.195) is determined in the following form:

$$D_\eta = \int_{-\infty}^{\infty} S_\eta(f)\,df = 2\int_{f_0-\frac{\Delta f}{2}}^{f_0+\frac{\Delta f}{2}} S_\xi(f)\,df \geq 0. \tag{1.197}$$

If the band filter is the narrow-band filter, i.e., the following condition

$$\Delta f \ll f_0 \tag{1.198}$$

is true and the spectral density $S_\xi(f)$ is the continuous function within the limits of the bandwidth Δf then we can write

$$D_\eta \simeq S_\xi \Delta f. \tag{1.199}$$

Next we list the main properties of the spectral density:

- *The first property*. The spectral density of the stationary stochastic process (real or complex-valued) is the non-negative value:

$$S(f) \geq 0. \tag{1.200}$$

  This property follows from Eqs. (1.197) and (1.199). If we suppose that the spectral density $S(f) < 0$ is within the limits of the bandwidth Δf, then the variance of the stochastic process at the output of the linear system is negative, but it is impossible. Thus, if the non-negative spectral density $S(f)$ is given, which is equivalent to the non-negative value determined before correlation function $R(\tau)$, then the stationary stochastic process $\xi(t)$ with the spectral density $S(f)$ and correlation function $R(\tau)$ exists.

- *The second property*. The spectral density of the stationary in a broad sense stochastic process is always the real function because

$$K(-\tau) = K^*(\tau). \tag{1.201}$$

Moreover, the spectral density is the even function with respect to frequency when the stochastic process is real. Since the correlation function $R(\tau)$ of the real stochastic process is the even function with respect to the argument then we can write

$$S_0(-f) = \int_{-\infty}^{\infty} R(\tau) \cdot e^{j2\pi f \tau}\, d\tau = S_0(f). \tag{1.202}$$

Taking into account an evenness of the spectral density, Eqs. (1.188) and (1.196) can be written in the following forms:

$$S_0(-f) = \int_{-\infty}^{\infty} R(\tau)\cos(2\pi f \tau)\, d\tau = 2\int_{0}^{\infty} R(\tau)\cos(2\pi f \tau)\, d\tau; \tag{1.203}$$

$$R(\tau) = \int_{-\infty}^{\infty} S_0(f)\cos(2\pi f \tau)\, df = 2\int_{0}^{\infty} S_0(f)\cos(2\pi f \tau)\, df. \tag{1.204}$$

Consequently, the spectral density and the correlation function of the real stationary in a broad sense stochastic process are related to each other by the mutual Fourier cosine-transform. Since the correlation function of the real stochastic process is the real function of argument then one can see from Eq. (1.204) that the spectral density is also the real function with respect to frequency.

- *The third property*. The correlation function $R(\tau)$ and the spectral density $S_0(\tau)$ of the stationary in a broad sense stochastic process possess all properties that are characteristic of the pair of the mutual Fourier transform. In particular, the spectrum $S_0(f)$ is wider, the correlation function $R(\tau)$ is narrower, and vice versa. This result is expressed quantitatively by the uncertainty principle or relationship.

Note that the spectral density $S_0(f)$ is defined under positive and negative values of frequency in all above-mentioned formulae. According to Eq. (1.202) for the real stochastic process the equality

$$S_0(f) = S_0(-f) \tag{1.205}$$

is true. Introduce the one-sided spectrum $S_0^+(f)$ that is not equal to zero only at the positive frequencies $f \geq 0$:

$$S_0^+(f) = S_0(f) + S_0(-f) = 2S_0(f). \tag{1.206}$$

Then in terms of Eqs. (1.203) and (1.204) we can write finally the Wiener–Heanchen formulae in the following form:

$$S^{+}(f) = 4\int_{0}^{\infty} K(\tau)\cos(2\pi f\tau)\,d\tau, \quad f \geq 0; \tag{1.207}$$

$$K(\tau) = \int_{0}^{\infty} S^{+}(f)\cos(2\pi f\tau)\,df, \quad f \geq 0. \tag{1.208}$$

Equations (1.185)–(1.189) are widely used in theory; however, in practice there is a need to use Eqs. (1.207) and (1.208).

In practice there is the concept of the effective spectrum bandwidth. There are some definitions of the effective spectrum bandwidth. The first definition is determined in the following form:

$$\Delta f_{ef}^{2} = \frac{\int_{-\infty}^{\infty} (f - \overline{f}) S_{0}^{2}(f)\,df}{\int_{-\infty}^{\infty} S_{0}^{2}(f)\,df}, \tag{1.209}$$

where

$$\overline{f} = \frac{\int_{-\infty}^{\infty} f S_{0}^{2}(f)\,df}{\int_{-\infty}^{\infty} S_{0}^{2}(f)\,df}. \tag{1.210}$$

In practice the second definition in the following form

$$\Delta f_{ef} = \frac{\int_{0}^{\infty} S_{0}^{2}(f)\,df}{S_{0}^{2}(f)\,df} \tag{1.211}$$

is widely used, where $S_0(f_0)$ is the spectral density under the characteristic frequency f_0, for example, the resonant frequency or central frequency of the filter or the linear system. The value $S_0(f_0)$ corresponds to the maximum of the spectral density.

The stationary stochastic process with the spectral density concentrated within the limits of the narrow-band interval Δf near the frequency

$$f_0 \gg \Delta f \tag{1.212}$$

is called the narrow-band stochastic process. Otherwise the stochastic process is not narrow-band. In the determination of the spectral density of the narrow-band stochastic process the following formula is very useful. If the spectral density $S_{0_\xi}(f)$ corresponds to the correlation function $R_\xi(\tau)$ then the spectral density $S_{0_\eta}(f)$ corresponding to the correlation function

$$R_\eta(\tau) = R_\xi(\tau)\cos(2\pi f_0 \tau) \tag{1.213}$$

is determined in the following form:

$$S_{0_\eta}(f) = 0.5\left[S_{0_\xi}(f + f_0) + S_{0_\xi}(f - f_0)\right]. \tag{1.214}$$

Henceforth we will use the following Wiener–Heanchen formulae:

$$S(\omega) = \int_{-\infty}^{\infty} K(\tau)\cdot e^{-j\omega\tau}\,d\tau; \tag{1.215}$$

$$K(\tau) = \frac{1}{2\pi}\int_{-\infty}^{\infty} S(\omega)\cdot e^{j\omega\tau}\,d\omega; \tag{1.216}$$

$$S_0(\omega) = \int_{-\infty}^{\infty} R(\tau)\cdot e^{-j\omega\tau}\,d\tau; \tag{1.217}$$

$$R(\tau) = \frac{1}{2\pi}\int_{-\infty}^{\infty} S_0(\omega)\cdot e^{j\omega\tau}\,d\omega. \tag{1.218}$$

In the process, we must bear in mind the following equality

$$S(\omega) = S_0(\omega) + 2\pi m_\xi^2 \delta(\omega) \tag{1.219}$$

that follows from Eq. (1.191) in terms of Eq. (I.34) (see Appendix I).

Instead of the spectral density $S_0(\omega)$ we can use the normalized spectral density in the following form:

$$s_0(\omega) = \frac{S_0(\omega)}{D}, \tag{1.220}$$

where D is the variance of the considered stochastic processes. The normalized spectral density and the correlation function of the stationary in a broad sense stochastic process are functionally related to each other by the pair of mutual Fourier transforms:

$$s_0(\omega) = \int_{-\infty}^{\infty} r(\tau)\cdot e^{-j\omega\tau}\,d\tau; \tag{1.221}$$

$$r(\tau) = \frac{1}{2\pi}\int_{-\infty}^{\infty} s_0(\omega)\cdot e^{j\omega\tau}\,d\omega. \tag{1.222}$$

According to Eqs. (1.207) and (1.208) we can write

$$s_0^+(f) = 4\int_{-\infty}^{\infty} r(\tau)\cos(2\pi f\tau)\,d\tau, \quad f \geq 0; \tag{1.223}$$

$$r(\tau) = \int_{-\infty}^{\infty} s_0^+(f)\cos(2\pi f\tau)\,df, \quad f \geq 0. \tag{1.224}$$

Note that the normalized spectral density satisfies the following condition:

$$\int_{-\infty}^{\infty} s_0(\omega)\,d\omega = 1, \quad s_0(\omega) \geq 0. \tag{1.225}$$

The mutual spectral density of two stationary stochastic processes $\xi(t)$ and $\eta(t)$ is determined by the Fourier transform of the mutual covariance functions in the following form:

$$S_{\xi\eta}(\omega) = \int_{-\infty}^{\infty} K_{\xi\eta}(\tau)\cdot e^{-j\omega\tau}\,d\tau = S_{\eta\xi}^*(\omega). \tag{1.226}$$

Using the inverse Fourier transform, we can write

$$K_{\xi\eta}(\tau) = \frac{1}{2\pi}\int_{-\infty}^{\infty} S_{\xi\eta}(\omega)\cdot e^{j\omega\tau}\,d\omega. \tag{1.227}$$

Under the condition $\tau = 0$ we can write

$$\frac{1}{2\pi}\int_{-\infty}^{\infty} S_{\xi\eta}(\omega)\,d\omega = K_{\xi\eta}(0) = M[\xi(t)\eta^*(t)]. \tag{1.228}$$

An example of the application of Eq. (1.228) is: if the stochastic process $\xi(t)$ is the random voltage and the stochastic process $\eta(t)$ is the random current, then the covariance function $K_{\xi\eta}(\tau)$ under the condition $\tau = 0$, i.e., $K_{\xi\eta}(0)$ can be considered as the average power.

For more detailed information regarding spectral density the reader should consult References 11 and 12.

1.6 Statistical Characteristics

1.6.1 Main Definitions

The probability characteristics of the random variables and the stochastic processes defined by experimental data are random and differ from true values of the probability characteristics, since the results observed during one or more experiments are random. From here onward, the probability characteristics defined by experimental data are called the statistical characteristics. Often the statistical characteristics are called the estimations of the probability characteristics: the mean, the variance, the correlation function, the spectral density, and so on. With an increase in a sample size or time interval, within the limits of which the stochastic process is observed, the statistical characteristics or estimations tend to approach the true probability characteristics of the stochastic processes or the random variables.

It is very important to know a precise definition of the statistical characteristics, i.e., a degree of possible deviations of the statistical characteristics from true values of the corresponding probability characteristics. The statistical characteristics are random variables, and possible deviations of the statistical characteristics from the true values of the corresponding probability characteristics decrease with an increase in a set of experimental data. For this reason, the precision of definition of the statistical characteristics increases with an increase in a set of experimental data.

As a degree of precision of the statistical characteristic definition we can take two values: the deviation Δ of the mean of the statistical characteristic from the true value of the corresponding probability characteristic and the variance D of the statistical characteristic. If α^* is the statistical value of the probability characteristic and α is the true value of the same probability characteristic then we can write

$$\Delta = M[\alpha^*] - \alpha; \tag{1.229}$$

$$D = \sigma^2 = M\{[\alpha^* - M[\alpha^*]]^2\}. \tag{1.230}$$

The root mean square error $\widetilde{A}$ determined in the following form

$$\widetilde{A} = \sqrt{M[(\alpha^* - \alpha)^2]} \tag{1.231}$$

is taken as a degree of precision in parallel with the values Δ and σ^2. It is easy to verify the correctness of the following equality

$$\widetilde{A} = \sqrt{\Delta^2 + \sigma^2}. \tag{1.232}$$

If the statistical characteristic or estimation is so defined that the value $\Delta = 0$, for this case the statistical characteristic is called unbiased. If $\Delta \neq 0$

the statistical characteristic or estimation is called biased. Note that the mean square deviation σ of the unbiased estimation is equal to the mean square error $\widetilde{A}$ of the unbiased estimation. If the following condition $\sigma = \sigma_{min}$ is true the statistical characteristic is called the best. Thus, the statistical characteristic is the best and unbiased if both conditions $\sigma = \sigma_{min}$ and $\Delta = 0$ are satisfied. If all information inherent in experimental data is used to define the statistical characteristic then this estimation (the statistical characteristic) is called sufficient.

The unbiased statistical characteristic (the unbiased estimation) is called consistent when the sample size or time interval, within the limits of which the stochastic process is observed, tends to approach ∞ ($N \to \infty$ or $T \to \infty$, where N is the sample size and $[0, T]$ is the time interval) and $\sigma \to 0$. The biased estimation is called consistent when $\sigma \to 0$ and $\Delta \to 0$ as $N \to \infty$ or $T \to \infty$. Otherwise the estimation is called inconsistent. The estimation defined on the basis of the maximum likelihood method is called the likelihood estimation.

Let m_Y be the mean and σ_Y be the variance of the random variable Y. If experimental data are independent of each other and obtained under the same experimental conditions, then according to the central limit theorem[1,2] the mean estimation or the statistical mean is equal to

$$m_Y^* = \frac{1}{N} \cdot \sum_{k=1}^{N} Y_k. \tag{1.233}$$

It is easy to prove the correctness of the following equality

$$\Delta = M[m_Y^*] - m_Y = 0, \tag{1.234}$$

i.e., the mean estimation m_Y^* is unbiased. It is easy to prove that the mean estimation m_Y^* is sufficient, too. Then

$$\sigma^2 = \frac{1}{N} \cdot \sigma_Y^2 \tag{1.235}$$

or

$$\sigma = \frac{1}{\sqrt{N}} \cdot \sigma_Y, \tag{1.236}$$

where σ_Y is the root mean square deviation of the random variable Y.

In addition to the precision it is often useful to determine certainty of definition of the statistical characteristic. As a degree of certainty either the probability that the true value of the probability characteristic lies within the limits of the interval given with respect to the statistical characteristic taken or the interval with respect to the statistical characteristic, within the limits of which the true value of the probability characteristic lies with the given probability, is taken. This probability and corresponding interval are called the fiducial probability and the confidence interval. It is obvious that if the fiducial probability at the given interval is high or the confidence interval at the given probability is less, the certainty of definition of the statistical characteristic is high.

1.6.2 Statistical Numerical Characteristics

During each experiment assume that variables X are random and considered as the random variables. Define these random variables $X_1, X_2, \ldots, X_n$ in accordance with the first, the second, ..., the n-th experiment. Owing to independence between experiments the random variables $X_1, X_2, \ldots, X_n$ are independent, too.

Define the statistical mean of the function $\Psi(X)$ of the random variable X. According to definitions in References 1, 3, and 5–7 the mean of the function $\Psi(X)$ is determined in the following form

$$M[\Psi(X)] = \int_{-\infty}^{\infty} \Psi(x) f_X(x)\, dx, \tag{1.237}$$

where $f_X(x)$ is the probability distribution density of the random variable X. The value $\frac{1}{n}\sum_{i=1}^{n} \Psi(X_i)$ is the average of the function $\Psi(X)$ during n independent experiments. According to the central limit theorem for all $\varepsilon > 0$, where ε is infinitesimal, the following equality

$$\lim_{n\to\infty} P\left\{\left|\frac{1}{n}\sum_{i=1}^{n} \Psi(X_i) - M[\Psi(X)]\right| \geq \varepsilon\right\} = 0 \tag{1.238}$$

is true. Because of this, the statistical mean of the function $\Psi(X)$ of the random variable X is determined in the following form:

$$M^*[\Psi(X)] = \frac{1}{n} \cdot \sum_{i=1}^{n} \Psi(X_i). \tag{1.239}$$

The statistical initial moment of the k-th order of the random variable X is determined in the following form:

$$m_k^*[X] = m_k^* = M^*[X^k] = \frac{1}{n} \cdot \sum_{i=1}^{n} X_i^k. \tag{1.240}$$

The statistical mean of the random variable X is determined in the following form:

$$m_X^* = m_1^*[X] = M^*[X] = \frac{1}{n} \cdot \sum_{i=1}^{n} X_i. \tag{1.241}$$

The statistical central moment of the k-th order of the random variable X is determined in the following form:

$$D_k^*[X] = D_k^* = M^*\left[(X - m_X)^k\right] = \frac{1}{n} \cdot \sum_{i=1}^{n} (X_i - m_X)^k. \tag{1.242}$$

The statistical variance of the random variable X is determined in the following form:

$$D_{XX}^* = D_2^*[X] = M^*\left[(X - m_X)^2\right] = \frac{1}{n} \cdot \sum_{i=1}^{n} (X_i - m_X)^2. \tag{1.243}$$

The statistical root mean square deviation of the random variable X is determined in the following form:

$$\sigma_X^* = \sqrt{D_{XX}^*}. \tag{1.244}$$

The statistical initial and central higher-order moments can be determined in an analogous way.

In the course of n experiments the random variables $X_1, X_2, \ldots, X_n$ take the non-random values $x_1, x_2, \ldots, x_n$, respectively, that can be used for definition of the statistical initial and central moments of the random variable X. For example,

$$\overline{m_k^*} = \frac{1}{n} \cdot \sum_{i=1}^{n} x_i^k; \tag{1.245}$$

$$\overline{D_k^*} = \frac{1}{n} \cdot \sum_{i=1}^{n} (x_i - m_X)^k. \tag{1.246}$$

In terms of Eqs. (1.240)–(1.244) and using Eqs. (1.245) and (1.246), we can define the statistical mean and the statistical variance of the random variable X on the basis of experimental values $x_1, x_2, \ldots, x_n$ in the following form

$$\overline{m_1^*} = \frac{1}{n} \cdot \sum_{i=1}^{n} x_i; \tag{1.247}$$

$$\overline{D_{XX}^*} = (\overline{\sigma_X^*})^2 = \frac{1}{n} \cdot \sum_{i=1}^{n} (x_i - m_X)^k. \tag{1.248}$$

The statistical moments $\overline{m_k^*}$ and $\overline{D_k^*}$ are the non-random variables and in a general case may differ from the corresponding moments m_k and D_k of the random variable X. As $n \to \infty$ we can think that $\overline{m_k^*} \to m_k$ and $\overline{D_k^*} \to D_k$.

In practice we are unable to ever fully define or we do not know at all the *a priori* true values of the mean m_X of the random variable X. However, we have to define, for example, the statistical central moments D_k^* of the random variable X. If we change the mean m_X by the statistical mean m_X^* in Eq. (1.242) we can determine the statistical central moment of the k-th order in the following form:

$$D_{k_s}^* = \frac{1}{n} \cdot \sum_{i=1}^{n} (X_i - m_X^*)^k. \tag{1.249}$$

It is obvious that the statistical central moment $D_{k_s}^*$ in Eq. (1.249) differs from the statistical central moment D_k^* in Eq. (1.242). As the value n gets higher, the difference between $D_{k_s}^*$ and D_k^* becomes less. But the mean of the statistical central moment of the k-th order is not equal to the true central moment D_k. It is not a great problem to show this. According to Eq. (1.249) and in terms of Eq. (1.241) we can write, for example,

$$D_{2_s}^* = \frac{1}{n} \cdot \sum_{i=1}^{n} \left(X_i - \frac{1}{n} \cdot \sum_{j=1}^{n} X_j \right)^2. \tag{1.250}$$

Add and subtract the mean m_X. Then we can write

$$D_{2_s}^* = \frac{1}{n} \cdot \sum_{i=1}^{n} (X_i - m_X)^2 - \frac{2}{n^2} \cdot \sum_{i=1}^{n} \sum_{j=1}^{n} (X_i - m_X)(X_j - m_X) + \frac{1}{n^2} \cdot \left\{ \sum_{j=1}^{n} (X_j - m_X) \right\}^2. \tag{1.251}$$

The mean of the statistical central moment of the second order $D_{2_s}^*$ is equal to

$$M\left[D_{2_s}^*\right] = \frac{1}{n} \cdot \sum_{i=1}^{n} M\{(X_i - m_X)^2\} - \frac{2}{n^2} \cdot \sum_{i=1}^{n} \sum_{j=1}^{n} M\{(X_i - m_X)(X_j - m_X)\} + \frac{1}{n^2} \cdot M\left\{ \left[\sum_{j=1}^{n} (X_j - m_X) \right]^2 \right\}. \tag{1.252}$$

Taking into account that the random variables $X_1, X_2, \ldots, X_n$ are independent, we can write that

$$M\left[D_{2_s}^*\right] = \frac{1}{n} \cdot \sum_{i=1}^{n} M\{(X_i - m_X)^2\} - \frac{2}{n^2} \cdot \sum_{i=1}^{n} \sum_{j=1}^{n} M\{(X_i - m_X)^2\} + \frac{1}{n^2} \cdot \sum_{j=1}^{n} M\{(X_j - m_X)^2\}. \tag{1.253}$$

But it is well known that

$$M\{(X_i - m_X)^2\} = M\{(X_j - m_X)^2\} = D_2, \tag{1.254}$$

where

$$D_2 = D_{XX} \tag{1.255}$$

is the central moment of the second order of the random variable X. Because of this, we obtain the following final result

$$M\left[D^*_{2_s}\right] = M\left\{\frac{1}{n} \cdot \sum_{i=1}^{n}(X_i - m_X)^2\right\} = \frac{n-1}{n} \cdot D_2. \tag{1.256}$$

Thus, under the finite value n the statistical central moment determined by Eq. (1.249) is the biased estimation of the central moment in the average.

The statistical central moment of the k-th order equal in average to the central moment of the k-th order is denoted by D^{**}_k. Then the statistical central moment of the second order D^{**}_2, which does not differ in average from the central moment D_2, is determined in the following form:

$$D^{**}_2 = \frac{n}{n-1} \cdot D^*_{2_s}. \tag{1.257}$$

In terms of Eq. (1.249) and using Eq. (1.257), we can write

$$D^{**}_2 = D^{**}_{XX} = (\sigma^{**}_X)^2 = \frac{1}{n-1} \cdot \sum_{i=1}^{n}(X_i - m^*_X)^2, \tag{1.258}$$

where D^{**}_{XX} is the statistical variance and σ^{**}_X is the statistical mean square deviation of the random variable X. The mean of the statistical variance D^{**}_2 is equal to the true value of the variance

$$M[D^{**}_2] = M[D^{**}_{XX}] = D_2 = D_{XX}. \tag{1.259}$$

The statistical variance of the random variable X, which is defined on the basis of n experiments and in terms of the statistical mean $\overline{m^*_X}$, is determined by

$$\overline{D^{**}_{XX}} = (\overline{\sigma^{**}_X})^2 = \frac{1}{n-1} \cdot \sum_{i=1}^{n}(X_i - m^*_X)^2. \tag{1.260}$$

1.6.3 Precision of the Statistical Characteristic Definition

We first consider the statistical initial moments. The statistical initial moment m^*_k of the k-th order of the random variable X is determined by Eq. (1.240). The mean of the statistical initial moment m^*_k is equal to the true value of the

k-th order initial moment

$$M[m_k^*] = m_k = M\{X^k\}. \tag{1.261}$$

The variance D_{m_k}, or the root mean square deviation σ_{m_k} of the statistical initial moment m_k^*, from its mean m_k, is determined in the following form:

$$D_{m_k} = \sigma_{m_k}^2 = M\left[m_k^{*^2}\right] - m_k^2. \tag{1.262}$$

Define the first term in Eq. (1.262):

$$\begin{aligned} M\left[m_k^{*^2}\right] &= \frac{1}{n^2} \cdot M\left\{\left(\sum_{i=1}^{n} X_i^k\right)^2\right\} = \frac{1}{n^2} \cdot M\left\{\sum_{i=1}^{n} X_i^k \sum_{j=1}^{n} X_j^k\right\} \\ &= \frac{1}{n^2} \cdot \sum_{i=1}^{n}\sum_{j=1}^{n} M\{X_i^k X_j^k\}. \end{aligned} \tag{1.263}$$

Taking into account that under the condition $i \neq j$ the random variables X_i and X_j are independent, we can write that

$$M\left[m_k^{*^2}\right] = \frac{1}{n^2} \cdot M\{X_i^{2k}\} + \frac{1}{n^2} \cdot \sum_{i=1}^{n}\sum_{j=1}^{n} M\{X_i^k\}M\{X_j^k\}. \tag{1.264}$$

In terms of Eq. (1.261) Eq. (1.264) takes the following form:

$$M\left[m_k^{*^2}\right] = \frac{1}{n^2} \cdot \left[n m_{2k} + n(n-1)m_k^2\right] = \frac{1}{n} \cdot \left[m_{2k} + (n-1)m_k^2\right]. \tag{1.265}$$

Substituting Eq. (1.265) in Eq. (1.264), we can write

$$D_{m_k} = \sigma_{m_k}^2 = \frac{m_{2k} - m_k^2}{n}. \tag{1.266}$$

Only the statistical initial moment of the first order given by Eq. (1.241) is widely used among other statistical initial moments. According to Eq. (1.241) we can write

$$M[m_X^*] = m_X. \tag{1.267}$$

The variance of the statistical mean m_X^* is determined in the following form:

$$\begin{aligned} D_{m_X} &= \sigma_{m_X}^2 = M\left\{(m_X^* - m_X)^2\right\} = \frac{m_2 - m_1^2}{n} \\ &= \frac{M[X^2] - \{M[X]\}^2}{n} = \frac{D_{XX}}{n}, \end{aligned} \tag{1.268}$$

where D_{XX} is the variance of the random variable X. The root mean square deviation is determined in the following form:

$$\sigma_{m_X} = \sqrt{\frac{D_{XX}}{n}} = \frac{1}{\sqrt{n}} \cdot \sigma_X. \tag{1.269}$$

Reference to Eq. (1.269) shows that the root mean square deviation of the statistical mean of the random variable X in $\sqrt{n}$ times less than the variance σ_X of the random value X (n is the number of experiments).

Consider the statistical central moments of the random variable X when the mean m_X is known. In this case the statistical central moment of the k-th order of the random variable X is determined by Eq. (1.242). The mean of the statistical central moment D_k^* is equal to D_k. Define the variance D_{D_k} and the mean square deviation σ_{D_k} in the following form:

$$D_{D_k} = \sigma^2_{D_k} = M\left[D_k^{*^2}\right] - D_k^2. \tag{1.270}$$

Determine the first term in Eq. (1.270):

$$\begin{aligned} M\left[D_k^{*^2}\right] &= \frac{1}{n^2} \cdot M\left\{\left[\sum_{i=1}^{n}(X_i - m_X)^k\right]^2\right\} \\ &= \frac{1}{n^2} \cdot M\left\{\sum_{i=1}^{n}(X_i - m_X)^k \sum_{j=1}^{n}(X_j - m_X)^k\right\} \\ &= \frac{1}{n^2} \cdot \sum_{i=1}^{n}\sum_{j=1}^{n} M\{(X_i - m_X)^k (X_j - m_X)^k\}. \end{aligned} \tag{1.271}$$

Taking into account that under the condition $i \neq j$ the random variables $X_i - m_X$ and $X_j - m_X$ are independent, we can write

$$M\left[D_k^{*^2}\right] = \frac{1}{n^2} \cdot \left[nD_{2k} + n(n-1)D_k^2\right] = \frac{1}{n} \cdot \left[D_{2k} + (n-1)D_k^2\right]. \tag{1.272}$$

In particular, the equality

$$\begin{aligned} M\left[D_{XX}^{*^2}\right] &= M\left[D_2^{*^2}\right] = \frac{1}{n} \cdot \left[D_4 + (n-1)D_2^2\right] \\ &= \frac{1}{n} \cdot \left[D_4 + (n-1)D_{XX}^2\right] = \frac{1}{n} \cdot \left[D_4 + (n-1)\sigma_X^4\right] \end{aligned} \tag{1.273}$$

is true, where D_{XX} and σ_X are the variance and the mean square deviation of the random variable X, respectively. Substituting Eq. (1.272) in Eq. (1.270), we can write

$$D_{D_k} = \sigma^2_{D_k} = \frac{D_{2k} - D_k^2}{n}, \tag{1.274}$$

where n is the number of experiments. In particular, it is very convenient to use the relative precision of definition of the statistical central moment of the k-th order that is determined in the following form:

$$\frac{\sigma_{D_k}}{D_k} = \frac{1}{\sqrt{n}} \cdot \sqrt{\frac{D_{2k}}{D_k^2} - 1}. \qquad (1.275)$$

Consider the case when the mean m_X of the random variable X is unknown. In this case the statistical central moment D_k^{**} of the k-th order is used for definition of the statistical characteristic precision. The statistical variance D_{XX}^{**} of the random variable X is determined by Eq. (1.258). The mean of the statistical variance D_{XX}^{**} is equal to D_{XX}. Define the variance and the mean square deviation of the statistical variance D_{XX}^{**} in the following form:

$$\begin{aligned} D_D = \sigma_D^2 &= M\left[D_{XX}^{**^2}\right] - \left\{M\left[D_{XX}^{**}\right]\right\}^2 \\ &= M\left[D_{XX}^{**^2}\right] - D_{XX}^2 = M\left[D_{XX}^{**^2}\right] - D_2^2. \end{aligned} \qquad (1.276)$$

Define the first term in Eq. (1.276). We can write the variance D_{XX}^{**} in the following form:

$$\begin{aligned} D_{XX}^{**} &= \frac{1}{n-1} \cdot \sum_{i=1}^{n} (X_i - m_X^*)^2 \\ &= \frac{1}{n-1} \cdot \left[\sum_{i=1}^{n} (X_i - m_X)^2 - 2(m_X^* - m_X) \sum_{i=1}^{n} (X_i - m_X) + n(m_X^* - m_X)^2\right]. \end{aligned} \qquad (1.277)$$

But

$$m_X^* - m_X = \frac{1}{n} \cdot \sum_{j=1}^{n} X_j - m_X = \frac{1}{n} \cdot \sum_{j=1}^{n} (X_j - m_X). \qquad (1.278)$$

Because of this, the following equality

$$D_{XX}^{**} = \frac{1}{n-1} \cdot \left\{\sum_{i=1}^{n} (X_i - m_X)^2 - \frac{1}{n} \cdot \left[\sum_{j=1}^{n} (X_j - m_X)\right]^2\right\} \qquad (1.279)$$

is true.

Then

$$\begin{aligned} M\left[D_{XX}^{**^2}\right] = \frac{1}{(n-1)^2} \cdot \Bigg\{ & M\left\{\left[\sum_{i=1}^{n} (X_i - m_X)^2\right]^2\right\} \\ & - \frac{2}{n} \cdot M\left\{\sum_{i=1}^{n} (X_i - m_X)^2 \left[\sum_{j=1}^{n} (X_j - m_X)\right]^2\right\} \\ & + \frac{1}{n^2} \cdot M\left\{\left[\sum_{j=1}^{n} (X_j - m_X)\right]^4\right\}\Bigg\}. \end{aligned} \qquad (1.280)$$

Define all terms in Eq. (1.280). Taking into consideration that the random variables $X_1, X_2, \ldots, X_n$ are independent under the condition $i \neq j$, we can write the first term in Eq. (1.280) in the following form:

$$M\left\{\left[\sum_{i=1}^{n}(X_i - m_X)^2\right]^2\right\} = \sum_{i=1}^{n} M\{(X_i - m_X)^4\} + \sum_{i=1}^{n}\sum_{j=1}^{n} M\{(X_i - m_X)^2\}M\{(X_j - m_X)^2\}. \quad (1.281)$$

Thus we can write

$$M\left\{\left[\sum_{i=1}^{n}(X_i - m_X)^2\right]^2\right\} = nD_4 + n(n-1)D_2^2. \quad (1.282)$$

The second term in Eq. (1.280) in terms of Eq. (1.282) can be written in the following form:

$$M\left\{\sum_{i=1}^{n}(X_i - m_X)^2\left[\sum_{j=1}^{n}(X_i - m_X)\right]^2\right\} = M\left\{\left[\sum_{i=1}^{n}(X_i - m_X)^2\right]^2\right\} = nD_4 + n(n-1)D_2^2. \quad (1.283)$$

The third term in Eq. (1.280) can be written in the following form:

$$\begin{aligned} M\left\{\left[\sum_{j=1}^{n}(X_j - m_X)\right]^4\right\} &= M\left\{\left[(X_n - m_X) + \sum_{j=1}^{n-1}(X_j - m_X)\right]^4\right\} \\ &= M\{(X_n - m_X)^4\} + 4M\left\{(X_n - m_X)^3\sum_{j=1}^{n-1}(X_j - m_X)\right\} \\ &\quad + 6M\left\{(X_n - m_X)^2\left[\sum_{j=1}^{n-1}(X_j - m_X)\right]^2\right\} \\ &\quad + 4M\left\{(X_n - m_X)\left[\sum_{j=1}^{n-1}(X_j - m_X)\right]^3\right\} + M\left\{\left[\sum_{j=1}^{n-1}(X_j - m_X)\right]^4\right\}. \end{aligned} \quad (1.284)$$

It is obvious that

$$M\{(X_n - m_X)^4\} = D_4; \quad (1.285)$$

$$M\left\{(X_n - m_X)^3\sum_{j=1}^{n-1}(X_j - m_X)\right\} = M\{(X_n - m_X)^3\}M\left\{\sum_{j=1}^{n-1}(X_j - m_X)\right\} = 0; \quad (1.286)$$

$$M\left\{(X_n - m_X)\left[\sum_{j=1}^{n-1}(X_j - m_X)\right]^3\right\} = M\{X_n - m_X\}M\left\{\left[\sum_{j=1}^{n-1}(X_j - m_X)\right]^3\right\} = 0, \tag{1.287}$$

$$M\left\{(X_n - m_X)^2\left[\sum_{j=1}^{n-1}(X_j - m_X)\right]^2\right\} = M\{(X_n - m_X)^2\}M\left\{\left[\sum_{j=1}^{n-1}(X_j - m_X)\right]^2\right\}$$
$$= (n-1)D_2^2. \tag{1.288}$$

Substituting Eqs. (1.285)–(1.288) in Eq. (1.284), we can write

$$M\left\{\left[\sum_{j=1}^{n}(X_j - m_X)\right]^4\right\} = M\left\{\left[\sum_{j=1}^{n-1}(X_j - m_X)\right]^4\right\} + D_4 + 6(n-1)D_2^2. \tag{1.289}$$

If the condition $n = 1$ is satisfied we can write

$$M\left\{\left[\sum_{j=1}^{n}(X_j - m_X)\right]^4\right\} = M\{(X_j - m_X)^4\}. \tag{1.290}$$

Using Eq. (1.289), we can write

$$M\left\{\left[\sum_{j=1}^{n}(X_j - m_X)\right]^4\right\} = nD_4 + 3n(n-1)D_2^2. \tag{1.291}$$

Substituting Eqs. (1.282), (1.283), and (1.291) in Eq. (1.280), we can write

$$M\left[D_{XX}^{**^2}\right] = \frac{1}{n}\cdot\left[D_4 + \frac{n^2 - 2n + 3}{n-1}\cdot D_2\right]. \tag{1.292}$$

Substituting Eq. (1.293) in Eq. (1.276), we can write finally

$$D_D = \sigma_D^2 = \frac{1}{n}\cdot\left[D_4 - \left(1 - \frac{2}{n-1}\right)D_2^2\right]. \tag{1.293}$$

In particular, if the random variable X is Gaussian, i.e., $D_4 = 3D_2^2$, we can write

$$D_D = \sigma_D^2 = \frac{2}{n-1}\cdot D_2^2. \tag{1.294}$$

1.6.4 Statistical Characteristics of the Ergodic Stationary Stochastic Processes

If the stationary stochastic process is ergodic with respect to the mean then an average value of this process converges in the variance and the probability to the mean within the limits of the time interval $[0, T]$ as time increases indefinitely. This feature of the ergodic stochastic process allows us to define the mean using a single realization during one experiment instead of multiple n realizations during multiple n experiments. By this reason, the statistical mean of the considered ergodic stationary stochastic process $X(t)$ can be determined within the limits of the time interval $[0, T]$ in the following form:

$$m_X^* = \frac{1}{T}\int_0^T X(t)\,dt. \tag{1.295}$$

The statistical mean m_X^* is the random variable that converges to the true mean m_X of the stationary stochastic process $X(t)$ as $T \to \infty$.

Assume that $x(t)$ is the realization of the stationary stochastic process $X(t)$ within the limits of the time interval $[0, T]$. Then the statistical mean $\overline{m_X^*}$ is determined in the following form:

$$\overline{m_X^*} = \frac{1}{T}\int_0^T x(t)\,dt. \tag{1.296}$$

If the stationary stochastic process is ergodic with respect to the correlation function then an average value of the product between the random variables $X(t) - m_X$ and $X(t+\tau) - m_X$ of the stationary stochastic process $X(t)$ converges in the variance and the probability to the correlation function $R_{XX}(\tau)$. This ergodic property of the stationary stochastic process $X(t)$ allows us to define the correlation function using a single realization during one experiment instead of multiple n realizations during multiple n experiments. The statistical correlation function is determined in the following form:

$$R_{XX}^*(\tau) = \frac{1}{T-\tau}\int_0^T [X(t) - m_X][X(t+\tau) - m_X]\,dt. \tag{1.297}$$

The statistical correlation function $R_{XX}^*(\tau)$ is the random function with the argument τ as one can see from Eq. (1.297) and converges to the correlation function $R_{XX}(\tau)$ within the limits of the time interval $[0, T]$ as time increases indefinitely. Assume that $x(t)$ is the realization of the stationary stochastic process $X(t)$ within the limits of the time interval $[0, T]$. Then the statistical correlation function is determined in the following form:

$$\overline{R_{XX}^*}(\tau) = \frac{1}{T-\tau}\int_0^T [x(t) - m_X][x(t+\tau) - m_X]\,dt. \tag{1.298}$$

For definition of the statistical correlation function there is a need to know the true mean m_X as it follows from Eqs. (1.297) and (1.298). Replacing the mean m_X by the statistical mean m_X^* or $\overline{m_X^*}$, the statistical correlation function is determined in the following form:

$$R_{XX}^{**}(\tau) = \frac{1}{T-\tau}\int_0^T [X(t) - m_X^*][X(t+\tau) - m_X^*]\,dt; \tag{1.299}$$

$$\overline{R_{XX}^{**}}(\tau) = \frac{1}{T-\tau}\int_0^T \left[x(t) - \overline{m_X^*}\right]\left[x(t+\tau) - \overline{m_X^*}\right] dt. \tag{1.300}$$

Note that for definition of the statistical correlation function by Eqs. (1.299) and (1.300) the stationary stochastic process $X(t)$ must be ergodic with respect to both the correlation function and the mean.

The immediate values $X(t)$ are often used for definition of the statistical correlation function. Reference to Eq. (1.299) shows that

$$R_{XX}^{**}(\tau) = \frac{1}{T-\tau}\int_0^T X(t)X(t+\tau)\,dt$$
$$- m_X^*\left[\frac{1}{T-\tau}\int_0^T X(t)\,dt + \frac{1}{T-\tau}\int_0^T X(t+\tau)\,dt - m_X^*\right]. \tag{1.301}$$

Introducing designations

$$m_{X_1}^* = \frac{1}{T-\tau}\int_0^T X(t)\,dt; \tag{1.302}$$

$$m_{X_2}^* = \frac{1}{T-\tau}\int_0^T X(t+\tau)\,dt = \frac{1}{T-\tau}\int_\tau^T X(t_1)\,dt_1, \tag{1.303}$$

we can write that

$$R_{XX}^{**}(\tau) = \frac{1}{T-\tau}\int_0^{T-\tau} X(t)X(t+\tau)\,dt - m_X^*\left(m_{X_1}^* + m_{X_2}^* - m_X^*\right). \tag{1.304}$$

It is obvious that the means $m_{X_1}^*$ and $m_{X_2}^*$ are the statistical means of the stationary stochastic process $X(t)$ within the limits of the time intervals $[0, T-\tau]$ and $[\tau, T]$ or within the limits of the time interval, a duration of which is equal to $[0, T]$. For this reason,

$$m_X^* \simeq m_{X_1}^* \simeq m_{X_2}^*. \tag{1.305}$$

Substituting Eq. (1.305) in Eq. (1.304), we can write the final approximate result

$$R^{**}_{XX}(\tau) \simeq \frac{1}{T-\tau}\int_0^T X(t)X(t+\tau)\,dt - m^{*^2}_X. \tag{1.306}$$

Knowing the statistical correlation function $R^*_{XX}(\tau)$ or $R^{**}_{XX}(\tau)$, we can define the statistical variance D^*_{XX} or D^{**}_{XX} or the statistical mean square deviation σ^*_X or σ^*_{XX} of the stationary stochastic process $X(t)$, which is ergodic with respect to the correlation function. We list the main relationships:

$$D^*_{XX} = \sigma^{*^2}_X = R^*_{XX}(0); \tag{1.307}$$

$$D^{**}_{XX} = \sigma^{**^2}_X = R^{**}_{XX}(0); \tag{1.308}$$

$$D^*_{XX} = \sigma^{*^2}_X = \frac{1}{T}\int_0^T [X(t) - m_X]^2\,dt; \tag{1.309}$$

$$D^{**}_{XX} = \sigma^{**^2}_X = \frac{1}{T}\int_0^T [X(t) - m^*_X]^2\,dt. \tag{1.310}$$

In terms of Eq. (1.295) Eq. (1.310) takes the following form:

$$D^{**}_{XX} = \sigma^{**^2}_X = \frac{1}{T}\int_0^T X^2(t)\,dt - m^2_X. \tag{1.311}$$

1.6.5 Precision of Statistical Characteristics of the Ergodic Stationary Stochastic Processes

Consider the precision of definition of the statistical mean m^*_X of the ergodic stationary stochastic process $X(t)$. According to Eq. (1.295) the mean of the statistical mean m^*_X is determined in the following form:

$$M[m^*_X] = \frac{1}{T}\int_0^T M[X(t)]\,dt = \frac{1}{T}\int_0^T m_X\,dt = m_X. \tag{1.312}$$

The mean square deviation σ_m is considered as a degree of definition for precision of the statistical mean

$$\sigma^2_m = M\left\{\left[\frac{1}{T}\int_0^T X(t)\,dt - m_X\right]^2\right\} = \frac{1}{T^2}\int_0^T\int_0^T R_{XX}(t_2 - t_1)\,dt_1\,dt_2, \tag{1.313}$$

where $R_{XX}(t_2 - t_1)$ is the correlation function of the stationary stochastic process $X(t)$. Introducing a change of variables $\tau = t_2 - t_1$, we can rewrite Eq. (1.313) in the following form:

$$\sigma_m^2 = \int_0^T dt_1 \int_0^T R_{XX}(\tau)\,d\tau. \tag{1.314}$$

Denote

$$\gamma(\varepsilon) = \int_0^{\varepsilon} R_{XX}(\tau)\,d\tau. \tag{1.315}$$

Integrating Eq. (1.314) by parts, in terms of Eq. (1.315) we can write

$$\sigma_m^2 = \frac{1}{T^2}\left\{[\gamma(T - t_1) - \gamma(t_1)]\|_0^T + \int_0^T [R_{XX}(T - t_1) - R_{XX}(t_1)]t_1\,dt_1\right\}. \tag{1.316}$$

Taking into account Eq. (1.315), we can write

$$\sigma_m^2 = \frac{2}{T}\int_0^T \left(1 - \frac{\tau}{T}\right) R_{XX}(\tau)\,d\tau. \tag{1.317}$$

Since the stationary stochastic process $X(t)$ is ergodic, then owing to Eq. (1.317) the mean square deviation $\sigma_m \to 0$ as $T \to \infty$, otherwise $\sigma_m \neq 0$. One can see from Eq. (1.317) that the mean square deviation σ_m depends on the correlation function $R_{XX}(\tau)$ and the time interval $[0, T]$.

Suppose that the mean m_X of the ergodic stationary stochastic process $X(t)$ is known. In this case the mean of the statistical correlation function $R_{XX}^{**}(\tau)$ determined by Eq. (1.297) is equal to the correlation function $R_{XX}(\tau)$:

$$M[R_{XX}^*(\tau)] = \frac{1}{T - \tau}\int_0^{T-\tau} R_{XX}(\tau)\,d\tau = R_{XX}(\tau). \tag{1.318}$$

Consequently, the statistical correlation function $R_{XX}^*(\tau)$ is equal in average to the correlation function $R_{XX}(\tau)$. Reference to Eq. (1.318) shows that the mean of the statistical variance D_{XX}^* determined by Eq. (1.309) is equal to the variance D_{XX}.

The mean square deviation $\sigma_R(\tau)$ is considered as a degree of definition precision of the statistical correlation function $R_{XX}^*(\tau)$

$$\sigma_R^2(\tau) = M\left[R_{XX}^{*^2}(\tau)\right] - R_{XX}^2(\tau). \tag{1.319}$$

But

$$M\left[R_{XX}^{*^2}(\tau)\right] = \frac{1}{(T-\tau)^2}\int_0^{T-\tau}\int_0^{T-\tau} M\{[X(t_1)-m_X][X(t_1+\tau)-m_X] \times [X(t_2)-m_X][X(t_2+\tau)-m_X]\}\,dt_1\,dt_2. \tag{1.320}$$

The integrand in Eq. (1.320) is the central moment of the fourth order. Let the stationary stochastic process $X(t)$ be the Gaussian process. Then, as was shown in Reference 5,

$$M\{[X(t_1)-m_X][X(t_1+\tau)-m_X][X(t_2)-m_X][X(t_2+\tau)-m_X]\} = R_{XX}^2(\tau) + R_{XX}^2(t_2-t_1) + R_{XX}(t_2-t_1+\tau)R_{XX}(t_2-t_1-\tau). \tag{1.321}$$

Consequently,

$$M\left[R_{XX}^{*^2}(\tau)\right] = R_{XX}^2(\tau) + \frac{1}{(T-\tau)^2}\int_0^{T-\tau}\int_0^{T-\tau}\left[R_{XX}^2(t_2-t_1) + R_{XX}(t_2-t_1+\tau)R_{XX}(t_2-t_1-\tau)\right]dt_1\,dt_2. \tag{1.322}$$

Substituting Eq. (1.322) in Eq. (1.319), we can write

$$\begin{aligned}\sigma_R^2(\tau) &= \frac{1}{(T-\tau)^2}\int_0^{T-\tau}\int_0^{T-\tau}\left[R_{XX}^2(t_2-t_1) + R_{XX}(t_2-t_1+\tau)R_{XX}(t_2-t_1-\tau)\right]dt_1\,dt_2\\ &= \frac{2}{T-\tau}\int_0^{T-\tau}\left(1-\frac{\tau_1}{T-\tau}\right)\left[R_{XX}^2(\tau_1) + R_{XX}(\tau_1+\tau)R_{XX}(\tau_1-\tau)\right]d\tau_1,\end{aligned} \tag{1.323}$$

where $\tau_1 = t_2 - t_1$.

Equation (1.323) can be used for definition of the statistical correlation function precision of the Gaussian ergodic stationary process in the case when the mean m_X is known. Under the condition $\tau = 0$ the variance

$$\sigma_R(\tau) = \sigma_R(0) \tag{1.324}$$

is the mean square deviation σ_D of the statistical variance D_{XX}^* of the Gaussian ergodic stationary process from the true value D_{XX}. According to Eq. (1.323) the following equality

$$\sigma_D^2 = \sigma_R^2(0) = \frac{4}{T}\int_0^{T}\left(1-\frac{\tau_1}{T}\right)R_{XX}^2(\tau_1)\,d\tau_1 \tag{1.325}$$

is true.

Consider the case when the mean m_X of the stationary stochastic process $X(t)$ is unknown and the statistical correlation function $R^{**}_{XX}(\tau)$ is determined by Eq. (1.299). In this case the following inequality

$$M[R^{**}_{XX}(\tau)] \neq R_{XX}(\tau) \tag{1.326}$$

is true.

For this reason, a degree of definition precision of the statistical correlation function is defined by both the mean square deviation $\sigma_R(\tau)$ and the difference $M[R^{**}_{XX}(\tau)] - R_{XX}(\tau)$. The equation for definition of the mean square deviation $\sigma_R(\tau)$ is very cumbersome and difficult to work with. Because of this, under high values of T we can use Eq. (1.323) for definition of the estimation $\sigma_R(\tau)$. To estimate a deviation of the mean $M[R^{**}_{XX}(\tau)]$ of the statistical correlation function $R^{**}_{XX}(\tau)$ from the correlation function $R_{XX}(\tau)$ define as a preliminary the mean $M[R^{**}_{XX}(\tau)]$.

1.6.5.1 Definition of the Mean $M[R^{**}_{XX}(\tau)]$

Using Eq. (1.299), we can write

$$M[R^{**}_{XX}(\tau)] = \frac{1}{T-\tau}\int_0^{T-\tau} M\{[X(t) - m^*_X][X(t+\tau) - m^*_X]\}\,dt \tag{1.327}$$

or

$$\begin{aligned} M[R^{**}_{XX}(\tau)] &= \frac{1}{T-\tau}\int_0^{T-\tau} M\{[X(t) - m_X][X(t+\tau) - m_X]\}\,dt \\ &- \frac{1}{T-\tau}\int_0^{T-\tau} M\{[m^*_X - m_X][X(t+\tau) - m_X]\}\,dt \\ &- \frac{1}{T-\tau}\int_0^{T-\tau} M\{[X(t) - m_X][m^*_X - m_X]\}\,dt \\ &+ \frac{1}{T-\tau}\int_0^{T-\tau} M\{(m^*_X - m_X)^2\}\,dt. \end{aligned} \tag{1.328}$$

Consider integrands in Eq. (1.328). It is obvious that

$$M\{[X(t) - m_X][X(t+\tau) - m_X]\} = R_{XX}(\tau). \tag{1.329}$$

In terms of Eq. (1.295) we can write

$$m^*_X - m_X = \frac{1}{T}\int_0^{T} [X(t') - m_X]\,dt'. \tag{1.330}$$

Using Eq. (1.330), we can write

$$M\{[m_X^* - m_X][X(t+\tau) - m_X]\} = \frac{1}{T}\int_0^T R_{XX}(t+\tau-t')\,dt'; \qquad (1.331)$$

$$M\{[X(t) - m_X][m_X^* - m_X]\} = \frac{1}{T}\int_0^T M\{[X(t') - m_X][X(t) - m_X]\}\,dt'$$

$$= \frac{1}{T}\int_0^T R_{XX}(t-t')\,dt'; \qquad (1.332)$$

$$M\{(m_X^* - m_X)^2\} = \frac{1}{T^2}\int_0^T\int_0^T R_{XX}(t_2 - t_1)\,dt_1\,dt_2. \qquad (1.333)$$

Substituting Eqs. (1.329)–(1.333) in Eq. (1.328), we can write

$$M[R_{XX}^{**}(\tau)] = \frac{1}{T-\tau}\int_0^{T-\tau} R_{XX}(\tau)\,d\tau$$

$$- \frac{1}{(T-\tau)T}\int_0^{T-\tau}\int_0^{T-\tau}[R_{XX}(t-t'+\tau) + R_{XX}(t-t')]\,dt\,dt'$$

$$+ \frac{1}{(T-\tau)T^2}\int_0^{T-\tau} dt \int_0^T\int_0^T R_{XX}(t_2 - t_1)\,dt_1\,dt_2. \qquad (1.334)$$

Integrating with respect to the variable t, we can write

$$M[R_{XX}^{**}(\tau)] = R_{XX}(\tau) - \frac{1}{(T-\tau)T}\int_0^{T-\tau}\int_0^T[R_{XX}(t-t'+\tau) + R_{XX}(t-t')]\,dt\,dt'$$

$$+ \frac{1}{T^2}\int_0^T\int_0^T R_{XX}(t_2 - t_1)\,dt_1\,dt_2. \qquad (1.335)$$

According to Eqs. (1.313) and (1.317) we can write

$$\frac{1}{T^2}\int_0^T\int_0^T R_{XX}(t_2 - t_1)\,dt_1\,dt_2 = \frac{2}{T}\int_0^T \left(1 - \frac{\tau_1}{T}\right) R_{XX}(\tau_1)\,d\tau_1. \qquad (1.336)$$

Using Eqs. (1.335) and (1.336), it is not a great problem to show that

$$M[R_{XX}^{**}(\tau)] = R_{XX}(\tau) - \frac{1}{T}\int_0^T [R_{XX}(\tau_1 - \tau) + R_{XX}(\tau_1)]\,d\tau_1$$
$$- \frac{2}{T-\tau}\int_0^T \left(1 - \frac{\tau_1}{T}\right)[R_{XX}(\tau_1 - \tau) + R_{XX}(\tau_1)]\,d\tau_1$$
$$+ \frac{2}{T}\int_0^T \left(1 - \frac{\tau_1}{T}\right) R_{XX}(\tau_1)\,d\tau_1. \qquad (1.337)$$

Taking into account that

$$\int_\tau^T \left(1 - \frac{\tau_1}{T}\right)[R_{XX}(\tau_1 - \tau) + R_{XX}(\tau_1)]\,d\tau_1$$
$$= \int_0^T \left(1 - \frac{\tau_1}{T}\right)[R_{XX}(\tau_1 - \tau) + R_{XX}(\tau_1)]\,d\tau_1$$
$$- \int_0^\tau \left(1 - \frac{\tau_1}{T}\right)[R_{XX}(\tau_1 - \tau) + R_{XX}(\tau_1)]\,d\tau_1, \qquad (1.338)$$

we can write

$$M[R_{XX}^{**}(\tau)] = R_{XX}(\tau)$$
$$- \frac{2}{(T-\tau)T}\int_0^T \left(1 - \frac{\tau_1}{T}\right)[\tau R_{XX}(\tau_1) + T R_{XX}(\tau_1 - \tau)]\,d\tau_1$$
$$+ \frac{1}{(T-\tau)T}\int_0^\tau (T + \tau - 2\tau_1)[R_{XX} + (\tau_1) + R_{XX}(\tau_1)]\,d\tau_1. \qquad (1.339)$$

Consequently, when the mean m_X of the stationary stochastic process $X(t)$ is unknown then the mean $M[R_{XX}^{**}(\tau)]$ is not equal to the correlation function $R_{XX}(\tau)$ of this process. This means that under definition of the statistical correlation function $R_{XX}^{**}(\tau)$ determined by Eq. (1.299) there is the error determined in the following form:

$$
\begin{aligned}
M[R^{**}_{XX}(\tau)] &- R_{XX}(\tau)\\
&= -\frac{2}{(T-\tau)T}\int_0^T\left(1-\frac{\tau_1}{T}\right)[\tau R_{XX}(\tau_1) + T R_{XX}(\tau_1-\tau)]\,d\tau_1\\
&+\frac{1}{(T-\tau)T}\int_0^{\tau}(T+\tau-2\tau_1)[R_{XX}(\tau_1)+R_{XX}(\tau_1-\tau)]\,d\tau_1. \qquad (1.340)
\end{aligned}
$$

1.6.5.2 The Mean of the Statistical Variance

Using Eq. (1.339), we can define the mean of the statistical variance D^{**}_{XX}. Actually,

$$M[D^{**}_{XX}] = M[R^{**}_{XX}(0)]. \qquad (1.341)$$

Taking into account Eq. (1.339) and in terms of the equality

$$D_{XX} = R_{XX}(0), \qquad (1.342)$$

we can write

$$M[D^{**}_{XX}] = D_{XX} - \frac{2}{T}\int_0^T\left(1-\frac{\tau_1}{T}\right)R_{XX}(\tau_1)\,d\tau_1. \qquad (1.343)$$

Using Eq. (1.342), we can define the mean of the absolute error $(D^{**}_{XX} - D_{XX})$ and the relative error $(\frac{D^{**}_{XX}-D_{XX}}{D_{XX}})$ under definition of the statistical variance D^{**}_{XX} and under definition of the mean of the ratio $\frac{D^{**}_{XX}}{D_{XX}}$:

$$M[D^{**}_{XX} - D_{XX}] = -\frac{2}{T}\int_0^T\left(1-\frac{\tau_1}{T}\right)R_{XX}(\tau_1)\,d\tau_1; \qquad (1.344)$$

$$M\left[\frac{D^{**}_{XX}-D_{XX}}{D_{XX}}\right] = -\frac{2}{T}\int_0^T\left(1-\frac{\tau_1}{T}\right)r_{XX}(\tau_1)\,d\tau_1; \qquad (1.345)$$

$$M\left[\frac{D^{**}_{XX}}{D_{XX}}\right] = 1-\frac{2}{T}\int_0^T\left(1-\frac{\tau_1}{T}\right)r_{XX}(\tau_1)\,d\tau_1, \qquad (1.346)$$

where $r_{XX}(\tau_1)$ is the normalized correlation function of the stationary stochastic process $X(t)$.

1.7 Conclusions

This chapter focused on the main concepts and definitions in the theory of probability and statistics. The concept of probability was defined and the main properties of probability were discussed. The main characteristics of the probability distribution function and the probability distribution density for the continuous and discrete variables were also considered. The initial and central moments of the random variables were also characterized.

Special attention was paid to the stochastic processes. Five main forms of the stochastic processes were considered and defined: the discrete random sequence (the discrete stochastic process with a discrete time), the random sequence (the continuous stochastic process with a discrete time), the discrete (discontinuous) stochastic process (the discrete stochastic process with a continuous time), the continuous stochastic process, and the stochastic point process.

The probability distribution function and the probability distribution density of the stochastic process were discussed. The cumulant and the correlation functions of the stochastic process were defined. The stochastic processes were classified into the stationary stochastic processes in both a narrow and a broad sense. Properties and characteristics of the stochastic processes in a narrow and broad sense were studied. The ergodic and non-ergodic stationary stochastic processes were considered. Properties of ergodicity of the stationary stochastic processes with respect to the mean and the correlation function were discussed.

The correlation function and the spectral density of the stationary stochastic process were considered in additional detail. The main features of the correlation function and the spectral density were discussed. The concept of the effective spectrum bandwidth of the stationary stochastic process was defined.

Considerable attention was paid to the estimations or the statistical characteristics of the mean, the variance, and the correlation function of the random variables and the stationary stochastic processes. These estimations are very useful in practice. The statistical characteristics of the random variables and the stationary stochastic processes such as the statistical mean, the statistical variance, and the statistical correlation function were discussed for cases in which the mean of the random variables and the stationary stochastic processes are both known and unknown. Very important formulae for definition of the statistical mean, the statistical variance, and the statistical correlation function and their estimations, which are widely used in practice, were presented for the random variables and the stationary stochastic processes.

The main purpose of this chapter was to review briefly the main concepts and definitions essential to an understanding of the theory of probability and statistics. Such a review is essential because this book is based on the statistical decision-making rules applied to signal processing in the presence of additive Gaussian noise and multiplicative noise. Information provided in this chapter is very useful for a better understanding of the subsequent chapters.

References

1. Crammer, H., *Mathematical Methods of Statistics*, Princeton University Press, Princeton, NJ, 1946.
2. Van der Waerden, B., *Mathematische Statistik*, Springer-Verlag, Berlin, 1957.
3. Papoulis, A., *Probability, Random Variables, and Stochastic Processes*, McGraw-Hill, New York, 1965.
4. Box, G. and Tiao, G., *Bayesian Inference in Statistical Analysis*, Addison-Wesley, Cambridge, MA, 1973.
5. Pugachev, V., *Probability Theory and Mathematical Statistics*, Nauka, Moscow, 1979 (in Russian).
6. Lehmann, E., *Testing Statistical Hypotheses*, 2nd ed., Wiley, New York, 1986.
7. Lifshiz, N. and Pugachev, V., *Probability Analysis of Automatic Control Systems*, Soviet Radio, Moscow, Parts 1 and 2, 1963 (in Russian).
8. Ventzel, E. and Ovcharov, L., *Probability Theory*, Nauka, Moscow, 1973 (in Russian).
9. Jazwinski, A., *Stochastic Processes and Filtering Theory*, Academic Press, New York, 1970.
10. Wong, E., *Stochastic Processes in Information and Dynamical Systems*, McGraw-Hill, New York, 1971.
11. Tikhonov, Y., *Statistical Radio Engineering*, Radio and Svyaz, Moscow, 1982 (in Russian).
12. Andersson, T., *An Introduction to Multivariate Statistical Analysis*, 2nd ed., Wiley, New York, 1984.

2

Classical and Modern Approaches to Signal Detection Theory

This chapter is devoted to a brief analysis of classical and modern signal detection theories. There are two avenues in classical signal detection theory, the Gaussian and Markov approaches, each of which differs in the investigative technique employed.

Modern signal detection theory is only concerned with specific problems of signal detection that are more closely parallel to real practice. The primary avenues of investigation in modern signal detection theory are:[1–45]

- Solution for signal detection problems under conditions of *a priori* parametric and non-parametric uncertainty
- Further use of sequential analysis in signal detection problems; extension of the main results of sequential analysis in signal detection problems with *a priori* uncertainty; the development of more precise techniques to compute the characteristics of sequential procedures
- Solution for signal detection problems for the non-Gaussian signals, clutter, and noise
- Solution for the problems of statistical design, construction, and study of the unified signal detection-measurement and control algorithm

2.1 Gaussian Approach

The Gaussian approach to signal detection in the presence of noise is based on the assumption that the signals, interference, and noise being studied are appropriately characterized by Gaussian processes with known statistical characteristics.

2.1.1 Gaussian Processes

The real stochastic process $\xi(t)$ is called the Gaussian process if for any limited set of the time instants $t_1, t_2, \ldots, t_n$ the random variables $\xi_1 = \xi(t_1), \ldots, \xi_n = \xi(t_n)$ have the joint normal probability distribution density determined in the following form:

$$f_n(x_1, \ldots, x_n; t_1, \ldots, t_n) = \frac{1}{\sqrt{(2\pi)^n |\vec{\mathbf{R}}|}} \exp\left[-\frac{1}{2}(\vec{\mathbf{x}} - \vec{\mathbf{m}})^{\mathrm{T}} \vec{\mathbf{R}}^{-1} (\vec{\mathbf{x}} - \vec{\mathbf{m}})\right] \quad (2.1)$$

and the characteristic function that is equal to

$$\Theta_n(j\vartheta_1, \ldots, j\vartheta_n; t_1, \ldots, t_n) = \exp\left(j\vec{\mathbf{m}}^{\mathrm{T}}\vec{\vartheta} - \frac{1}{2}\vec{\vartheta}^{\mathrm{T}}\vec{\mathbf{R}}\vec{\vartheta}\right), \quad (2.2)$$

where $\vec{\mathbf{R}}^{-1}$ is the reciprocal matrix with respect to the matrix $\vec{\mathbf{R}}$; the symbol $\mathbf{T}$ denotes the transpose;

$$\vec{\vartheta}^{\mathrm{T}} = [\vartheta_1, \vartheta_2, \ldots, \vartheta_n] \quad (2.3)$$

is the row vector; the matrix $\vec{\mathbf{R}}$ is determined in the following form:

$$\vec{\mathbf{R}} = \left\| \begin{matrix} R_{11} & R_{12} & \cdots & R_{1n} \\ R_{21} & R_{22} & \cdots & R_{2n} \\ \cdots & \cdots & \cdots & \cdots \\ R_{n1} & R_{n2} & \cdots & R_{nn} \end{matrix} \right\|; \quad (2.4)$$

and

$$|\vec{\mathbf{R}}| = \det \vec{\mathbf{R}} \quad (2.5)$$

is the determinant of the correlation matrix.

For the case in which we do not use the vector form, Eqs. (2.1) and (2.2) can be written in the following form:

$$f_n(x_1, \ldots, x_n; t_1, \ldots, t_n) = \frac{1}{\sqrt{(2\pi)^n |\vec{\mathbf{R}}|}} \exp\left[-\frac{1}{2|\vec{\mathbf{R}}|} \sum_{i,j=1}^{n} A_{ij}(x_i - m_i)(x_j - m_j)\right]; \quad (2.6)$$

$$\Theta_n(j\vartheta_1, \ldots, j\vartheta_n; t_1, \ldots, t_n) = \exp\left(j\sum_{i=1}^{n} m_i\vartheta_i - \frac{1}{2}\sum_{i,j=1}^{n} R_{ij}\vartheta_i\vartheta_j\right), \quad (2.7)$$

where A_{ij} is the cofactor of the element R_{ij} of the correlation matrix given by Eq. (2.4).

The mean, variance, and correlation function of the Gaussian process are determined in the following form:

$$\begin{cases} m_i = m_\xi(t_i) = M[\xi(t_i)]; \\ \sigma_i^2 = D_i = D_\xi(t_i) = M\{[\xi(t_i) - m_\xi(t_i)]^2\}; \\ R_{ij} = R_{ji} = R_\xi(t_i, t_j) = M\{[\xi(t_i) - m_\xi(t_i)][\xi(t_j) - m_\xi(t_j)]\} \\ \quad = \sqrt{D_\xi(t_i)D_\xi(t_j)}\, r_\xi(t_i, t_j), \qquad i, j = \overline{1, n}. \end{cases} \quad (2.8)$$

Let us recall the main properties of the Gaussian processes. We list the main properties below, in the absence of proofs:

- The Gaussian process is exhaustively defined by the mean $m_\xi(t)$ and the correlation function $R_\xi(t_1, t_2)$.
- If two values of the Gaussian process are uncorrelated then these values are independent of each other.
- The Gaussian process is the stationary stochastic process both in a narrow sense and in a broad sense. When defined as such, these concepts coincide for the Gaussian process.
- The conditional probability distribution density of the Gaussian process is normal.
- When the Gaussian process $\xi(t)$ acts at the input of a linear system with the pulse response $h(t, \tau)$, then the process $\eta(t)$ at the output of that same linear system is the Gaussian process, for example,

$$\eta(t) = \int_0^T h(t, \tau)\xi(\tau)\, d\tau, \tag{2.9}$$

  where $\eta(t)$ is the Gaussian process.
- If the Gaussian process $\xi(t)$ acts at the input of a non-linear system, the process $\eta(t)$ forming at the output of that same non-linear system is the non-Gaussian process. For example, if

$$\eta(t) = f[t, \xi(t)], \tag{2.10}$$

  where $f[t, \xi(t)]$ is the non-linear transformation with respect to the Gaussian process $\xi(t)$, and the process $\eta(t)$ is the non-Gaussian process.
- Under linear transformation the correlated values of the Gaussian process can be transformed to the uncorrelated values of the Gaussian process.
- Optimal estimation of values of the Gaussian process by criterion of the minimal mean square error is the linear estimation.
- The Gaussian process with the rational spectral density is simultaneously the Markov process.

2.1.2 Karhunen–Loeve Series Expansion Method

The Gaussian approach to signal detection theory was based on the investigations carried out by F. Lehan, R. Parks, D. Youla, J. Thomas, E. Wong, I. Bolshakow, V. Repin, H. Van Trees, and others.[46–52] Two methods are widely used for definition of the optimal signal detection algorithm in the Gaussian approach: the method based on the Karhunen–Loeve series expansion[53,54] and the variational method.[51,52,55] Consider briefly these methods under the following initial conditions.

The input stochastic process presented in the following form

$$\vec{X}(t) = \vec{m}[t, \vec{a}(t)] + \vec{n}(t) \tag{2.11}$$

acts at the input of a signal processing system, where $\vec{m}[t, \vec{a}(t)]$ is the known function; $\vec{a}(t)$ is the received signal obeying the Gaussian law; and $\vec{n}(t)$ is the additive Gaussian noise. Here $\vec{X}(t)$, $\vec{m}[t, \vec{a}(t)]$, and $\vec{n}(t)$ are the column vectors.

A full set of correlation functions can be represented in the following form:

$$R(t, \tau) = \begin{Vmatrix} R_{aa}(t, \tau) & R_{an}(t, \tau) \\ R_{na}(t, \tau) & R_{nn}(t, \tau) \end{Vmatrix}. \tag{2.12}$$

Under these conditions it is required to define the optimal algorithm of signal detection in the presence of additive Gaussian noise. The criterion of the maximal *a posteriori* probability distribution density for the vector $\vec{a}(t)$ is taken as a criterion of optimality.[47–49,52]

Suppose that the input stochastic process $X(t)$ is analyzed within the limits of the time interval $[0, T]$ and takes the form

$$X(t) = m[t, a(t)] + n(t), \tag{2.13}$$

where $m[t, a(t)]$ is the known function of two arguments; $a(t)$ is the Gaussian signal; and $n(t)$ is the additive Gaussian noise. As is well known,[53,54] the continuous Gaussian processes $a(t)$ and $n(t)$ can be represented by the Karhunen–Loeve series expansion within the limits of the time interval $[0, T]$:

$$a(t) = \sum_{i=1}^{\infty} \frac{A_i}{\sqrt{\lambda_i}} \Xi_i(t), \tag{2.14}$$

$$n(t) = \sum_{i=1}^{\infty} \frac{N_i}{\sqrt{\mu_i}} \Upsilon_i(t), \tag{2.15}$$

where $\Xi_i(t)$ and $\Upsilon_i(t)$ are the orthogonal eigenfunctions defined by the correlation functions $R_a(t, \tau)$ and $R_n(t, \tau)$, respectively; λ_i and μ_i are the eigenvalues.

Assume that the random variables A_i and N_i in Eqs. (2.14) and (2.15) are the independent Gaussian variables with zero mean and the unit value variance for simplicity. The essence of the considered method is reduced to a definition of the optimal estimation of the random parameters A_i of the Gaussian process $a(t)$, since the processes $a(t)$ and $n(t)$ are defined by a set of the independent random parameters A_i and N_i under the use of the Karhunen–Loeve series expansion. For definition of optimal estimation of the random parameters A_i there is a need to know the *a posteriori* probability distribution density $f_N(A_i \mid X_i)$, where N is the sample size.

Knowledge of maximal extreme peaks of the probability distribution density $f_N(A_i \mid X_i)$ allows us to define a structure of the optimal detector. Finally we obtain the following integral equation, and the optimal estimation of the signal (random parameters A_i) must satisfy this equation

$$\widehat{a}(t) = \int_0^T R_a(t,\tau) \cdot \frac{\partial m[\tau, \widehat{a}(\tau)]}{\partial a} \cdot g(\tau)\, d\tau, \tag{2.16}$$

where

$$\widehat{a}(t) = \sum_{\ell=1}^{\infty} \frac{\widehat{A}_\ell \Xi_\ell(t)}{\sqrt{\lambda_\ell}} \tag{2.17}$$

and

$$g(\tau) = \int_0^T Q_n(\tau, y)\{X(y) - m[y, \widehat{a}(y)]\}\, dy. \tag{2.18}$$

The function $Q_n(\tau, y)$, like the function $Q_a(\tau, y)$, is determined in the following form:

$$\begin{cases} \int_0^T R_a(t,y) Q_a(y,\tau)\, dy = \delta(t-\tau); \\ \\ \int_0^T R_n(t,y) Q_n(y,\tau)\, dy = \delta(t-\tau), \end{cases} \tag{2.19}$$

where $\delta(t)$ is the delta function. One can find more detailed information in References 53–55. For the realizable optimal detector in the case when $n(t)$ is the additive white Gaussian noise with the spectral density $\frac{N_0}{2}$ the function $Q_n(t, \tau)$ has the following physical sense:

$$Q_n(t,\tau) = \frac{2}{N_0}\delta(t-\tau). \tag{2.20}$$

2.1.3 Variational Method

In the variational method for definition of optimal estimation of the signal $a(t)$ (random parameters A_i) the following presentation of the signal $a(t)$ is used:

$$a(t) = \widehat{a}(t) + \epsilon\varrho(t), \tag{2.21}$$

where ϵ is the variable parameter and $\varrho(t)$ is the arbitrary function. In this case the probability distribution density $f[a(t) \mid X(t)]$ is the continuous function of ϵ and possesses the maximum at $\epsilon = 0$ by the initial conditions. Differentiating the probability distribution density $f[a(t) \mid X(t)]$ with respect to the parameter ϵ under the condition given in Eq. (2.21) and setting the end result equal to zero, we can obtain the optimal estimation $\widehat{a}(t)$ defined by Eq. (2.16). We can see that the end result is the same as in Section 2.1.2.

2.2 Markov Approach

The Markov approach to signal detection in the presence of additive Gaussian noise is based on the assumption that the signals, interference, and noise to be analyzed are appropriately characterized by the Markov processes.

2.2.1 Markov Processes

The Markov process is called the stochastic process, for which under the fixed value $\xi(u)$ the random variable $\xi(t), t > u$ does not depend on the parameter $\xi(s), s < u$. There are four main forms of the Markov processes:[56,57]

- The Markov chain is the Markov process, for which the range of values and domain of definition are the discrete sets.
- The Markov sequence is the Markov process, for which the range of values is a continuous set and the domain of definition is a discrete set.
- The discrete Markov process is the Markov process, for which the range of values is a discrete set and the domain of definition is a continuous set.
- The continuous Markov process is the Markov process, for which both the range of values and the domain of definition are continuous sets.

The main peculiarity for all forms of the Markov processes is the following. The stochastic process $\xi(t)$ is the Markov process if for any n time instants $t_1 < t_2 < \cdots < t_n$ within the limits of the time interval $[0, T]$ the conditional probability distribution function for the value $\xi(t_n)$ under the fixed values $\xi(t_1), \xi(t_2), \ldots, \xi(t_{n-1})$ depends only on the value $\xi(t_{n-1})$, i.e., under the given values $\xi_1, \xi_2, \ldots, \xi_n$ the following equality

$$P[\xi(t_n) \leq \xi_n \mid \xi(t_1) = \xi_1, \ldots, \xi(t_{n-1}) = \xi_{n-1}] = P[\xi(t_n) \leq \xi_n \mid \xi(t_{n-1}) = \xi_{n-1}] \tag{2.22}$$

is true.

For three time instants $t_i > t_j > t_k$ Eq. (2.22) takes the following form:

$$P[\xi(t_i) \leq \xi_i \mid \xi(t_k) = \xi_k, \ \xi(t_j) = \xi_j] = P[\xi(t_i) \leq \xi_i \mid \xi(t_j) = \xi_j]. \tag{2.23}$$

For this reason, every so often we can say that the characteristic peculiarity of the Markov processes is the following: if the Markov process is completely known at the instant t_j then the Markov process at the instant t_i does not depend on the Markov process at the instant t_k.

The following relationship that is symmetric with respect to time

$$\begin{aligned} &P[\xi(t_i) \leq \xi_i, \xi(t_k) \leq \xi_k \mid \xi(t_j) = \xi_j] \\ &\quad = P[\xi(t_i) \leq \xi_i \mid \xi(t_j) = \xi_j] P[\xi(t_k) \leq \xi_k \mid \xi(t_j) = \xi_j] \end{aligned} \tag{2.24}$$

can be used to define the Markov process. This written form means that under the fixed Markov process at the present instant t_j the Markov processes at the instants t_i and t_k are independent. From the above-mentioned statements it follows that the n-order probability distribution density of the Markov process can be determined in the following form:

$$f(\xi_1, \xi_2, \ldots, \xi_n) = f(\xi_1) \prod_{i=1}^{n-1} f(\xi_{i+1} \mid \xi_i). \tag{2.25}$$

There is one more general and very important property of the continuous Markov process. Consider an evolution of the probability

$$P[\xi(t) \leq \xi \mid \xi(t_0) = \xi_0], \tag{2.26}$$

which is determined in the following form:

$$\frac{d}{dt} = \mathcal{L}P, \tag{2.27}$$

where $\mathcal{L}$ is the linear operator (the matrix, differential operator, and so on). This peculiarity of the continuous Markov process allows us to study statistical characteristics using differential equations. The character of initial conditions and boundary conditions for Eq. (2.27) can take various forms and are defined by solving physical problems.

2.2.2 Basic Principles

The basic research results concerning the Markov approach can be found in References 58–64. Let the vector stochastic process be observed at the input of a signal processing system

$$\vec{\mathcal{X}}(t) = \vec{h}(t, \vec{X}) + \vec{n}(t), \tag{2.28}$$

where $\vec{n}(t)$ is the $(p \times 1)$-order column vector, the components of which are the additive Gaussian noise with zero mean and the correlation matrix

$$\overline{\vec{n}(t)\vec{n}^T(t)} = \mathcal{N}\delta(t - \tau), \tag{2.29}$$

where $\mathcal{N}$ is the $(p \times p)$-order positive defined matrix; the sign T implies the conjugation; the line over terms implies an averaging. The function $\vec{h}(t, \vec{X})$ is the known p-order non-linear function of the time t and vector $\vec{X}(t)$. The components of the vector $\vec{X}(t)$ must be determined and estimated at the output of the receiver of the signal processing system. The vector $\vec{X}(t)$ is the m-order Markov process, which can be written as a function of time in the following form:

$$\frac{d\vec{X}(t)}{dt} = \vec{F}(t, \vec{X}) + G_{\vec{\mathcal{X}}}(t), \tag{2.30}$$

under the initial condition

$$\vec{X}(t = 0) = \vec{X}_0, \tag{2.31}$$

where $\vec{F}(t, \vec{X})$ is the known m-order non-linear vector function in a general case; G is the known $(m \times m)$-order matrix; $\vec{\mathcal{X}}(t)$ is the m-order Gaussian process with zero mean and the correlation matrix

$$\overline{\vec{\mathcal{X}}(t)\vec{\mathcal{X}}^T(t)} = X\delta(t - \tau), \tag{2.32}$$

where X is the $(m \times m)$-order positive defined matrix. Note in the case when $\vec{F}(t, \vec{X})$ is the linear function, then $\vec{X}(t)$ is the Gaussian process.

As is well known[58,59] the multivariate *a priori* probability distribution density of components of the vector $\vec{X}(t)$ obeys the Fokker–Planck equation. All information regarding the process $\vec{X}(t)$ that can be obtained when the input stochastic process $\vec{\mathcal{X}}(t)$ is observed within the limits of the time interval $[0, T]$ is enclosed in the *a posteriori* probability distribution density $f[\vec{X}, t \mid \vec{\mathcal{X}}(t)]$ of components of the vector $\vec{X}(t)$ under the condition that the process $\vec{\mathcal{X}}(t)$ can be determined within the limits of the time interval $[0, T]$.

It is well known[65–67] that the optimal estimation of the vector $\vec{X}(t)$ by the criterion of the minimal mean square error is determined in the following form:

$$\vec{X}^*(t) = \int_{-\infty}^{\infty} \cdots \int_{-\infty}^{\infty} \vec{X} f[\vec{X}, t \mid \vec{\mathcal{X}}(t)] \prod_{\ell=1}^{m} d X_\ell. \tag{2.33}$$

Using results in References 58–60 and 63, we can see that definition of optimal estimation of the vector $\vec{X}(t)$ reduces to equations of optimal linear filtering of the Gaussian–Markov processes.[68]

When the signal-to-noise ratio at the input of the receiver is high, and the accuracy of the estimation $\vec{X}^*(t)$ (see Eq. (2.33)) is very high, the problem of non-linear filtering may be approximately reduced to the problem of linear filtering using the approximation of the non-linear functions $\vec{F}(t, \vec{X})$ and $\vec{h}(t, \vec{X})$ by the linear functions using the Taylor series expansion of the linear functions in the neighborhood of the estimation $\vec{X}^*(t)$.

The Markov approach may be applied to detection of the signals with the amplitude, phase, and frequency modulation laws. The use of the Markov approach for the detection of signals using the amplitude modulation law allows us to construct the optimal linear filter for the predetermined modulation law. In doing so, we match the filter responses with the signal parameters as a rule. It should be pointed out that, in contrast to the Gaussian approach, the Markov approach allows us to construct optimal detectors that are realizable in practice.

2.3 Bayes' Decision-Making Rule

Let the input stochastic process $X(t)$ be represented as both the noise $n(t)$ (the hypothesis H_0) and the arbitrary combination of the signal $a(t)$ and noise $n(t)$ (the hypothesis H_1):

$$X(t) = \begin{cases} a(t) \otimes n(t) & \Rightarrow \quad H_1; \\ n(t) & \Rightarrow \quad H_0, \end{cases} \tag{2.34}$$

where the symbol $\otimes$ signifies the arbitrary combination of the signal and noise. We must define a signal processing algorithm of the input stochastic process $X(t)$ and characteristics of the signal processing algorithm and, using the defined signal processing algorithm, make a decision: a "yes" or a "no" signal in the input stochastic process $X(t)$. The definition of the signal detection algorithm reduces to applying the decision-making rule to the observed input data $X(t)$ and deciding in favor of either of two hypotheses: H_0 or H_1. Bayes' risk, the probability of false alarm, and the probability of signal omission are the functions of the initial parameters of the signal and noise and can be used as characteristics of the defined signal detection algorithm, depending on the selection of the optimal criterion.

Bayes' criterion is used given *a priori* total information about the signal and noise. Bayes' optimal decision-making rule or the optimal partition of the n-order sample vector $\vec{X}(t)$ on two disjoint regions $\vec{X}_0(t)$ and $\vec{X}_1(t)$ is based on the minimization of the average risk[2,5,16,30,35,44,45,69–71]

$$\Re = \sum_{i,j=0}^{N} P_i \Pi_{ij} \int_{\vec{X}_j} f_N(\vec{X} \mid H_i)\, d\vec{X}. \tag{2.35}$$

Here $\|\Pi_{ij}\|$ $(i, j = 0, 1)$ is the loss matrix;

$$P_1 = 1 - P_0 \tag{2.36}$$

is the *a priori* probability of a "yes" signal in the input stochastic process; P_0 is the *a priori* probability of a "no" signal in the input stochastic process; $f_N(\vec{X} \mid H_i)$ is the likelihood function (the conditional probability distribution density) of the sample

$$\vec{X} = (X_1, X_2, \ldots, X_N) \tag{2.37}$$

under the assumption that the hypothesis H_1 is true, and N is the sample size.

Minimizing Eq. (2.35),[72] we can obtain the structure of the Bayes' detector, but we must use a wide range of *a priori* data. The loss matrix, *a priori* probabilities of a "yes" and a "no" signal in the input stochastic process, the model signal and noise, and techniques of combining the model signal and noise defining the likelihood function must be predetermined for this purpose. Because of this, criteria differing from the Bayes' criterion are used in solving problems of signal detection in the presence of additive Gaussian noise. So, when *a priori* probabilities of a "yes" or a "no" signal in the input stochastic process are unknown, the minimax criterion can be used. The minimax signal detection algorithm is a particular case of the Bayes' signal detection algorithm under the least preferential values of the probabilities P_0^* and P_1^*, in which the averaged Bayes' (minimal) risk is determined in the following form:

$$\Re(P_0^*, P_1^*) \geq \Re(P_0, P_1) \qquad \forall P_0 + P_1 = 1. \tag{2.38}$$

The sign $\forall$ signifies "under all."

When *a priori* probabilities P_0 and P_1 are known and the loss matrix is unknown, the criterion of maximum *a posteriori* probability may be used. In accordance with this criterion, the decision is made in favor of the hypothesis that has the maximum *a posteriori* probability:

$$P\{\vec{X} \mid H_i\} = \frac{P_1 f_N(\vec{X} \mid H_i)}{P_0 f_N(\vec{X} \mid H_0) + P_1 f_N(\vec{X} \mid H_1)}, \qquad i = 0, 1. \tag{2.39}$$

When *a priori* probabilities of a "yes" or a "no" signal in the input stochastic process and the loss matrix are unknown, the criterion of the maximal likelihood based on a comparison of the likelihood functions $f_N(\vec{X} \mid H_1)$ and $f_N(\vec{X} \mid H_0)$ is often used.

Apart from the decision-making criteria mentioned above, the Neyman–Pearson criterion is used widely.[1,2,5,44,45,71] For the Neyman–Pearson criterion the probability of false alarm,

$$P_F = \int_{\vec{X}_1} f_N(\vec{X} \mid H_0)\, d\vec{X}, \tag{2.40}$$

is fixed, and the probability of signal omission,

$$P_M = \int_{\vec{X}_0} f_N(\vec{X} \mid H_1)\, d\vec{X} = 1 - \int_{\vec{X}_1} f(\vec{X} \mid H_1)\, d\vec{X} = 1 - P_D, \tag{2.41}$$

is minimized.

The Neyman–Pearson criterion, as well as the criterion of the maximal likelihood ratio, does not require a knowledge of the *a priori* probabilities of a "yes" and a "no" signal in the input stochastic process, as well as a knowledge of the loss matrix. Note that the design and construction of the optimal detector of signals in the presence of additive Gaussian noise by any one of the above-mentioned criteria require the availability of *a priori* data for construction of the likelihood functions $f_N(\vec{X} \mid H_1)$ and $f_N(\vec{X} \mid H_0)$.

We suggest that the deterministic signal $a(t)$ in the presence of additive Gaussian noise $n(t)$ must be detected using the observed sample determined by Eq. (2.37). Then the signal detection algorithm reduces to a comparison of the likelihood ratio

$$\ell(\vec{X}) = \frac{f_N(\vec{X} \mid H_1)}{f_N(\vec{X} \mid H_0)} \tag{2.42}$$

with the threshold K.

The threshold K is determined by the chosen criterion of optimality of the signal detection in the presence of additive Gaussian noise. When Bayes' criterion is used, the threshold is determined in the following form:

$$K = \frac{P_0(\Pi_{10} - \Pi_{00})}{P_1(\Pi_{01} - \Pi_{11})}. \tag{2.43}$$

For the Neyman–Pearson criterion

$$K = K(P_F), \tag{2.44}$$

where $K(P_F)$ is chosen given the condition of providing the required probability of false alarm P_F (see Eq. (2.40)).

If the continuous input stochastic process $X(t)$ (but not the sample $\vec{X}$) is used for signal detection, then the likelihood ratio functional

$$\ell[X(t)] = \lim_{N \to \infty} \ell(X_1, X_2, \ldots, X_N) \tag{2.45}$$

is compared with the threshold.[10,11,15,73–76] However, the detection problem of the deterministic signal in the presence of additive Gaussian noise, all statistical parameters of which are known *a priori*, is rarely encountered in nature or practice.

In practice, signal detection in the presence of additive Gaussian noise requires solving the signal detection problems under conditions of *a priori* uncertainty. Since *a priori* uncertainty relative to the signal and the additive Gaussian noise may vary in kind, methods of overcoming *a priori* uncertainty in signal detection problems in the presence of additive Gaussian noise must take on various approaches as well.

2.4 Unbiased and Invariant Decision-Making Rules

2.4.1 Unbiased Rule

Detection of signals under conditions of *a priori* parametric uncertainty involves the testing of complex statistical hypotheses with respect to the probability distribution density of the observed process $\vec{X}(t)$. The hypothesis H_0—a "no" signal in the input stochastic process—indicates that this probability distribution density belongs to the class

$$\mathcal{P}_0 = \{f(\vec{X} \mid \vec{\vartheta}); \vec{\vartheta} \in \Theta_0\}. \tag{2.46}$$

The hypothesis H_1—a "yes" signal in the input stochastic process—indicates that the process $\vec{X}(t)$ obeys the probability distribution density in the class

$$\mathcal{P}_1 = \{f(\vec{X} \mid \vec{\vartheta}); \vec{\vartheta} \in \Theta_1\}. \tag{2.47}$$

Hereafter $f(\vec{X} \mid \vec{\vartheta})$ is the probability distribution density of samples of the observed input stochastic process $\vec{X}(t)$; $\vec{\vartheta}$ is the multivariate parameter of the probability distribution density in a general case; Θ_0 and Θ_1 are the disjoint sets of the parametric space Θ,

$$\Theta = \Theta_0 \bigcup \Theta_1. \tag{2.48}$$

Hypotheses testing rules are the decision functions $\varphi(\vec{X})$, which preassign the decision-making rule favoring one of two hypotheses concerning the input stochastic process $\vec{X}(t)$.[28,31,71] Non-randomized decision-making rules

$$\varphi(\vec{X}) = 1 \quad \text{at } \vec{X} \in \vec{X}_1 \tag{2.49}$$

and

$$\varphi(\vec{X}) = 0 \quad \text{at } \vec{X} \in \vec{X}_0 \tag{2.50}$$

are usually used in practice, where $\vec{X}_1$ and $\vec{X}_0$ are the disjoint sets of the space $\vec{X}$ of realizations of the observed input stochastic process

$$\vec{X} = \vec{X}_1 \bigcup \vec{X}_0. \tag{2.51}$$

The decision-making rule favoring the hypothesis H_1 is made if

$$\varphi(\vec{X}) = 1, \tag{2.52}$$

and the decision-making rule favoring the hypothesis H_0 is made if

$$\varphi(\vec{X}) = 0. \tag{2.53}$$

In the case of the randomized decision-making rule, the function $\varphi(\vec{X})$ defines the probability of the decision-making rule favoring the hypothesis H_1 with regard to the input stochastic process $\vec{X}(t)$. The probability of the decision-making rule favoring the hypothesis H_0 is equal to $1 - \varphi(\vec{X})$. Probability performance of the hypothesis testing rules is the power function

$$\beta(\varphi \mid \vec{\vartheta}) = M[\varphi \mid \vec{\vartheta}], \tag{2.54}$$

where $M[. \mid \vec{\vartheta}]$ is the mean of distribution of the observed process having the probability distribution density $f(\vec{X} \mid \vec{\vartheta})$. The power function is equal to the probability of false alarm P_F under the condition $\Theta \in \Theta_0$ and the probability of detection P_D under the condition $\Theta \in \Theta_1$.

The decision-making rule under the conditions of *a priori* uncertainty must satisfy the following requirements. First, this decision-making rule must be structurally stable. In other words, the decision function $\varphi(\vec{X})$ of the decision-making rule must not depend on unknown parameters of the probability distribution density of the observed input stochastic process $X(t)$. Second, the losses in detection effectiveness must be minimal in comparison with the losses of the decision-making rules, which are optimal under *a priori* total information; and the detection performance must be stable with respect to the variation of *a priori* indeterminate parameters in the signal detection problem.

The uniformly most powerful tests satisfy the listed requirements most efficiently. The characteristic peculiarity of the uniformly most powerful test is that this test ensures the maximal probability of detection P_D under all $\vec{\vartheta} \in \Theta_1$ given the same decision function and the predetermined probability of false alarm P_F on the set Θ_0. Stability of the detection performances of the uniformly most powerful test is ensured by the predetermined probability of false alarm P_F under all variations of the parameter $\vec{\vartheta} \in \Theta_0$, and minimal losses in effectiveness are ensured by maximization of the probability of detection P_D of the signal under all $\vec{\vartheta} \in \Theta_1$.

However, the uniformly most powerful tests are very rare in practice. Techniques based on the principles of unbiasedness, similarity, and invariance[34,42,71] are used effectively to solve problems of signal detection in the presence of additive Gaussian noise in practice. Consider the principles of unbiasedness and similarity. The class $\mathcal{F}_{US}$ of the decision-making rules is generated by synthesis of the optimal decision-making rule on the basis of the principle of unbiasedness. The power functions of these decision-making rules satisfy the conditions of unbiasedness:

$$\begin{cases} \beta(\varphi \mid \vec{\vartheta}) \leq P_F & \forall \vec{\vartheta} \in \Theta_0; \\ \\ \beta(\varphi \mid \vec{\vartheta}) \geq P_F & \forall \vec{\vartheta} \in \Theta_1. \end{cases} \tag{2.55}$$

The first condition in Eq. (2.55) ensures, as in the case of the uniformly most powerful test, the predetermined probability of false alarm P_F. The second condition in Eq. (2.55) increases the stability of the decision-making rule in the sense that the values of the probability of detection P_D, which are less than that of the probability of false alarm P_F, are eliminated. The decision-making rules belonging to the class $\mathcal{F}_{US}$ are called the unbiased decision-making rules. Note that the uniformly most powerful test belonging to the class of all unbiased decision-making rules is unbiased.[71,77,78] Because of this, transfer into the class $\mathcal{F}_{US}$ does not eliminate the uniformly most powerful test if the uniformly most powerful test exists. The optimal decision-making rule in the Neyman–Pearson criterion sense, for which the probability of detection P_D is maximal for each $\vec{\vartheta} \in \Theta_1$ from the class $\mathcal{F}_{US}$, is called the unbiased uniformly most powerful test.

The principle of unbiasedness allows us to construct the uniformly most powerful test if the parameter $\vec{\vartheta}$ is classified into the useful $\tilde{\gamma}$ and nuisance $\tilde{\pi}$ parameters. The parameter defining a fulfillment of the hypothesis H_0 or the alternative H_1 is called the useful parameter $\tilde{\gamma}$. The parameter whose modification does not influence on the fulfillment of any testing hypotheses is called the nuisance parameter $\tilde{\pi}$. As a rule, when solving signal detection problems in the presence of additive Gaussian noise, the parameters of the probability distribution density of the additive Gaussian noise are the nuisance parameters and the parameters depending on the energy characteristics of signals are the useful parameters.

For example, in detection of the deterministic signals in the presence of additive Gaussian noise the useful parameter is defined as

$$\tilde{\gamma} = \frac{S}{\sigma_n^2}, \tag{2.56}$$

where $S(t)$ is the amplitude of the signal, σ_n^2 is the variance of the additive Gaussian noise, and the nuisance parameter is defined as

$$\tilde{\pi} = \frac{1}{2\sigma_n^2}. \tag{2.57}$$

The sets Θ_0 and Θ_1, given the useful and nuisance parameters, are expressed by

$$\Theta_0 = \Gamma_0 \times \Pi \tag{2.58}$$

and

$$\Theta_1 = \Gamma_1 \times \Pi, \tag{2.59}$$

where Γ_0 and Γ_1 are the ranges of the useful parameter $\tilde{\gamma}$ at the hypotheses H_0 and H_1, respectively, and Π is the range of the nuisance parameter $\tilde{\pi}$.

The power function $\beta(\varphi \mid \vec{\vartheta})$ is the continuous function with respect to the parameter $\vec{\vartheta}$. Given this, the unbiased decision-making rules for this function are equal to the constant value of the probability of false alarm P_F at the boundary Δ of the sets Θ_0 and Θ_1. As a consequence, the class $\mathcal{F}_{SS}$ of the decision-making rules is generated by synthesis of the unbiased uniformly most powerful test. The class $\mathcal{F}_{SS}$ must satisfy the following condition:

$$\beta(\varphi \mid \vec{\vartheta}) = P_F \qquad \text{under all} \qquad \vec{\vartheta} \in \Delta. \tag{2.60}$$

Thereafter the uniformly most powerful test for the probability of false alarm P_F is found in the class $\mathcal{F}_{SS}$ and it is verified that this uniformly most powerful test is unbiased. The condition in Eq. (2.60) is called the condition of similarity of the decision-making rules in the set Δ. The decision-making rule satisfying the condition in Eq. (2.60) is called the similar decision-making rule in the set Δ. The uniformly most powerful test in the class $\mathcal{F}_{SS}$ is called the similar uniformly most powerful test.

The lemma, according to which the unbiased and similar uniformly most powerful tests at the same probability of false alarm P_F coincide if their power functions are continuous with respect to the parameter $\vec{\vartheta}$ and if the similarity set is the boundary of the sets Θ_0 and Θ_1, is the basis for transferring into the class of similar decision-making rules generated by synthesis of the unbiased uniformly most powerful test.[71,79] The problem of synthesis is essentially simplified when transferring similar decision-making rules if the nuisance parameter possesses the sufficient statistic. The use of the sufficient statistic of the nuisance parameter allows us to exclude the unknown nuisance parameter from consideration regarding synthesis of the unbiased uniformly most powerful test.[71,80]

Consider the detection problem of the signal $\sqrt{E_a}a(t)$ in the presence of additive Gaussian noise. The input stochastic process $X(t), t \in [0, T]$ is observed at the output of the preliminary filter (PF) with the Π-amplitude-frequency response and bandwidth ΔF. The PF is the linear section of the receiver or detector. The noise at the input of the PF is considered as the additive white Gaussian noise with the correlation function

$$\frac{N_0}{2}\delta(t_2 - t_1). \tag{2.61}$$

The signal $a(t)$ is the completely known signal with the unit energy. Assume that the energy

$$E_a \in [0, \infty) \tag{2.62}$$

of the received signal and the spectral power density

$$\frac{N_0}{2} \in [0, \infty) \tag{2.63}$$

of the additive white Gaussian noise are *a priori* indeterminate parameters. The coefficient of signal transmission of the PF is equal to 1.

Generate the sample

$$\dot{\vec{X}} = (\dot{X}_1, \dot{X}_2, \ldots, \dot{X}_N) \tag{2.64}$$

using the imaginary envelope $\dot{X}(t)$ of the observed input stochastic process at the instant $t_i = \frac{i}{\Delta F}$. The probability distribution density of the considered sample takes the following form

$$f\left(\dot{\vec{X}} \,|\, E_a; \frac{N_0}{2}\right) = k \exp\left\{-\frac{1}{2N_0\Delta F}\|\dot{\vec{X}} - \sqrt{E_a}\dot{\vec{a}}\|^2\right\}, \tag{2.65}$$

where

$$k = \frac{1}{2\pi N_0 \Delta F}; \tag{2.66}$$

$$\dot{\vec{a}} = (\dot{a}_1, \dot{a}_2, \ldots, \dot{a}_N) \tag{2.67}$$

is the signal vector formed by reading the imaginary envelope $\dot{a}(t)$ of the signal $a(t)$, and $\|.\|$ is the normalized determinant in the imaginary Euclidean space.

Let us introduce the useful

$$\tilde{\gamma} = \frac{\sqrt{E_a}}{N_0} \tag{2.68}$$

and the nuisance

$$\tilde{\pi} = -\frac{1}{2N_0} \tag{2.69}$$

parameters and the statistics:

$$U(\dot{\vec{X}}) = \frac{1}{T\Delta F} Re\langle \dot{\vec{X}}; \dot{\vec{a}} \rangle \tag{2.70}$$

and

$$T(\dot{\vec{X}}) = \frac{1}{T\Delta F} \|\dot{\vec{X}}\|^2, \tag{2.71}$$

where $\langle .;. \rangle$ denotes the scalar product in the imaginary Euclidean space. At high values of the $T\Delta F$ (signal base) the statistics can be approximated in the integral form

$$U(\dot{\vec{X}}) = Re \int_0^T \dot{X}(t)\dot{a}^*(t)\, dt; \tag{2.72}$$

$$T(\dot{\vec{X}}) = \int_0^T |\ \dot{X}(t)\ |^2\ dt, \tag{2.73}$$

where $\dot{a}^*(t)$ is the complex conjugate of the process $\dot{a}(t)$.

Using the introduced useful $\tilde{\gamma}$ and nuisance $\tilde{\pi}$ parameters and in terms of Eqs. (2.72) and (2.73), we can rewrite Eq. (2.65) in the form of the probability distribution density of the exponential class of distributions:

$$f(\dot{\vec{X}} \mid \tilde{\gamma}; \tilde{\pi}) = k \exp\left(\frac{\tilde{\gamma}^2}{2\tilde{\pi}}\right) \exp\left[\tilde{\pi} T(\dot{\vec{X}}) + \tilde{\gamma} U(\dot{\vec{X}})\right];\ \tilde{\gamma} \in [0, \infty); \tilde{\pi} \in (-\infty, 0]. \tag{2.74}$$

The problem of signal detection in the presence of additive Gaussian noise using the useful $\tilde{\gamma}$ and nuisance $\tilde{\pi}$ parameters is formulated as a test of the complex hypotheses:

$$\begin{cases} H_0 : \vec{\vartheta} = (\tilde{\gamma}, \tilde{\pi}) \in \Theta_0 = \Gamma_0 \times \Pi;\ \Gamma_0 = \{\tilde{\gamma} = 0\},\ \Pi = (-\infty, 0]; \\ H_1 : \vec{\vartheta} \in \Theta_1 = \Gamma_1 \times \Pi, \quad \Gamma_1 = [0, \infty). \end{cases} \tag{2.75}$$

One can see from Eq. (2.74) and the theorem of factorization[41,71] that the statistic determined by Eq. (2.72) is sufficient for the useful parameter $\tilde{\gamma}$ and the statistic determined by Eq. (2.73) is sufficient for the nuisance parameter $\tilde{\pi}$. According to the theorem about completeness of the class of the probability distribution densities, the statistic $T(\vec{X})$ generates the complete class at the boundary

$$\Delta = \Gamma_0 \times \Pi. \tag{2.76}$$

Using Eq. (2.74) and the properties of the exponential classes of the probability distribution densities,[71,81] we can show that the power function for any decision-making rules is the continuous power function with respect to the parameter $\vec{\vartheta}$, and the class of conditional distributions has the monotone likelihood with respect to the statistic $U(\vec{X})$ for all values of the statistic $T(\vec{X})$.

Thus, the unbiased uniformly most powerful test for the considered problem of signal detection in the presence of additive Gaussian noise has the decision function

$$\varphi(\dot{\vec{X}}) = \begin{cases} 1 & \text{at } U(\dot{\vec{X}}) \geq L\left[T(\dot{\vec{X}});\, P_F\right]; \\ 0 & \text{otherwise,} \end{cases} \tag{2.77}$$

where $L[T(\dot{\vec{X}}); P_F]$ is the threshold function.

As was shown in References 30 and 41, the threshold function can be determined in the following form:

$$L\left[T(\dot{\vec{X}});\, P_F\right] = K(P_F)V_1\left[T(\dot{\vec{X}})\right] + V_2\left[T(\dot{\vec{X}})\right], \tag{2.78}$$

where $K(P_F)$ is the constant defined only by the probability of false alarm P_F; the statistics $V_1[T(\vec{X})]$ and $V_2[T(\vec{X})]$ are monotone with respect to the statistic $U(\vec{X})$ and are independent of the statistic $T(\vec{X})$ in the probability sense.

In terms of Eq. (2.78) the threshold function takes the following form:

$$L\left[T(\dot{\vec{X}});\, P_F\right] = K(P_F)\sqrt{T(\dot{\vec{X}})}. \tag{2.79}$$

In terms of Eqs. (2.72), (2.77), and (2.79), the decision function takes the following form:

$$\varphi(\dot{\vec{X}}) = \begin{cases} 1 & \text{at } Re \int\limits_0^T \dot{X}(t)\dot{a}^*(t)\,dt \geq K(P_F)\sqrt{\int\limits_0^T |\, \dot{X}(t)\,|^2\, dt}; \\ 0 & \text{otherwise.} \end{cases} \tag{2.80}$$

2.4.2 Invariant Rule

The invariance principle is often useful when there is no unbiased uniformly most powerful test. This principle is based on the view of *a priori* uncertainty in the form of action of some arbitrary transformations from the fixed group $\vec{G}$

on the observed input stochastic process or on a sample of this observed input stochastic process. In this view of *a priori* uncertainty the following conditions are assumed:

- The symmetry of the class

$$\mathcal{P} = \{f(\vec{X} \mid \vec{\vartheta}); \vec{\vartheta} \in \Theta\} \tag{2.81}$$

  of the probability distribution densities of samples of the observed input stochastic process with respect to the group $\vec{G}$.
- The invariance of the sets Θ_0 and Θ_1 determining the hypotheses H_0 and H_1 in the parametric space Θ with respect to the induced group $\vec{G}_*$ of transformations of the space Θ.

The set $\Theta_i (i = 0, 1)$ is invariant with respect to the group $\vec{G}_*$ if

$$g_* \vec{\vartheta} \in \Theta_i \tag{2.82}$$

for all $\vec{\vartheta} \in \Theta$ and for all $g_* \in \vec{G}_*$, where g_* is the elementary induced transformation (element of the group $\vec{G}_*$).

The symmetry of the class $\mathcal{P}$ with respect to the group $\vec{G}$ ensures the membership of the probability distribution density of the transformed sample $g\vec{X}$ in the class $\mathcal{P}$ under all $g \in \vec{G}$. The invariance of the sets Θ_0 and Θ_1 demonstrates that the transformation of the sample $\vec{X}$ by the operators of the group $\vec{G}$ does not violate the stated hypotheses. Without fulfillment of these two conditions the group $\vec{G}$ cannot be used for a view of *a priori* uncertainty.

Problems with the symmetric classes $\mathcal{P}$ and invariant sets Θ_0 and Θ_1 are called symmetric with respect to the group $\vec{G}$. Since *a priori* uncertainty in symmetric problems reduces to a variation of the probability distribution densities of the observed input stochastic process during transformations of the induced group $\vec{G}_*$, it is natural to require that the power functions of hypothesis testing rules be independent of the variation of the probability distribution density. Thus, the decision-making rules with the invariant power functions $\beta(\varphi \mid \vec{\vartheta})$ with respect to the group $\vec{G}_*$ are generated. In other words, the decision-making rules, for which

$$\beta(\varphi \mid g_*\vec{\vartheta}) = \beta(\varphi \mid \vec{\vartheta}) \tag{2.83}$$

for all $\vec{\vartheta} \in \Theta$ and $g_* \in \vec{G}_*$, are obtained.

For symmetric problems the condition of the invariance of the power function is observed, if the decision function $\varphi(\vec{X})$ of the decision-making rule is invariant with respect to the group

$$\vec{G} : \varphi(g\vec{x}) = \varphi(\vec{x}) \tag{2.84}$$

for all $\vec{x} \in \vec{X}$ and $g \in \vec{G}$.

This statement follows from these equalities:

$$\beta(\varphi \mid g_*\vec{\vartheta}) = \int_{\vec{X}} \varphi(\vec{x}) f(\vec{x} \mid g_*\vec{\vartheta})\, d\vec{x}$$
$$= \int_{\vec{X}} \varphi(g\vec{x}) f(g\vec{x} \mid g_*\vec{\vartheta}) \mid J_g \mid d\vec{x}$$
$$= \int_{\vec{X}} \varphi(\vec{x}) f(\vec{x} \mid \vec{\vartheta})\, d\vec{x} = \beta(\varphi \mid \vec{\vartheta}), \tag{2.85}$$

where J_g is the Jacobian of the transformation g.[38,82]

Because of this, when the invariance principle of the class for all hypothesis testing rules is used, the class $\mathcal{F}_{IS}$ of the invariant decision-making rules possessing the invariant decision functions with respect to the group $\vec{G}$ is generated. The generation of the class $\mathcal{F}_{IS}$ ensures, on the one hand, the stability of the decision-making rules under conditions of *a priori* uncertainty and, on the other hand, the possibility of creating the necessary premises for including the optimal uniformly most powerful test in this class. Including the optimal decision-making rule in the class $\mathcal{F}_{IS}$ yields the invariant uniformly most powerful test.

Consider the detection problem of the signal $\sqrt{E_a}\, a(t, \varphi_0)$ with the arbitrary initial phase φ_0 in the presence of additive Gaussian noise. The additive Gaussian noise at the input of the PF is the supposed white Gaussian noise with the correlation function determined by Eq. (2.61)

Assume that the signal $a(t, \varphi_0)$ at the fixed phase φ_0 is known and has the unit energy. The observed input stochastic process $X(t),\ t \in [0, T]$ is the process at the output of the PF with the Π-amplitude-frequency response, unit coefficient of signal transmission, and bandwidth ΔF. The signal energy being within the limits of the interval determined by Eq. (2.62) and the initial phase

$$\varphi_0 \in [-\pi, \pi] \tag{2.86}$$

are considered *a priori* indeterminate parameters of the incoming signal.

The spectral power density of the additive Gaussian noise determined by Eq. (2.63) is considered an *a priori* indeterminate parameter of the additive Gaussian noise.

The sample of readings of the imaginary envelope $\dot{X}(t)$ of the process $X(t)$ at the instant $t_i = \frac{i}{\Delta F}$, which is determined by Eq. (2.64), is generated. The following value

$$N = T\Delta F \tag{2.87}$$

is the signal base.

The probability distribution density of this sample is determined in the following form:

$$f\left(\dot{\vec{X}} \middle| E_a; \varphi_0; \frac{N_0}{2}\right) = k \exp\left\{-\frac{1}{2N_0\Delta F}\left\|\dot{\vec{X}} - \sqrt{E_a}\exp(-i\varphi_0)\dot{\vec{a}}\right\|^2\right\}, \tag{2.88}$$

where the coefficient k is determined by Eq. (2.66), and the vector $\dot{\vec{a}}$ determined by Eq. (2.67) is the signal vector formed by readings of the imaginary envelope $\dot{a}(t)$ of the signal $a(t, \varphi_0)$ under the condition $\varphi_0 = 0$. The $\|.\|$ is the normalized determinant in the imaginary Euclidean space. Introduce the useful parameter

$$\vec{\vartheta} = (\vartheta_1, \vartheta_2), \tag{2.89}$$

where

$$\vartheta_1 = \frac{\sqrt{E_a}\cos\varphi_0}{N_0}; \tag{2.90}$$

$$\vartheta_2 = \frac{\sqrt{E_a}\sin\varphi_0}{N_0}, \tag{2.91}$$

and the nuisance parameter

$$\tilde{\pi} = \frac{1}{2N_0}, \tag{2.92}$$

and the statistics

$$U_1(\dot{\vec{X}}) = \frac{1}{\Delta F} Re\,\langle\, \dot{\vec{X}}; \dot{\vec{a}}\,\rangle; \tag{2.93}$$

$$U_2(\dot{\vec{X}}) = \frac{1}{\Delta F} Im\,\langle \dot{\vec{X}}; \dot{\vec{a}}\,\rangle; \tag{2.94}$$

$$T(\dot{\vec{X}}) = \frac{1}{\Delta F} \|\dot{\vec{X}}\|^2, \tag{2.95}$$

where $\langle\, .;.\,\rangle$ is the scalar product in the imaginary Euclidean space.

At high values of the signal base determined by Eq. (2.87) the statistics $U_1(\dot{\vec{X}})$, $U_2(\dot{\vec{X}})$, and $T(\dot{\vec{X}})$ can be approximated in the integral form:

$$U_1(\dot{\vec{X}}) = Re \int_0^T \dot{X}(t)\dot{a}^*(t)\,dt; \tag{2.96}$$

$$U_2(\dot{\vec{X}}) = Im \int_0^T \dot{X}(t)\dot{a}^*(t)\,dt; \tag{2.97}$$

$$T(\dot{\vec{X}}) = \int_0^T |\ \dot{X}(t)\ |^2\ dt. \tag{2.98}$$

Then the probability distribution density (see Eq. (2.88)) can be rewritten in the form of the exponential-type distribution:

$$f(\dot{\vec{X}} \mid \vec{\vartheta}) = k\exp\left(-\frac{\vartheta_1^2 + \vartheta_2^2}{4\tilde{\pi}}\right)\exp\left[-\tilde{\pi}T(\dot{\vec{X}}) + \vec{\vartheta}_1 U_1(\dot{\vec{X}}) + \vec{\vartheta}_2 U_2(\dot{\vec{X}})\right], \tag{2.99}$$

where

$$\vec{\vartheta} = (\vartheta_1, \vartheta_2, \tilde{\pi}). \tag{2.100}$$

Using the parameter $\vec{\vartheta}$, we can formulate the problem of signal detection in the presence of additive Gaussian noise as the problem of testing complex hypotheses:

$$\begin{cases} H_0 : \vec{\vartheta} \in \Theta_0 = \{\vec{\vartheta} : \quad \vartheta_1^2 + \vartheta_2^2 = 0; \quad 0 < \tilde{\pi} < \infty\}; \\ H_1 : \vec{\vartheta} \in \Theta_1 = \{\vec{\vartheta} : \quad 0 < \vartheta_1^2 + \vartheta_2^2 < \infty; \quad 0 < \tilde{\pi} < \infty\}. \end{cases} \tag{2.101}$$

It is impossible to use the principle of unbiasedness for solving this problem of signal detection, since the useful parameter $(\vartheta_1; \vartheta_2)$ is not univariate. For this reason, we will attempt to use the invariance principle.

The uncertainty in the initial phase, energy of the signal, and spectral power density of the additive Gaussian noise can be represented by the group

$$\vec{G} = \{g : \dot{\vec{X}} \to \delta \cdot e^{-i\varphi}\dot{\vec{X}}, \delta \in [0, \infty); \varphi \in [-\pi, \pi]\}. \tag{2.102}$$

Noting that the Jacobian

$$|J_g| = \delta \tag{2.103}$$

and

$$\frac{2}{N_0}\left\|\delta \cdot e^{-i\varphi}\dot{\vec{X}} - \sqrt{E_a} \cdot e^{-i\varphi_0}\dot{\vec{a}}\right\|^2 = \frac{1}{N_0'}\left\|\dot{\vec{X}} - \sqrt{E_a'} \cdot e^{-i\varphi_0'}\dot{\vec{a}}\right\|^2, \tag{2.104}$$

where

$$N_0' = \frac{N_0}{2\delta^2}; \tag{2.105}$$

$$E_a' = \frac{E_a}{\delta^2}; \tag{2.106}$$

$$\varphi_0' = \varphi_0 - \varphi, \tag{2.107}$$

there is no difficulty in defining the symmetry of the class $\mathcal{P}$ with the probability distribution density shown in Eq. (2.99) with respect to the group $\vec{G}$.

The group $\vec{G}$ is induced in a space of the statistics

$$\left(U_1\left(\dot{\vec{X}}\right); U_2\left(\dot{\vec{X}}\right); T\left(\dot{\vec{X}}\right)\right) \tag{2.108}$$

given by Eqs. (2.96)–(2.98) into the group of linear transformations with the matrices

$$\vec{C} = [C_{ik}], \tag{2.109}$$

where

$$C_{11} = C_{22} = \delta \cos\varphi; \tag{2.110}$$

$$C_{12} = -C_{21} = \delta \sin\varphi; \tag{2.111}$$

$$C_{33} = \delta^2 \tag{2.112}$$

and other elements are equal to zero. The statistics determined by Eqs. (2.96)–(2.98) according to the theorem of factorization are sufficient for the class $\mathcal{P}$.

For this reason, we shall consider the statistics determined by Eq. (2.108) instead of the sample $\vec{X}$ and present the group $\vec{G}$ by linear transformation with the matrices $\vec{C}$. Since the class of the probability distribution densities of the statistics determined by Eq. (2.108) is exponential, the group $\vec{G}_*$ induced in the space determined by Eq. (2.48) consists of linear transformations whose matrices are

$$\vec{C}_* = (\vec{C}^{-1}). \tag{2.113}$$

It is easy to verify that the sets Θ_0 and Θ_1 are invariant with respect to the group $\vec{G}_*$.

Thus, we can use the invariance principle for the present problem of signal detection in the presence of additive Gaussian noise. The maximal invariances of the groups $\vec{G}$ and $\vec{G}_*$ are determined, respectively, in the following form:

$$Z = \frac{U_1^2 + U_2^2}{T} \tag{2.114}$$

and

$$\tilde{\gamma} = \frac{\vartheta_1^2 + \vartheta_2^2}{\tilde{\pi}}. \tag{2.115}$$

The hypotheses H_0 and H_1 can be formulated as follows:

$$\begin{cases} H_0 : \tilde{\gamma} \in \Gamma_0 = \{\tilde{\gamma} = 0\}; \\ H_1 : \tilde{\gamma} \in \Gamma_1 = \{\tilde{\gamma} : 0 < \tilde{\gamma} < \infty\}. \end{cases} \tag{2.116}$$

The statistic Z has a non-central β-distribution law with the non-centralized parameter $\sqrt{\frac{\tilde{\gamma}}{2}}$.[42,71] The class of the β-distribution laws has the monotone likelihood ratio with respect to the statistic Z. Because of this, there is the invariant uniformly most powerful test of the hypotheses H_0 and H_1. The decision function is determined in the following form:

$$\varphi(\vec{X}) = \begin{cases} 1 & \text{at } U_1^2(\dot{\vec{X}}) + U_2^2(\dot{\vec{X}}) \geq K(P_F)T(\dot{\vec{X}}); \\ 0 & \text{otherwise.} \end{cases} \tag{2.117}$$

The decision-making rule in Eq. (2.117) has the stable probability of false alarm P_F for any variations of the spectral power density of the additive Gaussian noise, ensures the invariance of the probability of detection P_D with respect to the initial phase of the signal, and maximizes the probability of detection P_D for all values of the signal-to-noise ratio

$$q_T^2 = \frac{2E_a}{N_0} \tag{2.118}$$

in the class of the invariant decision-making rules. Since we have sufficient statistics in Eqs. (2.96)–(2.98), and their probability distribution densities generate the total class according to Eq. (2.99) and the theorem about completeness of the class, then the decision-making rule in Eq. (2.117) maximizes the probability of detection P_D not only in the class of the invariant decision-making rules, but in the more extended class of the decision-making rules as well, in which the power function has the above-mentioned peculiarities of invariance.

2.5 Mini-Max Decision-Making Rule

Consider the detection problem of the Gaussian signals in the presence of additive non-stationary and correlated Gaussian noise with the unknown spectral power density $\frac{N_0}{2}$. The signal is the discrete sample

$$\vec{a} = (a_1, a_2, \ldots, a_N) \tag{2.119}$$

obeying the Gaussian distribution law with zero mean

$$M[a_i] = 0, \qquad i = \overline{1, N}, \tag{2.120}$$

where N is the sample size.

Assume that the correlation matrix of the signal

$$\vec{R}_a = M[\vec{a}\vec{a}^+] = \|R_{a_{ij}}\|, \tag{2.121}$$

where

$$R_{a_{ij}} = M[a_i a_j], \tag{2.122}$$

is known with an accuracy of some unknown averaged energy E_a^{av} of the signal:

$$\vec{R}_a = E_a^{av}\vec{R}_{0a}; \tag{2.123}$$

$$\vec{R}_{0a}^T = 1; \tag{2.124}$$

$$E_a^{av} = M[\vec{a}^+\vec{a}] = \vec{R}_a^T, \tag{2.125}$$

where $\vec{R}_{0a}$ is the completely known correlation matrix of the signal. The matrix $\vec{R}_{0a}$ is normalized by the signal energy.

The additive non-stationary and correlated Gaussian noise is determined in the following form:

$$\vec{n} = (n_1, n_2, \ldots, n_N)^+ \tag{2.126}$$

with zero mean

$$M[n_i] = 0, \qquad i = \overline{1, N}. \tag{2.127}$$

The correlation matrix of the additive non-stationary and correlated Gaussian noise

$$\vec{R}_n = M[\vec{n}\vec{n}^+] = \|R_{n_{ij}}\|, \tag{2.128}$$

where

$$R_{n_{ij}} = M[n_i n_j], \tag{2.129}$$

is known with an accuracy of the spectral power density $\frac{N_0}{2}$ of the additive non-stationary and correlated Gaussian noise:

$$\vec{R}_n = \frac{N_0}{2}\vec{R}_{0n}; \tag{2.130}$$

$$\vec{R}_{0n}^T = 1; \tag{2.131}$$

$$\frac{N_0}{2} = M[\vec{n}^+\vec{n}] = \vec{R}_n^T, \tag{2.132}$$

where $\vec{R}_{0n}$ is the completely known normalized correlation matrix of the additive non-stationary and correlated Gaussian noise.

As before, we must define the optimal decision-making rule using the sample determined by Eq. (2.37), where

$$X_i = a_i + n_i \tag{2.133}$$

if there is a "yes" signal in the input stochastic process (the hypothesis H_1) and

$$X_i = n_i \tag{2.134}$$

if there is a "no" signal in the input stochastic process (the hypothesis H_0).

The decision-making rule must ensure the guaranteed probability of detection

$$P_D \geq P_{D_0} \tag{2.135}$$

at the minimal signal-to-noise ratio q_{T_0} and for all

$$q_T \geq q_{T_0}. \tag{2.136}$$

In doing so, there are two kinds of limitations of the probability of false alarm:[28,30,31,44,83]

$$P_F = P_{F_0} \tag{2.137}$$

and

$$P_F \leq P_{F_0}. \tag{2.138}$$

The conditional probability distribution densities of samples of the input stochastic process under the hypotheses of a "yes" and a "no" signal in the input stochastic process have the following form:

$$f_1\left(\vec{X} \mid q_T, \frac{N_0}{2}\right) = \frac{\exp\left[-N_0 \vec{X}^+ (q_T \vec{R}_{0a} + \vec{R}_{0n})^{-1} \vec{X}\right]}{(\pi N_0)^{\frac{N}{2}} \sqrt{|q_T \vec{R}_{0a} + \vec{R}_{0n}|}}; \tag{2.139}$$

$$f_0\left(\vec{X} \,\middle|\, \frac{N_0}{2}\right) = \frac{\exp\left[-N_0 \vec{X}^+ \vec{R}_{0n}^{-1} \vec{X}\right]}{(\pi N_0)^{\frac{N}{2}} \sqrt{|\vec{R}_{0n}|}}. \tag{2.140}$$

In terms of Eqs. (2.139) and (2.140) we can define the likelihood ratio of the maximal invariant for the condition

$$q_T = q_{T_0} \tag{2.141}$$

in the following form:[30,83–85]

$$\ell = \frac{\int f_1(\vec{X} \mid q_{T_0}, \frac{N_0}{2}) d(\frac{N_0}{2})}{\int f_0(\vec{X} | \frac{N_0}{2}) d(\frac{N_0}{2})} = \frac{\sqrt{|\vec{R}_{0n}|}}{\sqrt{|q_{T_0} \vec{R}_{0a} + \vec{R}_{0n}|}} \cdot \sqrt{\left\{\frac{\vec{X}^+ \vec{R}_{0n}^{-1} \vec{X}}{\vec{X}^T (q_{T_0} \vec{R}_{0a} + \vec{R}_{0n})^{-1} \vec{X}}\right\}^N}. \tag{2.142}$$

Consequently, the unique invariant most powerful test under the condition determined by Eq. (2.141) takes the following form:

$$\vec{X}^+ \vec{R}_{0n}^{-1} \vec{X} \geq K \vec{X}^+ \left(q_{T_0} \vec{R}_{0a} + \vec{R}_{0n}\right)^{-1} \vec{X}, \qquad K > 1, \tag{2.143}$$

where the threshold value K is identically defined by a predetermined value of the probability of false alarm P_F. A monotone increase in the probability of detection P_D with increasing the signal-to-noise ratio q_T proves that the decision-making rule is mini-max under two kinds of limitations of the probability of false alarm P_F.

The mini-max decision-making rule (see Eq. (2.143)) can be presented in the following form:[41,42,44,86]

$$\vec{X}^+ \left[\vec{R}_{0n}^{-1} - \left(q_{T_0} \vec{R}_{0a} + \vec{R}_{0n}\right)^{-1}\right] \vec{X} \geq K_1 \vec{X}_1 + \vec{R}_{0n}^{-1} \vec{X}, \tag{2.144}$$

where

$$K_1 = \frac{K-1}{K} > 0, \tag{2.145}$$

and in the equivalent form using the quadrature component

$$\vec{X}^+ \left[\vec{R}_{0n}^{-1} - K \left(q_{T_0} \vec{R}_{0a} + \vec{R}_{0n}\right)^{-1}\right] \vec{X} \geq 0. \tag{2.146}$$

The optimal mini-max decision-making rule (see Eq. (2.143)) can be transformed for the weak signals determined by the following condition

$$q_{T_0} \to 0 \tag{2.147}$$

into the following inequality:

$$\vec{X}^{+}\vec{R}_{0n}^{-1}\vec{R}_{0a}\vec{R}_{0n}^{-1}\vec{X} \geq K\vec{X}^{+}\vec{R}_{0n}^{-1}\vec{X} \tag{2.148}$$

and for the powerful signals determined by the following condition

$$q_{T_0} \to \infty \tag{2.149}$$

into the following inequality

$$\vec{X}^{+}\vec{R}_{0n}^{-1}\vec{X} > K\vec{X}^{+}\vec{R}_{0a}^{-1}\vec{X}. \tag{2.150}$$

2.6 Sequential Signal Detection

The decision-making rules or signal detection algorithms considered in the previous sections are based on the suggestion that the problem of testing statistical hypotheses is singly solved after observing the sample with the sample size N. There is another approach to solving this problem that is of some interest. This approach is based on the repeated testing of a possibility to make a decision in favor of the hypothesis H_0 or the alternative H_1, once a new element of the sample of the observed process becomes available. Such a large body of procedures is called sequential, and therefore the associated decision-making analysis is called sequential analysis.

In the sequential analysis of the statistical hypotheses for each k-th step, in which the attempt to make a decision is carried out, three ranges of the decision statistic must be defined: the admissible range $\mathcal{L}_0$, the critical range $\mathcal{L}_1$, and the range of uncertainty $\mathcal{L}_U$. If the decision statistic belongs to the range $\mathcal{L}_0$, the decision in favor of the hypothesis H_0 is made; if the decision statistic belongs to the range $\mathcal{L}_1$, the decision in favor of the hypothesis H_1 is made. If the decision statistic belongs to the range of uncertainty $\mathcal{L}_U$, it is assumed that the data on the k-th step are not sufficient for making a decision, and there is a need to analyze the next element of the sample.

Contrary to non-sequential procedures, in which the sample size N under the known probability distribution densities $f(\vec{X} \mid H_0)$ and $f(\vec{X} \mid H_1)$ is determined in advance, reasoning from a necessity to define the predetermined probability of false alarm P_F and the probability of detection P_D, under the sequential procedure the sample size N is a random variable depending on the input data.

One of the most effective sequential criteria was suggested and investigated by A. Wald.[87,88] The sequential Wald criterion is to compare the likelihood ratio

$$\Lambda(X_1, X_2, \ldots, X_N) \tag{2.151}$$

of the sample determined by Eq. (2.37) on the each k-th step with two fixed thresholds A_1 and B_1, where $A_1 > B_1$. If

$$\Lambda \leq B_1 \tag{2.152}$$

the decision in favor of the hypothesis H_0 is made. If

$$\Lambda \geq A_1 \tag{2.153}$$

the decision in favor of the hypothesis H_1 is made. If

$$B_1 < \Lambda < A_1 \tag{2.154}$$

the test must be continued.

The thresholds A_1 and B_1 of the Wald criterion can be defined by the predetermined probability of false alarm P_F and the probability of detection P_D. Let the sample determined by Eq. (2.37) be the vector of a set of the samples $\{X_i\}$.

The decision in favor of the hypothesis H_1 is made if

$$\Lambda = \frac{f(\vec{X} \mid H_1)}{f(\vec{X} \mid H_0)} \geq A_1 \tag{2.155}$$

or

$$f(\vec{X} \mid H_1) \geq A_1 f(\vec{X} \mid H_0). \tag{2.156}$$

This condition is correct for any sample of the set $\vec{X}_1$, $\vec{X}_1 \in \vec{X}$. Because of this, we can write

$$\int_{\vec{X}_1} f(\vec{X} \mid H_1)\, d\vec{X} \geq A_1 \int_{\vec{X}_1} f(\vec{X} \mid H_0)\, d\vec{X}. \tag{2.157}$$

In terms of Eqs. (2.40) and (2.41), the last inequality may be rewritten in the following form:

$$P_D \geq A_1 P_F \qquad \text{or} \qquad A_1 \leq \frac{P_D}{P_F}. \tag{2.158}$$

The inequality in Eq. (2.158) is the upper bound for the threshold A_1. In an analogous way consider the case when the decision in favor of the hypothesis H_0 is made. Then

$$\Lambda = \frac{f(\vec{X} \mid H_1)}{f(\vec{X} \mid H_0)} \leq B_1 \tag{2.159}$$

or

$$f(\vec{X} \mid H_1) \leq B_1 f(\vec{X} \mid H_0). \tag{2.160}$$

Integrating by the set $\vec{X}_0$, $\vec{X}_0 \in \vec{X}$, we obtain

$$1 - P_D \leq B_1(1 - P_F); \tag{2.161}$$

$$B_1 \geq \frac{1 - P_D}{1 - P_F}. \tag{2.162}$$

The inequalities in Eqs. (2.161) and (2.162) are the lower bound for the threshold B_1. Note that the bounds of the thresholds A_1 and B_1 are only defined by the probability of false alarm P_F and the probability of detection P_D and are independent of any kind of probability distribution densities.

In actual practice the logarithm of the likelihood ratio is convenient to use for independent samples. In doing so,

$$Z_N = \sum_{i=1}^{N} z_i = Z_{N-1} + z_N = \ln\{\lambda(X_1)\lambda(X_2)\cdots\lambda(X_N)\}. \tag{2.163}$$

The accumulated value of the decision statistic Z_N at each step is compared with the thresholds

$$A = \ln A_1 = \ln\frac{P_D}{P_F} \tag{2.164}$$

and

$$B = \ln B_1 = \ln\frac{1 - P_D}{1 - P_F}. \tag{2.165}$$

When

$$Z_N \leq B \tag{2.166}$$

the decision in favor of the hypothesis H_0 is made. If

$$Z_N \geq A \tag{2.167}$$

the decision in favor of the hypothesis H_1 is made. When

$$B < Z_N < A \tag{2.168}$$

the test must be continued for the next sequential sample.

It is well known[8,13,32,40,43,71,87,88] that the number of steps of the sequential procedures is finite with the probability equal to the unit value for the independent uniform sample. Thus, the sequential Wald decision-making rule is comprised of determination of the decision statistic z_N for each element of the sample, determination of the accumulated value of the decision statistic

$$Z_N = Z_{N-1} + z_N, \tag{2.169}$$

and a comparison of the accumulated value of the decision statistic Z_N at each step of the procedure with the thresholds A and B.

The Wald–Wolfowitz theorem[89,90] holds that the procedure of sequential analysis of the likelihood ratio with the fixed thresholds requires the minimal sample size in the statistical sense for all criteria, solving the problem of recognizing the statistical hypotheses with the probability of false alarm P_F and the probability of detection P_D, which do not exceed the predetermined values of the probability of false alarm P_F and the probability of signal omission

$$P_M = 1 - P_D. \tag{2.170}$$

The Wald–Wolfowitz theorem proves the optimality of the sequential Wald decision-making rule for time measure of the cost of observations at the input stochastic process. When one or more assumptions made under the Wald–Wolfowitz theorem are violated, difficulties in finding the optimal sequential decision-making rule arise. However, some investigations[8,12,20,23,91–100] show that there are effective modifications of the sequential Wald criterion in many important practical problems under conditions differing from the conditions of the Wald–Wolfowitz theorem and possessing characteristics that are very close to those of the optimal sequential Wald decision-making rule.

So, the optimal detector must be constructed on the basis of the vector sufficient statistic $\vec{Z}(t)$ to fit the detection problem of the correlated Markov signals in the presence of additive correlated Gaussian noise. Coordinates of the sufficient statistic $\vec{Z}(t)$ are the values of the input stochastic process $\vec{X}(t)$ and values of the process $m(t)$ at the output of the optimal filter in addition to the logarithm of the likelihood ratio $z(t)$.[94,97,101] A search for the optimal decision thresholds meets with serious mathematical difficulties.

At the same time the sequential decision-making rule based on the comparison of the scalar statistic $z(t)$ with constant thresholds is asymptotic (to the) optimal for this problem.[23] This detector is optimal for measuring the cost of observations of the input stochastic process with respect to the detection problem of the correlated signals in the presence of additive white Gaussian noise.[101–104] This detector is asymptotic optimal for the time measure of the cost too, but with the high signal-to-noise ratio, it is at a moderate disadvantage in comparison with the more complex optimal detector using the sufficient statistic $\vec{Z}\{z(t);m(t)\}$ and the thresholds that are variable in time.[97,105] Immediate application of the Wald criterion to the complex hypotheses is often ineffective,[21,28,71] but there are some techniques that allow us to take advantage of the sequential signal detection algorithms for problems with complex hypotheses.

One of the main characteristics of the optimal decision-making rule for sequential signal detection in the presence of additive Gaussian noise is the mean of the number of steps of the sequential procedure under the hypotheses H_0 and H_1 (the averaged sample size). Decreasing the average sample size for the sequential decision-making rule in comparison with the fixed sample size that is necessary for the signal detection in the presence of additive Gaussian noise by the Neyman–Pearson criterion is the main advantage of the sequential signal detection algorithm, which may be used in various automatic signal detection systems. For the independent uniform sample $\vec{X}$, the accumulated value of the decision statistic Z_N is the random sum of uniformly distributed components at the instant of decision-making.

Thus, the following relationships are true:[90]

$$M[N \mid H_{0,1}] = \frac{M[Z_N \mid H_{0,1}]}{M[z_i \mid H_{0,1}]}, \tag{2.171}$$

where

$$M[z_i \mid H_{0,1}] = \bar{z}_{0,1} = \int_{-\infty}^{\infty} \left[\ln \frac{f(X_i \mid H_1)}{f(X_i \mid H_0)} \right] f(X_i \mid H_{0,1})\, dX_i. \tag{2.172}$$

Thus,

$$\overline{N}_0 = M[N \mid H_0] = \frac{P_F A + (1 - P_F)B}{\bar{z}_0}; \tag{2.173}$$

$$\overline{N}_1 = M[N \mid H_1] = \frac{P_D A + (1 - P_D)B}{\bar{z}_1}. \tag{2.174}$$

As was shown in References 24, 93, and 95, Eq. (2.174) is true for the powerful signals if the additional term is introduced into the numerators. Moreover, this additional term must be equal to the mean of an event when the decision statistic exceeds the thresholds at the instant of finishing the procedure. When the probability of false alarm P_F and the probability of signal omission determined by Eq. (2.170) are the same and a disposition of the decision thresholds A and B is symmetric relative to zero, the gain in the average sample size of the sequential decision-making rule is about double in comparison with the Neyman–Pearson criterion that is equivalent in probabilities of error.[87–90]

In practice the requirements for the probability of false alarm P_F and the probability of signal omission P_M determined by Eq. (2.170) often differ in value.[17,36,38,42,43] As a consequence, the disposition of the thresholds A and B is not symmetric. So, for signal detection by radar systems the probability of false alarm that is between

$$P_F = 10^{-3} \quad \text{and} \quad P_F = 10^{-12} \tag{2.175}$$

and is much less than the probability of signal omission, that is between

$$P_M = 0.1 \quad \text{and} \quad P_M = 0.5. \tag{2.176}$$

In this case

$$A \gg B \tag{2.177}$$

and

$$\frac{\overline{N}_1}{\overline{N}_0} \approx \frac{A}{B} \tag{2.178}$$

as

$$\bar{z}_0 \approx \bar{z}_1. \tag{2.179}$$

The average sample size $\overline{N}_1$ of the sequential decision-making rule when a "yes" signal exists in the input stochastic process is approximately equal to the fixed sample size when the Neyman–Pearson criterion is applied that is equivalent in probabilities of error; and the average sample size of the sequential decision-making rule when a "no" signal exists in the input stochastic process is much less in comparison with the fixed sample size when the Neyman–Pearson criterion is applied.[106–109]

2.7 Signal Detection in the Presence of Non-Gaussian Noise

The signal detection problem is formulated in the following manner. The observed data

$$X_k = a_k + \xi_k \tag{2.180}$$

are the sample of the input stochastic process

$$X(t) = a(t) + \xi(t) \tag{2.181}$$

with the sample size N. The input stochastic process $X(t)$ is distributed within the limits of the time interval $[0, T]$; $a(t)$ is the signal; and $\xi(t)$ is the additive noise with the predetermined correlation function and the univariate probability distribution density. The additive noise $\xi(t)$ is a result of the composed linear and non-linear transformations of the white Gaussian noise n_l

$$\xi_k = V\left(\sum_{l=1}^{N} A_{kl} n_l\right). \tag{2.182}$$

The linear transformation may be fulfilled, for example, by the preliminary filter (linear system), at the input of the receiver or detector

$$\eta_k = \sum_{l=1}^{N} A_{kl} n_l, \tag{2.183}$$

where

$$M[\eta] = 0; \tag{2.184}$$

$$M[\eta^2] = 1, \tag{2.185}$$

and

$$M[\eta_k \eta_l] = R_{0\eta}(|k - l|). \tag{2.186}$$

The non-linear transformation possesses the function $V(\eta)$ with the reciprocal unambiguous function $Q(\xi)$. Moreover,

$$\frac{d\,Q(\xi)}{d\xi} > 0. \tag{2.187}$$

The operator in Eq. (2.182) is defined by two characteristics: the matrix $\| A_{kl} \|$ and the function $V(\eta)$. Using the input data X_k, we need to make a decision—a "yes" or a "no" signal in the input stochastic process. The additive noise ξ_k is the correlated non-Gaussian process in a general case.

Primary attention is given to the detection problem of the weak signals when the input signal-to-noise ratio is much less than 1. For example, radioengineering systems using complex signals, radar systems with moving-target selection in the presence of additive noise, radioengineering systems under conditions of radio-opposition, passive sonar systems—all of them operate under the conditions of the weak input signals.[9,10,12,21,27,28,36,37]

The Neyman–Pearson detector maximizing the probability of detection P_D at the fixed probability of false alarm P_F makes the decision a "yes" signal in the input stochastic process when the following inequality

$$\ell = \frac{f(\vec{X} \mid H_1)}{f(\vec{X} \mid H_0)} \geq K \tag{2.188}$$

is satisfied, where $f(\vec{X} \mid H_1)$ and $f(\vec{X} \mid H_0)$ are the probability distribution densities of the input data determined by Eq. (2.154) when a "yes" and a "no" signal exists in the input stochastic process, respectively; K is the threshold defined by the predetermined probability of false alarm P_F. For the additive non-Gaussian noise the signal detection algorithms resulting from Eq. (2.188) are expressed by cumbersome calculations, which are not often used in practice.[17,18,20,22,25,29,30,34,38,42]

When the input signal-to-noise ratio is much less than 1 there are other approaches for the optimization of signal detection in the presence of additive non-Gaussian noise. One is based on the idea of local optimality.[22,43] The energy parameter ϑ of the signal is introduced, and if the condition $\vartheta = 0$ is satisfied the signal disappears; and the probability distribution density $f(\vec{X} \mid H_1)$, when a "yes" signal exists in the input stochastic process, transforms to the probability distribution density $f(\vec{X} \mid H_0)$, when a "no" signal exists in the input stochastic process. The decision-making rule $\mathcal{R}_{\ell o}$, wherein

$$\left.\frac{\partial P_D(\vec{\vartheta}, \mathcal{R}_{\ell o})}{\partial \vartheta}\right|_{\vartheta=0} > \left.\frac{\partial}{\partial \vartheta} P_D(\vec{\vartheta}, \mathcal{R})\right|_{\vartheta=0} \tag{2.189}$$

at the fixed probability of false alarm P_F, is considered as the locally optimum decision-making rule, where $\mathcal{R}$ is the non-optimal decision-making rule.

Thus, if the Neyman–Pearson detector maximizes the probability of detection P_D, so the locally optimum detector maximizes a slope of the curve of the probability of detection P_D as a function of the parameter ϑ at the point $\vartheta = 0$. Following E. Lehmann [110], using the fundamental Neyman–Pearson lemma, we obtain the following statement: the derivative of the probability of detection P_D with respect to the parameter ϑ becomes maximal at the fixed probability of false alarm P_F, when the decision a "yes" signal in the input stochastic process is made if the inequality

$$\frac{\partial}{\partial \vartheta} \ell(\vartheta) \bigg|_{\vartheta=0} \geq K \tag{2.190}$$

is satisfied.

Note that

$$\frac{\partial}{\partial \vartheta} \ell(\vartheta) \bigg|_{\vartheta=0} = \frac{\partial}{\partial \vartheta} \ln \ell(\vartheta) \bigg|_{\vartheta=0}, \tag{2.191}$$

since

$$\ell(\vartheta)|_{\vartheta=0} = 1. \tag{2.192}$$

Equation (2.190) is the initial premise for the synthesis and construction of locally optimum detectors.

Another approach to synthesis and construction of the optimal detector of the weak signals is based on the criterion of asymptotic optimality.[17,37] The asymptotic optimal detector of the weak signals is designed so the detector, under the predetermined signal energy and other energy characteristics of the signal defining the probability of detection P_D, can ensure the same probability of detection P_D as the Neyman–Pearson criterion, in the limit, as the sample size $N \to \infty$. The condition $N \to \infty$ must not change the predetermined energy characteristics of the signal, which define the probability of detection P_D.

One approach to do this is to use the dependence of the signal amplitude on the sample size N. The asymptotic optimal detector can be synthesized and constructed on the basis of the likelihood ratio (see Eq. (2.188)) by the power-series expansion of the signal function, but dropping terms of the expansion, making a zero contribution to the mean and variance of statistics forming by ended accumulators, as $N \to \infty$. In specific cases the locally optimum detector may possess the peculiarities of the asymptotic optimal detector. The conditions, under which the locally optimum detector is the asymptotic optimal detector, are formulated in References 111–116. These conditions satisfy, for example, the detection problems of the deterministic and noise non-coherent signals in the presence of additive non-Gaussian noise with independent samples. In a general case the locally optimum detectors do not ensure the maximal probability of detection P_D, even asymptotically.

Consider the asymptotic optimal detector that converges to the Neyman–Pearson detector in the probability of detection P_D under a sufficiently large sample size N. The sufficient sample size N depends on the nature of the additive noise and the required probability of detection P_D. In many cases the sample size N can be considered as sufficient, if at this sample size N the required value of the probability of detection P_D can be reached, when the input signal-to-noise ratio is less than 10^{-3}. With increasing the input signal-to-noise ratio, the effectiveness of the asymptotic optimal detector of signals in the presence of additive non-Gaussian noise decreases; however, the effectiveness can be higher than that for the Neyman–Pearson detector of signals in the presence of additive Gaussian noise.[117–124]

The quality of the signal detection algorithms is estimated by the probability of detection P_D and the probability of false alarm P_F. However, a calculation of the probability of detection P_D and the probability of false alarm P_F requires a knowledge of the probability distribution density $f(Z)$ of the statistic Z at the detector output. The statistic Z is compared with the threshold and the calculation of the probability distribution density $f(Z)$ is a very difficult problem, especially for the additive non-Gaussian noise. In this particular case, when the statistic Z is the Gaussian statistic, the probability of detection P_D and the probability of false alarm P_F are the monotone functions of the signal-to-noise ratio at the detector output

$$q_{out} = \frac{\left(M_{Z_1} - M_{Z_0}\right)^2}{\sigma^2_{Z_0}}, \tag{2.193}$$

where M_{Z_1} and M_{Z_0} are the means of the statistic Z when a "yes" or a "no" signal exists in the input stochastic process, respectively; $\sigma^2_{Z_0}$ is the variance of the statistic Z. It is assumed that the variance of the statistic Z at the detector output for the events a "yes" and a "no" signal in the input stochastic process is the same. In a general case the probability of detection P_D and the probability of false alarm P_F are functions of the signal-to-noise ratio q_{out} at the output of the detector, and additionally are functions of other statistical characteristics of the signal and noise.

At the present time investigations in signal detection in the presence of additive non-Gaussian noise are carried out in two directions: the additive uncorrelated non-Gaussian noise and the additive correlated non-Gaussian noise.

The additive uncorrelated non-Gaussian noise is the stationary non-Gaussian process with a wide-range spectrum within the limits of the bandwidth interval $[0, \Delta F_n]$. If we approximate this spectrum by the right-angled spectrum, the correlation function of the additive non-Gaussian noise $R_\xi(\tau)$ will be equal to zero at the points

$$\tau = \frac{k}{2\Delta F_n}, \tag{2.194}$$

where $k = 1, 2, \ldots$. If we take the readings of the input stochastic process determined by Eq. (2.181) within the limits of the time interval

$$\Delta t = \frac{1}{2\Delta F_n}, \tag{2.195}$$

we obtain the sequence determined by Eq. (2.180), in which the additive non-Gaussian noise readings ξ_k are not correlated. If under this condition the effective spectrum bandwidth of the signal is less than the spectrum bandwidth of the additive non-Gaussian noise, therefore, according to the theorem of readings, the sample a_k of the signal will be equivalent to the continuous process $a(t)$.

In the case of the additive correlated non-Gaussian noise the noise ξ_k is the additive sum of the correlated component V_k and the uncorrelated Gaussian noise n_k:

$$\xi_k = V_k + n_k. \tag{2.196}$$

One can read more about results of signal detection in the presence of uncorrelated and correlated non-Gaussian noise in References 17, 18, 23, 37, 110, 112, 119, and 125–136.

2.8 Non-Parametric Signal Detection

At the present time the problem of signal detection in the presence noise is universally accepted as a statistical problem with *a priori* uncertainty. This can be expressed for a set of parameters on occasion in which the probability distribution function $F_0(X)$, when a "no" signal exists in the input stochastic process (the hypothesis H_0), and the probability distribution function $F_1(X)$, when a "yes" signal exists in the input stochastic process (the hypothesis H_1), are inexactly known and may vary during the observation and analysis of the input stochastic process.

Under these conditions, the classical signal detection algorithms specialized to, as a rule, the additive Gaussian noise may be ineffective. For signal detection in the presence of additive non-Gaussian noise, the classical signal detection algorithms lose optimality. When only one of the statistical parameters of the additive noise—for example, the variance—is changed, the classical signal detection algorithms that remain structurally optimal do not ensure correctness of detection performances calculated by computers.

One of the ways to overcome *a priori* uncertainty is to synthesize and construct adaptive signal detection algorithms, the structure and parameters of which can be changed in accordance with the results of the input stochastic data analysis. In the event that the comparatively easily controlled parameter of the signal or noise is unknown or changed, *a priori* uncertainty can be overcome by an adaptation of the detector parameters in the course of the observation and analysis of the input stochastic process. But the problem of adaptation of the detectors is essentially complicated when some parameters, or a type of the probability distribution functions $F_0(X)$ and $F_1(X)$, are unknown. For this reason, these signal detection algorithms are very complex and not practical for the real-time applications.[3,19,26]

Another way to overcome *a priori* uncertainty is to design and construct signal detection algorithms that are non-sensitive or weakly sensitive to the changing statistical characteristics of the signal and noise. When parameters of the probability distribution functions $F_0(X)$ and $F_1(X)$ are unknown, this leads to similar and invariant signal detection algorithms. When the type of the probability distribution functions $F_0(X)$ and $F_1(X)$ is unknown, this leads to non-parametric signal detection algorithms.[4,9,17,20,21,23]

In this relationship, non-parametric signal detection algorithms have been used more often in recent years for the detection of signals in the presence of noise. The statistical method is called a non-parametric method if its use does not assume any knowledge of the type of the probability distribution functions. It is customary to assume in signal detection theory that the detector is called a non-parametric detector if, when a "no" signal exists in the input stochastic process, the probability distribution density of the decision statistic at the detector output does not depend on the probability distribution function of the noise.[9,12,17,22,28] For this reason, we mean that the detector ensures the stable probability of false alarm P_F independently of the statistical characteristics of the noise. The quality of the detector is defined by the probability of false alarm P_F and by the probability of detection P_D (or the probability of signal omission P_M). Because of this, the problem of stabilization of the probability of false alarm P_F or the probability of detection P_D under changing statistical characteristics of the noise is a very important problem. Non-parametric signal detection algorithms are able to preserve the detection performances under changing statistical parameters of the noise.

The harnessing of the non-parametric signal processing is especially useful both from the viewpoint of stabilization of the probability of false alarm P_F and from the viewpoint of detection effectiveness or the probability of detection P_D in the case when the probability distribution density of the noise does not obey the Gaussian law. There are many examples of the presence of the additive non-Gaussian noise in real practice during signal detection: non-linear transformation of the additive Gaussian noise by the logarithmic receiver, radio-opposition, stochastic pulse clutter, signal reflection of the surface of the Earth and sea, and so on. For example, during signal reflection of the Earth's surface, the probability distribution density of the amplitude envelope of the signal obeys the logarithmic-normal Gaussian and Weibull laws. It is well known that the model of Nakagami law applies for the target return signals.[10,29–31,36,42]

In many instances there is a need to reject the Gaussian model of the noise. In fact, the accuracy of the approximation of the real probability distribution density by the Gaussian law is very high only for the medium part of the probability distribution density (within the limits of $3\sigma_n^2$). For the remainder of the probability distribution density, the accuracy of approximation decreases rapidly, as one moves farther and farther away from the center of the probability distribution density curve. For example, radar systems operate at the probability of false alarm between $P_F = 10^{-3}$ and $P_F = 10^{-13}$, which corresponds to the remainder of the distribution probability density, where the approximation by the Gaussian law is not accurate.[33,34,36,38,42,137]

Synthesis of the optimum non-parametric signal detection algorithms involves serious mathematical difficulties, which are not overcome in practice. Solving the problem of the synthesis of the optimum non-parametric decision-making rules is possible only for an asymptotic case as $N \to \infty$. For limited N, the well-known non-parametric decision-making rules were obtained by heuristic procedures.

Consider briefly the well-known non-parametric decision-making rules under limited N. Assume that there is the alternative

$$F_1(\vec{X}) = F_0(\vec{X} - \vec{a}), \tag{2.197}$$

where

$$F_1(\vec{X}) < F_0(\vec{X}). \tag{2.198}$$

Under the hypothesis H_0, the independent sample determined by Eq. (2.154) is the noise sample with zero mean. Under the hypothesis H_1, the sample $\vec{X}$ with nonzero mean allows us to generate sign testing based on the polarity of the sample

$$Z = \sum_{i=1}^{N} h(X_i); \tag{2.199}$$

$$h(X_i) = \begin{cases} 1, & X_i > 0; \\ 0, & X_i < 0. \end{cases} \tag{2.200}$$

Occasionally, the centered statistic[138–143]

$$\mathcal{Z} = \sum_{i=1}^{N} \operatorname{sgn}(X_i), \tag{2.201}$$

where

$$\operatorname{sgn}(X_i) = \frac{X_i}{|X_i|} = 1 \tag{2.202}$$

under the condition $X_i > 0$, and

$$\operatorname{sgn}(X_i) = \frac{X_i}{|X_i|} = -1 \tag{2.203}$$

under the condition $X_i < 0$,

$$\operatorname{sgn}(X) = 2h(X) - 1 \tag{2.204}$$

is considered instead of Eq. (2.200).

When the mean of the probability distribution function $F_0(X)$ is unknown, we obtain a two-sample sign testing based on the comparison of signs of the pairs $X_i - n_i$, where $(X_1, X_2, \ldots, X_N)$ is the analyzed sample and $(n_1, n_2, \ldots, n_N)$ is the noise sample (reference sample)

$$Z = \sum_{i=1}^{N} h(X_i - n_i), \tag{2.205}$$

$$h(X_i - n_i) = \begin{cases} 1, & X_i > 0; \\ 0, & X_i < 0. \end{cases} \tag{2.206}$$

Since the number of units in Eqs. (2.200) and (2.206) is equivalent to the number of positive results under the Bernoulli testing, the probability of exceeding the threshold K is equal to

$$P(Z > K) = \sum_{i=K+1}^{N} C_N^i p^i (1-p)^{N-i}, \tag{2.207}$$

where

$$p = P(X > n) = \int F_0(X)\, dF_1(X) \tag{2.208}$$

is the probability of the event $X > n\,(X > 0)$; C_N^i is the number of combinations of N elements taken i at a time. Under the hypothesis H_0

$$P = 0.5 \tag{2.209}$$

Because of this, the probability of false alarm

$$P_F = P(Z > K \mid H_0) = \frac{1}{2^N} \sum_{i=K+1}^{N} C_N^i \tag{2.210}$$

does not depend on the probability distribution function $F_0(X)$, which validates the non-parametric testing. The probability of detection P_D depends naturally on the probability distribution functions $F_0(X)$ and $F_1(X)$. The relation

$$P(Z > C \mid H_1) \tag{2.211}$$

shows how the probability of detection P_D depends on the parameter p.

Taking into consideration a deviation of sample elements from the elements of the noise sample, the rank algorithms are more powerful. Due to the truth of the hypothesis for a uniform independent sample, the rank value of any elements of the sample is equiprobable (the readings X_i and n_i are mixed uniformly at the variational sequence), the decision-making rules based on the any rank statistic $Z(\vec{R})$, where the function of the rank vector is determined in the form

$$\vec{R} = (R_1, R_2, \ldots, R_N), \tag{2.212}$$

are non-parametric independently of the probability distribution function $F_0(X)$.

The rank statistics are invariant with respect to the non-linear monotone transformations of sample readings taken. The harnessing of the rank procedures leads to information losses. However, these losses are decreased with increasing the sample size N, and the rank signal detection algorithms are as effective as the optimum signal detection algorithms. Thus, we can consider the rank signal detection algorithms as the asymptotic optimum algorithms.

The Wilcoxon rank decision-making rule uses a statistic based on the sum of ranks and can be applied for the detection problems of signals in the presence of additive noise with zero mean and the symmetric probability distribution density.[110,144] The decision a "yes" signal in the input stochastic process is made if the following condition

$$Z = \sum_{i=1}^{N} R_i^+ > K, \tag{2.213}$$

is satisfied, where R_i^+ is the value of the rank of the positive readings X_i at the variational sequence, where the readings X_i are ranked by an absolute value; K is the threshold.

If the probability distribution function $F_0(X)$ is unknown and

$$F_1(X) < F_0(X), \tag{2.214}$$

we obtain a two-sample Wilcoxon rank signal detection algorithm,[145]

$$Z = \sum_{i=1}^{N} R_i > K, \tag{2.215}$$

where R_i is the rank of the readings X_i at the variational sequence that consists of the independent readings of the noise sample (the reference sample) $(n_1, n_2, \ldots, n_N)$ and the independent readings of the analyzed sample determined by Eq. (2.154).

Non-parametric decision-making rules for constructing a detector of signals in the presence of additive noise must take into account the following considerations: first, the decision-making rule must incorporate the highest power; second, the detector of the signals in the presence of additive noise must operate in real time. The decision-making rules based on the sign statistics are the most simple. A quantity of N operations are needed to define the decision statistic when the sample size is equal to N. The decision-making rules based on goodness-of-fit tests are very difficult to calculate, even with a computer.[30]

Construction of the detectors on the basis of the goodness-of-fit tests functioning in real time is also very difficult. Moreover, the effectiveness of these detectors is worse than the effectiveness of the rank detectors. The asymptotic optimum rank signal detection algorithms are the most effective, but the realization of these detectors in practice is a very complex problem. Wilcoxon rank decision-making rules are difficult but realizable. In addition, Wilcoxon rank signal detection algorithms are comparable with the parametric signal detection algorithms when $N > 15$. To make the decision a "yes" or a "no" signal in the input stochastic process by the Wilcoxon rank signal detection algorithm it is necessary to carry out N^2 operations. For this reason, the rank detectors based on the Wilcoxon statistics are the most widely used in practice.

There is also a need to consider the problem of the number of detector inputs. The one-input detectors use a single sample for making a "yes" or a "no" signal decision in the input stochastic process and are sensitive to variation in the symmetry of the probability distribution density when a "yes" signal exists in the input stochastic process or the signal parameters are varied during observation and analysis. Hypotheses associated with symmetry of the probability distribution density and independence of random values of the samples correspond to particular cases of detection problems for signals in the presence of additive noise.

So, the hypothesis for symmetry of the probability distribution density holds only in the case of the detection of coherent signals in the presence of additive noise. Testing of the hypothesis with respect to independence of random values of the samples does not allow us to detect the mean of the signal. Effectiveness of one-input detectors is very poor.

The two-sample decision-making rules cover more general cases of signal detection in the presence of additive noise. These rules require fewer *a priori* data, because the reference (noise) sample is used. The decision a "yes" signal in the input stochastic process is made when there is a contrast between the observed samples and the reference samples. The rank statistic is the quantitative measure of this contrast.[146,147]

Consider the two-input Neyman–Pearson rank detector. The input sample determined by Eq. (2.154) is observed. For the each X_i the reference sample $(n_1, n_2, \ldots, n_N)$, relative to which the rank for X_i is computed, is observed. Next the rank for X_{i+1} is computed relative to M readings of the new reference sample $(n_{i+11}, n_{i+12}, \ldots)$ and so on. After that the statistic $Z(\vec{R})$ is determined by results of definition of the rank vector determined by Eq. (2.212).

Summing the ranks, we obtain the modified Wilcoxon testing statistic:

$$Z = \sum_{i=1}^{N} R_i . \tag{2.216}$$

In Eq. (2.216) the rank for X_i is determined relative to the i-th reference sample. Thus, the reference sample is made anew at each subsequent step.

Furthermore, modified testing, using more information about the additive noise, is more effective.

Another statistic that differs from the statistic in Eq. (2.216) by the value $0.5N(N+1)$ is introduced:

$$Z = \sum_{i=1}^{N} r_i = \sum_{i=1}^{N} \sum_{j=1}^{M} h(X_i - n_{ij}); \tag{2.217}$$

$$h(X_i - n_{ij}) = \begin{cases} 1, & X_i > n_{ij}; \\ 0, & X_i < n_{ij}, \end{cases} \tag{2.218}$$

where

$$r_i + 1 = R_i, \tag{2.219}$$

and $h(X_i - n_{ij})$ is the indicator of exceeding X_i over n_{ij}. The reference and observed samples and the results of the determination of ranks can be represented in the matrix form:

$$\begin{Vmatrix} n_{11} & n_{12} & \cdots & n_{1M} & X_1 \\ n_{21} & n_{22} & \cdots & n_{2M} & X_2 \\ \cdots & \cdots & \cdots & \cdots & \cdots \\ n_{N1} & n_{N2} & \cdots & n_{NM} & X_N \end{Vmatrix} \rightarrow \begin{Vmatrix} r_1 \\ r_2 \\ \vdots \\ r_N \end{Vmatrix}. \tag{2.220}$$

The signal detection algorithm has the following form when a "yes" signal exists in the input stochastic process:

$$Z > K. \tag{2.221}$$

With the rank vector

$$\vec{r} = (r_1, r_2, \ldots, r_N) \tag{2.222}$$

we may use another function of the vector $\vec{r}$ as the statistic $Z(\vec{r})$. This function cannot be a Wilcoxon-type function, but it can be the statistic based on the determination of the likelihood ratio of the rank vector

$$\ell = \frac{P(\vec{r} \mid H_1)}{P(\vec{r} \mid H_0)} = \prod_{i=1}^{N} \frac{P(r_i \mid H_1)}{P(r_i \mid H_0)}, \tag{2.223}$$

where $P(r_i \mid H_0)$ and $P(r_i \mid H_1)$ are the probabilities of the rank r_i under the hypotheses H_0 and H_1, respectively, or it can be the statistic based on the binary rank quantization:[112,148]

$$Z = \sum_{i=1}^{N} c_i = C, \tag{2.224}$$

where

$$c_i = \begin{cases} 1, & r_i > K_1; \\ 0, & r_i \leq K_1. \end{cases} \tag{2.225}$$

Here K_1 is the quantization threshold.

The model considered corresponds to the multi-channel detector (for example in radar systems) in which the range channels are used as independent channels. Here the use of fast Doppler channels is also possible.[149] In addition to radar systems, this model can also be used in sonar and communication systems.

The following avenues of investigation relevant to non-parametric signal detection are:

- Sequential rank signal detection[3,4,19,26,87,150–165]
- Adaptive sequential rank signal detection[20,87,88,133,166–174]
- Non-parametric signal detection in the presence of correlated Gaussian noise[158,175–190]
- Rank and sign-rank signal detection[42,144,166,191–196]
- Non-parametric signal detection based on the mixed statistics[151,166,197–203]

2.9 Conclusions

In this chapter we considered briefly the main features of classical and modern signal detection theories and various approaches to signal detection problems in the presence of additive noise. The main purpose of this chapter was to introduce readers to the problem of signal detection in the presence of additive noise and to present the main avenues of investigation in this area without detailed analysis. For this reason, the aspects considered in this chapter were presented in a concise form. The main features of classical and modern signal detection theories were summarized briefly.

The Gaussian and Markov approaches to problems of the optimal detection of signals in the presence of additive Gaussian noise are in conflict with each other in the sense that, although the techniques of these approaches are essentially different, their areas of application coincide. At present, it is impossible to give an unambiguous answer as to which approach (Gaussian or Markov) is more suitable for solving the problems presented by the construction of the optimal detector of signals in the presence of additive Gaussian noise. Because of this, it is of interest to compare the Gaussian and Markov approaches within the boundaries of classical signal detection theory.

The Markov approach has the following advantages over the Gaussian approach:

- The optimal detectors of signals in the presence of additive Gaussian noise constructed on the basis of the Markov approach are described by differential equations. These differential equations are more simply solved using numerical techniques than the integral equations used with the Gaussian approach.
- The Markov approach allows us to construct the optimal detectors of signals in the presence of additive Gaussian noise that are realizable in practice. Furthermore, these optimal detectors coincide structurally with the realizable optimal detectors constructed on the basis of the Gaussian approach when the heuristic procedures suggested by H. Van Trees[46,51,52] are used.
- The Markov approach may be used to solve the problems of signal detection in the presence of additive Gaussian noise in communication systems, in which the signals are non-Gaussian.

The Markov approach has the following disadvantages:

- The Markov approach can only be applied to communication systems in which the power spectrums of stochastic processes and frequency responses of linear filters are bilinear functions of frequency.
- The Markov approach cannot be applied to communication systems in which there are delay lines present at both the transmitter and receiver or if delay lines are present at the communication channels. In particular, it is impossible to solve the problems of optimal signal detection in the presence of additive Gaussian noise under multi-beam signal processing using the Markov approach.
- The Markov approach does not allow us to construct optimal detectors of signals in the presence of additive Gaussian noise when smoothing of signals or optimal filtering with a time lag is carried out.

The Gaussian approach is free from the disadvantages indicated above.

We should also note the following differences between the Gaussian and Markov approaches:

- Under the Gaussian approach the criterion of maximal *a posteriori* probability distribution density is used as the optimal decision-making criterion. Under the Markov approach, the criterion of minimum of the mean square error of filtering is used as the optimal decision-making criterion. Therefore, it is of interest to point out the results presented in Reference 63. H. Kushner derived the integral-differential equations for the maximal *a posteriori* probability distribution density $f[\vec{X}, t \mid \vec{\mathcal{X}}(t)]$ of components of the Markov vector process $\vec{X}(t)$. This equation might be used to construct the optimal detectors of analog signals by the Markov approach, but these detectors are optimal on the basis of the decision-making criterion used under the Gaussian approach.

- The mathematics used with the Markov approach is more difficult than that used with the Gaussian approach. The distinctive feature of the mathematics used with the Markov approach is that all the components of the Markov processes are included in integro-differential equations, even those components that are not detected by the receiver.

It is noteworthy that the Gaussian and Markov approaches have only been investigated for application to the over threshold functional domain of the optimal detectors. In the threshold and under threshold functional domains of the optimal detectors, neither the Gaussian nor Markov approach allows us to solve the problems associated with the construction of optimal detectors of signals in the presence of additive Gaussian noise. When the input signal-to-noise ratio is high, the difference between the optimal decision-making criteria under the Gaussian and Markov approaches mentioned above is not essential; and if both the Gaussian and Markov approaches are to be applied, the structures of the optimal realizable detectors are the same.

Analysis of the detection problems of signals in the presence of additive Gaussian noise under conditions of *a priori* parametric uncertainty is characterized by the fact that the probability distribution density of the input stochastic process is known, but ignorance of *a priori* knowledge about the signal and noise is a deficiency of the parameters of this probability distribution density. In doing so, the signal detection problem is formulated as the test of the complex hypotheses with respect to the probability distribution density of the observed sample. The hypothesis H_1—a "yes" signal in the input stochastic process—is based on the fact that the probability distribution density of the observed input stochastic process takes the form $f_1(\vec{X} \mid \vec{\vartheta})$, where $\vec{\vartheta}$ is the vector of the unknown parameters of the signal and the additive Gaussian noise. The hypothesis H_0—a "no" signal in the input stochastic process—is based on the fact that the probability distribution density of the observed input stochastic process takes the form $f_0(\vec{X} \mid \vec{\vartheta})$ and depends only on the unknown parameters of the additive Gaussian noise.

One of the approaches to solving the signal detection problem under conditions of *a priori* parametric uncertainty is the Bayes' technique. This approach is based on the fact that the unknown parameters of the signal are supposedly random with *a priori* probability distribution density $f(\vec{\vartheta})$. The knowledge of the *a priori* probability distribution density allows us to derive the unconditional probability distribution density

$$f_1(\vec{X}) = \int_{\vec{\Theta}} f_1(\vec{X} \mid \vec{\vartheta}) f(\vec{\vartheta})\, d\vec{\vartheta}. \tag{2.226}$$

One can eliminate the unknown parameters by averaging and thus make the hypothesis H_1 a simple hypothesis. The assumption of randomness of the unknown parameters is not necessarily correct, and this is the well-known limitation of the Bayes' approach. Consequently, an *a priori* probability distribution density $f(\vec{\vartheta})$ does not always exist. However, if an *a priori* probability distribution density $f(\vec{\vartheta})$ is given, and the Bayes' approach can be applied, the mathematical difficulties associated with averaging with respect to *a priori* probability distribution density $f(\vec{\vartheta})$ arise.

These difficulties can be effectively overcome in the event of a high signal-to-noise ratio. In this case, the probability distribution density $f_1(\vec{X} \mid \vec{\vartheta})$ has a clearly defined maximum in the neighborhood of the real value of the parameter $\vec{\vartheta}$, and the probability distribution density $f_1(\vec{\vartheta})$ can be determined in the following manner:

$$f_1(\vec{X}) \approx f_1\big(\vec{X} \,\big|\, \widehat{\vec{\vartheta}}\big), \tag{2.227}$$

where $\widehat{\vec{\vartheta}}$ is the estimation of the maximal likelihood ratio of the parameter $\vec{\vartheta}$. The relationship between the Bayes' approach and the maximal likelihood ratio is defined, and the structure of the optimal detector is demonstrated.[30,36,39,44,45]

Another approach to solving the signal detection problem under the conditions of *a priori* parametric uncertainty is to ignore the *a priori* probability distribution density of the unknown parameters. Various properties of signal detection must be considered individually in this event, and the problems of improving the properties of the signal detection are considerable. To solve these contradictions we must introduce limitations of various kinds, and a large number of approaches to the signal detection optimization problem is possible.

The formalization of the optimization problem under the conditions of *a priori* parametric uncertainty is very simple on the quality level. Let $\varphi(\vec{X})$ be the optimal decision-making function that must be defined (criterion). The non-randomized decision-making function takes the following form:

$$\varphi(\vec{X}) = \begin{cases} 1, & \vec{X} \in \vec{X}_1; \\ 0, & \vec{X} \in \vec{X}_0. \end{cases} \tag{2.228}$$

For this reason, we mean that the hypothesis H_1 is accepted when the sample $\vec{X}$ belongs to the set $\vec{X}_1$; otherwise, it is rejected. The randomized decision-making function takes any one of the values

$$0 \le \varphi(\vec{X}) \le 1, \tag{2.229}$$

and when the randomized decision-making function is equal to $\varphi(\vec{X})$, the hypothesis H_1 is accepted with the probability equal to $\varphi(\vec{X})$. Then the probability of false alarm (significance of criterion) is determined in the following form:

$$P_F(\vec{\vartheta}) = \int_{\vec{X}} \varphi(\vec{X}) f_0(\vec{X} \mid \vec{\vartheta})\, d\vec{X}, \tag{2.230}$$

and the probability of detection (power of criterion) is given by

$$P_D(\vec{\vartheta}) = \int_{\vec{X}} \varphi(\vec{X}) f_1(\vec{X} \mid \vec{\vartheta})\, d\vec{X}. \tag{2.231}$$

The problem of optimization lies in a choice of the decision-making function $\varphi(\vec{X})$ such that the probability of false alarm $P_F(\vec{\vartheta})$ is much lower and the probability of detection $P_D(\vec{\vartheta})$ is much higher. It is obvious that these two conditions are inconsistent. Moreover, since the probability of false alarm $P_F(\vec{\vartheta})$ and the probability of detection $P_D(\vec{\vartheta})$ are simultaneous functions, the minimization (or maximization) of these functions at one point may conflict with minimization (or maximization) of these functions at another point. For these reasons we need to use additional conditions and limitations. If the probability of false alarm $P_F(\vec{\vartheta})$ is fixed, then there is the problem of defining the criterion, for which the probability of detection $P_D(\vec{\vartheta})$ is maximal simultaneously for all values of the parameter $\vec{\vartheta}$. This criterion is called the uniformly most powerful criterion under the given condition with respect to the probability of false alarm $P_F(\vec{\vartheta})$. Determination of the uniformly most powerful criteria is the main problem of the optimization of signal detection under conditions of *a priori* parametric uncertainty.

It is simplest to fix the probability of false alarm $P_F(\vec{\vartheta})$, which is independent of the parameter $\vec{\vartheta}$. If this criterion exists, it is called the similar criterion. Conditions of similarity allow us often to limit the subset of criteria, and the existence of the uniformly most powerful criterion is ensured in this subset, therefore the way to determine this criterion is clear. However, to define the only true decision we must limit the criterion by the condition of unbiasedness

$$P_F(\vec{\vartheta}) \leq P_D(\vec{\vartheta}). \tag{2.232}$$

The technique of sequential analysis is widely used for signal detection in the presence of additive noise. It is well known that sequential analysis uses samples of non-fixed beforehand sizes. The technique of sequential analysis supposes a control of the experimental study of the observations based on the information received. This statement of the problem is adequate for signal detection in complex and non-stationary noise situations. It is known that the problem of definition of the characteristics of the sequential procedure is a most difficult one, involving the definition of the thresholds under the predetermined probability of false alarm P_F, the probability of signal omission P_M, and the computer calculation of the average observation time of the samples.

Employment of sequential analysis is very useful for solving signal detection problems under nuisance parameters. Because the sample size is controlled in the course of the observations of the input stochastic process during sequential analysis, the conditions of existence of the invariant uniformly most powerful criteria are attenuated. For the weak signals (i.e., the signal-to-noise ratio at the input of the receiver is much less than 1) there are the effective invariant or nearly invariant sequential decision-making rules. Problems of signal detection in the presence of additive Gaussian noise with unknown energy characteristics are examples of the use of sequential analysis.

Problems of signal detection in the additive non-Gaussian noise are very real. The additive non-Gaussian noise is widely found in nature. Improving detection performances by means of more exact consideration and analysis of the statistical characteristics of the additive non-Gaussian noise is a very important issue. Signal detection theory in the presence of additive non-Gaussian noise is constructed mainly for two noise models: the Markov processes and the non-linear transformation of the Gaussian noise. A great deal of attention is focused on the consideration and analysis of optimal signal detection algorithms and potential noise immunity of signal processing systems.

Non-parametric signal detection was considered with a special emphasis on signal detection problems, in which the probability distribution density of the signal and noise and the probability distribution density of the noise are unknown. In doing so, the signal detection problems are not reduced to a definition of the unknown parameters of the probability distribution density. Non-parametric techniques are not based on the knowledge of the functional type of the probability distribution density. As a rule, the way to overcome the non-parametric uncertainty is to examine the decision-making statistics, which are independent of the probability distribution density of the additive noise. This ensures the stabilization of the probability of false alarm P_F or the probability of signal omission P_M with a given accuracy under variable noise conditions. Synthesis of non-parametric detectors meets insurmountable mathematical difficulties in practice. This problem may be solved only in the asymptotic case, when the sample size of observations tends to approach ∞. If the sample size is limited, the non-parametric decision-making criteria are proposed. These criteria are based on the test of the sign statistic and the rank statistic. Obviously, the probability distribution density of the sign statistic does not depend on the type of additive noise if it is independent and symmetrically distributed with respect to zero. In a similar way, the probability distribution density of the rank statistic does not depend on the additive noise if it is independent and stationary.

References

1. Middleton, D., *An Introduction to Statistical Communication Theory*, McGraw-Hill, New York, 1960.
2. Selin, I., *Detection Theory*, Princeton University Press, Princeton, NJ, 1965.
3. Noether, G., *Elements of Non-Parametric Statistics*, Wiley, New York, 1967.
4. Hajek, J. and Sidak, Z., *Theory of Rank Tests*, Academic Press, New York, 1967.

5. Van Trees, H., *Detection, Estimation and Modulation Theory. Part I: Detection, Estimation, and Linear Modulation Theory*, Wiley, New York, 1968.
6. Helstrom, C., *Statistical Theory of Signal Detection*, 2nd ed., Pergamon Press, Oxford, London, 1968.
7. Gallager, R., *Information Theory and Reliable Communications*, Wiley, New York, 1968.
8. Fu, K., *Sequential Methods in Pattern Recognition and Machine Learning*, Academic Press, New York, 1968.
9. Thomas, J., *An Introduction to Statistical Communication Theory*, Wiley, New York, 1969.
10. Jazwinski, A., *Stochastic Processes and Filtering Theory*, Academic Press, New York, London, 1970.
11. Van Trees, H., *Detection, Estimation and Modulation Theory. Part II: Non-Linear Modulation Theory*, Wiley, New York, 1970.
12. Schwartz, M., *Information, Transmission, Modulation, and Noise*, 2nd ed., McGraw-Hill, New York, 1970.
13. Ghosh, B., *Sequential Tests of Statistical Hypotheses*, Addison-Wesley, Cambridge, MA, 1970.
14. Wong, E., *Stochastic Processes in Information and Dynamical Systems*, McGraw-Hill, New York, 1971.
15. Van Trees, H., *Detection, Estimation and Modulation Theory. Part III: Radar-Sonar Signal Processing and Gaussian Signals in Noise*, Wiley, New York, 1972.
16. Box, G. and Tiao, G., *Bayesian Inference in Statistical Analysis*, Addison-Wesley, Cambridge, MA, 1973.
17. Levin, B., *Theoretical Foundations of Statistical Radio Engineering. Parts I–III*, Soviet Radio, Moscow, 1974–1976 (in Russian).
18. Tikhonov, V. and Kulman, N., *Non-Linear Filtering and Quasideterministic Signal Processing*, Soviet Radio, Moscow, 1975 (in Russian).
19. Gibson, J. and Melsa, J., *Introduction to Non-Parametric Detection with Applications*, Academic Press, New York, 1975.
20. Repin, V. and Tartakovskiy, G., *Statistical Synthesis under a Priori Uncertainty and Adaptation of Information Systems*, Soviet Radio, Moscow, 1977 (in Russian).
21. Huber, P., *Robust Statistical Procedures*, SIAM, Philadelphia, 1977.
22. Kulikov, E. and Trifonov, A., *Estimation of Signal Parameters in Noise*, Soviet Radio, Moscow, 1978 (in Russian).
23. Sosulin, Yu., *Detection and Estimation Theory of Stochastic Signals*, Soviet Radio, Moscow, 1978 (in Russian).
24. Ibragimov, I. and Rozanov, Y., *Gaussian Random Processes*, Springer-Verlag, New York, 1978.
25. Anderson, B. and Moore, J., *Optimal Filtering*, Prentice-Hall, Englewood Cliffs, NJ, 1979.
26. Kassam, S. and Thomas, J., eds., *Non-Parametric Detection Theory and Applications*, Dowden, Hutchinson & Ross, Stroudsburg, PA, 1980.
27. Shirman, Y. and Manjos, V., *Theory and Methods in Radar Signal Processing*, Radio and Svyaz, Moscow, 1981 (in Russian).
28. Huber, P., *Robust Statistics*, Wiley, New York, 1981.
29. Blachman, N., *Noise and Its Effect in Communications*, 2nd ed., Krieger, Malabar, FL, 1982.

30. Bacut, P., et al., *Signal Detection Theory*, Radio and Svyaz, Moscow, 1984 (in Russian).
31. Anderson, T., *An Introduction to Multivariate Statistical Analysis*, 2nd ed., Wiley, New York, 1984.
32. Siegmund, D., *Sequential Analysis: Tests and Confidence Intervals*, Springer-Verlag, New York, 1985.
33. Silverman, B., *Density Estimation for Statistics and Data Analysis*, Chapman & Hall, London, 1986.
34. Blahut, R., *Principles of Information Theory*, Addison-Wesley, Reading, MA, 1987.
35. Weber, C., *Elements of Detection and Signal Design*, Springer-Verlag, New York, 1987.
36. Skolnik, M., *Radar Applications*, IEEE Press, New York, 1988.
37. Kassam, S., *Signal Detection in Non-Gaussian Noise*, Springer-Verlag, Berlin, 1988.
38. Poor, V., *Introduction to Signal Detection and Estimation*, Springer-Verlag, New York, 1988.
39. Scharf, L., *Statistical Signal Processing, Detection, Estimation, and Time Series Analysis*, Addison-Wesley, Reading, MA, 1991.
40. Ghosh, B. and Sen, P., Eds., *Handbook on Sequential Analysis*, Marcel Dekker, New York, 1991.
41. Snyder, D. and Miller, M., *Random Point Processes in Time and Space*, Springer-Verlag, New York, 1991.
42. Fisher, N., *Statistical Analysis of Circular Data*, Cambridge University Press, Cambridge, U.K., 1993.
43. Porat, B., *Digital Processing of Random Signals: Theory and Methods*, Prentice-Hall, Englewood Cliffs, NJ, 1994.
44. Helstrom, C., *Elements of Signal Detection and Estimation*, Prentice-Hall, Englewood Cliffs, NJ, 1995.
45. McDonough, R. and Whallen, A., *Detection of Signals in Noise*, 2nd ed., Academic Press, New York, 1995.
46. Van Trees, H., *Optimum Array Processing*, Wiley, New York, 1971.
47. Lehan, F. and Parks, R., Optimum demodulation, *IRE Natl. Conv. Rec.*, Pt. 8, 1953, pp. 101–103.
48. Youla, D., The use of maximum likelihood in estimating continuously modulated intelligence which has been corrupted by noise, *IRE Trans.*, Vol. IT-3, March 1954, pp. 90–105.
49. Thomas, J. and Wong, E., On the statistical theory of optimum demodulation, *IRE Trans.*, Vol. IT-6, No. 5, September 1960, pp. 420–425.
50. Bolshakov, I. and Repin, V., The problems of non-linear filtering, *Avtomatika and Telemechanika*, No. 4, 1961, pp. 47–55 (in Russian).
51. Van Trees, H., The structure of efficient demodulator for multi-dimensional phase modulated signals, *IEEE Trans.*, Vol. CS-11, No. 3, 1963, pp. 235–247.
52. Van Trees, H., Analog communication over randomly-time varying channels, *IEEE Trans.*, Vol. IT-12, No. 1, January 1966, pp. 51–63.
53. Loeve, M., Sur les fonctions aleatoires stationnaires de second order, *Rev. Sci.*, Vol. 83, 1945, pp. 297–310.
54. Karhunen, K., Über linearen Methoden in der Wahrscheinlichkeitsrechnung, *Ann. Acad. Sci. Fennical*, Ser. A., Vol. 1, No. 2, 1946, pp. 127–139.

55. Van Trees, H., Bounds on the accuracy attainable in the estimation of continuous random processes, *IEEE Trans.*, Vol. IT-12, July 1966, pp. 425–437.
56. Tikhonov, V. and Mironov, M., *Markov Processes*, Soviet Radio, Moscow, 1977 (in Russian).
57. Cox, D. and Miller, H., *The Theory of Stochastic Processes*, Wiley, New York, 1965.
58. Stratonovich, R., *Selected Problems of Fluctuation Theory in Radio Engineering*, Soviet Radio, Moscow, 1961 (in Russian).
59. Stratonovich, R., *Conditional Markov Processes and Applications to Theory of Optimal Control*, Moscow State University, Moscow, 1966 (in Russian).
60. Kushner, H., On the differential equations satisfied by conditional probability densities of Markov processes with applications, *J. SIAM on Control. Ser. A.*, Vol. 2, No. 1, 1964, pp. 37–45.
61. Snyder, D., Some useful expressions for optimum linear filtering in white noise, *Proc. IEEE*, Vol. 53, No. 6, 1965, pp. 532–539.
62. Snyder, D., Optimum linear filtering of an integrated signal in white noise, *IEEE Trans.*, Vol. AES-2, No. 2, 1966, pp. 153–161.
63. Kushner, H., Non-linear filtering. The exact dynamical equations satisfied by a conditional mode, *IEEE Trans.*, Vol. AC-12, No. 3, 1967, pp. 247–255.
64. Snyder, D., The state-variable approach to analog communication theory, *IEEE Trans.*, Vol. IT-14, No. 1, January 1968, pp. 57–65.
65. Middleton, D., Acoustic signal detection by simple correlator in the presence of non-Gaussian noise, *J. Acoust. Soc. Am.*, Vol. 34, October 1962, pp. 1598–1609.
66. Esposito, R., Middleton, D., and Mullen, J., Advantages of amplitude and phase adaptivity in the detection of signals subject to slow Rayleigh fading, *IEEE Trans.*, Vol. IT-11, No. 5, October 1965, pp. 473–482.
67. Groginskiy, H., Wilson, L., and Middleton, D., Adaptive detection of statistical signals in noise, *IEEE Trans.*, Vol. IT-12, No. 4, July 1966, pp. 357–363.
68. Kalman, R. and Bucy, R., New results in linear filtering and prediction theory, *J. Basic Engrg., Trans. ASME*, Vol. 83, March 1961, pp. 187–195.
69. Crammer, H., *Mathematical Methods of Statistics*, Princeton University Press, Princeton, NJ, 1946.
70. Rao, C., *Advanced Statistical Methods in Biometric Research*, Wiley, New York, 1952.
71. Lehmann, E., *Testing Statistical Hypotheses*, 2nd ed., Wiley, New York, 1986.
72. Lugosi, G. and Zeger, K., Non-parametric estimation via empirical risk minimization, *IEEE Trans.*, Vol. IT-41, No. 3, May 1995, pp. 677–687.
73. Middleton, D., On the detection of stochastic signals in additive normal noise, Part I, *IRE Trans.*, Vol. IT-13, No. 2, June 1957, pp. 343–359.
74. Middleton, D., On the detection of stochastic signals in additive normal noise, Part II, *IRE Trans.*, Vol. IT-6, No. 2, June 1960, pp. 349–360.
75. Swerling, P., Probability of detection for fluctuating targets, *IRE Trans.*, Vol. IT-6, No. 1, April 1960, pp. 269–308.
76. Middleton, D., On singular and non-singular optimum (Bayes) tests for the detection of normal stochastic signals in normal noise, *IRE Trans.*, Vol. IT-7, No. 1, April 1961, pp. 105–113.
77. Martin, R. and McGath, C., Robust detection of stochastic signals, *IEEE Trans.*, Vol. IT-20, No. 4, July 1974, pp. 537–541.
78. Sonalker, R. and Shen, C., Rapid estimation and detection scheme for unknown discretized rectangular inputs, *IEEE Trans.*, Vol. AC-20, No. 1, February 1975, pp. 142–144.

79. Martin, R. and Schwartz, S., Robust detection of a known signal in nearly Gaussian noise, *IEEE Trans.*, Vol. IT-17, No. 1, January 1971, pp. 50–56.
80. Pollak, M., Optimal detection of a change in distribution, *Ann. Stat.*, Vol. 13, 1985, pp. 206–227.
81. Verdu, S., Minimum probability of error for asynchronous Gaussian multiple-access channels, *IEEE Trans.*, Vol. IT-32, No. 1, January 1986, pp. 85–96.
82. Papadopoulos, H. and Wornell, G., Maximum-likelihood estimation of a class of chaotic signals, *IEEE Trans.*, Vol. IT-41, No. 1, January 1995, pp. 312–317.
83. Huber, P. and Strassen, V., Minimax tests and the Neyman–Pearson lemma for capacities, *Ann. Math. Statistics*, Vol. 1, 1973, pp. 251–263.
84. D'Appolito, J. and Hutchinson, C., Minimax approach to the design of low sensitivity state estimators, *Automatica*, Vol. 8, 1972, pp. 599–608.
85. Kelsey, P. and Haddad, A., A note on detectors for joint minimax detection-estimation schemes, *IEEE Trans.*, Vol. AC-18, No. 5, October 1973, pp. 558–559.
86. Abraham, D., A page test with nuisance parameter estimation, *IEEE Trans.*, Vol. IT-42, No. 6, November 1996, pp. 2242–2252.
87. Wald, A., *Sequential Analysis*, Wiley, New York, 1947.
88. Wald, A., *Statistical Decision Functions*, Wiley, New York, 1950.
89. Wald, A. and Wolfowitz, J., Optimum character of the sequential probability ratio test, *Ann. Math. Statistics*, Vol. 19, 1948, pp. 326–339.
90. Wald, A. and Wolfowitz, J., Bayes solutions of sequential decision problems, *Ann. Math. Statistics*, Vol. 21, 1950, pp. 82–99.
91. Kolmogorov, A. and Prokchorov, Yu., Sums of random values of random components, *UMN*, Vol. 4, No. 4, 1949, pp. 168–172 (in Russian).
92. Aivazian, S., Comparison of the optimal properties of the Wald and Neyman–Pearson criteria, *Theory of Probability and Its Applications*, Vol. 4, 1959, pp. 83–89.
93. Rozanov, B., Distribution of accumulated value of decision statistic under sequential analysis, *Radio Eng. Electron. Phys.*, Vol. 17, No. 10, 1972, pp. 1021–1034.
94. Sosulin, Yu. and Fishman, M., Sequential detection of signals in noise, *Eng. Cybern.*, No. 2, 1973, pp. 189–197.
95. Vlasov, I., Calculation of sequential analysis duration, *Radio Eng. Electron. Phys.*, Vol. 19, No. 1, 1974, pp. 187–189.
96. Tantarantana, S. and Thomas, J., Relative efficiency of the sequential probability ratio test in signal detection, *IEEE Trans.*, Vol. IT-24, No. 1, January 1978, pp. 22–31.
97. Sosulin, Yu., Tartakovskiy, A., and Fishman, M., Sequential detection of correlated Gaussian signal in white noise, *Radio Eng. Electron. Phys.*, Vol. 24, No. 4, 1979, pp. 720–732.
98. Vlasov, I., Kuzmina, E., and Solovyev, G., Sequential procedure with composed statistic for signal detection in multi-channel systems, *Radio Eng. Electron. Phys.*, Vol. 28, No. 9, 1983, pp. 965–973.
99. Tantarantana, S., Sequential detection of a positive signal, *Communications and Networks*, Black, I. and Poor, H., Eds., Springer-Verlag, New York, 1986.
100. Efremovich, S., Sequential non-parametric estimation with assigned risk, *Ann. Statist.*, Vol. 23, December 1995, pp. 1376–1392.
101. Sosulin, Yu. and Fishman, M., Optimal sequential detection under information degree of cost of observations, *Eng. Cybern.*, No. 3, 1974, pp. 169–176.
102. Koplowitz, J. and Roberts, R., Sequential estimation with a finite statistic, *IEEE Trans.*, Vol. IT-19, No. 3, May 1973, pp. 631–635.

103. Morris, J. and Van de Linde, D., Robust quantization of discrete-time signals with independent samples, *IEEE Trans.*, Vol. COM-22, No. 12, December 1974, pp. 1897–1902.
104. El-Sawy, A. and Van de Linde, D., Robust sequential detection of signals in noise, *IEEE Trans.*, Vol. IT-25, No. 3, May 1979, pp. 346–353.
105. Bussang, J., Sequential methods in radar detection, *Proc. IEEE*, Vol. 58, May 1970, pp. 731–743.
106. Lainiotis, D. and Park, S., On joint detection, estimation and system identification-discrete data case, *Int. J. Control*, Vol. 17, No. 3, 1973, pp. 609–633.
107. Spaulding, A., Optimum threshold signal detection in broad-band impulsive noise employing both time and spatial sampling, *IEEE Trans.*, Vol. COM-29, No. 2, February 1981, pp. 147–152.
108. Zhang, Q., Wong, K., Patrick, C., and Reilly, J., Statistical analysis of the performance of information theoretic criteria in the detection of the number of signals in array processing, *IEEE Trans.*, Vol. ASSP-37, No. 10, October 1989, pp. 1557–1567.
109. DeLucia, J. and Poor, H., Performance analysis of sequential tests between Poisson processes, *IEEE Trans.*, Vol. IT-43, No. 1, January 1997, pp. 221–238.
110. Lehmann, E., The power rank tests, *Ann. Math. Statistics*, Vol. 24, 1953, pp. 237–249.
111. Capon, J., On the asymptotic efficiency of locally optimum detectors, *IRE Trans.*, Vol. IT-7, No. 1, April 1961, pp. 67–71.
112. Antonov, O., Optimal signal detection in non-Gaussian noise, *Radio Eng. Electron. Phys.*, Vol. 12, No. 4, 1967, pp. 579–587.
113. Martinez, A., Swaszeck, P., and Thomas, J., Locally optimum detection in multivariate non-Gaussian noise, *IEEE Trans.*, Vol. IT-30, No. 6, November 1984, pp. 815–822.
114. Spaulding, A., Locally optimum and suboptimum detector performance in a non-Gaussian interference environment, *IEEE Trans.*, Vol. COM-33, No. 6, June 1985, pp. 509–517.
115. Blachman, N., Gaussian noise. Part I: The shape of large excursions, *IEEE Trans.*, Vol. IT-34, No. 6, November 1988, pp. 1396–1400.
116. Maras, A., Locally optimum detection in mixing average non-Gaussian noise, *IEEE Trans.*, Vol. COM-36, No. 8, August 1988, pp. 907–912.
117. Valeev, V. and Sosulin, Yu., Detection of weak coherent signals in correlated non-Gaussian noise, *Radio Eng. Electron. Phys.*, Vol. 14, No. 2, 1969, pp. 229–237.
118. Kassam, S. and Thomas, J., Asymptotically robust detection of a known signal in contaminated non-Gaussian noise, *IEEE Trans.*, Vol. IT-22, No. 1, January 1976, pp. 22–26.
119. Sheehy, J., Optimum detection of signals in non-Gaussian noise, *J. Acoust. Soc. Am.*, Vol. 63, No. 1, January 1978, pp. 81–90.
120. Farina, A. and Russo, A., Radar detection of correlated targets in clutter, *IEEE Trans.*, Vol. AES-22, No. 5, September 1982, pp. 513–536.
121. Middleton, D., Threshold detection in non-Gaussian interference environments: exposition and interpretation of new results for EMC applications, *IEEE Trans.*, Vol. EMC-26, No. 1, February 1984, pp. 19–28.
122. Middleton, D. and Spaulding, A., Elements of weak signal detection in non-Gaussian noise environments, *Advances in Statistical Signal Processing*, Vol. 2, Poor, H. and Thomas, J., Eds., JAI Press, Greenwich, CT, 1993, chap. 5.

123. Garth, L. and Poor, V., Detection of non-Gaussian signals: A paradigm for modern statistical signal processing, *Proc. IEEE*, Vol. 82, No. 7, 1994, pp. 1061–1095.
124. Middleton, D., Threshold detection in correlated non-Gaussian noise fields, *IEEE Trans.*, Vol. IT-41, No. 4, July 1995, pp. 976–1000.
125. Levin, B. and Kushnir, A., Asymptotically optimum rank algorithms of signal detection in noise, *Radio Eng. Electron. Phys.*, Vol. 14, No. 2, 1969, pp. 221–228.
126. Rappaport, S. and Kurz, L., An optimal non-linear detector for digital data transmission through non-Gaussian channels, *IEEE Trans.*, Vol. COM-14, No. 6, June 1966, pp. 266–274.
127. Miller, J. and Thomas, J., Detectors for discrete-time signals in non-Gaussian noise, *IEEE Trans.*, Vol. IT-18, No. 2, March 1972, pp. 241–250.
128. Modestino, J. and Ningo, A., Detection of weak signals in narrow-band non-Gaussian noise, *IEEE Trans.*, Vol. IT-25, No. 5, September 1979, pp. 592–600.
129. Halverson, D. and Wise, G., A detection scheme for dependent noise processes, *J. Franklin Inst.*, Vol. 309, May 1980, pp. 287–300.
130. Pakula, L. and Kay, S., Detection performance of the circular correlation coefficient receiver, *IEEE Trans.*, Vol. ASSP-34, No. 3, June 1986, pp. 399–404.
131. Conte, E., Lops, M., and Ullo, S., Detection of signals with unknown parameters in correlated K-distributed clutter, *Proc. IEEE*, Vol. 138, No. 2, April 1991, pp. 279–287.
132. Levin, B., Kushnir, A., and Pinskiy, A., Asymptotic optimal signal detection algorithms in correlated noise, *Radio Eng. Electron. Phys.*, Vol. 16, No. 5, 1971, pp. 719–732.
133. Menon, M., Estimation of the shape and scale parameters of the Weibull distribution, *Technometrics*, Vol. 5, No. 2, 1963, pp. 175–182.
134. Blum, R. and Kassam, S., Approximate analysis of the convergence of relative efficiency to ARE for known signal detection, *IEEE Trans.*, Vol. IT-37, No. 1, January 1991, pp. 199–206.
135. Hero, A. and Kim, J., Simultaneous signal detection and classification under a false alarm constraint, *Proc. IEEE ICASSP-90*, Albuquerque, 1990, pp. 457–463.
136. Ozturk, A., Chakravarthi, P., and Weiner, D., On determining the radar threshold for non-Gaussian processes from experimental data, *IEEE Trans.*, Vol. IT-42, No. 4, July 1996, pp. 1310–1316.
137. Cox, D., A penalty method for non-parametric estimation of the logarithmic derivative of a density function, *Ann. Inst. Statist. Math.*, Vol. 37, 1985, pp. 271–288.
138. Schwartz, M., A coincidence procedure for signal detection, *IRE Trans.*, Vol. IT-2, No. 6, December 1956, pp. 135–139.
139. Capon, J., Optimum coincidence procedures for detecting weak signals in noise, *IRE Int. Conv. Rec.*, Pt. 4, 1960, pp. 154–166.
140. Wolff, S., Thomas, J., and Williams, T., The polarity coincidence correlator: a non-parametric detection device, *IRE Trans.*, Vol. IT-8, No. 1, January 1962, pp. 97–115.
141. Kanefsky, M., Detection of weak signals with polarity coincidence arrays, *IEEE Trans.*, Vol. IT-12, No. 2, April 1966, pp. 260–268.
142. Shin, J. and Kassam, S., Multi-level coincidence correlators for random signal detection, *IEEE Trans.*, Vol. IT-25, No. 1, January 1979, pp. 47–53.
143. Masry, E. and Bullo, F., Convergence analysis of the sign algorithm for adaptive filtering, *IEEE Trans.*, Vol. IT-41, No. 2, March 1995, pp. 489–495.
144. Wilcoxon, F., Individual comparisons by ranking methods, *Biometrics*, Vol. 1, 1945, pp. 80–83.

145. Van der Waerden, B., *Mathematische Statistik*, Springer-Verlag, Berlin, 1957.
146. Morris, J., Optimal probability-of-error thresholds and performance for two versions of the sign detector, *IEEE Trans.*, Vol. COM-39, No. 12, December 1991, pp. 1726–1728.
147. McKellips, A. and Verdu, S., Worst case additive noise for binary-input channels and zero-threshold detection under constraints of power and divergence, *IEEE Trans.*, Vol. IT-43, No. 4, July 1997, pp. 1256–1264.
148. Middleton, D. and Esposito, R., Simultaneous optimum detection and estimation of signals in noise, *IEEE Trans.*, Vol. IT-14, No. 3, May 1968, pp. 434–444.
149. Verdu, S., Multiuser detection, *Advances in Statistical Signal Processing*, Vol. 2 (Signal Detection), JAI Press, Greenwich, CT, 1993, pp. 369–409.
150. Woinskiy, M., Non-parametric detection using spectral data, *IEEE Trans.*, Vol. IT-18, No. 1, January 1972, pp. 110–118.
151. Brikker, A., Guaranteed detection performances of non-parametric detectors of signals, *Telecommun. Eng.*, Vol. 37, No. 2, 1983, pp. 69–74.
152. Aazhang, B. and Poor, V., On optimum and nearly optimum data quantization for signal detection, *IEEE Trans.*, Vol. COM-32, No. 7, July 1984, pp. 745–751.
153. Hossjer, O. and Mettiji, M., Robust multiple classification of known signals in additive noise—an asymptotic weak signal approach, *IEEE Trans.*, Vol. IT-39, No. 2, March 1993, pp. 594–608.
154. Blachman, N., Gaussian noise: prediction based on its value and *N* derivatives, *Proc. IEEE. Part F: Radar and Signal Processing*, Vol. 140, No. 2, 1993, pp. 98–102.
155. Baygun, B. and Hero, O., Optimal simultaneous detection and estimation under a false alarm constraint, *IEEE Trans.*, Vol. IT-41, No. 3, May 1995, pp. 688–703.
156. Blum, S., Necessary conditions for optimum distributed sensor detectors under the Neyman–Pearson criterion, *IEEE Trans.*, Vol. IT-42, No. 3, May 1996, pp. 990–994.
157. Blum, S., Locally optimum distributed detection of correlated random signals based on ranks, *IEEE Trans.*, Vol. IT-42, No. 3, May 1996, pp. 931–942.
158. Ching, Y. and Kurz, L., Non-parametric detectors based on *m*-interval partitioning, *IEEE Trans.*, Vol. IT-18, No. 2, March 1972, pp. 251–257.
159. Al-Hussaini, E., Badran, F., and Turner, L., Modified average and modified rank squared non-parametric detectors, *IEEE Trans.*, Vol. AES-14, No. 2, March 1978, pp. 242–251.
160. Sanz-Gonzalez, J., Non-parametric rank detectors on quantized radar video signals, *IEEE Trans.*, Vol. AES-39, No. 5, November 1990, pp. 969–975.
161. Song, I. and Kassam, S., Locally optimum rank detection of correlated random signals in additive noise, *IEEE Trans.*, Vol. IT-38, No. 4, July 1992, pp. 1311–1322.
162. Blum, R. and Kassam, S., Distributed cell-averaging CFAR detection in dependent sensors, *IEEE Trans.*, Vol. IT-41, No. 2, March 1995, pp. 513–518.
163. Blum, R., Quantization in multi-sensor random signal detection, *IEEE Trans.*, Vol. IT-41, No. 1, January 1995, pp. 204–215.
164. Akimov, P., Yefremov, V., and Kubasov, A., Non-parametric binary Neyman–Pearson detector, *News of Univ., Ser.: Radio Electron.*, Vol. 21, No. 4, 1978, pp. 78–83 (in Russian).
165. Akimov, P. and Yefremov, V., Truncated sequential rank procedure under multichannel detection, *Radio Eng. Electron Phys.*, Vol. 21, No. 7, 1976, pp. 1452–1457.
166. Hajek, J., Asymptotically most powerful rank-order tests, *Ann. Math. Statistics*, Vol. 33, 1962, pp. 1124–1147.

167. Viterbi, A., The effect of sequential decision feedback on communication over the Gaussian channel, *Informat. Contr.*, Vol. 8, February 1965, pp. 212–225.
168. Kassam, S. and Thomas, J., Generalizations of the sign detector based on conditional tests, *IEEE Trans.*, Vol. COM-24, No. 5, May 1976, pp. 481–487.
169. Kassam, S. and Thomas, J., Improved non-parametric coincidence detectors, *J. Franklin Inst.*, Vol. 303, January 1977, pp. 75–84.
170. Bickel, P., On adaptive estimation, *Ann. Statist.*, Vol. 10, 1982, pp. 647–671.
171. Verhoeckx, N. and Claasen, T., Some considerations on the design of adaptive digital filters equipped with the sign algorithm, *IEEE Trans.*, Vol. COM-32, No. 3, March 1984, pp. 258–266.
172. Mathews, V. and Chu, S., Improved convergence analysis of stochastic gradient adaptive filters using the sign algorithm, *IEEE Trans.*, Vol. ASSP-35, No. 3, May 1987, pp. 450–454.
173. Akimov, P., Adaptation of binary rank detection procedure, *News of Univ., Ser.: Radio Electron.*, Vol. 22, No. 7, 1979, pp. 31–37 (in Russian).
174. Conte, E., Lops, M., and Ullo, S., A new model for coherent Weibull clutter, *Proc. Int. Conf. on Radar*, Paris, France, April 24–28, 1989, pp. 482–487.
175. Kassam, S. and Thomas, J., Dead-zone limiter: an application of conditional tests, *J. Acoust. Soc. Am.*, Vol. 60, No. 4, October 1976, pp. 857–862.
176. Poor, V. and Thomas, J., Asymptotically robust quantization for detection, *IEEE Trans.*, Vol. IT-24, No. 2, March 1978, pp. 222–229.
177. Poor, V., Robust decision design using a distance criterion, *IEEE Trans.*, Vol. IT-26, No. 5, September 1980, pp. 575–587.
178. Bath, W. and Van de Linde, D., Robust memoryless quantization for minimum signal distortion, *IEEE Trans.*, Vol. IT-28, No. 2, March 1982, pp. 296–306.
179. Poor, V., Signal detection in the presence of weakly dependent noise. Part II: Robust detection, *IEEE Trans.*, Vol. IT-28, No. 5, September 1982, pp. 744–752.
180. Picinbono, B. and Duvaut, P., Optimum quantization for detection, *IEEE Trans.*, Vol. COM-36, No. 11, November 1988, pp. 1254–1258.
181. Song, I. and Kassam, S., Locally optimum detection of signals in a generalized observation model: the random signal case, *IEEE Trans.*, Vol. IT-36, No. 3, May 1990, pp. 516–530.
182. Phamdo, N. and Favardin, N., Optimal detection of discrete Markov sources over discrete memoryless channels—applications to combined source–channel coding, *IEEE Trans.*, Vol. IT-40, No. 1, January 1994, pp. 186–193.
183. Alajaji, F., Phamdo, N., Favardin, N., and Fuja, T., Detection of binary Markov sources over channels with additive Markov noise, *IEEE Trans.*, Vol. IT-42, No. 1, January 1996, pp. 230–239.
184. Gustafsson, R., Hossjer, O., and Oberg, T., Adaptive detection of known signals in additive noise by means of kernel density estimators, *IEEE Trans.*, Vol. IT-43, No. 4, July 1997, pp. 1192–1204.
185. Akimov, P., Kubasov, A., and Litnovskiy, V., Binary rank detection of deterministic signal in Markov noise, *Radio Eng. Electron. Phys.*, Vol. 25, No. 7, 1980, pp. 1454–1459.
186. Poor, V. and Thomas, J., Memoryless quantizer-detectors for constant signals in *m*-dependent noise, *IEEE Trans.*, Vol. IT-26, No. 4, July 1980, pp. 423–432.
187. Kassam, S., Optimum quantization for signal detection, *IEEE Trans.*, Vol. COM-25, No. 5, May 1977, pp. 497–507.

188. Kassam, S., Locally robust array detectors for random signals, *IEEE Trans.*, Vol. IT-24, No. 2, March 1978, pp. 309–316.
189. Al-Hussaini, E. and Turner, L., The asymptotic performance of two-sample non-parametric detectors when detecting non-fluctuating signals in non-Gaussian noise, *IEEE Trans.*, Vol. IT-25, No. 1, January 1979, pp. 124–127.
190. Kassam, S. and Poor, V., Robust techniques for signal processing: a survey, *Proc. IEEE*, Vol. 73, 1985, pp. 433–481.
191. Morris, J., On single-sample robust detection for known signals and additive unknown-mean amplitude-bounded random interference, *IEEE Trans.*, Vol. IT-26, No. 2, March 1980, pp. 225–237.
192. Morris, J., On single-sample robust detection of known signals with additive unknown-mean amplitude-bounded random interference. Part II: The randomized decision rule solution, *IEEE Trans.*, Vol. IT-27, No. 1, January 1981, pp. 132–136.
193. Benitz, G. and Blucklew, J., Asymptotically optimum quantizers for detection of I.I.D. data, *IEEE Trans.*, Vol. IT-35, No. 2, March 1989, pp. 316–327.
194. Blum, R. and Kassam, S., Optimum distributed detection of weak signals in dependent sensors, *IEEE Trans.*, Vol. IT-38, No. 3, March 1992, pp. 1066–1079.
195. Kazakos, D., New error bounds and optimum quantization for multi-sensor distributed signal detection, *IEEE Trans.*, Vol. COM-40, No. 7, July 1992, pp. 1144–1151.
196. Willet, P. and Warren, D., The suboptimality of randomized tests in distributed and quantized detection systems, *IEEE Trans.*, Vol. IT-38, No. 2, March 1992, pp. 355–362.
197. Brikker, A., Non-parametric signal detection algorithms based on mixed statistics with linear transformation of input data, *Radio Eng. Electron Phys.*, Vol. 26, No. 10, 1981, pp. 2119–2128.
198. Akimov, P., Non-parametric detection of signals, *Telecommun. Radio Eng.*, Vol. 31/32, Pt. 1, No. 11, 1977, pp. 26–36.
199. Akimov, P. and Yefremov, V., Detection performances of rank radar detector, *Radio Eng. Electron. Phys.*, Vol. 29, No. 7, 1977, pp. 1527–1531.
200. Kay, S., Robust detection by autoregressive spectrum analysis, *IEEE Trans.*, Vol. ASSP-30, No. 2, April 1982, pp. 256–269.
201. Orsak, G. and Paris, B., On the relationship between measures of discrimination and the performance of suboptimal detectors, *IEEE Trans.*, Vol. IT-41, No. 1, January 1995, pp. 188–203.
202. Po-Ning, Chen and Papamarcon, A., Error bounds for parallel distributed detection under the Neyman–Pearson criterion, *IEEE Trans.*, Vol. IT-41, No. 2, March 1995, pp. 528–533.
203. Sun Yong, Kim and Iickho, Song, On the score functions of the two-sample locally optimum rank test statistic for random signals in additive noise, *IEEE Trans.*, Vol. IT-41, No. 3, May 1995, pp. 842–846.

3

Main Characteristics of Multiplicative Noise

At the present time, signal processing systems in various areas of application very intensively utilize signals with complex structure, pulse sequences, and continuous signals with large duration. Improvement in characteristics of signal processing systems is fundamentally contingent on a time interval (for example, the duration of signals) of the coherent signal processing system. If the time interval value in the coherent signal processing is high, the effect in energy characteristics, resolution, and noise immunity is high. The time interval of coherent signal processing is defined by the noise and interference of the environment.

Stochastic distortions of parameters in the transmitted signal, attributable to unforeseen changes in instantaneous values of the signal phase and amplitude as a function of time, can be considered as multiplicative noise. There are two main classes of the noise and interference: additive and multiplicative. The effect of the addition of noise and interference to the signal generates an appearance of false information in the case of additive noise. For this reason, the parameters of the received signal, which is an additive mixture of the transmitted signal, noise, and interference, differ from the parameters of the transmitted signal.

Under the stimulus of the multiplicative noise, a false information is a consequence of changed parameters of transmitted signals, for example, the parameters of transmitted signals are corrupted by the noise and interference. Thus, the impact of the additive noise and interference may be lowered by an increase in the signal-to-noise ratio. However, in the case of the multiplicative noise and interference, an increase in the signal-to-noise ratio does not produce any positive effects.[1–3]

Let us consider briefly a classification of the noise and interference.

3.1 Classification of the Noise and Interference

Quality and integrity of any signal processing system are defined by statistical characteristics of the noise and interference, which are caused by an electromagnetic field of the environment. Consider a classification of noise and interference in accordance with their physical nature and peculiarities of the stimulus on the parameters of transmitted signals. This classification is based on dividing all forms of noise and interference into three main classes: additive, multiplicative, and intersymbol noise and interference.

Multiplicative noise is called the stochastic distortions of parameters of the transmitted (or target return) signal at the input of the receiver or detector of a signal processing system. Multiplicative noise has a determining impact on the noise immunity of signal processing systems and must be studied very carefully. Under the stimulus of multiplicative noise, the received signal takes the following form:

$$a(t) = \mu(t)\, a_0(t), \tag{3.1}$$

where $a_0(t)$ is the transmitted signal.

Additive noise can act independently of a "yes" or a "no" signal in the stochastic process at the input of the receiver or detector of a signal processing system. Additive noise can be classified into both natural and man-made noise and interference. The natural noise is referred to as the Earth's electromagnetic radiation, radiation of atmosphere, cosmic radiation, set noise of linear tract of the receiver or detector, and so on. Additive noise is inherent over all frequency ranges and acts continuously at the input of the receiver or detector of signal processing systems in the form of composition of the fluctuating noise and pulse interference.[4–6]

Random changes in electromagnetic field at the input of the receiver or detector of signal processing systems, caused by composition of weak radiation of various sources, give rise to the fluctuating noise. A random sequence of non-overlapping in time power pulses of electromagnetic field caused by atmospherics (for example) is the pulse interference. Maximal spectral power density of the pulse interference is numerous factors of 10 dB greater than that of the fluctuating noise. As a consequence, natural noise and interference are close to the stationary Gaussian noise only within the limits of frequency range, where the pulse interference is absent, and differ sharply from the Gaussian noise when the pulse interference is present.[7]

Natural noise and interference are sufficiently well understood through experiment; however, an exhaustive mathematical statistical model of the natural noise and interference is not yet defined. To the contrary, fluctuating noise is defined by the Gaussian model, and composition of the fluctuating noise and pulse interference is defined by various mathematical models. Each mathematical model allows us to describe a composition of fluctuating noise and pulse interference only for particular cases of signal processing when natural noise and interference become dominant.

Man-made noise and interference are divided into two classes that differ from each other in principle. The first class, called inadvertent interference, possesses a stochastic character. Examples of inadvertent interference include radio radiation of the world network, various radio engineering systems, mobile systems, and so on. Composition of natural noise and inadvertent interference is wide-band, but the spectral density of this composition is non-uniform and, likewise, the change of instantaneous values is the non-stationary stochastic process. Composition of the natural and inadvertent interference can be represented in the form of the sum of three independent components: fluctuating noise, pulse interference, and lumped interference.[4,5,8–10]

The lumped interference is called a radiation, the spectrum of which is concentrated in the neighborhood of the central (or resonant) frequency and, with a spectral bandwidth that is larger in comparison to the spectral bandwidth of the signal at the input of the receiver or detector of signal processing systems.

Pulse interference is called a sharp short-term increase in the spectral power density of the additive noise that occurs within the spectrum bandwidth limits of the signal.

Composition of the natural and inadvertent interference can possess peculiarities of the stationary stochastic process and can be considered as the fluctuating noise when the lumped interference and pulse interference are absent. In this case the additive noise has the following form:

$$\xi(t) = n(t) + \eta(t) + \zeta(t), \tag{3.2}$$

where $n(t)$ is the fluctuating noise, $\eta(t)$ is the lumped interference, and $\zeta(t)$ is the pulse interference.

The deliberate interference is generated specifically for the purpose of reducing efficiency and noise immunity of the signal processing system. Special sources of interference (generators or transmitters) may be used for generation of the deliberate interference. In many cases the deliberate interference can be considered as the additive fluctuating noise, pulse interference, and lumped interference. In some cases the deliberate interference can be thought of as the multiplicative noise.

In practice the fluctuating noise, lumped interference, and pulse interference are usually considered for representation of a total composition of the additive noise. When doing so, the mathematical model of each form of the noise and interference must be correctly justified.

The intersymbol interference is a special class of the noise. The intersymbol interference is caused by an increase in speed of data transmission under the condition of limited spectrum bandwidth of the signal, especially for multi-beam and multi-channel signal processing. The intersymbol interference acts simultaneously as the additive and multiplicative noise.

Thus, the additive noise and the multiplicative noise are the main factors that define the noise immunity of signal processing systems in various areas of application.[11–16]

3.2 Sources of Multiplicative Noise

Impact of multiplicative noise on signal parameters and qualitative characteristics of signal processing systems essentially depends on the relationship between the time interval $[0, T]$, within the limits of which the signal is processed and analyzed, and the rate of changes in the signal phase and amplitude as a function of time. If the time of correlation (or period) of the multiplicative noise is more greater than the duration T of an incoming coherent signal, then changes in amplitude and phase of the signal caused by the multiplicative noise are slow. We may ignore these changes within the limits of the time interval $[0, T]$, assuming that the amplitude and phase of the signal are varied from one realization of the signal to another. In this case distortions of structure and form of the signal are practically absent, and the impact of the multiplicative noise diminishes to changes in the amplitude and initial phase of the signal.[17]

For cases in which the time of correlation of the multiplicative noise is a commensurate time interval $[0, T]$ within the limits of which the coherent signal is processed, or less, then the signal structure is distorted by the multiplicative noise. In other words, the modulation laws of amplitude and phase of the signal are randomly changed. This case is of additional interest to us.

Thus it follows that an initiation of the multiplicative noise in the sense of noise modulation of the signal is caused by both the characteristics of the source of multiplicative noise and the signal parameters. Under predetermined characteristics, the source of the noise can generate multiplicative noise for coherent signals with large duration and does not generate multiplicative noise for coherent signals with small duration—changing only the amplitude and initial phase of the signal.

The multiplicative noise can be generated by all elements of a data transmission channel, both during the propagation of radio signals and under signal processing by a receiver or detector of signal processing systems. Let us consider briefly the main sources of the multiplicative noise.

3.2.1 Propagation and Reflection of Radio Waves

Under propagation and reflection of radio waves, an initiation of the multiplicative noise is caused by the following reasons. Coefficient of refraction of the environment, in which the radio waves are propagated, is varied as a function of time by fluctuations in temperature, pressure, and humidity in the turbulent troposphere, and fluctuations of electron concentration in the ionosphere and cosmos. Fluctuations of the coefficient of refraction of the environment give rise to fluctuations in phase of the signal propagated in this environment.

During the process of propagation the radio waves can be scattered by local heterogeneities in both the troposphere and the ionosphere. In this case, the signal at the input of the receiver or detector of signal processing systems is the result of an action of interference by a set of signals generated by individual regions of scattering and shifted in time. The shift in time between the individual signals is distributed continuously within the limits of the time interval from 0.01 μsec to 10 μsec. As a consequence the resulting signal is distorted. The distortions of the signal are higher and the spectrum of the signal is wider under a definite duration of the signal. In accordance, the amplitude- and phase-modulated signals are distorted significantly. Stochastic variations in time of mutual locations of local scattering heterogeneities give rise to random distortions in signal structure.[18–23]

Analogous phenomena may take place under reflection of radio waves from the ionosphere or from the Earth's surface. In these cases, the signal at the input of the receiver or detector of signal processing systems is the sum of the mirror-reflected signal (the regular component) and the signals scattered by local heterogeneities (the scattered component). Under reflection of high-frequency radio waves from the ionosphere, the scattered component can be higher by power in comparison with the regular component; and the shift in time between elementary signals that form the scattered component is distributed continuously within the limits of the time interval that is defined by 10 to 100 μsec. Stochastic location of individual heterogeneities gives rise to a random character of distortions in structure of the resulting signal.

Under propagation of radio waves through solar and interplanetary plasmas, the random phase modulation of the signal is formed owing to moving a non-uniform plasma environment with a high velocity (approximately 500 km/h).

Space heterogeneities of environment, in which radio waves are propagated, can serve as sources of multiplicative noise when the transmitter and receiver (or detector) of signal processing systems move. For this reason, the distribution of heterogeneities is varied as a function of time.[22,24–26,34]

A probable source of multiplicative noise is the overlapping modulation of signals under propagation of radio waves in the non-linear environment. For example, a signal in the ionosphere is subjected to an additional parasitic modulation by virtue of stimulus by another signal (parasitic signal) propagated simultaneously with the signal, and possessing much higher spectral density. This effect is characteristic of the environment-frequency radio waves to a greater extent.[27–34]

The stimulus of multiplicative noise under propagation of radio waves can be defined as an effect of passing the signal through some equivalent linear four-terminal network with parameters varying as a function of time.

3.2.2 Transmitter

Background noise in the generator, which gives rise to fluctuations in amplitude and phase of the signal at the output of the generator, is one reason that multiplicative noise is formed. Frequency multiplication used by the transmitter of some signal processing systems causes the fluctuations in phase of the signal to increase.

Additional conditions traceable to the formation of multiplicative noise include: fluctuations and non-stabilities of voltage in the power supply; interference of the generator and power amplifier; and fluctuations of parameters in basic elements of signal processing systems.

3.2.3 Receiver or Detector

The main sources of multiplicative noise for the receiver or detector are the same as those identified for the transmitter. Additional sources of multiplicative noise are fluctuations in frequency of heterodyne, phase distortions, and so on. Radio interference generated by other receivers or detectors of signal processing systems is a very serious source of multiplicative noise. The parasitic (or overlapping modulation of the signal) is formed under interaction between the signal and interference of non-linear elements of the receiver or detector. Multiplicative noise as an interference, the spectrum bandwidth of which is out of the spectrum bandwidth of the signal, has a great impact on qualitative characteristics of signal processing systems.

The sources of multiplicative noise mentioned above are not, of course, exhaustive. For example, specific sources of multiplicative noise result from the operation of radar systems in the case of side scanning. However, even a brief analysis of sources of multiplicative noise shows that distortions of signal parameters under the stimulus of multiplicative noise are an integral peculiarity of operation of signal processing systems in practice, as the background noise is an integral peculiarity of the receiver or detector in practice.[35–38]

3.2.4 Stimulus of Multiplicative Noise

During the period of passing the input signal through various elements and blocks of a data transmission channel, distortions of the signal are increased because of a stimulus of the multiplicative noise that takes place at each element and block. Thus, there is an accumulation of distortions in the signal. For this reason, despite the fact that distortions of the signal generated by the multiplicative noise of each element and each block of a data transmission channel are low, resulting distortions of the signal can be high.

Impact of multiplicative noise on qualitative characteristics of signal processing systems is discussed in the literature much less often than the impact of additive noise. This fact can be explained in the following manner. A necessity to take into account an impact of the additive noise on qualitative characteristics of signal processing systems was revealed at the beginning of the advancement of the theory of signal processing and signal detection in radar, communications, navigation, and so on. The importance of taking into account multiplicative noise was established later, when signal processing systems in various areas had started to use coherent signals with large duration.

Let us consider the main effects caused by multiplicative noise during signal processing and detection. Distortions of structure of the signal caused by multiplicative noise impair a coherence of the signal. For this reason, a process of coherent accumulation of the signal by the receiver or detector of a signal processing system becomes less effective and the energy (or power) of the signal at the output of the receiver or detector decreases. This is a reason for the deterioration of all qualitative characteristics of signal processing systems (the probability of detection, resolution, precision of definition of signal parameters) which are defined by the signal-to-noise ratio at the output of the receiver or detector. Moreover, the parasitic modulation of the signal by multiplicative noise leads to an extension of signal spectrum that additionally impairs a possibility to select signals by frequency, and decreases a precision of measurements of signal frequency (for example, under a definition of the Doppler shift in frequency).[39–42]

Impact of the multiplicative noise on the wide-band pulse signals, i.e., the signals with the signal base

$$T\beta \gg 1 \quad \text{or} \quad T\Delta f \gg 1, \tag{3.3}$$

leads to deterioration in the quality of signal processing (the bandwidth compression of the signal) by the receiver or detector of signal processing systems. As is well known, during processing of such signals—the frequency-modulated or phase-modulated signals (for example), there are non-compensated components—the secondary peaks or overshoots of the decision statistic—in conjunction with the signal—the main peak of the decision statistic—at the output of the detector.

Under the stimulus of multiplicative noise, the main peak of the decision statistic at the output of the detector is decreased in magnitude, and a relative level of the secondary peaks (or overshoots) of the decision statistic—the non-compensated components—is increased as a consequence of distortions in phase and amplitude constituents among individual components of the signal. In the process, the probability to make a decision a "yes" signal in the input stochastic process, when one of the side overshoots—the non-compensated components—may be considered as the expected and detected signal, is decreased. Precision of definition of frequency of the signal and location of the signal on the time axis—the time when the signal appears—is impaired. Selection of the signals by frequency is impaired and resolution of signal processing systems using these signals is impaired, as well.

Under a high level of multiplicative noise, the level of the secondary overshoots—the non-compensated components—of the decision statistic at the output of the detector becomes proportionate to the level of the signal statistic. Thus, the signal at the output of the detector is divided and the process of the bandwidth compression of the signal at the output of the detector is destroyed.

In radar systems with the phased antenna and side scanning, the multiplicative noise can induce distortions in phase and amplitude relationships between signals forming by elementary antennas, i.e., errors in phasing arise due to distortions in the directional diagram of the antenna.[43–53]

Thus, the multiplicative noise impacts all main qualitative characteristics of signal processing systems in various areas of application and must be taken into serious account in analysis and study of any signal processing system. In the process, there is a need to consider the additive noise as the background noise of the receiver or detector, and in many cases there is a need to consider the additive noise as an external interference. Thus, the signal processing or signal detection problem is reduced to investigations into the additive noise and the multiplicative noise that are present simultaneously.

3.3 Classification and Main Properties of Multiplicative Noise

Let us consider some signal processing systems using radio signals for transmission and acquisition of information. We assume that qualitative characteristics of signal processing systems are defined by the probability of false alarm, probability of signal detection, precision of definition and measurement of signal parameters, resolution for one signal parameter or other, and so on. Then in the majority of cases we may not specify a type of signal processing system, for example, radar, communications, navigation, telecommunications, remote sensing, cosmic, optical radar, sonar, and so on, and we use a generalized concept called a *signal processing system*.

Any signal processing system has the following common blocks that are characteristic of signal processing system in any area of application. They are:

- Source of the information data
- Input transformer: the block transforming an input information in accordance with signal parameters, for example, the coder in communications, navigation systems, and telecommunications; the target in radar, sonar, and remote sensing
- Generator of signals
- Channel: for example, the communication channel or radio channel
- Receiver or detector
- Decision block: the block transforming the signal parameters into information, on the basis of which the decision is made, for example, the decoder in communications, navigation, and telecommunications; the display in radar, sonar, and remote sensing

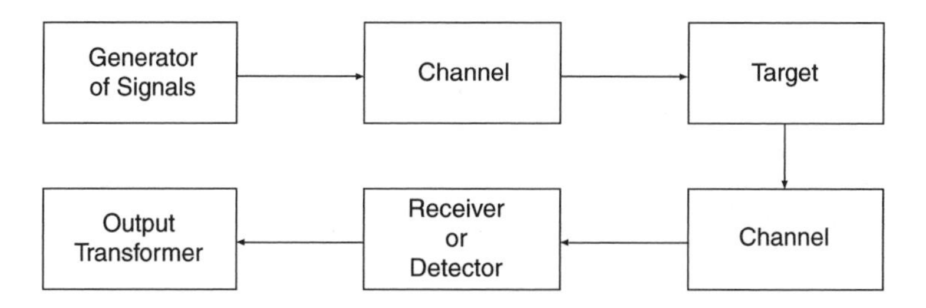

FIGURE 3.1
The information channel for radar systems.

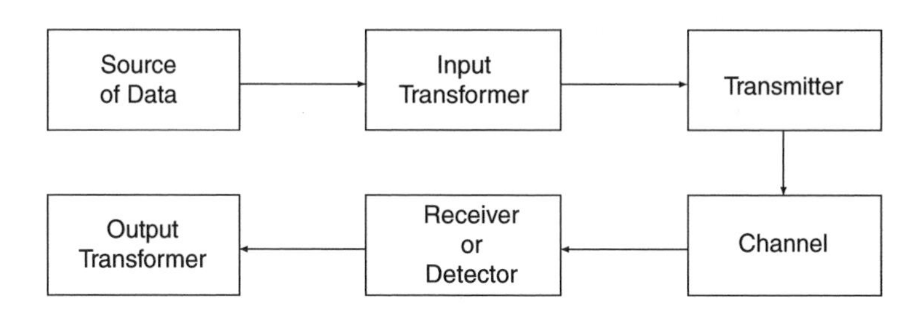

FIGURE 3.2
The information channel for communication and navigation systems.

Totality of these blocks will henceforth be called the *information channel*. The information channel in, for example, acoustics, optical communications, and radar has analogous blocks.

3.3.1 Main Peculiarities of Multiplicative Noise

All of these blocks may be arranged in an arbitrary way in the information channel for any signal processing system in various areas of application. So, for example, the information channel for radar systems is shown in Fig. 3.1 and the information channel for communications and navigation systems is shown in Fig. 3.2.

As we can see from Section 3.2, where the main sources of multiplicative noise are discussed, multiplicative noise can arise in the majority of blocks of the information channel. Some blocks can be considered as a variety of sources of multiplicative noise, for example, the high-frequency amplifier and frequency transformer of the receiver or detector are the sources of multiplicative noise.

As a rule, radio signal processing systems use narrow-band signals. The spectrum bandwidth of these signals is much less than the carrier frequency. The narrow-band signal formed by modulation in amplitude and phase or frequency takes the following form[54]

$$a(t) = S(t)\cos[\omega_0 t + \Psi(t) + \varphi_0], \tag{3.4}$$

where $S(t)$ is the amplitude envelope of the signal defined by the amplitude modulation law; $\Psi(t)$ is the phase modulation law of the signal, for example, under the frequency modulation law

$$\Psi(t) = \int \Omega(t)\,dt, \tag{3.5}$$

where $\Omega(t)$ is the frequency modulation law; ω_0 is the carrier frequency of the signal; and φ_0 is the initial phase of the signal. Information transferred by the signal is contained in the functions $S(t)$ and $\Psi(t)$. We will often use Eq. (3.4) in the complex form

$$a(t) = Re\left\{\dot{S}(t)\cdot e^{j(\omega_0 t + \varphi_0)}\right\}, \tag{3.6}$$

where

$$\dot{S}(t) = S(t)\cdot e^{j\Psi(t)} \tag{3.7}$$

is the complex envelope of the signal containing information transferred by the signal. The point at the top signifies the complex value.

Multiplicative noise causes distortions in amplitude and phase of the signal. The signal distorted by the multiplicative noise can be written in the following form:

$$a_M(t) = A(t)S(t)\cos[\omega_0 t + \Psi(t) + \varphi_0 + \varphi(t)], \tag{3.8}$$

where $\varphi(t)$ is the variation of the signal phase caused by multiplicative noise—distortions in phase of the signal; and

$$A(t) \geq 0$$

is the pure factor characterizing the changes in the amplitude envelope of the signal caused by multiplicative noise—distortions in amplitude of the signal. Assume that the spectrum bandwidth of the functions $\varphi(t)$ and $A(t)$ is much less than the carrier frequency of the signal, i.e., distortions have a character of parasitic modulation of the signal. The amplitude envelope of the signal $a_M(t)$ is equal to

$$S_M(t) = A(t)S(t). \tag{3.9}$$

The signal in Eq. (3.8) can be rewritten in the following form:

$$a_M(t) = Re\left\{\dot{S}_M(t)\cdot e^{j(\omega_0 t + \varphi_0)}\right\}, \tag{3.10}$$

where

$$\dot{S}_M(t) = \dot{S}(t)A(t) \cdot e^{j\varphi(t)} \tag{3.11}$$

is the complex amplitude envelope of the signal distorted by multiplicative noise.

Thus, the complex amplitude envelope of the signal containing the transferred information is changed under the stimulus of the multiplicative noise, resulting in distortions of this information. The function

$$\dot{M}(t) = A(t) \cdot e^{j\varphi(t)} \tag{3.12}$$

characterizes completely the parasitic modulation of the signal caused by the multiplicative noise. The function $\dot{M}(t)$ is called the *noise modulation function* of multiplicative noise.

If the signal $a(t)$ has the energy equal to unity, for example,

$$a(t) = \cos \omega_0 t \tag{3.13}$$

then the signal after being distorted by multiplicative noise takes the following form:

$$a_M(t) = A(t) \cos[\omega_0 t + \varphi(t)] = Re\left\{\dot{M}(t) \cdot e^{j\omega_0 t}\right\}. \tag{3.14}$$

It follows that the noise modulation function can be considered as the complex amplitude envelope of the signal forming under the same amplitude and phase modulation laws of the signal with the unit energy, according to which the multiplicative noise distorts the signal. This representation of the noise modulation function $\dot{M}(t)$ of multiplicative noise is very useful for definition of noise characteristics in some cases.

As can be seen from Section 3.2, an initiation of the multiplicative noise is caused by the parameters of individual blocks and elements of the information channel changing in time. In the majority of cases, an initiation of multiplicative noise can be defined as a result of passing the narrow-band signal through the equivalent linear four-terminal network with parameters varying as a function of time (see Fig. 3.3). Equivalence in a given sense can be thought of as equivalence with respect to a definite element or block of the information channel. This definition is useful for defining some properties of

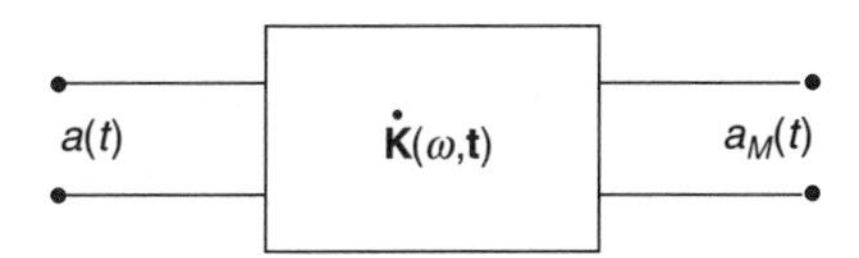

FIGURE 3.3
Linear four-terminal network with parameters varying as a function of time.

multiplicative noise and the essence of the noise modulation function $\dot{M}(t)$ of multiplicative noise.

The linear four-terminal network with parameters varying in time is characterized by the transfer characteristic:

$$\dot{K}(\omega, t) = A_k(\omega, t) \cdot e^{j\varphi_k(\omega,t)}, \tag{3.15}$$

where $A_k(\omega, t)$ is the module of the transfer characteristic (the amplitude characteristic of the linear four-terminal network) and $\varphi_k(\omega, t)$ is the argument of the transfer characteristic (the phase characteristic of the linear four-terminal network).

The signal at the output of the linear four-terminal network with varying in time parameters is determined in the following form:

$$a_M(t) = \frac{1}{2\pi} \int_{-\infty}^{\infty} \dot{S}_a(\omega)\dot{K}(\omega, t) \cdot e^{j\omega t}\, dt, \tag{3.16}$$

where $\dot{S}_a(\omega)$ is the spectrum (the Fourier transform) of the input signal $a(t)$.

Taking into account the following peculiarities of the signal spectrum and transfer characteristic

$$\dot{K}(-\omega, t) = K^*(\omega, t); \tag{3.17}$$

$$\dot{S}_a(-\omega) = S_a^*(\omega), \tag{3.18}$$

where the symbol $*$ signifies the complex conjugate value, and in terms of Eq. (3.16), the complex amplitude envelope of the signal at the output of the linear four-terminal network with varying in time parameters is determined in the following form:

$$\dot{S}_M(t) = \frac{1}{2\pi} \int_{0}^{\infty} \dot{S}_S(\Omega)\dot{K}(\omega_0 + \Omega, t) \cdot e^{j\Omega t}\, d\Omega, \tag{3.19}$$

where $\dot{S}_S(\Omega)$ is the spectrum of the complex amplitude envelope of the signal and ω_0 is the carrier frequency of the signal.

As is well known, the signal at the output of the linear four-terminal network with parameters varying in time possesses the delay τ_0 that is not varied in time, and the structure of the signal is not distorted when the following conditions

$$A_k(\omega, t) = A_k = const \tag{3.20}$$

and

$$\frac{\partial}{\partial\omega}\varphi_k(\omega, t) = \tau_0 = const \tag{3.21}$$

are satisfied within the limits of the signal bandwidth.

If

$$A_k(\omega, t) \quad \text{and} \quad \frac{\partial}{\partial \omega}\varphi_k(\omega, t) \tag{3.22}$$

are the functions of frequency and time within the limits of the signal bandwidth then the structure of the signal at the output of the linear four-terminal network with parameters varying in time is distorted in parallel with the delay. The functions indicated in Eq. (3.22) cause distortions of the signal in the frequency domain. Dependence of these characteristics on time leads to initiation of the multiplicative noise.[55–57]

Consider an initiation of the multiplicative noise during passing the signal determined by Eq. (3.4) through the linear four-terminal network with variable parameters. Using the Taylor series expansion for the functions indicated in Eq. (3.22) by the argument ω in the neighborhood of the carrier frequency ω_0 of the signal

$$A_k(\omega, t) = A_k(\omega_0, t) + \frac{\partial}{\partial \omega} A_k(\omega, t)\bigg|_{\omega=\omega_0} (\omega - \omega_0) + \cdots, \tag{3.23}$$

$$\varphi_k(\omega, t) = \varphi_k(\omega_0, t) + \frac{\partial}{\partial \omega}\varphi_k(\omega, t)\bigg|_{\omega=\omega_0} (\omega - \omega_0) + \cdots. \tag{3.24}$$

Usually the spectrum bandwidth of the narrow-band signal is chosen such that there are no distortions of the signal in frequency. Reference to Eq. (3.15) demonstrates that this occurs if the function $A_k(\omega, t)$ does not depend on frequency within the limits of the signal bandwidth and, consequently, we can neglect all terms in Eq. (3.23) except the first term. Moreover, the function $\varphi_k(\omega, t)$ is the linear function of frequency and we can neglect all terms in Eq. (3.24) except the first and second terms.

Under these conditions the transfer function of the linear four-terminal network with parameters varying in time in Eq. (3.15) is determined within the limits of the narrow-band signal bandwidth by the following equation

$$\dot{K}(\omega, t) = A_k(\omega_0, t) \cdot e^{j\varphi_k(\omega_0, t)} \cdot e^{-j(\omega-\omega_0)\tau} = \dot{K}(\omega_0, t) \cdot e^{-j(\omega-\omega_0)\tau}, \quad \omega \geq 0, \tag{3.25}$$

where

$$\tau = \tau(t) = \frac{\partial}{\partial \omega}\varphi_k(\omega, t)\bigg|_{\omega=\omega_0} \tag{3.26}$$

is the rate of changes in the phase-frequency response of the linear four-terminal network with parameters varying in time under the frequency equal to the carrier frequency of the signal.

Substituting Eq. (3.25) in Eq. (3.19) in terms of Eqs. (3.23) and (3.24), the complex amplitude envelope of the signal at the output of the linear four-terminal network with parameters varying in time takes the following form:

$$\dot{S}_M(t) = \dot{K}(\omega_0, t)\frac{1}{2\pi}\int_0^\infty \dot{S}_S(\Omega)\cdot e^{j\Omega(t-\tau)}\,d\Omega = \dot{K}(\omega_0, t)\dot{S}(t-\tau). \tag{3.27}$$

The signal at the output of the linear four-terminal network with parameters varying in time under these conditions takes the following form:

$$\begin{aligned} a_M(t) &= Re\left\{\dot{S}_M(t)\cdot e^{j(\omega_0 t+\varphi_0)}\right\} \\ &= A_k(\omega_0, t)S(t-\tau)\cos[\omega_0 t + \Psi(t-\tau) + \varphi_k(\omega_0, t) + \varphi_0]. \end{aligned} \tag{3.28}$$

If we compare Eqs. (3.3) and (3.28) we can see that the signal is subjected to an additional modulation law in accordance with the function $A_k(\omega_0, t)$ and the phase modulation law $\varphi_k(\omega_0, t)$ during passing through the linear four-terminal network with parameters varying in time. Furthermore, the complex amplitude envelope of the signal has the delay τ varying in time. In the process, distortions of the signal are caused by both an additional modulation and the delay varying in time.

We proceed to determine the noise modulation function defining these distortions of the signal. In a general case the delay τ in Eq. (3.26) consists of the component τ_0 that does not vary in time and the component defined by the rate of changing the phase-frequency response as a function of time:

$$\tau = \tau_0 + \Delta\tau(t). \tag{3.29}$$

We use the Taylor series expansion for the functions

$$\Psi(t-\tau) \quad \text{and} \quad S(t-\tau). \tag{3.30}$$

Suppose that a maximal variation of the variable $\Delta\tau(t)$ during the time equal to the duration of the signal is less than the time of correlation of the signal. Because of this, we can limit by the first and second terms of expansion:

$$\Psi(t-\tau) \simeq \Psi(t-\tau_0) - \Psi'(t-\tau_0)\Delta\tau(t), \tag{3.31}$$

$$S(t-\tau) \simeq S(t-\tau_0) - S'(t-\tau_0)\Delta\tau(t). \tag{3.32}$$

We introduce designations

$$\xi_\tau = \frac{S'(t-\tau_0)}{S(t-\tau_0)}\cdot\Delta\tau(t), \tag{3.33}$$

$$\varphi_\tau = \Psi'(t-\tau_0)\Delta\tau(t). \tag{3.34}$$

Substituting Eqs. (3.31)–(3.34) in Eq. (3.27), we can write

$$\dot{S}_M(t) = \dot{K}(\omega_0, t)[1-\xi_\tau(t)]\cdot e^{-j\varphi_\tau(t)}\dot{S}(t-\tau_0). \tag{3.35}$$

We can see from Eq. (3.35) that during passing the signal through the linear four-terminal network with parameters varying in time and the transfer characteristic given by Eq. (3.25), the complex amplitude envelope has the delay τ_0 that does not vary in time and, furthermore, the signal has distortions in amplitude and phase, which are determined by the noise modulation function of the multiplicative noise

$$\dot{M}(t) = \dot{K}(\omega_0, t)[1 - \xi_\tau(t)] \cdot e^{-j\varphi_\tau(t)}. \tag{3.36}$$

Reference to Eqs. (3.33), (3.34), and (3.36) shows that the magnitude of distortions in amplitude and phase of the signal does not depend on the energy of the signal. Because of this, under the stimulus of the multiplicative noise, as opposed to the stimulus of additive noise, an increase in the energy of the signal does not lead to a decrease in the influence of the multiplicative noise.

The noise modulation function $\dot{M}(t)$ of the multiplicative noise in Eq. (3.36) is the product of two components. The first component $\dot{K}(\omega_0, t)$ is only defined by characteristics of the linear four-terminal network with parameters varying in time, which is equivalent to an element or block of the information channel. The second component containing the functions $\xi_\tau(t)$ and $\varphi_\tau(t)$ is defined by both characteristics of the linear four-terminal network with parameters varying in time and the amplitude modulation law $\dot{S}(t)$ (see Eqs. (3.33) and (3.34)) in a general case.

In practice, the noise modulation function does not depend on parameters of the signal and is only defined by parameters of the equivalent linear four-terminal network with parameters varying in time. Consider the following cases.

Case 1. Under distortions of the signal in amplitude the transfer characteristic of the equivalent linear four-terminal network with parameters varying in time can be determined in the following form:

$$\dot{K}(\omega_0, t) = \dot{K}_1(\omega_0) K_2(t). \tag{3.37}$$

Substituting Eq. (3.37) in Eq. (3.19) under the condition that distortions in frequency of the signal are absent, we obtain the complex amplitude envelope of the signal at the output of the linear four-terminal network with parameters varying in time in the following form:

$$\dot{S}_M(t) = K_2(t)\dot{S}(t - \tau_0), \tag{3.38}$$

where τ_0 is the delay of the signal at the output of the linear four-terminal network with parameters varying in time with the transfer characteristic $\dot{K}_1(\omega)$. In doing so, the delay τ_0 does not vary in time. In this case the noise modulation function of the multiplicative noise is determined in the following form

$$\dot{M}(t) = K_2(t) \tag{3.39}$$

and does not depend on parameters of the signal.

Case 2. If the condition

$$|\Delta\tau_{\max}(t)| \leq \frac{1}{\Delta F_a} \tag{3.40}$$

is satisfied within the limits of the time interval $[0, T]$, where ΔF_a is the spectrum bandwidth of the signal, we can believe with sufficient precision that

$$\dot{S}(t - \tau_0 - \Delta\tau(t)) \simeq \dot{S}(t - \tau_0) \tag{3.41}$$

and Eq. (3.27) can be rewritten in the following form

$$\dot{S}_M(t) \simeq \dot{K}(\omega_0, t)\dot{S}(t - \tau_0). \tag{3.42}$$

The noise modulation function $\dot{M}(t)$ of the multiplicative noise follows from Eq. (3.42)

$$\dot{M}(t) = \dot{K}(\omega_0, t) \tag{3.43}$$

and does not depend on parameters of the signal.

For example,[58] under propagation of radio waves with

$$\tau_0 = 2 \times 10^{-3} sec \tag{3.44}$$

the mean square deviation of propagation time $\Delta\tau(t)$ caused by random heterogeneities in atmosphere, when the multi-beam propagation is absent, is equal to 10^{-9} sec. As we can see from Eq. (3.40), during these values of $\Delta\tau(t)$, the noise modulation function $\dot{M}(t)$ of the multiplicative noise does not depend on parameters of the signal for any duration of the signal if the spectrum bandwidth of the signal is less than 100 MHz. When the spectrum bandwidth of the signal is less, the impact of parameters of the signal on the noise modulation function $\dot{M}(t)$ of the multiplicative noise under the given values of $\Delta\tau(t)_{\max}$ is less.

Consider the limiting case—the harmonic signal

$$a(t) = S_0 \cos \omega_0 t, \tag{3.45}$$

the amplitude envelope spectrum of which is defined by the delta function

$$\dot{S}_S(\Omega) = 2\pi S_0 \delta(\Omega). \tag{3.46}$$

Substituting Eq. (3.46) in Eq. (3.19), we can write

$$\dot{S}_M(t) = S_0 \dot{K}(\omega_0, t), \tag{3.47}$$

$$\dot{M}(t) = \dot{K}(\omega_0, t). \tag{3.48}$$

Thus, in this case the noise modulation function $\dot{M}(t)$ of the multiplicative noise does not depend on parameters of the signal for all values of $\Delta\tau_{\max}(t)$.

So, the process of initiation of the multiplicative noise, in some elements or blocks of the information channel, can be defined as a result of passing the signal through the equivalent linear four-terminal network with parameters varying in time. Moreover, the equivalent linear four-terminal network with parameters varying in time does not form distortions in frequency of the signal within the limits of the spectrum bandwidth of the signal. If the condition in Eq. (3.40) is satisfied or if there are distortions in amplitude of the signal, then the noise modulation function $\dot{M}(t)$ of the multiplicative noise is the transfer characteristic of the equivalent linear four-terminal network with parameters varying in time at the frequency equal to the carrier frequency of the signal.

Under these conditions, the noise modulation function $\dot{M}(t)$ of the multiplicative noise defines the properties of the information channel, or some individual elements or blocks of the information channel. For this reason, the noise modulation function $\dot{M}(t)$ of the multiplicative noise under these conditions can be thought of as a function that is independent of a "yes" or a "no" signal in the input stochastic process. The multiplicative noise, as opposed to the additive noise, exists if—and only if—the signal is present in the input stochastic process.[59–64]

If the signals have the same carrier frequency and spectrum bandwidth, that is, there are no essential distortions in frequency of the signal and the condition determined by Eq. (3.40) is satisfied, then during passing through the given information channel these signals are subjected to the same multiplicative noise. Under the conditions mentioned, the character and extent of distortions in amplitude and phase of the signal do not depend on the structure of the signal. They are entirely defined by the properties of the information channel, through which the signal is transferred. This does not mean that the extent of impact of these distortions of the signal on qualitative characteristics of signal processing systems does not depend on the structure of the signal used by signal processing systems. Since multiplicative noise is completely characterized by the noise modulation function $\dot{M}(t)$ of the multiplicative noise, it is worthwhile to carry out the classification of the multiplicative noise in accordance with the properties of the noise modulation function $\dot{M}(t)$ of the multiplicative noise.

3.3.2 Classification of Multiplicative Noise

Multiplicative noise can be classified according to the following features:

- In accordance with the form of distortions of the signal, we can distinguish:[65–68]
  - — Phase distortions or phase multiplicative noise

$$\dot{M}(t) = e^{j\varphi(t)} \tag{3.49}$$

  - — Amplitude distortions or amplitude multiplicative noise

$$\dot{M}(t) = A(t) \tag{3.50}$$

— Amplitude-phase distortions or a general case of the multiplicative noise

$$\dot{M}(t) = A(t) \cdot e^{j\varphi(t)} \tag{3.51}$$

- According to the form of process defining distortions of the signal, we can distinguish:
  - — Deterministic multiplicative noise—$\dot{M}(t)$ is the deterministic function—and quasideterministic multiplicative noise—$\dot{M}(t)$ is the definite function with one or more random parameters. The particular case of deterministic and quasideterministic multiplicative noise, which is widely used in practice, is the periodic multiplicative noise—$\dot{M}(t)$ is the periodic function.
  - — Stochastic multiplicative noise—$\dot{M}(t)$ is the stochastic function. The particular case of stochastic multiplicative noise, which is widely used in practice, is the fluctuating multiplicative noise—$\dot{M}(t)$ is the fluctuating process. If the noise modulation function is the stationary stochastic process—$A(t)$ and $\varphi(t)$ are the stationary and stationary related in a general case stochastic processes—this multiplicative noise is called the stationary multiplicative noise.
- According to the relative rate of changes in amplitude and phase of the signal, caused by a stimulus of multiplicative noise, we can distinguish the following particular cases:[69–74]
  - — Rapid multiplicative noise—the time of correlation or the period of the noise modulation function $\dot{M}(t)$ is much less than the duration of the coherent signal
  - — Slow multiplicative noise—the time of correlation or the period of the noise modulation function $\dot{M}(t)$ is much greater than the duration of the coherent signal

3.4 Correlation Function and Energy Spectrum of Multiplicative Noise

As we can see from Section 3.3, multiplicative noise is completely characterized by the noise modulation function $\dot{M}(t)$. We will now study the main characteristics of the noise modulation function, which are widely used in practice in analysis of the impact of multiplicative noise on parameters of the signals and qualitative characteristics of signal processing systems. These main characteristics of the noise modulation function are the correlation function $\dot{R}(\tau)$ and the energy spectrum $G_M(\Omega)$.

In the case of fluctuating multiplicative noise, the following characteristics of fluctuations of the noise modulation function $\dot{M}(t)$ of the multiplicative noise are also widely used in practice: the correlation function $\dot{R}_V(\tau)$ and the energy spectrum $G_V(\Omega)$. In the case of deterministic and quasideterministic multiplicative noise, the complex spectrum $\dot{S}_M(\Omega)$ of the noise modulation function $\dot{M}(t)$ of the multiplicative noise also is often used in practice.

As was discussed in Section 3.3, the noise modulation function $\dot{M}(t)$ of multiplicative noise can be considered as the complex amplitude envelope of the signal formed under modulation of the harmonic signal, with the unit amplitude (or unit energy of the signal), by the multiplicative noise. Using the well-known relationships between the spectrum of the narrow-band signal and the spectrum of complex amplitude envelope of the narrow-band signal,[69,75,76] the spectrum $\dot{S}_M(\Omega)$ and the energy spectrum $G_M(\Omega)$ of the noise modulation function $\dot{M}(t)$ of the multiplicative noise, and also the spectrum $G_V(\Omega)$ of fluctuations of the noise modulation function $\dot{M}(t)$ of the multiplicative noise can be determined by the corresponding spectra $S_0(\omega)$ and $G_1(\omega)$ of the harmonic signal modulated in accordance with the corresponding law and by the energy spectrum $G_0(\omega)$ of fluctuations of this harmonic signal in the following form:

$$\dot{S}_M(\Omega) = 2\dot{S}_0(\omega_0 + \Omega); \tag{3.52}$$

$$G_M(\Omega) = 4G_1(\omega_0 + \Omega); \tag{3.53}$$

$$G_V(\Omega) = 4G_0(\omega_0 + \Omega), \tag{3.54}$$

where ω_0 is the carrier frequency of the harmonic signal and

$$\Omega > -\omega_0. \tag{3.55}$$

The spectrum of the harmonic signal, modulated on amplitude and phase by deterministic and quasideterministic functions and the stationary wide-band fluctuation process, is discussed in more detail in References 2, 3, 54, 77, and 78. For this reason, we briefly consider the spectrum of the noise modulation function $\dot{M}(t)$ of the multiplicative noise for these cases and discuss only the main results. Appendix II provides an additional detailed discussion of the following cases: when the changes in amplitude and phase of the signal are mutually correlated and are narrow-band stochastic processes and when the changes in phase of the signal are non-stationary pulse-fluctuation processes. In this section we consider only the main results for these cases.

3.4.1 Deterministic and Quasideterministic Multiplicative Noise

Periodic character of the changes in amplitude and phase of the signal can be caused by the following: rotation of the target reflecting radio waves around its own center of gravity; interferences of feed circuits; and others. Let

$$\varphi(t) = \varphi(t + T_M) \tag{3.56}$$

be the periodic deterministic or quasideterministic function satisfying the Dirichlet boundary conditions, and T_M be the period of the multiplicative noise.

In the case of fluctuation-related changes in amplitude and phase of the signal, the noise modulation function of the multiplicative noise takes the following form:

$$\dot{M}(t) = A[\varphi(t)] \cdot e^{j\varphi(t)}. \tag{3.57}$$

We proceed to define the complex spectrum of the noise modulation function $\dot{M}(t)$ of the multiplicative noise. Using the Fourier series expansion of the noise modulation function $\dot{M}(t)$ of the multiplicative noise in Eq. (3.57), we can write

$$\dot{M}(t) = \sum_{n=-\infty}^{\infty} \dot{C}_n \cdot e^{jn\Omega_M t}, \tag{3.58}$$

where

$$\dot{C}_n = \frac{1}{T_M} \int_0^{T_M} A[\varphi(t)] \cdot e^{j[\varphi(t) - n\Omega_M t]}\, dt \tag{3.59}$$

and

$$\Omega_M = \frac{2\pi}{T_M}. \tag{3.60}$$

Taking into account Eq. (3.58), the complex spectrum of the noise modulation function $\dot{M}(t)$ of the multiplicative noise can be written in the following form:

$$\dot{S}_M(\Omega) = \sum_{n=-\infty}^{\infty} \dot{C}_n \delta(\Omega - n\Omega_M). \tag{3.61}$$

Usually the totality of values $|\dot{C}_n|$ is called the spectrum of amplitudes and the totality of values argument $\dot{C}_n$ is called the spectrum of phases.[1,2,69,79]

In the case of the quasideterministic multiplicative noise, when the function $\varphi(t)$ possesses the random initial phase uniformly distributed within the limits of the interval $[0, 2\pi]$, the noise modulation function $\dot{M}(t)$ of the multiplicative noise is the stochastic function. The correlation function of the stochastic noise modulation function $\dot{M}(t)$ of the multiplicative noise is determined in the following form:

$$\dot{R}_M(\tau) = m_1[\dot{M}(t)M^*(t-\tau)] = \sum_{n=-\infty}^{\infty} \overline{|\dot{C}_n|^2} \cdot e^{jn\Omega_M \tau}, \tag{3.62}$$

where

$$\overline{|\dot{C}_n|^2} = m_1\left[|\dot{C}_n|^2\right] \tag{3.63}$$

and $m_1[.]$ is the mean (see Chapter 1).

Taking into consideration the fact that the energy spectrum of the noise modulation function $\dot{M}(t)$ of the multiplicative noise is the Fourier transform of the correlation function,[2] we can write

$$G_M(\Omega) = \sum_{n=-\infty}^{\infty} \overline{|\dot{C}_n|^2}\delta(\Omega - n\Omega_M). \tag{3.64}$$

The power of each discrete component in Eq. (3.64) at the frequency $n\Omega_M$ is defined by the mean of squared module of the Fourier series expansion coefficient of the noise modulation function $\dot{M}(t)$ of the multiplicative noise within the limits of the interval $[0, T_M]$.

Let, for example, the function of distortions in amplitude and phase of the signal be determined in the following form:

$$A[\varphi(t)] = e^{-k_1\varphi(t)}, \tag{3.65}$$

and the phase of the signal be subjected to the function

$$\varphi(t) = \varphi_m \sin[\Omega_M t + \varphi_0], \tag{3.66}$$

where k_1 is the constant factor; φ_0 is the random initial phase of the signal, which is distributed uniformly within the limits of the interval $[0, 2\pi]$. Then, taking into account Eqs. (3.58), (3.64), and (3.65) and in terms of the well-known expansion[80]

$$e^{j\dot{x}\sin\varphi} = \sum_{n=-\infty}^{\infty} J_n(\dot{x}) \cdot e^{jn\varphi}, \tag{3.67}$$

where $J_n(\dot{x})$ is the Bessel function of complex argument, we can write

$$\dot{S}_M(\Omega) = \sum_{n=-\infty}^{\infty} J_n(\varphi_m\dot{\psi}) \cdot e^{jn\varphi_0}\delta(\Omega - n\Omega_M), \tag{3.68}$$

$$\dot{G}_M(\Omega) = \sum_{n=-\infty}^{\infty} |J_n(\varphi_m\dot{\psi})|^2\delta(\Omega - n\Omega_M), \tag{3.69}$$

where

$$\dot{\psi} = 1 + jk_1. \tag{3.70}$$

The functions

$$|\dot{C}_0|^2 = |J_0(\varphi_m\dot{\psi})|^2 \tag{3.71}$$

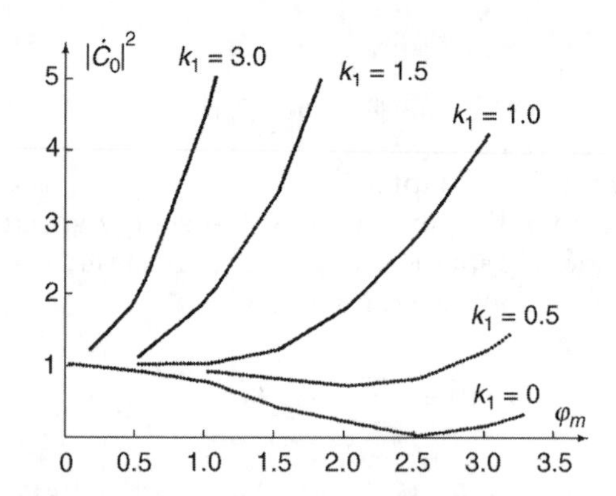

FIGURE 3.4
Function $|\dot{C}_0|^2 = |J_0(\varphi_m \dot{\psi})|^2$.

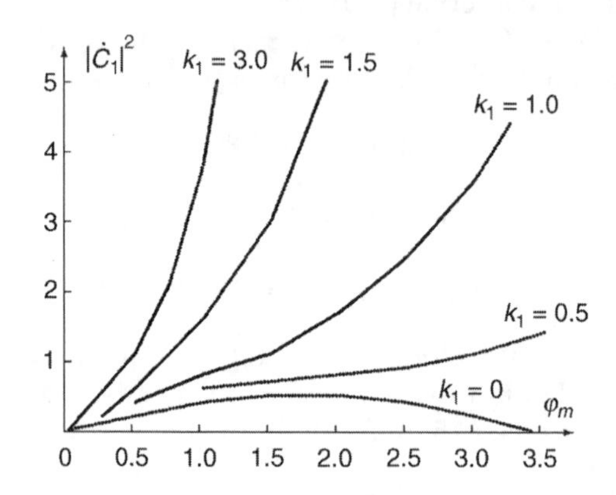

FIGURE 3.5
Function $|\dot{C}_1|^2 = |J_1(\varphi_m \dot{\psi})|^2$.

and

$$|\dot{C}_1|^2 = |J_1(\varphi_m \dot{\psi})|^2 \tag{3.72}$$

define the degree of distortions in phase of the signal under various values k_1, and are shown in Figs. 3.4 and 3.5 when distortions in phase of the signal are modulated by the sinusoidal law and are subjected to the exponential dependence between the changes in amplitude and phase of the signal (see Eq. (3.65)). It is easy to see, when distortions in amplitude of the signal are absent and

$$k_1 = 0 \tag{3.73}$$

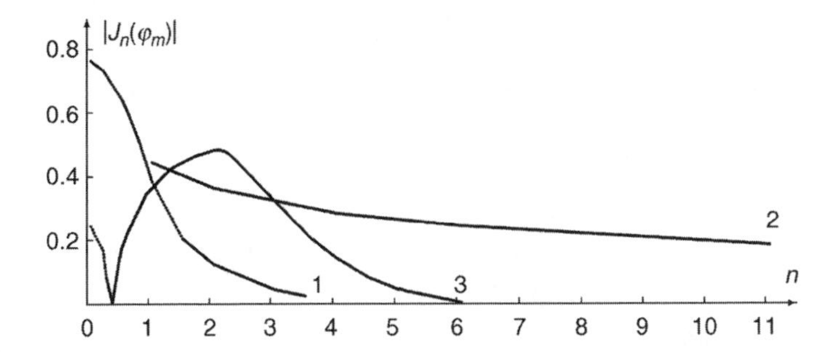

FIGURE 3.6
Function $|J_m(\varphi_m)|$ at φ_m = const; for 1, $\varphi_m = 1$; for 2, $J_n(n)$; for 3, $\varphi_m = 3$.

and only the distortions modulated by sinusoidal law in phase of the signal are present, Eq. (3.69) can be transformed into the well-known equation

$$\dot{G}_M(\Omega) = \sum_{n=-\infty}^{\infty} |J_n(\varphi_m)|^2 \delta(\Omega - n\Omega_M), \tag{3.74}$$

where $J_n(\varphi_m)$ is the Bessel function of the real argument.

The function $|J_n(\varphi_m)|$ vs. n at the degree of distortions in phase of the signal

$$\varphi_m = const \tag{3.75}$$

is shown in Fig. 3.6. Curve 2 corresponds to the case when the degree of distortions in phase of the signal is equal to the order n of the Bessel function. Curves 1 and 3 correspond to the case when

$$\varphi_m = 1 \tag{3.76}$$

and

$$\varphi_m = 3, \tag{3.77}$$

respectively. Dependences shown in Fig. 3.6 characterize the energy spectrum of the noise modulation function $\dot{M}(t)$ of the multiplicative noise during quasideterministic sinusoidal distortions in phase of the signal.

Table 3.1 represents the squared module of the Fourier coefficients $|\dot{C}_n|^2$ for the phase modulation functions $\varphi(t)$ of the signal, which are widely used in practice. Table 3.1 contains the function-related amplitude-phase distortions of the signal given by Eq. (3.65) in the second column and the only phase distortions case of the signal ($k_1 = 0$) in the third column.

The spectrum of the noise modulation function $\dot{M}(t)$ of the multiplicative noise under the parabolic function of changes in phase of the signal for the conditions

$$\varphi_m = 2\pi, 5\pi, 8\pi, 20\pi, \tag{3.78}$$

TABLE 3.1

Squared Module of the Fourier Coefficients $|\dot{C}_n|^2$

No.	$\varphi(t)$	$\lvert\dot{C}_n\rvert^2 \quad (\dot{\psi} = 1 + jk_1)$	$\lvert\dot{C}_n\rvert^2 \quad (\dot{\psi} = 1; k_1 = 0)$
1	$\varphi_m \sin[\Omega_M t + \varphi_0]$	$\left\lvert J_n(\varphi_m\dot{\psi})\right\rvert^2$	$\left\lvert J_n(\varphi_M)\right\rvert^2$
2	$\beta t, \quad 0 < t \leq T_M$	$\left\lvert \frac{\mathrm{sh}\left[\frac{T_M}{2}(\beta\dot{\psi} - jn\Omega_M)\right]}{\frac{T_M}{2}(\beta\dot{\psi} - jn\Omega_M)} \right\rvert^2$	$\left\lvert \frac{\sin\left[\frac{T_M}{2}(\beta - n\Omega_M)\right]}{\frac{T_M}{2}(\beta - n\Omega_M)} \right\rvert^2$
3	$\varphi_m\left[\frac{(t-b)^2}{b^2} - 1\right]$, $0 < t \leq T_M$	$\frac{1}{4\varphi_m\lvert\dot{\psi}\rvert}\lvert\Phi(\dot{y}_1) - \Phi(\dot{y}_2)\rvert^2$, where $\dot{y}_1 = (1 - 2\Delta)\sqrt{\varphi_m\dot{\psi}} - j\frac{n\pi}{2(1-\Delta)\sqrt{\varphi_m\dot{\psi}}}$; $\Delta = 1 - \frac{T_M}{2b}$; $\dot{y}_2 = \sqrt{\varphi_m\dot{\psi}} - \frac{n\pi}{2(1-\Delta)\sqrt{\varphi_m\dot{\psi}}}$; $\Phi(\dot{y})$ is the error integral[80]	$\frac{\pi}{8\varphi_m(1-\Delta)^2}\left\{[C(x_1) - C(x_2)]^2 + [S(x_1) - S(x_2)]^2\right\}$, where $C(x)$ and $S(x)$ are the Fresnel integrals;[80] $x_1 = \sqrt{\frac{2}{\pi}}\left[(1 - 2\Delta)\sqrt{\varphi_m} - \frac{n\pi}{2(1-\Delta)\sqrt{\varphi_m}}\right]$; $x_2 = \sqrt{\frac{2}{\pi}}\left[\sqrt{\varphi_m} - \frac{n\pi}{2(1-\Delta)\sqrt{\varphi_m}}\right]$; $\Delta = 1 - \frac{T_M}{2b}$.

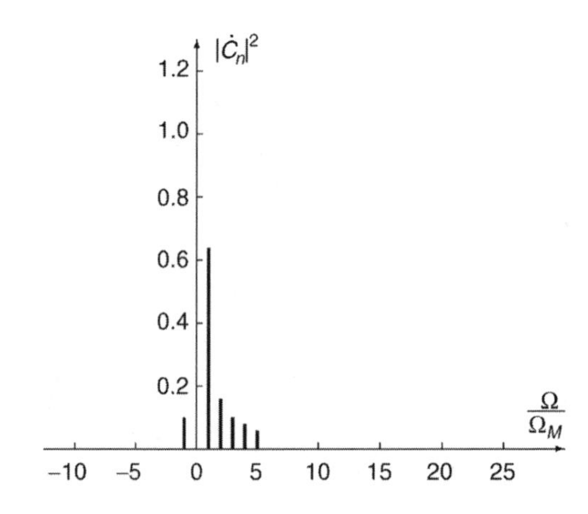

FIGURE 3.7
Spectrum of the noise modulation function at $\varphi_m = 2\pi$.

and

$$\Delta = 0.4 \tag{3.79}$$

(Table 3.1, third row) is shown in Figs. 3.7–3.10, respectively.

If distortions in amplitude and phase of the signal are function-related by the relationship

$$A[\varphi(t)] = [1 + k_2\varphi(t - \tau)], \tag{3.80}$$

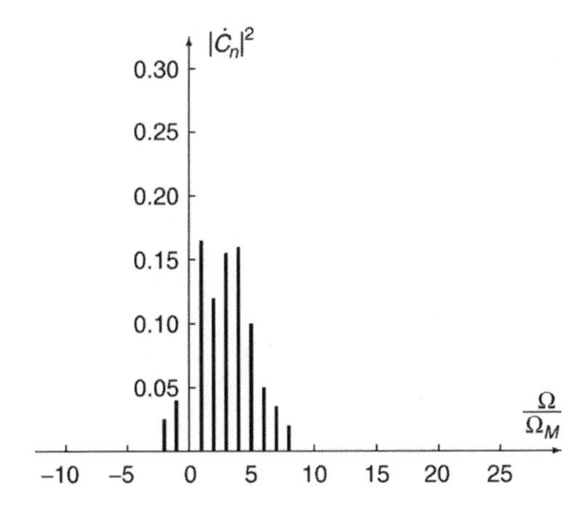

FIGURE 3.8
Spectrum of the noise modulation function at $\varphi_m = 5\pi$.

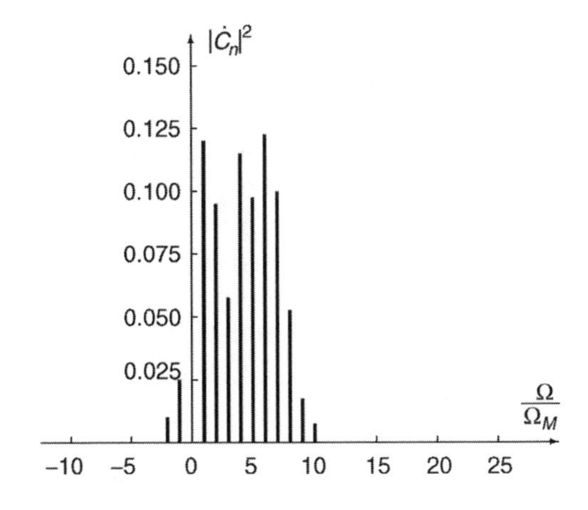

FIGURE 3.9
Spectrum of the noise modulation function at $\varphi_m = 8\pi$.

where τ is the delay and k_2 is the modulation index, then, taking into account results in Reference 4, by relation to Eq. (3.61), we can write

$$\dot{S}(\Omega) = \sum_{n=-\infty}^{\infty} \dot{C}_n \left(1 + k_2 \sum_{m=-\infty}^{\infty} \dot{R}_m \cdot e^{jm\Omega_M \tau}\right) \delta(\Omega - n\Omega_M), \tag{3.81}$$

where

$$\dot{R}_m = \frac{1}{T_M} \int_0^{T_M} \varphi(t) \cdot e^{jm\Omega_M t}\, dt; \tag{3.82}$$

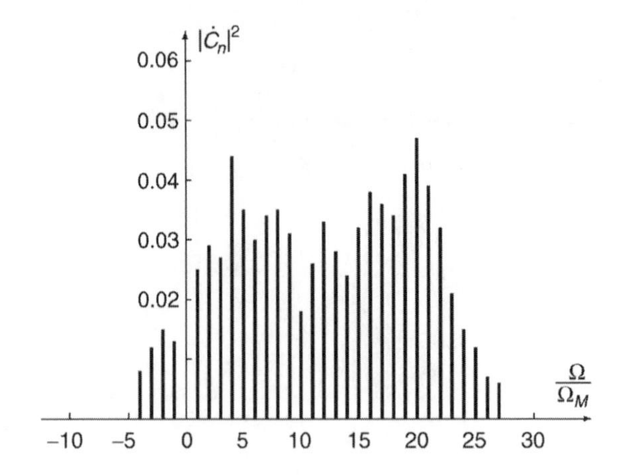

FIGURE 3.10
Spectrum of the noise modulation function at $\varphi_m = 20\pi$.

$\dot{C}_n$ is the coefficient under the Fourier series expansion of the noise modulation function $\dot{M}(t)$ of the multiplicative noise at the only distortions in phase of the signal case (see the third column in Table 3.1).

As is well known,[4] the energy spectrum under the function-related amplitude-phase distortions of the signal given by Eq. (3.80) is asymmetric. If the modulation index k_2 is high, then the asymmetry of the energy spectrum is high. The main part of the asymmetric energy spectrum of the noise modulation function $\dot{M}(t)$ of the multiplicative noise can be greater or less than the zero frequency as a function of the modulation index k_2 and delay τ.[2,81]

3.4.2 Stationary Fluctuating Multiplicative Noise—Correlation Function of the Noise Modulation Function

If distortions in amplitude and phase of the signal caused by the multiplicative noise are the stationary stochastic processes and the amplitude and phase of the signal are function-related, we can consider these distortions as the stationary function-related processes. Then the noise modulation function $\dot{M}(t)$ of the multiplicative noise is the stationary function. In this case the noise modulation function takes the following form:

$$\dot{M}(t) = \overline{\dot{M}} + \dot{V}_0(t), \tag{3.83}$$

where

$$\overline{\dot{M}} = m_1[\dot{M}(t)] \tag{3.84}$$

is the mean of the noise modulation function $\dot{M}(t)$ of the multiplicative noise, which does not vary in time, and $\dot{V}_0(t)$ is the fluctuation of the noise modulation function $\dot{M}(t)$ of the multiplicative noise.

The correlation function $\dot{R}_M(\tau)$ of the noise modulation function $\dot{M}(t)$ of the multiplicative noise and the correlation function $\dot{R}_V(\tau)$ of fluctuations of the noise modulation function $\dot{M}(t)$ of the multiplicative noise can be defined using the characteristic function of distortions in amplitude and phase of the signal. We proceed to introduce the four-dimensional characteristic function of distortions in amplitude and phase of the signal at the instants $t_1 = t$ and $t_2 = t - \tau$:

$$\Theta_4^{A(t)\varphi(t)}(x_1, x_2, x_3, x_4) = m_1\left[e^{j(x_1 A_1 + x_2 A_2 + x_3\varphi_1 + x_4\varphi_2)}\right], \tag{3.85}$$

where

$$A_1 = A(t_1 = t); \tag{3.86}$$

$$A_2 = A(t_2 = t - \tau); \tag{3.87}$$

$$\varphi_1 = \varphi(t_1 = t); \tag{3.88}$$

and

$$\varphi_2 = \varphi(t_2 = t - \tau). \tag{3.89}$$

Using the characteristic function determined by Eq. (3.85), it is simple to show that the correlation function $\dot{R}_M(\tau)$ of the noise modulation function $\dot{M}(t)$ of the multiplicative noise can be written in the following form:

$$\begin{aligned}\dot{R}_M(\tau) &= m_1[\dot{M}(t)M^*(t-\tau)]\\ &= -\left[\frac{\partial^2}{\partial x_1 \partial x_2}\Theta_4^{A(t)\varphi(t)}(x_1, x_2, 1, -1)\right]_{x_1=0;\ x_2=0}.\end{aligned} \tag{3.90}$$

In terms of Eq. (3.83), the correlation function of fluctuations of the noise modulation function $\dot{M}(t)$ of the multiplicative noise takes the following form:

$$\dot{R}_V(\tau) = m_1[\dot{V}_0(t)V_0^*(t-\tau)] = \dot{R}_M(\tau) - \overline{|\dot{M}|^2}. \tag{3.91}$$

The mean $\overline{\dot{M}}$ of the noise modulation function $\dot{M}(t)$ of the multiplicative noise can be defined using the two-dimensional characteristic function of distortions in amplitude and phase of the signal at the same coinciding instants—$\Theta_2^{A(t)\varphi(t)}(x_1, x_2)$. Actually

$$\Theta_2^{A(t)\varphi(t)}(x_1, x_2) = m_1\left[e^{j[x_1 A(t) + x_2\varphi(t)]}\right]. \tag{3.92}$$

Referring to Eq. (3.92) shows that

$$\overline{\dot{M}} = m_1\left[A(t)e^{j\varphi(t)}\right] = -j\left[\frac{\partial}{\partial x_1}\Theta_2^{A(t)\varphi(t)}(x_1, x_2)\right]_{x_1=0}. \tag{3.93}$$

In terms of Eqs. (3.90), (3.91), and (3.93) the correlation function of fluctuations of the noise modulation function $\dot{M}(t)$ of the multiplicative noise is determined in the following form:

$$\dot{R}_V(\tau) = -\left[\frac{\partial^2}{\partial x_1 \partial x_2}\Theta_4^{A(t)\varphi(t)}(x_1, x_2, 1, -1)\right]_{x_1=0;\ x_2=0} - \left|\left[\frac{\partial}{\partial x_1}\Theta_2^{A(t)\varphi(t)}(x_1, 1)\right]_{x_1=0}\right|^2. \tag{3.94}$$

In many cases the pure factor $A(t)$ defining distortions in amplitude of the signal can be written in the following form:

$$A(t) = A_0[1 + \xi(t)], \tag{3.95}$$

where A_0 is the mean of the pure factor $A(t)$, $\xi(t)$ is the stationary stochastic process with zero mean, and

$$1 + \xi(t) \geq 0. \tag{3.96}$$

In the process, the correlation function $\dot{R}_M(\tau)$ of the noise modulation function $\dot{M}(t)$ of the multiplicative noise is determined in the following form:

$$\dot{R}_M(\tau) = A_0^2\left\{\Theta_4^{\xi(t)\varphi(t)}(0, 0, 1, -1) - j\left[\frac{\partial}{\partial x_1}\Theta_4^{\xi(t)\varphi(t)}(x_1, 0, 1, -1)\right]_{x_1=0} - j\left[\frac{\partial}{\partial x_2}\Theta_4^{\xi(t)\varphi(t)}(0, x_2, 1, -1)\right]_{x_2=0} - \left[\frac{\partial^2}{\partial x_1 \partial x_2}\Theta_4^{\xi(t)\varphi(t)}(x_1, x_2, 1, -1)\right]_{x_1=0;\ x_2=0}\right\}, \tag{3.97}$$

where

$$\Theta_4^{\xi(t)\varphi(t)}(x_1, x_2, x_3, x_4) = m_1\left[e^{j(x_1\xi_1 + x_2\xi_2 + x_3\varphi_1 + x_4\varphi_2)}\right] \tag{3.98}$$

is the four-dimensional characteristic function of the processes $\xi(t)$ and $\varphi(t)$ at the instants $t_1 = t$ and $t_2 = t - \tau$, i.e.,

$$\xi_1 = \xi(t_1 = t); \tag{3.99}$$

$$\xi_2 = \xi(t_2 = t - \tau); \tag{3.100}$$

$$\varphi_1 = \varphi(t_1 = t); \tag{3.101}$$

and

$$\varphi_2 = \varphi(t_2 = t - \tau). \tag{3.102}$$

In this case the mean of the noise modulation function $\dot{M}(t)$ of the multiplicative noise is determined in the following form:

$$\overline{\dot{M}} = A_0 \Theta_2^{\xi(t)\varphi(t)}(0, 1) - jA_0 \left[\frac{\partial}{\partial x_1} \Theta_2^{\xi(t)\varphi(t)}(x_1, 1) \right]_{x_1=0}, \tag{3.103}$$

where

$$\Theta_2^{\xi(t)\varphi(t)}(x_1, x_2) = m_1 \left[e^{j[x_1\xi(t)+x_2\varphi(t)]} \right] \tag{3.104}$$

is the two-dimensional characteristic function of the processes $\xi(t)$ and $\varphi(t)$ at the coinciding instants.

Substituting Eqs. (3.97) and (3.103) in Eq. (3.91), we can be certain that the correlation function of fluctuations of the noise modulation function $\dot{M}(t)$ of the multiplicative noise takes the following form:

$$\begin{aligned} \dot{R}_V(\tau) = A_0^2 \Bigg\{ & \Theta_4^{\xi(t)\varphi(t)}(0, 0, 1, -1) - j \left[\frac{\partial}{\partial x_1} \Theta_4^{\xi(t)\varphi(t)}(x_1, 0, 1, -1) \right]_{x_1=0} \\ & - j \left[\frac{\partial}{\partial x_2} \Theta_4^{\xi(t)\varphi(t)}(0, x_2, 1, -1) \right]_{x_2=0} \\ & - \left[\frac{\partial^2}{\partial x_1 \partial x_2} \Theta_4^{\xi(t)\varphi(t)}(x_1, x_2, 1, -1) \right]_{x_1=0;\ x_2=0} \\ & - \left| \Theta_2^{\xi(t)\varphi(t)}(0, 1) - j \left[\frac{\partial}{\partial x_1} \Theta_2^{\xi(t)\varphi(t)}(x_1, 1) \right]_{x_1=0} \right|^2 \Bigg\}. \end{aligned} \tag{3.105}$$

Therefore, we can easily define the correlation functions $\dot{R}_M(\tau)$ and $\dot{R}_V(\tau)$ of the noise modulation function $\dot{M}(t)$ of the multiplicative noise, if the characteristic functions corresponding to the distribution law of distortions in amplitude and phase of the signal are known.

By way of example let us consider the case widely used in practice in which the phase $\varphi(t)$ and amplitude $\xi(t)$ distortions of the signal obey the distribution law that is close to the Gaussian distribution law.[82–84] In doing so, it must be emphasized that the distribution law of distortions in amplitude $\xi(t)$ of the signal can be approximated by the Gaussian distribution law if and only if the condition determined by Eq. (3.96) is satisfied, i.e., the degree of amplitude distortions of the signal is not so high.

If the mean square deviation σ_ξ of distortions in amplitude $\xi(t)$ of the signal is much less than unity, then the probability

$$P[1 + \xi(t) < 0] \tag{3.106}$$

is negligible, and the Gaussian distribution law can be a very good approximation to define distortions in amplitude of the signal. For example, if $\sigma_\xi \leq 0.25$, then

$$P[1 + \xi(t) < 0] < 0.023 \tag{3.107}$$

and if $\sigma_\xi \leq 0.17$, then the probability

$$P[1 + \xi(t) < 0] < 0.00015. \tag{3.108}$$

This limitation does not impact on distortions in phase of the signal.

For the Gaussian distribution law the characteristic functions $\Theta_4^{\xi(t)\varphi(t)}(x_1, x_2, x_3, x_4)$ and $\Theta_2^{\xi(t)\varphi(t)}(x_1, x_2)$ are determined in the following form:[54]

$$\begin{aligned}\Theta_4^{\xi(t)\varphi(t)}(x_1, x_2, x_3, x_4) = \exp\big\{ -0.5\big[&\sigma_\xi^2 x_1^2 + \sigma_\xi^2 x_2^2 + \sigma_\varphi^2 x_3^2 + \sigma_\varphi^2 x_4^2 \\ &+ 2\sigma_\xi^2 r_\xi(\tau) x_1 x_2 + 2\sigma_\varphi^2 r_\varphi(\tau) x_3 x_4 + 2\sigma_\xi \sigma_\varphi r_{\xi\varphi}(0) x_1 x_3 \\ &+ 2\sigma_\xi \sigma_\varphi r_{\xi\varphi}(0) x_2 x_4 + 2\sigma_\xi \sigma_\varphi r_{\xi\varphi}(\tau) x_1 x_4 \\ &+ 2\sigma_\xi \sigma_\varphi r_{\xi\varphi}(-\tau) x_2 x_3\big]\big\};\end{aligned} \tag{3.109}$$

$$\Theta_2^{\xi(t)\varphi(t)}(x_1, x_2) = \exp\big\{ -0.5\big[\sigma_\xi^2 x_1^2 + \sigma_\varphi^2 x_2^2 + 2\sigma_\xi \sigma_\varphi r_{\xi\varphi}(0) x_1 x_2\big]\big\}, \tag{3.110}$$

where σ_φ^2 is the variance of distortions in phase of the signal; σ_ξ^2 is the variance of distortions in amplitude of the signal; $r_\varphi(\tau)$ is the coefficient of correlation during the period of distortions in phase of the signal; $r_\xi(\tau)$ is the coefficient of correlation during the period of distortions in amplitude of the signal; and $r_{\xi\varphi}(\tau)$ is the coefficient of mutual correlation between distortions in amplitude $\xi(t)$ of the signal and distortions in phase $\varphi(t)$ of the signal.

Substituting Eqs. (3.109) and (3.110) in Eqs. (3.97) and (3.107), respectively, we can write

$$\begin{aligned}\dot{R}_M(\tau) = A_0^2\{&1 + \sigma_\xi^2 r_\xi(\tau) + j\sigma_\xi\sigma_\varphi[r_{\xi\varphi}(-\tau) - r_{\xi\varphi}(\tau)] \\ &+ \sigma_\xi^2\sigma_\varphi^2[r_{\xi\varphi}(0) - r_{\xi\varphi}(\tau)][r_{\xi\varphi}(0) - r_{\xi\varphi}(-\tau)]\} \cdot e^{-\sigma_\varphi^2[1 - r_\varphi(\tau)]};\end{aligned} \tag{3.111}$$

$$\begin{aligned}\dot{R}_V(\tau) = A_0^2\{&1 + \sigma_\xi^2 r_\xi(\tau) + j\sigma_\xi\sigma_\varphi[r_{\xi\varphi}(-\tau) - r_{\xi\varphi}(\tau)] \\ &+ \sigma_\xi^2\sigma_\varphi^2[r_{\xi\varphi}(0) - r_{\xi\varphi}(\tau)][r_{\xi\varphi}(0) - r_{\xi\varphi}(-\tau)]\} \cdot e^{-\sigma_\varphi^2[1 - r_\varphi(\tau)]} \\ &- A_0^2 \cdot e^{-\sigma_\varphi^2}\big[1 + \sigma_\xi^2\sigma_\varphi^2 r_{\xi\varphi}^2(0)\big].\end{aligned} \tag{3.112}$$

If distortions in amplitude and phase of the signal are not correlated between each other, then the correlation functions of the noise modulation function $\dot{M}(t)$ of the multiplicative noise can be determined in the following form:

$$\dot{R}_M(\tau) = A_0^2\big[1 + \sigma_\xi^2 r_\xi(\tau)\big] \cdot e^{-\sigma_\varphi^2[1 - r_\varphi(\tau)]}; \tag{3.113}$$

$$\dot{R}_V(\tau) = A_0^2\big[1 + \sigma_\xi^2 r_\xi(\tau)\big] \cdot e^{-\sigma_\varphi^2[1 - r_\varphi(\tau)]} - A_0^2 \cdot e^{-\sigma_\varphi^2}. \tag{3.114}$$

If there are only distortions in phase of the signal, then the correlation functions of the noise modulation function $\dot{M}(t)$ of the multiplicative noise take the following form:

$$\dot{R}_M(\tau) = e^{-\sigma_\varphi^2[1-r_\varphi(\tau)]}; \tag{3.115}$$

$$\dot{R}_V(\tau) = e^{-\sigma_\varphi^2}\left[e^{\sigma_\varphi^2 r_\varphi(\tau)} - 1\right]. \tag{3.116}$$

It follows that the correlation functions $\dot{R}_M(\tau)$ and $\dot{R}_V(\tau)$, of the noise modulation function $\dot{M}(t)$ of multiplicative noise, are the complex functions if distortions in amplitude and phase of the signal are interdependent

$$r_{\xi\varphi}(\tau) \neq 0. \tag{3.117}$$

In the process, the corresponding energy spectra ($G_M(\Omega)$ and $G_V(\Omega)$) are asymmetric with respect to zero. If distortions in amplitude and phase of the signal are independent, or if there are distortions only in phase or only in amplitude of the signal, then the correlation functions $\dot{R}_M(\tau)$ and $\dot{R}_V(\tau)$ of the noise modulation function $\dot{M}(t)$ of multiplicative noise are the real functions, and the energy spectra $G_M(\Omega)$ and $G_V(\tau)$ are symmetric with respect to zero.

Consider the case in which the time of correlation of the multiplicative noise is much higher than the duration of the signal distorted by the multiplicative noise. This multiplicative noise is called the slow multiplicative noise (see Section 3.3.2), since the amplitude and phase of the signal are slowly changed under the stimulus of the multiplicative noise under the given conditions. Let the stochastic processes $A(t)$ or $\xi(t)$ and $\varphi(t)$, defining the multiplicative noise, be differentiated in the mean square sense that is true if the average power of their derivatives is finite.[54,85,86] Using the Maclaurin series expansion for the correlation functions $\dot{R}_M(\tau)$ and $\dot{R}_V(\tau)$ under the condition of slow multiplicative noise, we can reach more simplified approximate mathematical expressions for the correlation functions $\dot{R}_M(\tau)$ and $\dot{R}_V(\tau)$ in comparison with the expressions mentioned above (see Appendix II).

If distortions in amplitude $A(t)$ or $\xi(t)$ and in phase $\varphi(t)$ of the signal are independent, then the correlation functions $\dot{R}_M(\tau)$ and $\dot{R}_V(\tau)$ of the noise modulation function $\dot{M}(t)$ of the multiplicative noise are defined under the condition of the slow multiplicative noise in the following form:

$$\begin{aligned}\dot{R}_M(\tau) &\simeq \overline{A^2} - 0.5\tau^2\left(\sigma_{A'}^2 + \overline{A^2}\sigma_\omega^2\right)\\ &= A_0^2\left(1+\sigma_\xi^2\right) - 0.5\tau^2 A_0^2\left[\sigma_{\xi'}^2 + \left(1+\sigma_\xi^2\right)\sigma_\omega^2\right];\end{aligned} \tag{3.118}$$

$$\begin{aligned}\dot{R}_V(\tau) &\simeq \overline{A^2} - \alpha_0^2 - 0.5\tau^2\left(\sigma_{A'}^2 + \overline{A^2}\sigma_\omega^2\right)\\ &= A_0^2\left(1+\sigma_\xi^2\right) - \alpha_0^2 - 0.5\tau^2 A_0^2\left[\sigma_{\xi'}^2 + \left(1+\sigma_\xi^2\right)\sigma_\omega^2\right],\end{aligned} \tag{3.119}$$

where

$$\overline{A^2} = m_1[A^2(t)]; \tag{3.120}$$

$\sigma^2_{A'}$ is the variance of the derivative of the function $A(t)$; σ^2_ξ is the variance of the stochastic process $\xi(t)$; $\sigma^2_{\xi'}$ is the variance of derivative of the stochastic function $\xi(t)$; σ^2_ω is the variance of derivative of the phase distortions $\varphi(t)$ of the signal (the variance of instantaneous frequency of the signal caused by the multiplicative noise); and $\alpha_0 = |\overline{\dot{M}}|$ is the module of the mean of the noise modulation function $\dot{M}(t)$ of the multiplicative noise.

If there are distortions only in phase of the signal caused by the multiplicative noise, then the correlation functions $\dot{R}_M(\tau)$ and $\dot{R}_V(\tau)$ have the following form:

$$\dot{R}_M(\tau) \simeq 1 + 0.5\tau^2\sigma^2_\omega; \tag{3.122}$$

$$\dot{R}_V(\tau) \simeq 1 - \alpha_0^2 - 0.5\tau^2\sigma^2_\omega. \tag{3.123}$$

When distortions in amplitude and phase of the signal caused by the multiplicative noise are correlated and obey the Gaussian distribution law, then the following approximate formulae are true under the condition of the slow multiplicative noise:

$$\begin{aligned}\dot{R}_M(\tau) \simeq A_0^2\{1 + \sigma^2_\xi + 2j\tau\sigma_\xi\sigma_\omega r_{\xi\omega}(0)\\ - 0.5\tau^2\left[\sigma^2_{\xi'} + 2\sigma^2_\xi\sigma^2_\omega r_{\xi\omega}(0) + \left(1 + \sigma^2_\xi\right)\sigma^2_\omega\right]\};\end{aligned} \tag{3.124}$$

$$\begin{aligned}\dot{R}_V(\tau) \simeq A_0^2\{1 + \sigma^2_\xi - \left[1 + \sigma^2_\xi\sigma^2_\varphi r_{\xi\varphi}(0)\right] \cdot e^{-\sigma^2_\varphi} + 2j\tau\sigma_\xi\sigma_\omega r_{\xi\omega}(0)\\ - 0.5\tau^2\left[\sigma^2_{\xi'} + 2\sigma^2_\xi\sigma^2_\omega r_{\xi\omega}(0) + \left(1 + \sigma^2_\xi\right)\sigma^2_\omega\right]\},\end{aligned} \tag{3.125}$$

where $r_{\xi\omega}(0)$ is the coefficient of mutual correlation between the stochastic function $\xi(t)$ and the derivative $\varphi'(t)$ of phase distortions of the signal caused by the multiplicative noise at the coinciding instants.

The approximate relationships for the correlation functions $\dot{R}_M(\tau)$ and $\dot{R}_V(\tau)$ of the noise modulation function $\dot{M}(t)$ of the multiplicative noise mentioned above allow us to define these functions under the condition in which the multiplicative noise is slow. This is for the cases in which the correlation functions of distortions in amplitude and phase of the signal are unknown but their characteristics are simpler—the variances of distortions in amplitude and phase of the signal and derivatives of distortions in amplitude and phase of the signal.

We have just considered the case in which distortions in amplitude and phase of the signal caused by multiplicative noise are related statistically. Below we consider the correlation function $\dot{R}_M(\tau)$, of the noise modulation function $\dot{M}(t)$ of the multiplicative noise, when the functional relationship between distortions in the amplitude and phase of the signal is determined by Eq. (3.65).

Using Eq. (II.45) (see Appendix II) and the two-dimensional characteristic function,[54] the correlation function $\dot{R}_M(\tau)$, of the noise modulation function

$\dot{M}(t)$ of the multiplicative noise, can be represented in the following form:

- The stochastic distortions in phase $\varphi(t)$ of the signal obey the Gaussian distribution law:

$$\dot{R}_M(\tau) = e^{-\sigma_\varphi^2(1+k_1^2)[1-r_\varphi(\tau)]}. \tag{3.126}$$

- The stochastic distortions in phase $\varphi(t)$ of the signal obey the exponential distribution law:

$$\dot{R}_M(\tau) = \frac{1}{1 - 2k_1\sigma_\varphi^2 + 4\sigma_\varphi^4(1+k_1^2)\left[1 - r_\varphi^2(\tau)\right]}. \tag{3.127}$$

3.4.3 Stationary Fluctuating Multiplicative Noise—Energy Spectrum of the Noise Modulation Function

The energy spectrum of the noise modulation function $\dot{M}(t)$ of the multiplicative noise is defined by the Fourier transform of the correlation function $\dot{R}_M(\tau)$:

$$G_M(\Omega) = \int_{-\infty}^{\infty} \dot{R}_M(\tau) \cdot e^{-j\omega\tau}\, d\tau. \tag{3.128}$$

The energy spectrum of fluctuations of the noise modulation function $\dot{M}(t)$ of the multiplicative noise is defined by the Fourier transform of the correlation function $\dot{R}_V(\tau)$:

$$G_V(\Omega) = \int_{-\infty}^{\infty} \dot{R}_V(\tau) \cdot e^{-j\omega\tau}\, d\tau. \tag{3.129}$$

Since the noise modulation function $\dot{M}(t)$ of the multiplicative noise is the pure function, the correlation functions $\dot{R}_M(\tau)$ and $\dot{R}_V(\tau)$ are pure functions, as well. The functions $G_M(\Omega)$ and $G_V(\Omega)$ have the dimensional representation of time. As previously mentioned, the energy spectra $G_M(\Omega)$ and $G_V(\Omega)$ can be defined by the energy spectra $G_1(\omega)$ and $G_0(\omega)$, respectively, of the harmonic oscillation modulated by corresponding law (see Eqs. (3.53) and (3.54), respectively).

3.4.3.1 *Distortions of the Signal Attributable to the Wide-Band Gaussian Stationary Process*

The problem of analysis of the energy spectrum of harmonic oscillation modulated in phase or amplitude by the wide-band Gaussian stationary process is discussed in greater detail in References 1–3, 54, and 69. This fact allows us to find mathematical expressions of the energy spectra $G_M(\Omega)$ and $G_V(\Omega)$. The concepts—*the wide-band stationary stochastic process* and *the narrow-band stationary stochastic process*—are used here in the same sense as in Reference 54: that is, the bandwidth of energy spectrum of the narrow-band stationary stochastic process is much less than the carrier frequency, and this condition is not satisfied for the wide-band stationary stochastic process.

Continuing, it is appropriate to use results discussed in Reference 54. Let $\dot{R}_\varphi(\tau)$ be the correlation function of distortions in phase of the signal. The correlation function $\dot{R}_\varphi(\tau)$ is differentiated under the condition $\tau = 0$. If $\sigma_\varphi^2 \gg 1$, then the energy spectrum $G_M(\Omega)$ can be determined in the following form:

$$G_M(\Omega) \simeq 2\pi \cdot e^{-\sigma_\varphi^2}\delta(\Omega) + \frac{\sqrt{2\pi}}{\sigma_\varphi \Omega_{1_\varphi}} \cdot e^{-\frac{\Omega^2}{2\sigma_\varphi^2 \Omega_{1_\varphi}^2}} \left(1 - e^{-\sigma_\varphi^2}\right), \tag{3.130}$$

where

$$\Omega_{1_\varphi}^2 = -r''_\varphi(0) = \frac{R_\omega(0)}{\sigma_\varphi^2} = \frac{\sigma_\omega^2}{\sigma_\varphi^2}. \tag{3.131}$$

The energy spectrum $G_V(\Omega)$ differs from the energy spectrum $G_M(\Omega)$ in that the discrete component or the first term on the right side of Eq. (3.130) is absent.

In this case the bandwidth of the energy spectrum is determined:

$$\Delta\Omega_M = \sqrt{2\pi}\sigma_\varphi \Omega_{1_\varphi} = \sqrt{2\pi}\sigma_\varphi. \tag{3.132}$$

Under the condition $\sigma_\varphi^2 \ll 1$, the energy spectra are determined in the following form:

$$G_M(\Omega) \simeq 2\pi \cdot e^{-\sigma_\varphi^2}\delta(\Omega) + 2G_\varphi(\Omega); \tag{3.133}$$

$$G_V(\Omega) \simeq 2G_\varphi(\Omega), \tag{3.134}$$

where $G_\varphi(\Omega)$ is the energy spectrum of distortions in phase of the signal when the coefficient of correlation

$$r_\varphi(\tau) = \exp\left\{-\frac{\Delta\Omega_\varphi}{\pi} \cdot |\tau|\right\} \tag{3.135}$$

is not differentiated under the condition $\tau = 0$, and the energy spectrum of the noise modulation function $\dot{M}(t)$ of the multiplicative noise has the following form:[2]

$$G_M(\Omega) = 2\pi \cdot e^{-\sigma_\varphi^2}\left(\delta(\Omega) + \frac{1}{\Delta\Omega_\varphi}\sum_{n=1}^{\infty}\frac{\sigma_\varphi^{2n}}{(n-1)!\left[n^2 + \left(\frac{\Omega}{\Delta\Omega_\varphi}\right)^2\right]}\right), \tag{3.136}$$

where $\Delta\Omega_\varphi$ is the equivalent bandwidth of the energy spectrum of distortions in phase of the signal.

3.4.3.2 Distortions of the Signal Attributable to the Narrow-Band Gaussian Stationary Process

Assume that distortions in phase $\varphi(t)$ of the signal are the narrow-band Gaussian stationary process with the correlation function

$$R_{\varphi}(\tau) = \sigma_{\varphi}^{2} A_{\varphi}(\tau) \cos \Omega_{M} \tau, \tag{3.137}$$

where σ_{φ}^{2} is the variance of phase fluctuations of the signal; $A_{\varphi}(\tau)$ is the envelope of the coefficient of correlation of the phase fluctuations $\varphi(t)$ of the signal; and Ω_{M} is the central (or resonant) frequency of the energy spectrum of narrow-band fluctuations.

The energy spectrum of the noise modulation function $\dot{M}(t)$ of the multiplicative noise, when the correlation function of distortions in phase of the signal is determined by Eq. (3.137) and distortions in amplitude of the signal are function-related with distortions in phase of the signal by dependence given in Eq. (3.65), is discussed in Appendix II.

During the period in which the degree of distortions in phase of the signal is determined by

$$\sigma_{\varphi} < \frac{1}{|\dot{\psi}|}, \tag{3.138}$$

where

$$|\dot{\psi}| = \sqrt{1 + k_{1}^{2}}, \tag{3.139}$$

the energy spectrum of the noise modulation function $\dot{M}(t)$ of the multiplicative noise takes the following form:

$$G_{M}(\Omega) = C_{0} \left\{ \delta(\Omega) + \frac{\sigma_{\varphi_{eq}}^{4}}{4} \cdot G_{2\varphi}(\Omega) + \frac{\sigma_{\varphi_{eq}}^{4}}{8} \cdot G_{2\varphi}(\Omega \pm 2\Omega_{M}) + \frac{\sigma_{\varphi_{eq}}^{2}}{4} \cdot G_{\varphi}(\Omega \pm \Omega_{M}) + \cdots \right\}, \tag{3.140}$$

where

$$G_{n\varphi}(\Omega \pm m\Omega_{M}) = \int_{-\infty}^{\infty} A_{\varphi}^{n}(\tau) \cdot e^{j(\Omega \pm \Omega_{M})\tau}\, d\tau; \tag{3.141}$$

$$\sigma_{\varphi_{eq}}^{2} = |\dot{\psi}| \sigma_{\varphi}^{2} \quad \text{at } k_{1} \neq 0; \tag{3.142}$$

$$\sigma_{\varphi_{eq}}^{2} = \sigma_{\varphi}^{2} \quad \text{at } k_{1} = 0; \tag{3.143}$$

$$C_{0} = \dot{\Theta}_{1}^{\varphi}(\dot{\psi}) \dot{\Theta}_{1}^{\varphi}(-\dot{\psi}^{*}). \tag{3.144}$$

TABLE 3.2

Functions $G_{n\varphi}(\Omega \pm m\Omega_M)$ for the Envelope $A_\varphi(\tau)$ of the Coefficient of Correlation of Phase Distortions of the Signal

No.	$A_\varphi(t)$	$G_{n\varphi}(\Omega)$
1	$e^{-\beta^2\tau^2}, \quad \beta = \frac{\Delta\Omega_\varphi}{\sqrt{\pi}}$	$\sqrt{\frac{\pi}{n}}\frac{1}{\beta}e^{-\frac{(\Omega \pm m\Omega_M)^2}{4n\beta^2}}, \quad n \neq 0$
2	$\frac{\sin\left[\frac{\tau}{2}\Delta\Omega_\varphi\right]}{\frac{\tau}{2}\Delta\Omega_\varphi}$	$\frac{\pi(n\Delta\Omega_\varphi)+(\lvert\Omega \pm m\Omega_M\rvert)^{n-1}}{2^{n-2}(n-1)!(\Delta\Omega_\varphi)^n}, \quad n \neq 0, \lvert\Omega \pm m\Omega_M\rvert < n\Omega_\varphi$
3	$e^{-\alpha\lvert\tau\rvert}, \quad \alpha = \frac{2\Delta\Omega_\varphi}{\pi}$	$\frac{2n\alpha}{n^2\alpha^2+(\Omega \pm m\Omega_M)^2}, \quad n \neq 0$

Table 3.2 represents the functions $G_{n\varphi}(\Omega \pm m\Omega_M)$ for the envelope $A^n_\varphi(\tau)$ of the coefficient of correlation that is widely used in practice.[54] In Table 3.2, the value $\Delta\Omega_\varphi$ signifies the equivalent bandwidth of the energy spectrum of phase fluctuations of the signal. Reference to Eq. (3.140) shows that the continuous component of the energy spectrum at the frequency $\Omega \pm \Omega_M$ is $\frac{2}{\sigma^2_{\varphi_{eq}}}$ times more than that of the continuous component of the energy spectrum at the frequency $\Omega = 0$.

It is well known[54,87–89] that under the condition of distortions only in phase of the signal, when $\varphi(t)$ is the wide-band Gaussian stationary process, the energy spectrum $G_M(\Omega)$ contains a discrete component at the frequency $\Omega = 0$ and the continuous component of energy spectrum in the neighborhood of the frequency $\Omega = 0$. We can see from Eq. (3.140) that during the period of narrow-band distortions in phase of the signal the continuous component of the energy spectrum is concentrated in the neighborhood of the frequency $\Omega = m\Omega_M$, where $m = 0, \pm 1, \pm 2, \ldots$. Under the degree of distortions in phase of the signal equal to

$$\sigma_\varphi \gg \frac{1}{|\dot{\psi}|}, \tag{3.145}$$

as we can see from Appendix II, the energy spectrum of the noise modulation function $\dot{M}(t)$ of the multiplicative noise takes the following form:

$$G_M(\Omega) = \sum_{m=0}^{\infty} C_0 I_m\left(\frac{\sigma^2_{\varphi_{eq}}}{2}\right) G_{n_0\varphi}(\Omega \pm m\Omega_M), \tag{3.146}$$

where

$$n_0 = 0.5\left(\sqrt{\sigma^2_{\varphi_{eq}} + 4m^2} - 1\right). \tag{3.147}$$

Thus, reference to Eq. (3.146) shows that under the condition

$$\sigma^2_{\varphi_{eq}} \gg 1 \tag{3.148}$$

a character of envelope of the energy spectrum depends essentially on the ratio between the average frequency Ω_M of the narrow-band Gaussian stationary process and the bandwidth $\Delta G_{n_0\varphi}$ of the energy spectrum defined by the envelope $G_{n_0}(\Omega \pm m\Omega_M)$. In the process, there are the following cases:

Case 1: $\Omega_M > \Delta G_{n_0\varphi}$. In this case, the continuous component of the energy spectrum is concentrated in the neighborhood of frequencies $m\Omega_M$ and the envelope of the continuous component is a set of the non-overlapping envelopes $G_{n_0}(\Omega \pm m\Omega_M)$, shifted in the frequency Ω_M with respect to each other (see Fig. 3.11, curve 1).

Case 2: $\Omega_M \ll \Delta G_{n_0\varphi}$. In this case, the continuous component of the energy spectrum $G_M(\Omega)$ that is symmetric with respect to the frequency $\omega = 0$ has the envelope

$$G_M(\Omega) \simeq G_{n_0\varphi}(\Omega), \tag{3.149}$$

taking into account the fact that $m\Omega_M \ll G_{n_0\varphi}$ and, consequently, an extension of the continuous component of the energy spectrum under the frequencies $m\Omega_M$ is negligible. Because of this, we can believe that the total power of the energy spectrum is concentrated in the neighborhood of the frequency $\Omega = 0$ (see Fig. 3.11, curve 3).

Case 3: $\Omega_M \leq \Delta G_{n_0\varphi}$. In this case, the envelope of the energy spectrum $G_M(\Omega)$ at each point is defined by summing the envelopes $G_{n_0\varphi}(\Omega \pm m\Omega_M)$ at the associated point. Probable behavior of the energy spectrum $G_M(\omega)$ is shown in Fig. 3.11, curve 2.

As we can see from the preceding discussion, if there is the narrow-band Gaussian stationary multiplicative noise in which the energy spectrum of the noise modulation function $\dot{M}(t)$ of the multiplicative noise has a discrete component, then the power depends on the degree of distortions in phase of

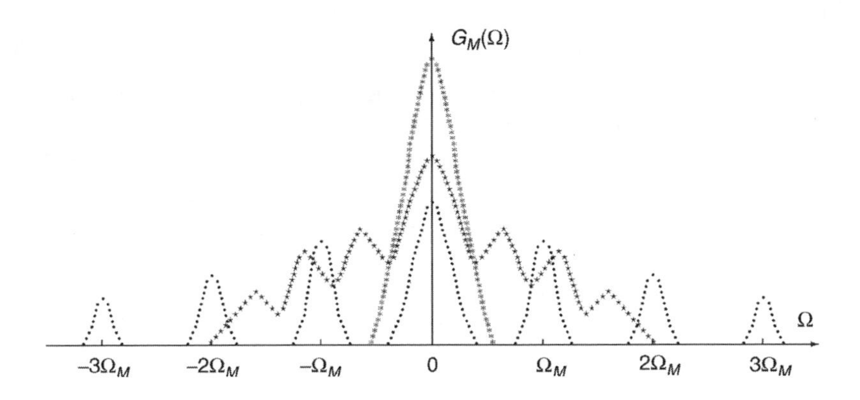

FIGURE 3.11
Spectrum of the noise modulation function: for $(\ldots)$, $\Omega_m > \Delta G_{n_0\varphi}$; for $(* * *)$, $\Omega_m \leq \Delta G_{n_0\varphi}$; for $(\star\star\star)$, $\Omega_m \ll \Delta G_{n_0\varphi}$.

the signal. Character of the probability distribution density of the continuous component of the energy spectrum by frequency depends essentially on the ratio between the bandwidth of the energy spectrum of distortions in phase of the signal and the average frequency of the energy spectrum at the constant degree of distortions in phase of the signal.

3.4.4 Pulse-Fluctuating Multiplicative Noise— Energy Spectrum of the Noise Modulation Function

In many cases distortions in phase of the signal are pulses, the structure of which is known. However, amplitudes of pulses and their location on the time axis are random. Consider the energy spectrum of the noise modulation function $\dot{M}(t)$, of the multiplicative noise when the pulse multiplicative noise is fluctuating, and the function $\varphi(t)$ (distortions in phase of the signal) is a sequence of pulses with the stochastic amplitude φ_n and the clock time interval T_M that is constant in time. The structure of the pulses is defined by the deterministic function $y(t)$, for $0 < t \leq T_M$. In practice this case can be found, for example, when there is pulse interference in feed circuits of amplifiers of a receiver or detector of signal processing systems.

Pulses of any k-th realization of the multiplicative noise are generated by the function $y(t)$ under the product of values of the function $y(t)$ on $\varphi_n^{(k)}$, and shift of the function $y(t)$ in the value nT_M on the time axis. In Reference 54 the reader can find the procedure of definition of the energy spectrum of the non-stationary pulse process, the stochastic parameters of which are the amplitude, time of appearance of the pulse on the time axis, and duration of the pulse. In the case of the pulse-fluctuating phase multiplicative noise, the phase of the signal is the non-stationary stochastic parameter of the noise modulation function $\dot{M}(t)$ of the multiplicative noise.

Taking into consideration that the function $y(t_n - nT_M)$ being outside the time interval

$$nT_M < t \leq (n+1)T_M \tag{3.150}$$

is equal to zero under the fixed value n, the noise modulation function $\dot{M}(t)$ of the multiplicative noise can be written in the following form:

$$\dot{M}(t) = \sum_{n=-\infty}^{\infty} \dot{M}_n(t), \tag{3.151}$$

where

$$\dot{M}_n(t) = \begin{cases} e^{j\varphi_n^{(k)} y(t-nT_M)} & \text{at } nT_M < t \leq (n+1)T_M; \\ 0 & \text{at } t \leq nT_M,\ t > (n+1)T_M \end{cases} \tag{3.152}$$

is the noise modulation function of the multiplicative noise with the duration T_M corresponding to the n-th elementary pulse of the stochastic process $\varphi(t)$.

The energy spectrum of the noise modulation function $\dot{M}(t)$ of the multiplicative noise in Eq. (3.152) is discussed in Appendix II under the condition when the amplitudes $\varphi_1^{(k)}, \varphi_2^{(k)}, \ldots, \varphi_n^{(k)}$ are the stationary uncorrelated pulse sequence and can be written in the following form:

$$G_M(\Omega) = \frac{2}{T_M}\left\{K_0(\Omega) - K_\infty(\Omega) + \frac{2\pi}{T_M}K_\infty(\Omega)\sum_{m=-\infty}^{\infty}\delta(\Omega - m\Omega_M)\right\}, \quad (3.153)$$

where

$$K_0(\Omega) = m_1\left\{\left|\int_0^{T_M} e^{j\varphi_m y(t)} \cdot e^{-j\Omega t}dt\right|^2\right\}; \quad (3.154)$$

$$K_\infty(\Omega) = \left|m_1\left\{\int_0^{T_M} e^{j\varphi_m y(t)} \cdot e^{-j\Omega t}dt\right\}\right|^2; \quad (3.155)$$

and

$$\Omega_M = \frac{2\pi}{T_M}. \quad (3.156)$$

Reference to Eq. (3.153) shows that the energy spectrum of the noise modulation function $\dot{M}(t)$ of the multiplicative noise is the sum of the continuous and discrete components.

For example, the energy spectrum of the noise modulation function $\dot{M}(t)$ of the multiplicative noise is defined in Eq. (3.153) when distortions in phase of the signal are the pulse sequence, the structure of which is determined by the following function

$$y(t) = \frac{\varphi_m}{T_M} \cdot t, \quad 0 < t \leq T_M. \quad (3.157)$$

Suppose that the random variables φ_m obey the Gaussian distribution law with the mean φ_0 and variance σ_φ^2. Under definition of the functions $K_\infty(\Omega)$ and $K_0(\Omega)$ in terms of Eq. (3.153), the continuous component $G_c(\Omega)$ and discrete component $G_d(\Omega)$ of the energy spectrum $G_M(\Omega)$ of the noise modulation function $\dot{M}(t)$ of the multiplicative noise are determined in the following form:

$$G_c(\Omega) = \frac{2\sqrt{2\pi}T_M}{\sigma_\varphi} \cdot e^{-\frac{(\Omega_0-\Omega)^2T_M^2}{2\sigma_\varphi^2}} - \frac{4\pi T_M}{\sigma_\varphi^2} \cdot e^{-\frac{(\Omega_0-\Omega)^2T_M^2}{\sigma_\varphi^2}}; \quad (3.158)$$

$$G_d(\Omega) = \frac{8\pi^2}{\sigma_\varphi^2} \cdot e^{-\frac{(\Omega_0-\Omega)^2T_M^2}{2\sigma_\varphi^2}} \sum_{m=-\infty}^{\infty}\delta(\Omega - m\Omega_M), \quad (3.159)$$

where

$$\Omega = \frac{\varphi_0}{T_M}. \tag{3.160}$$

The ratio between the continuous component and the discrete component of the energy spectrum of the noise modulation function $\dot{M}(t)$ of the multiplicative noise under the condition $\sigma_\varphi^2 \gg 1$ is equal to

$$\frac{G_c(\Omega)}{G_d(\omega)} = \frac{\sigma_\varphi}{2\sqrt{2\pi}\,\Omega_M} \cdot \exp\left\{-\frac{4\pi^2\Omega^2}{\sigma_\varphi^2\Omega_M^2}\right\} \tag{3.161}$$

and

$$\mathcal{L} = \frac{\int\limits_{-\infty}^{\infty} G_c(\Omega)\,d\Omega}{\int\limits_{-\infty}^{\infty} G_d(\Omega)\,d\Omega} = \frac{\sigma_\varphi}{\sqrt{\pi}} - 1. \tag{3.162}$$

The power component of the energy spectrum at the frequency $\Omega = 0$, which defines the relative level of constant component of the noise modulation function $\dot{M}(t)$ of the multiplicative noise, can be written in the following form:

$$\alpha_0^2 = \frac{8\pi^2}{\sigma_\varphi^2} \cdot \exp\left\{-\frac{\varphi_0^2}{\sigma_\varphi^2}\right\}. \tag{3.163}$$

We can determine that the energy spectrum of the noise modulation function $\dot{M}(t)$ of the multiplicative noise consists of only discrete components under the condition $\sigma_\varphi^2 \ll 1$:

$$G_M(\Omega) = 2\pi \left[\frac{\sin\frac{(\Omega-\Omega_0)T_M}{2}}{\frac{(\Omega-\Omega_0)T_M}{2}}\right] \sum_{m=-\infty}^{\infty} \delta(\Omega - m\Omega_M) = G_d(\Omega). \tag{3.164}$$

The correlation functions and energy spectrum discussed in this section allow us to define parameters of the signal under the stimulus of the multiplicative noise and parameters of the process at the output of any linear system.

3.5 Generalized Statistical Model of Multiplicative Noise

Signal propagation in real radio channels with anisotropic peculiarities has a multi-beam character. Because of this, the received signal has a complex angular spectrum. The number of beams is random. Therefore, every beam is subjected to absorptive fading caused by changing radiolucent or reflected features, interlocation, and dimensions or configurations of heterogeneities, and also by changes in radio wave-length. Superposition of radio signals leads to interference fading of the resulting signal, which in totality with radiolucent can be considered as multiplicative noise generating distortions in parameters (amplitude and phase) of the incoming signal at the input of the receiver or detector of the signal processing system.[90]

Henceforth we assume that the maximal delay of the received signals by each beam (or radio channel) is determined in the following form:

$$\Delta t_{\max} = \max \Delta t_j \qquad \forall j = \overline{1, M}, \tag{3.165}$$

where M is the number of the observed received signals (radio channels or beams). The maximal delay of the received signals must satisfy the following condition

$$\Delta t_{\max} \ll T, \tag{3.166}$$

where T is the duration of the elementary received signal. When the condition in Eq. (3.166) is not satisfied, then the interference fading is considered as the intersymbol interference. The number of received signals is defined by the aggregate and location of heterogeneities that scatter the signal under its propagation. It is impossible to separate entirely the received signals in radio channels. For this reason, the signal fading possesses both the absorptive and interference character.

Suppose that the resulting signal at the input of the receiver of the signal processing system takes the following form:

$$\dot{a}(t) = S(t) \cdot e^{j(\omega_0 t - \Delta\varphi)}, \tag{3.167}$$

where

$$S(t) = \sqrt{\left(\sum_{j=1}^{M} S_j(t) \cos \Delta\varphi_j\right)^2 + \left(\sum_{j=1}^{M} S_j(t) \sin \Delta\varphi_j\right)^2}; \tag{3.168}$$

$$\Delta\varphi = \operatorname{arctg} \frac{\sum_{j=1}^{M} S_j(t) \sin \Delta\varphi_j}{\sum_{j=1}^{M} S_j(t) \cos \Delta\varphi_j}; \tag{3.169}$$

and M is the number of beams of angular spectrum of the received signal. The multiplicative noise shows itself as the stochastic amplitude envelope $S(t)$ and random phase $\Delta\varphi$ of the incoming signal at the input of the receiver or detector of the signal processing system.

Statistical characteristics of the received signal $\dot{a}(t)$ depend on various agents: the disposition of the receiver; distance of radio wave propagation; frequency range and features of radio wave propagation in communication channels; frequency-time parameters of radio signals; and directional diagram and polarization of the transmitting and receiving antennas. In this association, statistical properties of the amplitude envelope $S(t)$ and phase $\Delta\varphi$ of the received signal $\dot{a}(t)$ are essentially different for radio channels, radio relay channels, cosmic satellite channels, radar channels, communication channels, mobile and telecommunication channels, and so on.

Experimental investigations and subjective choice of mathematical models for statistical representation of radio signals in various channels, each of which is in a good agreement with data of the corresponding experiment, lead to a set of models of the multiplicative noise, which has a common physical nature in any communication channel in effect. At the present time, there is a need to design the generalized model of the multiplicative noise, allowing us to define the amplitude envelope and phase of the received signal for any kind of communication channel on the unified background basis.

The model of multiplicative noise extending the set of models for various real communication channels used in practice (as mentioned above) is discussed in Reference 27. The received signal is the pair of orthogonal components that are the independent narrow-band Gaussian processes. Conditions, in which this assumption is true, can be realized using an adaptive phase shifter eliminating the mutual correlation of orthogonal components of the received signal. However, this principle cannot be used in signal processing systems of any kind absolutely.

Consider the generalized statistical model of the signal, taking into account the mutual correlation of components of the received signal.[91] This model is generated on the basis of the model discussed in Reference 27. Analysis of theoretical and experimental study allows us to assume that the transformation of the received signal under its propagation is linear; for example, the communication channel for each received signal (beam) $j \in [1, M]$ can be presented as the linear four-terminal network with the characteristic of transmission $\mu_j(\omega, t)$. The signal at the input of the linear four-terminal network takes the following form:

$$\dot{a}(t) = S(t) \cdot e^{j[\omega_0 t + \Psi(t)]}. \tag{3.170}$$

The received signal has a limited spectrum. This fact allows us to carry out the Fourier series expansion of the received signal

$$\dot{a}(t) = \sum_{k=k_1}^{k_m} S_k(t) \cdot e^{j[k\Omega_0 t + \Psi_k(t)]}, \tag{3.171}$$

where $S_k(t)$ is the amplitude of the k-th harmonic component of the received signal; $\Psi_k(t)$ is the phase of the k-th harmonic component of the received signal; and Ω_0 is the frequency satisfying the conditions

$$(k_m - k_1)\Omega_0 = 2\pi \Delta F \tag{3.172}$$

and

$$\frac{(k_m + k_1)\Omega_0}{2} = \omega_0 = 2\pi f_0 \tag{3.173}$$

as long as the spectrum of the received signal is symmetric.

The multi-beam communication channel is characterized by the number M of beams or channels and by the characteristic of transmission of the received signal. The characteristic of transmission of the received signal by the j-th channel is determined by

$$\dot{\mu}_j(\omega, t) = \mu_j(\omega, t) \cdot e^{j\varphi_j(\omega,t)} = \mu_\Delta \cdot e^{j\varphi_j}, \tag{3.174}$$

where μ_j is the amplitude characteristic—the module of the characteristic of transmission of the received signal and φ_j is the phase characteristic—the argument of the characteristic of transmission of the received signal.

The signal propagated by the j-th channel is connected to the generated signal $a_0(t)$ by the relationship

$$\dot{a}_j(t) = \dot{\mu}_j(\omega, t) a_0(t) \cdot e^{-j\omega_0 t_j}, \tag{3.175}$$

where t_j is the time of propagation of the received signal by the j-th channel. The time t_j is the same for all frequency components of the received signal. So, the received signal transmitted by the j-th channel takes the following form:

$$\dot{a}_j(t) = \mu_j(\omega, t)\, a(t) \cdot e^{j[\omega_0 t + \Psi(t) + \varphi_j(\omega,t)]}. \tag{3.176}$$

The multi-beam communication channel can be represented as (quantity) M four-terminal networks, parallel to one another. Since the communication channel can be considered linear, in accordance with the principle of superposition, the received signal in analytical form can be written as the sum of M signals:

$$\dot{a}_j(t) = \sum_{j=1}^{M} \mu_j(\omega, t)\, a(t) \cdot e^{j[\omega_0 t + \Psi(t) + \varphi_j(\omega,t)]}. \tag{3.177}$$

Using a representation of the received signal on the basis of harmonic functions, we can write

$$a(t) = \sum_{j=1}^{M} \sum_{k=k_1}^{k_m} \mu_j(\omega, t) S_k(t) \cos[k\Omega_0 t + \Psi_k(t) + \varphi_j(\omega, t)] \tag{3.178}$$

or

$$\begin{aligned} a(t) &= \sum_{j=1}^{M} \mu_j(\omega, t) \cos \varphi_j(\omega, t) \sum_{k=k_1}^{k_m} S_k(t) \cos[k\Omega_0 t + \Psi_k(t)] \\ &\quad - \sum_{j=1}^{M} \mu_j(\omega, t) \sin \varphi_j(\omega, t) \sum_{k=k_1}^{k_m} S_k(t) \sin[k\Omega_0 t + \Psi_k(t)] \\ &= X \sum_{k=k_1}^{k_m} S_k(t) \cos[k\Omega_0 t + \Psi_k(t)] - Y \sum_{k=k_1}^{k_m} S_k(t) \sin[k\Omega_0 t + \Psi_k(t)] \\ &= E \sum_{k=k_1}^{k_m} S_k(t) \cos[k\Omega_0 t + \Psi_k(t) + \vartheta(t)], \end{aligned} \tag{3.179}$$

where

$$E = \sqrt{X^2 + Y^2}; \tag{3.180}$$

$$\vartheta(t) = \operatorname{arctg} \frac{Y}{X}; \tag{3.181}$$

$$X = \sum_{j=1}^{M} \mu_j(\omega, t) \cos \varphi_j(\omega, t) = \sum_{j=1}^{M} x_j; \tag{3.182}$$

$$Y = \sum_{j=1}^{M} \mu_j(\omega, t) \sin \varphi_j(\omega, t) = \sum_{j=1}^{M} y_j. \tag{3.183}$$

To analyze the statistical properties of the received signal $a(t)$, there is a need to know the statistical characteristics of the Hilbert components X and Y that are the sum of random variables, where M is also the random variable. Each random variable distinguishes the characteristic of transmission by the j-th channel

$$j \in [1, M]. \tag{3.184}$$

As was shown in Reference 28, under the condition $M > 5$ the probability distribution density of the sum of X and Y can be approximated by the Gaussian law—even under the assumption that there is a correlation among the Hilbert components.[54] Thus, the joint probability distribution density of the Hilbert components X and Y can be determined in the following form:

$$\begin{aligned} f(X; Y) = \frac{1}{2\pi\sigma_X\sigma_Y\sqrt{1 - r_{XY}^2}} \cdot \exp\left\{ -\frac{1}{2(1 - r_{XY}^2)} \left[\frac{(X - m_X)^2}{\sigma_X^2} \right.\right. \\ \left.\left. - 2r_{XY} \frac{(X - m_X)(Y - m_Y)}{\sigma_X\sigma_Y} + \frac{(Y - m_Y)^2}{\sigma_Y^2} \right] \right\}, \end{aligned} \tag{3.185}$$

where m_X and σ_X^2 are the mean and variance of the Hilbert component X of the received signal, respectively; m_Y and σ_Y^2 are the mean and variance of the Hilbert component Y of the received signal, respectively; and r_{XY} is the coefficient of correlation between the Hilbert components X and Y of the received signal.

We define the mean, variance, and coefficient of correlation for the Hilbert components of the received signal:[54]

$$m_X = \sum_{j=1}^{M} \int_0^{\infty} \int_{-\infty}^{\infty} \mu_j \cos\varphi_j f_2(\mu_j; \varphi_j)\, d\mu_j\, d\varphi_j; \tag{3.186}$$

$$m_Y = \sum_{j=1}^{M} \int_0^{\infty} \int_{-\infty}^{\infty} \mu_j \sin\varphi_j f_2(\mu_j; \varphi_j)\, d\mu_j\, d\varphi_j; \tag{3.187}$$

$$\sigma_X^2 = \sum_{j=1}^{M} \int_0^{\infty} \int_{-\infty}^{\infty} \mu_j^2 \cos^2\varphi_j f_2(\mu_j; \varphi_j)\, d\mu_j\, d\varphi_j - m_X^2; \tag{3.188}$$

$$\sigma_Y^2 = \sum_{j=1}^{M} \int_0^{\infty} \int_{-\infty}^{\infty} \mu_j^2 \sin^2\varphi_j f_2(\mu_j; \varphi_j)\, d\mu_j\, d\varphi_j - m_Y^2; \tag{3.189}$$

$$\begin{aligned} r_{XY} = \sigma_X\sigma_Y \Bigg[&\frac{1}{2} \sum_{j=1}^{M} \int_0^{\infty} \int_{-\infty}^{\infty} \mu_j^2 \sin 2\varphi_j f_2(\mu_j; \varphi_j)\, d\mu_j\, d\varphi_j \\ &+ \sum_{j=1}^{M} \sum_{\gamma=1}^{M} \int_0^{\infty} \int_{-\infty}^{\infty} \mu_j \cos\varphi_j f(\mu_j; \varphi_j)\, d\mu_j\, d\varphi_j \\ &\times \int_0^{\infty} \int_{-\infty}^{\infty} \mu_\gamma \sin\varphi_\gamma f(\mu_\gamma; \varphi_\gamma)\, d\mu_\gamma\, d\varphi_\gamma - m_X m_Y \Bigg]. \end{aligned} \tag{3.190}$$

We can see from Eqs. (3.185)–(3.190) that the joint probability distribution density of the Hilbert components X and Y (of the received signal in Eq. (3.185)) is related to the joint probability distribution density of the module; the argument of the characteristic of transmission by the j-th channel determined by Eq. (3.184); and the number of channels. In turn, these characteristics depend on physical properties of the environment of radio wave propagation and also depend on characteristics and parameters of the transmitting and receiving antennas.

Note the following three distinctive characteristics of signal propagation by the communication channels: one-beam propagation, two-beam propagation, and clear-cut multi-beam propagation.

The first case is of practical significance for communication channels in the cosmos and atmosphere for any range of frequencies. The second case is most common in radio relay channels; communication channels, for example, between sea-ships; satellite communication channels for meter and decimeter radio waves given off by reflections from the Earth's surface; and aircraft communication channels. The third case is characteristic of channels based on the troposphere and ionosphere radio wave propagation, and in satellite communication channels under the use of millimeter radio waves.

In the first case

$$M = 1 \quad \text{and} \quad f_2(X; Y) = \delta(X_0; Y_0), \tag{3.191}$$

for example, the received signal is known completely. In the second case, the first signal (the signal propagated by the first channel) can be considered as the deterministic signal, and the second signal (the signal propagated by the second channel) is formed by mirror reflection—for example, from the Earth's surface. The third case is of prime interest for simulation of the multiplicative noise because the third case generalizes both the first and the second cases.

Assuming

$$M = 1; \tag{3.192}$$

$$m_X = X_0; \tag{3.193}$$

$$m_Y = Y_0; \tag{3.194}$$

$$\sigma_X = \sigma_Y = r_{XY} = 0, \tag{3.195}$$

we obtain, in particular, the model for the first case. Omitting from consideration the first and second channels, we obtain the model for the second case. Taking into account this fact, we suppose that the signals are commensurately statistically independent.

Using Eqs. (3.124)–(3.128) and proceeding to apply the polar coordinates, we can write

$$\begin{aligned} f_2(E; \vartheta) = {} & \frac{E}{2\pi\sigma_X\sigma_Y\sqrt{1 - r_{XY}^2}} \cdot \exp\left\{ -\frac{1}{2\sigma_X\sigma_Y\sqrt{1 - r_{XY}^2}} \right. \\ & \times \left\{ E^2 \cos^2\vartheta \left(\frac{\sigma_X}{\sigma_Y} + \frac{\sigma_X}{\sigma_Y} \cdot \frac{\sin^2\vartheta}{\cos^2\vartheta} - 2r_{XY} \cdot \frac{\sin\vartheta}{\cos\vartheta} \right) \right. \\ & + E_0^2 \cos^2\varphi_0 \left(\frac{\sigma_X}{\sigma_Y} + \frac{\sigma_X}{\sigma_Y} \cdot \frac{\sin^2\varphi_0}{\cos^2\varphi_0} - 2r_{XY} \cdot \frac{\sin\varphi_0}{\cos\varphi_0} \right) \\ & - 2EE_0 \cos\vartheta \cos\varphi_0 \left[\frac{\sigma_X}{\sigma_Y} + \frac{\sigma_X}{\sigma_Y} \cdot \frac{\sin\vartheta \sin\varphi_0}{\cos\vartheta \cos\varphi_0} \right. \\ & \left.\left.\left. - r_{XY} \left(\frac{\sin\vartheta}{\cos\vartheta} - \frac{\sin\varphi_0}{\cos\varphi_0} \right) \right] \right\} \right\}, \end{aligned} \tag{3.196}$$

where

$$E_0 = \sqrt{m_X^2 + m_Y^2} \tag{3.197}$$

and

$$\varphi_0 = \text{arctg}\frac{m_X}{m_Y}. \tag{3.198}$$

The statistical models presented in Eqs. (3.185) and (3.198) define a stochastic character of changes in the characteristic of transmission of the multi-beam communication channel. It should be pointed out that these models are true for multi-beam communication channels of any physical nature.

The random character of changing the characteristic of transmission of the communication channel leads to random distortions in parameters (amplitude and phase) of the received signal (see Eq. (3.177)). In a general case, the received signal $\dot{a}(t)$ is the non-stationary stochastic process.

Actually, reference to Eq. (3.179) shows that the Hilbert components X and Y are multiplying out with harmonic deterministic functions possessing a period that is proportional to the frequency Ω_0:

$$m'_X = m_X \sum_{k=k_1}^{k_m} S_k(t)\cos[k\Omega_0 t + \Psi_k(t)]; \tag{3.199}$$

$$m'_Y = m_Y \sum_{k=k_1}^{k_m} S_k(t)\sin[k\Omega_0 t + \Psi_k(t)]; \tag{3.200}$$

$$\sigma_X^{\prime 2} = \sigma_X^2 \left\{\sum_{k=k_1}^{k_m} S_k(t)\cos[k\Omega_0 t + \Psi_k(t)]\right\}^2; \tag{3.201}$$

$$\sigma_Y^{\prime 2} = \sigma_Y^2 \left\{\sum_{k=k_1}^{k_m} S_k(t)\sin[k\Omega_0 t + \Psi_k(t)]\right\}^2; \tag{3.202}$$

$$r'_{XY} = r_{XY}\sigma_X\sigma_Y \sum_{k=k_1}^{k_m} S_k(t)\sin 2[k\Omega_0 t + \Psi_k(t)]. \tag{3.203}$$

Reference to Eq. (3.134) shows that the received signal is the periodic non-stationary stochastic process. Considering the case in practice, and showing that the communication channel is stationary only under the conditions

$$\frac{\sigma_X}{\sigma_Y} = 1 \tag{3.204}$$

and

$$r_{XY} = 0. \tag{3.205}$$

Before proceeding on to the generalized model of multiplicative noise, we introduce the following assumption, taking into consideration the following peculiarities of radio wave propagation:

- The module μ_j and argument φ_j of the characteristic of transmission of communication channel are statistically independent, for example,

$$f_2(\mu_j; \varphi_j) = f(\mu_j) f(\varphi_j). \tag{3.206}$$

- The probability distribution densities $f(\mu_j)$ and $f(\varphi_j)$ are the same for various ways of radio wave propagation, for example,

$$f(\mu_j) = f(\mu), \qquad f(\varphi_j) = f(\varphi), \qquad \forall j \in [1, M]. \tag{3.207}$$

The first assumption is true since μ depends essentially on a scattering volume and φ_j depends on the signal delay—for example, on disposition in a space of the reflecting point. Legitimacy of the second assumption is based on the same physical nature of a communication channel for any method of signal propagation.

Then in terms of Eqs. (3.206) and (3.207) Eqs. (3.186)–(3.190) take the following form:

$$m_X = M \int_0^\infty \mu f(\mu)\, d\mu \int_{-\infty}^\infty \cos\varphi f(\varphi)\, d\varphi; \tag{3.208}$$

$$m_Y = M \int_0^\infty \mu f(\mu)\, d\mu \int_{-\infty}^\infty \sin\varphi f(\varphi)\, d\varphi; \tag{3.209}$$

$$\sigma_X^2 = M \int_0^\infty \mu^2 f(\mu)\, d\mu \int_{-\infty}^\infty \cos^2\varphi f(\varphi)\, d\varphi - m_X^2; \tag{3.210}$$

$$\sigma_Y^2 = M \int_0^\infty \mu^2 f(\mu)\, d\mu \int_{-\infty}^\infty \sin^2\varphi f(\varphi)\, d\varphi - m_Y^2; \tag{3.211}$$

$$r_{XY} = \frac{1}{\sigma_X \sigma_Y} \left\{ \frac{\mu}{2} \int_0^\infty \mu^2 f(\mu)\, d\mu \int_{-\infty}^\infty \sin 2\varphi f(\varphi)\, d\varphi \right.$$
$$+ M(M-1) \int_0^\infty \mu^2 f(\mu)\, d\mu \int_{-\infty}^\infty \cos\varphi f(\varphi)\, d\varphi$$
$$\left. \times \int_{-\infty}^\infty \sin\varphi f(\varphi)\, d\varphi - m_X m_Y \right\}. \tag{3.212}$$

The means and variances of the Hilbert components, X and Y of the received signal, are different. The coefficient of their mutual correlation does not equal zero. Because of this, for an unambiguous definition of the statistical model of the multiplicative noise it is necessary and sufficient to define only five parameters in Eqs. (3.208)–(3.212). In the process, an *a priori* uncertainty in statistics is the parametric *a priori* uncertainty. Taking into consideration a physical nature of processes that take place in communication channels, we are able to generalize a set of parameters and define them by new parameters, the number of which is less than the original number of parameters.

Eq. (3.196) is transformed, introducing the normalized variables

$$A = \frac{E}{\sqrt{\sigma_X \sigma_Y}} \tag{3.213}$$

and

$$a = \frac{E_0}{\sqrt{\sigma_X \sigma_Y}}, \tag{3.214}$$

and the parameter of non-stationary state

$$\alpha = \sigma_X^2 - \sigma_Y^2. \tag{3.215}$$

Then Eq. (3.196) takes the following form:[18]

$$\begin{aligned} f_2(A, \vartheta, \alpha, r_{XY}, a, \varphi_0) &= \frac{A}{2\pi\sqrt{1 - r_{XY}^2}} \\ &\times \exp\left\{ -\frac{a^2}{2(1 - r_{XY}^2)\sqrt{1-\alpha^2}} \cdot \{1 - B(\alpha, r_{XY}) \cos[2\varphi_0 - \beta(\alpha, r_{XY})]\} \right. \\ &- \frac{A^2}{2(1 - r_{XY}^2)\sqrt{1-\alpha^2}} \cdot \{1 - B(\alpha, r_{XY}) \cos[2\vartheta - \beta(\alpha, r_{XY})]\} \\ &\left. - \frac{Aa}{1 - r_{XY}^2} \cdot \{r_{XY} \sin(\vartheta + \varphi_0) - C(\alpha, \varphi_0) \cos[\vartheta - \gamma(\alpha, \varphi_0)]\} \right\}, \end{aligned} \tag{3.216}$$

where

$$B(\alpha, r_{XY}) = \sqrt{\alpha^2 + r_{XY}^2(1 - \alpha^2)}; \tag{3.217}$$

$$\beta(\alpha, r_{XY}) = \mathrm{arctg} \frac{r_{XY}\sqrt{1-\alpha^2}}{\alpha}; \tag{3.218}$$

$$C(\alpha, \varphi_0) = \sqrt{\frac{1-\alpha}{1+\alpha} \cos^2 \varphi_0 + \frac{1+\alpha}{1-\alpha} \sin^2 \varphi_0}; \tag{3.219}$$

and

$$\gamma(\alpha, \varphi_0) = \mathrm{arctg} \left[\mathrm{tg}\varphi_0 \frac{1+\alpha}{1-\alpha} \right]. \tag{3.220}$$

The parameters α and r_{XY} in Eq. (3.216) characterize the non-stationary state of the communication channel. The parameter φ_0 defines the initial phase of the phase characteristic of the communication channel. The parameter α defines a power of the regular component of the communication channel.

Equation (3.216) is the initial condition to define the one-dimensional probability distribution density of amplitude envelope and phase of the received signal:

$$f(A,\alpha,r_{XY},a,\varphi_0) = \int_0^{2\pi} f_2(A,\vartheta,\alpha,r_{XY},a,\varphi_0)\,d\varphi$$

$$= \frac{A}{\sqrt{1-r_{XY}^2}}\cdot\exp\left\{-\frac{A^2+a^2\{1-B(\alpha,r_{XY})\cos[2\varphi_0-\beta(\alpha,r_{XY})]\}}{2(1-r_{XY}^2)\sqrt{1-\alpha^2}}\right\}$$

$$\times\sum_{n=0}^{\infty}\varepsilon_n J_n\left[\frac{A^2B(\alpha,r_{XY})}{2(1-r_{XY}^2)\sqrt{1-\alpha^2}}\right]$$

$$\times J_{2n}\left[\frac{Aa}{1-r_{XY}^2}\sqrt{r_{XY}^2+C^2(\alpha,\varphi_0)-\frac{2r_{XY}\sin 2\varphi_0}{\sqrt{1-\alpha^2}}}\right]\cos 2n\mathcal{V}(\alpha,r_{XY},\varphi_0), \quad (3.221)$$

where

$$\varepsilon_n = 1 \quad \text{at } n=0; \qquad (3.222)$$

$$\varepsilon_n = 2 \quad \text{at } n\neq 0; \qquad (3.223)$$

$J_n(x)$ and $J_{2n}(x)$ are the Bessel functions of the first kind;

$$\mathcal{V}(\alpha,r_{XY},\varphi_0) = \Delta(\alpha,r_{XY},\varphi_0) - \frac{\beta(\alpha,r_{XY})}{2}; \qquad (3.224)$$

$$\Delta(\alpha,r_{XY},\varphi_0) = \operatorname{arctg}\frac{\sqrt{\frac{1+\alpha}{1-\alpha}}\cdot\sin\varphi_0 - r_{XY}\cos\varphi_0}{\sqrt{\frac{1-\alpha}{1+\alpha}}\cdot\cos\varphi_0 - r_{XY}\sin\varphi_0}; \qquad (3.225)$$

$$f(\vartheta,\alpha,r_{XY},a,\varphi_0) = \int_0^{\infty} f_2(A,\vartheta,\alpha,r_{XY},a,\varphi_0)\,dA$$

$$= \frac{\sqrt{(1-r_{XY}^2)(1-\alpha^2)}}{2\pi\{1-B(\alpha,r_{XY})\cos[2\vartheta-\beta(\alpha,r_{XY})]\}}$$

$$\times\exp\left\{-\frac{a^2}{2(1-r_{XY}^2)\sqrt{1-\alpha^2}}\{1-B(\alpha,r_{XY})\cos[2\varphi_0-\beta(\alpha,r_{XY})]\}\right\}$$

$$\times\left\{1+a\sqrt{\frac{\pi}{2}}\cdot\frac{\sqrt[4]{1-\alpha^2}}{\sqrt{1-r_{XY}^2}}\cdot\frac{C(\alpha,\varphi_0)\cos[\vartheta-\gamma(\alpha,\varphi_0)]-r_{XY}\sin[\vartheta+\varphi_0]}{\sqrt{1-\beta(\alpha,r_{XY})\cos[2\vartheta-\beta(\alpha,r_{XY})]}}\right\}$$

$$\times\exp\left\{-\frac{a^2\sqrt{1-\alpha^2}}{2(1-r_{XY}^2)}\cdot\frac{\{C(\alpha,\varphi_0)\cos[\vartheta-\gamma(\alpha,\varphi_0)]-r_{XY}\sin[\vartheta-\varphi_0]\}^2}{1-B(\alpha,r_{XY})\cos[2\vartheta-\beta(\alpha,r_{XY})]}\right\}$$

$$\times\left\{1+\frac{\Phi\{a\sqrt[4]{1-\alpha^2}\{C(\alpha,\varphi_0)\cos[\vartheta-\gamma(\alpha,\varphi_0)]-r_{XY}\sin[\vartheta-\varphi_0]\}\}}{\sqrt{2(1-r_{XY}^2)\{1-B(\alpha,r_{XY})\cos[2\vartheta-\beta(\alpha,r_{XY})]\}}}\right\}, \qquad (3.226)$$

where

$$\Phi(x) = \frac{2}{\sqrt{\pi}} \int_{0}^{x} e^{t^2} dt \tag{3.227}$$

is the error integral.

On account of the non-stationary state of the communication channel, the amplitude envelope and phase of the received signal are interdependent. The probability distribution density of the amplitude envelope of the received signal is very sensitive to variations in its numerical parameters.

Variation in parameters allows us to use the suggested generalized model to define the multiplicative noise in communication channels with various peculiarities of signal propagation. The generalized model of the multiplicative noise has been obtained based on general assumptions setting minimal limits under the use of the generalized model of the multiplicative noise.

The generalized model of multiplicative noise allows us to define various forms of signal fading when the sampling data of the amplitude envelope of a signal have two absolute values, for example, there is a bimodal probability distribution density of the amplitude envelope of the signal.

Analysis of communication channels with the use of geometric representation of a signal shows that the bimodal probability distribution density of the amplitude envelope of the signal can take place in real communication channels. This peculiarity is characteristic of the probability distribution densities, possessing σ_X and σ_Y, that are much less in comparison with other parameters of the probability distribution density. To prove this statement, there is a need to study Eq. (3.226) under the condition

$$|\alpha| = 1. \tag{3.228}$$

Physical sources of the bimodal probability distribution density of the amplitude envelope of the characteristic of transmission of a communication channel, which are related to the Doppler frequency shift in the spectrum of individual signals under the multi-beam radio wave propagation, are explained in Reference 19.

We can see from Eqs. (3.208)–(3.212) that parameters of the probability distribution density determined by Eqs. (3.216)–(3.226) depend on the probability distribution density of the module $f(\mu)$ and the probability distribution density of the argument $f(\varphi)$ of the characteristic of transmission of a communication channel.

Analysis of interdependency of the initial parameters $m_X, m_Y, \sigma_X^2, \sigma_Y^2, r_{XY}$, and the parameters α, r_{XY}, a, and φ_0 when the probability distribution density of the argument φ obeys the uniform and Gaussian distribution laws, showed that the parameters α and r_{XY} do not depend on the number M of the received signals and are only defined by numerical characteristics of the probability distribution densities $f(\mu)$ and $f(\varphi)$. The parameter φ_0 also does not depend on the number M of the received signals and the form of the probability distribution densities $f(\mu)$ and $f(\varphi)$.

The particular statistical models following from Eqs. (3.216)–(3.226) are in good agreement with known statistical models. The number of probable particular models can be determined in the following form:

$$N = \sum_{k=0}^{d} C_d^k, \tag{3.229}$$

where C_d^k is the binomial coefficient of parameters of the probability distribution density. In the case $d = 4$, then $N = 16$.

Analysis of Eq. (3.212) shows that under the condition $a = 0$ the argument φ cannot be defined, for example, the value φ_0 can take any value within the limits of the interval $[0, 2\pi]$ with the probability equal to zero. Thus, twelve particular models are completely defined, and four particular models (for which $a = 0$) cannot be completely defined and coincide with corresponding particular models under the condition

$$\varphi \in [0, \pi] \cup [\pi, 2\pi] \tag{3.230}$$

and form new particular models at the points

$$\varphi_0 = n\pi, \tag{3.231}$$

where $n = 0, 1, 2, \ldots$.

It is appropriate to consider the particular models when $a = 0$ only at $\varphi_0 = 0$ in conjunction with the angular disposition vector (the length of which is equal to zero) cannot impact on the form of the model.

The well-known Rayleigh,[1,2,54,69] Rice,[1,54] Hoyt,[92] and Nakagami[93–95] models are the particular cases of the generalized model of the multiplicative noise in Eqs. (3.216)–(3.226). Table 3.3 shows values of parameters of the generalized model of the multiplicative noise, which correspond to these particular cases.

As discussed previously, the multiplicative noise is the periodic nonstationary process in a general case. In this case, the necessary conditions

TABLE 3.3

Parameters of Generalized Model of the Multiplicative Noise

No.	Parameters	Particular Models
1	$\alpha = 0, r_{XY} = 0, a = 0, \varphi_0 = 0$	Rayleigh Model
2	$\alpha = 0, r_{XY} = 0, a \neq 0, \varphi_0 \neq 0$	Rice Model
3	$\alpha \neq 0, r_{XY} = 0, a \neq 0, \varphi_0 = 0$	Hoyt Model
4	$\alpha \neq 0, r_{XY} = 0, a \neq 0, \varphi_0 \neq 0$	Generalized Hoyt Model
5	$\alpha = 0, r_{XY} \neq 0, a = 0, \varphi_0 = 0$	ρ-Vector Model
6	$\alpha = 0, r_{XY} \neq 0, a \neq 0, \varphi_0 \neq 0$	Generalized ρ-vector Model

of stationary state of the stochastic process defining the received signal are the following:[20]

- Quasimonochromaticity of the received signal
- Symmetry of spectrum of the received signal
- Equality to zero of the first moment of phase of the received signal
- Constancy of the first and second moments of amplitude envelope and phase of the received signal

The first condition of stationary state is satisfied with sufficient accuracy for signals propagated by any real communication channels. The second condition of stationary state is not ever satisfied. For example, this condition is not satisfied under the use of one-band signals; when selective fading in the communication channels are present; when the frequency characteristics of filters of the transmitter are non-uniform and asymmetric; and so on. The most common criterion of stationary state of the multiplicative noise is the third condition that essentially combines the first and second conditions.

We proceed to estimate a stationary state of the received signal at the output of the multi-beam communication channel (see Eq. (3.196)). Consider the properties of the multiplicative noise, assuming that the received signal takes the following form:

$$a(t) = X\cos\omega_0 t - Y\sin\omega_0 t = E\cos[\omega_0 t + \vartheta(t)]. \tag{3.232}$$

Define the first (central) and second (initial) moments of the stochastic process $a(t)$ determined by Eq. (3.232) on the basis of the generalized model of the multiplicative noise in Eq. (3.216):

$$M[a(t)] = \frac{a}{\sqrt{2}} \cdot \sigma\sqrt[4]{1-\alpha^2}\cos[\omega_0 t + \varphi_0]; \tag{3.233}$$

$$D[a(t)] = \frac{\sigma^2}{2}\left(1 + \alpha\cos 2\omega_0 t - r_{XY}\sqrt{1-\alpha^2}\sin 2\omega_0 t\right); \tag{3.234}$$

$$m_2[a(t)] = \frac{\sigma^2}{2}\left[1 + \alpha\cos 2\omega_0 t - r_{XY}\sqrt{1-\alpha^2}\sin 2\omega_0 t + a^2\sqrt{1-\alpha^2}\cos^2[\omega_0 t + \varphi_0]\right], \tag{3.235}$$

where

$$\sigma^2 = \sigma_X^2 + \sigma_Y^2. \tag{3.236}$$

Reference to Eqs. (3.233)–(3.236) shows that only the Rayleigh channel is stationary under the condition in which various parameters of the probability distribution density are taken into account.

The considered generalized statistical model of the received signal defines a general case for communication channels, and allows us to reduce an *a priori* uncertainty in statistical characteristics of the multiplicative noise to a parametric *a priori* uncertainty that is easily overcome under determination of sampling estimations of parameters of the generalized Gaussian four-parametric model of the received signal.

3.6 Conclusions

Discussion in this chapter allows us to draw the following conclusions.

- Classification of the noise and interference in accordance with their physical nature and peculiarities as a stimulus on parameters of transmitted signals is considered. This classification is based on partitioning all forms of the noise and interference into three main classes: additive, multiplicative, and intersymbol noise and interference. An exhaustive mathematical statistical model of the natural noise and interference is not yet defined. Each mathematical statistical model allows us to describe a composition of the fluctuating noise and pulse interference, only for particular cases of signal processing when the natural noise and interference become dominant. The additive noise and the multiplicative noise are the main factors defining the noise immunity of signal processing systems in various areas of application.
- The stimulus of multiplicative noise on signal parameters and qualitative characteristics of signal processing systems depends essentially on the relationship between the time interval $[0, T]$, within the limits of which the signal is processed, observed, and analyzed, and the rate of changes in the signal phase and amplitude as a function of time. For cases in which the time of correlation of the multiplicative noise is a commensurate time interval $[0, T]$ within the limits of which the coherent signal is processed, or less, then the signal structure is distorted by the multiplicative noise. In other words, the modulation laws of amplitude and phase of the signal are changed randomly.
- The multiplicative noise can be generated by all elements of a data transmission channel both during radio wave propagation and under signal processing by a receiver or detector of signal processing systems. The main sources of the multiplicative noise are random fluctuations of parameters in the troposphere, ionosphere (temperature, humidity, electron concentrations, and so on); presence of local heterogeneities in the troposphere and ionosphere; reflection of electromagnetic waves from the ionosphere or the Earth's surface; movement of non-uniform solar and interplanetary plasmas; movement of the receiver or detector caused by mobile signal processing systems; overlapping modulation of signals under propagation of electromagnetic

waves in the non-linear environment; the background noise of the generator of signals giving rise to fluctuations of amplitude and phase of the signal; fluctuations and non-stabilities of voltage in the power supply, interference of the generator and power amplifier, and fluctuations of parameters in basic elements of signal processing systems; radio interference generated by other receivers or detectors of signal processing systems; and so on. Specific sources of the multiplicative noise take place, for example, under functioning radar systems in the case of side scanning.

- In the majority of cases, an initiation of multiplicative noise can be defined as a result of passing the narrow-band signal through the equivalent linear four-terminal network with parameters varying as a function of time.
- Magnitude of distortions in amplitude and phase of the signal, caused by the multiplicative noise, do not depend on the energy of the signal. Because of this, under the stimulus of multiplicative noise as opposed to the stimulus of additive noise, an increase in the energy of the signal does not lead to a decrease in the influence of the multiplicative noise.
- In accordance with the form of distortions of the signal, there are the phase distortions, amplitude distortions, and amplitude-phase distortions. According to the form of process defining distortions of the signal, there are the deterministic (or quasideterministic) and stochastic multiplicative noise. According to the relative rate of changes in amplitude and phase of the signal, caused by the multiplicative noise, there are rapid and slow multiplicative noise.
- The multiplicative noise is completely characterized by the noise modulation function, $\dot{M}(t)$. The main characteristics of the noise modulation function of the multiplicative noise are the correlation function and the energy spectrum. The correlation function and energy spectrum of the deterministic and quasideterministic multiplicative noise, stationary fluctuating multiplicative noise, and pulse-fluctuating multiplicative noise were discussed. The correlation function of the noise modulation function of the multiplicative noise can be easily defined if the characteristic functions corresponding to the distribution laws of distortions in amplitude and phase of the signal are known. The correlation function of the noise modulation function $\dot{M}(t)$ of the multiplicative noise is the complex function if the amplitude and phase distortions of the signal are interdependent, and the energy spectrum in this case is asymmetric with respect to zero frequency. If the amplitude and phase distortions of the signal are independent, or if there are distortions only in phase or only in amplitude of the signal, then the correlation function of the noise modulation function of the multiplicative noise is real, and the energy spectrum in this case is symmetric with respect to zero frequency. When the multiplicative noise is the narrow-band Gaussian stationary process the

energy spectrum of the noise modulation function $\dot{M}(t)$ of the multiplicative noise has a discrete component, the power of which depends on the degree of distortions in phase of the signal. The character of distortions of the continuous component of the energy spectrum by frequency depends essentially on the ratio between the bandwidth of the energy spectrum of distortions in phase of the signal and the average frequency of the energy spectrum at the constant degree of distortions in phase of the signal.

- The generalized statistical model of the multiplicative noise was discussed. This model allows us to define the amplitude envelope and phase of the received signal for any kind of communication channel on the unified basis. All well-known models, for example, the Rayleigh model, Rice model, Hoyt model, and Nakagami model, are the particular cases of the suggested generalized statistical model of multiplicative noise.

References

1. Helstrom, C., *Statistical Theory of Signal Detection*, 2nd ed., Pergamon Press, Oxford, U.K. 1968.
2. Middleton, D., *An Introduction to Statistical Communication Theory*, McGraw-Hill, New York, 1960.
3. Skolnik, M., *Radar Applications*, IEEE Press, New York, 1988.
4. Middleton, D., Statistical-physical models of man-made radio noise. Part 1: First-order probability models of the instantaneous amplitude, Office of Telecommunications, Report 74-36, April 1974, pp. 1–76.
5. Middleton, D., Statistical-physical models of man-made radio noise. Part 2: First-order probability models of the envelope and phase, Office of Telecommunications, Report 76-86, April 1976, pp. 76–124.
6. Middleton, D., Man-made noise in urban environments and transportation systems: models and measurements, *IEEE Trans.*, Vol. IT-22, No. 4, 1973, pp. 25–77.
7. Spaulding, A. and Middleton, D., Optimum reception in an impulsive interference environment. Part 1: Coherent detection; Part 2: Incoherent detection, *IEEE Trans.*, Vol. COM-25, No. 9, 1977, pp. 910–934.
8. Ziemer, R., Error probabilities due to impulsive noise, *IEEE Trans.*, Vol. COM-15, No. 6, 1967, pp. 471–474.
9. Hyghn, H. and Lecours, M., Impulsive noise in non-coherent M-ary digital systems, *IEEE Trans.*, Vol. COM-23, No. 2, 1975, pp. 31–38.
10. Bello, F. and Esposito, R., Error probabilities due to impulsive noise in linear and hard-limited DPS systems, *IEEE Trans.*, Vol. COM-19, No. 2, 1971, pp. 14–20.
11. Barness, Y., Steered beam and LMS interference canceller comparison, *IEEE Trans.*, Vol. AES-11, No. 1, 1983, pp. 30–39.
12. Friedlander, B., System identification techniques for adaptive noise canceling, *IEEE Trans.*, Vol. SSP-10, No. 5, 1982, pp. 699–709.

13. Black, W., An impulsive noise canceller, *IEEE Trans.*, Vol. COM-13, No. 12, 1963, pp. 506–518.
14. Schwartz, M., Bennett, W., and Stein, S., *Communication Systems and Techniques*, McGraw-Hill, New York, 1966.
15. Di Franco, J. and Rubin, W., *Radar Detection*, Artech House, Norwood, MA, 1980.
16. Conte, E., De Maio, A., and Galdi, C., Signal detection in compound-Gaussian noise: Neyman–Pearson and CFAR detectors, *IEEE Trans.*, Vol. SP-48, No. 2, 2000, pp. 419–428.
17. Zhou, G. and Giannakis, G., Harmonics in multiplicative and additive noise: performance analysis of cyclic estimators, *IEEE Trans.*, Vol. SP-43, No. 6, 1995, pp. 1445–1460.
18. Pozdnyak, S. and Melititzkiy, V., *Introduction to Statistical Theory of Radio Wave Propagation*, Soviet Radio, Moscow, 1984 (in Russian).
19. Alpert, Ya., *Propagation of Electromagnetic Waves and Troposphere*, Nauka, Moscow, 1972 (in Russian).
20. Rytov, S., *Introduction to Statistical Radio Physics*, Nauka, Moscow, 1966 (in Russian).
21. Gallager, R., *Information Theory and Reliable Communication*, Wiley, New York, 1968.
22. Jakes, W., *Microwave Mobile Communications*, IEEE Press, Piscataway, NJ, 1993.
23. Cox, D. and Leck, R., Correlation bandwidth and delay spread multi-path propagation statistics for 910 MHz urban mobile radio, *IEEE Trans.*, Vol. COM-23, No. 11, 1975, pp. 1271–1280.
24. Rappaport, T., *Wireless Communications: Principles and Practice*, Prentice-Hall, Englewood Cliffs, NJ, 1996.
25. Yacoub, M., *Foundations of Mobile Radio Engineering*, CRC Press, Boca Raton, FL, 1993.
26. Greenwood, D. and Hanzo, L., Characterization of mobile radio channels, *Mobile Radio Communications*, Steele, R., Ed., IEEE Press, Piscataway, NJ, 1992.
27. Klovskiy, D., *Transmission of Discrete Signals Using Radio Channels*, Svyaz, Moscow, 1969 (in Russian).
28. Phink, L., *Transmission Theory of Discrete Signals*, Soviet Radio, Moscow, 1970 (in Russian).
29. Anderson, B. and Moore, J., *Optimal Filtering*, Prentice-Hall, Englewood Cliffs, NJ, 1979.
30. Cover, T. and Thomas, J., *Elements of Information Theory*, Wiley, New York, 1991.
31. Proakis, J., *Digital Communications*, 3rd ed., McGraw-Hill, New York, 1995.
32. Varshney, P., *Distributed Detection and Data Fusion*, Springer-Verlag, New York, 1996.
33. Kennedy, R., *Fading Dispersive Communications Channels*, Wiley–Interscience, New York, 1969.
34. Stüber, G., *Principles of Mobile Communications*, Kluwer Academic Publishers, Boston, 1996.
35. Hudson, J., *Adaptive Array Principles*, Peter Peregrinus, London, 1981.
36. Compton, R., *Adaptive Antennas*, Prentice-Hall, Englewood Cliffs, NJ, 1988.
37. Ulaby, F. and Elachi, C., *Radar Polarimetry for Geoscience Applications*, Artech House, Norwood, MA, 1990.
38. Monzingo, R. and Miller, T., *Introduction to Adaptive Arrays*, Wiley, New York, 1980.

39. Baird, C. and Rassweiler, G., Adaptive side lobe nulling using digitally controlled phase-shifters, *IEEE Trans.*, Vol. AP-24, No. 5, 1976, pp. 638–649.
40. Brennan, L. and Reed, I., Theory of adaptive radars, *IEEE Trans.*, Vol. AES-9, No. 2, 1973, pp. 237–252.
41. Scharf, L., *Statistical Signal Processing: Detection, Estimation and Time Series Analysis*, Addison-Wesley, New York, 1991.
42. Biglier, E., Proakis, J., and Shamai, S., Fading channels: information-theoretic and communication aspects, *IEEE Trans.*, Vol. IT-44, No. 5, 1998, pp. 2619–2692.
43. Salwen, H., Differential phase-shift keying performance under time-selective multi-path fading, *IEEE Trans.*, Vol. COM-23, No. 3, 1975, pp. 371–379.
44. Huang, T., Omura, J., and Biederman, L., Bit error rate comparison of repeater and regenerative communication satellites, *IEEE Trans.*, Vol. COM-28, No. 7, 1980, pp. 1088–1097.
45. White, W., Cascade preprocessors for adaptive antennas, *IEEE Trans.*, Vol. AP-24, No. 5, 1976, pp. 670–684.
46. Hackett, C., Adaptive arrays can be used to separate communication signal, *IEEE Trans.*, Vol. AP-17, No. 2, 1981, pp. 234–247.
47. Al-Khatib and Compton, R., Gain optimizing algorithm for adaptive arrays, *IEEE Trans.*, Vol. AP-26, No. 2, 1978, pp. 228–235.
48. Compton, R., The tripole antenna and adaptive array with full polarization flexibility, *IEEE Trans.*, Vol. AP-29, No. 6, 1981, pp. 944–952.
49. Davis, R., Brennan, L., and Reed, L., Angle estimation with adaptive arrays in external noise field, *IEEE Trans.*, Vol. AES-19, No. 2, 1976, pp. 179–184.
50. Thompson, P., Adaptation by direct phase-shift adjustment in narrow-band adaptive antenna systems, *IEEE Trans.*, Vol. AP-24, No. 5, 1976, pp. 756–760.
51. Hochwald, B. and Marzetta, T., Unitary space-time modulation for multiple-antenna communications in Rayleigh flat fading, *IEEE Trans.*, Vol. IT-46, No. 2, 2000, pp. 453–564.
52. Marzetta, T. and Hochwald, B., Capacity of a mobile multiple-antenna communication link in Rayleigh flat fading, *IEEE Trans.*, Vol. IT-45, No. 1, 1999, pp. 139–157.
53. Seshadri, N. and Ninters, J., Two signalling schemes for improving the error performance of frequency-division-duplex (FDD) transmission systems using transmitter antenna diversity, *Int. J. Wire. Inform. Net.*, Vol. 1, No. 1, 1994, pp. 49–59.
54. Levin, B., *Theoretical Foundations of Statistical Radio Engineering*, Parts 1–3, Soviet Radio, Moscow, 1974–1976 (in Russian).
55. Eyuboglu, M. and Qureshi, S., Detection of code modulation signals on linear severely distorted channels using decision–feedback noise prediction with interleaving, *IEEE Trans.*, Vol. COM-36, No. 4, 1988, pp. 401–409.
56. Foschini, G., Layered space-time architecture for wireless communications in a fading environment when using multi-element antennas, *Bell Labs. Tech. J.*, Vol. 1, No. 2, 1966, pp. 41–59.
57. Hayes, J., Adaptive feedback communications, *IEEE Trans.*, Vol. COM-16, No. 2, 1968, pp. 29–34.
58. Matsumoto, S., Sango, J., and Segawa, J., 200 Mb/S 16-QAM digital radio-relay system operating in 4 and 5 GHz hands, *Telecommun. Rev. Jap.*, Vol. 24, 1982, pp. 65–73.

59. Widrow, B. and McCool, J., Comparison of adaptive algorithms based on the methods of steepest descent and random search, *IEEE Trans.*, Vol. AP-24, No. 5, 1976, pp. 615–637.
60. Mueller, K., A new fast-converging mean-square algorithm for adaptive equalizers with partial-response signalling, *BSTJ*, Vol. 54, No. 1, 1975, pp. 143–153.
61. Gitlin, R. and Magee, F., Self-orthogonalizing adaptive equalization algorithms, *IEEE Trans.*, Vol. COM-25, No. 7, 1977, pp. 666–672.
62. Sari, H., Simplified algorithms for adaptive channel equalization, *Philips J. Res.*, Vol. 37, No. 1–2, 1982, pp. 56–77.
63. Brandwood, B., A complex gradient operator and its application in adaptive array theory, in *Proc. IEE*, Part F, Vol. 130, No. 1, 1983, pp. 11–16.
64. Buttler, P. and Cantoni, A., Non-iterative automatic equalization, *IEEE Trans.*, Vol. COM-23, No. 6, 1975, pp. 621–633.
65. Hsu, F., Square root Kalman filtering for high-speed data received over fading dispersive HF channels, *IEEE Trans.*, Vol. IT-28, No. 5, 1982, pp. 753–763.
66. Goldsmith, A., Variable-rate variable-power MQAM for fading channels, in *Proc. IEEE Vehicular Technology Conf.*, Atlanta, GA, April 1996, pp. 815–819.
67. Viterbi, A. and Omura, J., *Principles of Digital Communications and Coding*, McGraw-Hill, New York, 1979.
68. Pillai, S., Oh, H., Youla, D., and Gnerci, J., Optimum transmit-receiver design in the presence of signal-dependent interference and channel noise, *IEEE Trans.*, Vol. IT-46, No. 2, 2000, pp. 577–584.
69. Helstrom, C., *Elements of Signal Detection and Estimation*, Prentice-Hall, Englewood Cliffs, NJ, 1995.
70. Blahut, R., Computation of channel capacity and rate-distortion functions, *IEEE Trans.*, Vol. IT-18, No. 4, 1972, pp. 460–473.
71. Ligdas, P. and Farvardin, N., Optimizing the transmit power for slow fading channels, *IEEE Trans.*, Vol. IT-46, No. 2, 2000, pp. 565–576.
72. Gallager, R., Residual noise after interference cancellation on fading multi-path channels, in *Proc. Communications, Computation, Control, and Signal Processing*, Kluwer Academic Publishers, Boston, 1997, pp. 67–77.
73. Lee, E. and Messerschmitt, D., *Digital Communications*, Kluwer Academic Publishers, Boston, 1992.
74. Marzetta, T. and Hochwald, B., Multiple antenna communications when nobody knows the Rayleigh fading coefficients, in *Proc. 35th Allerton Conf. Communications, Control and Computation*, Urbana, IL, 1997.
75. Porat, B., *Digital Processing of Random Signals: Theory and Methods*, Prentice-Hall, Englewood Cliffs, NJ, 1994.
76. Kay, S., *Fundamentals of Statistical Signal Processing*, Prentice-Hall, Englewood Cliffs, NJ, 1993.
77. Oppenheim, A., Schafer, R., and Buck, J., *Discrete-Time Signal Processing*, 2nd ed., Prentice-Hall, Upper Saddle River, NJ, 1999.
78. Quazi, A., Array beam response in the presence of amplitude and phase fluctuations, *J. Acoust. Soc. Am.*, Vol. 72, July 1982, pp. 171–180.
79. Zhou, G., Ginnakis, G., and Swami, A., On polynomial phase signals with time-varying amplitudes, *IEEE Trans.*, Vol. SP-44, No. 4, 1996, pp. 848–861.
80. Gradshteyn, I. and Ryzhik, I., *Table of Integrals, Series, and Products*, 5th ed., Academic Press, New York, 1994.

81. Compton, R., The power inversion array concept and performance, *IEEE Trans.*, Vol. AES-15, No. 11, 1979, pp. 803–810.
82. Falcorner, D. and Ljung, L., Application of fast Kalman estimation to adaptive equalization, *IEEE Trans.*, Vol. COM-26, No. 10, 1978, pp. 1439–1446.
83. Godard, D., Channel equalization using a Kalman filter for fast data transmission, *IBM J. Res. Develop.*, May 1974, pp. 267–273.
84. Conte, E. and Ricci, G., Sensitivity study of the GLRT detection in compound-Gaussian clutter, *IEEE Trans.*, Vol. AES-34, No. 1, 1998, pp. 6308–6316.
85. Medard, M., The effect upon channel capacity in wireless communications of perfect and imperfect knowledge of the channel, *IEEE Trans.*, Vol. IT-46, No. 3, 2000, pp. 933–946.
86. Lapidoth, A. and Shamai, S., Fading channels: how perfect need "perfect side information" be, in *Proc. 1999 IEEE Information Theory Workshop*, Kruger National Park, South Africa, June 20–25, 1999, pp. 36–39.
87. Kavcic, A. and Moura, J., The Viterbi algorithm and Markov noise memory, *IEEE Trans.*, Vol. IT-46, No. 1, 2000, pp. 291–301.
88. Appelbaum, S. and Chapman, D., Adaptive arrays with main beam constraints, *IEEE Trans.*, Vol. AP-24, No. 9, 1976, pp. 650–662.
89. Haimovich, A. and Bar-Ness, Y., An eigenanalysis interference canceller, *IEEE Trans.*, Vol. SP-39, No. 1, 1991, pp. 76–84.
90. Wei, L., Coded M–DPSK with built-in time diversity for fading channels, *IEEE Trans.*, Vol. IT-39, No. 6, 1993, pp. 1820–1839.
91. Kamnev, E. et al., *Signal Processing Noise in Communication Channels*, Radio and Svyaz, Moscow, 1985 (in Russian).
92. Hoyt, R., Probability functions for the modulus and angle of the normal complex variate, *BSTJ*, No. 2, 1947, pp. 223–237.
93. Nakagami, M., Statistical characteristics of short-wave fading, *J. Inst. Elec. Commun. Engrs.*, Japan, February 1943, pp. 239–245.
94. Nakagami, M., Wada, S., and Fujimura, S., Some considerations on random phase problems from the standpoint of fading, *J. Inst. Elec. Commun. Engrs.*, Japan, November 1953.
95. Nakagami, M., *Statistical Methods in Radio Wave Propagation*, Hofman, W., Ed., Pergamon Press, New York, 1960.

4

Statistical Characteristics of Signals under the Stimulus of Multiplicative Noise

This chapter discusses the main characteristics of signals distorted by multiplicative noise and the functional relationship between statistical characteristics of the signal and the noise modulation function $\dot{M}(t)$ of the multiplicative noise. As was shown in Chapter 3, the noise modulation function $\dot{M}(t)$ characterizes the multiplicative noise completely.

The noise modulation function $\dot{M}(t)$ of the multiplicative noise is the complex amplitude envelope of the signal formed under modulation of the harmonic oscillation by the same law, by which the multiplicative noise distorts the signal. Because of this, the characteristics of the complex amplitude envelopes of the signals distorted by the multiplicative noise can be defined using the main characteristics of the harmonic oscillation modulated by the same distortions. Both the theoretical and experimental characteristics of the harmonic oscillation are more easily defined than the characteristics of the complex amplitude envelopes of the signals distorted by the multiplicative noise.[1,2]

The signal distorted by the multiplicative noise takes the following form:

$$a_M(t) = Re\{\dot{S}_M(t) \cdot e^{j(\omega_0 t + \varphi_0)}\}, \tag{4.1}$$

where

$$\dot{S}_M(t) = \dot{M}(t)\dot{S}(t) \tag{4.2}$$

is the complex amplitude envelope of the signal $a_M(t)$ and $\dot{S}(t)$ is the complex amplitude envelope of the undistorted signal $a(t)$.

4.1 Deterministic and Quasideterministic Multiplicative Noise

4.1.1 Periodic Multiplicative Noise

The discussion of periodic multiplicative noise begins with a consideration of the functional relationship between the following: the structure of the signal $a_M(t)$ distorted by the multiplicative noise; the structure of the undistorted

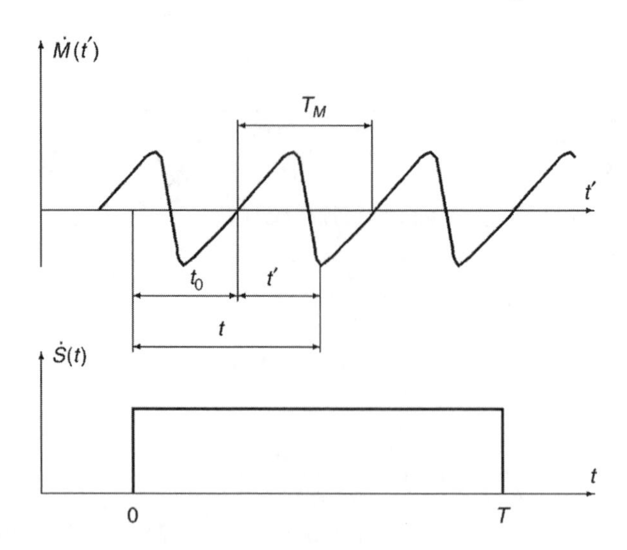

FIGURE 4.1
Domain of the noise modulation function definition.

signal $a(t)$; and characteristics of the noise modulation function $\dot{M}(t)$ of the multiplicative noise.

Let the noise modulation function $\dot{M}(t)$ of the multiplicative noise be the periodic function with the period

$$T_M = \frac{2\pi}{\Omega_M}. \tag{4.3}$$

In the majority of cases, the sources generating distortions in parameters of the signal—the sources of the multiplicative noise—exist independently of a "yes" or a "no" signal in the stochastic process at the input of the receiver or detector of signal processing system. Because of this, the function defining the multiplicative noise is valid for both periods within the limits of the time interval $[0, T]$ in which the signal is observed and periods outside of this interval as well (see Fig. 4.1).

When a "yes" signal exists in the input stochastic process, the complex amplitude envelope $\dot{S}(t)$ of the signal can take any position on the time axis with respect to the noise modulation function $\dot{M}(t)$ of the multiplicative noise. If, for any realizations of the signal and the multiplicative noise, the shift in time t_0 (of the complex amplitude envelope of the signal on the time axis (see Fig. 4.1) with respect to the noise modulation function $\dot{M}(t)$ of the multiplicative noise) is constant, then the multiplicative noise is considered as deterministic multiplicative noise. If the shift in time t_0 (of the complex amplitude envelope of the signal on the time axis, with respect to the noise modulation function $\dot{M}(t)$ of the multiplicative noise) is a random variable, then the multiplicative noise is considered as quasideterministic multiplicative noise. These definitions are in accordance with the classifications introduced in Section 3.3.

If the noise modulation function, $\dot{M}(t)$ of the multiplicative noise, satisfies the Dirichlet boundary conditions, then these conditions as a rule are satisfied for the multiplicative noise. The noise can be presented using the Fourier series expansion in the following form:[3–5]

$$\dot{M}(t') = \sum_{n=-\infty}^{\infty} \dot{C}_n \cdot e^{jn\Omega_M t'}, \tag{4.4}$$

where

$$\dot{C}_n = \frac{1}{T_M} \int_0^{T_M} \dot{M}(t) \cdot e^{-jn\Omega_M t}\, dt. \tag{4.5}$$

Going to a new instant (see Fig. 4.1),

$$t = t' + t_0, \tag{4.6}$$

Equation (4.4) can be rewritten in the following form:

$$\dot{M}(t) = \sum_{n=-\infty}^{\infty} \dot{C}_n \cdot e^{jn(\Omega_M t - \vartheta)}, \tag{4.7}$$

where

$$\vartheta = \Omega_M t_0 \tag{4.8}$$

is the initial phase of the noise modulation function $\dot{M}(t)$ of the multiplicative noise with respect to the complex amplitude envelope of the signal. If the multiplicative noise is the deterministic multiplicative noise, then the initial phase ϑ is constant. If the multiplicative noise is the quasideterministic multiplicative noise the initial phase ϑ is a random variable.

Substituting Eq. (4.7) in Eq. (4.1), we can write

$$a_M(t) = \sum_{n=-\infty}^{\infty} Re\{\dot{C}_n \cdot e^{-jn\vartheta}\, \dot{S}(t) \cdot e^{j(\omega_0 - n\Omega_M)t}\}. \tag{4.9}$$

We next introduce the following designations:

- $a(t, \beta)$ is the signal that differs from the signal $a(t)$ by a shift in the initial phase by the value β.
- $a(t, \beta, \Omega)$ is the signal that differs from the signal $a(t)$ by a shift in the carrier frequency by the value Ω, and a shift in the initial phase by the value β.

Then Eq. (4.9) in terms of Eq. (4.2) can be written in the following form:

$$a_M(t) = |\dot{C}_0| a(t, \beta_0) + \sum_{m \neq 0} |\dot{C}_m| a(t, \beta_m, m\Omega_M), \tag{4.10}$$

where

$$\beta_0 = \arg \dot{C}_0; \tag{4.11}$$

$$\beta_m = \arg \dot{C}_m - m\vartheta. \tag{4.12}$$

The first term on the right side of Eq. (4.10) differs from the signal $a(t)$ by only the scale factor $|\dot{C}_0|$ and the initial phase β_0, an unessential parameter. Thus, this component of the signal distorted by the multiplicative noise duplicates the undistorted signal with an accuracy of the initial phase in the changed scale. This component of the signal is called the undistorted component of the signal, and is denoted by the undistorted component of the signal $m(t)$.

The scale factor

$$\alpha_0 = |\dot{C}_0| = \frac{m(t)}{a(t, \beta)} \tag{4.13}$$

is called the relative level of the undistorted component of the signal. The second term on the right side of Eq. (4.10) is the component of the signal distorted by the multiplicative noise, which is absent in the undistorted component of the signal. This component is caused by a stimulus of the multiplicative noise. The second term on the right side of Eq. (4.10) is called the noise component of the signal and is denoted by $\upsilon(t)$.

Thus, according to Eq. (4.10) a stimulus of the periodic multiplicative noise on the signal reduces to changes in amplitude and initial phase of the signal—the scale factor determined by

$$\alpha_0 = |\dot{C}|, \tag{4.14}$$

and the initial phase β_0—and the appearance of some equivalent additive component $\upsilon(t)$—the noise component of the distorted signal.

Analysis of parameters of the signal distorted by the periodic multiplicative noise has just been carried out in time space. Evidently, the same results can be obtained under an analysis of parameters of the signal in frequency-spectral space.

The first component—the undistorted component of the signal—is the spectral components of the signal distorted by the multiplicative noise, which are present in the undistorted signal.

The second component—the noise component of the signal—is the spectral components of the signal distorted by the multiplicative noise, which are absent in the undistorted signal and appear only as a result of the stimulus of multiplicative noise.[6,7]

Taking into consideration the statements mentioned above, we can rewrite Eq. (4.10) in the following form:

$$a_M(t) = \alpha_0\, a(t, \beta_0) + \upsilon(t) = m(t) + \upsilon(t). \tag{4.15}$$

The relative level of the undistorted component of the signal α_0 and its initial phase β_0 are defined by the module and argument of the constant coefficient

$\dot{C}_0$ under the use of the Fourier series expansion of the noise modulation function $\dot{M}(t)$ of the multiplicative noise:

$$\alpha_0 = |\dot{C}_0| = \frac{1}{T_M}\left|\int_0^{T_M} \dot{M}(t)\,dt\right|; \tag{4.16}$$

$$\beta_0 = \arg \dot{C}_0 = \arg\left\{\int_0^{T_M} \dot{M}(t)\,dt\right\}. \tag{4.17}$$

Taking into account the fact that variables α_0 and β_0 do not depend on the structure of the signal $a(t)$ distorted by the multiplicative noise, and in terms of Eq. (4.10) when

$$a(t) = \cos \omega_0 t, \tag{4.18}$$

it is easy to define the values α_0 and β_0 that are equal to the amplitude and phase of the residual signal, respectively, at the carrier frequency under modulation of the harmonic oscillation with the unit amplitude by the same law, by which the signal is distorted by the multiplicative noise.

The noise component of the distorted signal is a composition of signals that are similar in structure to the undistorted signal, but are shifted in frequency. Moreover, the relative levels of the individual signals and their initial phases are defined by corresponding coefficients using the Fourier series expansion of the noise modulation function $\dot{M}(t)$ of the multiplicative noise:

$$v(t) = \sum_{m=-\infty}^{\infty} \alpha_m a(t, \beta_m, m\Omega_M), \tag{4.19}$$

where

$$\alpha_m = |\dot{C}_m|, \tag{4.20}$$

$$\beta_m = \arg \dot{C}_m - m\vartheta. \tag{4.21}$$

Substituting Eq. (4.18) in Eq. (4.10), it is easy to verify that the values α_m and β_m are the amplitude and phase, respectively, of the m-th harmonics of the signal formed under modulation of the harmonic oscillation with the unit amplitude by the same law, by which the signal is distorted by the multiplicative noise.

If the frequency of the periodic multiplicative noise

$$f_M = \frac{1}{T_M} \tag{4.22}$$

is greater than the spectrum bandwidth ΔF_a of the undistorted signal $a(t)$ the individual terms of the sum in Eq. (4.19) are separated in frequency by intervals that exceed their spectrum bandwidths, i.e., they are not overlapped in frequency and can be separated. In this case, the multiplicative noise gives rise to multiplication of the signal in frequency space.

In a similar fashion, the undistorted component of the signal is not overlapped by the noise component of the signal if the condition

$$f_M > \Delta F_a \tag{4.23}$$

is satisfied. These two components of the signal distorted by the periodic multiplicative noise can be separated.

Let us now consider an example of periodic multiplicative noise. Assume that the multiplicative noise has a character of sinusoidal distortions in phase of the signal:

$$\varphi(t') = \varphi_m \sin \Omega_M t', \tag{4.24}$$

where φ_m is the maximal deviation of the phase and

$$t' = t + t_0. \tag{4.25}$$

In doing so, the noise modulation function $\dot{M}(t)$ of the multiplicative noise is determined in the following form:

$$\dot{M}(t) = \exp\{j\varphi_m \sin[\Omega_m t - \vartheta]\}, \tag{4.26}$$

where the parameter ϑ is determined by Eq. (4.8).

The coefficients of the Fourier series expansion of the noise modulation function $\dot{M}(t)$ of the multiplicative noise are determined in the following form:[8]

$$\dot{C}_n = \frac{1}{T_M}\int\limits_0^{T_M} e^{j\varphi_m \sin \Omega_M \tau} \cdot e^{-jn\Omega_M \tau} = J_n(\varphi_m), \tag{4.27}$$

where $J_n(x)$ is the Bessel function of the n-th order.

Hence it follows that

$$\alpha_n = |\dot{C}| = J_n(\varphi_m); \tag{4.28}$$

$$\arg \dot{C}_n = 0; \tag{4.29}$$

$$\beta_n = -n\vartheta. \tag{4.30}$$

The signal distorted by this multiplicative noise can be written on the basis of Eq. (4.10) in the following form:

$$a_M(t) = J_0(\varphi_m)a(t) + \sum_{m=-\infty}^{\infty} J_m(\varphi_m)a(t, -m\vartheta, m\Omega_M), \quad m \neq 0 \tag{4.31}$$

and the components of this signal are determined by

$$m(t) = J_0(\varphi_m)a(t); \tag{4.32}$$

$$v(t) = \sum_{m=-\infty}^{\infty} J_m(\varphi_m)a(t, -m\vartheta, m\Omega_M), \quad m \neq 0. \tag{4.33}$$

If the initial phase ϑ is constant (as in the case of deterministic multiplicative noise), then the phase relationship between the individual terms on the right side of Eq. (4.19) are the same for all realizations of the signal and the multiplicative noise. The noise component $v(t)$ of the signal, as it follows from Eq. (4.19), is the deterministic function of time, which is given within the limits of the interval, in which the signal $a(t)$ exists.[9]

If the initial phase ϑ is the random variable that is case of quasideterministic multiplicative noise, then the initial phases of the individual terms on the right side of Eq. (4.19)

$$\beta_m = \arg \dot{C}_m - m \tag{4.34}$$

are the random variables. The phase relations between the individual components are changed randomly from realization to realization.

In the process, the noise component $v(t)$ of the signal is changed randomly. In many cases it would appear reasonable that the initial phase ϑ is distributed uniformly within the limits of the interval $[0, 2\pi]$. Then the mean of the noise component of the signal in terms of Eqs. (4.9) and (4.19) takes the following form:

$$m_1[v(t)] = \sum_{m=-\infty}^{\infty} Re\{m_1[e^{-jm\vartheta}]\dot{C}_m\dot{S}(t) \cdot e^{j(\omega_0 t + m\Omega_M t + \varphi_0)}\} = 0, \quad m \neq 0, \tag{4.35}$$

as

$$m_1[e^{-jm\vartheta}] = 0 \tag{4.36}$$

if the initial phase ϑ is distributed uniformly within the limits of the interval $[0, 2\pi]$.

Thus, in this case the mean of the noise component of the signal is equal to zero. Consequently, the undistorted component of the signal is the mean of the signal distorted by the periodic quasideterministic multiplicative noise:

$$m_1[a_M(t)] = \alpha_0\, a(t, \beta_0). \tag{4.37}$$

We proceed to define the variance of the noise component of the signal under the periodic quasideterministic multiplicative noise. In terms of Eqs. (4.9) and (4.19), and the relationship

$$Re\,\dot{a}\,Re\,\dot{b} = \frac{1}{2}Re\,\dot{a}\dot{b}^* + \frac{1}{2}Re\,\dot{a}\dot{b} \tag{4.38}$$

the variance of the noise component of the signal can be determined in the following form:

$$\begin{aligned}\sigma_v^2(t) &= D[v(t)] \\ &= \frac{1}{2}m_1\left\{Re\left\{\sum_{m=-\infty\,(m\neq 0)}^{\infty} \dot{C}_m \cdot e^{-jm\vartheta}\,\dot{S}(t)\cdot e^{j(\omega_0 t + m\Omega_M t + \varphi_0)}\right.\right. \\ &\quad \times \left.\left.\sum_{k=-\infty\,(k\neq 0)}^{\infty} C_k^* \cdot e^{jk\vartheta_0}\,S^*(t)\cdot e^{-j(\omega_0 t + k\Omega_M t + \varphi_0)}\right\}\right\} \\ &\quad + \frac{1}{2}m_1\left\{Re\left\{\sum_{m=-\infty\,(m\neq 0)}^{\infty} \dot{C}_m \cdot e^{-jm\vartheta}\,\dot{S}(t)\cdot e^{j(\omega_0 t + m\Omega_M t + \varphi_0)}\right.\right. \\ &\quad \times \left.\left.\sum_{k=-\infty\,(k\neq 0)}^{\infty} \dot{C}_k \cdot e^{-jk\vartheta}\,\dot{S}(t)\cdot e^{j(\omega_0 t + k\Omega_M t + \varphi_0)}\right\}\right\}. \end{aligned} \tag{4.39}$$

If the initial phase is distributed uniformly within the limits of the interval $[0, 2\pi]$, then the following equalities

$$m_1\left[e^{-j(m+k)\vartheta}\right] = 0; \tag{4.40}$$

$$m_1\left[e^{-j(m-k)\vartheta}\right] = \begin{cases} 0 & \text{at } m \neq k; \\ 1 & \text{at } m = k \end{cases} \tag{4.41}$$

are true.

Taking into account Eqs. (4.40) and (4.41), Eq. (4.39) has the following form:

$$\sigma_v^2(t) = \frac{1}{2}S^2(t) \sum_{m=-\infty\,(m\neq 0)}^{\infty} C_m^2, \tag{4.42}$$

where

$$C_m = |\dot{C}_m|. \tag{4.43}$$

Reference to Eq. (4.5) shows that

$$\sum_{n=-\infty}^{\infty} C_n^2 = \frac{1}{T_M}\int_0^{T_M} |\dot{M}(t)|^2\,dt = \frac{1}{T_M}\int_0^{T_M} A^2(t)\,dt, \tag{4.44}$$

where $A(t)$ is the factor characterizing distortions in amplitude of the signal.

Consequently,

$$\sum_{m=-\infty\,(m\neq 0)}^{\infty} C_m^2 = \frac{1}{T_M}\int_0^{T_M} A^2(t)\,dt - \alpha_0^2 \tag{4.45}$$

and the variance of the noise component of the signal is determined in the following form:

$$\sigma_v^2(t) = \frac{1}{2} S^2(t) \left[\frac{1}{T_M} \int_0^{T_M} A^2(t)\, dt - \alpha_0^2 \right]. \tag{4.46}$$

Equation (4.46) shows that the variance of the noise component of the signal distorted by the periodic quasideterministic multiplicative noise is proportional to the squared amplitude envelope of the signal.

The value

$$\frac{1}{T_M} \int_0^{T_M} A^2(t)\, dt \tag{4.47}$$

is approximately equal to unity in the majority of cases. When distortions in amplitude of the signal are absent, this value is equal to unity. In this case

$$\sum_{m=-\infty\,(m\neq 0)}^{\infty} C_m^2 = 1 - \alpha_0^2 \tag{4.48}$$

and the variance of the noise component of the signal is determined in the following form:

$$\sigma_v^2(t) = \frac{1}{2} \left(1 - \alpha_0^2\right) S^2(t). \tag{4.49}$$

Thus, during the condition of the periodic multiplicative noise the distorted signal can be represented as the sum of two components: the undistorted signal and the noise component of the signal. The first component of the signal duplicates the undistorted signal in the changed scale and, in a general case, does so with some constant shift in phase of the signal.

In the case of periodic deterministic multiplicative noise, the second component is the deterministic function of time, which is given within the limits of the signal duration. In the case of periodic quasideterministic multiplicative noise, the second component is the stochastic function that is also given within the limits of the signal duration. If the frequency of the periodic multiplicative noise is greater than the spectrum bandwidth of the signal, then the first and second components of the signal can be separated by frequency selection.[10,11]

4.1.2 Non-Periodic Deterministic Multiplicative Noise

In a general case, when the deterministic multiplicative noise is not periodic we can use the Fourier series expansion of the noise modulation function $\dot{M}(t)$ of the multiplicative noise within the limits of the signal duration

$$\dot{M}(t) = \sum_{k=-\infty}^{\infty} \dot{C}_k^T \cdot e^{jk\frac{2\pi}{T}t}, \quad 0 \leq t \leq T, \tag{4.50}$$

where

$$\dot{C}_k^T = \frac{1}{T}\int_0^T \dot{M}(t) \cdot e^{-jk\frac{2\pi}{T}t}\,dt. \tag{4.51}$$

Contrary to Eq. (4.4), Eq. (4.50) defines correctly the noise modulation function $\dot{M}(t)$ of the multiplicative noise only within the limits of the signal duration $[0, T]$. The noise modulation function $\dot{M}(t)$ of the multiplicative noise is defined incorrectly by Eq. (4.50) outside of the interval $[0, T]$. Moreover, since the amplitude factor $S(t)$ is equal to zero outside the interval $[0, T]$ (see Eqs. (4.1) and (4.2) that define the signal distorted by the multiplicative noise), the character of the noise modulation function $\dot{M}(t)$ of the multiplicative noise in Eq. (4.50) is unessential outside the interval $[0, T]$ and the noise modulation function $\dot{M}(t)$ determined by Eq. (4.50) can be used in Eq. (4.2).

Substituting Eq. (4.50) in Eq. (4.2) and using the designations introduced above, we obtain that the signal distorted by deterministic multiplicative noise is determined in the following form:

$$a_M(t) = |\dot{C}_0^T| a\left(t, \beta_0^T\right) + \sum_{m=-\infty\ (m\neq 0)}^{\infty} |\dot{C}_m^T| a\left(t, \beta_m^T, m \cdot \frac{2\pi}{T}\right), \tag{4.52}$$

where

$$\beta_m^T = \arg \dot{C}_m^T. \tag{4.53}$$

The first term on the right side of Eq. (4.52) is the undistorted component of the signal. The second term on the right side of Eq. (4.52) is the noise component of the signal.

The undistorted component of the signal is determined in the following form:

$$m(t) = Re\{\dot{C}_0^T \dot{S}(t) \cdot e^{j\omega_0 t}\} = \alpha_0^T \cdot a\left(t, \beta_0^T\right), \tag{4.54}$$

where the relative level of the undistorted component of the signal is determined by

$$\alpha_0^T = |\dot{C}_0^T| = \left|\int_0^T \dot{M}(t)\,dt\right|. \tag{4.55}$$

The noise component of the signal is determined by

$$\begin{aligned} v(t) &= \sum_{m=-\infty\ (m\neq 0)}^{\infty} |\dot{C}_m^T| a\left(t, \beta_m^T, m \cdot \frac{2\pi}{T}\right) \\ &= \sum_{m=-\infty\ (m\neq 0)}^{\infty} \alpha_m^T \cdot a\left(t, \beta_m^T, m \cdot \frac{2\pi}{T}\right). \end{aligned} \tag{4.56}$$

In a general case, Eqs. (4.52), (4.54), and (4.56) define the signal distorted by the multiplicative noise and its components when there is deterministic multiplicative noise. Equations (4.52), (4.54), and (4.56) are also true in the particular case of periodic multiplicative noise. Thus, the signal distorted by the periodic multiplicative noise can be determined correctly both by Eq. (4.10) and by Eq. (4.52).

This phenomenon is explained by the following statements of fact. Equations (4.52), (4.54), and (4.56) are based on the Fourier series expansion of the noise modulation function $\dot{M}(t)$ of the multiplicative noise. However, this expansion is not unique within the limits of the finite interval.[12] In addition, it can be shown that when the duration of the signal is much greater than the period T_M of the multiplicative noise—the case in which the multiplicative noise is thought of as periodic multiplicative noise—Eqs. (4.10) and (4.52) become the same.

In an effort to prove this determination, compare the coefficients

$$\alpha_k^T = \left|\dot{C}_k^T\right| \tag{4.57}$$

and

$$\alpha_m = |\dot{C}_m| \tag{4.58}$$

along with the corresponding shifts in frequency

$$\frac{2\pi k}{T} \qquad \text{and} \qquad m\Omega_M \tag{4.59}$$

in Eqs. (4.10) and (4.52), and also define the coefficients $\dot{C}_k^T$ (see Eq. (4.51)) using the coefficients $\dot{C}_m$.

Substituting the noise modulation function $\dot{M}(t)$ from Eq. (4.7) that is true for the periodic multiplicative noise in Eq. (4.51), which defines the arbitrary deterministic multiplicative noise, we can write

$$\begin{aligned}\dot{C}_k^T &= \frac{1}{T}\sum_{m=-\infty}^{\infty} \dot{C}_m \cdot e^{-jm\vartheta} \int_0^T e^{j\left(m\Omega_M - \frac{2\pi k}{T}\right)t}\, dt \\ &= \sum_{m=-\infty}^{\infty} \dot{C}_m \cdot e^{-jm\vartheta} \cdot \frac{e^{jm\Omega_M t} - 1}{j(m\Omega_M T - 2\pi k)}\end{aligned} \tag{4.60}$$

when the condition

$$T > T_M$$

is satisfied.

In a general case, the interval $[0, T]$ can be represented in the following form:

$$T = nT_M + \Delta T, \tag{4.61}$$

where n is an integer and

$$\Delta T < T_M.$$

In this case, Eq. (4.60) can be written in the following form:

$$\dot{C}_k^T = \sum_{m=-\infty}^{\infty} \dot{C}_m \cdot e^{-jm\vartheta} \cdot e^{jm\Omega_M \frac{T}{2}} \cdot \frac{\sin\left[\frac{m\Omega_M \Delta T}{2}\right]}{\frac{m\Omega_M \Delta T}{2} + \pi(mn - k)}. \tag{4.62}$$

Next we define the coefficients

$$\left(\alpha_k^T\right)^2 = \left|\dot{C}_k^T\right|^2. \tag{4.63}$$

Taking into account the fact that the initial phase ϑ (determined by Eq. (4.8)) characterizes the initial phase of the noise modulation function $\dot{M}(t)$ of the multiplicative noise with respect to the complex amplitude envelope $\dot{S}(t)$ of the signal, then it follows that the noise modulation function $\dot{M}(t)$ of the multiplicative noise is the periodic function in Eq. (4.42).

If an instant of appearance of the signal on the time axis (see Fig. 4.1) can take any position within the limits of the interval $[0, T_M]$ with the same probabilities, where T_M is the period of the multiplicative noise, then the initial phase ϑ is distributed uniformly within the limits of the interval $[0, 2\pi]$.[13]

We define the mean of the value $(\alpha_k^T)^2$ for all possible values t_0:

$$\overline{\left(\alpha_k^T\right)^2} = \int_{-\infty}^{\infty} \left(\alpha_k^T\right)^2 f(\vartheta)\, d\vartheta = \frac{1}{2\pi} \int_0^{2\pi} \dot{C}_k^T C_k^{*T}\, d\vartheta, \tag{4.64}$$

where $f(\vartheta)$ is the probability distribution density of the initial phase ϑ.

There is the following equality for the probability distribution density $f(\vartheta)$ of the initial phase ϑ:

$$m_1\left[e^{j(m_1 - m_2)\vartheta}\right] = \begin{cases} 0 & \text{at } m_1 \neq m_2; \\ 1 & \text{at } m_1 = m_2. \end{cases} \tag{4.65}$$

Substituting the value $\dot{C}_k^T$ from Eq. (4.62) in Eq. (4.64), and taking into account Eq. (4.65), we can write

$$\overline{\left(\alpha_k^T\right)^2} = \sum_{m=-\infty}^{\infty} \alpha_m^2 \cdot \frac{\sin^2 \frac{\pi m \Delta T}{T_M}}{\pi^2\left(m \cdot \frac{\Delta T}{T_M} + mn - k\right)^2}. \tag{4.66}$$

Equation (4.66) can be written in the following form:

$$\overline{\left(\alpha_k^T\right)^2} = \alpha_{m=\frac{k}{n}}^2 \cdot \frac{\sin^2\left[\pi \cdot \frac{k}{n} \cdot \frac{\Delta T}{T_M}\right]}{\left[\pi \cdot \frac{k}{n} \cdot \frac{\Delta T}{T_M}\right]^2} + \sum_{m=-\infty\, (m \neq \frac{k}{n})}^{\infty} \alpha_m^2 \cdot \frac{\sin^2 \frac{\pi m \Delta T}{T_M}}{\pi^2\left(m \cdot \frac{\Delta T}{T_M} + mn - k\right)^2}. \tag{4.67}$$

During the condition

$$\Delta T = 0$$

Equation (4.67) takes the following form:

$$\overline{\left(\alpha_k^T\right)^2} = \begin{cases} \alpha^2_{m=\frac{k}{n}} & \text{at } k = mn; \\ 0 & \text{at } k \neq mn. \end{cases} \tag{4.68}$$

Reference to Eq. (4.68) shows that during the condition

$$\Delta T = 0 \qquad \text{or} \qquad T = nT_M \tag{4.69}$$

all terms in Eq. (4.52) are equal to zero except the terms that have a shift in frequency determined in the following form

$$k \cdot \frac{2\pi}{T} = mn \cdot \frac{2\pi}{nT_M} = m\Omega_M. \tag{4.70}$$

Other terms in Eq. (4.52) are equal to zero because the shift in frequency for each term in Eqs. (4.10) and (4.52) is the same. The mean square deviation of the coefficients in Eq. (4.52) is equal to the coefficients in Eq. (4.10). Thus, both terms and the relative levels of the terms in Eq. (4.52) are the same as in Eq. (4.10) during the condition that is determined by Eq. (4.69).

For the condition

$$\Delta T \neq 0 \tag{4.71}$$

the equality in Eq. (4.68) is approximated if the following condition

$$T \gg T_M, \tag{4.72}$$

is satisfied, i.e., the value n is sufficiently high.

Actually, reference to Eq. (4.66) shows that

$$\lim_{n\to\infty} \overline{\left(\alpha_k^T\right)^2} = \begin{cases} \alpha^2_{m=\frac{k}{n}} & \text{at } k = mn; \\ 0 & \text{at } k \neq mn. \end{cases} \tag{4.73}$$

The ratio $\frac{\Delta T}{T_M}$ is low and the number of terms of expansion in Eq. (4.7), for which the coefficients

$$\alpha_m = |\dot{C}_m| \tag{4.74}$$

have a great value, is low, the value $\overline{(\alpha_k^T)^2}$ tends to approach the limit in Eq. (4.73) more rapidly.

For example, consider the case of sinusoidal distortions in phase of the signal (see Eq. (4.31)). For this case, the coefficients α_m under the condition

$$\varphi_m = 0.5 \tag{4.75}$$

TABLE 4.1

The Values of the Coefficients $\widetilde{\alpha}_k^T = \sqrt{\left(\alpha_k^T\right)^2}$

		k								
n	$\frac{\Delta T}{T_M}$	0	1	2	3	4	5	6	7	8
2	0.00	0.90	0.00	0.28	0.00	0.03				
	0.10	0.90	0.03	0.28	0.03	0.03				
	0.30	0.90	0.06	0.24	0.10	0.04				
	0.50	0.90	0.07	0.18	0.18	0.06				
4	0.00	0.90	0.00	0.00	0.00	0.28	0.00	0.00	0.00	0.03
	0.10	0.90	0.01	0.01	0.03	0.28	0.03	0.01	0.01	0.03
	0.30	0.90	0.03	0.03	0.06	0.24	0.10	0.04	0.03	0.03
	0.50	0.90	0.03	0.04	0.06	0.18	0.06	0.04	0.04	0.03

have the following values:

$$m = 0, \qquad \alpha_m = 0.90; \tag{4.76}$$

$$m = \pm 1, \qquad \alpha_m = 0.28; \tag{4.77}$$

$$m = \pm 2, \qquad \alpha_m = 0.03. \tag{4.78}$$

Table 4.1 shows the values of the coefficients, which are determined in the following form:

$$\widetilde{\alpha}_k^T = \sqrt{\left(\alpha_k^T\right)^2}, \tag{4.79}$$

given by Eq. (4.67) for this case at two values of n, and four values (per each value of n) of the ratio $\frac{\Delta T}{T_M}$ from zero to the maximal value equal to 0.5.

Reference to Table 4.1 shows that both at $n = 2$ and at $n = 4$ the relative level of the undistorted component of the signal is determined in the following form:

$$\alpha_0^T = \alpha_0. \tag{4.80}$$

For other values of k at $n = 4$ the coefficients $\widetilde{\alpha}_k^T$ are close to the coefficients α_m at $k = mn$, and the values $\widetilde{\alpha}_k^T$ are negligible in comparison with the coefficients α_m under the condition $k \neq mn$.

Thus, if the duration of the signal is a multiple of the period of the multiplicative noise, then the presentation of the signal, distorted by the multiplicative noise in the form of two components based on the use of the Fourier series expansion of the noise modulation function $\dot{M}(t)$ of the multiplicative noise, gives the same results regardless of whether this expansion is carried out within the limits of the interval equal to the signal duration or this expansion is carried out within the limits of the interval equal to the period of the multiplicative noise.

If the signal duration is not a multiple of the period of the multiplicative noise, then the results are close to those that occur when the duration of the signal is several times greater than the period of the multiplicative noise.[14,15]

4.2 Stationary Fluctuating Multiplicative Noise

Under the stimulus of fluctuating multiplicative noise, changes in amplitude and phase of the signal as a function of source of initiation of the multiplicative noise can be functionally related or are in statistical dependence—the independent changes are the particular case.

Assume that the changes in amplitude and phase of the signal are realizations of the stationary and stationary related stochastic processes in the case of statistical dependence. Also assume that the noise modulation function $\dot{M}(t)$ of the multiplicative noise is the stationary stochastic process.

Suppose that the time of correlation of the noise modulation function $\dot{M}(t)$ of the multiplicative noise is much more than the period of the carrier frequency of the signal. As a rule, this relationship is always satisfied.

By a stimulus of multiplicative noise, the deterministic or quasideterministic signal $a(t)$ is transformed into the realization of the fluctuating nonstationary stochastic process $a_M(t)$. An ensemble of realizations of the signal $a_M(t)$ distorted by the multiplicative noise corresponds to an ensemble of realizations of the noise modulation function $\dot{M}(t)$ of the multiplicative noise.[16]

In this section we discuss the structure of the signal distorted by the fluctuating multiplicative noise and the main characteristics of this signal including the mean, the correlation function, and the energy spectrum. More complete characteristics of this signal are discussed in Sections 4.4 and 4.5.

In order to define the structure of the signal distorted by the stationary fluctuating multiplicative noise we consider the mean of this signal. Since the stochastic character of the signal $a_M(t)$ is caused by fluctuations in amplitude and phase of the signal, the mean of the signal $a_M(t)$ is determined in the following form:

$$\overline{a_M(t)} = m_1[a_M(t)] = \int_{-\infty}^{\infty}\int_{-\infty}^{\infty} a_M(t) f(A, \varphi)\, dA\, d\varphi, \tag{4.81}$$

where $f(A, \varphi)$ is the joint probability distribution density of distortions in amplitude $A(t)$ and phase $\varphi(t)$ of the signal at the coinciding instants.

Taking into account Eqs. (3.12) and (4.2), the mean of the signal distorted by the multiplicative noise can be written in the following form:

$$\begin{aligned}\overline{a_M(t)} &= Re\left\{\dot{S}(t) \cdot e^{j(\omega_0 t + \varphi_0)} \int_{-\infty}^{\infty}\int_{-\infty}^{\infty} A(t) \cdot e^{j\varphi(t)} f(A, \varphi)\, dA\, d\varphi\right\}\\ &= Re\left\{\overline{\dot{M}(t)}\dot{S}(t) \cdot e^{j(\omega_0 t + \varphi_0)}\right\},\end{aligned} \tag{4.82}$$

where

$$\overline{\dot{M}(t)} = \int_{-\infty}^{\infty}\int_{-\infty}^{\infty} A(t) \cdot e^{j\varphi(t)} f(A, \varphi)\, d A d\varphi \tag{4.83}$$

is the mean of the noise modulation function $\dot{M}(t)$ of the multiplicative noise. Since the noise modulation function $\dot{M}(t)$ of the multiplicative noise is the stationary function, the mean of the noise modulation function $\dot{M}(t)$ of the multiplicative noise does not depend on the time.

Denoting

$$\left|\overline{\dot{M}(t)}\right| = \alpha_0 \tag{4.84}$$

and

$$\arg \overline{\dot{M}(t)} = \beta_0 \tag{4.85}$$

and taking into account Eq. (4.1), Eq. (4.82) can be rewritten in the following form:

$$\overline{a_M(t)} = \alpha_0\, Re\{\dot{S}(t) \cdot e^{j(\omega_0 t + \varphi_0 + \beta_0)}\} = \alpha_0 \cdot a\,(t,\, \beta_0), \tag{4.86}$$

where $a\,(t,\, \beta_0)$ is the signal that differs from the signal $a\,(t)$ only by the constant shift in phase of the signal by the value β_0.

Thus, if distortions in amplitude $A(t)$ and phase $\varphi(t)$ of the signal are stationary, then the mean of the signal distorted by the multiplicative noise faithfully copies the undistorted signal in the changed scale α_0 determined by Eq. (4.84), this occurs with a precision of the unessential parameter—the constant shift in the phase β_0. Because of this, the mean of the signal $\overline{a_M(t)}$ can be considered as the undistorted component of the signal for an ensemble of realizations of the signal distorted by the stationary fluctuating multiplicative noise:[17]

$$m(t) = \overline{a_M(t)} = \alpha_0 \cdot a\,(t,\, \beta_0). \tag{4.87}$$

The coefficient α_0 determined by Eq. (4.84) is the relative level of the undistorted component of the signal. As follows from Eq. (4.85), the relative level α_0 of the undistorted component of the signal and the shift in phase β_0 do not depend on the structure of the signal and are defined only by the noise modulation function $\dot{M}(t)$ of the multiplicative noise.

Another component of the signal $a_M(t)$

$$\upsilon(t) = a_M(t) - \overline{a_M(t)} \tag{4.88}$$

is fluctuations with a zero mean, i.e., the pure stochastic function that does not contain the deterministic and quasideterministic components of the signal. This component of the signal $a_M(t)$, which is absent in the undistorted signal and appears as a result of the stimulus of the multiplicative noise, is called the noise component of the signal.

Thus, the ensemble of realizations of the signal distorted by the stationary fluctuating multiplicative noise can be presented in the form of a sum of two components—the undistorted component of the signal and the noise component of the signal:

$$a_M(t) = m(t) + v(t) = \alpha_0 \cdot a(t, \beta_0) + v(t). \qquad (4.89)$$

The first component is analogous to the signal, i.e., this component is the deterministic or quasideterministic function. The second component is the stochastic function.

Under this assumption, an impact of the stationary fluctuating multiplicative noise reduces to the following: changes in the amplitude of the signal by α_0 times; a shift in phase of the signal by the value β_0; and the initiation of some equivalent additive component $v(t)$—the noise component of the signal, related to the signal. This is analogous to the case considered in the previous section, with the deterministic or quasideterministic multiplicative noise (see Section 4.1).

However, contrary to Section 4.1, the given expansion of the signal $a_M(t)$ on the deterministic and stochastic components is unique, since the noise component of the signal under this expansion does not contain any components that are similar to the undistorted signal.

This notion of the signal distorted by the multiplicative noise (see Eq. (4.89)) has just been introduced in the discussion of characteristics of the signal in time space. Analogous notion of the signal can be obtained by analysis of characteristics of the signal in frequency space.

Let, for example, the signal $a(t)$ be periodic—the continuous periodic signal or a periodic sequence of pulses. Then the energy spectrum of the signal $a_M(t)$, distorted by the stationary fluctuating multiplicative noise, contains both the discrete components during the same frequencies that are used for the energy spectrum of the undistorted signal $a(t)$ and the continuous component that is absent from the energy spectrum of the undistorted signal $a(t)$.

These discrete components of the energy spectrum of the signal $a_M(t)$ correspond to the undistorted component of the signal, and the continuous component of the energy spectrum of the signal corresponds to the noise component of the signal.[18]

The notion of the signal distorted by the stationary fluctuating multiplicative noise in the form of two additive components considered above, has been introduced for the ensemble of realizations of the signal. Characteristics of the ensemble of realizations of the signal, distorted by the stationary fluctuating multiplicative noise, are required for statistical estimation of the impact of multiplicative noise on the functionality of signal processing systems in various areas of applications.

However, in some cases the estimation of degree of distortions of individual realizations of the signal during the condition of stationary fluctuating multiplicative noise is of definite interest to us. Relationships that permit us to carry out this estimation are discussed and presented in Section 4.3. Section 4.3 discusses the problem as devoted to the determination of both the extent and the conditions during which characteristics defining the ensemble of realizations of the signal can characterize distortions of individual realizations of the signal, along with methods for defining characteristics of the ensemble using only the realization of the signal.

In Section 4.3, the reader can also find that the characteristics given for the ensemble of realizations of the signal, distorted by the stationary fluctuating multiplicative noise, can provide us a sufficient notion—with respect to individual realizations of this signal—when the time of correlation of distortions is much less than the signal duration. For these cases, we can define the characteristics of the ensemble of realizations using only the realization of the signal.

Let us consider the main characteristics of two components of the signal distorted by the multiplicative noise given by Eq. (4.89).

4.2.1 Characteristics of Undistorted Component of the Signal

Using Eqs. (3.93), (4.83), and (4.85), the relative level α_0 of the undistorted component of the signal and the shift in the phase β_0 of the signal can be determined on the basis of the characteristic function of the stochastic processes $A(t)$ and $\varphi(t)$:

$$\alpha_0 = \left|\overline{\dot{M}(t)}\right| = \left|\frac{\partial}{\partial x_1}\Theta_2^{A(t)\varphi(t)}(x_1, 1)\right|_{x_1=0}; \tag{4.90}$$

$$\beta_0 = \arg \overline{\dot{M}(t)} = \arg\left[\frac{\partial}{\partial x_1}\Theta_2^{A(t)\varphi(t)}(x_1, 1)\right]_{x_1=0} - \frac{\pi}{2}, \tag{4.91}$$

where

$$\Theta_2^{A(t)\varphi(t)}(x_1, x_2) = m_1\left[e^{j[x_1 A(t)+x_2\varphi(t)]}\right] \tag{4.92}$$

is the two-dimensional characteristic function of the stochastic processes $A(t)$ and $\varphi(t)$ at coinciding instants.

If the pure amplitude factor $A(t)$ of the signal can be determined in the form of Eq. (3.95)

$$A(t) = A_0[1 + \xi(t)] \tag{4.93}$$

then in terms of Eq. (3.103), the values α_0 and β_0 have the following form:

$$\alpha_0 = A_0\left|\Theta_2^{\xi(t)\varphi(t)}(0, 1) - j\left[\frac{\partial}{\partial x_1}\Theta_2^{\xi(t)\varphi(t)}(x_1, 1)\right]_{x_1=0}\right|; \tag{4.94}$$

$$\beta_0 = \arg\left\{\Theta_2^{\xi(t)\varphi(t)}(0, 1) - j\left[\frac{\partial}{\partial x_1}\Theta_2^{\xi(t)\varphi(t)}(x_1, 1)\right]_{x_1=0}\right\}, \tag{4.95}$$

where $\Theta_2^{\xi(t)\varphi(t)}(x_1, x_2)$ is the characteristic function of the stochastic processes $\xi(t)$ and $\varphi(t)$ at coinciding instants.

If the changes in amplitude and phase of the signal are independent of each other, then we can write

$$|\overline{\dot{M}(t)}| = \overline{A(t)} \cdot e^{j\overline{\varphi(t)}} \tag{4.96}$$

and

$$\alpha_0 = A_0 \left|\Theta_1^{\varphi(t)}(1)\right| ; \tag{4.97}$$

$$\beta_0 = \arg\left\{\Theta_1^{\varphi(t)}(1)\right\}, \tag{4.98}$$

where $\Theta_1^{\varphi(t)}(x)$ is the one-dimensional characteristic function of the changes in the phase $\varphi(t)$ of the signal.

When there are distortions only in phase of the signal we can write

$$\alpha_0 = \left|\Theta_1^{\varphi(t)}(1)\right|; \tag{4.99}$$

$$\beta_0 = \arg\left\{\Theta_1^{\varphi(t)}(1)\right\}. \tag{4.100}$$

Since the relative level α_0 determined by Eq. (4.84) of the undistorted component of the signal is defined only by the noise modulation function $\dot{M}(t)$ of the multiplicative noise and does not depend on the structure of the signal $a(t)$, the value α_0^2 can be defined as the relative level of the discrete component—the residual signal on the carrier frequency—of the energy spectrum of the signal formed during modulation of harmonic oscillation by the same law by which the signal is distorted by the multiplicative noise.

As an example, consider the case that is often used in practice, in which distortions in amplitude $\xi(t)$ and phase $\varphi(t)$ of the signal obey the probability distribution density that is close to the Gaussian distribution law. When the stochastic processes $\xi(t)$ and $\varphi(t)$ obey the Gaussian probability distribution density under the use of Eqs. (3.110), (4.94), and (4.95) we can write[19,20]

$$\alpha_0^2 = A_0^2 \cdot e^{-\sigma_\varphi^2}\left[1 + r_{\xi\varphi}(0)\sigma_\xi^2\sigma_\varphi^2\right]; \tag{4.101}$$

$$\beta_0 = \text{arctg}\,[r_{\xi\varphi}(0)\sigma_\xi\sigma_\varphi], \tag{4.102}$$

where σ_ξ^2 is the variance of distortions in amplitude of the signal; σ_φ^2 is the variance of distortions in phase of the signal; and $r_{\xi\varphi}(0)$ is the coefficient of mutual correlation between the stochastic functions $\xi(t)$ and $\varphi(t)$ at coinciding instants.

When changes in amplitude and phase of the signal are independent at coinciding instants, then we can write

$$\alpha_0^2 = A_0^2 \cdot e^{-\sigma_\varphi^2}, \quad \beta_0 = 0. \tag{4.103}$$

When there are distortions only in phase of the signal caused by the multiplicative noise, then we can write

$$\alpha_0^2 = e^{-\sigma_\varphi^2}, \quad \beta_0 = 0. \tag{4.104}$$

Reference to Eqs. (4.102)–(4.104) shows that, as a rule, the relative level of the undistorted component of the signal is defined by the degree of distortions in phase of the signal: the degree of distortions in phase of the signal is high, the relative level α_0^2 is low. Comparing Eqs. (4.102) and (4.103), we can see that the relative level α_0^2 of the undistorted component of the signal is higher when there is a functional relationship between the changes in amplitude and phase of the signal as opposed to the case in which distortions in amplitude and phase of the signal are independent.[21]

In some cases instantaneous values of the changes in phase of the signal, caused by the multiplicative noise, are approximated very well by the Gaussian probability distribution density of random variable bounded within the limits of the interval,

$$[-\varphi_m, \varphi_m], \tag{4.105}$$

the symmetric bilateral limitation.

Dependence of the relative level α_0^2 of the undistorted component of the signal on the limitation φ_m and variance σ^2 of random variable before limitation[22] is shown in Fig. 4.2 for the case in which there are distortions only in phase of the signal. Reference to Fig. 4.2 shows that the relative level α_0^2 of the undistorted component of the signal is decreased with an increase in the variance σ^2. When the variance σ^2 is constant, then the limitation φ_m is low and the

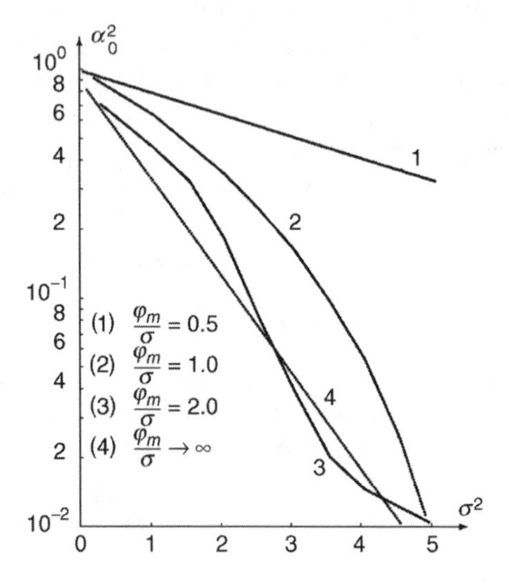

FIGURE 4.2
Dependence α_0^2 vs. the values φ_m and σ^2.

relative level α_0^2 of the undistorted component of the signal is high. At the given maximal value φ_m, the ratio $\frac{\varphi_m}{\sigma}$ is low and the relative level α_0^2 (of the undistorted component of the signal) is low.

We proceed to consider the relative level of the undistorted component of the signal—when there are distortions only in phase of the signal—for other forms of the probability distribution density of the phase that differ from the Gaussian distribution law. Using Eqs. (4.99)–(4.100) and Table II.1 (see Appendix II), we can obtain the following equalities:

- Under the uniform probability distribution density of deviations in phase of the signal, within the limits of the interval determined by Eq. (4.105)

$$\alpha_0 = \frac{\sin \varphi_m}{\varphi_m}. \tag{4.106}$$

- Under the Laplace probability distribution density of deviations in phase of the signal

$$\alpha_0 = \frac{1}{1 - 0.5\sigma_{\varphi_m}^2}. \tag{4.107}$$

Equations (4.90)–(4.100), and related examples considered above, show that the relative level of the undistorted component of the signal can be defined in a general case by the joint probability distribution density of the changes in amplitude and phase of the signal. If there are distortions only in phase of the signal, then the relative level of the undistorted component of the signal can be defined by the one-dimensional probability distribution density of changes in phase and does not depend on spectral characteristics of the multiplicative noise.

When the joint probability distribution density of the changes in amplitude and phase of the signal is given, then the relative level of the undistorted component of the signal is defined by a degree of distortions. In doing so, distortions in phase of the signal have a great impact.

In a general case, when the degree of distortions in phase of the signal is high, then the relative level of the undistorted component of the signal is low. However, for some forms of the probability distribution density, for example, the uniform probability distribution density, the limited Gaussian probability distribution density, and so on, this function cannot be monotone.[23]

4.2.2 Characteristics of the Noise Component of the Signal

Since the first component of the signal distorted by the stationary fluctuating multiplicative noise (see Eq. (4.89)) is the mean of the distorted signal, the noise component of the signal is the stochastic process with a zero mean. This stochastic process exists within the limits of the interval, where the undistorted signal $a(t)$ is present.

The correlation function and the energy spectrum of the noise component of the signal are defined using corresponding characteristics of the noise modulation function $\dot{M}(t)$ of the multiplicative noise and characteristics of the signal formed during modulation of the harmonic oscillation with the unit amplitude by the same law, by which the signal is distorted by the multiplicative noise.

Fluctuations of the noise modulation function $\dot{M}(t)$ of the multiplicative noise are denoted in the following manner:

$$\dot{V}_0(t) = \dot{M}(t) - \overline{\dot{M}(t)}. \tag{4.108}$$

So, the noise component of the distorted signal $a_M(t)$, in accordance with Eqs. (4.1), (4.2), (4.82), and (4.88), can be written in the following form:

$$\begin{aligned} v(t) &= a_M(t) - \overline{a_M(t)} \\ &= Re\{\dot{M}(t)\dot{S}(t) \cdot e^{j(\omega_0 t+\varphi_0)}\} - Re\{\overline{\dot{M}(t)}\, S(t) \cdot e^{j(\omega_0 t+\varphi_0)}\} \\ &= Re\{\dot{V}_0(t)\dot{S}(t) \cdot e^{j(\omega_0 t+\varphi_0)}\} = Re\{\dot{V}(t) \cdot e^{j(\omega_0 t+\varphi_0)}\}, \end{aligned} \tag{4.109}$$

where

$$\dot{V}(t) = \dot{V}_0(t)\dot{S}(t) \tag{4.110}$$

is the complex amplitude envelope of the noise component of the signal $a_M(t)$.

In the case of the harmonic oscillation with the unit amplitude

$$\dot{S}(t) = 1 \tag{4.111}$$

and during modulation of this signal by the similar multiplicative noise, the noise component of the signal is determined in the following form:

$$v_0(t) = Re\{\dot{V}_0(t) \cdot e^{j(\omega_0 t+\varphi_0)}\}. \tag{4.112}$$

Hence, it follows that the fluctuations $\dot{V}_0(t)$ of the noise modulation function $\dot{M}(t)$ of the multiplicative noise can be considered as the complex amplitude envelope of the noise component of the harmonic oscillation with the unit amplitude distorted by the multiplicative noise.

In a general case, the stochastic process $v_0(t)$ is the non-stationary process, even though the function $\varphi(t)$ (that is defined by distortions in phase of the signal) and the function $A(t)$ (that is defined by distortions in amplitude of the signal) are the stationary functions and the function $\dot{V}_0(t)$ is also the stationary stochastic process.[3] The exception is the case for which instantaneous values of the stochastic function $\varphi(t)$ obey the uniform probability distribution density within the limits of the interval $[0, 2\pi]$ during distortions in phase of the signal or during independent distortions in amplitude and phase of the signal. This problem is discussed in greater detail in Section 4.4.

Since the time of correlation of the noise modulation function $\dot{M}(t)$ of the multiplicative noise is much greater than the period of the carrier frequency of the signal, the complex amplitude envelope $\dot{V}_0(t)$ of fluctuations of the noise modulation function $\dot{M}(t)$ of the multiplicative noise, as well as the complex amplitude envelope $\dot{S}(t)$ of the signal $a_M(t)$, is a slowly varying function.

The correlation function of the noise component $v(t)$ of the signal is determined in the following form:

$$R_v(t, t-\tau) = m_1\{Re\{\dot{V}(t)\cdot e^{j(\omega_0 t+\varphi_0)}\}Re\{\dot{V}(t-\tau)\cdot e^{j[\omega_0(t-\tau)+\varphi_0]}\}\}. \quad (4.113)$$

Taking into account Eq. (4.38), Eq. (4.113) can be rewritten in the following form:

$$R_v(t, t-\tau) = \frac{1}{2} m_1\{Re\{\dot{V}(t)V^*(t-\tau)\cdot e^{j\omega_0\tau}\} + Re\{\dot{V}(t)\dot{V}(t-\tau)\cdot e^{j[\omega_0(2t-\tau)+2\varphi_0]}\}\}. \quad (4.114)$$

We proceed to introduce the correlation function of fluctuations of the noise modulation function $\dot{M}(t)$ of the multiplicative noise

$$\dot{R}_V(\tau) = m_1[\dot{V}_0(t)V_0^*(t-\tau)]. \quad (4.115)$$

Since the function $\dot{V}_0(t)$, as well as the noise modulation function $\dot{M}(t)$ of the multiplicative noise, is the stationary stochastic process, the correlation function $\dot{R}_V(\tau)$ depends on the value τ only.

We can also see that when the process $\dot{V}_0(t)$ is the stationary stochastic function, the mean of the product

$$\dot{V}_0(t)\dot{V}_0(t-\tau)$$

is the function of the shift in time τ only:

$$m_1[\dot{V}_0(t)\dot{V}_0(t-\tau)] = \dot{D}_V(\tau). \quad (4.116)$$

As was mentioned above, the function $\dot{V}_0(t)$ is the complex amplitude envelope of the stochastic process $v_0(t)$. If the stochastic process $v_0(t)$ is the stationary process, then the following equality for the complex envelope of amplitude of this process

$$\dot{D}_V(\tau) \equiv 0 \quad (4.117)$$

is true.[24,25]

As mentioned above, the stochastic process $v_0(t)$ is the stationary process only in particular cases. However, in a general case the stochastic process is the non-stationary process, and the following condition

$$\dot{D}_V(\tau) \neq 0 \quad (4.118)$$

is true.

Consider two cases:

- The initial phase φ_0 of the signal is the constant value that is the same for all realizations of the signal.
- The initial phase φ_0 of the signal is the random variable distributed uniformly within the limits of the interval $[0, 2\pi]$.

In the first case the correlation function defined by Eq. (4.114) in terms of Eqs. (4.110), (4.115), and (4.116) takes the following form:

$$R_\upsilon(t, t-\tau) = \frac{1}{2}\{Re\{\dot{R}_V(\tau)\dot{S}(t)S^*(t-\tau)\cdot e^{j\omega_0\tau}\}$$
$$+ Re\{\dot{D}_V(\tau)\,\dot{S}(t)\,\dot{S}(t-\tau)\cdot e^{j[\omega_0(2t-\tau)+2\varphi_0]}\}\}. \quad (4.119)$$

Reference to Eq. (4.119) shows that the stochastic process $\upsilon(t)$ is the non-stationary process: that is, its correlation function depends on both the value τ and the value t.

Dependence on the value t is defined by two factors:

- Dependence of the complex amplitude envelope $\dot{S}(t)$ of the signal causes relatively slow changes in both the first and the second terms in Eq. (4.119) as a function of time.
- The presence of the factor in the second term in Eq. (4.119). In the process, this factor is oscillated with the frequency $2\omega_0$. In a general case, this circumstance leads to the non-stationary state of the noise component of the signal, even though the dependence on the complex amplitude envelope of the signal is absent (see Eq. (4.112)).

In the second case, when the initial phase φ_0 of the signal is distributed uniformly within the limits of the interval $[0, 2\pi]$, it follows from Eq. (4.114), taking into account the independence of the initial phase φ_0 of the signal and the complex amplitude envelope $\dot{V}(t)$ of the noise component of the signal $a_M(t)$, that

$$R_\upsilon(t, t-\tau) = \frac{1}{2}Re\{\dot{R}_V(\tau)\dot{S}(t)S^*(t-\tau)\cdot e^{j\omega_0\tau}\}$$
$$+ Re\{\dot{D}_V(\tau)m_1[e^{2j\varphi_0}]\dot{S}(t)\dot{S}(t-\tau)\cdot e^{j\omega_0(2t-\tau)}\}. \quad (4.120)$$

Under the given probability distribution density

$$m_1[e^{2j\varphi_0}] = 0 \quad (4.121)$$

and Eq. (4.120) takes the following form:

$$R_\upsilon(t, t-\tau) = \frac{1}{2}Re\{\dot{R}_V(\tau)\dot{S}(t)S^*(t-\tau)\cdot e^{j\omega_0\tau}\}. \quad (4.122)$$

In this case, it is easy to show that the non-stationary state of the noise component $\upsilon(t)$ of the signal $a_M(t)$ is caused by the first factor mentioned above, but the stochastic process $\upsilon_0(t)$ is the stationary function.

There is a need to note that under definite conditions that are often used in practice the inequality

$$\dot{D}_V(\tau) \ll \dot{R}_V(\tau)$$

is carried out. In doing so, under an approximate definition of estimation of parameters of the signal distorted by the multiplicative noise, we can neglect the second term on the right side of Eq. (4.119) and use Eq. (4.122) for the first case.[26]

Since the stochastic function $\upsilon(t)$ is the non-stationary function, there is a need to define the mean of the correlation function of the stochastic process $\upsilon(t)$ [27], for the purpose of defining the energy spectrum in the sense of the spectral power density averaged within the limits of the signal duration T:

$$R_{\upsilon}^{av}(\tau) = < R_{\upsilon}(t, t-\tau) > = \frac{1}{T}\int_{0}^{T} R_{\upsilon}(t, t-\tau)\,dt. \qquad (4.123)$$

The designation $< \ldots >$ signifies the averaging with respect to time.

Substituting the value of the correlation function from Eqs. (4.119) and (4.122) into Eq. (4.123), and neglecting a contribution of the fast oscillating second term in Eq. (4.119) during averaging with respect to time, we obtain the average correlation function for the two cases considered above:

$$R_{\upsilon}^{av}(\tau) = \frac{E_a}{T}\cdot Re\{\dot{\rho}(\tau, \Omega)\dot{R}_V(\tau)\cdot e^{j\omega_0\tau}\}, \qquad (4.124)$$

where

$$\dot{\rho}(\tau, \Omega) = \frac{1}{2E_a}\int_{0}^{T} \dot{S}(t)S^*(t-\tau)\cdot e^{j\Omega t}dt \qquad (4.125)$$

is the normalized complex autocorrelation function of the signal $a(t)$, and E_a is the energy of the signal $a(t)$.

In an analogous way, we can define the mean of the correlation function of the stochastic process $\upsilon_0(t)$ determined by Eq. (4.112)

$$R_{0}^{av}(\tau) = \frac{1}{2}Re\{\dot{R}_V(\tau)\cdot e^{j\omega_0\tau}\}. \qquad (4.126)$$

Thus, the average correlation function of the noise component of the signal $a_M(t)$ distorted by the multiplicative noise can be defined using the auto-correlation function $\dot{\rho}(\tau, \Omega)$ of the signal $a(t)$ and the correlation function of fluctuations $\dot{R}_V(\tau)$ of the noise modulation function $\dot{M}(t)$ of the multiplicative noise.[28]

Next, we define the energy spectrum of the noise component $G_{\upsilon}(\omega)$ of the signal. Based on Eq. (4.124), the energy spectrum of the noise component of the signal can be represented in the following form:

$$G_{\upsilon}(\omega) = \int_{-\infty}^{\infty} R_{\upsilon}^{av}(\tau)\cdot e^{-j\omega_0\tau}d\tau$$

$$= \frac{2E_a}{T}\int_{0}^{\infty} Re\{\dot{\rho}(\tau, \Omega)\dot{R}_V(\tau)\cdot e^{j\omega_0\tau}\}\cos\omega\tau\,d\tau. \qquad (4.127)$$

Transforming the integrand in Eq. (4.127), neglecting the fast oscillating terms, taking into account that the bandwidth of the energy spectrum of the stochastic function $\upsilon(t)$ is much less than the carrier frequency, and recognizing the energy spectrum is in the neighborhood of the carrier frequency, we can write

$$G_\upsilon(\omega) = \frac{E_a}{T} \cdot Re \int_{-\infty}^{\infty} \dot{\rho}(\tau, 0)\dot{R}_V(\tau) \cdot e^{j(\omega_0-\omega)\tau} d\tau. \tag{4.128}$$

Equation (4.128) defines the energy spectrum of the noise component of the signal distorted by the stationary fluctuating multiplicative noise, using the autocorrelation function of the signal and the correlation function of fluctuations of the noise modulation function $\dot{M}(t)$ of the multiplicative noise.

Based on Eq. (4.126), the energy spectrum of the stochastic process $\upsilon_0(t)$ can be determined in the following form:

$$G_0(\omega) = \frac{1}{2} Re \int_{0}^{\infty} \dot{R}_V(\tau) \cdot e^{j(\omega_0-\omega)\tau} d\tau. \tag{4.129}$$

Denoting the energy spectrum of fluctuations of the noise modulation function $\dot{M}(t)$ of the multiplicative noise by $G_V(\Omega)$

$$G_V(\Omega) = \int_{-\infty}^{\infty} \dot{R}_V(\tau) \cdot e^{-j\Omega\tau} d\tau = 2Re \int_{0}^{\infty} \dot{R}_V(\tau) \cdot e^{-j\Omega\tau} d\tau \tag{4.130}$$

and comparing Eqs. (4.129) and (4.130), it is easy to show that

$$G_V(\Omega) = 4G_0(\omega_0 + \Omega) \tag{4.131}$$

and the energy spectrum $G_0(\omega)$ can be determined in the following form:[24,25]

$$G_0(\omega) = \frac{1}{4} G_V(\Omega - \omega_0) + \frac{1}{4} G_V(-\Omega - \omega_0). \tag{4.132}$$

The complex autocorrelation function of the signal can be defined using the energy spectrum $G_S(\omega)$ of the complex amplitude envelope $\dot{S}(t)$ of the signal:[27]

$$\dot{\rho}(\tau, 0) = \frac{T}{2E_a} \int_{-\infty}^{\infty} G_S(\omega) \cdot e^{j\omega\tau} d\omega = \frac{1}{2E_a} \int_{-\infty}^{\infty} |\dot{S}_S(\omega)|^2 \cdot e^{j\omega\tau} d\omega, \tag{4.133}$$

where $\dot{S}_S(\omega)$ is the Fourier transform of the complex amplitude envelope $\dot{S}(t)$ of the signal $a_M(t)$.

In terms of Eqs. (4.129) and (4.133), Eq. (4.128) can be written in the following form:

$$G_v(\omega) = \frac{1}{2\pi}\int_{-\infty}^{\infty} G_0(\Omega)G_S(\Omega-\omega)\,d\Omega$$

$$= \frac{1}{2\pi T}\int_{-\infty}^{\infty} G_0(\Omega)|\dot{S}_S(\Omega-\omega)|^2\,d\Omega. \tag{4.134}$$

Equation (4.134) represents the energy spectrum of the noise component of the signal $a_M(t)$ distorted by the multiplicative noise as a convolution of the energy spectrum of complex amplitude envelope of the undistorted signal and the energy spectrum of fluctuations formed under modulation of the harmonic oscillation by the same law, by which the signal is distorted by the multiplicative noise.

The analysis considered above shows that the bandwidth of the energy spectrum of the noise component of the signal distorted by the multiplicative noise is defined by the energy spectrum of the undistorted signal and the bandwidth of energy spectrum of the noise modulation function $\dot{M}(t)$ of the multiplicative noise. The energy spectrum of the noise component of the signal is wider than the energy spectrum of the undistorted signal.

The bandwidth of the energy spectrum of the noise modulation function $\dot{M}(t)$ of the multiplicative noise is wider, and the extension of the energy spectrum of the signal caused by the fluctuating multiplicative noise is also wider. The bandwidth of the energy spectrum of the noise modulation function $\dot{M}(t)$ of the multiplicative noise depends on both the degree of distortions in phase and amplitude of the signal and the bandwidth of their energy spectrum.

4.3 Ensemble and Individual Realizations of the Signal

Characteristics of the ensemble of realizations of the signal distorted by the fluctuating multiplicative noise were just discussed in the previous section. These characteristics are the relative level of the undistorted component of the signal that defines the mean of the distorted signal and the correlation function or the energy spectrum of the noise component of the signal that characterizes fluctuations of the distorted signal.

However, in some cases an estimation of degree of distortions of individual realizations of the signal by the fluctuating multiplicative noise is of great interest to us. For this reason, it is worthwhile to consider the characteristics of individual realizations of the signal that are required for this estimation and compare them to the characteristics of the ensemble of realizations of the signals.

This action allows us to define how the characteristics defined for the ensemble of realizations of the signal, distorted by the multiplicative noise, give us a notion regarding distortions of individual realizations of the signal, and how we can use only the realization of the signal to define the characteristics of the ensemble of the signal.

Distortions in the individual i-th realization of the signal under the stimulus of fluctuating multiplicative noise can be characterized by the undistorted component $m_i(t)$ and the noise component $v_i(t)$ of the signal. Each individual realization of the signal distorted by the multiplicative noise is the deterministic function.[29,30]

Because of this, both the undistorted component $m_i(t)$ and the noise component $v_i(t)$ of the signal for the individual i-th realization of the signal can be defined using the known realization of the noise modulation function $\dot{M}(t)$ of the multiplicative noise as in Section 4.1.

4.3.1 Undistorted Component of the Signal in Individual Realizations and Ensemble of the Signals

Using Eqs. (4.50)–(4.54) for the undistorted component of the i-th realization of the signal distorted by the fluctuating multiplicative noise, we can write

$$m_i(t) = \alpha_{0_i}^T \cdot a\left(t, \beta_{0_i}^T\right); \tag{4.135}$$

$$\alpha_{0_i}^T = \left|\dot{C}_{0_i}^T\right|; \tag{4.136}$$

$$\beta_{0_i}^T = \arg \dot{C}_{0_i}^T; \tag{4.137}$$

$$\dot{C}_{0_i}^T = \frac{1}{T}\int_0^T \dot{M}_i(t)\,dt. \tag{4.138}$$

The value $\alpha_{0_i}^T$ is the relative level of the undistorted component of the signal for the i-th realization. Contrary to the values α_0 and β_0 that are defined by the ensemble of realizations of the signal distorted by the multiplicative noise, the values $\alpha_{0_i}^T$ and $\beta_{0_i}^T$ are the random variables, since the noise modulation function $\dot{M}_i(t)$ of the multiplicative noise is a realization of the stochastic process.

It is easy to show that the mean of the random variable $\dot{C}_{0_i}^T$ that defines the undistorted component of the signal for individual realizations of the signal is equal to the value $\overline{\dot{M}(t)}$ that defines the undistorted component of the ensemble of realizations of the signal:

$$\overline{\dot{C}_{0_i}^T} = m_1\left[\dot{C}_{0_i}^T\right] = \frac{1}{T}\int_0^T m_1[\dot{M}(t)]\,dt = \overline{\dot{M}(t)}. \tag{4.139}$$

The mean $\overline{m_i(t)}$, of the undistorted component of individual realizations of the signal under the stimulus of stationary fluctuating multiplicative noise is equal to the undistorted component of the ensemble of realizations of the signal:

$$\overline{m_i(t)} = m_1\left[\alpha_{0_i}^T \cdot a\left(t, \beta_{0_i}^T\right)\right] = Re\left\{\overline{\dot{C}_{0_i}^T}\,\dot{S}(t) \cdot e^{j\omega_0 t}\right\}$$
$$= Re\left\{\overline{\dot{M}(t)}\dot{S}(t) \cdot e^{j\omega_0 t}\right\} = \alpha_0 \cdot a(t, \beta_0) = m(t). \tag{4.140}$$

Since there is the equation

$$\overline{\dot{C}_{0_i}^T} = \overline{\dot{M}(t)}, \tag{4.141}$$

a difference between the relative level of the undistorted component of individual realizations of the signal and the relative level of the undistorted component of the ensemble of realizations of the signal can be defined by the variance of the value $\dot{C}_{0_i}^T$:

$$\sigma_{C_0}^2 = m_1\left[\dot{C}_{0_i}^T C_{0_i}^{*T}\right] - \left|m_1\left[\dot{C}_{0_i}^T\right]\right|^2$$
$$= m_1\left[\left(\alpha_{0_i}^T\right)^2\right] - \left|\overline{\dot{M}(t)}\right|^2 = \overline{\left(\alpha_{0_i}^T\right)^2} - \alpha_0^2. \tag{4.142}$$

It is easy to show that the value $(\alpha_{0_i}^T)^2$ is the relative level of power of the undistorted component of individual realizations of the signal $a_{M_i}(t)$ and that α_0^2 is the relative level of power of the undistorted component of the ensemble of realizations of the signal $a_M(t)$. Thus, the variance $\sigma_{C_0}^2$ is equal to a difference between the relative level of power of the undistorted component of individual realizations of the signal and the relative level of power of the undistorted component of the ensemble of realizations of the signal.

We define the mean $\overline{(\alpha_{0_i}^T)^2}$:

$$\overline{\left(\alpha_{0_i}^T\right)^2} = m_1\left[\dot{C}_{0_i}^T C_{0_i}^{*T}\right] = \frac{1}{T^2}\int_0^T\!\!\int_0^T m_1[\dot{M}_i(t_1)\dot{M}_i(t_2)]\,dt_1\,dt_2 \tag{4.143}$$

or in terms of Eqs. (4.50), (4.85), and (4.115) we can write

$$\overline{\left(\alpha_{0_i}^T\right)^2} = \alpha_0^2 + \frac{1}{T^2}\int_0^T\!\!\int_0^T \dot{R}_V(t_1 - t_2)\,dt_1\,dt_2. \tag{4.144}$$

Hence it follows

$$\sigma_{C_0}^2 = \overline{\left(\alpha_{0_i}^T\right)^2} - \alpha_0^2 = \frac{1}{T^2}\int_0^T\!\!\int_0^T \dot{R}_V(t_1 - t_2)\,dt_1\,dt_2. \tag{4.145}$$

We use the well-known inequality:[31]

$$\frac{1}{T^2}\int_0^T\!\!\int_0^T \dot{R}_V(t_1 - t_2)\,dt_1\,dt_2 < \frac{2\tau_c}{T} \cdot \dot{R}_V(0) + \epsilon, \tag{4.146}$$

where τ_c is the time of correlation of fluctuations of the noise modulation function $\dot{M}(t)$ of the multiplicative noise, and ϵ is the initial point of reading of the time of correlation.

Comparing Eqs. (4.145) and (4.146), it is easy to show that the time of correlation of fluctuations of the noise modulation function $\dot{M}(t)$ of the multiplicative noise is less in comparison to the signal duration, and the mean of the relative level of power of the undistorted component of the signal of individual realizations of the signal is close to the relative level of power of the undistorted component of the signal of the ensemble of realizations of the signal.

Taking into account that, as a rule,

$$\dot{R}_V(0) \leq 1, \tag{4.147}$$

reference to Eq. (4.146) shows that the difference between the values $\overline{(\alpha_{0_i}^T)^2}$ and α_0^2 does not exceed the value $\frac{2\tau_c}{T}$. Thus, the value of the relative level of the undistorted component of the signal defined in Section 4.2 provides the sufficiently completed notion, regarding the degree of distortions of individual realizations of the signal, for the case in which the time of correlation of the noise modulation function $\dot{M}(t)$ of the multiplicative noise is much less than the signal duration.[32]

The difference between the values $\overline{(\alpha_{0_i}^T)^2}$ and α_0^2 is defined using the parameters of the energy spectrum of the noise modulation function $\dot{M}(t)$ of the multiplicative noise.

Changing the variables

$$\tau = t_1 - t_2 \tag{4.148}$$

in Eq. (4.145) and following the procedure of integration, we can write

$$\sigma_{C_0}^2 = \overline{\left(\alpha_{0_i}^T\right)^2} - \alpha_0^2 = \frac{2}{T}\int_0^T \left(1 - \frac{\tau}{T}\right) Re\{\dot{R}_V(\tau)\}\, d\tau. \tag{4.149}$$

Assuming that the normalized central moment (the center of gravity) of the curve $Re\{\dot{R}_V(\tau)\}$ during the condition

$$\tau \geq 0$$

is denoted as τ_{cg}:

$$\tau_{cg} = \frac{\int_0^\infty \tau Re\{\dot{R}_V(\tau)\}\, d\tau}{\int_0^\infty Re\{\dot{R}_V(\tau)\}\, d\tau} \tag{4.150}$$

and, taking into account that according to Eqs. (3.54) and (4.130) the following equality

$$\int_{-\infty}^{\infty} \dot{R}_V(\tau)\, d\tau = G_V(0) = 4G_0(\omega_0) \tag{4.151}$$

is true, Eq. (4.149) can be rewritten in the following form:

$$\sigma_{C_0}^2 = \frac{G_V(0)}{T}\left(1 - \frac{\tau_{cg}}{T}\right) = \frac{4G_0(\omega_0)}{T}\left(1 - \frac{\tau_{cg}}{T}\right). \tag{4.152}$$

The bandwidth of the complex amplitude envelope of the signal is approximately equal to $\frac{1}{T}$. Reference to Eq. (4.152) shows that during the condition

$$\tau_{cg} \ll T \tag{4.153}$$

the value $\overline{(\alpha_{0_i}^T)^2}$ exceeds the relative level of the undistorted component of the signal α_0^2 by the value that is approximately equal to a power of fluctuations of the noise modulation function $\dot{M}(t)$ of the multiplicative noise within the limits of the bandwidth of the amplitude envelope of the signal.

As follows from Eqs. (4.145) and (4.152), the difference between the values $\overline{(\alpha_{0_i}^T)^2}$ and α_0^2 is caused by the following fact. The value $\overline{(\alpha_{0_i}^T)^2}$ consists of power of the undistorted component of the ensemble of realizations of the signal—the term α_0^2—and the part of power of the noise component $\upsilon(t)$ of the ensemble of realizations of the signal, which is defined by fluctuations of the undistorted component of individual realizations of the signal. It is defined by the right side of Eqs. (4.145) and (4.152).

As an example, consider the mean of the relative level of power of the undistorted component of individual realizations of the signal during distortions only in phase of the signal and the Gaussian probability distribution density of phase of the signal.

Based on Eq. (4.129), in this case the correlation function $\dot{R}_V(\tau)$ can be determined by[33]

$$\dot{R}_V(\tau) = e^{-\sigma_\varphi^2} \cdot \left(e^{\sigma_\varphi^2 r_\varphi(\tau)} - 1\right) = \alpha_0^2\left(e^{\sigma_\varphi^2 r_\varphi(\tau)} - 1\right), \tag{4.154}$$

where σ_φ^2 is the variance of phase of the signal; $r_\varphi(\tau)$ is the coefficient of correlation of the changes in phase of the signal according to the stochastic process $\varphi(t)$.

Consider the value $\overline{(\alpha_{0_i}^T)^2}$ for two forms of approximation of the coefficient of correlation: triangular and Gaussian.[34] For the triangular approximation, the coefficient of correlation is determined in the following form

$$r_\varphi(\tau) = \begin{cases} 1 - \frac{|\tau|}{2\tau_\varphi} & \text{at } |\tau| \leq \tau_\varphi; \\ 0 & \text{at } |\tau| > \tau_\varphi, \end{cases} \tag{4.155}$$

where τ_φ is the time of correlation of the stochastic process $\varphi(t)$ determined by

$$\tau_\varphi = \frac{1}{2}\int\limits_{-\infty}^{\infty} r_\varphi(\tau)\, d\tau. \tag{4.156}$$

Substituting the value $\dot{R}_V(\tau)$ from Eq. (4.154) in terms of the given value $r_\varphi(\tau)$ in Eq. (4.155), and carrying out all required transformations, we can write

$$\sigma^2_{\hat{C}_0} = \left(a^2 - 2a + \frac{2a^2}{\sigma^2_\varphi}\right) \cdot e^{-\sigma^2_\varphi} + 2\left(1 - e^{-\sigma^2_\varphi}\right) \cdot \left(\frac{a}{\sigma^2_\varphi} - \frac{a^2}{\sigma^4_\varphi}\right)$$
$$= \alpha_0^2\left(a^2 - 2a + \frac{2a^2}{\sigma^2_\varphi}\right) + 2\left(1 - \alpha_0^2\right) \cdot \left(\frac{a}{\sigma^2_\varphi} - \frac{a^2}{\sigma^4_\varphi}\right), \tag{4.157}$$

where

$$a = \frac{2\tau_\varphi}{T}. \tag{4.158}$$

Dependence of the value $\sigma^2_{\hat{C}_0}$ vs. the values a and σ_φ is shown in Fig. 4.3.

Under the Gaussian approximation, the coefficient of correlation is determined in the following form:

$$r_\varphi(\tau) = \exp\left(-\frac{\pi T^2}{4\tau^2_\varphi}\right), \tag{4.159}$$

where the time of correlation τ_φ is defined by Eq. (4.156).

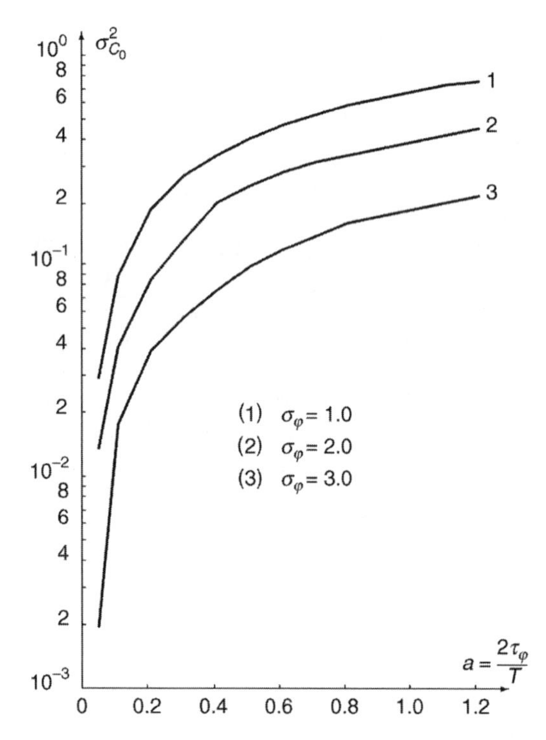

FIGURE 4.3
Dependence $\sigma^2_{\hat{C}_0}$ vs. the values a and σ_φ. Triangular approximation.

Using the series expansion discussed in Reference 35, the value $\sigma^2_{C_0}$ can be written in the following form:

$$\sigma^2_{C_0} = \frac{1}{4} \cdot e^{-\sigma^2_\varphi} \sum_{m=1}^{\infty} \frac{\sigma_\varphi^{2m}}{m!} \cdot \mathcal{L}\left(\frac{1.13a}{\sqrt{m}}\right) = \frac{\alpha_0}{4} \cdot \sum_{m=1}^{\infty} \frac{\sigma_\varphi^{2m}}{m!} \cdot \mathcal{L}\left(\frac{1.13a}{\sqrt{m}}\right), \tag{4.160}$$

where

$$\mathcal{L}(x) = \sqrt{\pi} x^2 \left[\frac{2}{x} erf\left(\frac{2}{x}\right) - x^2 \left(1 - e^{\frac{4}{x^2}}\right)\right] \tag{4.161}$$

and

$$erf(y) = \frac{2}{\sqrt{\pi}} \int_0^y e^{-t^2}\, dt. \tag{4.162}$$

The series converges very rapidly. The function $\mathcal{L}(x)$ is discussed in Reference 35 in greater detail. Dependence of the value $\sigma^2_{C_0}$ vs. the values a and σ_φ is shown in Fig. 4.4. Reference to Figs. 4.3 and 4.4 shows that with a decrease

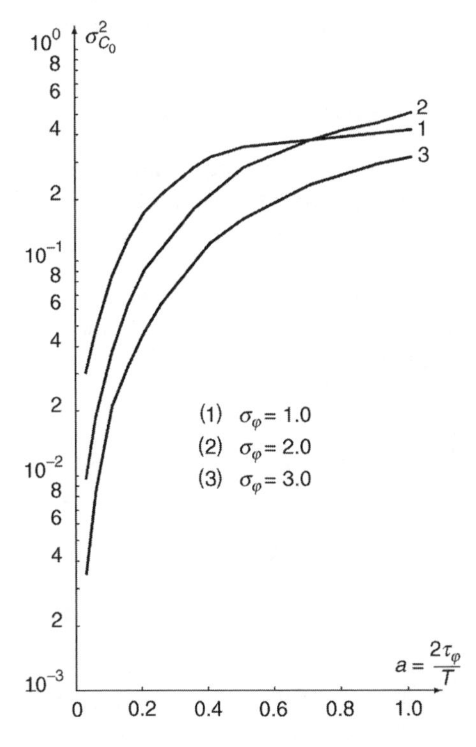

FIGURE 4.4
Dependence $\sigma^2_{C_0}$ vs. the values a and σ_φ. Gaussian approximation.

in the value τ_φ and an increase in the value σ_φ—both of these factors lead to a decrease in the time of correlation τ_c of the noise modulation function $\dot{M}(t)$ of the multiplicative noise—the value $\sigma^2_{C_0}$ is decreased and the value $\overline{(\alpha^T_{0_i})^2}$ tends to approach the relative level of the undistorted component of the signal α^2_0.

4.3.2 Noise Component of the Signal in Individual Realizations and Ensemble of the Signals

As was shown in the previous sections, during the condition of stationary fluctuating multiplicative noise the mean of the undistorted component of individual realizations of the signal coincides with the undistorted component of the ensemble of realizations of the signal:

$$\overline{m_i(t)} = m(t). \tag{4.163}$$

Since

$$m(t) = m_1[a_M(t)], \tag{4.164}$$

the mean of the noise component of individual realizations of the signal, during the condition of stationary fluctuating multiplicative noise, is equal to zero:

$$\overline{\upsilon_i(t)} = m_1[\upsilon_i(t)] = 0. \tag{4.165}$$

Next we compare the correlation function and the energy spectrum of the noise component of individual realizations of the signal with the correlation function and the energy spectrum of the ensemble of realizations of the signal distorted by the multiplicative noise.

The noise component $\upsilon_i(t)$ of individual realizations of the signal is the deterministic function with the finite duration T and its energy spectrum is determined in the following form:[27]

$$G_{\upsilon_i}(\omega) = \int_{-\infty}^{\infty} K_{\upsilon_i}(\tau) \cdot e^{-j\omega\tau}\, d\tau, \tag{4.166}$$

where

$$K_{\upsilon_i}(\tau) = <\upsilon_i(t)\upsilon_i(t-\tau)> = \frac{1}{T}\int_0^T \upsilon_i(t)\upsilon_i(t-\tau)\, dt \tag{4.167}$$

is the time correlation function of the noise component $\upsilon_i(t)$ of the signal.

It is easy to show that the mean $\overline{K_{\upsilon_i}(\tau)}$ of the time correlation function and the mean $\overline{G_{\upsilon_i}(\omega)}$ of the energy spectrum of individual realizations of the signal are the mean of the correlation function $R_{\upsilon}^{av}(\tau)$ and the mean of the energy spectrum of the noise component $\upsilon(t)$ of the signal, respectively.

Actually,

$$\overline{K_{\upsilon_i}(\tau)} = m_1[< \upsilon_i(t)\upsilon_i(t-\tau) >]; \tag{4.168}$$

$$\overline{G_{\upsilon_i}(\omega)} = \int_{-\infty}^{\infty} m_1[< \upsilon_i(t)\upsilon_i(t-\tau) >] \cdot e^{-j\omega\tau} d\tau. \tag{4.169}$$

Interchanging the procedure of averaging with respect to time and the procedure of statistical averaging, and taking into consideration Eqs. (4.123) and (4.127), we can write

$$\begin{aligned} \overline{K_{\upsilon_i}(\tau)} &= < m_1[\upsilon_i(t)\upsilon_i(t-\tau)] > \\ &= < R_{\upsilon}(t, t-\tau) > = R_{\upsilon}^{av}(\tau); \end{aligned} \tag{4.170}$$

$$\begin{aligned} \overline{G_{\upsilon_i}(\omega)} &= \int_{-\infty}^{\infty} < m_1[\upsilon_i(t)\upsilon_i(t-\tau)] > \cdot e^{-j\omega\tau} d\tau \\ &= \int_{-\infty}^{\infty} R_{\upsilon}^{av}(\tau) \cdot e^{-j\omega\tau} d\tau = G_{\upsilon}(\omega). \end{aligned} \tag{4.171}$$

The time correlation function $K_{\upsilon_i}(\tau)$ and the energy spectrum $G_{\upsilon_i}(\omega)$ are the random variables under the fixed values of τ and ω. The means of these random variables, as illustrated above, are equal to $R_{\upsilon}^{av}(\tau)$ and $G_{\upsilon}(\omega)$, respectively, i.e., they are equal to corresponding characteristics of the ensemble of realizations of the signal distorted by the multiplicative noise.

But the time correlation function $K_{\upsilon_i}(\tau)$ of individual realization of the signal can differ from the correlation function $R_{\upsilon}^{av}(\tau)$, and the energy spectrum $G_{\upsilon_i}(\omega)$ of each individual realization of the signal can differ from the energy spectrum $G_{\upsilon}(\omega)$. The extent of these differences can be characterized by the variance of the random variables $K_{\upsilon_i}(\tau)$ and $G_{\upsilon_i}(\omega)$.[36]

The definition of the variance of the time correlation function $K_{\upsilon_i}(\tau)$ is very cumbersome. Because of this, we estimate a dependence between the variance of the time correlation function $K_{\upsilon_i}(\tau)$, the ratio between the time of correlation of the multiplicative noise and the signal duration we must analyze, and the variance of the time correlation function $K_{M_i}(\tau)$ of the individual i-th realization $a_{M_i}(t)$ of the signal $a_M(t)$ distorted by the multiplicative noise:

$$K_{M_i}(\tau) = < a_{M_i}(t)\, a_{M_i}(t-\tau) > = \frac{1}{T}\int_{0}^{T} a_{M_i}(t)\, a_{M_i}(t-\tau)\, dt. \tag{4.172}$$

Since the individual i-th realization $a_{M_i}(t)$ of the signal $a_M(t)$ is the sum of the undistorted component $m_i(t)$ and the noise component $\upsilon_i(t)$ of the signal and peculiarities of the undistorted component $m_i(t)$ of the signal that were discussed above, the analysis of the i-th realization $a_{M_i}(t)$ of the signal $a_M(t)$ allows us to define the main characteristics of the noise component $\upsilon_i(t)$ of the i-th realization of the signal.[37]

It is easy to show (see Eq. (4.170)) that the mean of the time correlation function $K_{M_i}(\tau)$ of the individual i-th realization $a_{M_i}(t)$ of the signal $a_M(t)$ is equal to the mean of the correlation function of the signal $a_M(t)$:

$$\overline{K_{M_i}(\tau)} = < R_m(t, t-\tau) > = R_m^{av}(\tau), \tag{4.173}$$

where

$$R_m(t, t-\tau) = m_1[a_M(t)a_M(t-\tau)]. \tag{4.174}$$

The variance of the correlation function $K_{M_i}(\tau)$ is equal to

$$\begin{aligned} D\big[K_{M_i}(\tau)\big] &= m_1\big[K_{M_i}^2(\tau)\big] - \big[\overline{K_{M_i}(\tau)}\big]^2 \\ &= m_1\big[K_{M_i}^2(\tau)\big] - \big[R_m^{av}(\tau)\big]^2. \end{aligned} \tag{4.175}$$

We proceed to determine the terms on the right side of Eq. (4.175). Substituting the value $a_{M_i}(t)$ defined by Eq. (4.2) in Eq. (4.172), and carrying out simple transformations as in Eq. (4.38), we can write

$$K_{M_i}(\tau) = \frac{1}{2T} Re\left\{ e^{j\omega_0 t} \int_0^T \dot{M}_i(t) M_i^*(t-\tau) \dot{S}(t) S^*(t-\tau)\, dt \right\}, \tag{4.176}$$

where $\dot{M}_i(t)$ is the i-th realization of the noise modulation function $\dot{M}(t)$ of the multiplicative noise.

Hence it follows

$$\begin{aligned} m_1\big[K_{M_i}^2(\tau)\big] &= \frac{1}{8T^2} Re \int_0^T\!\!\int_0^T m_1\big[\dot{M}_i(t) M_i^*(t_1-\tau) M_i^*(t_2) \dot{M}_i(t_2-\tau)\big] \\ &\quad \times \dot{S}(t_1) S^*(t_1-\tau) S^*(t_2) \dot{S}(t_2-\tau)\, dt_1\, dt_2 \\ &\quad + \frac{1}{8T^2} Re\left\{ e^{2j\omega_0\tau} \int_0^T\!\!\int_0^T m_1\big[\dot{M}(t_1) M_i^*(t_1-\tau) \dot{M}_i(t_2) M_i^*(t_2-\tau)\big] \right. \\ &\quad \left. \times \dot{S}(t_1) S^*(t_1-\tau) \dot{S}(t_2) S^*(t_2-\tau)\, dt_1\, dt_2 \right\}. \end{aligned} \tag{4.177}$$

Carrying out the same transformations that were fulfilled in Section 4.2 for the function $\upsilon(t)$ (see Eq. (4.124)) with respect to the function $a_M(t)$, the correlation function $R_m^{av}(\tau)$ takes the following form:

$$R_m^{av}(\tau) = \frac{1}{2T} Re\left\{ \dot{R}_M(\tau) \cdot e^{j\omega_0 \tau} \int_0^T \dot{S}(t) S^*(t-\tau)\, dt \right\}. \tag{4.178}$$

Hence it follows that

$$\begin{aligned}
\left[R_m^{av}(\tau)\right]^2 &= \frac{1}{8T^2} Re\left\{ |\dot{R}_M(\tau)|^2 \int_0^T\int_0^T \dot{S}(t_1) S^*(t_1-\tau) S^*(t_2) \dot{S}(t_2-\tau)\, dt_1\, dt_2 \right\} \\
&+ \frac{1}{8T^2} Re\left\{ \dot{R}_M^2(\tau) \cdot e^{2j\omega_0 \tau} \int_0^T\int_0^T \dot{S}(t_1) S^*(t_1-\tau) \dot{S}(t_2) S^*(t_2-\tau)\, dt_1\, dt_2 \right\}.
\end{aligned} \tag{4.179}$$

Substituting Eqs. (4.177) and (4.179) in Eq. (4.175), we can write

$$\begin{aligned}
D\left[K_{M_i}(\tau)\right] &= \frac{1}{8T^2} Re \int_0^T\int_0^T \left[X_1(t_2-t_1,\tau) - |\dot{R}_M(\tau)|^2\right] \\
&\times \dot{S}(t_1) S^*(t_1-\tau) S^*(t_2) \dot{S}(t_2-\tau)\, dt_1\, dt_2 \\
&+ \frac{1}{8T^2} Re\left\{ e^{2j\omega_0\tau} \int_0^T\int_0^T \left[X_2(t_2-t_1,\tau) - \dot{R}_M^2(\tau)\right] \right. \\
&\left. \times \dot{S}(t_1) S^*(t_1-\tau) \dot{S}(t_2) S^*(t_2-\tau)\, dt_1\, dt_2 \right\},
\end{aligned} \tag{4.180}$$

where

$$X_1(t_2-t_1,\tau) = m_1[\dot{M}_i(t_1) M_i^*(t_1-\tau) M_i^*(t_2) \dot{M}_i(t_2-\tau)] \tag{4.181}$$

and

$$X_2(t_2-t_1,\tau) = m_1[\dot{M}_i(t_1) M_i^*(t_1-\tau) \dot{M}_i(t_2) M_i^*(t_2-\tau)] \tag{4.182}$$

are the fourth mixed initial moments of the noise modulation function $\dot{M}(t)$ of the multiplicative noise.

Since the noise modulation function $\dot{M}(t)$ of the multiplicative noise is the stationary process according to initial conditions, the moments defined by Eqs. (4.181) and (4.182) depend on the difference $t_2 - t_1$ and do not depend on the instant of the time t_1.

After changing the variables

$$t_2 = t_1 + t \tag{4.183}$$

and using the procedure of integration with respect to the variables t_1 and t_2, Eq. (4.180) is transformed into the following form:

$$D\big[K_{M_i}(\tau)\big] = \frac{1}{8T^2} Re\left\{\int_0^T \big[X_1(t,\tau) - |\dot{R}_M(\tau)|^2\big]\,dt \right.$$
$$\left. \times \int_0^{T-t} \dot{S}(t_1)S^*(t_1-\tau)S^*(t_1+t)\dot{S}(t_1+t-\tau)\,dt_1\right\}$$
$$+ \frac{1}{8T^2} Re\left\{\int_{-T}^0 \big[X_1(t,\tau) - |\dot{R}_M(\tau)|^2\big]\,dt \right.$$
$$\left. \times \int_{-t}^{T} \dot{S}(t_1)S^*(t_1-\tau)S^*(t_1+t)\dot{S}(t_1+t-\tau)\,dt_1\right\}$$
$$+ \frac{1}{8T^2} Re\left\{e^{2j\omega_0\tau}\int_0^T \big[X_2(t,\tau) - \dot{R}_M^2(\tau)\big]\,dt \right.$$
$$\left. \times \int_0^{T-t} \dot{S}(t_1)S^*(t_1-\tau)\dot{S}(t_1+t)S^*(t_1+t-\tau)\,dt_1\right\}$$
$$+ \frac{1}{8T^2} Re\left\{e^{2j\omega_0\tau}\int_{-T}^0 \big[X_2(t,\tau) - \dot{R}_M^2(\tau)\big]\,dt \right.$$
$$\left. \times \int_{-t}^{T} \dot{S}(t_1)S^*(t_1-\tau)\dot{S}(t_1+t)S^*(t_1+t-\tau)\,dt_1\right\}. \qquad (4.184)$$

Taking into account limitations in changing the variable t for each term on the right side of Eq. (4.184), and assuming that the amplitude $S(t)$ of the signal $a_M(t)$ is the square wave-form, it is easy to show that

$$\int_0^T S^4(t)\,dt = \frac{4E_a^2}{T}, \qquad (4.185)$$

where E_a is the energy of the signal $a(t)$.

Taking into account Eq. (4.185) and based on Eq. (4.184), we obtain that the estimation of the variance of the time correlation function of individual realizations of the signal, distorted by stationary fluctuating multiplicative noise, is determined in the following form:

$$D\big[K_{M_i}(\tau)\big] \le \frac{E_a^2}{2T^3}\cdot Re\int_{-T}^{T}\big[X_1(t,\tau) + X_2(t,\tau)\cdot e^{2j\omega_0\tau}\big]\,dt$$
$$- \frac{E_a^2}{T^2}|\dot{R}_M(\tau)|^2 - \frac{E_a^2}{T^2}\cdot Re\big\{\dot{R}_M^2(\tau)\cdot e^{2j\omega_0\tau}\big\}, \qquad (4.186)$$

where $X_1(t, \tau)$ and $X_2(t, \tau)$ are defined by Eqs. (4.181) and (4.182) during the condition determined by Eq. (4.183).

To estimate the relative value of the variance $D[K_{M_i}(\tau)]$, it is worthwhile to use the squared maximal mean of the time correlation function $K_{M_i}(\tau)$.

According to Eqs. (4.173) and (4.178), we can write

$$\overline{K_{M_i}(0)} = R_m^{av}(0) = \frac{E_a}{T} \cdot \dot{R}_M(0) = \frac{E_a}{T} \cdot \overline{A^2(t)}, \tag{4.187}$$

where $\overline{A^2(t)}$ is the second initial moment of instantaneous values of the factor $A(t)$ of distortions in amplitude of the signal.

Then, using Eqs. (4.186) and (4.187), we can write

$$\frac{D\big[K_{M_i}(\tau)\big]}{\big[\overline{K_{M_i}(0)}\big]^2} \leq \frac{1}{2T[\overline{A^2(t)}]^2} \cdot Re \int_{-T}^{T} [X_1(t, \tau) + X_2(t, \tau) \cdot e^{2j\omega_0\tau}]\, dt$$
$$- \frac{1}{[\overline{A^2(t)}]^2} \cdot \big\{|\dot{R}_M(\tau)|^2 + Re\big\{\dot{R}_M^2(\tau) \cdot e^{2j\omega_0\tau}\big\}\big\}. \tag{4.188}$$

Equation (4.188) allows us to estimate a difference between the time correlation function of individual realizations of the signal by the multiplicative noise and the correlation function of the ensemble of realizations of the signal.

As an example, consider the case that is very commonly used in practice in which the fluctuating multiplicative noise causes the distortions in phase of the signal. Moreover, the deviations in phase obey the Gaussian probability distribution density.

In this case we can write

$$\dot{M}(t) = e^{j\varphi(t)}; \tag{4.189}$$

$$\overline{A^2(t)} = 1; \tag{4.190}$$

$$\dot{R}_M(\tau) = e^{-\sigma_\varphi^2[1 - r_\varphi(\tau)]}, \tag{4.191}$$

where $r_\varphi(\tau)$ is the coefficient of correlation of distortions in phase $\varphi(t)$ of the signal (see Eq. (3.116)).

We define the functions $X_1(t, \tau)$ and $X_2(t, \tau)$. During distortions in phase of the signal, these functions can be defined using the characteristic function of distortions in phase of the signal:

$$X_1(t, \tau) = m_1\left[e^{j[\varphi(t_1) - \varphi(t_1 - \tau) - \varphi(t_1 + t) + \varphi(t_1 + t - \tau)]}\right]$$
$$= \Theta_4^{\varphi(t)}(1, -1, -1, 1); \tag{4.192}$$

$$X_2(t, \tau) = m_1\left[e^{j[\varphi(t_1) - \varphi(t_1 - \tau) + \varphi(t_1 + t) - \varphi(t_1 + t - \tau)]}\right]$$
$$= \Theta_4^{\varphi(t)}(1, -1, 1, -1), \tag{4.193}$$

where $\Theta_4^{\varphi(t)}(x_1, x_2, x_3, x_4)$ is the four-dimensional characteristic function of distortions in phase $\varphi(t)$ of the signal at the instants $t_1, t_1 - \tau, t_1 + t, t_1 + t - \tau$.

Under the Gaussian probability distribution density of distortions in phase $\varphi(t)$ using the general formula of the four-dimensional characteristic function[27] and Eqs. (4.192) and (4.193), we can write

$$X_1(t,\tau) = e^{-\sigma_\varphi^2[2-2r_\varphi(\tau)-2r_\varphi(t)+r_\varphi(t+\tau)+r_\varphi(t-\tau)]}; \tag{4.194}$$

$$X_2(t,\tau) = e^{-\sigma_\varphi^2[2-2r_\varphi(\tau)+2r_\varphi(t)-r_\varphi(t+\tau)-r_\varphi(t-\tau)]}. \tag{4.195}$$

Reference to Eqs. (4.194) and (4.195) shows that the functions $X_1(t,\tau)$ and $X_2(t,\tau)$ are the real functions. Substituting Eqs. (3.141), (4.194), and (4.195) in Eq. (4.188), we obtain that the estimation of the relative level of the variance of the time correlation function of individual realizations of the signal, when distortions in phase of the signal obey the Gaussian probability distribution density, is determined in the following form:

$$\frac{D\left[K_{M_i}(\tau)\right]}{\left[K_{M_i}(0)\right]^2} \le y(\tau), \tag{4.196}$$

where

$$y(\tau) = \left\{\frac{1}{2T}\int_{-T}^{T} e^{\sigma_\varphi^2[2r_\varphi(t)-r_\varphi(t+\tau)-r_\varphi(t-\tau)]}\,dt - 1 + \frac{\cos 2\omega_0\tau}{2T}\right. \\ \left. \times \int_{-T}^{T} e^{-\sigma_\varphi^2[2r_\varphi(t)-r_\varphi(t+\tau)-r_\varphi(t-\tau)]}\,dt - \cos 2\omega_0\tau\right\} \cdot e^{-2\sigma_\varphi^2[1-r_\varphi(\tau)]}. \tag{4.197}$$

The function $y(\tau)$ is the rapid oscillating function with a frequency equal to $2\omega_0$ that is maximal under the condition

$$\cos 2\omega_0\tau = \pm 1. \tag{4.198}$$

The amplitude envelope of this function is determined in the following form:

$$Y(\tau) = \left\{\frac{1}{2T}\int_{-T}^{T} e^{\sigma_\varphi^2[2r_\varphi(t)-r_\varphi(t+\tau)-r_\varphi(t-\tau)]}\,dt - 1\right. \\ \left. \pm \frac{1}{2T}\int_{-T}^{T} e^{-\sigma_\varphi^2[2r_\varphi(t)-r_\varphi(t+\tau)-r_\varphi(t-\tau)]}\,dt \mp 1\right\} \cdot e^{-2\sigma_\varphi^2[1-r_\varphi(\tau)]}. \tag{4.199}$$

The sign in Eq. (4.199) must be kept in the following manner such that the function $Y(\tau)$ would be maximal. In the process, the function $Y(\tau)$ defines the upper estimation of the relative level of the variance of the time correlation function of individual realizations of the signal at the given value τ.

Let the coefficient of correlation of distortions in phase of the signal be approximated in the following form:

$$r_\varphi(\tau) = \begin{cases} 1 - \frac{|\tau|}{2\tau_\varphi} & \text{at } |\tau| \leq 2\tau_\varphi; \\ 0 & \text{at } |\tau| > 2\tau_\varphi, \end{cases} \tag{4.200}$$

where τ_φ is the time of correlation of distortions in phase of the signal.

Substituting Eq. (4.200) in Eq. (4.199), and integrating under the following conditions

$$0 \leq \tau \leq \tau_\varphi \tag{4.201}$$

and

$$T \geq 2\tau_\varphi + \tau, \tag{4.202}$$

we can write

$$Y(\tau) = \frac{2\tau_\varphi}{\sigma_\varphi^2 T}\left[\text{ch}\, 2\sigma_\varphi^2 - \text{ch}\, \sigma_\varphi^2\left(2 - \frac{\tau}{2\tau_\varphi}\right) + \text{ch}\frac{\tau}{\tau_\varphi} - \text{ch}\,\frac{\tau}{2\tau_\varphi}\sigma_\varphi^2\right] \cdot e^{-\sigma_\varphi^2 \frac{\tau}{\tau_\varphi}}. \tag{4.203}$$

Reference to Eq. (4.203) shows that the function $Y(\tau)$ depends linearly on $\frac{\tau_\varphi}{T}$, the ratio between the time of correlation of distortions in phase of the signal and the signal duration. As was shown above, the function $Y(\tau)$ is the upper estimation of the relative value of the variance of the time correlation function of individual realizations of the signal.

Hence, it follows that with an increase in the ratio $\frac{T}{\tau_\varphi}$ this variance is decreased and the time correlation function $K_{M_i}(\tau)$ of individual realizations of the signal tends to approach the mean of the correlation function $R_m^{av}(\tau)$ of the ensemble of individual realizations of the signal distorted by the multiplicative noise. An analogous relationship exists for the correlation function of the noise component $\upsilon(t)$ of the signal.

Dependence of the function $Y(\tau)$ vs. the ratio $\frac{T}{\tau_\varphi}$ at some values σ_φ^2 and $\frac{\tau}{\tau_\varphi}$ determined by Eq. (4.203) is shown in Figs. 4.5 and 4.6. Reference to Figs. 4.5 and 4.6 shows that during the condition

$$\frac{\tau}{\tau_\varphi} = 5 \tag{4.204}$$

the relative value of the variance of the time correlation function of individual realizations of the signal is very low, and the time correlation function defined in the case of individual realizations of the signal distorted by the multiplicative noise is very close to the correlation function of the ensemble of individual realizations of the signal distorted by the multiplicative noise.

The discussion in this section shows that the characteristics defined for the ensemble of individual realizations of the signal distorted by the multiplicative noise, provide a very good notion regarding distortions of individual

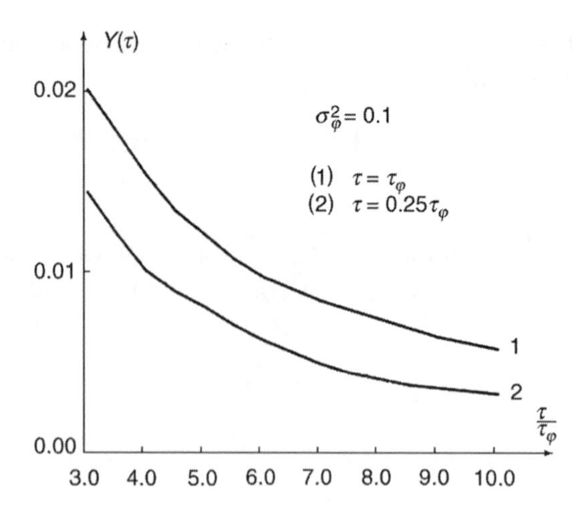

FIGURE 4.5
Dependence $Y(\tau)$ vs. the ratio $\frac{T}{\tau_\varphi}$ at $\sigma_\varphi^2 = 0.1$.

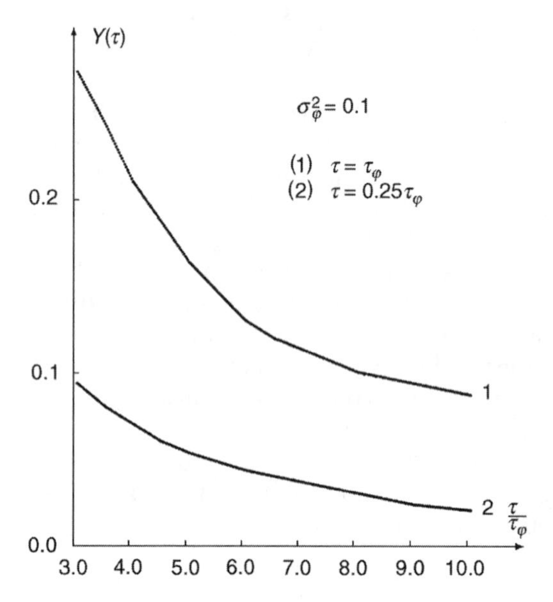

FIGURE 4.6
Dependence $Y(\tau)$ vs. the ratio $\frac{T}{\tau_\varphi}$ at $\sigma_\varphi^2 = 0.5$.

realizations of the signal when the time of correlation of distortions is much less than the signal duration.

Thus, as discussed in this section, the characteristics of the signal that is distorted by the multiplicative noise possess features similar to features of characteristics of the ergodic stochastic processes in a definite sense, even though these signals are not the ergodic processes.

4.4 Probability Distribution Density of the Signal in the Additive Gaussian Noise under the Stimulus of Multiplicative Noise

4.4.1 Probability Distribution Density of the Signal under the Stimulus of Multiplicative Noise

As was shown in Chapter 3, the normalized signal under the stimulus of multiplicative noise within the limits of the interval

$$0 \leq t \leq T \tag{4.205}$$

takes the following form:

$$\widetilde{a}(t) = \frac{a_M(t)}{S(t)} = A(t)\cos[\omega_0 t + \Psi_a(t) + \varphi(t)] = A(t)\cos\psi_{\tilde{a}}(t), \tag{4.206}$$

where $A(t)$ is the stochastic function that defines distortions in amplitude of the signal and $\varphi(t)$ is the stochastic function that defines distortions in phase of the signal. In a general case, the stochastic functions $A(t)$ and $\varphi(t)$ are functionally related. $S(t)$ is the known modulation law of amplitude of the signal; $\Psi_a(t)$ is the known modulation law of phase of the signal; and $\psi_{\tilde{a}}(t)$ is the total instantaneous phase of the signal $\widetilde{a}(t)$.

The mean of the value $\psi_{\tilde{a}}(t)$ is determined in the following form:

$$\overline{\psi_{\tilde{a}}(t)} = m_1[\psi_{\tilde{a}}(t)] = \omega_0 t + \Psi_a(t) + \varphi_\varphi, \tag{4.207}$$

where φ_φ is the mean of distortions in phase of the signal.

We define the one-dimensional probability distribution density $f[\widetilde{a}(t), t]$ of the signal $\widetilde{a}(t)$. The multivariate probability distribution density of the signal $\widetilde{a}(t)$ is discussed in Section 4.5. The joint probability distribution density of instantaneous values of the signal $\widetilde{a}(t)$, amplitude envelope factor $A(t)$, and phase $\varphi(t)$ in terms of Eq. (4.206) and using the main property of the delta function (see Appendix I) can be written in the following form:[27]

$$\begin{aligned} f[\widetilde{a}(t), A(t), \varphi(t)] &= f[A(t), \varphi(t)]\delta[\widetilde{a}(t) - A(t)\cos\psi_{\tilde{a}}(t)] \\ &= \frac{1}{2\pi} f[A(t), \varphi(t)] \int_{-\infty}^{\infty} e^{jn[\tilde{a}(t) - A(t)\cos\psi_{\tilde{a}}(t)]}\, dn. \end{aligned} \tag{4.208}$$

Taking into account the periodic character of the function $\cos\psi_{\tilde{a}}(t)$, and using the series expansion discussed in Reference 38,

$$e^{jnA(t)\cos\psi_{\tilde{a}}(t)} = \sum_{\ell=-\infty}^{\infty} J_\ell[nA(t)] \cdot e^{j\ell\psi_{\tilde{a}}(t)}, \tag{4.209}$$

where $J_\ell[nA(t)]$ is the Bessel function of real argument and integer order, during the condition

$$\widetilde{a}(t) \leq A(t) \tag{4.210}$$

we can write:[38]

$$f[\widetilde{a}(t), A(t), \varphi(t)] = \frac{f[A(t), \varphi(t)]}{\pi\sqrt{A^2(t) - \widetilde{a}^2(t)}} \cdot \sum_{\ell=-\infty}^{\infty} e^{j\ell[\psi_{\widetilde{a}}(t) + \arcsin\frac{\widetilde{a}(t)}{A(t)}]}. \tag{4.211}$$

Equation (4.211) allows us to define the probability distribution density $f[\widetilde{a}(t)]$ of instantaneous values of the signal $\widetilde{a}(t)$ during the condition of distortions in amplitude and phase of the signal with the probability distribution density $f[A(t), \varphi(t)]$.

4.4.1.1 Amplitude and Phase Distortions of the Signal as Independent

Consider the case in which distortions in amplitude and phase of the signal are independent only at coinciding instants. Then the probability distribution density of instantaneous values of the signal during the condition

$$|\widetilde{a}(t)| \leq A(t) \tag{4.212}$$

is determined in the following form:

$$\begin{aligned} f[\widetilde{a}(t)] &= \int_0^{\infty}\int_{-\infty}^{\infty} f[\widetilde{a}(t), A(t), \varphi(t)]d[A(t)]d[\varphi(t)] \\ &= \frac{1}{\pi}\int_{|\widetilde{a}(t)|}^{\infty} \frac{f[A(t)]}{\sqrt{A^2(t) - \widetilde{a}^2(t)}} d[A(t)] + \frac{1}{\pi}\sum_{\ell=-\infty}^{\infty} \Theta_1^{\varphi(t)}(\ell) \cdot e^{j\ell\overline{\psi_{\widetilde{a}}(t)}} \\ &\quad \times \int_{|\widetilde{a}(t)|}^{\infty} \frac{f[A(t)]}{\sqrt{A^2(t) - \widetilde{a}^2(t)}} \cdot e^{j\ell \arcsin\frac{\widetilde{a}(t)}{A(t)}} d[A(t)], \end{aligned} \tag{4.213}$$

where $\Theta_1^{\varphi(t)}(\ell)$ is the one-dimensional characteristic function of distortions in phase of the signal $\widetilde{a}(t)$ (see Eq. (4.206)). Table I.1 in Appendix I shows some examples of the one-dimensional characteristic function of distortions in phase of the signal.

If there are distortions only in phase of the signal, and the condition

$$f[A(t)] = \delta[A(t) - A_0(t)] \tag{4.214}$$

is satisfied, then under the use of Eq. (4.212) and if the condition determined by Eq. (4.212) is satisfied, we can write

$$\begin{aligned} f[\widetilde{a}(t)] = \frac{1}{\pi\sqrt{A_0^2(t) - \widetilde{a}^2(t)}}\Bigg\{1 + 2\sum_{\ell=1}^{\infty}\left|\Theta_1^{\varphi(t)}(\ell)\right| \cos\Bigg\{\ell[\omega_0 t + \Psi_a(t)] \\ + \arg\Theta_1^{\varphi(t)}(\ell) + \ell\arcsin\frac{\widetilde{a}(t)}{A_0(t)}\Bigg\}\Bigg\}. \end{aligned} \tag{4.215}$$

The first, second, third, and fourth initial moments of the probability distribution density given by Eq. (4.215) take the following forms:

$$m_1[\tilde{a}(t)] = A_0(t)\left|\Theta_1^{\varphi(t)}(1)\right| \cos\left[\omega_0 t + \Psi_a(t) + \arg \Theta_1^{\varphi(t)}(\ell)\right]; \tag{4.216}$$

$$m_2[\tilde{a}(t)] = \frac{A_0^2(t)}{2}\left\{1 + \left|\Theta_1^{\varphi(t)}(2)\right| \cos\left[2\omega_0 t + 2\Psi_a(t) + \arg \Theta_1^{\varphi(t)}(2)\right]\right\}; \tag{4.217}$$

$$m_3[\tilde{a}(t)] = \frac{A_0^3(t)}{4}\left\{3\left|\Theta_1^{\varphi(t)}(1)\right| \cos\left[\omega_0 t + \Psi_a(t) + \arg \Theta_1^{\varphi(t)}(1)\right] + \left|\Theta_1^{\varphi(t)}(3)\right| \cos\left[3\omega_0 t + 3\Psi_a(t) + \arg \Theta_1^{\varphi(t)}(3)\right]\right\}; \tag{4.218}$$

$$m_4[\tilde{a}(t)] = \frac{3}{8}A_0^4(t)\left\{1 + \frac{4}{3}\left|\Theta_1^{\varphi(t)}(2)\right| \cos\left[2\omega_0 t + 2\Psi_a(t) + \arg \Theta_1^{\varphi(t)}(2)\right] + \frac{1}{3}\left|\Theta_1^{\varphi(t)}(4)\right| \cos\left[4\omega_0 t + 4\Psi_a(t) + \arg \Theta_1^{\varphi(t)}(4)\right]\right\}. \tag{4.219}$$

Reference to Eqs. (4.213) and (4.215) shows that the first terms are the function that defines the probability distribution density of the signal with the random initial phase distributed uniformly within the limits of the interval $[0, 2\pi]$.

The contribution of other terms in Eqs. (4.213) and (4.215) is defined by the characteristic function of distortions in phase of the signal, which tends to approach zero in the limit:

$$\lim_{\ell\to\infty} \Theta_1^{\varphi(t)}(\ell) \to 0. \tag{4.220}$$

The characteristic function $\Theta_1^{\varphi(t)}(\ell)$, when the variance σ_φ^2 is thought of as a parameter, for the case of distortions in phase of the signal, which are distributed according to the Gaussian probability distribution density with a zero mean, are shown in Fig. 4.7.

Reference to Eqs. (4.85), (4.86), and (4.216) shows that the mean of the signal, under the stimulus of multiplicative noise, is defined by the relative level α_0^2 of the undistorted component of the signal (see Sections 4.2 and 4.3).

When the variance of phase of the signal is satisfied by the following condition

$$\sigma_\varphi^2 > 1$$

and if the following condition

$$\Theta_1^{\varphi(t)}(\ell > 1) \ll \Theta_1^{\varphi(t)}(1) \tag{4.221}$$

is true, we can be limited by two terms in Eqs. (4.213) and (4.215) that define the probability distribution density of the signal $\tilde{a}(t)$.

For example, in the case of Eq. (4.215) during the condition

$$\sigma_\varphi^2 > 1$$

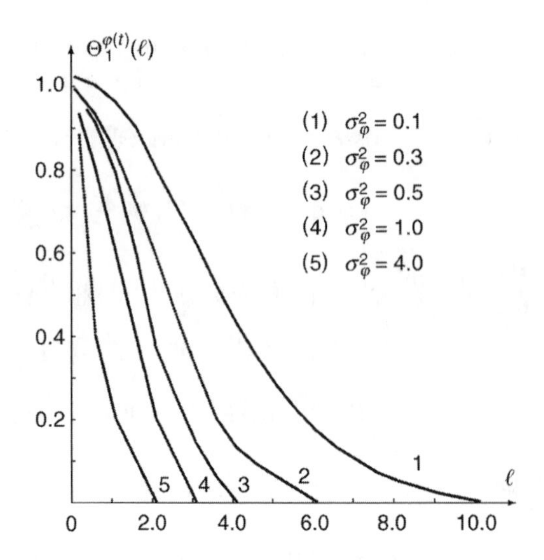

FIGURE 4.7
Characteristic function $\Theta_1^{\varphi(t)}(\ell)$. Gaussian distortions in phase.

we can write

$$f[\widetilde{a}(t)] = \frac{1}{\pi\sqrt{A_0^2(t) - \widetilde{a}^2(t)}}\left\{1 + 2\left|\Theta_1^{\varphi(t)}(1)\right| \times \cos\left[\omega_0 t + \Psi_a(t) + \arcsin\frac{\widetilde{a}(t)}{A_0(t)} + \arg\Theta_1^{\varphi(t)}(1)\right]\right\}. \tag{4.222}$$

The probability distribution densities determined by Eqs. (4.213) and (4.222), when the phase of the signal is not distributed uniformly within the limits of the interval $[0, 2\pi]$, depend on the time and the signal is the non-stationary stochastic function for this case.

When the variance of phase of the signal is much greater than unity, i.e.,

$$\sigma_\varphi^2 \gg 1$$

the second term in Eq. (4.222) is much less than the first term (see Fig. 4.7), and we can believe that the probability distribution density of the signal, independently of the probability distribution density of distortions in phase of the signal $f[\varphi(t)]$, is defined by the probability distribution density of the signal with the random initial phase distributed uniformly within the limits of the interval $[0, 2\pi]$ during the conditions

$$|\widetilde{a}(t)| \leq A_0(t) \qquad \text{and} \qquad \sigma_\varphi^2 \gg 1. \tag{4.223}$$

So, we can write

$$f[\widetilde{a}(t)] \simeq \frac{1}{\pi\sqrt{A_0^2(t) - \widetilde{a}^2(t)}}. \tag{4.224}$$

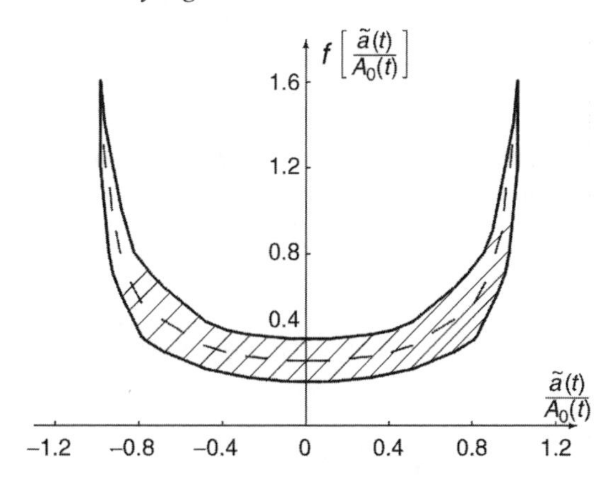

FIGURE 4.8
The probability distribution density of the signal during Gaussian distortions in phase.

Taking into account the well-known formula[27] for the probability distribution density of the random variable

$$\widetilde{a}_f(t) = \widetilde{a}(t) - \alpha_0 \cdot \widetilde{a}_0(t) \tag{4.225}$$

that is linearly related with another random variable $\widetilde{a}(t)$, we can obtain the probability distribution density of the fluctuating component of the signal under the stimulus of multiplicative noise:

$$f[\widetilde{a}_f(t)] = f_{\widetilde{a}}[\widetilde{a}_f(t) + \alpha_0 \cdot \widetilde{a}_0(t)], \tag{4.226}$$

where

$$\widetilde{a}_0(t) = \frac{a(t)}{S(t)} \tag{4.227}$$

is the normalized undistorted signal.

The probability distribution density of the signal with the amplitude that is not varied in time and the random phase distributed in accordance with the Gaussian probability distribution density with a zero mean and the variance $\sigma_\varphi^2 = 4$ radian is shown in Fig. 4.8.

The shaded region is the domain of definition of the probability distribution density of the signal. With an increase in the variance

$$\sigma_\varphi^2 \to \infty$$

the area of this domain tends to approach zero, the dotted line in Fig. 4.8. Consequently, the signal given by Eq. (4.206) can be considered as the stationary stochastic function during distortions in phase of the signal if the phase of the signal is distributed uniformly within the limits of the interval $[0, 2\pi]$.

This statement is true in the case of other probability distribution densities of phase of the signal in which the variance of distortions in phase of the signal is much greater than unity,[39,40] i.e.,

$$\sigma_\varphi^2 \gg 1.$$

If the amplitude and phase of the signal are independent stochastic functions, introducing the variable

$$A(t) = |\widetilde{a}(t)| \operatorname{ch} n \tag{4.228}$$

into Eq. (4.213), we can write

$$f[\widetilde{a}(t)] = f_0[\widetilde{a}(t)] + f_f[\widetilde{a}(t)], \tag{4.229}$$

where

$$f_0[\widetilde{a}(t)] = \frac{1}{\pi}\int_0^\infty f_A[\widetilde{a}(t) \operatorname{ch} n]\, dn; \tag{4.230}$$

$$\begin{aligned} f_f[\widetilde{a}(t)] &= \frac{1}{\pi}\sum_{\ell=-\infty\,(\ell\neq 0)}^{\infty} \Theta_1^{\varphi(t)}(\ell)\cdot e^{j\ell\overline{\psi_{\widetilde{a}}(t)}} \\ &\quad\times \int_{|\widetilde{a}(t)|}^{\infty} \frac{f[A(t)]}{\sqrt{A^2(t)-\widetilde{a}^2(t)}}\cdot e^{j\ell \arcsin\frac{\widetilde{a}(t)}{A(t)}}\, d[A(t)] \\ &= \frac{1}{\pi}\sum_{\ell=-\infty\,(\ell\neq 0)}^{\infty} \Theta_1^{\varphi(t)}(\ell)\cdot e^{j\ell\overline{\psi_{\widetilde{a}}(t)}}\cdot\frac{1}{j\ell} \\ &\quad\times \int_{|\widetilde{a}(t)|}^{\infty} \frac{d}{d[\widetilde{a}(t)]}\left\{f[A(t)]\cdot e^{j\ell \arcsin\frac{\widetilde{a}(t)}{A(t)}}\right\} d[A(t)]. \end{aligned} \tag{4.231}$$

The term $f_0[\widetilde{a}(t)]$ defines the probability distribution density of instantaneous values of the signal in Eq. (4.206) during independent distortions in amplitude and phase of the signal, when distortions in phase of the signal are distributed uniformly within the limits of the interval $[0, 2\pi]$, or under arbitrary probability distribution density of phase of the signal when the variance of phase of the signal satisfies the condition

$$\sigma_\varphi^2 \gg 1.$$

The initial moments of the probability distribution density determined by Eqs. (4.229)–(4.231) take the following form:

$$m_1[\widetilde{a}(t)] = 0; \tag{4.232}$$

$$m_2[\widetilde{a}(t)] = \frac{1}{2} m_2[A(t)]; \tag{4.233}$$

$$m_3[\widetilde{a}(t)] = 0; \tag{4.234}$$

$$m_4[\widetilde{a}(t)] = \frac{3}{8} m_4[A(t)]. \tag{4.235}$$

Taking into account that[41]

$$\int_{|\tilde{a}(t)|}^{\infty} \frac{d}{d[\tilde{a}(t)]} \{ f[A(t)] \cdot e^{j\ell \arcsin \frac{\tilde{a}(t)}{A(t)}} \} d[A(t)]$$

$$= \frac{d}{d[\tilde{a}(t)]} \left\{ \int_{|\tilde{a}(t)|}^{\infty} f[A(t)] \cdot e^{j\ell \arcsin \frac{\tilde{a}(t)}{A(t)}} d[A(t)] \right\} + f_A[|\tilde{a}(t)|] \cdot e^{j\ell \frac{\pi}{2}} \quad (4.236)$$

and in terms of the condition determined by Eq. (4.214), we can write that the characteristic function can be determined in the following form:

$$\Theta_1^{\varphi(t)}(\ell) \simeq \int_{|\tilde{a}(t)|}^{\infty} f[A(t)] \cdot e^{j\ell \arcsin \frac{\tilde{a}(t)}{A(t)}} d[A(t)]. \quad (4.237)$$

The probability distribution density in Eq. (4.213) during the condition determined by Eq. (4.214) takes the following form:

$$f[\tilde{a}(t)] = f_0[\tilde{a}(t)] + \frac{2}{\pi} \sum_{\ell=1}^{\infty} \frac{1}{\ell} \left| \Theta_1^{\varphi(t)}(\ell) \right| f_A[|\tilde{a}(t)|]$$

$$\times \sin \left\{ \ell \left[\omega_0 t + \Psi_a(t) + \frac{\pi}{2} \right] + \arg \Theta_1^{\varphi(t)}(\ell) \right\}. \quad (4.238)$$

Reference to Eq. (4.238) shows that under the independent distortions in amplitude and phase of the signal and during the condition

$$\sigma_\varphi^2 \gg 1,$$

when the following condition

$$\left| \Theta_1^{\varphi(t)}(\ell \geq 1) \right| \gg 1 \quad (4.239)$$

is satisfied, all terms of the series are much less than the term $f_0[\tilde{a}(t)]$ and, consequently, the probability distribution density of the signal depends very weakly on the probability distribution density $f[\varphi(t)]$ and is defined by Eqs. (4.230) and (4.231).

In other words, we can state with confidence that the probability distribution density depends only on the probability distribution density of the amplitude envelope factor $f[A(t)]$ of the signal.

It is very important to note the functional relationship between the characteristic function of the instantaneous values of the signal and the probability distribution density of the complex amplitude envelope factor of the signal, when the phase of the signal is distributed uniformly within the limits of the interval $[0, 2\pi]$, or when the condition

$$\sigma_\varphi^2 \gg 1$$

is satisfied, and for the case of other probability distribution densities of the signal.

We can see from Eq. (4.211) that during the condition determined by Eq. (4.214) and in terms of the equality[31,41]

$$\frac{1}{\sqrt{A^2(t) - \tilde{a}^2(t)}} \cdot e^{j\ell \arcsin\frac{\tilde{a}(t)}{A(t)}} = \int_{-\infty}^{\infty} J_\ell[nA(t)] \cdot e^{jn\tilde{a}(t)}\, dn \tag{4.240}$$

we can write

$$f[\tilde{a}(t)] = \int_0^{\infty}\int_{-\infty}^{\infty} f[A(t)]J_0[nA(t)] \cdot e^{jn\tilde{a}(t)} d[A(t)]\, dn. \tag{4.241}$$

Taking the product between both sides of Eq. (4.241) and the function $e^{jv\tilde{a}(t)}$, and integrating with respect to the variable $\tilde{a}(t)$ in terms of properties of the delta function, we can write[42]

$$\Theta_1^{\tilde{a}(t)}(v) = \int_0^{\infty} J_0[vA(t)]f[A(t)]d[A(t)] = H_0\left\{\frac{f[A(t)]}{A(t)}\right\}, \tag{4.242}$$

where

$$H_0\left\{\frac{f[A(t)]}{A(t)}\right\} \tag{4.243}$$

is the Hancel transform.

In terms of the Hancel transform, we can write

$$\frac{f[A(t)]}{A(t)} = \int_0^{\infty} \Theta_1^{\tilde{a}(t)}(v)J_0[vA(t)]v\, dv. \tag{4.244}$$

The k-th initial moment of the probability distribution density of the signal given by Eq. (4.206) is determined in the following form:

$$m_k[\tilde{a}(t)] = \frac{1}{jk}\left[\frac{d^k}{dv^k}\Theta_1^{\tilde{a}(t)}(v)\right]_{v=0} = \frac{1}{jk}\left\{\frac{d^k}{dv^k}H_0\left\{\frac{f[A(t)]}{A(t)}, v\right\}\right\}_{v=0}. \tag{4.245}$$

When the variance of distortions in phase of the signal is satisfied to the condition

$$\sigma_\varphi^2 > 1,$$

reference to Eq. (4.238) shows that

$$f[\tilde{a}(t)] = f_0[\tilde{a}(t)] + \frac{2}{\pi}\left|\Theta_1^{\varphi(t)}(1)\right| f_A[|\tilde{a}(t)|] \times \sin\left[\omega_0 t + \Psi_a(t) + \frac{\pi}{2} + \arg\Theta_1^{\varphi(t)}(1)\right]. \tag{4.246}$$

The first and second moments of the probability distribution density determined by Eq. (4.246) during the condition

$$\sigma_\varphi^2 > 1$$

have the following forms:

$$m_1[\widetilde{a}(t)] = \frac{2}{\pi} m_1[A(t)] \left|\Theta_1^{\varphi(t)}(1)\right| \sin\left[\omega_0 t + \Psi_a(t) + \frac{\pi}{2} + \arg \Theta_1^{\varphi(t)}(1)\right]; \quad (4.247)$$

$$m_2[\widetilde{a}(t)] = \frac{1}{2} m_2[A(t)] \left\{1 + \frac{4}{\pi} \left|\Theta_1^{\varphi(t)}(1)\right| \sin\left[\omega_0 t + \Psi_a(t) + \frac{\pi}{2} + \arg \Theta_1^{\varphi(t)}(1)\right]\right\}, \quad (4.248)$$

where $m_1[A(t)]$ is the mean of the amplitude envelope factor of the signal, and $m_2[A(t)]$ is the second initial moment of the amplitude envelope factor of the signal.

Table 4.2 shows the probability distribution densities of the signal determined by Eq. (4.206) when distortions in phase of the signal are distributed uniformly within the limits of the interval $[0, 2\pi]$, and in the case of high distortions in phase of the signal

$$\sigma_\varphi^2 \gg 1$$

for the probability distribution densities of the complex amplitude envelope factor of the signal $f[A(t)]$ of some types, which are often used in practice.

4.4.1.2 Amplitude and Phase Distortions of the Signal as Functionally Related

Consider the case in which there is a functional relationship between fluctuations of amplitude and phase of the signal

$$\varphi(t) = g[A(t)]. \quad (4.249)$$

The two-dimensional probability distribution density $f[A(t), \varphi(t)]$ determined by Eq. (4.211) can be written in the following form:

$$f[A(t), \varphi(t)] = f[A(t)]\delta\{\varphi(t) - g[A(t)]\}. \quad (4.250)$$

Substituting Eq. (4.250) in Eq. (4.211), after some transformations we can write

$$\begin{aligned} f[\widetilde{a}(t)] &= \frac{1}{\pi} \sum_{\ell=-\infty}^{\infty} \int_{|\widetilde{a}(t)|}^{\infty} \int_{-\infty}^{\infty} \frac{f[A(t)]\delta\{\varphi(t) - g[A(t)]\}}{\sqrt{A^2(t) - \widetilde{a}^2(t)}} \\ &\quad \times e^{j\ell\left[\arcsin \frac{\widetilde{a}(t)}{A(t)} + \psi_{\widetilde{a}}(t)\right]}\, d[A(t)] d[\varphi(t)] \\ &= \frac{1}{\pi} \int_0^{\infty} f[|\widetilde{a}(t)| \operatorname{ch} n]\, dn + \frac{2}{\pi} f_A[|\widetilde{a}(t)|] \\ &\quad \times \sum_{\ell=1}^{\infty} \frac{\sin\{\ell[\omega_0 t + \Psi_a(t) + \frac{\pi}{2} + g[|\widetilde{a}(t)|]]\}}{\ell}. \end{aligned} \quad (4.251)$$

TABLE 4.2

Probability Distribution Densities

$f[A(t)]$	$\delta[A(t)-A_0]$	$\frac{1}{A_0-B_0}$, $B_0 \leq A_0$	$\frac{A(t)}{\sigma_A} \cdot e^{-\frac{A^2(t)}{2\sigma_A^2}}$, $A(t) > 0$	$\frac{1}{\sigma_A} \cdot e^{-\frac{A(t)}{\sigma_A}}$, $A(t) > 0$	$\frac{1}{\pi\sqrt{A_0^2-A^2(t)}}$, $\lvert A(t)\rvert \leq A_0$
$\Theta_1^{A(t)}(x)$	e^{jxA_0}	$\frac{1}{jx}\frac{e^{jxA_0}-e^{jxB_0}}{A_0-B_0}$	$1+\sqrt{2\pi}\,jx\sigma_A \times\Phi(jx\sigma_A)\cdot e^{-\frac{\sigma_A^2x^2}{2}}$	$\frac{2}{1-jx\sigma_A}$	$J_0[xA(t)]$
$f_0(x)$	$\frac{1}{\pi\sqrt{A_0^2-x^2}}$	$\frac{1}{2\pi(A_0-B_0)} \times\ln\frac{A_0+\sqrt{A_0^2-x^2}}{A_0-\sqrt{A_0^2-x^2}}$	$\frac{1}{\sqrt{2\pi}\sigma_A}\cdot e^{-\frac{x^2}{2\sigma_A^2}}$	$\frac{1}{\pi\sigma_A}K_0\left(\frac{x}{\sigma_A}\right)$	$\frac{1}{A_0}F\left(\frac{\pi}{2}, Z\right)$, $Z=\sqrt{1-\frac{x^2}{A_0^2}}$
$\Gamma_0=\int_\beta^\infty \frac{f[A(t)]}{A(t)}d[A(t)]$	$\frac{1}{A_0}$	$\frac{1}{A_0-B_0}\ln\frac{A_0}{B_0}$	$\sqrt{2\pi}\left[1-\Phi\left(\frac{\beta}{\sigma_A}\right)\right]$, $\beta > 0$	$-\frac{1}{\sigma_A}Ei\left(-\frac{\beta}{\sigma_A}\right)$, $\beta > 0$	$\frac{1}{2\pi A_0} \times\ln\frac{A_0+\sqrt{A_0^2-\beta^2}}{A_0-\sqrt{A_0^2-\beta^2}}$, $0 < \beta < A_0$

Note: $\Phi(x)$ is the integral of error; $K_0(x)$ is the McDonald function of the zero order; $F(\varphi, k)$ is the elliptic integral of the first order; $Ei(x)$ is the integral exponential function.

Reference to Eq. (4.251) shows that if distortions in amplitude and phase of the signal are functionally related, then the probability distribution density of the signal can be defined using the statistical parameters of the complex amplitude envelope factor of the signal.

The initial moment of the k-th order of the probability distribution density, determined by Eq. (4.251), has the following form:

$$m_k[\widetilde{a}(t)] = \frac{1}{\pi} \int_{-\infty}^{\infty} \widetilde{a}^k(t) f[\widetilde{a}(t)] d[\widetilde{a}(t)] = m_{k_u}[\widetilde{a}(t)] + m_{k_f}[\widetilde{a}(t)], \tag{4.252}$$

where $m_{k_u}[\widetilde{a}(t)]$ is the component defined by the uniform probability distribution density of phase of the signal given by Eq. (4.245), and $m_{k_f}[\widetilde{a}(t)]$ is the component defined by the probability distribution density of phase of the signal, which differs from the uniform probability distribution density.

Using the Maclaurin series expansion for the function

$$g[\widetilde{a}(t)] = \sum_{k=0}^{\infty} \frac{\widetilde{a}^k(t)}{k!} \cdot g^{(k)}(0) \tag{4.253}$$

and limiting to the first and second terms of the series, we can write

$$m_{k_f}[\widetilde{a}(t)] = \frac{4}{\pi} \cdot Im \sum_{\ell=1}^{\infty} \frac{1}{\ell[jg'(0)]^k} \cdot \frac{d^k}{d\ell^k} \{\Theta_1^{A(t)}[\ell g'(0)]\} \cdot e^{j\ell[\omega_0 t + \Psi_a(t) + g(0)]}, \tag{4.254}$$

where

$$\Theta_1^{A(t)}[\ell g'(0)] \tag{4.255}$$

is the characteristic function of distortions in amplitude of the signal (see Table I.1, Appendix I).

Reference to Eq. (4.254) shows that the mean of the signal given by Eq. (4.206) depends essentially on a character of distortions in amplitude of the signal, and is defined by the component contingent on a derivative of function of distortions in amplitude of the signal, in addition to the component defined by the uniform probability distribution density of phase of the signal within the limits of the interval $[0, 2\pi]$.

4.4.2 Probability Distribution Density of the Signal in Additive Gaussian Noise under the Stimulus of Multiplicative Noise

In the majority of practical problems the signal is considered in the additive Gaussian noise $n(t)$. The probability distribution density of the additive Gaussian noise is determined in the following form:

$$f[n(t)] = \frac{1}{\sqrt{2\pi}\sigma_n} \cdot e^{-\frac{n^2(t)}{2\sigma_n^2}}, \tag{4.256}$$

where σ_n^2 is the variance of the additive Gaussian noise.

Taking into account the probability distribution density of instantaneous values of the signal given by Eq. (4.206) and discussed in Section 4.4.1, we define the probability distribution density of the signal in the additive Gaussian noise

$$a(t) = a_M(t) + n(t), \quad 0 \leq t \leq T, \tag{4.257}$$

where

$$a_M(t) = S(t)\tilde{a}(t). \tag{4.258}$$

As is well known, the probability distribution density of the sum of random variables is determined in the following form:

$$f[a(t)] = \int_{-\infty}^{\infty} f[a_M(t)] f[a(t) - a_M(t)] d[a_M(t)]. \tag{4.259}$$

Substituting Eq. (4.257) in terms of Eq. (4.206) and changing variables

$$s = \arcsin \frac{a_M(t)}{A(t)}, \tag{4.260}$$

we can write

$$f[a(t), A(t), \varphi(t)] = \frac{f[A(t), \varphi(t)]}{\pi\sqrt{2\pi}\,\sigma_n} \cdot e^{-\frac{2a^2(t)+A^2(t)S^2(t)}{4\sigma_n^2}} \sum_{\ell=-\infty}^{\infty} e^{j\ell\psi_{\tilde{a}}(t)}$$
$$\times \int_{-\frac{\pi}{2}}^{\frac{\pi}{2}} e^{-\left[\frac{A^2(t)S^2(t)}{4\sigma_n^2}\cos 2s + \frac{A(t)S(t)a(t)}{\sigma_n^2}\cos s - j\ell s\right]} ds. \tag{4.261}$$

Taking into account the series expansion[38]

$$e^{-a\cos x} = \sum_{n=-\infty}^{\infty} (-1)^n I_n(a) \cdot e^{jnx}, \tag{4.262}$$

where $I_n(a)$ is the modified Bessel function, Eq. (4.261) takes the following form:

$$f[a(t), A(t), \varphi(t)] = \frac{f[A(t), \varphi(t)]}{\pi\sqrt{2\pi}\,\sigma_n} \cdot e^{-\frac{2a^2(t)+A^2(t)S^2(t)}{4\sigma_n^2}}$$
$$\times \sum_{\ell=-\infty}^{\infty} \sum_{k=-\infty}^{\infty} I_{2k+\ell}\left[\frac{A(t)S(t)a(t)}{\sigma_n^2}\right] I_{2k}\left[\frac{A^2(t)S^2(t)}{4\sigma_n^2}\right] \cdot e^{j\ell\psi_{\tilde{a}}(t)}; \tag{4.263}$$

$$f[a(t)] = \frac{\pi}{\pi\sqrt{2\pi}\,\sigma_n} \cdot e^{-\frac{a^2(t)}{2\sigma_n^2}}$$
$$\times \sum_{\ell=-\infty}^{\infty} \sum_{k=-\infty}^{\infty} \Theta_1^{\varphi(t)}(\ell) F[a(t), k, \ell] \cdot e^{j\ell[\omega_0 t + \Psi_a(t)]}, \tag{4.264}$$

where

$$F[a(t), k, \ell] = \int_0^\infty I_{2k+\ell}\left[\frac{A(t)S(t)a(t)}{\sigma_n^2}\right] I_{2k}\left[\frac{A^2(t)S^2(t)}{4\sigma_n^2}\right] \cdot e^{-\frac{A^2(t)S^2(t)}{4\sigma_n^2}} f[A(t)]d[A(t)] \tag{4.265}$$

and

$$\psi_{\tilde{a}}(t) = \omega_0 t + \Psi_a(t) + \varphi(t). \tag{4.266}$$

We proceed to introduce the signal-to-noise ratio in the following form:

$$q^2 = \frac{A^2(t)S^2(t)}{2\sigma_n^2} \tag{4.267}$$

and consider Eq. (4.263) for two cases:

- The signal-to-noise ratio is very high—the powerful signals
$$q^2 \gg 1.$$
- The signal-to-noise ratio is very low—the weak signals
$$q^2 \ll 1.$$

4.4.2.1 Amplitude and Phase Distortions of the Signal as Independent

Consider the case in which distortions in amplitude and phase of the signal are independent under the stimulus of multiplicative noise

$$f[A(t), \varphi(t)] = f[A(t)]f[\varphi(t)]. \tag{4.268}$$

Assume that the signal-to-noise ratio given by Eq. (4.267) is equal to

$$q^2 \ll 1,$$

the case of the weak signals. Limiting in this case by terms with the number $k = 0$ and assuming

$$I_0(x) \cdot e^{-x} \simeq 1 \tag{4.269}$$

in Eq. (4.263), we can write

$$f[a(t), A(t), \varphi(t)] = \frac{f[A(t), \varphi(t)]}{\pi\sqrt{2\pi}\sigma_n} \cdot e^{-\frac{a^2(t)}{2\sigma_n^2}} \sum_{\ell=-\infty}^{\infty} I_\ell\left[\frac{A(t)S(t)a(t)}{\sigma_n^2}\right] \cdot e^{j\ell\psi_{\tilde{a}}(t)}. \tag{4.270}$$

If distortions in amplitude of the signal are absent under the stimulus of multiplicative noise

$$f[A(t)] = \delta[A(t) - 1] \tag{4.271}$$

and distortions in phase of the signal are distributed in an arbitrary way, then the probability distribution density of the signal during the condition

$$q^2 \ll 1$$

takes the following form:

$$f[a(t)] = \frac{1}{\pi\sqrt{2\pi}\sigma_n} \cdot e^{-\frac{a^2(t)}{2\sigma_n^2}} \sum_{\ell=-\infty}^{\infty} \Theta_1^{\varphi(t)}(\ell) I_\ell \left[\frac{S(t)a(t)}{2\sigma_n^2}\right] \cdot e^{j\ell[\omega_0 t + \Psi_a(t)]}. \quad (4.272)$$

When the following condition

$$\sigma_\varphi^2 \gg 1$$

is satisfied, and distortions in phase of the signal are distributed uniformly within the limits of the interval $[0, 2\pi]$, and during the condition

$$q^2 \ll 1$$

we can write

$$f[a(t)] \simeq \frac{1}{\pi\sqrt{2\pi}\sigma_n} \cdot e^{-\frac{a^2(t)}{2\sigma_n^2}} I_0\left[\frac{S(t)a(t)}{\sigma_n^2}\right]. \quad (4.273)$$

If the amplitude envelope factor $A(t)$ obeys the arbitrary probability distribution density, then the probability distribution density of the signal given by Eq. (4.257) during the condition

$$q^2 \ll 1$$

is determined in the following form:

$$f[a(t)] = \frac{1}{\pi\sqrt{2\pi}\sigma_n} \cdot e^{-\frac{a^2(t)}{2\sigma_n^2}} \sum_{\ell=-\infty}^{\infty} \Theta_1^{\varphi(t)}(\ell) F[a(t), \ell] \cdot e^{j\ell[\omega_0 t + \Psi_a(t)]}, \quad (4.274)$$

where

$$F[a(t), \ell] = \int_0^\infty I_\ell\left[\frac{A(t)S(t)a(t)}{\sigma_n^2}\right] f[A(t)]d[A(t)] \quad (4.275)$$

is the integral transform of the probability distribution density $f[A(t)]$ of the complex amplitude envelope factor of the signal.[38]

As a first approximation, for signal-to-noise ratio values

$$q^2 \ll 1$$

we can believe that

$$I_\ell\left[\frac{A(t)S(t)a(t)}{\sigma_n^2}\right] \simeq \frac{1}{\ell!}\left[\frac{A(t)S(t)a(t)}{2\sigma_n^2}\right]^\ell. \quad (4.276)$$

Taking into consideration that

$$I_\ell(x) = I_{-\ell}(x), \quad (4.277)$$

we can write

$$F[a(t), \ell] = \frac{1}{\ell!}\left[\frac{S(t)a(t)}{2\sigma_n^2}\right]^{\ell} m_\ell[A(t)], \tag{4.278}$$

where $m_\ell[A(t)]$ is the initial moment of the ℓ-th order.

Equation (4.274) in terms of Eq. (4.278) during the condition

$$q^2 \ll 1$$

can be written in the following form:

$$f[a(t)] = \frac{1}{\pi\sqrt{2\pi}\sigma_n}\cdot e^{-\frac{a^2(t)}{2\sigma_n^2}}\left\{1 + 2\sum_{\ell=1}^{\infty}\left[\frac{S(t)a(t)}{2\sigma_n^2}\right]^{\ell}\right.$$
$$\left.\times\frac{m_\ell[A(t)]}{\ell!}\cdot Re\left\{\Theta_1^{\varphi(t)}(\ell)e^{j\ell[\omega_0 t + \Psi_a(t)]}\right\}\right\}. \tag{4.279}$$

For conditions in which the signal-to-noise ratio is equal to

$$q^2 \gg 1,$$

the case of the powerful signals, taking into account that

$$I_{2k}\left[\frac{A^2(t)S^2(t)}{4\sigma_n^2}\right] \simeq \sqrt{\frac{2}{n}}\cdot\frac{\sigma_n}{A(t)S(t)}\cdot e^{\frac{A^2(t)S^2(t)}{4\sigma_n^2}} \tag{4.280}$$

and

$$\sum_{k=1}^{\infty} I_{2k}(x) = \frac{1}{2}\,\mathrm{ch}\,x; \tag{4.281}$$

$$\sum_{k=1}^{\infty} I_{2k+1}(x) = \frac{1}{2}\,\mathrm{sh}\,x, \tag{4.282}$$

the probability distribution density determined by Eq. (4.263) during the conditions

$$\sigma_\varphi^2 > 1$$

and

$$q^2 \gg 1$$

has the following form:

$$f[a(t), A(t), \varphi(t)] = \frac{1}{\pi^2 A(t)S(t)}\cdot e^{-\frac{a^2(t)}{2\sigma_n^2}} f[A(t), \varphi(t)]$$
$$\times\left\{\mathrm{ch}\frac{A(t)S(t)a(t)}{\sigma_n^2} + \mathrm{sh}\frac{A(t)S(t)a(t)}{\sigma_n^2}\cdot\cos\psi_{\tilde{a}}(t)\right\}. \tag{4.283}$$

Under the arbitrary probability distribution densities of the complex amplitude envelope factor $f[A(t)]$ and phase $f[\varphi(t)]$ of the signal and the condition

$$q^2 \gg 1$$

we can write

$$f[a(t)] = \frac{1}{\pi^2} \cdot e^{-\frac{a^2(t)}{2\sigma_n^2}} \{\Gamma_{ch}[a(t)S(t)] + \Gamma_{sh}[a(t)S(t)] Re\{\Theta_1^{\varphi(t)}(1) e^{j[\omega_0 t + \Psi_a(t)]}\}\}, \tag{4.284}$$

where

$$\Gamma_{ch}[a(t)S(t)] = \int_0^\infty \frac{f[A(t)]}{A(t)} \cdot \text{ch}\left[\frac{A(t)a(t)S(t)}{\sigma_n^2}\right] d[A(t)]; \tag{4.285}$$

$$\Gamma_{sh}[a(t)S(t)] = \int_0^\infty \frac{f[A(t)]}{A(t)} \cdot \text{sh}\left[\frac{A(t)a(t)S(t)}{\sigma_n^2}\right] d[A(t)]. \tag{4.286}$$

Determination of the arbitrary probability distribution density $f[A(t)]$ in Eqs. (4.285) and (4.286) is very difficult.

For this reason, using the series expansions

$$\text{ch}\, x = \sum_{k=1}^{\infty} \frac{x^{2k}}{(2k)!} \tag{4.287}$$

and

$$\text{sh}\, x = \sum_{k=1}^{\infty} \frac{x^{2k+1}}{(2k+1)!}, \tag{4.288}$$

we can define the functions

$$\Gamma_{ch}[a(t)S(t)]$$

and

$$\Gamma_{sh}[a(t)S(t)],$$

using the initial moments of the amplitude envelope factor of the signal determined by Eq. (4.206):

$$\Gamma_{ch}[a(t)S(t)] = \int_0^\infty \frac{f[A(t)]}{A(t)} d[A(t)] + \sum_{k=1}^{\infty} \frac{1}{(2k)!} \left[\frac{a(t)S(t)}{\sigma_n^2}\right]^{2k} m_{2k-1}[A(t)]; \tag{4.289}$$

$$\Gamma_{sh}[a(t)S(t)] = \frac{a(t)S(t)}{\sigma_n^2} + \sum_{k=1}^{\infty} \frac{1}{(2k+1)!} \left[\frac{a(t)S(t)}{\sigma_n^2}\right]^{2k+1} m_{2k}[A(t)], \tag{4.290}$$

where $m_{2k-1}[A(t)]$ is the initial moment of the $(2k-1)$-th order of the probability distribution density $f[A(t)]$ and $m_{2k}[A(t)]$ is the initial moment of the $2k$-th order of the probability distribution density $f[A(t)]$.

Table 4.2 shows the values of the coefficient

$$\Gamma_0 = \int_0^\infty \frac{f[A(t)]}{A(t)} d[A(t)] \tag{4.291}$$

for some forms of the probability distribution density of the amplitude envelope factor $f[A(t)]$.

Substituting Eq. (4.290) in Eq. (4.284), we define the series expansion of the probability distribution density of the signal determined by Eq. (4.257) with respect to the probability distribution density of the Gaussian stochastic process, with the weight coefficients given by Eqs. (4.289) and (4.290) and defined by the initial moments of the amplitude envelope factor of the signal.

When the following condition

$$\sigma_\varphi^2 \gg 1$$

is satisfied, see Fig. 4.7, and distortions in phase of the signal obey the uniform probability distribution density within the limits of the interval $[0, 2\pi]$, Eq. (4.284) can be written in the following form:

$$f[a(t)] = \frac{\Gamma_{ch}[a(t)]}{\pi^2} \cdot e^{-\frac{a^2(t)}{2\sigma_n^2}}. \tag{4.292}$$

4.4.2.2 Amplitude and Phase Distortions of the Signal as Functionally Related

We define the probability distribution density of the signal determined by Eq. (4.257) when distortions in amplitude and phase of the signal $a_M(t)$ are functionally related, i.e., the conditions in Eqs. (4.249) and (4.250) are satisfied. If the signal-to-noise ratio satisfies the condition

$$q^2 \ll 1$$

Equation (4.257), after integration with respect to the complex amplitude envelope factor $A(t)$ and phase $\varphi(t)$ of the signal, takes the following form:

$$f[a(t)] = \frac{1}{\pi\sqrt{2\pi}\sigma_n} \cdot e^{-\frac{a^2(t)}{\sigma_n^2}} \sum_{\ell=-\infty}^{\infty} e^{j\ell[\overline{\psi_a(t)}+g(0)]}$$
$$\times \int_0^\infty I_\ell\left[\frac{A(t)S(t)a(t)}{\sigma_n^2}\right] \cdot e^{j\ell A(t)g'(0)} f[A(t)]\, d[A(t)]. \tag{4.293}$$

Using the power series expansion for the Bessel function $I_\ell(x)$ and limiting during the condition

$$q^2 \ll 1$$

by the term

$$I_\ell\left[\frac{A(t)S(t)a(t)}{\sigma_n^2}\right] \simeq \frac{1}{\ell!}\left[\frac{A(t)S(t)a(t)}{\sigma_n^2}\right]^\ell \tag{4.294}$$

and taking into account that

$$A(t) \cdot e^{j\ell A(t)g'(0)} = \left\{ \frac{d^n}{[jg'(0)]^n d\ell^n} \cdot e^{j\ell A(t)g'(0)} \right\}_{n=\ell}, \tag{4.295}$$

under the condition mentioned above, we can finally write

$$f[a(t)] = \frac{1}{\pi\sqrt{2\pi}\sigma_n} \cdot e^{-\frac{a^2(t)}{2\sigma_n^2}} \sum_{\ell=-\infty}^{\infty} \frac{1}{[jg'(0)]^\ell} \times \left[\frac{S(t)a(t)}{\sigma_n^2}\right]^\ell \left\{ \frac{d^n}{d\ell^n} \Theta_1^{A(t)}[\ell g'(0)] \right\}_{n=\ell} \cdot e^{j\ell\overline{\psi_{\tilde{a}}(t)}}, \tag{4.296}$$

where

$$\overline{\psi_{\tilde{a}}(t)} = \omega_0 t + \Psi_a(t). \tag{4.297}$$

When the conditions

$$q^2 \gg 1$$

and

$$\sigma_\varphi^2 > 1$$

are jointly satisfied, and

$$f[A(t), \varphi(t)] = f[A(t)]\delta[\varphi(t) - g[A(t)]] \tag{4.298}$$

Equation (4.284) has the following form:

$$f[a(t)] = \frac{1}{\pi^2} \cdot e^{-\frac{a^2(t)}{2\sigma_n^2}} \left\{ \Gamma_{ch}[a(t)S(t)] + Re\left\{ e^{j\overline{\psi_{\tilde{a}}(t)}} \int_0^\infty \mathrm{sh}\frac{A(t)a(t)S(t)}{\sigma_n^2} f[A(t)] \cdot e^{jg(A)} d[A(t)] \right\} \right\}. \tag{4.299}$$

When the condition

$$q^2 \ll 1$$

is satisfied we can write

$$f[a(t)] = \frac{1}{\pi^2} \cdot e^{-\frac{a^2(t)}{2\sigma_n^2}} \left\{ \Gamma_{ch}[a(t)S(t)] + \sum_{k=1}^{\infty} C_k[a(t)] \cos[\omega_0 t + \Psi_a(t) + g(0)] \right\}, \tag{4.300}$$

where

$$C_k[a(t)] = \frac{1}{[jg'(0)]^{2k+1}} \left[\frac{a(t)S(t)}{2\sigma_n^2}\right]^{2k+1} \frac{1}{(2k+1)!} \left\{\frac{d^{2k+1}}{d\ell^{2k+1}} \Theta_1^{A(t)}[\ell g'(0)]\right\}_{\ell=1}. \tag{4.301}$$

Contrary to Eq. (4.284), during distortions in amplitude and phase of the signal, which are functionally related, the weight factors in Eq. (4.300), under the use of series expansion of the probability distribution density of the signal $a(t)$ determined by Eq. (4.206) according to the Gaussian distribution law, are defined by the derivative with respect to the characteristic function of fluctuations of the amplitude envelope factor $A(t)$. The non-stationary term in Eq. (4.284) differs from the non-stationary term in Eq. (4.300) only by the initial phase defined by the function given by Eq. (4.249) at the zero point.

Thus, using the probability distribution density of the signal determined by Eq. (4.206) and distorted by the multiplicative noise and the signal and the additive Gaussian noise determined by Eq. (4.257), we can see that if the probability distribution density of phase of the signal differs from the uniform probability distribution density of phase of the signal within the limits of the interval $[0, 2\pi]$, the signal is the non-stationary stochastic process that approaches the stationary stochastic function with an increase in the degree of phase distortions

$$\sigma_\varphi^2 \gg 1$$

of the signal.

4.5 Multivariate Probability Distribution Density of Instantaneous Values of the Signal under the Stimulus of Fluctuating Multiplicative Noise

The signal under the stimulus of multiplicative noise takes the following form:

$$a_M(t) = A(t)S_0(t)\cos[\omega_0 t + \Psi_a(t) + \varphi(t)], \tag{4.302}$$

where $A(t)$ is the stochastic amplitude envelope factor of the signal; $\varphi(t)$ is the random distortions in phase of the signal; $S_0(t)$ is the known modulation law of amplitude of the signal; and $\Psi_a(t)$ is the known modulation law of phase of the signal.

For the sake of convenience, we use the following form of the signal:

$$\hat{a}(t) = \frac{a_M(t)}{S_0(t)} = A(t)\cos\psi_{\hat{a}}(t), \tag{4.303}$$

where

$$\psi_{\hat{a}}(t) = \omega_0 t + \Psi_a(t) + \varphi(t) \tag{4.304}$$

is the random instantaneous phase of the signal with the mean

$$\overline{\psi_{\hat{a}}(t)} = \omega_0 t + \Psi_a(t). \tag{4.305}$$

Consider the case in which there are distortions only in phase of the signal

$$A(t) = 1.$$

As is well known,[43] the N-dimensional probability distribution density of the signal determined by Eq. (4.303) can be written in the following form:

$$f_0[\widehat{a}_1(t), \widehat{a}_2(t), \ldots, \widehat{a}_N(t)] = \underbrace{\int_{-\infty}^{\infty} \cdots \int_{-\infty}^{\infty}}_{N} f\left[\psi_{\hat{a}_1}(t), \psi_{\hat{a}_2}(t), \ldots, \psi_{\hat{a}_N}(t)\right]$$
$$\times \delta\left[\widehat{a}_1(t) - \cos\psi_{\hat{a}_1}(t)\right] \delta\left[\widehat{a}_2(t) - \cos\psi_{\hat{a}_2}(t)\right] \ldots \delta\left[\widehat{a}_N(t) - \cos\psi_{\hat{a}_N}(t)\right]$$
$$\times d\left[\psi_{\hat{a}_1}(t)\right] d\left[\psi_{\hat{a}_2}(t)\right] \ldots d\left[\psi_{\hat{a}_N}(t)\right], \tag{4.306}$$

where the transformation operator

$$\delta\left[\widehat{a}_N(t) - \cos\psi_{\hat{a}_N}(t)\right] \tag{4.307}$$

owing to a multi-formity of the function $\cos\psi_{\hat{a}_N}(t)$ defines the values

$$\psi_{\hat{a}_{N_k}}(t) = \left\{\psi_{\hat{a}_{N_1}}(t), \psi_{\hat{a}_{N_2}}(t), \ldots, \psi_{\hat{a}_{N_k}}(t)\right\} \tag{4.308}$$

with respect to each value $\widehat{a}_N(t)$. Moreover, the values $\psi_{\hat{a}_{N_k}}(t)$ differ from each other by the period $b = 2\pi$.

Taking into account the property of the delta function (see Appendix I), the transformation operator determined by Eq. (4.307) can be written in the following form:

$$\delta\left[\widehat{a}_N(t) - \cos\psi_{\hat{a}_N}(t)\right] = \sum_{k=-\infty}^{\infty} \left|\sin\psi_{\hat{a}_N}(t)\right|^{-1}_{\psi_{\hat{a}_N}(t)=\psi_{\hat{a}_{N_k}}(t)} \delta\left[\psi_{\hat{a}_N}(t) - \psi_{\hat{a}_{N_k}}(t)\right], \tag{4.309}$$

where

$$\psi_{\hat{a}_{N_k}}(t) = \left\{\arccos\widehat{a}_{N_0}(t), \arccos\widehat{a}_{N_0}(t) + 2\pi, \ldots, \arccos\widehat{a}_{N_0}(t) + 2k\pi\right\} \tag{4.310}$$

are the zeros of the function

$$r\left[\psi_{\hat{a}_N}(t)\right] = \widehat{a}_N(t) - \cos\psi_{\hat{a}_N}(t). \tag{4.311}$$

By virtue of periodicity of the function $\cos \psi_{\hat{a}_N}(t)$ and its derivative, Eq. (4.309) can be written in the following form:

$$\delta\left[\widehat{a}_N(t) - \cos \psi_{\hat{a}_N}(t)\right] = \left|\sin\left[\arccos \widehat{a}_{N_0}(t)\right]\right|^{-1} \times \sum_{k=-\infty}^{\infty} \delta\left[\psi_{\hat{a}_N}(t) - \arccos \widehat{a}_{N_0}(t) + 2k\pi\right]. \quad (4.312)$$

Substituting Eq. (4.312) in Eq. (4.306) and taking into account that the probability distribution density

$$f\left[\psi_{\hat{a}_1}(t), \psi_{\hat{a}_2}(t), \ldots, \psi_{\hat{a}_N}(t)\right] \quad (4.313)$$

and the characteristic function

$$\Theta_N^{\psi_{\hat{a}}(t)}(\vartheta_1, \vartheta_2, \ldots, \vartheta_N) \quad (4.314)$$

are the pair of functions related by the multivariate Fourier transform, we can write

$$f_0[\widehat{a}_1(t), \widehat{a}_2(t), \ldots, \widehat{a}_N(t)] = \sum_{k_1=-\infty}^{\infty} \cdots \sum_{k_N=-\infty}^{\infty} \underbrace{\int_{-\infty}^{\infty} \cdots \int_{-\infty}^{\infty}}_{N} \Theta_N^{\psi_{\hat{a}}(t)}(\vartheta_1, \vartheta_2, \ldots, \vartheta_N)$$

$$\times \left\{\prod_{r=1}^{N} \frac{1}{|2\pi \sin[\arccos \widehat{a}_r(t)]|} \cdot e^{j\vartheta_r[\arccos \hat{a}_{N_0}(t) + 2k_r\pi]}\right\} d\vartheta_1\, d\vartheta_2 \ldots d\vartheta_N. \quad (4.315)$$

Taking into consideration the equalities

$$\sum_{k=-\infty}^{\infty} e^{j\vartheta_k k 2\pi} = \sum_{\ell=-\infty}^{\infty} \delta(\vartheta - \ell) \quad (4.316)$$

and

$$|\sin[\arccos \widehat{a}_r(t)]|^{-1} = \frac{1}{\sqrt{1 - \widehat{a}_r^2(t)}}, \qquad |\widehat{a}_r(t)| \leq 1, \quad (4.317)$$

the N-dimensional probability distribution density of the signal if there are distortions only in phase of the signal during the condition

$$|\widehat{a}_r(t)| \leq 1 \quad (4.318)$$

has the following form after simple transformations:

$$f_0[\widehat{a}_1(t), \widehat{a}_2(t), \ldots, \widehat{a}_N(t)] = \sum_{\ell_1=-\infty}^{\infty} \cdots \sum_{\ell_N=-\infty}^{\infty} \left\{\Theta_N^{\psi_{\hat{a}}(t)}(\ell_1, \ell_2, \ldots, \ell_N)\right.$$

$$\left. \times \prod_{r=1}^{N} \frac{1}{2\pi\sqrt{1 - \widehat{a}_r^2(t)}} \cdot \{T_r[\widehat{a}_r(t)] + jU_r[\widehat{a}_r(t)]\} \cdot e^{j\ell_r[\omega_0 t + \Psi_a(t)]}\right\}, \quad (4.319)$$

where

$$T_r[\widehat{a}_r(t)] = \cos\{r \arccos [\widehat{a}_r(t)]\} \tag{4.320}$$

and

$$U_r[\widehat{a}_r(t)] = \sin\{r \arccos [\widehat{a}_r(t)]\} \tag{4.321}$$

are the Chebyshev polynomials.[38]

Consider the case in which there are distortions in amplitude and phase of the signal. Assume that distortions in amplitude and phase of the signal are the independent stochastic processes. In this case, the probability distribution density of product between the independent stochastic functions given by Eq. (4.303) is determined in the following form:

$$f[\widehat{a}_1(t), \widehat{a}_2(t), \ldots, \widehat{a}_N(t)] = \underbrace{\int_0^\infty \ldots \int_0^\infty}_{N} A_1(t) A_2(t) \ldots A_N(t)$$
$$\times f_A[A_1(t), A_2(t), \ldots A_N(t)] f_0\left[\frac{\widehat{a}_1(t)}{A_1(t)}, \frac{\widehat{a}_2(t)}{A_2(t)}, \ldots, \frac{\widehat{a}_N(t)}{A_N(t)}\right]$$
$$\times d[A_1(t)] d[A_2(T)] \ldots d[A_N(t)]. \tag{4.322}$$

Taking into account Eq. (4.319), the N-dimensional probability distribution density of the signal, if there are distortions only in amplitude and phase of the signal, can be written in the following form after simple transformations:

$$f[\widehat{a}_1, \widehat{a}_2(t), \ldots, \widehat{a}_N(t)] = \sum_{\ell_1=-\infty}^{\infty} \cdots \sum_{\ell_N=-\infty}^{\infty} \left\{ \Theta_N^{\psi_a(t)}(\ell_1, \ell_2, \ldots, \ell_N) \right.$$
$$\times \underbrace{\int_0^\infty \ldots \int_0^\infty}_{N} f_A\{\widehat{a}_1(t)\,\mathrm{ch}[\widehat{a}_1(t)], \widehat{a}_2(t)\,\mathrm{ch}[\widehat{a}_2(t)], \ldots, \widehat{a}_N(t)\,\mathrm{ch}[\widehat{a}_N(t)]\}$$
$$\left. \times \prod_{r=1}^{N} e^{j\ell_r\{\gamma[\widehat{a}_r(t)] - \omega_0 t - \Psi_a(t)\}} \right\} d\,[\widehat{a}_1(t)]\, d\,[\widehat{a}_2(t)] \ldots d[\widehat{a}_N(t)], \tag{4.323}$$

where

$$f_A\{\widehat{a}_1(t)\,\mathrm{ch}[\widehat{a}_1(t)], \ldots, \widehat{a}_N(t)\,\mathrm{ch}[\widehat{a}_N(t)]\} \tag{4.324}$$

is the N-dimensional probability distribution density of distortions in amplitude of the signal, and

$$\gamma[\widehat{a}_r(t)] = \arccos \frac{1}{\mathrm{ch}[\widehat{a}_r(t)]} \tag{4.325}$$

is the hyperbolic amplitude of the signal.

As an example, consider the two-dimensional probability distribution density of the signal determined by Eq. (4.303) during the condition

$$A(t) = 1 \tag{4.326}$$

that means an absence of distortions in amplitude of the signal, and distortions in phase of the signal either obey the Gaussian probability distribution density with a zero mean and the variance σ_φ^2 and the coefficient of correlation $r_\varphi(\tau)$ or are the quasideterministic process

$$\varphi(t) = \varphi_m(t)\sin[\Omega_M t + \varphi_0], \tag{4.327}$$

where φ_0 is the random initial phase of the signal, which is distributed uniformly within the limits of the interval $[0, 2\pi]$.

Taking into account the well-known relationships[27] and Eq. (4.319), we can write

$$\begin{aligned} f_0[a_{M_1}(t), a_{M_2}(t)] &= \frac{1}{\pi^2\sqrt{S_0^2(t) - a_{M_1}^2(t)}\cdot\sqrt{S_0^2(t) - a_{M_2}^2(t)}} \\ &\times \left\{1 + \sum_{\ell_1=-\infty}^{\infty}\sum_{\ell_2=-\infty}^{\infty} e^{-\frac{\sigma_\varphi^2}{2}[\ell_1^2+\ell_2^2+2r_\varphi(\tau)\ell_1\ell_2]}\right. \\ &\cdot\left\{T_{\ell_1}\left[\frac{a_{M_1}(t)}{S_0(t)}\right] + jU_{\ell_1}\left[\frac{a_{M_1}(t)}{S_0(t)}\right]\right\} \\ &\times \left.\left\{T_{\ell_2}\left[\frac{a_{M_2}(t)}{S_0(t)}\right] + jU_{\ell_2}\left[\frac{a_{M_2}(t)}{S_0(t)}\right]\right\}\cdot e^{-j\psi_{eq}(\ell_1,\ell_2)}\right\}; \end{aligned} \tag{4.328}$$

$$\begin{aligned} f_0[a_{M_1}(t), a_{M_2}(t)] &= \frac{1}{\pi^2\sqrt{S_0^2(t) - a_{M_1}^2(t)}\cdot\sqrt{S_0^2(t) - a_{M_2}^2(t)}} \\ &\times \left\{1 + \sum_{\ell_1=-\infty}^{\infty}\sum_{\ell_2=-\infty}^{\infty} J_0\left[\varphi_m(t)\sqrt{\ell_1^2 + \ell_2^2 + 2\ell_1\ell_2\cos\Omega_M\tau}\right]\right. \\ &\times \left\{T_{\ell_1}\left[\frac{a_{M_1}(t)}{S_0(t)}\right] + jU_{\ell_1}\left[\frac{a_{M_1}(t)}{S_0(t)}\right]\right\}\left\{T_{\ell_2}\left[\frac{a_{M_2}(t)}{S_0(t)}\right]\right. \\ &\left.\left. + jU_{\ell_2}\left[\frac{a_{M_2}(t)}{S_0(t)}\right]\right\}\cdot e^{-j\psi_{eq}(\ell_1,\ell_2)}\right\}, \end{aligned} \tag{4.329}$$

where

$$\psi_{eq}(\ell_1, \ell_2) = \omega_0(\ell_1 t_1 - \ell_2 t_2) + \Psi_a(t_1)\ell_1 - \Psi_a(t_2)\ell_2; \tag{4.330}$$

$$\tau = t_2 - t_1; \tag{4.331}$$

$$|a_M(t)| \leq S_0(t); \tag{4.332}$$

and $J_0(x)$ is the Bessel function of zero order.

Reference to Eqs. (4.328) and (4.329) shows that with an increase in the variance of distortions in phase of the signal, i.e., when the condition

$$\sigma_\varphi^2 \gg 1$$

is satisfied, or with an increase in the degree of distortions in phase of the signal

$$\lim_{m\to\infty} \varphi_m(t) \to \infty \tag{4.333}$$

under the stimulus of quasideterministic multiplicative noise, the two-dimensional probability distribution density tends to approach the probability distribution density defined by the product of the one-dimensional probability distribution densities of the signal, distortions in phase of the signal, which are distributed uniformly within the limits of the interval $[0, 2\pi]$:

$$f_0[a_{M_1}(t), a_{M_2}(t)] = \frac{1}{\pi^2\sqrt{S_0^2(t) - a_{M_1}^2(t)}\cdot\sqrt{S_0^2(t) - a_{M_2}^2(t)}}; \tag{4.334}$$

$$|a_{M_1}(t)| \leq S_0(t); \tag{4.335}$$

and

$$|a_{M_2}(t)| \leq S_0(t). \tag{4.336}$$

When the condition

$$\sigma_\varphi^2[1 - r_\varphi(t_2 - t_2)] > 3 \tag{4.337}$$

defining an uncorrelated relationship between the signals $A_{M_1}(t)$ and $A_{M_2}(t)$ is satisfied, the two-dimensional probability distribution density in Eq. (4.328) can be considered a product of the one-dimensional probability distribution densities determined by Eq. (4.334).

In practice, it is very useful to use the probability distribution density

$$f_0[a_{M_1}(t), a_{M_2}(t), \ldots, a_{M_N}(t)] \tag{4.338}$$

in Eqs. (4.319) and (4.323) for the case of independent samples of the signal when

$$f_0[a_{M_1}(t), a_{M_2}(t), \ldots, a_{M_N}(t)] = \prod_{r=1}^{N} f[a_{M_r}(t)]. \tag{4.339}$$

The characteristic function of the random variable

$$\varphi_1(t), \varphi_2(t), \ldots, \varphi_N(t) \tag{4.340}$$

obeying the N-dimensional Gaussian probability distribution density is determined by[44]

$$\Theta_N^{\varphi(t)}(\ell_1, \ell_2, \ldots, \ell_N) = \exp\left\{j \sum_{k=1}^{N} \overline{\psi_{\hat{a}}(t)\ell_k}\right\} \cdot \exp\left\{-\frac{1}{2}\sigma_\varphi^2 \sum_{k=1}^{N}\sum_{m=1}^{N} r_\varphi(t_k - t_m)\ell_k\ell_m\right\}. \tag{4.341}$$

Using the Taylor series expansion for the function of correlation between distortions in phase of the signal and taking into consideration that

$$\sqrt{-r''_\varphi(0)} = \Delta G_\varphi, \tag{4.342}$$

where ΔG_φ is the equivalent spectrum bandwidth of distortions in phase of the signal, Eq. (4.337) can be written in the following form

$$\Delta = t_2 - t_1 \geq \frac{1.7}{\sigma_\varphi \Delta G_\varphi}. \tag{4.343}$$

Equation (4.343) defines the condition wherein the readings t_k and t_m, of instantaneous values of the signal given by Eq. (4.303), can be thought of as independent and the probability distribution densities determined by Eqs. (4.319) and (4.323) can be written in the form as in Eq. (4.339).

It is well known that two uncorrelated Gaussian random variables are independent of each other or, in other words, the time of correlation of the Gaussian process coincides with the interval of independence.

Reference to Eqs. (4.328), (4.329), (4.339), and (4.343) shows that if readings of the signal distorted by the multiplicative noise are uncorrelated we can believe that these readings are independent.

4.6 Conclusions

Analysis of statistical characteristics of the signal distorted by multiplicative noise shows that under the deterministic, quasideterministic, and stationary stochastic multiplicative noise this signal can be considered as a sum of two components—the undistorted component of the signal and the additive noise component of the signal caused by the stimulus of multiplicative noise.

This form of notion of impact of the multiplicative noise on statistical characteristics of the signal offers the following advantages.

- Stimulus of the multiplicative noise on the signal is considered as a decrease in the amplitude of the signal and an appearance of the equivalent additive noise component of the signal. This consideration allows us to study an impact of the multiplicative noise on the main qualitative characteristics of signal processing systems in various areas of application—radar, communications, wireless communications, telecommunications, mobile communications, underwater signal processing, acoustics, sonar, remote sensing, and so on—based on the well-known methods of estimation of these characteristics in the additive noise. This fact allows us to obtain some formulae that are very useful in practice.

- The notion of the signal distorted by the multiplicative noise enables us to analyze the main qualitative characteristics of signal processing systems if the additive noise and the multiplicative noise act simultaneously.
- This notion of the signal distorted by the multiplicative noise helps us estimate the stimulus of the multiplicative noise on operation of signal processing systems.

Statistical characteristics of the undistorted component and the noise component of the signal $a_M(t)$ distorted by the multiplicative noise are defined using the statistical characteristics of the undistorted signal $a(t)$, the noise modulation function $\dot{M}(t)$ of the multiplicative noise, and the harmonic oscillation distorted by the same multiplicative noise as the signal.

In comparison to complex signals, statistical characteristics of the harmonic oscillation are more easily studied both by a theory and by an experiment. In the process, there is no need to define the statistical characteristics for each form of the signals, since by using statistical characteristics of the harmonic oscillation we are able to define them for any form of the signals.

References

1. Parsons, J., *The Mobile Radio Propagation Channel*, Wiley, New York, 1996.
2. Rappaport, T., *Wireless Communications Principles and Practice*, Prentice-Hall, Englewood Cliffs, NJ, 1996.
3. Brillinger, D., *Time Series: Data Analysis and Theory*, Holden Day, San Francisco, 1981.
4. Stark, H. and Woods, J., *Probability Random Processes and Estimation Theory for Engineers*, Prentice-Hall, Englewood Cliffs, NJ, 1986.
5. Proakis, J., *Digital Communications*, 3rd ed., McGraw-Hill, New York, 1995.
6. Stoica, P. and Moses, R., *Introduction to Spectral Analysis*, Prentice-Hall, Englewood Cliffs, NJ, 1997.
7. Porat, B., *Digital Processing of Random Signals: Theory and Methods*, Prentice-Hall, Englewood Cliffs, NJ, 1994.
8. Cover, T. and Thomas, J., *Elements of Information Theory*, Wiley, New York, 1991.

9. Kay, S., *Fundamentals of Statistical Signal Processing, Estimation Theory*, Prentice-Hall, Englewood Cliffs, NJ, 1993.
10. Poor, H., *An Introduction to Signal Detection and Estimation*, Springer-Verlag, New York, 1988.
11. Anderson, T., *An Introduction to Multivariate Statistical Analysis*, Wiley, New York, 1984.
12. Middleton, D., Statistical-physical models of electromagnetic interference, *IEEE Trans.*, Vol. EC-19, No. 2, 1977, pp. 106–127.
13. Carter, G., *Coherence and Time Delay Estimation*, IEEE Press, New York, 1993.
14. Kiefer, J., *Introduction to Statistical Inference*, Springer-Verlag, New York, 1987.
15. Haykin, S., *Adaptive Filter Theory*, 3rd ed., Prentice-Hall, Englewood Cliffs, NJ, 1996.
16. Shanmugan, K. and Breipohl, A., *Random Signals: Detection, Estimation and Data Analysis*, Wiley, New York, 1988.
17. Ghogho, M., Detection and estimation of signals in multiplicative and additive noise, Ph.D. dissertation, National Polytechnic Institute Toulouse, Toulouse, France, 1997.
18. Swami, A., Cramér-Rao bounds for deterministic signals in additive and multiplicative noise, *Signal Process.*, Vol. 53, August 1996, pp. 231–244.
19. Berger, O., *Statistical Decision Theory and Bayesian Analysis*, 2nd ed., Springer-Verlag, New York, 1985.
20. Gelman, A., Carlin, J., Stern, H., and Rubin, D., *Bayesian Data Analysis*, Chapman & Hall, London, 1995.
21. Wang, X. and Chen, R., Adaptive Bayesian multi-user detection for synchronous CDMA with Gaussian and impulsive noise, *IEEE Trans.*, Vol. SP-48, No. 7, 2000, pp. 2013–2028.
22. Sergievskiy, B. and Oganesyantz, L., The power on carrier frequency of oscillations with random initial phase at various probability distribution density of phase, *Radio Eng. Electron. Phys.*, Vol. 11, No. 6, 1966, pp. 72–78.
23. Oppenheim, A. and Schafer, R., *Digital Signal Processing*, Prentice-Hall, Englewood Cliffs, NJ, 1975.
24. Helstrom, C., *Statistical Theory of Signal Detection*, 2nd ed., Pergamon Press, Oxford, U.K., 1968.
25. Helstrom, C., *Elements of Signal Detection and Estimation*, Prentice-Hall, Englewood Cliffs, NJ, 1995.
26. Wicker, S., *Error Control Systems for Digital Communication and Storage*, Prentice-Hall, Upper Saddle River, NJ, 1995.
27. Levin, B., *Theoretical Foundations of Statistical Radio Engineering*, Parts I–III, Soviet Radio, Moscow, 1974 (in Russian).
28. Gallant, R., *Non-Linear Statistical Models*, Wiley, New York, 1987.
29. McDonald, K. and Blum, R., A statistical and physical mechanisms–based interference and noise model for array observations, *IEEE Trans.*, Vol. SP-48, No. 7, 2000, pp. 2044–2056.
30. Blackard, K., Rappaport, T., and Bostian, C., Measurements and models of radio frequency impulsive noise for indoor wireless communications, *IEEE J. Select. Areas Commun.*, Vol. 11, Sept. 1993, pp. 991–1001.
31. Lifshiz, N. and Pugachev, V., *Probability Analysis of Automatic Control Systems*, Parts 1 and 2, Soviet Radio, Moscow, 1963 (in Russian).

32. Middleton, D. and Spaulding, A., Elements of weak signal detection in non-Gaussian noise environments, in *Advances in Statistical Signal Processing, Vol. 2—Signal Detection*, Poor, H. and Thomas, J., Eds., JAI Press, Greenwich, CT, 1993.
33. Verdu, S., *Multi-User Detection*, Cambridge University Press, Cambridge, U.K., 1998.
34. Farina, A., *Antenna–Based Signal Processing Techniques for Radar Systems*, Artech House, Norwood, MA, 1992.
35. Shifrin, Ya., *Problems of Statistical Theory of Antennas*, Soviet Radio, Moscow, 1970 (in Russian).
36. Conte, E., Lops, M., and Ricci, G., Asymptotically optimum radar detection in compound–Gaussian clutter, *IEEE Trans.*, Vol. AES-31, No. 2, 1995, pp. 617–625.
37. Farina, A. and Gini, F., Interference blanking probabilities for SLB in correlated Gaussian clutter plus noise, *IEEE Trans.*, Vol. SP-48, No. 5, 2000, pp. 1481–1485.
38. Gradsteyn, I. and Ryzhik, I., *Table of Integrals, Series, and Products*, 5th ed., Academic Press, New York, 1994.
39. Kassam, S., *Signal Detection in Non-Gaussian Noise*, Springer-Verlag, New York, 1989.
40. Wang, X. and Poor, V., Robust multi-user detection in non-Gaussian channels, *IEEE Trans.*, Vol. SP-47, No. 2, 1999, pp. 289–305.
41. Ventzel, E. and Ovcharov, L., *Probability Theory*, Nauka, Moscow, 1973 (in Russian).
42. Rytov, S., *Introduction to Statistical Radio Physics*, Nauka, Moscow, 1966 (in Russian).
43. Middleton, D., *An Introduction to Statistical Communication Theory*, McGraw-Hill, New York, 1960.
44. Tichonov, V., *Statistical Radio Engineering*, Radio and Svyaz, Moscow, 1982 (in Russian).

5

Main Theoretical Principles of Generalized Approach to Signal Processing under the Stimulus of Multiplicative Noise

The generalized approach to signal processing in the presence of noise is based on a seemingly abstract idea: the introduction of an additional noise source, which does not carry any information about the signal, for the purpose of improving the detection performances and noise immunity of complex signal processing systems. The proposed generalized approach to signal processing in the presence of noise allows us to formulate decision-making rules based on the determination of the jointly sufficient statistics of the mean and variance of the likelihood function (or functional).

The classical and modern signal detection theories allow us to define only the sufficient statistic of the mean of the likelihood function (or functional). The presence of additional information about the statistical parameters of the likelihood function (or functional) leads to better qualitative characteristics of signal detection in comparison to the optimal signal detection algorithms of classical and modern theories.

The basic concepts of classical and modern signal detection theories[1–50] are briefly reviewed to help the reader understand the generalized approach to signal processing in the presence of noise.

5.1 Basic Concepts

The simplest signal detection problem is the problem of binary detection in the presence of additive Gaussian noise with a zero mean and the spectral power density $\frac{N_0}{2}$. The optimal detector can be realized as the matched filter or the correlation receiver. Detection quality depends on the normalized distance between two signal points of the decision-making space. This distance is characterized by signal energies, the coefficient of correlation between the signals, and the spectral power density of the additive Gaussian noise. If the signal energies are equal, then the optimal coefficient of correlation is equal to −1.

In addition, the signal waveform is of no consequence. In spite of the fact that the classical signal detection theory is very orderly and smooth, it cannot provide the most complete answer to the questions posed below. Let us consider briefly the results discussed in References 1–50.

The hypothesis H_0 must be chosen so that the input stochastic process is normal and has a zero mean vs. the alternative H_1, which is also normal, but has a mean that varies according to the known law $a(t)$. In a statistical context, this problem is solved as follows.

Let $X(t)$ be the input stochastic process, which is observed within the limits of the time interval $[0, T]$; $a(t)$, the signal; and $\xi(t)$, the additive Gaussian noise with a zero mean and the known variance σ_n^2:

$$X(t) = \begin{cases} a(t) + \xi(t) \Rightarrow H_1; \\ \xi(t) \qquad\quad \Rightarrow H_0. \end{cases} \tag{5.1}$$

As elements of the observed input stochastic sample, we take the uncorrelated coordinates

$$X_i = \sqrt{\lambda_i} \int_0^T X(t)\Xi_i(t)\,dt, \tag{5.2}$$

where $X(t)$ is the realization of the input stochastic process within the limits of the time interval $[0, T]$, and λ_i and $\Xi_i(t)$ are the eigenvalues and eigenfunctions of the integral equation

$$F(t) = \lambda \int_0^T R(y-t)\Xi(y)\,dy, \quad 0 < t < T, \tag{5.3}$$

where $R(t)$ is the known correlation function of the additive Gaussian noise.

As a rule, we take only the first N coordinates. Thus, for the hypothesis H_0 the likelihood function of the observed input stochastic sample $X_1, \ldots, X_N$ has the following form (note: for simplicity, we set the noise variance to be equal to unity):

$$f_{X\,|\,H_0}(X\,|\,H_0) = \frac{1}{(2\pi)^{\frac{N}{2}}} \cdot \exp\left\{-\frac{1}{2}\sum_{i=1}^{N} X_i^2\right\}. \tag{5.4}$$

This notation corresponds to a "no" signal in the observed input stochastic sample

$$X_1, \ldots, X_N.$$

For the observed input stochastic sample with a nonzero mean $a(t)$, for example, when considering the hypothesis H_1, we take for observed coordinates the values

$$X_i = a_i + \xi_i = \sqrt{\lambda_i} \int_0^T [a(t) + \xi(t)] \Xi_i(t)\, dt, \tag{5.5}$$

where $\xi(t)$ is the additive Gaussian noise with a zero mean and known variance.

Then the likelihood function in the presence of a signal in the observed input stochastic sample $X_1, \ldots, X_N$ has the following form

$$f_{X \mid H_1}(X \mid H_1) = \frac{1}{(2\pi)^{\frac{N}{2}}} \exp\left\{ -\frac{1}{2} \sum_{i=1}^{N} (X_i - a_i)^2 \right\}. \tag{5.6}$$

This notation corresponds to a "yes" signal in the observed input stochastic sample

$$X_1, \ldots, X_N.$$

Using Eqs. (5.4) and (5.6), we can write the likelihood function ratio in the following form:

$$\begin{aligned} \frac{f_{X \mid H_1}(X \mid H_1)}{f_{X \mid H_0}(X \mid H_0)} &= \frac{\exp\left\{-\frac{1}{2}\sum_{i=1}^{N}(X_i - a_i)^2\right\}}{\exp\left\{-\frac{1}{2}\sum_{i=1}^{N} X_i^2\right\}} \\ &= \exp\left\{ \sum_{i=1}^{N} X_i a_i - \frac{1}{2} \sum_{i=1}^{N} a_i^2 \right\} \\ &= \ell(X_1, \ldots, X_N) = C, \end{aligned} \tag{5.7}$$

where C is the constant, which is determined by the performance criterion of the decision-making rule.

Taking the logarithm in Eq. (5.7), we can write

$$\sum_{i=1}^{N} X_i a_i > K_{op} \Rightarrow H_1; \tag{5.8}$$

$$\sum_{i=1}^{N} X_i a_i \leq K_{op} \Rightarrow H_0; \tag{5.9}$$

$$K_{op} = \ln C + \frac{1}{2} \sum_{i=1}^{N} a_i^2, \tag{5.10}$$

where

$$\sum_{i=1}^{N} a_i^2 = E_a \tag{5.11}$$

is the signal energy and K_{op} is the threshold.

Letting $N \to \infty$ and transitioning to the integral form, and using the Parseval theorem,[8] we can maintain generality and write

$$\int_0^T X(t)a(t)\,dt > K_{op} \Rightarrow H_1; \tag{5.12}$$

$$\int_0^T X(t)a(t)\,dt \leq K_{op} \Rightarrow H_0; \tag{5.13}$$

$$K_{op} = \ln C + \frac{1}{2}\int_0^T a^2(t)\,dt, \tag{5.14}$$

where

$$\int_0^T a^2(t)\,dt = E_a \tag{5.15}$$

is the signal energy, and $[0, T]$ is the time interval, within the limits of which the input stochastic process is observed.

It is asserted that the signal detection algorithm determined by Eqs. (5.8)–(5.14) reduces to a calculation of the value

$$\sum_{i=1}^{N} X_i a_i$$

or

$$\int_0^T X(t)a(t)\,dt$$

and comparison to the threshold K_{op}.

This signal detection algorithm is optimal for any of the following chosen performance criteria—the Bayesian criterion, including as particular cases the *a posteriori* probability maximum and maximal likelihood; the Neyman–Pearson criterion; and the mini-max criterion—and is called the correlation signal detection algorithm, since the mutual correlation function between the input stochastic process $X(t)$ and signal $a(t)$ is defined.

Analysis of the signal detection algorithm determined by Eqs. (5.8)–(5.14) yields a property that, in combination with other factors, defines the noise immunity. The essence of the analysis reduces to substituting the actual values

$$X_i = a_i + \xi_i \tag{5.16}$$

and

$$X(t) = a(t) + \xi(t) \quad — \tag{5.17}$$

the hypothesis H_1—or

$$X_i = \xi_i \tag{5.18}$$

and

$$X(t) = \xi(t) \quad — \tag{5.19}$$

the hypothesis H_0—into Eqs. (5.8)–(5.14):

$$\sum_{i=1}^{N} X_i a_i = \sum_{i=1}^{N} a_i^2 + \sum_{i=1}^{N} a_i \xi_i \Rightarrow H_1; \tag{5.20}$$

$$\sum_{i=1}^{N} X_i a_i = \sum_{i=1}^{N} a_i \xi_i \Rightarrow H_0 \tag{5.21}$$

or

$$\int_0^T X(t) a(t)\, dt = \int_0^T a^2(t)\, dt + \int_0^T a(t) \xi(t)\, dt \Rightarrow H_1; \tag{5.22}$$

$$\int_0^T X(t) a(t)\, dt = \int_0^T a(t) \xi(t)\, dt \Rightarrow H_0, \tag{5.23}$$

where the terms

$$\sum_{i=1}^{N} a_i^2$$

and

$$\int_0^T a^2(t)\, dt$$

are the signal energy, and the terms

$$\sum_{i=1}^{N} a_i \xi_i$$

and

$$\int_0^T a(t)\xi(t)\,dt$$

are the noise component with a zero mean and the finite variance defined by

$$\lim_{N\to\infty} \overline{\left\{\sum_{i=1}^{N} a_i \xi_i\right\}^2} = \frac{E_a N_0}{2} \quad \text{as } N \to \infty \tag{5.24}$$

or as $T \to \infty$

$$\begin{aligned} \overline{\left[\int_{\infty} a(t)\xi(t)\,dt\right]^2} &= \int_0^{\infty} dt \int_0^{\infty} a(t)a(s)\overline{\xi(t)\xi(s)}\,ds \\ &= \frac{E_a N_0}{2}. \end{aligned} \tag{5.25}$$

$\frac{N_0}{2}$ is the spectral power density of the additive Gaussian noise.

The detection parameter

$$q = \sqrt{\frac{2E_a}{N_0}} \tag{5.26}$$

is taken as a qualitative characteristic of the signal detection algorithm determined by Eqs. (5.8)–(5.14). This parameter may also be called the voltage signal-to-noise ratio (SNR). This parameter is very important and, together with other factors, defines the noise immunity.

5.2 Criticism

Let us consider those factors generating questions in the synthesis of the signal detection algorithm determined by Eqs. (5.8)–(5.14). It is known that

$$\sum_{i=1}^{N} X_i,$$

which is the sufficient statistic of the mean, and

$$\sum_{i=1}^{N} X_i^2,$$

which is the sufficient statistic of the variance, are the jointly sufficient statistics characterizing the distribution law of the random value X_i.[8,19]

The sufficient statistics

$$\sum_{i=1}^{N} X_i^2$$

of the likelihood functions

$$f_{X\mid H_1}(X \mid H_1)$$

and

$$f_{X\mid H_0}(X \mid H_0)$$

are reduced in the synthesis of the signal detection algorithm determined by Eqs. (5.8)–(5.14). This is indeed the case in regard to the form of the expressions and assumptions of the statistical theory of decision making. However, in the physical sense, it causes a specific perplexity.

The point is that a "yes" signal—the mean a_i of the observed input stochastic sample

$$X_1, \ldots, X_N$$

is not zero—is indicated in the numerator of Eq. (5.7), and a "no" signal is indicated in the denominator of Eq. (5.7) under observation of the same coordinates. It would be difficult to imagine another approach for the same input stochastic sample

$$X_1, \ldots, X_N$$

in both the numerator and denominator of the likelihood function ratio.

The first question that arises is: Might a signal detection algorithm be constructed without loss of the sufficient statistic of the variance, which is one of the characteristics of the distribution law?

Another factor generating questions is that the signal detection is performed against the background of the noise component

$$\sum_{i=1}^{N} a_i \xi_i$$

or

$$\int_{0}^{T} a(t)\xi(t)\,dt$$

caused by the interaction between the signal and noise.

The variance of the noise component is proportional to the signal energy, which follows from Eqs. (5.24) and (5.25). The fact that the signal-to-noise ratio for the signal detection algorithm, which is determined by Eqs. (5.8)–(5.14), defined by Eq. (5.25) is not proportional to

$$\frac{2E_a}{N_0}$$

but rather proportional to the square root of the value

$$\frac{2E_a}{N_0}$$

is a consequence of this. A resulting question would be: Is this good or bad?

One would believe it is good, if the following condition

$$\frac{2E_a}{N_0} < 1$$

is satisfied. But if there is the condition

$$q < 1$$

and the probability of false alarm P_F is equal to 10^{-3}, for example, the probability of detection P_D does not exceed 0.1, which is a practically inoperative region for signal detection. If the conditions

$$\frac{2E_a}{N_0} > 1$$

and

$$q = \sqrt{\frac{2E_a}{N_0}}$$

are satisfied, then the probability of detection P_D is smaller in comparison to the proportional dependence

$$q = \frac{E_a}{N_0}.$$

This conclusion seems unusual, but it is real and is shown in References 51 and 52.

Analyzing Eqs. (5.8)–(5.14), we may note that this signal detection algorithm is considered to be optimal during the following conditions.

- The likelihood function (or functional) ratio is formed, using the same input stochastic sample, where the numerator assumes a "yes" signal, and the denominator assumes a "no" signal in the input stochastic process. In this case the standard quadratic statistic is reduced, and the additional information is lost—the sufficient statistic of the variance of the likelihood function. The expression obtained ensures the calculation of the sufficient statistic of the mean of the likelihood function only:

$$\sum_{i=1}^{N} X_i a_i$$

or

$$\int_0^T X(t)a(t)\,dt,$$

where a_i or $a(t)$ is the known signal.

- Theoretically speaking, the signal detection algorithm determined by Eqs. (5.8)–(5.14) is not realizable for the following reasons:
  - The mutual correlation function between the input stochastic process X_i or $X(t)$ and the signal a_i or $a(t)$ is defined by the left side of Eqs. (5.8)–(5.14), respectively;
  - The left side of Eqs. (5.8)–(5.14) vanishes given a "no" signal in the input stochastic process X_i or $X(t)$:

$$\sum_{i=1}^{N} X_i a_i$$

or

$$\int_0^T X(t)a(t)\,dt,$$

where $a_i = 0$ or $a(t) = 0$, *and any physical sense is lost.*

In practice, the signal detection algorithm determined by Eqs. (5.8)–(5.14) is realized if the signal structure a_i or $a(t)$ is replaced by its model a_{m_i} or $a_m(t)$ at the receiver, as a_i or $a(t)$ is the completely known signal—

$$a_{m_i} = ka_i \tag{5.27}$$

or

$$a_m(t) = ka(t), \tag{5.28}$$

where k is the coefficient of proportionality:

$$\sum_{i=1}^{N} X_i a_{m_i}$$

or

$$\int_0^T X(t)a_m(t)\,dt.$$

In this case, the left side of Eqs. (5.8)–(5.14) has a specific physical sense.

- If the signal structure a_i or $a(t)$ is replaced by its model a_{m_i} or $a_m(t)$, then the noise component

$$\sum_{i=1}^{N} a_{m_i}\xi_i$$

or

$$\int_0^T a_m(t)\xi(t)\,dt$$

arises, caused by the interaction between the model signal and noise, and always exists independently of what hypothesis (H_0 or H_1) is considered.

- The variance of the noise component noted above is proportional to the energy of the model signal, i.e.,

$$\frac{E_{a_m} N_0}{2},$$

where E_{a_m} is the energy of the model signal, and

$$\frac{N_0}{2}$$

is the spectral power density of the noise.

- The signal detection algorithm determined by Eqs. (5.8)–(5.14) does not allow us to obtain the ratio between the energy characteristics of the signal and noise in pure form, for example, in the form

$$\frac{2E_a}{N_0}.$$

It causes the probability of detection P_D to be a function of the square root of the ratio of the signal and noise energy characteristics, i.e., the voltage signal-to-noise ratio is proportional to

$$\sqrt{\frac{2E_a}{N_0}}.$$

- The signal detection algorithm determined by Eqs. (5.8)–(5.14) does not afford detection of a signal, whose structure does not correspond to that of the model signal at the receiver.

- In general, a detector constructed according to the signal detection algorithm determined by Eqs. (5.8)–(5.14) must be a tracker, not a true detector, because the instant of signal appearance on the time axis is unknown relative to the origin.

Considering the conditions of optimality of the signal detection algorithm determined by Eqs. (5.8)–(5.14) as briefly outlined above, if the same input stochastic sample is observed in the numerator and denominator of the likelihood function ratio, it is the author's opinion that it is necessary to undertake a critical review of the initial premises lying at the basis of the classical and modern signal detection theories.

5.3 Initial Premises

The signal detection algorithm determined by Eqs. (5.8)–(5.14) is based on the assumption that there is the frequency time region Z of the noise, where a signal may be present; for example, there is an observed stochastic sample from this region, relative to which there is a need to make the decision a "yes" signal—the hypothesis H_1—or a "no" signal—the hypothesis H_0.

We now proceed to modify the initial premises of the classical and modern signal detection theories. Suppose there are two independent frequency time noise regions Z and Z^* belonging to the space A (see Fig. 5.1). Noise from these regions obeys the same probability distribution densities with the same statistical parameters. The same probability distribution density and equality of the statistical parameters have been chosen for simplicity of this analysis. In general, the probability distribution densities and statistical parameters may not be equal.

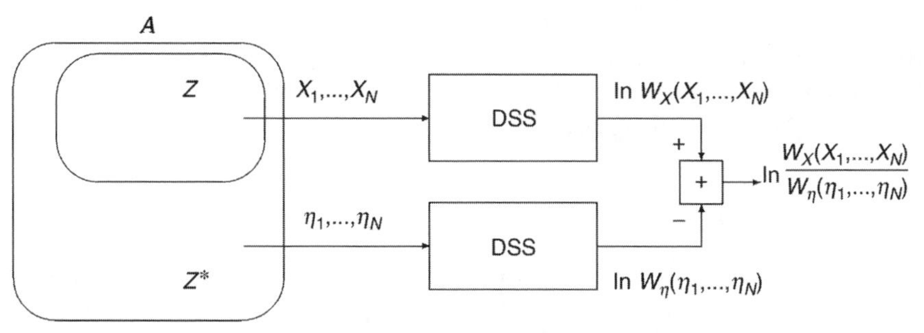

FIGURE 5.1
Definition of jointly sufficient statistics.

A "yes" signal is possible in the noise region Z as before. *It is known a priori that a "no" signal is obtained in the noise region* Z^*. Below we will call the noise region Z^* the reference region, and consequently, the observed sample from this region is called the reference sample.

There is a need to make the decision a "yes" signal—the hypothesis H_1—or a "no" signal—the hypothesis H_0—in the observed input stochastic sample from the region Z, by comparing the statistical parameters of the probability distribution density of this observed input stochastic sample with those of the observed input stochastic sample from the reference region Z^*.

The problem posed in the preceding sections of this chapter must be solved using the statistical decision-making theory. Thus, it is necessary to accumulate and compare statistical data defining the statistical parameters of the probability distribution densities of the observed input stochastic samples from two independent frequency time regions Z and Z^*.

If the probability distribution density statistical parameters for two samples are equal, or agree with each other within the limits of a given accuracy, then the decision a "no" signal in the observed input stochastic sample from the region Z is made—the hypothesis H_0. If the probability distribution density statistical parameters of the observed input stochastic sample from the region Z differ from those of the reference sample from the region Z^* by a value that exceeds the prescribed error limit, then the decision a "yes" signal in the region Z is made—the hypothesis H_1.

5.4 Likelihood Ratio

Now the problem is to obtain jointly sufficient statistics to define the statistical parameters of the probability distribution densities. For this purpose let us avail ourselves of one of the well-known results.[7,18,24,29,31,45,53–55]

It is known that a sufficient statistic is determined from the condition that the likelihood function has an extremum. In general, the condition of an extremum of the likelihood function, relative to the parameter to be determined with a prescribed accuracy, is determined in the following form:

$$\frac{\partial f_X(X_1, \ldots, X_N \mid \vartheta)}{\partial \vartheta} = 0, \tag{5.29}$$

where N is the sample size determining the prescribed accuracy, and ϑ is the parameter to be determined. However, this equation is not used in practice.

A simple mathematical procedure simplifies the representation of this equation. Since the logarithm is a monotonic function, the extrema of the functions

$$f_X(X_1, \ldots, X_N)$$

and

$$\ln f_X(X_1, \ldots, X_N)$$

are reached at the same values of the parameter ϑ. Therefore, the likelihood function equation is usually written in the following form

$$\frac{\partial \ln f_X(X_1, \ldots, X_N \mid \vartheta)}{\partial \vartheta} = 0. \tag{5.30}$$

As was shown in References 19 and 31 using Eqs. (5.4), (5.6), and (5.30), it is easy to prove that the values

$$\sum_{i=1}^{N} X_i a_i$$

and

$$\sum_{i=1}^{N} X_i^2$$

are the jointly sufficient statistics of the likelihood function parameters in Eqs. (5.4) and (5.6) for the observed input stochastic sample

$$X_1, \ldots, X_N.$$

The likelihood function for the reference sample

$$\eta_1, \ldots, \eta_N$$

for unit variance is determined in the following form:

$$f_\eta(\eta_1, \ldots, \eta_N) = \frac{1}{(2\pi)^{\frac{N}{2}}} \exp\left\{-\frac{1}{2}\sum_{i=1}^{N} \eta_i^2\right\}, \tag{5.31}$$

where

$$\sum_{i=1}^{N} \eta_i^2$$

is the sufficient statistic of the likelihood function parameters for the reference sample

$$\eta_1, \ldots, \eta_N.$$

In the definition of the sufficient statistics using the input stochastic samples

$$X_1, \ldots, X_N$$

and

$$\eta_1, \ldots, \eta_N$$

the problem of their comparison arises.

Usually for this purpose, a difference is used (see Fig. 5.1). The resulting sufficient statistics are observed at the output of the difference device:

$$\ln f_X(X_1, \ldots, X_N) - \ln f_\eta(\eta_1, \ldots, \eta_N) = \frac{1}{2}\left\{\sum_{i=1}^{N} 2X_i a_i - \sum_{i=1}^{N} X_i^2 + \sum_{i=1}^{N} \eta_i^2 - \sum_{i=1}^{N} a_i^2\right\}. \tag{5.32}$$

It is customary to reference the last term on the right side of Eq. (5.32) to a threshold independent of the observed input stochastic sample, as in Eqs. (5.8)–(5.10). Equation (5.32), obtained by definition of the resulting sufficient statistics, is the logarithm of the likelihood function.

The signal detection algorithm based on two independent observed input stochastic samples, one of which is the reference sample with *a priori* information a "no" signal, follows from Eq. (5.32):

$$\ln f_X(X_1, \ldots, X_N) - \ln f_\eta(\eta_1, \ldots, \eta_N) = \ln\left\{\frac{f_X(X_1, \ldots, X_N)}{f_\eta(\eta_1, \ldots, \eta_N)}\right\} = \frac{1}{2}\left\{\sum_{i=1}^{N} 2X_i a_i - \sum_{i=1}^{N} X_i^2 + \sum_{i=1}^{N} \eta_i^2 - \sum_{i=1}^{N} a_i^2\right\} = \ln C \tag{5.33}$$

or

$$\sum_{i=1}^{N} 2X_i a_i - \sum_{i=1}^{N} X_i^2 + \sum_{i=1}^{N} \eta_i^2 = K_g, \tag{5.34}$$

where K_g is the threshold.

Proceeding from generally accepted concepts, it follows that the hypothesis H_1—a "yes" signal in the observed input stochastic sample

$$X_1, \ldots, X_N \quad —$$

is assumed if the following inequality is fulfilled:

$$\sum_{i=1}^{N} 2X_i a_i - \sum_{i=1}^{N} X_i^2 + \sum_{i=1}^{N} \eta_i^2 > K_g, \tag{5.35}$$

and the hypothesis H_0—a "no" signal in the observed input stochastic sample

$$X_1, \ldots, X_N \quad —$$

is assumed if the opposite inequality is fulfilled.

The first term on the left side of Eq. (5.35) is the signal detection algorithm determined by Eqs. (5.8)–(5.14) with the factor 2.

The more rigorous form of Eq. (5.35), based on the analysis performed in Sections 3.1 and 3.2, is the following:

$$\sum_{i=1}^{N} 2X_i a_{m_i} - \sum_{i=1}^{N} X_i^2 + \sum_{i=1}^{N} \eta_i^2 > K_g, \tag{5.36}$$

where a_{m_i} is the model signal.

Letting $N \to \infty$ and transitioning to the integral form, and using the Parseval theorem,[8] we maintain generality and can write

$$2\int_0^T X(t)a_m(t)\,dt - \int_0^T X^2(t)\,dt + \int_0^T \eta^2(t)\,dt > K_g, \tag{5.37}$$

where $[0, T]$ is the time interval, within the limits of which the input stochastic process is observed.

Analysis of the signal detection algorithm in Eqs. (5.36) and (5.37), performed by the same procedure as in Sections 3.1 and 3.2, shows that when considering the hypothesis H_1:

$$X_i = a_i + \xi_i \tag{5.38}$$

or

$$X(t) = a(t) + \xi(t) \tag{5.39}$$

given

$$a_i = a_{m_i} \tag{5.40}$$

or

$$a(t) = a_m(t), \tag{5.41}$$

and the left side of Eqs. (5.36) and (5.37) has the following form

$$2\sum_{i=1}^{N}[a_i + \xi_i]a_{m_i} - \sum_{i=1}^{N}[a_i + \xi_i]^2 + \sum_{i=1}^{N}\eta_i^2 = \sum_{i=1}^{N}a_i^2 + \sum_{i=1}^{N}\eta_i^2 - \sum_{i=1}^{N}\xi_i^2 \tag{5.42}$$

or

$$2\int_0^T [a(t) + \xi(t)]a_m(t)\,dt - \int_0^T [a(t) + \xi(t)]^2\,dt + \int_0^T \eta^2(t)\,dt$$
$$= \int_0^T a^2(t)\,dt + \int_0^T \eta^2(t)\,dt - \int_0^T \xi^2(t)\,dt, \tag{5.43}$$

respectively, where the terms

$$\sum_{i=1}^{N} a_i^2 = E_a \qquad \text{and} \qquad \int_0^T a^2(t)\,dt \tag{5.44}$$

are the signal energy, and the terms

$$\sum_{i=1}^{N} \eta_i^2 - \sum_{i=1}^{N} \xi_i^2 \tag{5.45}$$

and

$$\int_0^T \eta^2(t)\,dt - \int_0^T \xi^2(t)\,dt \tag{5.46}$$

are the background noise.

When considering the hypothesis H_0:

$$X_i = \xi_i \tag{5.47}$$

or

$$X(t) = \xi(t) \tag{5.48}$$

and the conditions

$$a_i = 0 \tag{5.49}$$

or

$$a(t) = 0 \tag{5.50}$$

given

$$a_{m_i} = a_i \tag{5.51}$$

or

$$a(t) = a_m(t), \tag{5.52}$$

the left side of Eqs. (5.36) and (5.37) has the following form

$$\sum_{i=1}^{N} \eta_i^2 - \sum_{i=1}^{N} \xi_i^2, \tag{5.53}$$

or

$$\int_0^T \eta^2(t)\,dt - \int_0^T \xi^2(t)\,dt. \tag{5.54}$$

Subsequent analysis of the signal detection algorithm determined by Eqs. (5.36) and (5.37) will only be performed under the conditions determined by Eqs. (5.51) and (5.52). This statement is very important for further understanding of the generalized approach to signal processing in the presence of noise. How we do this becomes clear in the discussion of the experimental results presented in greater detail in References 51 and 52.

It must be emphasized that

$$\sum_{i=1}^{N} \eta_i^2 - \sum_{i=1}^{N} \xi_i^2 \to 0 \quad \text{as } N \to \infty \tag{5.55}$$

or

$$\int_0^T \eta^2(t)\,dt - \int_0^T \xi^2(t)\,dt \to 0 \quad \text{as } T \to \infty \tag{5.56}$$

in the statistical sense, since the processes ξ_i and η_i, or $\xi(t)$ and $\eta(t)$, are uncorrelated and have the same spectral power density of the additive Gaussian noise $\frac{N_0}{2}$ according to the initial conditions.

In this way it has been shown that both signal detection algorithms based on the observed input stochastic sample

$$X_1, \ldots, X_N$$

and the two independently observed input stochastic samples

$$X_1, \ldots, X_N$$

and

$$\eta_1, \ldots, \eta_N$$

have the same approach and are defined by the likelihood function using the statistical theory of decision making.

The difference is that the numerator and denominator of the likelihood function used, for synthesis of the signal detection algorithm determined by Eqs. (5.8)–(5.14), involve the same observed input stochastic sample (see Eqs. (5.4) and (5.6)), but a "yes" signal is assumed in the numerator and a "no" signal is assumed in the denominator.

The numerator of the likelihood function used for synthesis of the signal detection algorithm determined by Eqs. (5.36) and (5.37) involves the observed input stochastic sample, where a "yes" signal may be present, and the denominator involves the reference sample, which is known *a priori* to contain a "no" signal.

On this basis, it may be stated that only the sufficient statistic

$$\sum_{i=1}^{N} X_i a_i$$

or

$$\int_0^T X(t) a(t)\, dt$$

has been applied to define the mean of the likelihood function in the signal detection algorithm determined by Eqs. (5.8)–(5.14), respectively.

In the signal detection algorithm determined by Eqs. (5.36) and (5.37), the jointly sufficient statistics

$$\sum_{i=1}^{N} 2 X_i a_i$$

and

$$\sum_{i=1}^{N} (\eta_i^2 - X_i^2)$$

or

$$2 \int_0^T X(t)\, a(t)\, dt$$

and

$$\int_0^T [\eta^2(t) - X^2(t)]\, dt$$

are used to define the mean and variance of the likelihood function.

This fact permits us to obtain more complete information in the decision-making process in comparison to the signal detection algorithm determined by Eqs. (5.8)–(5.14).

The signal detection algorithm determined by Eqs. (5.36) and (5.37) is free from a number of conditions unique to the signal detection algorithm determined by Eqs. (5.8)–(5.14).

As the signal detection algorithm determined by Eqs. (5.8)–(5.14) is a component of the signal detection algorithm determined by Eqs. (5.36) and (5.37), the last of which has been called the generalized signal detection algorithm.

5.5 Engineering Interpretation

The technical realization of independent sampling from the regions Z and Z^* obeying the same probability distribution density with the same statistical parameters is not difficult. The solution of the problem of detecting the signal $a(t)$ with the additive Gaussian noise $n(t)$ is well known.[1–50,56]

The observed input stochastic process $X(t)$ is examined at the output of the linear section of the receiver, which has an ideal amplitude-frequency response and the bandwidth ΔF. It is supposed that the noise at the input of the linear section of the receiver is the additive white Gaussian noise, having the correlation function

$$\frac{N_0}{2} \cdot \delta(t_2 - t_1),$$

where $\delta(x)$ is the delta function (see Appendix I).

The signal $a(t)$ is assumed to be completely known, and the signal energy is taken to be equal to 1. The spectral power density $\frac{N_0}{2}$ is considered an *a priori* indeterminate parameter. The gain of the linear section of the receiver is equal to 1.

Through analysis, the problem is reduced to testing the complex hypothesis with the decision function

$$Re \int_0^T \dot{X}(t)\dot{a}^*(t)\,dt > K(P_F)\sqrt{\int_0^T |\dot{X}(t)|^2\,dt}, \tag{5.57}$$

where $\dot{a}^*(t)$ is the filter matched with the signal; $K(P_F)$ is the threshold defined by the probability of false alarm P_F; and

$$\int_0^T |\dot{X}(t)|^2\,dt$$

is the statistic defining the decision function.

It turns out that the signal detector constructed in accordance with the above decision function renders the probability of false alarm P_F stable, given an unknown noise power, and has the greatest probability of detection P_D for any signal-to-noise ratio.[29]

Let us interpret this problem. We use two linear sections of the receiver (instead of one section) for our set of statistics. These linear sections will be called the preliminary (PF) and additional (AF) filters. The amplitude-frequency responses of the PF and AF must obey the same law.

The resonant frequencies of the PF and AF must be detuned, relative to each other by a value determined from the well-known results,[57,58] for the purpose of providing uncorrelated statistics at the outputs of the PF and AF. The detuning value between the resonant frequencies of the PF and AF exceeds the effective signal bandwidth ΔF_a. As is well known,[57,58] if this value reaches $4\Delta F_a$ and $5\Delta F_a$, the coefficient of correlation between the statistics at the outputs of the PF and AF tends to approach zero.

In practice, these statistics may be regarded as uncorrelated. The effective bandwidth of the PF is equal to that of the signal frequency spectrum and can be even greater, but this is undesirable since the noise power at the output of the PF is proportional to the effective bandwidth.

The effective bandwidth of the AF may be smaller than the effective bandwidth of the PF; however, for simplicity of analysis, in this chapter the effective bandwidth of the AF is assumed to be the same as the effective bandwidth of the PF.

Thus, we can assume that uncorrelated samples of the observed input stochastic processes are formed at the outputs of the PF and AF. These samples obey the same probability distribution density with the same statistical parameters given that the same process is present at the inputs of the PF and AF, even if this process is the additive white Gaussian noise having the correlation function

$$\frac{N_0}{2} \cdot \delta(t_2 - t_1).$$

The physicotechnical interpretation of the signal detection algorithm determined by Eqs. (5.36) and (5.37) is the following.

- The AF may serve as the source of the observed reference sample

$$\eta_1, \ldots, \eta_N$$

  from the interference region Z^*. The resonant frequency of the AF is detuned relative to the carrier frequency of the signal by a value that can be determined on the basis of well-known results,[20,57,58] depending on the specific practical situation.
- The PF serves as the source of the sample

$$X_1, \ldots, X_N$$

  of the observed input stochastic process from the interference region Z. The bandwidth of the PF is matched with the effective bandwidth of the signal. The value of the bandwidth of the PF is matched with the value of the bandwidth of the AF.
- The first term of the generalized signal detection algorithm, determined by Eqs. (5.36) and (5.37), corresponds to synthesis of the correlation channel with twice the gain.

- The second term of the generalized signal detection algorithm, determined by Eqs. (5.36) and (5.37), corresponds to synthesis of the autocorrelation channel coupled with the PF.
- The third term of the generalized signal detection algorithm, determined by Eqs. (5.36) and (5.37), corresponds to synthesis of the autocorrelation channel coupled with the AF.
- The statistic of the autocorrelation channel coupled with the PF is subtracted from the statistic of the autocorrelation channel coupled with the AF. As a result,

$$\sum_{i=1}^{N} \eta_i^2 - \sum_{i=1}^{N} \xi_i^2 \to 0 \quad \text{as } N \to \infty \tag{5.58}$$

  or

$$\int_0^T \eta^2(t)\, dt - \int_0^T \xi^2(t)\, dt \to 0 \quad \text{as } T \to \infty \tag{5.59}$$

  in the statistical sense.
- The statistic of the autocorrelation channel coupled with the PF is subtracted from the statistic of the correlation channel. As a result, a complete compensation of the noise component

$$\sum_{i=1}^{N} a_{m_i} \xi_i$$

  or

$$\int_0^T a_m(t)\xi(t)\, dt$$

  of the signal detection algorithm determined by Eqs. (5.8)–(5.14) is achieved in the statistical sense if the conditions determined by Eqs. (5.51) and (5.52) are satisfied, where a_{m_i} or $a_m(t)$ is the model signal, and a_{1_i} or $a_1(t)$ is the signal at the output of the PF.

The detector shown in Fig. 5.2 is based on the physicotechnical interpretation of the generalized approach to signal processing in noise[51,52,59–62] stated above.

It is of special interest to compare these statements to the statements and analysis in Reference 29. Some opponents of the generalized approach to signal processing in the presence of noise erroneously believe that this approach is the same as the one-input two-sample signal detection approach in Reference 29.

For this purpose, we briefly recall the main statements of the one-input two-sample signal detection approach.

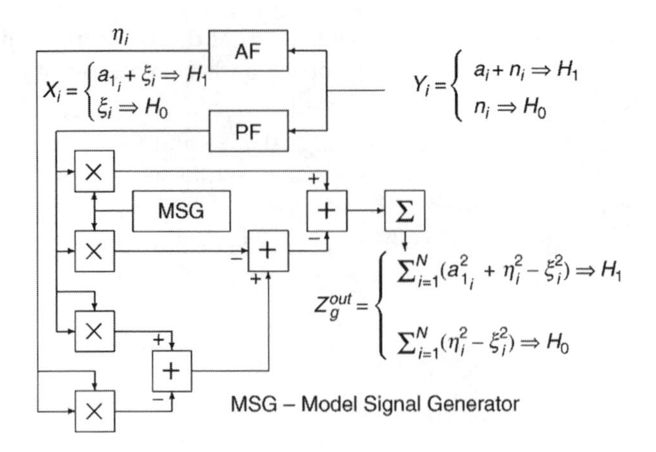

FIGURE 5.2
Physicotechnical interpretation of the generalized approach.

- First, the signal sample is generated only at the time instants that correspond to the expected signal. In other words, the signal sample is generated only at the time instants when the expected signal may appear in the time frequency space.
- Second, the noise channel is formed within the limits of the time intervals, in which it is *a priori* known that the signal is absent.

The generalized approach to signal processing in the presence of noise is based on the following statements.

- The first sample is generated independently of the time instants that correspond to the emergence of the signal in the time frequency space.
- The reference sample—the second sample—is formed simultaneously with the first sample at the same time intervals as the first sample and exists without any limitations in time or readings, but it is known *a priori* that a "no" signal exists in the reference sample owing to the conditions of generating the reference sample.
- The sample sizes of the first (signal) and second (reference) samples are the same.

These differences between the generalized approach to signal processing in the presence of noise and the one-input two-sample signal detection approach are very important.

In addition, we can see that the engineering interpretation of the generalized approach to signal processing in the presence of noise differs greatly from the one-input two-sample signal detection approach in Reference 29.

5.6 Generalized Detector

Consider the problem of specific interest in which the signal has the stochastic amplitude and random initial phase. The necessity of considering this problem stems from the fact that, in practice, some satellite signal processing systems use channels with an ionosphere mechanism of propagation and operate using frequencies that are higher than the maximal allowable frequency. Other satellite signal processing systems use channels with tropospheric scattering. These problems arise in radar during detection of fluctuating targets when the target return signal is a sequence of pulses of unknown amplitude and phase.

A signal with the stochastic amplitude and random initial phase can be written in the following form:

$$a(t, \varphi_0, A) = A(t)S(t)\cos[\omega_0 t + \Psi_a(t) - \varphi_0], \tag{5.60}$$

where ω_0 is the carrier frequency of the signal $a(t, \varphi_0, A)$; $S(t)$ is the known modulation law of amplitude of the signal $a(t, \varphi_0, A)$; $\Psi_a(t)$ is the known modulation law of phase of the signal $a(t, \varphi_0, A)$; φ_0 is the random initial phase of the signal $a(t, \varphi_0, A)$, which is uniformly distributed within the limits of the interval $[-\pi, \pi]$ and is time invariant within the limits of the time interval $[0, T]$; and $A(t)$ is the amplitude factor, which is a random value and a function of time in the general case.

Consider the generalized approach to signal processing in the presence of noise for the signals with the stochastic amplitude and random initial phase.[63–66] According to the generalized approach to signal processing in the presence of noise there is a need to form the following:

- First, the reference sample with *a priori* information a "no" signal in the input stochastic process
- Second, the autocorrelation channel for the purpose of compensating for the noise component of the correlation channel, which is caused by the interaction between the model signal and noise

The main principles of construction of the generalized detector for the signals with stochastic amplitude and random initial phase using the Neyman–Pearson criterion are in the following.[67–71]

The input stochastic process $Y(t)$ must pass through the preliminary filter (PF). The effective bandwidth of the PF is equal to ΔF_a, where ΔF_a is the effective spectrum bandwidth of the signal.

$$X(t) = a_1(t, \varphi_0, A) + \xi(t) \tag{5.61}$$

is the process at the output of the PF, if a "yes" signal exists in the input stochastic process—the hypothesis H_1.

$$X(t) = \xi(t) \tag{5.62}$$

is the process at the output of the PF, if a "no" signal exists in the input stochastic process—the hypothesis H_0; $a_1(t, \varphi_0, A)$ is the signal at the output of the PF determined by Eq. (5.60); and $\xi(t)$ is the noise at the output of the PF.

We need to form the reference sample for generation of the jointly sufficient statistics of the mean and variance of the likelihood function. For this purpose, the additional filter (AF) is formed in a parallel way to the PF. The amplitude-frequency response of the AF is analogous over the entire range of parameters to the amplitude-frequency response of the PF, but it is detuned in the resonant frequency relative to the PF for the purpose of providing uncorrelated statistics at the outputs of the PF and AF. The detuning value must be larger than the effective spectrum bandwidth of the signal so that the processes at the outputs of the PF and AF will be uncorrelated.

As was shown in References 57 and 58, if this detuning value reaches a value between $4\Delta F_a$ and $5\Delta F_a$, the processes at the outputs of the PF and AF are not correlated practically. For this condition, a coefficient of correlation between the statistics at the outputs of the PF and AF is not more than 0.05 for all practical purposes. The coefficient of correlation may be considered as a value tending to approach zero.

Thus, the process $\eta(t)$ is formed at the output of the AF:

$$\eta(t) = \xi_2(t)\cos[\omega'_n t + v'_i(t)], \tag{5.63}$$

where $\xi_2(t)$ is the random envelope of amplitude of the noise at the output of the AF; $v'_i(t)$ is the random phase of the noise at the output of the AF; and ω'_n is the medium frequency of the noise at the output of the AF.

The generalized detector for the signals with the stochastic amplitude and random initial phase is shown in Fig. 5.3. There are two filters with the non-overlapping amplitude-frequency responses, which must obey the same law: the preliminary filter (PF) and the additional filter (AF).

The PF is matched with the effective spectrum bandwidth of the signal with the carrier frequency ω_0. The AF does not pass the frequency ω_0. By this means, there is the following requirement for the AF: the resonant frequency of the AF must be detuned with respect to the resonant frequency of the PF to ensure the uncorrelated statistics at the outputs of both the AF and PF.[67–73] This requirement is necessary to ensure the complete compensation of the constant component of the background noise at the output of the generalized detector in the statistical sense.

For generation of the jointly sufficient statistics of the mean and variance of the likelihood function, in accordance with the generalized approach to signal processing in the presence of noise, there is a need to form the autocorrelation channel. These actions allow us to compensate the total noise component in the statistical sense. We proceed to show this.

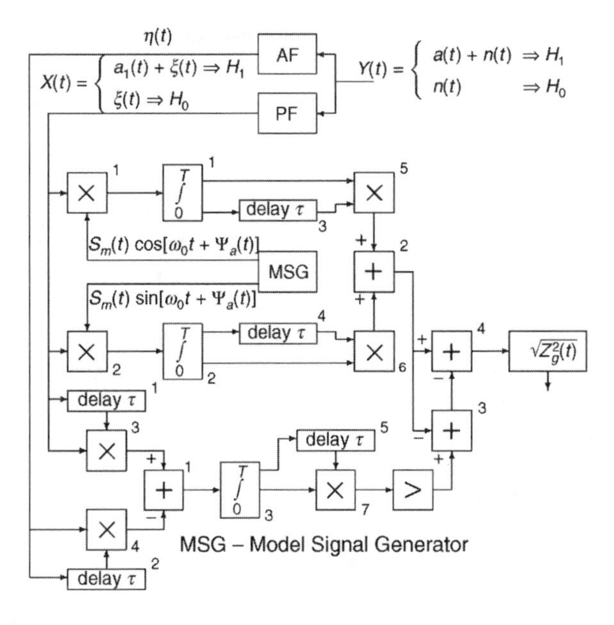

FIGURE 5.3
Generalized detector.

The generalized detector for the signals with the stochastic amplitude and random initial phase shown in Fig. 5.3 consists of:

- The correlation channel: the multipliers 1 and 2; the integrators 1 and 2; the square-law function generators 5 and 6; the summator 2; the model signal generator MSG
- The autocorrelation channel: the multipliers 3 and 4; the summator 1; the integrator 3; the square-law function generator 7; the amplifier (>)
- The compensating channel: the summators 3 and 4; the compensation of the total noise component in the statistical sense is carried out by the summators 3 and 4
- The delay blocks 1–5 are only used for specific technical problems, and are not taken into consideration during the analysis of theoretical principles of functionality

The compensating channel of the generalized detector allows us to compensate the noise component of the correlation channel of the generalized detector and the random component of the autocorrelation channel of the generalized detector in the statistical sense.

The noise component of the correlation channel of the generalized detector is created by the interaction between the model signal and noise. The random component of the autocorrelation channel of the generalized detector, which will be described below, is caused by the interaction between the signal and noise.

Let us analyze the generalized detector shown in Fig. 5.3 for the condition

$$S_m(t) = S(t), \tag{5.64}$$

i.e., the model signal $a_m(t)$ is completely matched with the signal $a_1(t, \varphi_0, A)$ at the output of the PF, and $\tau = 0$. How we are able to do this becomes clear in the discussion of experimental and application results presented in References 51 and 52.

We must take into consideration that the frequencies $2\omega_0, 2\omega_n$, or $2\omega'_n$ cannot pass through the PF and AF, respectively. The analysis is based on the results discussed in References 63–87 and the hypothesis H_1—a "yes" signal in the input stochastic process.

In terms of Eq. (5.60), the processes at the outputs of the multipliers 1–4 take the following form

$$\begin{aligned} y_1(t) &= [a_1(t, \varphi_0, A) + \xi(t)]a_{m_1}(t) \\ &= \{A(t)S(t)\cos[\omega_0 t + \Psi_a(t) - \varphi_0] \\ &\quad + \xi_1(t)\cos[\omega_n t + \upsilon(t)]\}S_m(t)\cos[\omega_0 t + \Psi_a(t)] \\ &= \frac{1}{2}\cos\varphi_0 A(t)S(t)S_m(t) + \frac{1}{2}S_m(t)\xi_1(t)\cos[\Psi_a(t) - \upsilon(t)]; \end{aligned} \tag{5.65}$$

$$\begin{aligned} y_2(t) &= [a_1(t, \varphi_0, A) + \xi(t)]a_{m_2}(t) \\ &= \{A(t)S(t)\cos[\omega_0 t + \Psi_a(t) - \varphi_0] \\ &\quad + \xi_1(t)\cos[\omega_n t + \upsilon(t)]\}S_m(t)\sin[\omega_0 t + \Psi_a(t)] \\ &= \frac{1}{2}\sin\varphi_0 A(t)S(t)S_m(t) \\ &\quad + \frac{1}{2}S_m(t)\xi_1(t)\sin[\Psi_a(t) - \upsilon(t)]; \end{aligned} \tag{5.66}$$

$$\begin{aligned} y_3(t) &= \frac{1}{2}A^2(t)S^2(t) \\ &\quad + A(t)S(t)\xi_1(t)\cos[\Psi_a(t) - \upsilon(t) - \varphi_0] + \frac{1}{2}\xi_1^2(t); \end{aligned} \tag{5.67}$$

$$\begin{aligned} y_4(t) &= \eta^2(t) \\ &= \xi_2(t)\cos[\omega'_n t + \upsilon'(t)]\xi_2(t)\cos[\omega'_n t + \upsilon'(t)] = \frac{1}{2}\xi_2^2(t), \end{aligned} \tag{5.68}$$

where

$$a_{m_{1,2}}(t) = S_m(t)\,{}^{\cos}_{\sin}[\omega_0 t + \Psi_a(t)] \tag{5.69}$$

is the model signal, which is formed at the output of the model signal generator MSG (Fig. 5.3). The notation $S_m(t)$ is retained to emphasize the interaction between the model signal and noise.

The process at the output of summator 1 takes the following form

$$
\begin{aligned}
y_5(t) &= \frac{1}{2}A^2(t)S^2(t) \\
&+ A(t)S(t)\xi_1(t)\cos[\Psi_a(t) - \upsilon(t) - \varphi_0] \\
&- \frac{1}{2}\left[\xi_2^2(t) - \xi_1^2(t)\right]. \qquad (5.70)
\end{aligned}
$$

The processes at the outputs of integrators 1–3 take the following form:

$$
\begin{aligned}
Z_1(t) &= \frac{1}{2}\cos\varphi_0 \int_0^T A(t)S(t)S_m(t)\,dt \\
&+ \frac{1}{2}\int_0^T S_m(t)\xi_1(t)\cos[\Psi_a(t) - \upsilon(t)]\,dt; \qquad (5.71)
\end{aligned}
$$

$$
\begin{aligned}
Z_2(t) &= \frac{1}{2}\sin\varphi_0 \int_0^T A(t)S(t)S_m(t)\,dt \\
&+ \frac{1}{2}\int_0^T S_m(t)\xi_1(t)\sin[\Psi_a(t) - \upsilon(t)]\,dt; \qquad (5.72)
\end{aligned}
$$

$$
\begin{aligned}
Z_3(t) &= \frac{1}{2}\int_0^T A^2(t)S^2(t)\,dt \\
&+ \int_0^T A(t)S(t)\xi_1(t)\cos[\Psi_a(t) - \upsilon(t) - \varphi_0]\,dt \\
&- \frac{1}{2}\int_0^T \left[\xi_2^2(t) - \xi_1^2(t)\right]dt. \qquad (5.73)
\end{aligned}
$$

It should be particularly emphasized that all integration operations must be read in the statistical sense.

Consider the process $Z_3(t)$ at the output of integrator 3 determined by Eq. (5.73). The first term of the process $Z_3(t)$, determined by Eq. (5.73), is proportional to the energy of the signal within the limits of the time interval $[0, T]$. The second term of the process $Z_3(t)$, determined by Eq. (5.73), is the random component of the autocorrelation channel of the generalized detector, which is caused by the interaction between the signal and noise. The third term of the process $Z_3(t)$, determined by Eq. (5.73), is the difference between the powers of the noise, which are formed at the outputs of the PF and AF, respectively.

The third term of the process $Z_3(t)$, determined by Eq. (5.73), does not participate in compensation of the noise components of the correlation channel of the generalized detector

$$\int_0^T S(t)S_m(t)\,dt \int_0^T A(t)S_m(t)\xi_1(t)\cos[\Psi_a(t) - \upsilon(t) - \varphi_0]\,dt \tag{5.74}$$

and

$$\left\{\int_0^T S_m(t)\xi_1(t)\,dt\right\}^2, \tag{5.75}$$

which are caused by the interaction between the model signal and noise. The third term of the process $Z_3(t)$, determined by Eq. (5.73), has the following physical sense.

The integral

$$\int_0^T \left[\xi_2^2(t) - \xi_1^2(t)\right]dt \tag{5.76}$$

is the background noise at the output of the generalized detector and tends to approach zero as $T \to \infty$ in the statistical sense (see Section 5.4). The background noise is only used for definition of the threshold K_g during decision making.

Based on this statement, the third term of the process $Z_3(t)$ determined by Eq. (5.73) can be discarded in the following analysis, but we will take it into account in the end result.

The processes at the outputs of the square-law function generators 1 and 2 take the following form

$$\begin{aligned}
Z_1^2(t) &= \frac{1}{4}\cos^2\varphi_0 \int_0^T A^2(t)S(t)S_m(t)\,dt \int_0^T S(t)S_m(t)\,dt \\
&\quad + \frac{1}{4}\left\{\int_0^T S_m(t)\xi_1(t)\cos[\Psi_a(t) - \upsilon(t)]\,dt\right\}^2 \\
&\quad + \frac{1}{2}\cos\varphi_0 \int_0^T S(t)S_m(t)\,dt \int_0^T A(t)S_m(t)\xi_1(t)\cos[\Psi_a(t) - \upsilon(t)]\,dt;
\end{aligned} \tag{5.77}$$

$$\begin{aligned}
Z_2^2(t) &= \frac{1}{4}\sin^2\varphi_0 \int_0^T A^2(t)S(t)S_m(t)\,dt \int_0^T S(t)S_m(t)\,dt \\
&\quad + \frac{1}{4}\left\{\int_0^T S_m(t)\xi_1(t)\sin[\Psi_a(t) - \upsilon(t)]\,dt\right\}^2 \\
&\quad + \frac{1}{2}\sin\varphi_0 \int_0^T S(t)S_m(t)\,dt \int_0^T A(t)S_m(t)\xi_1(t)\sin[\Psi_a(t) - \upsilon(t)]\,dt.
\end{aligned} \tag{5.78}$$

Using the straightforward mathematical transformations,[26,34] it is not difficult to show that

$$\frac{1}{4}\left\{\int_0^T S_m(t)\xi_1(t)\cos[\Psi_a(t)-\upsilon(t)]\,dt\right\}^2+\frac{1}{4}\left\{\int_0^T S_m(t)\xi_1(t)\sin[\Psi_a(t)-\upsilon(t)]\,dt\right\}^2$$
$$=\frac{1}{4}\left\{\int_0^T S_m(t)\xi_1(t)\,dt\right\}^2. \tag{5.79}$$

The process at the output of summator 2 in terms of Eq. (4.79) has the following form

$$Z_\Sigma^2(t)=\frac{1}{4}\int_0^T A^2(t)S(t)S_m(t)\,dt\int_0^T S(t)S_m(t)\,dt$$
$$+\frac{1}{4}\left\{\int_0^T S_m(t)\xi_1(t)\,dt\right\}^2$$
$$+\frac{1}{2}\int_0^T S(t)S_m(t)\,dt\int_0^T A(t)S_m(t)\xi_1(t)\cos[\Psi_a(t)-\upsilon(t)-\varphi_0]\,dt. \tag{5.80}$$

The process at the output of the square-law function generator 7 in terms of Eq. (5.79) takes the following form

$$Z_3^2(t)=\frac{1}{4}\int_0^T A^2(t)S^2(t)\,dt\int_0^T A^2(t)S^2(t)\,dt$$
$$+\frac{1}{2}\left\{\int_0^T A(t)S(t)\xi_1(t)\,dt\right\}^2$$
$$+\int_0^T A^2(t)S^2(t)\,dt\int_0^T A(t)S(t)\xi_1(t)\cos[\Psi_a(t)-\upsilon(t)-\varphi_0]\,dt. \tag{5.81}$$

Considering the second and third terms in Eqs. (5.80) and (5.81), respectively, we can see that they differ by the factor $A^2(t)$ under the condition determined by Eq. (5.64), and the second and third terms in Eqs. (5.80) and (5.81) agree within a factor of 2.

Before proceeding to questions of compensation of these terms in the statistical sense, let us consider two plausible cases.

The first case implies that the amplitude envelope of the signal is not the stochastic function of time within the limits of the time interval $[0, T]$ for a single sample and can be stochastic from sample to sample—the case of slow fluctuations. This case is very important for certain radar systems, where the signal may occur over and over.

The second case is based on the fact that the envelope of amplitude of the signal is the stochastic function of time within the limits of the time interval $[0, T]$ for a single sample, or in other words, the case of rapid fluctuations of the envelope of amplitude of the signal.

We consider these two cases in detail.

5.6.1 The Case of Slow Fluctuations

In this case, the compensation between the second and third terms in Eqs. (5.80) and (5.81) in the statistical sense is conceivable by averaging on a set M of realizations of the input stochastic process $X(t)$. All statistical characteristics of the input stochastic process $X(t)$ are invariant within the limits of the time interval $[0, T]$.

Then Eq. (5.81) may be written in the following form:

$$\sum_{j=1}^{M} Z_{3_j}^2(t) = \frac{1}{4}\sum_{j=1}^{M}\left\{\int_0^T A_j^2(t)S_j^2(t)\,dt \int_0^T A_j^2(t)S_j^2(t)\,dt\right\} + \frac{1}{2}\sum_{j=1}^{M}\left\{\int_0^T A_j(t)S_j(t)\xi_{1_j}(t)\,dt\right\}^2 + \sum_{j=1}^{M}\left\{\int_0^T A_j^2(t)S_j^2(t)\,dt \times \int_0^T A_j(t)S_j(t)\xi_{1_j}(t)\cos\left[\Psi_{a_j}(t) - \upsilon_j(t) - \varphi_{0_j}\right]dt\right\}. \quad (5.82)$$

Eq. (5.80) may be written in the identical form:

$$\sum_{j=1}^{M} Z_{\Sigma_j}^2(t) = \frac{1}{4}\sum_{j=1}^{M}\left\{\int_0^T A_j^2(t)S_j(t)S_{m_j}(t)\,dt \int_0^T S_j(t)S_{m_j}(t)\,dt\right\} + \frac{1}{4}\sum_{j=1}^{M}\left\{\int_0^T S_{m_j}(t)\xi_{1_j}(t)\,dt\right\}^2 + \frac{1}{2}\sum_{j=1}^{M}\left\{\int_0^T S_j(t)S_{m_j}(t)\,dt \times \int_0^T A_j(t)S_{m_j}(t)\xi_{1_j}(t)\cos\left[\Psi_{a_j}(t) - \upsilon_j(t) - \varphi_{0_j}\right]dt\right\}. \quad (5.83)$$

We introduce the designations for the condition determined by Eq. (5.64)

$$E_{a_{1_j}} = \int_0^T S_j(t) S_{m_j}(t)\,dt \tag{5.84}$$

and

$$\sigma^2_{A_j} E_{a_{1_j}} = \int_0^T A_j^2(t) S_j(t) S_{m_j}(t)\,dt, \tag{5.85}$$

where $E_{a_{1_j}}$ is the energy of the signal within the limits of the time interval $[0, T]$ in the j-th realization of the input stochastic process; and $\sigma^2_{A_j}(t)$ is the variance of the amplitude envelope factor $A_j(t)$ of the signal within the limits of the time interval $[0, T]$ in the j-th realization of the input stochastic process. An analogous representation of the variance σ^2_A is used in References 8 and 16.

Let us consider the second term in Eq. (5.82). This term can be represented as a product of two integrals:

$$\frac{1}{2}\sum_{j=1}^{M}\left\{\int_0^T A_j(t) S_j(t) \xi_{1_j}(t)\,dt\right\}^2$$
$$= \frac{1}{2}\sum_{j=1}^{M}\left\{\int_0^T dt \int_0^T A_j^2(t) S_j(t) S_j(\tau) \xi_{1_j}(t) \xi_{1_j}(\tau)\,d\tau\right\}. \tag{5.86}$$

Averaging the integrand with respect to the amplitude envelope factor $A_j(t)$ in the j-th realization for a set M of realizations of the input stochastic process, we can write

$$\int_0^T dt \int_0^T \overline{A_j^2(t)} S_j(t) S_j(\tau) \xi_{1_j}(t) \xi_{1_j}(\tau)\,d\tau$$
$$= \sigma^2_{A_j} \int_0^T dt \int_0^T S_j(t) S_j(\tau) \xi_{1_j}(t) \xi_{1_j}(\tau)\,d\tau. \tag{5.87}$$

In terms of Eqs. (5.85) and (5.87), Eqs. (5.82) and (5.83) can take the following form:

$$\sum_{j=1}^{M} Z^2_{3_j}(t) = \frac{1}{4}\sum_{j=1}^{M}\sigma^2_{A_j} E_{a_{1_j}} \int_0^T A_j^2(t) S_j^2(t)\,dt$$
$$+ \frac{1}{2}\sum_{j=1}^{M}\sigma^2_{A_j}\left\{\int_0^T S_j(t)\xi_{1_j}(t)\,dt\right\}^2$$
$$+ \sum_{j=1}^{M}\sigma^2_{A_j} E_{a_{1_j}} \int_0^T A_j(t) S_j(t) \xi_{1_j}(t) \cos\left[\Psi_{a_j}(t) - \upsilon_j(t) - \varphi_{0_j}\right] dt;$$
$$\tag{5.88}$$

$$\sum_{j=1}^{M} Z_{\Sigma_j}^2(t) = \frac{1}{4}\sum_{j=1}^{M} E_{a_{1_j}} \int_0^T A_j^2(t) S_j(t) S_{m_j}(t)\,dt + \frac{1}{4}\sum_{j=1}^{M}\left\{\int_0^T S_{m_j}(t)\xi_{1_j}(t)\,dt\right\}^2 + \frac{1}{2}\sum_{j=1}^{M} E_{a_{1_j}} \int_0^T A_j(t) S_{m_j}(t)\xi_{1_j}(t)\cos\left[\Psi_{a_j}(t) - \upsilon_j(t) - \varphi_{0_j}\right]dt. \tag{5.89}$$

Comparing Eqs. (5.88) and (5.89), we can see that the second and third terms differ by the value $\sum_{j=1}^{M}\sigma_{A_j}^2$ and by the factor 2. Because of this, the amplifier (>) of the autocorrelation channel of the generalized detector has the amplification factor $\frac{1}{\sigma_A^2}$ and is connected to the input of the compensating channel of the generalized detector.

Taking this fact into account, we may write Eq. (5.88) in the following form:

$$\sum_{j=1}^{M} Z_{3_j}^2(t) = \frac{1}{4}\sum_{j=1}^{M} E_{a_{1_j}} \int_0^T A_j^2(t) S_j^2(t)\,dt + \frac{1}{2}\sum_{j=1}^{M}\left\{\int_0^T S_j(t)\xi_{1_j}(t)\,dt\right\}^2 + \sum_{j=1}^{M} E_{a_{1_j}} \int_0^T A_j(t) S_j(t)\xi_{1_j}(t)\cos\left[\Psi_{a_j}(t) - \upsilon_j(t) - \varphi_{0_j}\right]dt. \tag{5.90}$$

Taking into consideration the condition determined by Eq. (5.64) and the discarded third term of the process $Z_3(t)$ in Eq. (5.73), the process at the output of the compensating channel of the generalized detector—the output of summator 4—takes the following form:

$$Z_g^{out2}(t) = \frac{1}{4}\sum_{j=1}^{M} E_{a_{1_j}} \int_0^T A_j^2(t) S_j^2(t)\,dt + \frac{1}{4}\sum_{j=1}^{M}\left\{\int_0^T \left[\xi_{2_j}^2(t) - \xi_{1_j}^2(t)\right]dt\right\}^2. \tag{5.91}$$

Reference to Eq. (5.91) shows that the compensation between the second and third terms in Eqs. (5.80) and (5.81), in the statistical sense, is performed at the output of the compensating channel of the generalized detector—the output of summator 4—during averaging on a set M of realizations of the input stochastic processes $X(t)$. It should be considered that all integration operations are implied in the statistical sense.

In this conjunction, the process at the output of the generalized detector takes the following form:

$$Z_g^{out}(t) = \frac{1}{2}\sqrt{\sum_{j=1}^{M} E_{a_{1_j}} \int_0^T A_j^2(t) S_j^2(t)\,dt + \sum_{j=1}^{M}\left\{\int_0^T \left[\xi_{2_j}^2(t) - \xi_{1_j}^2(t)\right] dt\right\}^2}. \quad (5.92)$$

The first term in Eq. (5.92) is the energy of the signal at the output of the generalized detector, and the second term in Eq. (5.92) is the background noise at the output of the generalized detector.

5.6.2 Case of Rapid Fluctuations

In this case, the envelope of amplitude of the signal is the stochastic function of a time within the limits of the time interval $[0, T]$ for a single realization of the input stochastic process $X(t)$.

The process at the output of summator 2—the output of the correlation channel of the generalized detector—takes the following form

$$\begin{aligned} Z_\Sigma^2(t) &= \frac{1}{4}\int_0^T S(t) S_m(t)\,dt \int_0^T A^2(t) S(t) S_m(t)\,dt \\ &+ \frac{1}{4}\left\{\int_0^T S_m(t)\xi_1(t)\,dt\right\}^2 \\ &+ \frac{1}{2}\int_0^T S(t) S_m(t)\,dt \int_0^T A(t) S_m(t)\xi_1(t)\cos[\Psi_a(t) - \upsilon(t) - \varphi_0]\,dt. \end{aligned} \quad (5.93)$$

The process at the output of the autocorrelation channel of the generalized detector—the output of square-law function generator 7—takes the following form

$$\begin{aligned} Z_3^2(t) &= \frac{1}{4}\int_0^T A^2(t) S^2(t)\,dt \int_0^T A^2(t) S^2(t)\,dt \\ &+ \frac{1}{2}\left\{\int_0^T A(t) S(t)\xi_1(t)\,dt\right\}^2 \\ &+ \int_0^T A^2(t) S^2(t)\,dt \int_0^T A(t) S(t)\xi_1(t)\cos[\Psi_a(t) - \upsilon(t) - \varphi_0]\,dt. \end{aligned} \quad (5.94)$$

We proceed to introduce the designations in accordance with the results discussed in References 72–75 for the condition determined by Eq. (5.64):

$$E_{a_1} = \int_0^T S(t) S_m(t)\,dt \quad (5.95)$$

and

$$\sigma_A^2 E_{a_1} = \int_0^T A^2(t) S(t) S_m(t)\, dt, \tag{5.96}$$

where E_{a_1} is the energy of the signal within the limits of the time interval $[0, T]$; σ_A^2 is the variance of the amplitude envelope factor $A(t)$ within the limits of the time interval $[0, T]$.

Let us consider the second term in Eq. (5.94) that can be presented as the product of two integrals.[26] Consider the double integral

$$\frac{1}{2}\left\{\int_0^T A(t) S(t) \xi_1(t)\, dt\right\}^2 = \frac{1}{2}\int_0^T dt \int_0^T A(t) A(\tau) S(t) S(\tau) \xi_1(t) \xi_1(\tau)\, d\tau. \tag{5.97}$$

Averaging the integrand with respect to the amplitude envelope factor $A(t)$ within the limits of the time interval $[0, T]$, we can write

$$\int_0^T dt \int_0^T \overline{A(t) A(\tau)} S(t) S(\tau) \xi_1(t) \xi_1(\tau)\, d\tau = \sigma_A^2 \int_0^T dt \int_0^T S(t) S(\tau) \xi_1(t) \xi_1(\tau)\, d\tau. \tag{5.98}$$

Equations (5.93) and (5.94) take the following form in terms of Eqs. (5.96) and (5.98):

$$\begin{aligned} Z_\Sigma^2(t) = {} & \frac{1}{4} E_{a_1} \int_0^T A^2(t) S(t) S_m(t)\, dt \\ & + \frac{1}{4}\left\{\int_0^T S_m(t) \xi_1(t)\, dt\right\}^2 \\ & + \frac{1}{2} E_{a_1} \int_0^T A(t) S_m(t) \xi_1(t) \cos[\Psi_a(t) - \upsilon(t) - \varphi_0]\, dt; \end{aligned} \tag{5.99}$$

$$\begin{aligned} Z_3^2(t) = {} & \frac{1}{4} \sigma_A^2 E_{a_1} \int_0^T A^2(t) S^2(t)\, dt \\ & + \frac{1}{2} \sigma_A^2 \left\{\int_0^T S(t) \xi_1(t)\, dt\right\}^2 \\ & + \sigma_A^2 E_{a_1} \int_0^T A(t) S(t) \xi_1(t) \cos[\Psi_a(t) - \upsilon(t) - \varphi_0]\, dt. \end{aligned} \tag{5.100}$$

Comparing Eqs. (5.99) and (5.100), we can see that the second and third terms differ by the variance σ_A^2 of the amplitude envelope factor and a factor of 2. Because of this, the amplifier (>) of the autocorrelation channel of the generalized detector has the amplification factor $\frac{1}{\sigma_A^2}$ and is connected to the input of the compensating channel of the generalized detector.

Taking this fact into consideration, Eq. (5.100) can be written in the following form:

$$Z_3^2(t) = \frac{1}{4} E_{a_1} \int_0^T A^2(t) S^2(t)\, dt + \frac{1}{2} \left\{ \int_0^T S(t) \xi_1(t)\, dt \right\}^2 + E_{a_1} \int_0^T A(t) S(t) \xi_1(t) \cos[\Psi_a(t) - \upsilon(t) - \varphi_0]\, dt. \qquad (5.101)$$

For the condition determined by Eq. (5.64) and in terms of the discarded third term of the process $Z_3(t)$ in Eq. (5.73), the process at the output of the compensating channel of the generalized detector—the output of summator 4—takes the following form:

$$Z_g^{out^2}(t) = 2Z_\Sigma^2(t) - Z_3^2(t) = \frac{1}{4} E_{a_1} \int_0^T A^2(t) S^2(t)\, dt + \frac{1}{4} \left\{ \int_0^T \left[\xi_2^2(t) - \xi_1^2(t)\right] dt \right\}^2. \qquad (5.102)$$

Equation (5.102) shows that the compensation between the second and third terms in Eqs. (5.80) and (5.81), in the statistical sense, is performed at the output of the compensating channel of the generalized detector—the output of summator 4—during averaging of the input stochastic processes $X(t)$ within the limits of the time interval $[0, T]$. It should be noted that all integration operations are implied in the statistical sense.

The process at the output of the generalized detector takes the following form:

$$Z_g^{out}(t) = \frac{1}{2} \sqrt{ E_{a_1} \int_0^T A^2(t) S^2(t)\, dt + \left\{ \int_0^T \left[\xi_2^2(t) - \xi_1^2(t)\right] dt \right\}^2 }. \qquad (5.103)$$

The first term in Eq. (5.103) is the energy of the signal at the output of the generalized detector, and the second term in Eq. (5.103) is the background noise at the output of the generalized detector.

Thus, if the envelope of amplitude of the signal is the stochastic function of a time within the limits of the time interval $[0, T]$, it is possible to perform a compensation between the noise component of the correlation channel of the generalized detector (which is created by the interaction between the model signal and noise—the second and third terms in Eq. (5.93)) and the random component of the autocorrelation channel of the generalized detector (which is caused by the interaction between the signal and noise—the second and third terms in Eq. (5.94)), without averaging of the input stochastic process $X(t)$ on a set of realizations.

This effect of compensation takes place within the limits of the time interval $[0, T]$. To attain these ends, the amplification factor of the amplifier (>) of the autocorrelation channel of the generalized detector must differ by $\frac{1}{\sigma_A^2}$ in comparison with the amplification factor of the correlation channel of the generalized detector.

Generation of the jointly sufficient statistics of the mean and variance of the likelihood function during the use of the generalized detector for the signals with stochastic amplitude and random initial phase allows us, in principle, to compensate the noise component of the correlation channel of the generalized detector (which is created by the interaction between the model signal and noise) and the random component of the autocorrelation channel of the generalized detector (which is created by the interaction between the signal and noise), using the compensating channel of the generalized detector.

The generalized detector allows us to increase the signal-to-noise ratio at the output of the detector in comparison to the optimal detectors of the classical and modern signal detection theories during the same input conditions.

5.7 Distribution Law

As shown in References 60, 62, and 85–89, and following from Eqs. (5.36)–(5.46), the background noise

$$\sum_{i=1}^{N} \eta_i^2 - \sum_{i=1}^{N} \xi_i^2 \quad \text{or} \quad \int_0^T \eta^2(t)\,dt - \int_0^T \xi^2(t)\,dt \tag{5.104}$$

is formed at the output of the generalized detector under the hypothesis H_0 and the conditions determined by Eqs. (5.51) and (5.52).

Assuming that the process at the output of the generalized detector is averaged, the background noise at the output of the generalized detector can be represented in the following form:

$$Z_g^n = \frac{1}{N}\sum_{i=1}^{N}\eta_i^2 - \frac{1}{N}\sum_{i=1}^{N}\xi_i^2 \tag{5.105}$$

or

$$Z_g^n(t) = \frac{1}{T}\int_0^T \eta^2(t)\,dt - \frac{1}{T}\int_0^T \xi^2(t)\,dt, \tag{5.106}$$

where $[0, T]$ is the time interval, within the limits of which the input stochastic process is observed; ξ_i or $\xi(t)$ is the noise at the output of the preliminary filter (PF) obeying the Gaussian probability distribution density with a zero mean and the finite variance σ_n^2; and η_i or $\eta(t)$ is the noise at the output of the additional filter (AF) obeying the Gaussian probability distribution density with a zero mean and the finite variance σ_n^2.

It is useful to determine the probability distribution density of the background noise at the output of the generalized detector during these conditions. For this purpose, consider the two cases below.

5.7.1 Process at the Input Integrator

Let

$$y = \left(\xi_1^2, \xi_2^2, \ldots, \xi_N^2\right) \tag{5.107}$$

and

$$y^* = \left(\eta_1^2, \eta_2^2, \ldots, \eta_N^2\right). \tag{5.108}$$

It is known[88,89] that the probability distribution density for the random values y and y^* is defined by the χ^2- distribution law with one degree of freedom:

$$f_N(y) = \frac{1}{\sqrt{2\pi y}\sigma_n}\cdot\exp\left(-\frac{y}{2\sigma_n^2}\right), \quad y > 0; \tag{5.109}$$

$$f_N(y^*) = \frac{1}{\sqrt{2\pi y^*}\sigma_n}\cdot\exp\left(-\frac{y^*}{2\sigma_n^2}\right), \quad y^* > 0. \tag{5.110}$$

The mean, or the central moment of the first order, can be determined in the following form:

$$M[y] = \int_0^\infty y f_N(y)\,dy; \tag{5.111}$$

$$M[y^*] = \int_0^\infty y^* f_N(y^*)\,dy^*. \tag{5.112}$$

Substituting Eqs. (5.109) and (5.110) into Eqs. (5.111) and (5.112), respectively, we can write

$$M[y] = \sigma_n^2 \tag{5.113}$$

and

$$M[y^*] = \sigma_n^2, \tag{5.114}$$

where σ_n^2 is the variance of the noise $\xi(t)$ and $\eta(t)$ at the outputs of the PF and AF, respectively.

The mean of the background noise

$$\eta^2(t) - \xi^2(t)$$

(see Fig. 5.2) at the input of the integrator of the generalized detector, in the statistical sense, is determined in the following form:

$$M[y^* - y] = \sigma_n^2 - \sigma_n^2 = 0. \tag{5.115}$$

We define the probability distribution density of the random value

$$y^* - y.$$

It is well known[89] that the probability distribution density for the independent random values $\eta^2(t)$ and $\xi^2(t)$ is given by

$$f_{\eta^2-\xi^2}(Z) = \int_0^\infty f_{\xi^2}(y) f_{\eta^2}(Z + y)\, dy, \tag{5.116}$$

where

$$Z = y^* - y. \tag{5.117}$$

Using Eqs. (5.109)–(5.112), and the tabulated integral[94]

$$\int_0^\infty \frac{\exp(-ax)\, dx}{\sqrt{x^2 + bx}} = \exp\left(\frac{ab}{2}\right) K_0\left(\frac{ab}{2}\right), \quad a > 0, b > 0, \tag{5.118}$$

we can write

$$f_{\eta^2-\xi^2}(Z) = \frac{1}{2\pi\sigma_n^2} \cdot K_0\left(\frac{Z}{2\sigma_n^2}\right), \tag{5.119}$$

where $K_0(x)$ is the modified second-order Bessel function of an imaginary argument or, as it is also called, McDonald's function. The probability distribution density in Eq. (5.119) is fully discussed in References 60, 80, and 90 (see Fig. 5.4).

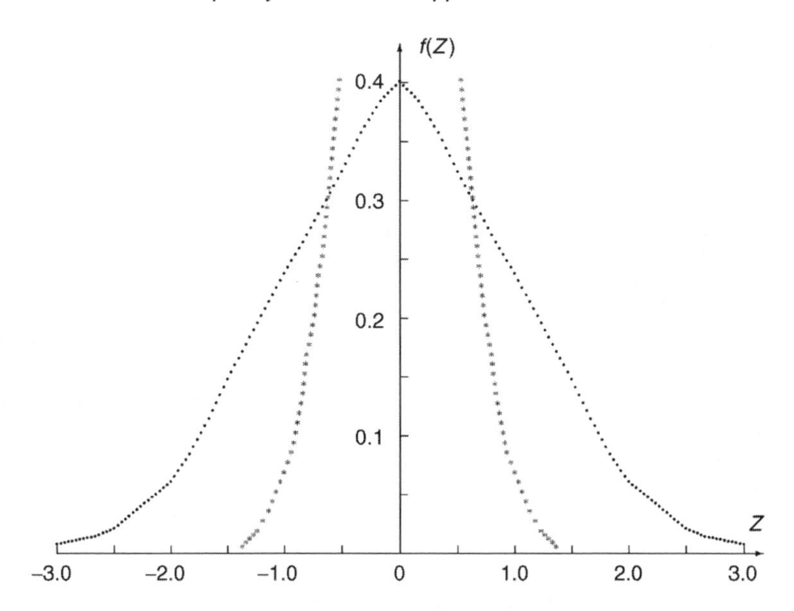

FIGURE 5.4
Probability distribution density: · · ·—Gaussian law; $* * *$—law defined by formula (5.119).

Taking into consideration Eq. (5.119), we can write

$$D[y^* - y] = \frac{1}{\pi \sigma_n^2} \cdot \int_0^\infty (y^* - y)^2 K_0\left(\frac{y^* - y}{2\sigma_n^2}\right) d(y^* - y), \tag{5.120}$$

where $D[.]$ is the central moment of the second order—the variance.

Using the tabulated integral[91]

$$\int_0^\infty x^b K_0(ax)\, dx = 2^{b-1} a^{-b-1} G[0.5(1+b)]^2, \tag{5.121}$$

where $G[.]$ is the gamma function, we can write

$$D[y^* - y] = 4\sigma_n^4. \tag{5.122}$$

5.7.2 Process at the Output Integrator

Assume that the process at the output of the integrator of the generalized detector is averaged within the limits of the time interval $[0, T]$. Then the background noise at the output of the integrator of the generalized detector takes the following form

$$Z_g^n(t) = \frac{1}{T}\int_0^T \eta^2(t)\,dt - \frac{1}{T}\int_0^T \xi^2(t)\,dt. \tag{5.123}$$

Let the process $n(t)$ at the input of the preliminary and additional filters (PF and AF, see Fig. 5.2) be the Gaussian process with a zero mean and the covariance function

$$K_n(t_1, t_2) = M[n(t_1)n(t_2)]. \tag{5.124}$$

When the properties of the process $n(t)$ are studied and the process $n(t)$ is a known function within the limits of the time interval $[0, T]$, it is convenient to represent the process $n(t)$ in the form of a set of the orthogonal functions $\varphi_k(t), k = 0, 1, 2, \ldots$.

Thus,

$$n(t) = \sum_{j=0}^{\infty} n_j\varphi_j(t), \quad t \in [0, T], \tag{5.125}$$

where

$$n_j = \int_0^T n(t)\varphi_j(t)\,dt, \quad j = 0, 1, 2, \ldots \tag{5.126}$$

and

$$\int_0^T \varphi_i(t)\varphi_j(t)\,dt = \delta_{ij}, \quad i, j = 0, 1, 2, \ldots, \tag{5.127}$$

δ_{ij} is the Kronecker symbol.

As is well known,[92] the solution of the uniform Fredholm equation with the kernel $K_n(t_1, t_2)$ is conveniently used by way of terms of a set of the orthogonal functions φ_i under an analogous representation of the process $n(t)$. In this case, the terms of a set $\varphi_n(t)$ are determined in the following form:

$$\lambda_j\varphi_j(t) = \int_0^T K_n(t, t_2)\varphi_j(t_2)\,dt_2, \quad t \in [0, T], \quad j = 0, 1, 2, \ldots. \tag{5.128}$$

It can easily be shown that n_j are the independent random variables obeying the Gaussian probability distribution density with a zero mean and the variance

$$D[n_j] = \lambda_j \quad j = 0, 1, 2, \ldots. \tag{5.129}$$

In the subsequent discussion we assume that

$$\lambda_0 \geq \lambda_1 \geq \lambda_2 \geq \ldots . \tag{5.130}$$

Consider the average power of the process $n(t)$ within the limits of the time interval $[0, T]$:

$$y = \frac{1}{T}\int_0^T n^2(t)\,dt = \frac{1}{T}\sum_{j=0}^{\infty} n_j^2 = \sum_{j=0}^{\infty} C_j X_j^2, \tag{5.131}$$

where

$$X_j = \frac{n_j}{\sqrt{\lambda_j}} \quad \text{and} \quad C_j = \frac{\lambda_j}{T}. \tag{5.132}$$

In Eq. (5.131) y is the linear combination of the random variables X_j. The values X_j are the independent random values obeying the Gaussian probability distribution density with a zero mean and variance equal to 1.

The characteristic function of the random value y is determined in the following form:[55]

$$\begin{aligned}\Theta(v) &= M[\exp(ivy)] = M\left[\exp\left(iv\sum_{j=0}^{\infty} C_j X_j^2\right)\right] \\ &= \prod_{j=0}^{\infty} M\left[\exp\left(ivC_j X_j^2\right)\right] = \prod_{j=0}^{\infty}\frac{1}{\sqrt{1-2ivC_j}}.\end{aligned} \tag{5.133}$$

The reader should bear in mind that only a positive sign ahead of the square root is allowed for any mathematical expressions containing the square root of an imaginary value. This allows us to bound the study of the phase angles within the limits of the phase interval

$$\left[-\frac{\pi}{2}, \frac{\pi}{2}\right].$$

The semi-invariants of the random value y can be determined using the characteristic function in Eq. (5.133). In accordance with the definition of the semi-invariants C_j,[93] we can write

$$\log\Theta(v) = \sum_{\nu=1}^{\infty}\frac{C_\nu}{\nu!}\cdot(i\nu)^\nu. \tag{5.134}$$

Equation (5.133) and the expression

$$\log(1-X) = -\sum_{n=1}^{\infty}\frac{X^n}{n} \tag{5.135}$$

show that

$$\log \Theta(v) = -\frac{1}{2}\sum_{j=0}^{\infty}\log(1-2ivC_j)$$

$$= \frac{1}{2}\sum_{k=1}^{\infty}\sum_{j=0}^{\infty}\frac{(2ivC_j)^k}{k} = \sum_{k=1}^{\infty}\frac{(iv)^k C_k}{k!}; \quad (5.136)$$

$$C_k = (k-1)!\,2^{k-1}\sum_{j=0}^{\infty} C_j^k. \quad (5.137)$$

When the semi-invariants are known, the moments of the random value y can be easily determined.[53] Equation (5.136) can be represented in the manner, where C_j are not found in an explicit form.

Taking into account the well-known expression[54,55]

$$K_n(t_1, t_2) = \sum_{j=0}^{\infty}\lambda_j\varphi_j(t_1)\varphi(t_2), \quad (5.138)$$

and the orthogonal properties of the terms φ_j, we can determine

$$C_k = \frac{(k-1)!\,2^{k-1}}{T^{(k)}}\int_0^T K_n^{(k)}(t_1, t_2)\,dt_1, \quad (5.139)$$

where $K_n^{(k)}(t_1, t_2)$ are the iteration kernels, which are determined by the formula

$$K_n^{(1)}(t_1, t_2) = K_n(t_1, t_2); \quad (5.140)$$

$$K_n^{(L)}(t_1, t_2) = \int_0^T K_n(t_1, x)K_n^{(L-1)}(x, t_2)\,dx, \quad L = 2, 3, \ldots. \quad (5.141)$$

The determination of the higher-order iteration kernels is a very difficult problem. Equation (5.139) has been written in the form suggested by Rice[93,94] as a result of computation of the first four semi-invariants of the random value y.

The probability distribution density of the random value y, using the Fourier transform with respect to the characteristic function $\Theta(v)$, takes the following form:

$$f(y) = \left(\frac{1}{2T}\right)^{\int_{-\infty}^{\infty}\frac{\exp(-ivy)dy}{\prod_{j=0}^{\infty}\frac{1}{\sqrt{1-2ivC_j}}}}, \quad (5.142)$$

and the probability distribution function can be written in the following form:

$$F(y) = 1 - \int_{y}^{\infty} f(x)\, dx. \tag{5.143}$$

We proceed to use the results in Eqs. (5.131)–(5.143). Consider the random process $\xi^2(t)$ within the limits of the time interval $[0, T]$. It is known from the initial conditions that the random process $\xi(t)$ obeys the Gaussian probability distribution density with a zero mean and the correlation function $R_\xi(\tau)$.

Let

$$\zeta_T = \frac{1}{T}\int_{0}^{T} \xi^2(t)\, dt, \tag{5.144}$$

where the integral is understood in the statistical sense.

The characteristic function of the random value ζ_T can be determined using Eq. (5.133). On this basis Eq. (5.133) can be written in the following form:

$$\Theta(v) = \prod_{j} \frac{1}{\sqrt{1 - \frac{2iv\sigma_n^2}{\lambda_j}}}. \tag{5.145}$$

The characteristic values λ_j in Eq. (5.145) must be determined using the equation

$$f(Z) = \frac{\lambda}{\sigma_n^2}\int_{-\infty}^{\infty} R(\tau - Z) h(\tau) f(\tau)\, d\tau, \tag{5.146}$$

where

$$h(\tau) = \frac{1}{T}, \quad \tau \in [0, T] \tag{5.147}$$

and

$$h(\tau) = 0, \quad \tau \notin [0, T], \tag{5.148}$$

and the correlation function $R(\tau - Z)$ in Eq. (5.106) is the same as the correlation function $R_\xi(\tau)$.

If the random process $\xi(t)$ is the result of passing the stochastic process $n(t)$ through the RLC oscillatory circuit with the resonant amplitude-frequency response (the PF or AF), then

$$R_\xi(\tau) = \sigma_n^2 \exp(-\beta|\tau|), \quad \beta = \frac{R}{2L} > 0. \tag{5.149}$$

Thus, in the case being considered the characteristic values defining the probability distribution density of the random process ζ_T can be determined using the integral equation

$$\lambda \int_0^T R_\xi(Z-\tau) f(\tau)\, d\tau = T f(Z). \tag{5.150}$$

There are no difficulties involved in transforming Eq. (5.150) into a differential equation of the second order. This differential equation is linear.

Actually, representing Eq. (5.150) in terms of Eq. (5.149) in the following form

$$\frac{T}{\lambda\sigma_n^2} \cdot f(Z) = \int_0^Z \exp[\beta(Z-\tau)] f(\tau)\, d\tau + \int_Z^T \exp[-\beta(\tau - Z)] f(\tau)\, d\tau \tag{5.151}$$

and differentiating both sides of this equation twice with respect to Z, we can write

$$\frac{T}{\lambda\sigma_n^2} \cdot f''(Z) = \beta^2 \int_0^Z \exp[-\beta(Z-\tau)] f(\tau)\, d\tau + \beta^2 \int_Z^T \exp[-\beta(\tau - Z)] f(\tau)\, d\tau - 2\beta f(Z). \tag{5.152}$$

Then

$$f''(Z) = \frac{\lambda\sigma_n^2}{T} \cdot \left[\beta^2 \int_0^T \exp[-\beta|Z-\tau|] f(\tau)\, d\tau - 2\beta f(Z)\right] \tag{5.153}$$

or

$$f''(Z) + \left(\frac{2\lambda\sigma_n^2}{\beta T} - 1\right)\beta^2 f(Z) = 0. \tag{5.154}$$

The general solution of Eq. (5.154), as is well known, takes the following form

$$f(Z) = C_1 \exp(ibZ\beta) + C_2 \exp(-ibZ\beta), \tag{5.155}$$

where

$$b^2 = \frac{2\lambda\sigma_n^2}{T\beta} - 1. \tag{5.156}$$

Substituting Eq. (5.155) into Eq. (5.150), we can see that this solution is the eigenfunction of the integral equation only if the following condition

$$b^2 > 0, \tag{5.157}$$

is satisfied, and b must be satisfied by one of the transcendental equations

$$\begin{cases} b\,\mathrm{tg}\,(\beta T b) = 1; \\ b\,\mathrm{ctg}\,(\beta T b) = -1. \end{cases} \tag{5.158}$$

Thus, the characteristic values in Eq. (5.150), for the random process with the correlation function determined in Eq. (5.149), are equal to

$$\lambda_k = \frac{T\beta}{2\sigma_n^2} \cdot \left(1 + b_k^2\right), \tag{5.159}$$

as long as the kernel in Eq. (5.142) is positive, and all eigenvalues in Eq. (5.142) are positive in magnitude.

The mean and variance of the random process ζ_T are equal to

$$M[\zeta_T] = \sigma_n^2; \tag{5.160}$$

$$D[\zeta_T] = \frac{\sigma_n^4}{4\kappa^2} \cdot [4\kappa - 1 + \exp(-4\kappa)], \tag{5.161}$$

where

$$\kappa = \frac{T\beta}{2}. \tag{5.162}$$

As $T \to \infty$, the variance is decreased by the formula

$$D[\zeta_T] \sim 2 \cdot \frac{\sigma_n^4}{T\beta}.$$

As a result, the probability distribution density of the process ζ_T is the asymptotic Gaussian probability distribution density

$$f_{\zeta_T}(Z) \sim \frac{1}{\sigma_n^2} \cdot \sqrt{\frac{\kappa}{2\pi}} \cdot \exp\left(-\kappa \cdot \frac{Z - \sigma_n^2}{2\sigma_n^4}\right) \tag{5.163}$$

with the parameters

$$\left(\sigma_n^2, \frac{2\sigma_n^4}{T\beta}\right).$$

The background noise at the output of the generalized detector is the difference between the random processes $\eta^2(t)$ and $\xi^2(t)$ within the limits of the time interval $[0, T]$. It is also known[88,89] that the probability distribution density of the difference between the random Gaussian values obeys the Gaussian probability distribution density.

Therefore, at the high value $T\beta$ the probability distribution density of the background noise at the output of the integrator of the generalized detector is the asymptotic Gaussian probability distribution density

$$f_{\eta^2-\xi^2}(Z) = \frac{1}{2\sigma_n^2} \cdot \sqrt{\frac{\kappa}{\pi}} \cdot \exp\left(-\frac{\kappa Z^2}{4\sigma_n^4}\right) \tag{5.164}$$

with the parameters

$$\left(0, \frac{4\sigma_n^4}{T\beta}\right).$$

Thus, the probability distribution density of the background noise at the output of the integrator of the generalized detector has been determined as $T \to 0$ and $T \to \infty$. For intervening values within the limits of the time interval $[0, T]$, the probability distribution density of the background noise can be determined using results in References 93 and 94 in terms of Reference 55. The referenced methods introduce errors of no more than 1%.

Consider the particular case when the noise $n(t)$ is the narrow-band stochastic process

$$n(t) = \xi_n(t)\cos[\omega_n t + \upsilon(t)], \tag{5.165}$$

where $\xi_n(t)$ is the random envelope of amplitude of the noise $n(t)$; ω_n is the medium frequency of the noise $n(t)$; and $\upsilon(t)$ is the random phase of the noise $n(t)$.

In accordance with References 51, 52, and 98 and the foregoing statements, the narrow-band stochastic processes

$$\xi(t) = \xi_1(t)\cos[\omega_n t + \upsilon(t)] \tag{5.166}$$

and

$$\eta(t) = \xi_2(t)\cos[\omega_n' t + \upsilon'(t)] \tag{5.167}$$

are formed at the outputs of the PF and AF, respectively, where $\xi_1(t)$ and $\xi_2(t)$ are the random envelopes of amplitudes.

Note that the processes $\xi_1(t)$ and $\xi_2(t)$ are the random envelopes of amplitudes of the noise $\xi(t)$ and $\eta(t)$ that are formed at the outputs of the PF and AF, respectively. In doing so, the noise $\xi(t)$ and $\eta(t)$ obey the Gaussian probability distribution density, but the envelopes $\xi_1(t)$ and $\xi_2(t)$ of the amplitudes of the noise $\xi(t)$ and $\eta(t)$ at the outputs of the PF and AF, respectively, obey the Rayleigh probability distribution density.

We define the probability distribution density of the background noise at the output of the generalized detector, during detection of the quasideterministic signal with a random initial phase in the presence of additive Gaussian noise, using the techniques and conditions discussed in References 79–81: a "no" signal exists in the input stochastic process $Y(t)$, and the model signal generator (MSG) (see Fig. 5.3) is turned off:

$$a_1(t) = 0 \qquad \text{and} \qquad a_m(t) = 0. \tag{5.168}$$

The background noise at the output of the generalized detector, determined by Eqs. (5.105) and (5.106) during detection of the stochastic signals in the presence of additive Gaussian noise, is the difference between squares of the envelopes $\xi_1(t)$ and $\xi_2(t)$ of amplitudes of the noise $\xi(t)$ and $\eta(t)$ that are formed at the outputs of the PF and AF, respectively.

Using the results discussed in References 74–81, we can express the background noise at the output of the generalized detector in the following form—the factor 0.5 can be dropped:

$$Z_g^{out}(t) = \frac{1}{T}\int_0^T \xi_2^2(t)dt - \frac{1}{T}\int_0^T \xi_1^2(t)\,dt. \tag{5.169}$$

As $T \to 0$, we can write

$$\xi_1^2(t) = \frac{1}{T}\int_0^T \xi_1^2(t)\,dt; \tag{5.170}$$

$$\xi_2^2(t) = \frac{1}{T}\int_0^T \xi_2^2(t)\,dt. \tag{5.171}$$

It is well known[88,89] that the probability distribution density of the Rayleigh random value x^2 with a zero mean and the finite variance σ_n^2 has the following form:

$$f(y) = \frac{1}{2\sigma_n^2} \cdot \exp\left(-\frac{y}{2\sigma_n^2}\right), \tag{5.172}$$

where $y = x^2$.

Thus, the probability distribution densities of the envelopes $\xi_1^2(t)$ and $\xi_2^2(t)$ of amplitudes of the noise $\xi(t)$ and $\eta(t)$ that are formed at the outputs of the PF and AF, respectively, are equal to

$$f_{\xi_1^2}(y_1) = \frac{1}{2\sigma_n^2} \cdot \exp\left(-\frac{y_1}{2\sigma_n^2}\right); \tag{5.173}$$

$$f_{\xi_2^2}(y_2) = \frac{1}{2\sigma_n^2} \cdot \exp\left(-\frac{y_2}{2\sigma_n^2}\right) \tag{5.174}$$

on the assumption that the envelopes $\xi_1(t)$ and $\xi_2(t)$ have a zero mean and the same finite variances σ_n^2.

Let

$$|z(t)| = y_2 - y_1, \tag{5.175}$$

where

$$y_1 = \left(\xi_1^2(t_1), \xi_1^2(t_2), \ldots, \xi_1^2(t_N)\right); \tag{5.176}$$

$$y_2 = \left(\xi_2^2(t_1), \xi_2^2(t_2), \ldots, \xi_2^2(t_N)\right). \tag{5.177}$$

Since the envelopes $\xi_1^2(t)$ and $\xi_2^2(t)$ of amplitudes of the noise $\xi(t)$ and $\eta(t)$ at the outputs of the PF and AF, respectively, are uncorrelated owing to the choice of the amplitude-frequency responses of the PF and AF, the probability distribution density of the random value $|z(t)|$ can be determined in the following form:[65]

$$f_{\xi_2^2-\xi_1^2}(|z|) = \int_0^\infty f_{\xi_1^2}(y_1) f_{\xi_2^2}(|z| + y_1)\, dy_1. \tag{5.178}$$

In this case,

$$\begin{aligned} f_{\xi_2^2-\xi_1^2}(|z|) &= \frac{1}{4\sigma_n^4} \int_0^\infty \exp\left(-\frac{y_1}{2\sigma_n^2}\right) \cdot \exp\left(-\frac{|z| + y_1}{2\sigma_n^2}\right) dy_1 \\ &= \frac{1}{4\sigma_n^4} \cdot \exp\left(-\frac{|z|}{2\sigma_n^2}\right) \int_0^\infty \exp\left(-\frac{y_1}{\sigma_n^2}\right) dy_1. \end{aligned} \tag{5.179}$$

Using the tabulated integral[91]

$$\int_0^\infty x^n \exp(-ax)\, dx = \frac{\Gamma(n+1)}{a^{n+1}}, \quad n > -1, \quad a > 0, \tag{5.180}$$

where $\Gamma(x)$ is the gamma function, we can write

$$f_{\xi_2^2-\xi_1^2}(|z|) = \frac{1}{4\sigma_n^2} \cdot \exp\left(-\frac{|z|}{2\sigma_n^2}\right), \tag{5.181}$$

where

$$\frac{1}{2\sigma_n^2} \cdot \exp\left(-\frac{|z|}{2\sigma_n^2}\right) \tag{5.182}$$

is the exponential-type probability distribution density.

Thus, as $T \to 0$, the probability distribution density of the background noise at the output of the generalized detector during detection of the signals with the stochastic amplitude and random initial phase in the presence of additive Gaussian noise obeys the exponential-type probability distribution density with the constant factor 0.5.

An analogous result is obtained in References 79–81. This very important fact must be taken into account for determination of the detection performances of the narrow-band stochastic signal with the random initial phase in the presence of additive Gaussian noise.

We consider a square of the envelope $\xi_1(t)$ of amplitude of the noise $\xi(t)$, which is formed at the output of the PF within the limits of the time interval $[0, T]$. Let

$$\zeta_{1_T} = \frac{1}{T}\int_0^T \xi_1^2(t)\,dt. \tag{5.183}$$

The characteristic function of the random value ζ_{1_T} can be determined using the general formula[93,94]

$$\Theta_1'(v) = \prod_j \frac{1}{\sqrt{1 - \frac{2iv\sigma_n^2}{\lambda_j}}}. \tag{5.184}$$

The characteristic values λ_j, which are included in the characteristic function, must be determined using the equation

$$\lambda \int_0^T R_\xi(z-\tau) f(\tau)\,d\tau = Tf(z), \tag{5.185}$$

where $R_\xi(\tau)$ is the correlation function of the quadrature component of the noise $\xi(t)$ at the output of the PF.

If the noise $\xi(t)$ is the result of passing the noise $n(t)$ through the PF or AF with the resonant amplitude-frequency response, then we can write

$$R_\xi(\tau) = \sigma_n^2 \exp(-\beta|\tau|), \quad \beta = \frac{R}{2L} > 0. \tag{5.186}$$

In this case, the characteristic values λ_j can be determined using the transcendental equation[92]

$$f''(z) + \left(\frac{2\lambda\sigma_n^2}{T\beta} - 1\right)\beta^2 f(z) = 0. \tag{5.187}$$

The mean and variance of the random value ζ_{1_T} are equal to

$$M[\zeta_{1_T}] = \sigma_n^2; \tag{5.188}$$

$$D[\zeta_{1_T}] = \frac{\sigma_n^4}{2\kappa^2} \cdot [4\kappa - 1 + \exp(-4\kappa)], \tag{5.189}$$

where the parameter κ is determined by Eq. (5.161).

As $T \to \infty$, the variance decreases according to the formula

$$D\left[\zeta_{1_T}\right] \sim \frac{2\sigma_n^4}{T\beta}. \tag{5.190}$$

In the process, the probability distribution density of square of the envelope $\xi_1(t)$ of amplitude of the noise $\xi(t)$ at the output of the PF obeys the asymptotic Gaussian probability distribution density with the parameters

$$\left(\sigma_n^2, \frac{2\sigma_n^4}{T\beta}\right).$$

The background noise of the generalized detector, during detection of the quasideterministic signal with the random initial phase in the presence of additive Gaussian noise, is determined by Eq. (5.169). Therefore, the probability distribution density of the background noise at the output of the generalized detector obeys the asymptotic Gaussian probability distribution density with the parameters

$$\left(0, \frac{4\sigma_n^4}{T\beta}\right)$$

as $T \to \infty$.

The analysis performed allows us to draw the following conclusions.

- The probability distribution density of the background noise at the output of the generalized detector has been determined for limiting values of the time interval $[0, T]$. The probability distribution density of the background noise at the output of the generalized detector is determined by Eq. (5.119) with the parameters

  $$(0, 4\sigma_n^4)$$

  when the time interval $[0, T]$ is infinitesimal. The probability distribution density of the background noise at the output of the generalized detector tends to approach the asymptotic Gaussian probability distribution density with the parameters

  $$\left(0, \frac{4\sigma_n^4}{T\beta}\right)$$

  at high values of the time interval $[0, T]$.
- If the noise $n(t)$ is the narrow-band stochastic process, then the probability distribution density of the background noise at the output of the generalized detector is defined by the exponential-type probability distribution density when the time interval $[0, T]$ is infinitesimal, and by the asymptotic Gaussian probability distribution density at high values of the time interval $[0, T]$.

5.8 Conclusions

We summarize briefly the main results discussed in this chapter.

- The proposed modification of the initial premises of the classical and modern signal detection theories assumes that there exists a frequency time region of the noise, where a "yes" signal may be found, and there exists a frequency time region of the noise, where it is known *a priori* that a "no" signal exists. This modification allows us to perform the theoretical synthesis of the generalized signal detection algorithm. Two uncorrelated samples are used, one of which is the reference sample, since it is known *a priori* that a "no" signal is found in this sample. This fact allows us to obtain the jointly sufficient statistics of the mean and variance of the likelihood function. The optimal signal detection algorithms of classical and modern theories, for the signals with known and unknown amplitude-phase-frequency structure, allow us to obtain only the sufficient statistic of the mean of the likelihood function and are components of the generalized signal detection algorithm.
- The physicotechnical interpretation of the generalized approach to signal processing in the presence of noise is a composite combination of optimal signal detection approaches of the classical and modern signal detection theories for the signals with both known and unknown amplitude-phase-frequency structure. The additional filter (AF) is the source of the reference sample. The resonant frequency of the AF is detuned relative to that of the preliminary filter (PF). The value of the detuning is greater than the effective spectral bandwidth of the signal. The use of the AF jointly with the PF forms the background noise at the output of the generalized detector. The background noise is the difference between the energy characteristics of the noise at the outputs of the PF and AF and tends to approach zero in the statistical sense. In other words, the background noise at the output of the generalized detector is formed as a result of generation of the jointly sufficient statistics of the mean and variance of the likelihood function for the generalized approach to signal processing in the presence of noise. The background noise is caused by both the noise at the output of the PF and the noise at the output of the AF. The background noise at the output of the generalized detector is independent of both the signal and the model signal.
- The correlation between the noise component

$$\sum_{i=1}^{N} a_i^* \xi_i$$

or

$$\int_0^T a_m(t)\xi(t)\,dt$$

of the correlation channel of the generalized detector and the random component

$$\sum_{i=1}^{N} a_{1_i}\xi_i$$

or

$$\int_0^T a_1(t)\xi(t)\,dt$$

of the autocorrelation channel of the generalized detector allows us to generate the jointly sufficient statistics of the mean and variance of the likelihood function. The noise component of the correlation channel of the generalized detector is caused by the interaction between the model signal and noise. The random component of the autocorrelation channel of the generalized detector is caused by the interaction between the signal and noise. The effect of compensation, between the noise component of the correlation channel of the generalized detector and the random component of the autocorrelation channel of the generalized detector, is caused by the generation of the jointly sufficient statistics of the mean and variance of the likelihood function for the generalized approach to signal processing in the presence of noise under employment of the generalized detectors in various complex signal processing systems. The effect of this compensation is carried out within the limits of the sample size $[1, N]$ or within the limits of the time interval $[0, T]$, for which the input stochastic process is observed.

- The use of generalized detectors for the signals with the stochastic amplitude and random initial phase in various complex signal processing systems has the following peculiarity. If the envelope of amplitude of the signal is not a stochastic function of time within the limits of the time interval $[0, T]$ for a single realization of the input stochastic process and can be stochastic from realization to realization, the generation of the jointly sufficient statistics of the mean and variance of the likelihood function is carried out by averaging on a set M of realizations of the input stochastic processes. If the envelope of amplitude of the signal is a stochastic function of time within the limits of the time interval $[0, T]$, then the generation of the jointly sufficient statistics of the mean and variance of the likelihood function is possible within the limits of the time interval $[0, T]$ without averaging the input stochastic processes on a set of realizations.

- The probability distribution density of the background noise at the output of the generalized detector has been defined for limiting values of the time interval $[0, T]$. The probability distribution density of the background noise, at the output of the generalized detector, is defined using McDonald's function with the parameters

$$(0, 4\sigma_n^4)$$

when the time interval $[0, T]$ is infinitesimal. The probability distribution density of the background noise at the output of the generalized detector tends to approach the asymptotic Gaussian probability distribution density with the parameters

$$\left(0, \frac{4\sigma_n^4}{T\beta}\right)$$

at high values of the time interval $[0, T]$. If the noise $n(t)$ is the narrowband stochastic process, then the probability distribution density of the background noise at the output of the generalized detector is determined by the exponential-type probability distribution density when the time interval $[0, T]$ is infinitesimal, and by the asymptotic Gaussian probability distribution density with the parameters

$$\left(0, \frac{4\sigma_n^4}{T\beta}\right)$$

at high values of the time interval $[0, T]$.

References

1. Kotelnikov, V., *Potential Noise Immunity Theory*, Soviet Radio, Moscow, 1956 (in Russian).
2. Wiener, N., *Non-Linear Problems in Stochastic Process Theory*, McGraw-Hill, New York, 1959.
3. Middleton, D., *An Introduction to Statistical Communication Theory*, McGraw-Hill, New York, 1960.
4. Shannon, K., *Research on Information Theory and Cybernetics*, McGraw-Hill, New York, 1961.
5. Wiener, N., *Cybernetics or Control and Communication in the Animal and the Machine*, 2nd ed., Wiley, New York, 1961.

6. Selin, I., *Detection Theory*, Princeton University Press, Princeton, NJ, 1965.
7. Miller, R., *Simultaneous Statistical Inference*, McGraw-Hill, New York, 1966.
8. Van Trees, H., *Detection, Estimation and Modulation Theory. Part I: Detection, Estimation, and Linear Modulation Theory*, Wiley, New York, 1968.
9. Helstrom, C., *Statistical Theory of Signal Detection*, 2nd ed., Pergamon Press, Oxford, 1968.
10. Gallager, R., *Information Theory and Reliable Communication*, Wiley, New York, 1968.
11. Thomas, J., *An Introduction to Statistical Communication Theory*, Wiley, New York, 1969.
12. Jazwinski, A., *Stochastic Processes and Filtering Theory*, Academic Press, New York, 1970.
13. Van Trees, H., *Detection, Estimation and Modulation Theory. Part II: Non-Linear Modulation Theory*, Wiley, New York, 1970.
14. Schwartz, M., *Information, Transmission, Modulation, and Noise*, 2nd ed., McGraw-Hill, New York, 1970.
15. Wong, E., *Stochastic Processes in Information and Dynamical Systems*, McGraw-Hill, New York, 1971.
16. Van Trees, H., *Detection, Estimation and Modulation Theory. Part III: Radar-Sonar Signal Processing and Gaussian Signals in Noise*, Wiley, New York, 1972.
17. Box, G. and Tiao, G., *Bayesian Inference in Statistical Analysis*, Addison-Wesley, Cambridge, MA, 1973.
18. Stratonovich, R., *Principles of Adaptive Processing*, Soviet Radio, Moscow, 1973 (in Russian).
19. Levin, B., *Theoretical Foundations of Statistical Radio Engineering. Parts I–III*, Soviet Radio, Moscow, 1974–1976 (in Russian).
20. Tikhonov, V. and Kulman, N., *Non-Linear Filtering and Quasideterministic Signal Processing*, Soviet Radio, Moscow, 1975 (in Russian).
21. Repin, V. and Tartakovskiy, G., *Statistical Synthesis under a priori Uncertainty and Adaptation of Information Systems*, Soviet Radio, Moscow, 1977 (in Russian).
22. Kulikov, E. and Trifonov, A., *Estimation of Signal Parameters in Noise*, Soviet Radio, Moscow, 1978 (in Russian).
23. Sosulin, Yu., *Detection and Estimation Theory of Stochastic Signals*, Soviet Radio, Moscow, 1978 (in Russian).
24. Ibragimov, I. and Rozanov, Y., *Gaussian Random Processes*, Springer-Verlag, New York, 1978.
25. Anderson, B. and Moore, J., *Optimal Filtering*, Prentice-Hall, Englewood Cliffs, NJ, 1979.
26. Shirman, Y. and Manjos, V., *Theory and Methods in Radar Signal Processing*, Radio and Svyaz, Moscow, 1981 (in Russian).
27. Huber, P., *Robust Statistics*, Wiley, New York, 1981.
28. Blachman, N., *Noise and Its Effect in Communications*, 2nd ed., Krieger, Malabar, FL, 1982.
29. Bacut, P. et al. *Signal Detection Theory*, Radio and Svyaz, Moscow, 1984 (in Russian).
30. Anderson, T., *An Introduction to Multivariate Statistical Analysis*, 2nd ed., Wiley, New York, 1984.
31. Lehmann, E., *Testing Statistical Hypotheses*, 2nd ed., Wiley, New York, 1986.

32. Silverman, B., *Density Estimation for Statistics and Data Analysis*, Chapman & Hall, London, 1986.
33. Bassevillee, M. and Benveniste, A., *Detection of Abrupt Changes in Signals and Dynamical Systems*, Springer-Verlag, Berlin, 1986.
34. Trifonov, A. and Shinakov, Yu., *Joint Signal Differentiation and Estimation of Signal Parameters in Noise*, Radio and Svyaz, Moscow, 1986 (in Russian).
35. Thomas, A., *Adaptive Signal Processing: Theory and Applications*, Wiley, New York, 1986.
36. Blahut, R., *Principles of Information Theory*, Addison-Wesley, Reading, MA, 1987.
37. Weber, C., *Elements of Detection and Signal Design*, Springer-Verlag, New York, 1987.
38. Skolnik, M., *Radar Applications*, IEEE Press, New York, 1988.
39. Kassam, S., *Signal Detection in Non-Gaussian Noise*, Springer-Verlag, Berlin, 1988.
40. Poor, V., *Introduction to Signal Detection and Estimation*, Springer-Verlag, New York, 1988.
41. Brook, D. and Wynne, R., *Signal Processing: Principles and Applications*, Pentech Press, London, 1988.
42. Porter, W. and Kak, S., *Advances in Communications and Signal Processing*, Springer-Verlag, Berlin, 1989.
43. Adrian, C., *Adaptive Detectors for Digital Modems*, Pentech Press, London, 1989.
44. Scharf, L., *Statistical Signal Processing, Detection, Estimation, and Time Series Analysis*, Addison-Wesley, Reading, MA, 1991.
45. Cover, T. and Thomas, J., *Elements of Information Theory*, Wiley, New York, 1991.
46. Basseville, M. and Nikiforov, I., *Detection of Abrupt Changes*, Prentice-Hall, Englewood Cliffs, NJ, 1993.
47. Dudgeon, D. and Johnson, D., *Array Signal Processing: Concepts and Techniques*, Prentice-Hall, Englewood Cliffs, NJ, 1993.
48. Porat, B., *Digital Processing of Random Signals: Theory and Methods*, Prentice-Hall, Englewood Cliffs, NJ, 1994.
49. Helstrom, C., *Elements of Signal Detection and Estimation*, Prentice-Hall, Englewood Cliffs, NJ, 1995.
50. McDonough, R. and Whallen, A., *Detection of Signals in Noise*, 2nd ed., Academic Press, New York, 1995.
51. Tuzlukov, V., *Signal Detection Theory*, Springer-Verlag, New York, 2001.
52. Tuzlukov, V., *Signal Processing in Noise: A New Methodology*, IEC, Minsk, 1998.
53. Crammer, H., *Mathematical Methods of Statistics*, Princeton University Press, Princeton, NJ, 1946.
54. Grenander, U., *Stochastic Processes and Statistical Inference*, Arkiv Mat, Uppsala, Sweden, 1950.
55. Rao, C., *Advanced Statistical Methods in Biometric Research*, Wiley, New York, 1952.
56. Kolmogorov, A., Theory of data transmission, *Reports of the Academy of Sciences of the USSR*, No. 2, 1957, pp. 66–99 (in Russian).
57. Maximov, M., Joint correlation of fluctuating noise at the outputs of frequency filters, *Radio Eng.*, No. 9, 1956, pp. 28–38 (in Russian).
58. Chernyak, Y., Joint correlation of noise voltage at the outputs of amplifiers with non-overlapping responses, *Radio Phys. and Elec.*, No. 4, 1960, pp. 551–561 (in Russian).

59. Tuzlukov, P. and Tuzlukov, V., Reliability increasing in signal processing in noise in communications, *Automatized Systems in Signal Processing*, Vol. 7, 1983, pp. 80–87 (in Russian).
60. Tuzlukov, V., Detection of deterministic signal in noise, *Radio Eng.*, No. 9, 1986, pp. 57–60.
61. Tuzlukov, V., Signal detection in noise in communications, *Radio Phys. and Elec.*, Vol. 15, 1986, pp. 6–12.
62. Tuzlukov, V., Detection of deterministic signal in noise, *Telecomm. and Radio Eng.*, Vol. 41, No. 10, 1987, pp. 128–131.
63. Tuzlukov, V., Detection of signals with stochastic parameters by employment of generalized algorithm, in *Proc. SPIE's 1997 International Symposium on AeroSense: Aerospace/Defense Sensing, Simulations and Controls*, Orlando, FL, 20–25 April, 1997, Vol. 3079, pp. 302–313.
64. Tuzlukov, V., Noise reduction by employment of generalized algorithm, in *Proc. 13th IEEE International Conference on Digital Signal Processing (DSP97)*, Santorini, Greece, 2–4 July, 1997, pp. 617–620.
65. Tuzlukov, V., Detection of signals with random initial phase by employment of generalized algorithm, in *Proc. SPIE's 1997 International Symposium on Optical Science, Engineering and Instrumentation*, San Diego, CA, 27 July–1 August, 1997, Vol. 3162, pp. 61–72.
66. Tuzlukov, V., Generalized detection algorithm for signals with stochastic parameters, in *Proc. 1997 IEEE International Geoscience and Remote Sensing Symposium (IGARSS '97)*, 4–8 August, 1997, Singapore, pp. 139–141.
67. Tuzlukov, V., Tracking systems for stochastic signal processing by employment of generalized algorithm, in *Proc. IEEE First International Conference on Information, Communications and Signal Processing (ICICS '97)*, Singapore, 9–12 September, 1997, pp. 311–315.
68. Tuzlukov, V., A new approach to signal detection theory, *Digital Signal Processing: A Review Journal*, Vol. 8, No. 3, 1998, pp. 166–184.
69. Tuzlukov, V., Signal-to-noise improvement under detection of stochastic signals using generalized detector, in *Proc. 1998 International Conference on Applications of Photonics Technology (ICAPT '98)*, Ottawa, Canada, 27–30 July, 1998.
70. Tuzlukov, V., Signal processing in noise in communications: a new approach. Tutorial No. 3, in *Proc. 6th IEEE International Conference on Electronics, Circuits and Systems*, Paphos, Cyprus, 5–8 September, 1999, pp. 5–128.
71. Tuzlukov, V., Detection of stochastic signals using generalized detector, in *Proc. 1999 IASTED International Conference on Signal and Image Processing*, Nassau, Bahamas, 18–21 October, 1999, pp. 95–99.
72. Tuzlukov, V., New remote sensing algorithms under detection of minefields in littoral waters, in *Proc. 3rd International Conference on Remote Sensing Technologies for Minefield Detection and Monitoring*, Easton, Washington, D.C., 17–20 May, 1999, pp. 182–241.
73. Tuzlukov, V., New remote sensing algorithms on the basis of generalized approach to signal processing in noise. Tutorial No. 2, in *Proc. Second International ICSC Symposium on Engineering of Intelligent Systems (EIS'2000)*, University of Paisley, Paisley, Scotland, 27–30 June, 2000.

74. Tuzlukov, V., Detection of quasideterministic signals in additive Gaussian noise, *News of the Belarussian Academy of Sciences. Ser. Phys.-Tech. Sci.*, No. 4, 1985, pp. 98–104 (in Russian).
75. Tuzlukov, V., Detection of signals with stochastic amplitude and random initial phase in additive Gaussian noise, *Radio Eng.*, No. 9, 1988, pp. 59–61 (in Russian).
76. Tuzlukov, V., Interference compensation in signal detection for a signal of arbitrary amplitude and initial phase, *Telecomm. and Radio Eng.*, Vol. 44, No. 10, 1989, pp. 131–132.
77. Tuzlukov, V., The generalized algorithm of detection in statistical pattern recognition, *Pattern Recognition and Image Analysis*, Vol. 3, No. 4, 1993, pp. 474–485.
78. Tuzlukov, V., Signal-to-noise improvement in video signal processing, in *Proc. SPIE's 1993 International Symposium on High-Definition Video*, Berlin, Germany, 5–9 April, 1993, Vol. 1976, pp. 346–358.
79. Tuzlukov, V., Probability distribution density of background noise at the output of generalized detector, *News of the Belarussian Academy of Sciences. Ser. Phys.-Tech. Sci.*, No. 4, 1993, pp. 63–70 (in Russian).
80. Tuzlukov, V., Distribution law at the generalized detector output, in *Proc. PRIA'95*, Minsk, Belarus, 19–21 September, 1995, pp. 145–150.
81. Tuzlukov, V., Statistical characteristics of process at the generalized detector output, in *Proc. PRIA'95*, Minsk, Belarus, 19–21 September, 1995, pp. 151–156.
82. Tuzlukov, V., Signal fidelity in radar processing by employment of generalized algorithm under detection of mines and mine-like targets, in *Proc. SPIE's 1998 International Symposium on AeroSense: Aerospace/Defense Sensing, Simulations and Controls*, Orlando, FL, 13–17 April, 1998, Vol. 3392, pp. 1206–1217.
83. Tuzlukov, V., Employment of the generalized detector for noise signals in radar systems, in *Proc. SPIE's 2000 International Symposium on AeroSense: Aerospace/Defense Sensing, Simulations and Controls*, Orlando, FL, 24–28 April, 2000, Vol. 4048.
84. Tuzlukov, V., Generalized detector for noise signals in radar systems, in *Proc. 10th Mediterranean Electrotechnical Conference (MELECON 2000)*, Nicosia, Cyprus, 29–31 May, 2000.
85. Tuzlukov, V., Generalized signal detection algorithm in additive noise, *News of the Belarussian Academy of Sciences. Ser. Phys.-Tech. Sci.*, No. 3, 1991, pp. 101–109 (in Russian).
86. Tuzlukov, V., Signal detection algorithm based on jointly sufficient statistics, *Problems of Efficiency Increasing in Military*, Vol. 3, 1992, pp. 48–55 (in Russian).
87. Tuzlukov, V., The generalized methodology of signal detection in noise, in *Proc. 1992 Korean Automatic Control Conference*, Seoul, South Korea, October 19–21, 1992, pp. 255–260.
88. Ventzel, E. and Ovcharov, L., *Probability Theory*, Nauka, Moscow, 1973 (in Russian).
89. Pugachev, V., *Probability Theory and Mathematical Statistics*, Nauka, Moscow, 1979 (in Russian).
90. Tuzlukov, V., Probability distribution density of background noise at the output of generalized detector, *News of the Belarussian Academy of Sciences. Ser. Phys.-Tech. Sci.*, No. 4, 1993, pp. 63–70 (in Russian).

91. Gradshteyn, I. and Ryzhik, I., *Table of Integrals, Series, and Products*, 5th ed., Academic Press, New York, 1994.
92. Slepian, D., Fluctuations of random noise, *BSTJ*, Vol. 37, No. 1, 1958, pp. 163–184.
93. Rice, S., Mathematical analysis of random noise, *BSTJ*, Vol. 23, No. 3, 1944, pp. 232–282.
94. Rice, S., Mathematical analysis of random noise, *BSTJ*, Vol. 24, No. 1, 1945, pp. 46–156.

6

Generalized Approach to Signal Processing under the Stimulus of Multiplicative Noise and Linear Systems

This chapter deals with analysis of signals distorted by the multiplicative noise at the output of linear systems with constant parameters, in particular, at the output of the preliminary filter (PF) at the input linear tract of the generalized detector.

As we know from Chapter 5, signals do not pass over the additional filter (AF) at the input linear tract of the generalized detector. For this reason, we only consider the PF at the input linear tract of the generalized detector.

The output of the model signal generator (MSG) of the generalized detector also can be considered as a linear system with constant parameters because the signal at the output of the PF at the input linear tract of the generalized detector must be matched with the model signal, for example, the reference voltage, at the output of the MSG. This is one of the main development principles of the properly functioning generalized detector.[1,2]

Statistical characteristics of the signals at the output of linear systems, with constant parameters of the generalized detector distorted by multiplicative noise, are defined using the statistical characteristics of the noise modulation function $\dot{M}(t)$ of the multiplicative noise and the same signals at the input of linear systems, when the multiplicative noise does not act.

6.1 Signal Characteristics at the Output of Linear System of the Generalized Detector under the Stimulus of Multiplicative Noise

Let us consider the signal at the output of the linear system with constant parameters—the PF at the input linear tract of the generalized detector. Let the bandwidth of the PF be $\Delta\omega_{PF}$. Assume that

$$\Delta\omega_{PF} \ll \omega_0, \tag{6.1}$$

where ω_0 is the central frequency of the bandwidth of the PF.

The pulse transient response of the PF can be written in the following form:

$$h(t) = Re\{\dot{H}(t) \cdot e^{j\omega_0 t}\}, \tag{6.2}$$

where $\dot{H}(t)$ is the complex envelope of the pulse transient response of the PF.

The undistorted signal at the input of the PF takes the following form

$$a(t) = Re\{\dot{S}(t) \cdot e^{j(\omega_a t + \varphi_0)}\}, \tag{6.3}$$

where ω_a is the carrier frequency of the signal; $\dot{S}(t)$ is the complex envelope of amplitude of the signal; and φ_0 is the initial phase of the signal.

The signal at the output of the PF takes the following form

$$\begin{aligned} a_1(t) &= \int_{-\infty}^{\infty} a(\tau)h(t-\tau)d\tau \\ &= \int_{-\infty}^{\infty} Re\{\dot{S}(\tau) \cdot e^{j(\omega_a \tau + \varphi_0)}\} Re\{\dot{H}(t-\tau) \cdot e^{j\omega_a (t-\tau)}\}\, d\tau. \end{aligned} \tag{6.4}$$

Using Eq. (4.38) and neglecting the fast oscillating term that contains the frequency $2\omega_0$ factor, we obtain that the signal at the output of the PF takes the following form:

$$a_1(t, \Omega) = \frac{1}{2} \cdot Re\left\{ e^{j\omega_0 t} \cdot e^{j\varphi_0} \int_{-\infty}^{\infty} \dot{S}(\tau)\dot{H}(t-\tau) \cdot e^{j\Omega\tau}\, d\tau \right\}, \tag{6.5}$$

where

$$\Omega = \omega_a - \omega_0 \tag{6.6}$$

is the detuning of the signal $a(t)$ with respect to the frequency of tuning the PF and $a_1(t, \Omega)$ is the signal at the output of the PF during detuning Ω.

The complex envelope of amplitude of the signal at the output of the PF as a function of time t and detuning Ω, as follows from Eq. (6.5), has the following form:

$$\dot{S}_1(t, \Omega) = \frac{1}{2} \int_{-\infty}^{\infty} \dot{S}(\tau)\dot{H}(t-\tau) \cdot e^{j\Omega\tau} d\tau. \tag{6.7}$$

Under the stimulus of the multiplicative noise characterized by the noise modulation function $\dot{M}(t)$, the complex envelope of amplitude of the signal at the input of the PF takes the following form (see Eq. (4.2)):

$$\dot{S}_M(t) = \dot{M}(t)\dot{S}(t). \tag{6.8}$$

Substituting Eq. (6.8) in Eqs. (6.5) and (6.7), the signal $a_1(t, \Omega)$ and the complex envelope of amplitude $\dot{S}_{1_M}(t, \Omega)$ of the signal at the output of the PF under the stimulus of the multiplicative noise take the following form:

$$a_{1_M}(t, \Omega) = \frac{1}{2} \cdot Re\left\{ e^{j\omega_0 t} \cdot e^{j\varphi_0} \int_{-\infty}^{\infty} \dot{M}(\tau)\dot{S}(\tau)\dot{H}(t-\tau) \cdot e^{j\Omega\tau}\, d\tau \right\}$$

$$= Re\left\{ \dot{S}_{1_M}(t, \Omega) \cdot e^{j\omega_0 t} \cdot e^{j\varphi_0} \right\}; \tag{6.9}$$

$$\dot{S}_{1_M}(t, \Omega) = \frac{1}{2} \int_{-\infty}^{\infty} \dot{M}(\tau)\dot{S}(\tau)\dot{H}(t-\tau) \cdot e^{j\Omega\tau} d\tau. \tag{6.10}$$

6.1.1 Deterministic and Quasideterministic Multiplicative Noise

Under the stimulus of deterministic multiplicative noise we can use the Fourier series expansion for the noise modulation function $\dot{M}(t)$ within the limits of the interval

$$0 \leq t \leq T, \tag{6.11}$$

where the signal exists:[3–5]

$$\dot{M}(t) = \sum_{k=-\infty}^{\infty} \dot{C}_k^T \cdot e^{jk\frac{2\pi}{T}t}. \tag{6.12}$$

Substituting Eq. (6.12) in Eq. (6.9), we obtain that the signal at the output of the PF as a function of time t and detuning Ω takes the following form:

$$a_{1_M}(t, \Omega) = \sum_{k=-\infty}^{\infty} Re\left\{ \dot{C}_k^T \cdot e^{j(\omega_0 t + \varphi_0)} \int_{-\infty}^{\infty} \dot{S}(\tau)\dot{H}(t-\tau) \cdot e^{j(\Omega + k\frac{2\pi}{T})\tau} d\tau \right\}. \tag{6.13}$$

In terms of Eq. (6.5), Eq. (6.13) can be rewritten in the following form:

$$a_{1_M}(t, \Omega) = \alpha_0^T \cdot a_1\left(t, \beta_0^T, \Omega\right) + s(t, \Omega), \tag{6.14}$$

where

$$s(t, \Omega) = \sum_{i=-\infty}^{\infty} \alpha_i^T \cdot a_1\left(t, \beta_i^T, \Omega + i \cdot \frac{2\pi}{T}\right), \quad i \neq 0; \tag{6.15}$$

$$\alpha_i^T = \left|\dot{C}_i^T\right|; \tag{6.16}$$

$$\beta_i^T = \arg \dot{C}_i^T; \tag{6.17}$$

$a_1(t, \beta, \Omega)$ is the signal at the input of the PF that differs from the signal $a_1(t)$ by the shift in initial phase by the value β and in frequency by the value Ω.

Thus, the signal at the output of the PF under the stimulus of multiplicative noise can be presented as the sum of the undistorted portion of the signal—the first term on the right side of Eq. (6.14), recurring from the undistorted signal with an accuracy of the initial phase in the scale α_0^T, and the signal noise component—the second term on the right side of Eq. (6.14).[6,7] The same result can be obtained using Eq. (4.10), taking into account that the principle of superposition is correct for the linear system (PF) of the generalized detector.

Reference to Eqs. (6.14) and (6.15) shows that in a general case of deterministic multiplicative noise the undistorted portion of the signal and individual components of the signal noise component $s(t)$ at the output of the PF cannot be separated by time or frequency selection, since individual components of the signal noise component and the undistorted portion of the signal at the output of the PF are overlapped by both time and frequency. This selection of individual components of the signal noise component is only possible in some particular cases.[8–10]

When the acting multiplicative noise is the periodic stochastic process with the period T_M, the noise modulation function $\dot{M}(t)$ of the multiplicative noise can be presented by the Fourier series expansion[11,12] in the following form:

$$\dot{M}(t) = \sum_{k=-\infty}^{\infty} \dot{C}_k \cdot e^{jk(\Omega_M t + \vartheta)}, \tag{6.18}$$

where

$$\Omega_M = \frac{2\pi}{T_M} \tag{6.19}$$

and ϑ is the initial phase of the noise modulation function $\dot{M}(t)$ of the multiplicative noise with respect to the signal $a(t)$.

In this case, using transformations that are analogous to the relationships mentioned above, we obtain

$$\begin{aligned} a_{1_M}(t, \Omega) &= \alpha_0 \cdot a_1(t, \beta_0, \Omega) + \sum_{\ell=-\infty}^{\infty} \alpha_\ell \cdot a_1(t, \beta_\ell, \Omega + \ell\Omega_M) \\ &= \alpha_0 \cdot a_1(t, \beta_0, \Omega) + s(t, \Omega), \quad \ell \neq 0, \end{aligned} \tag{6.20}$$

where

$$\alpha_\ell = |\dot{C}_\ell| \tag{6.21}$$

and

$$\beta_\ell = \arg \dot{C}_\ell - \ell\vartheta. \tag{6.22}$$

For some cases, under the stimulus of periodic multiplicative noise the undistorted portion of the signal and individual components of the signal noise component at the output of the PF can be separated by frequency selection. If the frequency Ω_M of the multiplicative noise is greater than the spectrum bandwidth $\Delta\Omega_a$ of the undistorted signal $a(t)$, then individual

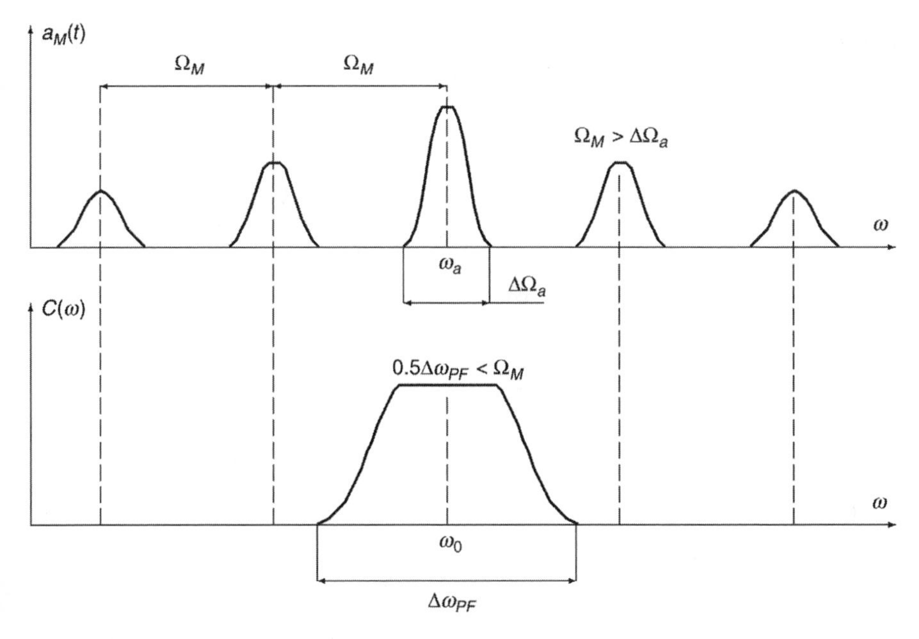

FIGURE 6.1
Spectrum of the signal $a_M(t)$ and amplitude-frequency response of the PF: the PF bandwidth $\Delta\omega_{PF} \ll 2\Omega_M$.

components of the signal noise component on the right side of Eq. (6.15) are not overlapped in frequency.

In the process, two cases are possible:

- *The first case*: The bandwidth of the PF satisfies the condition
$$\Delta\omega_{PF} < 2\Omega_M. \tag{6.23}$$
In this case only the undistorted portion of the signal is formed at the output of the PF (see Fig. 6.1).
- *The second case*: The bandwidth of the PF satisfies the condition
$$\Delta\omega_{PF} > 2\Omega_M. \tag{6.24}$$
In this case the undistorted portion of the signal and individual components of the signal noise component are formed at the output of the PF. Spectra of the undistorted portion of the signal and individual components of the signal noise component at the output of the PF are not overlapped, respectively, and can be separated by frequency selection (see Fig. 6.2).

Consider the case in which the shift in time between the signal and the multiplicative noise has a random character and the initial phase ϑ is a random variable—the case of the quasideterministic multiplicative noise.

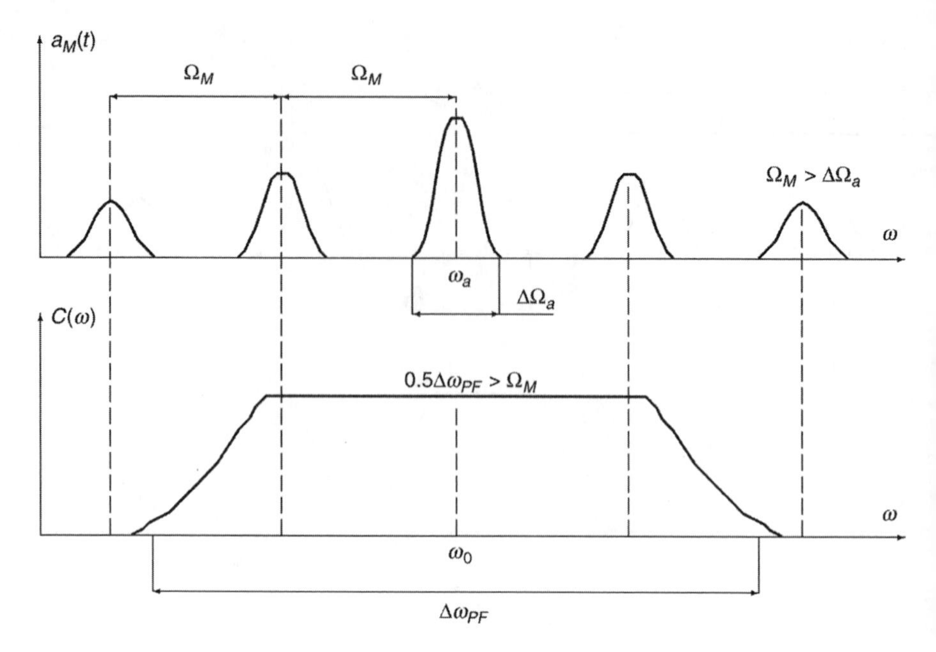

FIGURE 6.2
Spectrum of the signal $a_M(t)$ and amplitude-frequency response of the PF: the PF bandwidth $\Delta\omega_{PF} > 2\Omega_M$.

Assume that the shift in time is uniformly distributed within the limits of the period T_M of the multiplicative noise and, consequently, the initial phase ϑ is uniformly distributed within the limits of the interval $[0, 2\pi]$. It is easy to verify that for this case the mean of the signal $a_{1_M}(t, \Omega)$ at the output of the PF is defined by the undistorted portion of the signal, i.e.,

$$m_1\left[a_{1_M}(t, \Omega)\right] = \alpha_0 \cdot a_1(t, \beta_0, \Omega) \tag{6.25}$$

and the mean of the signal noise component at the output of the PF is equal to zero.

We proceed to define the variance of the signal noise component $\sigma_s^2(t, \Omega)$ at the output of the PF. Reference to Eq. (6.20) shows that

$$s(t, \Omega) = \sum_{\ell=-\infty}^{\infty} \alpha_\ell \cdot Re\left\{\dot{S}_1(t, \Omega + \ell\Omega_M) \cdot e^{j\omega_0 t} \cdot e^{j(\varphi_0 + \beta_\ell)}\right\}, \quad \ell \neq 0; \tag{6.26}$$

$$\sigma_s^2(t, \Omega) = m_1[(s(t, \Omega))^2] = \frac{1}{2} \sum_{\ell=-\infty}^{\infty} |\dot{S}_1(t, \Omega + \ell\Omega_M)|^2, \quad \ell \neq 0. \tag{6.27}$$

Thus, in this case the variance of the signal noise component at the output of the PF is the sum of powers of the individual components of the signal noise component.

6.1.2 Stationary Fluctuating Multiplicative Noise

During the stationary fluctuating multiplicative noise the noise modulation function $\dot{M}(t)$ can be written in the following form (see Eqs. (4.85) and (4.108)):

$$\dot{M}(t) = \alpha_0 \cdot e^{j\beta_0} + \dot{V}_0(t), \tag{6.28}$$

where the first term in Eq. (6.28) is the mean of the noise modulation function $\dot{M}(t)$ of the multiplicative noise, and the second term in Eq. (6.28) defines fluctuations of the noise modulation function $\dot{M}(t)$ of the multiplicative noise.

Substituting Eq. (6.28) in Eq. (6.9) and taking into account Eq. (6.5), we obtain that the signal at the output of the PF under the stimulus of the stationary fluctuating multiplicative noise takes the following form:

$$\begin{aligned} a_{1_M} &= \frac{\alpha_0}{2} \cdot Re\left\{ e^{j\omega_0 t} \cdot e^{j\varphi_0} \int\limits_{-\infty}^{\infty} \dot{S}(\tau)\dot{H}(t-\tau) \cdot e^{j\Omega\tau}\, d\tau \right\} \\ &+ \frac{1}{2} \cdot Re\left\{ e^{j\omega_0 t} \cdot e^{j\varphi_0} \int\limits_{-\infty}^{\infty} \dot{V}_0(\tau)\dot{S}(\tau)\dot{H}(t-\tau) \cdot e^{j\Omega\tau}\, d\tau \right\}. \end{aligned} \tag{6.29}$$

Reference to Eq. (6.29) shows that the first term on the right side is the signal that differs from the signal $a_1(t)$, which is formed at the output of the PF, by the constant factor α_0 and the constant shift in initial phase by the value β_0 (see Eq. (6.5)):

$$\alpha_0 \cdot a_1(t, \beta_0) = \frac{\alpha_0}{2} \cdot Re\left\{ e^{j\omega_0 t} \cdot e^{j(\varphi_0+\beta_0)} \int\limits_{-\infty}^{\infty} \dot{S}(\tau)\dot{H}(t-\tau) \cdot e^{j\Omega\tau}\, d\tau \right\}. \tag{6.30}$$

This is the undistorted portion of the signal at the output of the PF.

The second term

$$\begin{aligned} s(t, \Omega) &= \frac{1}{2} \cdot Re\left\{ e^{j\omega_0 t} \cdot e^{j\varphi_0} \int\limits_{-\infty}^{\infty} \dot{V}_0(\tau)\dot{S}(\tau)\dot{H}(t-\tau) \cdot e^{j\Omega\tau}\, d\tau \right\} \\ &= \int\limits_{-\infty}^{\infty} v(\tau)h(t-\tau)\, d\tau \end{aligned} \tag{6.31}$$

is the signal noise component at the output of the PF.

Thus,

$$a_{1_M}(t) = \alpha_0 \cdot a_1(t, \beta_0) + s(t, \Omega) = \alpha_0 \cdot a_1(t, \beta_0) + \int\limits_{-\infty}^{\infty} v(\tau)h(t-\tau)d\tau, \tag{6.32}$$

where $v(\tau)$ is the noise component of the signal distorted by the multiplicative noise.

These equations mentioned above can be obtained when the signal is the sum of two components (see Eq. (4.89)) and by using the principle of superposition that is true for the linear systems, in particular, for the PF of the generalized detector.

We define the statistical characteristics of the signal noise component $s(t, \Omega)$ at the output of the PF. The mean of the signal noise component $s(t, \Omega)$ at the output of the PF is equal to zero as the mean of the noise component of the signal at the input of the PF is equal to zero (see Section 4.2) and the PF is the linear system.

Consider the correlation function of the signal noise component $s(t, \Omega)$ at the output of the PF in time and frequency spaces:

$$R_s(t_1, t_2, \Omega_1, \Omega_2) = m_1[s(t_1, \Omega_1)s(t_2, \Omega_2)]. \tag{6.33}$$

Equation (6.33) characterizes a correlation between the values $s(t, \Omega)$ at the different instants t_1 and t_2 and during various values of frequency of the signal at the input of the PF: ω_{a_1} and ω_{a_2}. The values ω_{a_1} and ω_{a_2} are defined by the detuning in frequency

$$\Omega_1 = \omega_{a_1} - \omega_0 \tag{6.34}$$

and

$$\Omega_2 = \omega_{a_2} - \omega_0. \tag{6.35}$$

This correlation function allows us to estimate distortions of the signal both in the time space—distortions in the signal shape, and in the frequency space—distortions in spectrum of the signal.

Taking into consideration Eqs. (4.44) and (6.31) and given that $\dot{S}(t)$ and $\dot{H}(t)$ are the deterministic functions, the correlation function in Eq. (6.33) can be written in the following form:

$$\begin{aligned} R_s(t_1, t_2, \Omega_1, \Omega_2) = \frac{1}{8} \cdot Re\Bigg\{ e^{j\omega_0(t_1-t_2)} \int_{-\infty}^{\infty}\int_{-\infty}^{\infty} m_1[\dot{V}_0(\tau_1)V_0^*(\tau_2)]\dot{S}(\tau_1)S^*(\tau_2) \\ \times \dot{H}(t_1-\tau_1)H^*(t_2-\tau_2)\cdot e^{j(\Omega_1\tau_1-\Omega_2\tau_2)}\,d\tau_1\,d\tau_2\Bigg\} \\ + \frac{1}{8}\cdot Re\Bigg\{ e^{j\omega_0(t_1+t_2)} \int_{-\infty}^{\infty}\int_{-\infty}^{\infty} m_1\left[\dot{V}_0(\tau_1)\dot{V}_0(\tau_2)\cdot e^{2j\varphi_0}\right]\dot{S}(\tau_1)\dot{S}(\tau_2) \\ \times \dot{H}(t_1-\tau_1)\dot{H}(t_2-\tau_2)\cdot e^{j(\Omega_1\tau_1+\Omega_2\tau_2)}d\tau_1\,d\tau_2\Bigg\}. \end{aligned} \tag{6.36}$$

Since the initial phase of the signal and the noise modulation function $\dot{M}(t)$ of the multiplicative noise are independent, we can write that

$$\begin{aligned} m_1\left[\dot{V}_0(\tau_1)\dot{V}_0(\tau_2)\cdot e^{2j\varphi_0}\right] &= m_1[\dot{V}_0(\tau_1)\dot{V}_0(\tau_2)]m_1[e^{2j\varphi_0}] \\ &= \dot{D}_V(\tau_1-\tau_2)\Theta_1^{\varphi_0}(2), \end{aligned} \tag{6.37}$$

where $\dot{D}_V(\tau_1 - \tau_2)$ is defined by Eq. (4.116) and $\Theta_1^{\varphi_0}(2)$ is the one-dimensional characteristic function of the random phase φ_0.

Taking into account Eqs. (4.115) and (6.37), we obtain that the correlation function of the signal noise component at the output of the PF takes the following form:

$$R_s(t_1, t_2, \Omega_1, \Omega_2) = \frac{1}{8} \cdot Re\left\{ e^{j\omega_0(t_1-t_2)} \int_{-\infty}^{\infty}\int_{-\infty}^{\infty} \dot{R}_V(\tau_1-\tau_2)\dot{S}(\tau_1)S^*(\tau_2) \right.$$

$$\left. \times \dot{H}(t_1-\tau_1)H^*(t_2-\tau_2) \cdot e^{j(\Omega_1\tau_1-\Omega_2\tau_2)}\, d\tau_1\, d\tau_2 \right\}$$

$$+ \frac{1}{8} \cdot Re\left\{ \Theta_1^{\varphi_0}(2) \cdot e^{j\omega_0(t_1+t_2)} \int_{-\infty}^{\infty}\int_{-\infty}^{\infty} \dot{D}_V(\tau_1-\tau_2)\dot{S}(\tau_1)\dot{S}(\tau_2) \right.$$

$$\left. \times \dot{H}(t_1-\tau_1)\dot{H}(t_2-\tau_2) \cdot e^{j(\Omega_1\tau_1+\Omega_2\tau_2)}\, d\tau_1\, d\tau_2 \right\}$$

$$= R_{s_1} + R_{s_2}. \tag{6.38}$$

Reference to Eq. (6.38) shows that the correlation function of the signal noise component at the output of the PF takes two components. Each component depends on both the difference of instants

$$\Delta t = t_1 - t_2 \tag{6.39}$$

and on the time

$$t = t_1. \tag{6.40}$$

It is easy to verify that the first component determined by

$$R_{s_1} = \frac{1}{8} \cdot Re\left\{ e^{j\omega_0\Delta t} \int_{-\infty}^{\infty}\int_{-\infty}^{\infty} \dot{R}_V(\tau_1-\tau_2)\dot{S}(\tau_1)S^*(\tau_2) \right.$$

$$\left. \times \dot{H}(t-\tau_1)H^*(t-\Delta t-\tau_2) \cdot e^{j(\Omega_1\tau_1-\Omega_2\tau_2)}\, d\tau_1\, d\tau_2 \right\} \tag{6.41}$$

is varied with respect to the time t at a low rate and we can neglect these changes during the period of the carrier frequency of the signal.

The second component determined by

$$R_{s_2} = \frac{1}{8} \cdot Re\left\{ \Theta_1^{\varphi_0}(2) \cdot e^{2j\omega_0 t} \cdot e^{j\omega_0\Delta t} \int_{-\infty}^{\infty}\int_{-\infty}^{\infty} \dot{D}_V(\tau_1-\tau_2)\dot{S}(\tau_1)\dot{S}(\tau_2) \right.$$

$$\left. \times \dot{H}(t-\tau_1)\dot{H}(t-\Delta t-\tau_2) \cdot e^{j(\Omega_1\tau_1+\Omega_2\tau_2)}\, d\tau_1\, d\tau_2 \right\} \tag{6.42}$$

oscillates rapidly with the frequency equal to a doubled frequency of the signal.

We transform equations defining R_{s_1} and R_{s_2}. Making changes in variables in Eq. (6.41), we can write

$$R_{s_1} = \frac{1}{8} \cdot Re\left\{ e^{j\omega_0(t_1-t_2)} \int_{-\infty}^{\infty} \dot{R}_V(\tau_1) \cdot e^{j\Omega_2\tau_1} d\tau_1 \int_{-\infty}^{\infty} \dot{S}(\tau_2) S^*(\tau_2 - \tau_1) \right.$$

$$\left. \times \dot{H}(t_1 - \tau_2) H^*(t_2 - \tau_2 + \tau_1) \cdot e^{j(\Omega_1 - \Omega_2)\tau_2} \, d\tau_2 \right\}$$

$$= \frac{1}{8} \cdot Re\left\{ e^{j\omega_0(t_1-t_2)} \int_{-\infty}^{\infty} \dot{R}_V(\tau_1) \dot{\mathcal{X}}(t_1, t_2, \Omega_1, \Omega_2, \tau_1) \cdot e^{j\Omega_2\tau_1} \, d\tau_1 \right\}, \qquad (6.43)$$

where

$$\dot{\mathcal{X}}(t_1, t_2, \Omega_1, \Omega_2, \tau_1) = \int_{-\infty}^{\infty} \dot{S}(\tau_2) S^*(\tau_2 - \tau_1) \dot{H}(t_1 - \tau_2)$$

$$\times H^*(t_2 - \tau_2 + \tau_1) \cdot e^{j(\Omega_1 - \Omega_2)\tau_2} \, d\tau_2. \qquad (6.44)$$

The right side of Eq. (6.43) contains the Fourier transform of product between the two functions:

$$\dot{R}_V(\tau_1)$$

and

$$\dot{\mathcal{X}}(t_1, t_2, \Omega_1, \Omega_2, \tau_1),$$

which can be presented in the form of convolution of the Fourier transform of the factors:

$$\int_{-\infty}^{\infty} \dot{R}_V(\tau_1) \dot{\mathcal{X}}(\tau_1) \cdot e^{j\Omega_2\tau_1} d\tau_1 = \frac{1}{2\pi} \int_{-\infty}^{\infty} G_R(\omega) G_{\mathcal{X}}(-\Omega_2 - \omega) \, d\omega. \qquad (6.45)$$

The Fourier transform of the first factor is determined in the following form:

$$G_R(\omega) = \int_{-\infty}^{\infty} \dot{R}_V(\tau_1) \cdot e^{-j\omega\tau_1} d\tau_1 = G_V(\omega), \qquad (6.46)$$

where $G_V(\omega)$ is the energy spectrum of fluctuations of the noise modulation function $\dot{M}(t)$ of the stationary fluctuating multiplicative noise (see Eq. (4.130)).

The Fourier transform of the second factor is determined in the following form:

$$G_{\mathcal{X}}(\omega) = \int_{-\infty}^{\infty} \dot{\mathcal{X}}(\tau_1) \cdot e^{-j\omega\tau_1} d\tau_1 = \int_{-\infty}^{\infty}\int_{-\infty}^{\infty} \dot{S}(\tau_2) S^*(\tau_2 - \tau_1) \dot{H}(t_1 - \tau_2) \times H^*(t_2 - \tau_2 + \tau_1) \cdot e^{j(\Omega_1 - \Omega_2)\tau_2} \cdot e^{-j\omega\tau_1} \, d\tau_1 \, d\tau_2. \tag{6.47}$$

Reference to Eq. (6.47) shows that

$$G_{\mathcal{X}}(-\Omega_2 - \omega) = \int_{-\infty}^{\infty} \dot{S}(\tau_1) \dot{H}(t_1 - \tau_1) \cdot e^{j(\Omega_1 + \omega)\tau_1} \, d\tau_1 \times \int_{-\infty}^{\infty} S^*(\tau_2) H^*(t_2 - \tau_2) \cdot e^{-j(\Omega_2 + \omega)\tau_2} \, d\tau_2 = 4\dot{S}_1(t_1, \Omega_1 + \omega) S_1^*(t_2, \Omega_2 + \omega), \tag{6.48}$$

where $\dot{S}_1(t, \Omega)$ is the complex envelope of amplitude of the signal at the output of the PF when the multiplicative noise does not act (see Eq. (6.7)).

Representing the Fourier transform of product between the functions in the form of convolution of the Fourier transform for the factors,[13–15] we can write

$$\int_{-\infty}^{\infty} \dot{R}_V(\tau_1) \dot{\mathcal{X}}(\tau_1) \cdot e^{j\Omega_2\tau_1} d\tau_1 = \frac{2}{\pi} \int_{-\infty}^{\infty} G_V(\omega) \dot{S}_1(t_1, \Omega_1 + \omega) S_1^*(t_2, \Omega_2 + \omega) \, d\omega. \tag{6.49}$$

Substituting Eq. (6.49) in Eq. (6.43), we can write

$$R_{s_1} = \frac{1}{4\pi} \cdot Re\left\{ e^{j\omega_0(t_1 - t_2)} \int_{-\infty}^{\infty} G_V(\omega) \dot{S}_1(t_1, \Omega_1 + \omega) S_1^*(t_2, \Omega_2 + \omega) \, d\omega \right\}. \tag{6.50}$$

Making analogous transformations in Eq. (6.42), we can write

$$R_{s_2} = \frac{1}{4\pi} \cdot Re\left\{ \Theta_1^{\varphi_0}(2) \cdot e^{j\omega_0(t_1 + t_2)} \int_{-\infty}^{\infty} \dot{G}_D(\omega) \dot{S}_1(t_1, \Omega_1 + \omega) \dot{S}_1(t_2, \Omega_2 - \omega) \, d\omega \right\}, \tag{6.51}$$

where

$$\dot{G}_D(\omega) = \int_{-\infty}^{\infty} \dot{D}_V(\tau_1) \cdot e^{-j\omega\tau_1} \, d\tau_1 \tag{6.52}$$

is the Fourier transform of the function $\dot{D}_V(\tau_1)$.

In terms of Eqs. (6.50) and (6.51), the correlation function of the signal noise component at the output of the PF is determined in the following form

$$R_s(t_1, t_2, \Omega_1, \Omega_2) = \frac{1}{4\pi} \cdot Re\left\{ e^{j\omega_0(t_1 - t_2)} \int_{-\infty}^{\infty} G_V(\omega)\dot{S}_1(t_1, \Omega_1 + \omega) S_1^*(t_2, \Omega_2 + \omega)\, d\omega \right\} + \frac{1}{4\pi} \cdot Re\left\{ \Theta_1^{\varphi_0}(2) \cdot e^{j\omega_0(t_1 + t_2)} \int_{-\infty}^{\infty} \dot{G}_D(\omega)\dot{S}_1(t_1, \Omega_1 + \omega)\dot{S}_1(t_2, \Omega_2 - \omega)\, d\omega \right\}. \tag{6.53}$$

When the following conditions

$$t_1 = t_2 = t \tag{6.54}$$

and

$$\Omega_1 = \Omega_2 = \Omega \tag{6.55}$$

are satisfied, the variance of the signal noise component at the output of the PF on the basis of Eq. (6.53) takes the following form:

$$\sigma_s^2(t, \Omega) = \frac{1}{4\pi} \int_{-\infty}^{\infty} G_V(\omega) |\dot{S}_1(t, \Omega + \omega)|^2 d\omega + \frac{1}{4\pi} \cdot Re\left\{ \Theta_1^{\varphi_0}(2) \cdot e^{2j\omega_0 t} \int_{-\infty}^{\infty} \dot{G}_D(\omega)\dot{S}_1(t, \Omega + \omega)\dot{S}_1(t, \Omega - \omega)\, d\omega \right\}. \tag{6.56}$$

For many cases the initial phase of the signal, for example on the carrier frequency, can be considered as a random variable distributed uniformly within the limits of the interval $[0, 2\pi]$.[16–18]

For this case the one-dimensional characteristic function of the random initial phase φ_0 is equal to

$$\Theta_1^{\varphi_0}(2) = 0 \tag{6.57}$$

and Eq. (6.53) takes the following form:

$$R_s(t_1, t_2, \Omega_1, \Omega_2) = \frac{1}{4\pi} \cdot Re\left\{ e^{j\omega_0(t_1 - t_2)} \int_{-\infty}^{\infty} G_V(\omega)\dot{S}_1(t_1, \Omega_1 + \omega) S_1^*(t_2, \Omega_2 + \omega)\, d\omega \right\}, \tag{6.58}$$

and Eq. (6.38) takes the following form

$$R_s(t_1, t_2, \Omega_1, \Omega_2) = \frac{1}{8} \cdot Re\left\{ e^{j\omega_0(t_1 - t_2)} \int_{-\infty}^{\infty}\int_{-\infty}^{\infty} \dot{R}_V(\tau_1 - \tau_2)\dot{S}(\tau_1)S^*(\tau_2) \right.$$
$$\left. \times \dot{H}(t_1 - \tau_1)H^*(t_2 - \tau_2) \cdot e^{j(\Omega_1\tau_1 - \Omega_2\tau_2)} d\tau_1\, d\tau_2 \right\}. \quad (6.59)$$

For the conditions determined by Eqs. (6.54) and (6.55), the variance of the signal noise component at the output of the PF on the basis of Eq. (6.58) takes the following form:

$$\sigma_s^2(t, \Omega) = \frac{1}{4\pi} \int_{-\infty}^{\infty} G_V(\omega)|\dot{S}_1(t, \Omega + \omega)|^2\, d\omega$$
$$= \frac{1}{4\pi} \int_{-\infty}^{\infty} G_V(\Omega - \omega)|\dot{S}_1(t, \omega)|^2\, d\omega \quad (6.60)$$

when the initial phase of the signal is distributed uniformly. Equation (6.60) defines the relationship between the variance of the signal noise component at the output of the PF and time and detuning in frequency of the signal.

Reference to Eq. (6.60) shows that the signal noise component can generate a voltage at the output of the PF even during values of detuning in frequency Ω, when the undistorted portion of the signal does not generate a voltage at the output of the PF. Total frequency range of detuning, wherein the signal noise component generates a voltage at the output of the PF, is approximately equal to the sum of the spectrum bandwidth of undistorted portion of the signal at the output of the PF and the energy spectrum of the noise modulation function $\dot{M}(t)$ of the multiplicative noise.

If the bandwidth Ω_M, of the energy spectrum of the noise modulation function $\dot{M}(t)$ of the multiplicative noise, is much greater than the bandwidth $\Delta\omega_{PF}$ of the PF at the input linear tract of the generalized detector and the bandwidth $\Delta\Omega_a$ of the signal, and the function $G_V(\omega)$ is sufficiently smooth (see Fig. 6.3), then Eq. (6.60) can be written in a simpler form.

Actually, taking into consideration that

$$\dot{S}_1(t, \Omega) = 0 \quad (6.61)$$

when

$$|\Omega| > 0.5\Delta\omega_{PF} + \Delta\Omega_a \quad (6.62)$$

and assuming that for the condition

$$\Delta\Omega_M \gg \Delta\omega_{PF} + \Delta\Omega_a \quad (6.63)$$

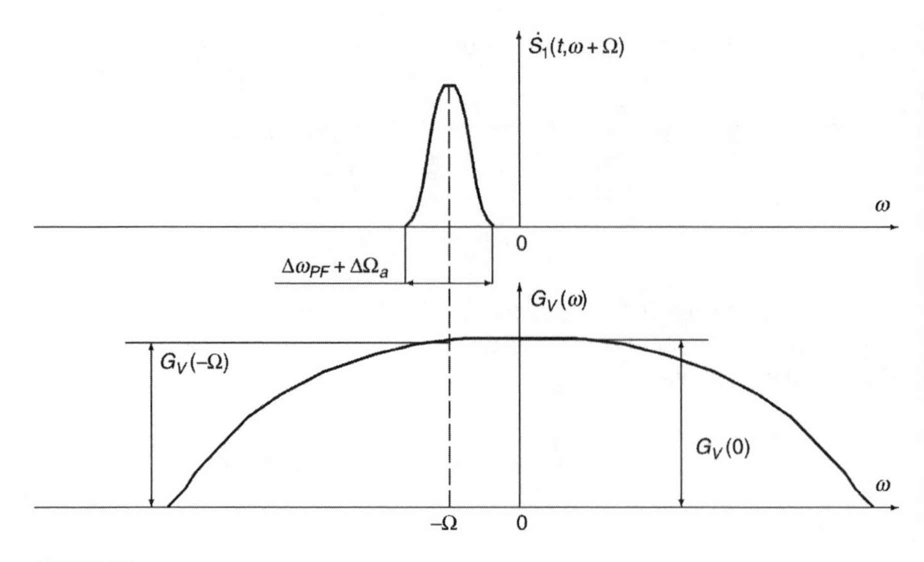

FIGURE 6.3
Energy spectrum of the noise modulation function $\dot{M}(t)$ of the multiplicative noise; the PF bandwidth $\Delta\omega_{PF}$ and bandwidth of the signal $\Delta\Omega_a$.

the variations in the function $G_V(\omega)$ within the limits of the interval

$$\Delta\omega_{PF} + \Delta\Omega_a \tag{6.64}$$

are relatively small, we can use the value of the function $G_V(\omega)$ in the middle point of the interval determined by Eq. (6.64) instead of the moving value of the function $G_V(\omega)$ at the moving instant.

Then we can write

$$\begin{aligned}\sigma_s^2(t, \Omega) &\simeq \frac{1}{4\pi} \cdot G_V(-\Omega) \int_{-\infty}^{\infty} |\dot{S}_1(t, \Omega + \omega)|^2 \, d\omega \\ &= \frac{1}{4\pi} \cdot G_V(-\Omega) \int_{-\infty}^{\infty} |\dot{S}_1(t, \omega)|^2 \, d\omega. \end{aligned} \tag{6.65}$$

The transition from Eq. (6.60) to the approximate expression in Eq. (6.65) corresponds to an exchange of the function $G_V(\omega)$ by its step approximation (see the dotted line in Fig. 6.3).

The correlation function and variance of the signal noise component at the output of the PF can be defined by the function $G_0(\omega)$, where $G_0(\omega)$ is the energy spectrum of fluctuations forming during modulation of the harmonic oscillation with the frequency ω_a and unit amplitude by the same modulation law, by which the multiplicative noise acts on the signal.[19–21]

Substituting Eqs. (4.131) and (6.6) in Eqs. (6.56) and (6.60) and making changes of variables, we can write

$$R_s(t_1, t_2, \omega_{a_1}, \omega_{a_2}) = \frac{1}{\pi} \cdot Re\left\{ e^{j\omega_0(t_1-t_2)} \int_{-\infty}^{\infty} G_0(\omega)\dot{S}_1(t_1, \omega_{a_1} + \omega)S_1^*(t_2, \omega_{a_2} + \omega)\, d\omega \right\}, \tag{6.66}$$

$$\sigma_s^2(t, \omega_a) = \frac{1}{\pi} \int_{-\infty}^{\infty} G_0(\omega)|\dot{S}_1(t, \omega_a + \omega)|^2\, d\omega, \tag{6.67}$$

where the energy spectrum $G_0(\omega)$ is defined by Eqs. (4.129) and (4.132).

The approximate formula for the variance of the signal noise component at the output of the PF, which is true for the condition

$$\Delta\Omega_M \gg \Delta\omega_{PF}, \tag{6.68}$$

is defined using the energy spectrum $G_0(\omega)$ in the following manner:

$$\sigma_s^2(t, \omega_a) \simeq \frac{1}{\pi} \cdot G_0(\omega_a) \int_{-\infty}^{\infty} |\dot{S}_1(t, \omega)|^2\, d\omega. \tag{6.69}$$

Pursuance of research of the signals distorted by the multiplicative noise at the output of the linear systems, in particular, the PF of the generalized detector, shows that the signal at the output of the PF, for example, is the sum of two components: the undistorted portion of the signal at the output of the PF, which recurs in the scale α_0 of the signal formed at the output of the PF by an action of the undistorted signal at the input of the generalized detector, and the signal noise component that arises as a consequence of the stimulus of the multiplicative noise.

Statistical characteristics of these two components can be defined using the complex envelope of amplitude of the signal formed at the output of the PF by an action of the undistorted signal at the input of the PF and owing to characteristics and parameters of the noise modulation function $\dot{M}(t)$ of the multiplicative noise.

6.2 Signal Characteristics at the Generalized Detector Output under the Stimulus of Multiplicative Noise

The model signal generator (MSG, see Fig. 5.2) is a very important block of the generalized detector. The use of the MSG allows us to define the mutual correlation function between the input stochastic process and the model signal at the output of the generalized detector and to solve the problem: a "yes" or a "no" signal at the input stochastic process (see Chapter 5).

The process at the output of the generalized detector is determined in the following form:

$$Z_g^{out}(t) = \int_0^T a_m(t)\xi(t)\,dt \quad \Rightarrow \quad \text{a "no" signal} \tag{6.70}$$

and

$$Z_g^{out}(t) = \int_0^T a_1(t)a_m(t)\,dt + \int_0^T a_m(t)\xi(t)\,dt \quad \Rightarrow \quad \text{a "yes" signal,} \tag{6.71}$$

where $a_m(t)$ is the model signal (the signal at the MSG output of the generalized detector). Other designations are the same as in Chapter 5.

In this chapter we consider the case of a "yes" signal at the input stochastic process and analyze the signal component

$$Z_a(t) = \int_0^T a_1(t)a_m(t)\,dt \tag{6.72}$$

at the output of the generalized detector under the stimulus of the multiplicative noise. Analysis is carried out for the condition

$$a_1(t) = a_m(t), \tag{6.73}$$

the main condition of functionality for the generalized detector.

The pulse transient response of the MSG matched with the signal $a_1(t)$ at the output of the PF at the input linear tract of the generalized detector is determined in the following form:

$$h_m(t) = Ca_1(t_0 - t), \tag{6.74}$$

where C is the constant coefficient depending on the factor of amplification of the MSG and t_0 is the delay of the model signal $a_m(t)$ by the MSG with respect to the signal $a_1(t)$.

In practice, the condition $t_0 \geq T$ always applies, where T is the duration of the signal $a_1(t)$ at the output of the PF at the input linear tract of the generalized detector.

Assume that

$$t_0 = T.$$

Using the complex envelope of amplitude of the signal $a_1(t)$, Eq. (6.74) can be written in the following form:[22]

$$h_m(t) = Re\{\dot{H}_m(t) \cdot e^{j\omega_0 t}\} = C Re\{S_m^*(t_0 - t) \cdot e^{-j\omega_0(t_0 - t)}\}, \tag{6.75}$$

where

$$\dot{H}_m(t) = C S_m^*(t_0 - t) \cdot e^{-j\omega_0 t_0} \tag{6.76}$$

is the complex envelope of pulse transient response of the MSG of the generalized detector.

Taking into consideration Eqs. (6.5) and (6.76), we obtain that the signal component at the output of the generalized detector takes the following form:

$$Z_a(t, \Omega) = \frac{C}{2} \cdot Re\left\{e^{j\omega_0(t_0 - t)} \cdot e^{j\varphi_0} \int_{-\infty}^{\infty} \dot{S}(\tau) S_m^*(\tau - t + t_0) \cdot e^{j\Omega\tau} d\tau\right\}, \tag{6.77}$$

where

$$\Omega = \omega_a - \omega_0 \tag{6.78}$$

is the detuning in frequency of the signal with respect to the tuning frequency of the MSG and $Z_a(t, \Omega)$ is the signal component at the output of the generalized detector during the detuning Ω.

We introduce a new variable

$$\widehat{\tau} = t - t_0. \tag{6.79}$$

Then the signal component at the output of the generalized detector has the following form:

$$Z_a(\widehat{\tau}, \Omega) = \frac{C}{2} \cdot Re\left\{e^{j\omega_0 \widehat{\tau}} \cdot e^{j\varphi_0} \int_{-\infty}^{\infty} \dot{S}(\tau) S_m^*(\tau - \widehat{\tau}) \cdot e^{j\Omega\tau} d\tau\right\} \tag{6.80}$$

and the complex envelope of amplitude of the signal component at the output of the generalized detector is determined in the following form:

$$\dot{S}_{Z_a}(\widehat{\tau}, \Omega) = \frac{C}{2} \int_{-\infty}^{\infty} \dot{S}(\tau) S_m^*(\tau - \widehat{\tau}) \cdot e^{j\Omega\tau} d\tau. \tag{6.81}$$

Taking into consideration that

$$\int_{-\infty}^{\infty} \dot{S}(\tau) S_m^*(\tau - \widehat{\tau}) \cdot e^{j\Omega\tau} d\tau = 2E_{a_1} \dot{\rho}(\widehat{\tau}, \Omega), \tag{6.82}$$

where $\dot{\rho}(\widehat{\tau}, \Omega)$ is the normalized autocorrelation function of the signal $a_1(t)$ and E_{a_1} is the energy of the signal $a_1(t)$, the complex envelope of amplitude of the signal component at the output of the generalized detector can be written in the following form:

$$\dot{S}_{Z_a}(\widehat{\tau}, \Omega) = C E_{a_1} \dot{\rho}(\widehat{\tau}, \Omega). \tag{6.83}$$

We proceed to define the signal component at the output of the generalized detector during a stimulus of the multiplicative noise. Substituting Eq. (6.76) in Eq. (6.9) and carrying out transformations that are analogous to the relationships mentioned above, we can write

$$Z_{a_M}(\widehat{\tau}, \Omega) = \frac{C}{2} \cdot Re\left\{ e^{j\omega_0 \widehat{\tau}} \cdot e^{j\varphi_0} \int_{-\infty}^{\infty} \dot{M}(\tau)\dot{S}(\tau)S_m^*(\tau - \widehat{\tau}) \cdot e^{j\Omega\tau}\, d\tau \right\}. \tag{6.84}$$

The complex envelope of amplitude of the signal component at the output of the generalized detector is determined in the following form:

$$\dot{S}_{Z_{a_M}}(\widehat{\tau}, \Omega) = \frac{C}{2} \int_{-\infty}^{\infty} \dot{M}(\tau)\dot{S}(\tau)S_m^*(\tau - \widehat{\tau}) \cdot e^{j\Omega\tau}\, d\tau. \tag{6.85}$$

6.2.1 Periodic Multiplicative Noise

Under the stimulus of periodic multiplicative noise the noise modulation function $\dot{M}(t)$ can be presented using the Fourier series expansion in the following form:[23–25]

$$\dot{M}(t) = \sum_{k=-\infty}^{\infty} \dot{C}_k \cdot e^{jk(\Omega_M t - \vartheta)}. \tag{6.86}$$

Substituting Eq. (6.86) in Eq. (6.85) and taking into account Eqs. (6.81) and (6.83), we obtain that the complex envelope of amplitude of the signal component at the output of the generalized detector can be written in the following manner:

$$\begin{aligned}\dot{S}_{Z_{a_M}}(\tau, \Omega) &= \sum_{k=-\infty}^{\infty} \dot{C}_k \cdot e^{-jk\vartheta}\, \dot{S}_{Z_a}(t, \Omega + k\Omega_M) \\ &= C E_{a_1} \sum_{k=-\infty}^{\infty} \alpha_k \cdot e^{j\beta_k}\, \dot{\rho}(\tau, \Omega + k\Omega_M),\end{aligned} \tag{6.87}$$

where

$$\alpha_k = |\dot{C}_k|, \tag{6.88}$$

and

$$\beta_k = \arg \dot{C}_k - k\vartheta. \tag{6.89}$$

In Eq. (6.87) the term corresponding to $k = 0$ is the undistorted portion of the signal at the output of the generalized detector. Other terms in Eq. (6.87) at $k \neq 0$ are the noise component of the total signal at the output of the generalized detector.

Since the autocorrelation function of the signal $\dot{\rho}(\tau, \Omega)$ has significant magnitude under the values of $|\Omega|$ that do not exceed the spectrum bandwidth $\Delta\Omega_a$ of the signal, we must only take into account those terms in Eq. (6.87), for which the following conditions

$$\alpha_k \neq 0 \tag{6.90}$$

and

$$|k\Omega_M| < \Delta\Omega_a \tag{6.91}$$

are true.

6.2.1.1 *The Case* $T_M < \frac{T}{2}$

As is well known, the presence of one or more main peaks jointly with side lobes (residuals), the level of which is much less than the level of the main peak, is characteristic of the autocorrelation functions. For the fixed value of τ lying near the main peak of the autocorrelation function $\dot{\rho}(\tau, \Omega)$ the frequency bandwidth overlapping by this peak is equal to $\frac{2}{T}$, where T is the duration of the signal. Because of this, when the period of the multiplicative noise satisfies the condition

$$T_M < \frac{T}{2} \tag{6.92}$$

the terms in Eq. (6.87) can be separated by frequency.

Thus, in this case the undistorted portion of the signal at the output of the generalized detector can be separated from the signal noise component at the output of the generalized detector and the signal noise component is divided into individual components, each of which repeats the undistorted signal at the output of the generalized detector in the scale α_k with an accuracy of the initial phase under detuning

$$\Omega + k\Omega_M. \tag{6.93}$$

For signals, the ambiguity function of which represents dependence between the shift in time and frequency—the frequency-modulated signals, a separation of the undistorted portion of the signal from the individual components of the signal noise component at the output of the generalized detector is possible by both frequency and time.[26–28]

For example, in the case of the frequency-modulated signals with a linear frequency modulation law the detuning of the signal in frequency by the value Ω_M generates the shift in time of the main peak of the autocorrelation function by the value

$$\tau_M = \frac{\Omega_M}{\Delta\omega_d} \cdot T, \tag{6.94}$$

where $\Delta\omega_d$ is the frequency deviation and T is the pulse duration for pulse signals or the period of modulation for continuous signals. Since for the fixed value Ω, the bandwidth of the main peak of the autocorrelation function $\dot{\rho}(\tau, \Omega)$ on the time axis is equal to

$$\frac{4\pi}{\Delta\omega_d},$$

the individual terms in Eq. (6.87) do not overlap in time for the condition determined by Eq. (6.92).

Neglecting an effect caused by the side lobes of the autocorrelation function, the complex envelope of amplitude of the signal component at the output of the generalized detector can be represented in the following form:

$$\dot{S}_{Z_{a_M}}(t) \simeq \sum_{k=-\infty}^{\infty} \alpha_k \cdot \dot{S}_{Z_a}(t, \Omega + k\Omega_M)$$
$$= CE_{a_1} \sum_{k=-\infty}^{\infty} \alpha_k \cdot |\dot{\rho}(\tau, \Omega + k\Omega_M)|. \tag{6.95}$$

In this case, the signal component at the output of the generalized detector is a sequence of particular signal components corresponding to individual terms in Eq. (6.95) with the period

$$\tau_M = \frac{\Omega_M}{\Delta\omega_d} \cdot T. \tag{6.96}$$

Particular signal components representing the individual components of the signal noise component can both be ahead of, and broken away from, the undistorted portion of the signal depending on the sign of the value k.

The total size of the whole sequence of particular signal components at the output of the generalized detector does not exceed $2T$, where T is the duration of the signal at the output of the PF at the input linear tract of the generalized detector. The amplitude of the k-th particular signal component at the output of the generalized detector is equal to

$$A_k(t) = CE_{a_{1_k}} \left| \dot{\rho}\left(\frac{k\Omega_M}{\Delta\omega_d} \cdot T, k\Omega_M \right) \right|. \tag{6.97}$$

As an example, consider the frequency-modulated signal component at the output of the generalized detector under the stimulus of multiplicative noise. For this signal component the autocorrelation function is determined in the following form:[22]

$$|\dot{\rho}(\tau, \Omega)| = \begin{cases} \left| \frac{\sin\left[\frac{1}{2}(\Omega + \Delta\omega_d \cdot \frac{\tau}{T})(T - |\tau|)\right]}{\frac{1}{2}(\Omega + \Delta\omega_d \cdot \frac{\tau}{T})} \right|, & |\tau| \leq T; \\ 0, & |\tau| > T. \end{cases} \tag{6.98}$$

For this case, in accordance with Eq. (6.97), the amplitudes of the particular signal components at the output of the generalized detector are determined in the following form:

$$A_k(t) = \begin{cases} CE_{a_{1_k}} \left(1 - \frac{|k\Omega_M|}{\Delta\omega_d}\right), & \left|\frac{k\Omega_M}{\Delta\omega_d}\right| \leq 1; \\ 0, & \left|\frac{k\Omega_M}{\Delta\omega_d}\right| > 1. \end{cases} \tag{6.99}$$

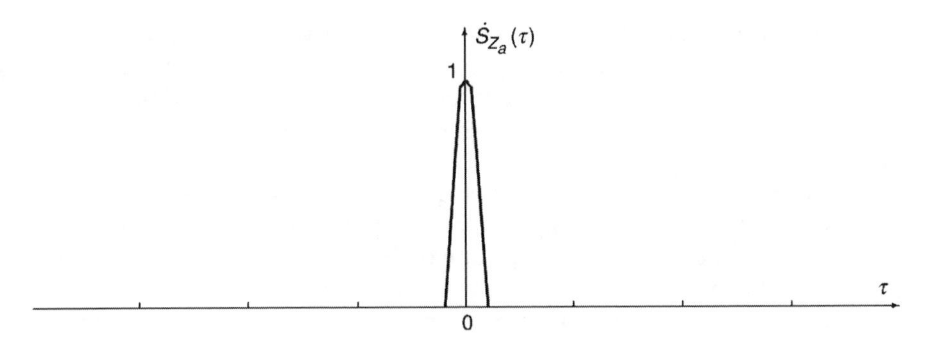

FIGURE 6.4
Signal at the output of the generalized detector: a "no" multiplicative noise.

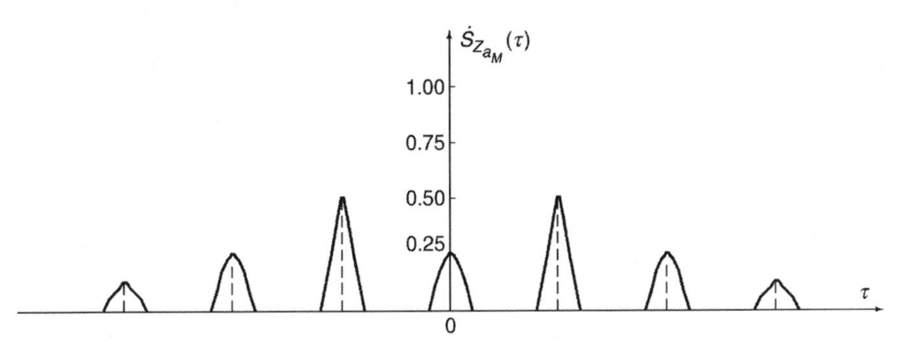

FIGURE 6.5
Signal at the output of the generalized detector: a "yes" multiplicative noise ($\varphi_m = 2\,radian$; $\frac{T}{T_M} = 9$; $\frac{\Delta\omega_d}{\Omega_M} = 7$).

The signal component at the output of the generalized detector is shown in Fig. 6.4 for the case of the rectangular frequency-modulated signal, when the multiplicative noise does not act. Fig. 6.5 shows the signal component when the multiplicative noise is periodic, in the form of sinusoidal distortions in phase of the signal. The side lobes of the autocorrelation function are not shown in Fig. 6.5.

6.2.1.2 *The Case* $T_M > \frac{T}{2}$

For this case the peaks of the autocorrelation functions

$$\dot{\rho}(\tau, \Omega + k\Omega_M) \tag{6.100}$$

and

$$\dot{\rho}[\tau, \Omega + (k+1)\Omega_M] \tag{6.101}$$

are overlapped and the individual terms in Eq. (6.87) cannot be separated.

Under frequency-modulated signals the duration of the signal component at the output of the generalized detector is increased because of the stimulus of the multiplicative noise. However, maximal value of duration of the signal component at the output of the generalized detector cannot exceed the value $2T$.

If the initial phase ϑ has a random character, the quasideterministic multiplicative noise, then relationships in phase between the individual terms of Eq. (6.87) are varied from one realization of the signal component at the output of the generalized detector to the other.[29,30] Because of this, the resulting signal component at the output of the generalized detector is varied from one realization to the other.

It is natural to assume that the initial phase is distributed uniformly within the limits of the interval $[0, 2\pi]$. For this case, the mean of the signal component at the output of the generalized detector is equal to the undistorted portion of the signal and is determined in the following form:

$$\begin{aligned} m_1\left[Z_{a_M}(\tau, \Omega)\right] &= \alpha_0 \cdot a_1(\tau, \beta_0, \Omega) \\ &= \alpha_0 \cdot C E_{a_1} Re\left\{\dot{\rho}(\tau, \Omega) \cdot e^{j(\omega_0 \tau + \varphi_0 + \beta_0)}\right\}. \end{aligned} \tag{6.102}$$

The variance of the noise component of the total signal $Z_{a_M}(\tau, \omega)$ at the output of the generalized detector is determined in the following form:

$$\sigma_s^2(\tau, \omega) = \frac{C^2 E_{a_1}^2}{2} \sum_{k=-\infty}^{\infty} \alpha_k^2 \cdot |\dot{\rho}(\tau, \Omega + k\Omega_M)|^2, \quad k \neq 0. \tag{6.103}$$

6.2.2 Fluctuating Multiplicative Noise

Let the noise modulation function $\dot{M}(t)$ of the multiplicative noise be the stationary stochastic function. For this case, the signal component at the output of the generalized detector according to Eqs. (6.29) and (6.32) can be written as the sum of two components. The first component is the undistorted portion of the total signal $Z_{a_M}(\tau, \Omega)$ at the output of the generalized detector and, in terms of Eq. (6.76), takes the following form:

$$\alpha_0 \cdot a_1(\hat{\tau}, \beta_0) = \frac{C}{2} \cdot Re\left\{e^{j(\omega_0 \hat{\tau} + \varphi_0 + \beta_0)} \int_{-\infty}^{\infty} \dot{S}(\tau) S_m^*(\tau - \hat{\tau}) \cdot e^{j\Omega\tau} d\tau\right\}. \tag{6.104}$$

The second component is the noise component of the total signal $Z_{a_M}(\tau, \Omega)$ at the output of the generalized detector and, in terms of Eq. (6.76), has the following form:

$$s(\hat{\tau}, \Omega) = \frac{C}{2} \cdot Re\left\{e^{j(\omega_0 \hat{\tau} + \varphi_0)} \int_{-\infty}^{\infty} \dot{V}_0(\tau) \dot{S}(\tau) S_m^*(\tau - \hat{\tau}) \cdot e^{j\Omega\tau} d\tau\right\}. \tag{6.105}$$

We next define the statistical characteristics of the signal noise component at the output of the generalized detector. As was shown in Section 6.1 the mean of the signal noise component is equal to zero.

The correlation function of the signal noise component at the output of the generalized detector can be defined on the basis of Eq. (6.53) using Eq. (6.83) and taking into account a new variable τ:

$$R_s(\tau_1, \tau_2, \Omega_1, \Omega_2)$$
$$= \frac{C^2 E_{a_1}^2}{4\pi} \cdot Re\left\{ e^{j\omega_0(\tau_1 - \tau_2)} \int_{-\infty}^{\infty} G_V(\omega)\dot{\rho}(\tau_1, \Omega_1 + \omega)\rho^*(\tau_2, \Omega_2 + \omega)\, d\omega \right\}$$
$$+ \frac{C^2 E_{a_1}^2}{4\pi} \cdot Re\left\{ \Theta_1^{\varphi_0}(2) \cdot e^{j\omega_0(\tau_1 + \tau_2)} \int_{-\infty}^{\infty} \dot{G}_D(\omega)\dot{\rho}(\tau_1, \Omega_1 + \omega)\dot{\rho}(\tau_2, \Omega_2 - \omega)\, d\omega \right\}. \tag{6.106}$$

The variance of the signal noise component at the output of the generalized detector can be defined using Eq. (6.54) in the following manner:

$$\sigma_s^2(\tau, \Omega) = \frac{C^2 E_{a_1}^2}{4\pi} \int_{-\infty}^{\infty} G_V(\omega)|\dot{\rho}(\tau, \Omega + \omega)|^2\, d\omega$$
$$+ \frac{C^2 E_{a_1}^2}{4\pi} \cdot Re\left\{ \Theta_1^{\varphi_0}(2) \cdot e^{2j\omega_0\tau} \int_{-\infty}^{\infty} \dot{G}_D(\omega)\dot{\rho}(\tau, \Omega + \omega)\dot{\rho}(\tau, \Omega - \omega)\, d\omega \right\}. \tag{6.107}$$

When the initial phase φ_0 of the signal is distributed uniformly within the limits of the interval $[0, 2\pi]$, then the one-dimensional characteristic function of the initial phase φ_0 of the signal $\Theta_1^{\varphi_0}(2)$ is determined by Eq. (6.55) and the correlation function (see Eq. (6.106)) and the variance (see Eq. (6.107)) of the signal noise component at the output of the generalized detector have the following form:

$$R_s(\tau_1, \tau_2, \Omega_1, \Omega_2)$$
$$= \frac{C^2 E_{a_1}^2}{4\pi} \cdot Re\left\{ e^{j\omega_0(\tau_1 - \tau_2)} \int_{-\infty}^{\infty} G_V(\omega)\dot{\rho}(\tau_1, \Omega_1 + \omega)\rho^*(\tau_2, \Omega_2 + \omega)\, d\omega \right\}; \tag{6.108}$$

$$\sigma_s^2(\tau, \Omega) = \frac{C^2 E_{a_1}^2}{4\pi} \int_{-\infty}^{\infty} G_V(\omega)|\dot{\rho}(\tau, \Omega + \omega)|^2\, d\omega$$
$$= C^2 E_{a_1}^2 \delta_1^2(\tau, \Omega), \tag{6.109}$$

where

$$\delta_1^2(\tau, \Omega) = \frac{1}{4\pi} \int_{-\infty}^{\infty} G_V(\omega)|\dot{\rho}(\tau, \Omega + \omega)|^2 \, d\omega \tag{6.110}$$

is the normalized variance. Henceforth in this section we assume that the condition mentioned above is satisfied.

The correlation function and variance of the signal noise component at the output of the generalized detector can be defined using the energy spectrum $G_0(\omega)$ of fluctuations of the harmonic oscillation with the unit amplitude distorted by the same multiplicative noise:

$$R_s(\tau_1, \tau_2, \omega_{a_1}, \omega_{a_2}) = \frac{C^2 E_{a_1}^2}{\pi} \cdot Re\left\{ e^{j\omega_0(\tau_1 - \tau_2)} \int_{-\infty}^{\infty} G_0(\omega)\dot{\rho}(\tau_1, \omega_{a_1} + \omega)\rho^*(\tau_2, \omega_{a_2} + \omega) \, d\omega \right\}; \tag{6.111}$$

$$\sigma_s^2(\tau, \omega) = \frac{C^2 E_{a_1}^2}{\pi} \int_{-\infty}^{\infty} G_0(\omega)|\dot{\rho}(\tau, \omega_a + \omega)|^2 \, d\omega. \tag{6.112}$$

The equations mentioned above allow us to make definite conclusions regarding peculiarities of the signal component at the output of the generalized detector under the stimulus of the multiplicative noise. Since the autocorrelation function differs from zero only within the limits of the interval

$$-T \leq \tau \leq T,$$

the signal noise component at the output of the generalized detector can only exist within the limits of this just mentioned interval as follows from Eqs. (6.107) and (6.109).

It should be pointed out that the just mentioned interval, which can be overlapped by the signal noise component at the output of the generalized detector, is much greater than the duration τ_y of the main peak of the autocorrelation function of the signal component at the output of the generalized detector for some types of the signals when the multiplicative noise does not act.

So, for example, for the wide-band signals—the signals with frequency or phase modulation—a duration of the main peak of the autocorrelation function of the signal component at the output of the generalized detector is Q_y times less approximately the duration T of the signal at the input of the generalized detector, where

$$Q_y = \Delta F_a T \tag{6.113}$$

is the coefficient and ΔF_a is the spectrum bandwidth of the signal $a(t)$ at the input of the generalized detector.

The signal noise component can overlap the interval $2T$. Because of this, if the signal noise component possesses an essential part of energy of the signal distorted by the multiplicative noise, that takes place if the parameter α_0 is very small, the multiplicative noise can "smear out" the signal component at the output of the generalized detector.

Physical reasons for the phenomenon mentioned above can be explained in the following way. During the undistorted wide-band signal processing by the generalized detector, a summation of individual components of the signal component is carried out at the instant of trailing edge of the pulse signal and a mutual compensation of components is carried out at other instants. For this reason, for the condition

$$\tau = 0$$

we have the main peak of the autocorrelation function of the signal component at the output of the generalized detector.

At other instants

$$\tau \neq 0$$

there are uncompensated residuals of the signal component at the output of the generalized detector, the amplitude of which are much less than the amplitude of the main peak of the autocorrelation function of the signal component at the output of the generalized detector.

The interval $2T$ consists of the signal duration T and the delay $t_0 = T$ caused by the MSG of the generalized detector (see Fig. 6.6a and b). For the stimulus of multiplicative noise the amplitude and phase relationships between individual components of the signal component at the output of the generalized detector are destroyed. For this reason, the summation of individual components of the signal component for the condition

$$\tau = 0$$

and their compensation at other instants are impaired.

As a result, the main peak of the autocorrelation function of the signal component at the output of the generalized detector decreases in amplitude and the amplitude of residuals increases (see Fig. 6.6c). Since phase and amplitude relationships between individual components of the signal component at the output of the generalized detector fluctuate because of the stimulus of the multiplicative noise, the main peak and residuals of the autocorrelation function of the signal component at the output of the generalized detector fluctuate.

Reference to Eq. (6.107) shows that the range of values of detuning Ω determined by Eqs. (6.6) and (6.78), during which there is the signal noise component at the output of the generalized detector, is equal to

$$\pm\left(\Delta\Omega_a + \frac{\Delta\Omega_M}{2}\right). \tag{6.114}$$

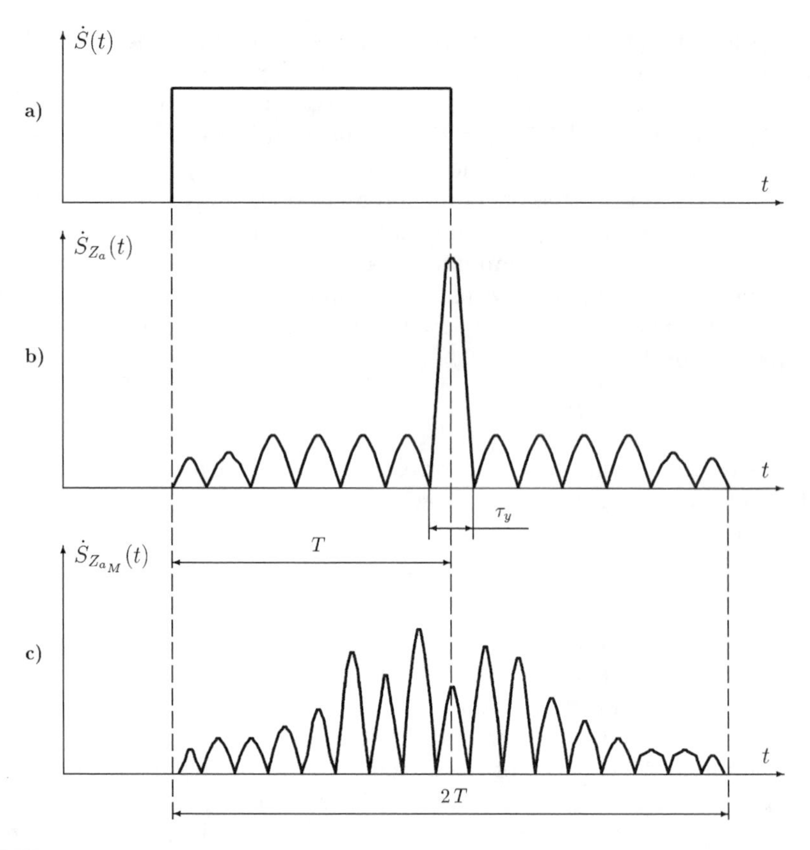

FIGURE 6.6
(a) Signal at the input of the generalized detector; (b) signal at the output of the generalized detector—a "no" multiplicative noise; (c) signal at the output of the generalized detector—a "yes" multiplicative noise.

Only a qualitative explanation of this phenomenon is discussed in this section. The degree of "smearing out" the signal component at the output of the generalized detector depends on the relative level α_0 of the undistorted portion of the signal component and on the shape and bandwidth of the energy spectrum $G_V(\omega)$ of the noise modulation function $\dot{M}(t)$ of the multiplicative noise. Our theoretical analysis is continued in the next section.

6.3 Signal Noise Component for Various Types of Signals

In this section we study relationships between the parameters of the noise modulation function $\dot{M}(t)$ of the multiplicative noise and statistical characteristics of the signal noise component at the output of the generalized detector under the stimulus of the stationary fluctuating multiplicative noise for various types of signals at the input of the generalized detector: the signals with a constant radio frequency carrier and with a square wave-form or bell-shaped amplitude envelope; the frequency-modulated signals with a bell-shaped amplitude envelope; and the signals with a phase-manipulated code.

The variance of the signal noise component at the output of the generalized detector is discussed. The correlation function of time and frequency of the signal noise component at the output of the generalized detector is also analyzed.

6.3.1 Frequency-Modulated Signals with Bell-Shaped Envelope and Signals with Bell-Shaped Envelope and Constant Radio Frequency Carrier

Definition of statistical characteristics of the signal noise component at the output of the generalized detector can be carried out simultaneously for the case of the frequency-modulated signals with a bell-shaped amplitude envelope and the signals with a bell-shaped amplitude envelope and constant radio frequency carrier at the input of the generalized detector.

Actually, in the case of the frequency-modulated signals with the unity amplitude with a bell-shaped amplitude envelope

$$a_1(t) = Re\left\{e^{j\omega_a t} \cdot e^{-\frac{\pi t^2}{T^2}} \cdot e^{-j\frac{\Delta\omega_d}{2T} \cdot t^2}\right\} \tag{6.115}$$

the complex autocorrelation function takes the following form:[31–34]

$$\dot{\rho}(\tau, \Omega) = \exp\left(-\frac{\pi\tau^2}{2T^2}\right) \cdot \exp\left[-\frac{\Delta\omega_d^2 T^2}{8\pi}\left(\frac{\tau}{T} - \frac{\Omega}{\Delta\omega_d}\right)^2\right] \cdot \exp\left(j \cdot \frac{\Omega\tau}{2}\right), \tag{6.116}$$

where T is the equivalent duration of the signal; $\Delta\omega_d$ is the frequency deviation within the limits of the signal duration T—for the condition

$$Q_y \gg 1 \tag{6.117}$$

the energy spectrum bandwidth of the signal $a_1(t)$ is approximately equal to the frequency deviation; and

$$Q_y = \frac{1}{2\pi} \cdot \Delta\omega_d T \tag{6.118}$$

is the coefficient defining a degree of signal truncation by the generalized detector during signal processing.

If we assume that the condition

$$\Delta\omega_d = 0 \tag{6.119}$$

is satisfied in Eq. (6.116), then the complex autocorrelation function of the signal with a bell-shaped amplitude envelope and constant radio frequency carrier has the following form:[35–37]

$$\dot{\rho}(\tau, \Omega) = \exp\left(-\frac{\pi\tau^2}{2T^2} - \frac{\Omega^2 T^2}{8\pi}\right) \cdot \exp\left(j \cdot \frac{\Omega\tau}{2}\right). \tag{6.120}$$

Assume that the energy spectrum of fluctuations of the noise modulation function $\dot{M}(t)$ of the multiplicative noise has a bell-shaped form:

$$G_V(\Omega) = 2\pi \cdot \frac{\overline{A^2(t)} - \alpha_0^2}{\Delta\Omega_M} \cdot e^{-\frac{\pi\Omega^2}{\Delta\Omega_M^2}}, \tag{6.121}$$

where the energy spectrum is defined for positive and negative values of frequency and $\Delta\Omega_M$ is the equivalent bandwidth of the energy spectrum $G_V(\Omega)$ determined in the following form:

$$\Delta\Omega_M = \frac{1}{G_V(0)} \int_{-\infty}^{\infty} G_V(\Omega)\, d\Omega. \tag{6.122}$$

The variance of fluctuations of the frequency-modulated signal with a bell-shaped envelope, which is distorted by the multiplicative noise at the output of the generalized detector, is defined by substitution of Eqs. (6.116) and (6.121) in Eq. (6.109).

After some mathematical transformations, using the tabulated integral in the form[38]

$$\int_{-\infty}^{\infty} e^{-px^2 \pm qx} dx = \sqrt{\frac{\pi}{p}} \cdot e^{\frac{q^2}{4p}}, \tag{6.123}$$

we can write

$$\sigma_s^2(\tau, \Omega) = \frac{C^2 E_{a_1}^2 \left(\overline{A^2(t)} - \alpha_0^2\right)}{2\sqrt{1+\xi^2}} \times \exp\left[-\frac{\pi\tau^2}{T^2(1+\xi^2)}\left(1 + \xi^2 + Q_y^2 - \frac{\Omega^2 T^2}{4\pi} - \Omega\tau Q_y\right)\right], \tag{6.124}$$

where

$$\xi = \frac{1}{2\pi} \cdot \Delta\Omega_M T = \Delta F_M T \tag{6.125}$$

is the parameter defining a ratio between the energy spectrum bandwidth of the noise modulation function $\dot{M}(t)$ of the multiplicative noise and the energy spectrum bandwidth of amplitude envelope of the signal.

In the case of the signal with a bell-shaped amplitude envelope and constant radio frequency carrier under substitution of the condition

$$Q_y = 0 \quad \text{or} \quad \Delta\omega_d = 0 \tag{6.126}$$

in Eq. (6.124) we can write

$$\sigma_s^2(\tau, \Omega) = \frac{C^2 E_{a_1}^2 \left(\overline{A^2(t)} - \alpha_0^2\right)}{2\sqrt{1+\xi^2}} \cdot \exp\left(-\frac{\pi\tau^2}{T^2}\right) \cdot \exp\left[-\frac{\Omega^2 T^2}{4\pi(1+\xi^2)}\right]. \tag{6.127}$$

Reference to Eqs. (6.124) and (6.127) shows that the presence of the multiplicative noise brings to the same extension in the frequency region on the frequency axis for the signals considered for the condition

$$\tau = 0.$$

However, in the case of the frequency-modulated signal the multiplicative noise gives rise to extension in the time region for the signal at the output of the generalized detector. An extent of the non-modulated signal does not vary on the time axis.

We define the normalized coefficient of correlation of the correlation function of the signal noise component at the output of the generalized detector. By definition the coefficient of correlation can be determined in the following form:[39,40]

$$r_s(\tau, \tau + \Delta\tau, \Omega, \Omega + \Delta\Omega) = \frac{R_s(\tau, \tau + \Delta\tau, \Omega, \Omega + \Delta\Omega)}{\sigma_s(\tau, \Omega)\sigma_s(\tau + \Delta\tau, \Omega + \Delta\Omega)}. \tag{6.128}$$

Substituting Eqs. (6.120) and (6.121) in Eq. (6.108) and further in Eq. (6.128), we can write

$$\begin{aligned} r_s(\tau, \tau + \Delta\tau, \Omega, \Omega + \Delta\Omega) &= \exp\left[-\frac{1}{4(1+\xi^2)}\left(\frac{\pi\xi^2(1+Q_y^2)}{T^2}\cdot\Delta\tau^2 + \frac{\xi^2 T^2}{4\pi}\cdot\Delta\Omega^2 + \xi^2 Q_y \Delta\tau\Delta\Omega\right)\right] \\ &\times \cos\left[\frac{1}{2(1+\xi^2)}\left(\frac{\pi\xi^2 Q_y}{T^2}\cdot(2\tau\Delta\tau + \Delta\tau^2)\right.\right. \\ &\left.\left. -\frac{2+\xi^2}{2}\cdot\Delta\tau\cdot\Delta\Omega + \xi^2\tau\Delta\Omega + \Omega\Delta\tau\right) + \omega_a\Delta\tau\right]. \end{aligned} \tag{6.129}$$

We limit a further analysis considering the coefficient of correlation in frequency for the condition

$$\tau = \Delta\tau = 0 \tag{6.130}$$

and in time during zero detuning

$$\Omega = \Delta\Omega = 0. \tag{6.131}$$

Then in terms of Eq. (6.129) we can write

$$r_{s_\Omega}(\Omega, \Omega + \Delta\Omega) = r_{s_\Omega}(\Delta\Omega) = \exp\left[-\frac{\xi^2 T^2}{16\pi(1+\xi^2)}\cdot\Delta\Omega^2\right]; \tag{6.132}$$

$$\begin{aligned} r_{s_\tau}(\tau, \tau + \Delta\tau) &= \exp\left[-\frac{\pi\xi^2(1+Q_y^2)}{4T^2(1+\xi^2)}\cdot\Delta\tau^2\right] \\ &\times \cos\left[\frac{\pi\xi^2 Q_y^2}{2T^2(1+\xi^2)}\cdot(2\tau\Delta\tau + \Delta\tau^2) + \omega_a\Delta\tau\right]. \end{aligned} \tag{6.133}$$

Reference to Eqs. (6.132) and (6.133) shows that the coefficient of correlation of the correlation function of the frequency-modulated signal with a bell-shaped amplitude envelope, which is distorted by the multiplicative noise, at the output of the generalized detector under the condition determined by Eq. (6.130) only depends on the difference $\Delta\Omega$ of arguments and does not depend on the coefficient Q_y.

In other words, we can say that the coefficient of correlation of the frequency-modulated signal with a bell-shaped amplitude envelope coincides with the corresponding coefficient of correlation of the signal with a bell-shaped amplitude envelope and the constant radio frequency carrier.

Correlation intervals in frequency $\Delta\Omega_{rs}$ in the cases of the non-modulated signal and the frequency-modulated signal with a bell-shaped amplitude envelope are equal to each other:

$$\Delta\Omega_{rs} = \frac{1}{2\pi}\int_0^\infty r_{s_\Omega}(\Delta\Omega)d(\Delta\Omega) = \frac{2\pi\sqrt{1+\xi^2}}{\xi T}. \tag{6.134}$$

The coefficient of correlation in time

$$r_{s_\tau}(\tau, \tau + \Delta\tau),$$

as can be seen from Eq. (6.133), in the case of the frequency-modulated signal depends on two parameters: τ and $\Delta\tau$. However, the amplitude envelope of the coefficient of correlation is defined by the difference $\Delta\tau$ between arguments.

In the case of the signal with a bell-shaped amplitude envelope and constant radio frequency carrier the coefficient of correlation in time can be obtained on the basis of Eq. (6.133) by substitution of the condition $Q_y = 0$:

$$r_{s_\tau}(\Delta\tau) = \exp\left[-\frac{\pi\xi^2\Delta\tau^2}{4T^2(1+\xi^2)}\right]\cdot\cos\omega_a\Delta\tau. \tag{6.135}$$

Reference to Eq. (6.135) shows that the coefficient of correlation in time $r_{s_\tau}(\Delta\tau)$ depends only on the difference $\Delta\tau$ of arguments in the case of modulation absence in frequency. The correlation interval in time for the frequency-modulated signal is defined, using the envelope of the coefficient of correlation (see Eq. (6.133)):

$$\Delta\tau_{rs} = \int_0^\infty \exp\left[-\frac{\pi\xi^2(1+Q_y^2)}{4T^2(1+\xi^2)}\cdot\Delta\tau^2\right]d(\Delta\tau) = \frac{T\sqrt{1+\xi^2}}{\xi\sqrt{1+Q_y^2}}. \tag{6.136}$$

For the case of the signal with a bell-shaped amplitude envelope and constant radio frequency carrier, using Eq. (6.136), we can write

$$\Delta\tau_{rs} = \frac{T}{\xi}\cdot\sqrt{1+\xi^2}. \tag{6.137}$$

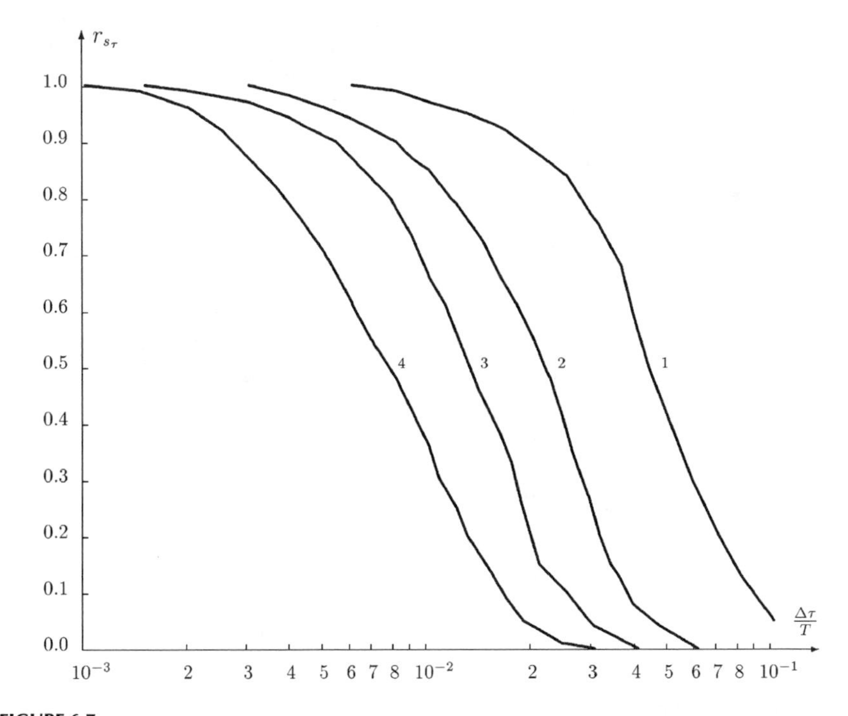

FIGURE 6.7
Dependence of envelope of the coefficient of correlation for the frequency-modulated signal with a bell-shaped amplitude envelope on the parameter $\frac{\Delta\tau}{T}$ at some values of the parameter ξ: for 1, $\xi = 0.2$; for 2, $\xi = 0.5$; for 3, $\xi = 1.0$; for 4, $\xi = 5.0$; $Q_y = 100$.

Equations (6.136) and (6.137) show us that the correlation interval in time decreases inversely proportionally to the energy spectrum bandwidth of the noise modulation function $\dot{M}(t)$ of the multiplicative noise and, when the condition

$$\Delta\Omega_M \geq \frac{6\pi}{T} \tag{6.138}$$

is satisfied, becomes practically equal to the equivalent duration of the amplitude envelope of the signal at the output of the generalized detector.

Dependence of envelope of the coefficient of correlation in time on the parameter

$$\frac{\Delta\tau}{T}$$

as can be seen from Eq. (6.97) in the case of the frequency-modulated signal with a bell-shaped amplitude envelope is shown in Fig. 6.7 for some values of the parameter

$$\xi = \frac{1}{2\pi} \cdot \Delta\Omega_M T \tag{6.139}$$

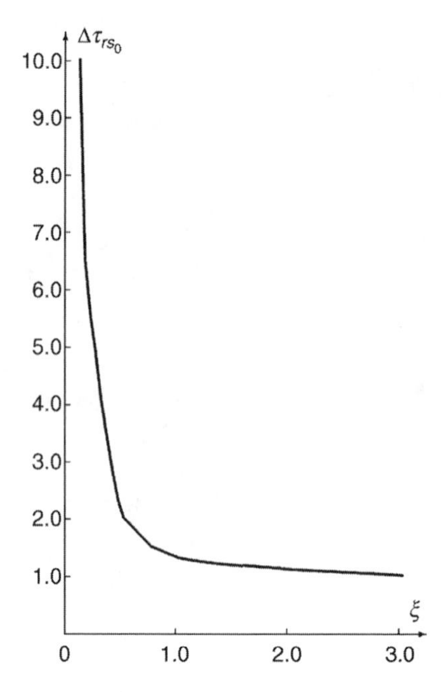

FIGURE 6.8
Dependence $\Delta\tau_{rs_0}$ on the parameter ξ.

for the condition

$$Q_y = 100. \tag{6.140}$$

Comparison of Eqs. (6.136) and (6.137) shows us that the dependences of the correlation intervals in time on the parameters of the noise modulation function $\dot{M}(t)$ of the multiplicative noise are the same in the case of the frequency-modulated signal with a bell-shaped amplitude envelope and in the case of the signal with a bell-shaped amplitude envelope and constant radio frequency carrier.

The function $\Delta\tau_{rs_0}$ vs. the parameter ξ is shown in Fig. 6.8. Moreover, the value $\Delta\tau_{rs_0}$ characterizes a ratio between the time of correlation and the duration of the undistorted signal component at the output of the generalized detector and is determined by:

- For the case of the frequency-modulated signal with a bell-shaped amplitude envelope:

$$\Delta\tau_{rs_0} = \frac{\Delta\tau_{rs}}{T} \cdot \sqrt{1 + Q_y^2}. \tag{6.141}$$

- For the case of the signal with a bell-shaped amplitude envelope and constant radio frequency carrier:

$$\Delta\tau_{rs_0} = \frac{\Delta\tau_{rs}}{T}. \tag{6.142}$$

One can see, from Eqs. (6.136) and (6.137) and Fig. 6.8, that during variation in the energy spectrum bandwidth of the noise modulation function $\dot{M}(t)$ of the multiplicative noise—the parameter ξ—from zero to infinity, the correlation interval in time is varied from infinity to the value equal to the duration of the undistorted signal component at the output of the generalized detector. The infinite value of the correlation interval $\Delta\tau_{rs}$ corresponds to the infinite correlation interval of the noise modulation function $\dot{M}(t)$ of the multiplicative noise.

Finiteness of the correlation interval $\Delta\tau_{rs}$ with an infinite decrease in the correlation interval of fluctuations of the noise modulation function $\dot{M}(t)$ of the multiplicative noise is caused by the limited bandwidth of the PF of the linear tract of the generalized detector. Note, that under the same durations of the signal with a bell-shaped amplitude envelope and constant radio frequency carrier and the frequency-modulated signal with a bell-shaped amplitude envelope, the correlation interval of the signal of the second type decreases rapidly in Q_y times with an increase in the parameter ξ under the stimulus of the multiplicative noise in comparison with the signal of the first type.

This fact is caused by a de-correlation process of the signal that is distorted by the multiplicative noise in the course of signal processing by the generalized detector and the amplitude-frequency response of the PF at the input linear tract of the generalized detector that is matched with the modulated signal using a complex law.

6.3.2 Signal with Square Wave-Form Envelope and Constant Radio Frequency Carrier

The autocorrelation function in time and frequency of the signal with the duration T and the square wave-form amplitude envelope and constant radio frequency carrier takes the following form:[41,42]

$$|\dot{\rho}(\tau, \Omega)|^2 = \left(\frac{2}{\Omega T}\right)^2 \cdot \sin^2\left[\frac{\Omega T}{2}\left(1 - \frac{|\tau|}{T}\right)\right]. \tag{6.143}$$

When the noise modulation function $\dot{M}(t)$ of the multiplicative noise has a bell-shaped energy spectrum, then substituting Eqs. (6.121) and (6.143) in Eq. (6.109) and introducing a new variable

$$x = \Omega + \omega, \tag{6.144}$$

the variance of fluctuations of the signal noise component at the output of the generalized detector takes the following form:

$$\sigma_s^2(\tau, \Omega) = \frac{C^2 E_{a_1}^2 \left(\overline{A^2(t)} - \alpha_0^2\right)}{2\Delta\Omega_M} \int\limits_{-\infty}^{\infty} \frac{\sin^2\left[\frac{xT}{2}\left(1 - \frac{|\tau|}{T}\right)\right]}{\left(\frac{xT}{2}\right)^2} \cdot e^{-\frac{\pi(x-\Omega)^2}{\Delta\Omega_M^2}}\, dx. \tag{6.145}$$

Henceforth, for simplicity assume that

$$C^2 = 1. \tag{6.146}$$

To determine the integral we change the first factor in Eq. (6.145) by the Fourier transform that is equal to:[43]

$$f(s) = \frac{1}{2\pi}\int_{-\infty}^{\infty} \frac{\sin^2\left[\frac{xT}{2}\left(1-\frac{|\tau|}{T}\right)\right]}{\left(\frac{xT}{2}\right)^2}\cdot \cos sx\,dx = \begin{cases} \frac{2}{T^2}\left[T\left(1-\frac{|\tau|}{T}\right)-s\right], & s < T\left(1-\frac{|\tau|}{T}\right); \\ 0, & s > T\left(1-\frac{|\tau|}{T}\right). \end{cases} \tag{6.147}$$

Taking into account Eq. (6.147), we can write

$$\sigma_s^2(\tau,\Omega) = \frac{E_{a_1}^2\left(\overline{A^2(t)}-\alpha_0^2\right)}{T^2\Delta\Omega_M}\int_{-\infty}^{\infty} e^{-\frac{\pi(x-\Omega)^2}{\Delta\Omega_M^2}}\,dx \times \int_0^{T\left(1-\frac{|\tau|}{T}\right)}\left[T\left(1-\frac{|\tau|}{T}\right)-s\right]\cdot\cos sx\,ds. \tag{6.148}$$

Integration with respect to the variable x in Eq. (6.148) gives us

$$\sigma_s^2(\tau,\Omega) = \frac{E_{a_1}^2\left(\overline{A^2(t)}-\alpha_0^2\right)}{T^2} \times \int_0^{T\left(1-\frac{|\tau|}{T}\right)}\left[T\left(1-\frac{|\tau|}{T}\right)-s\right]\cdot e^{-\frac{s^2\Delta\Omega_M^2}{4\pi}}\cos\Omega s\,ds. \tag{6.149}$$

Using the Taylor series expansion with respect to the variable Ω for the term $\cos\Omega s$ and carrying out an integration in terms of the tabulated integral,[44] we can write

$$\sigma_s^2(\tau,\Omega) = \frac{E_{a_1}^2\left(\overline{A^2(t)}-\alpha_0^2\right)}{2\sqrt{\pi}\xi}\cdot\sum_{n=-\infty}^{\infty}\frac{(-1)^n(4\pi)^n}{(2n)!}\cdot\left(\frac{\Omega}{\Delta\Omega}\right)^{2n} \times\left\{\left(1-\frac{|\tau|}{T}\right)\widetilde{G}\left[n+0.5;\pi\xi^2\left(1-\frac{|\tau|}{T}\right)^2\right] - \frac{1}{\xi\sqrt{\pi}}\cdot\widetilde{G}\left[n+1;\pi\xi^2\left(1-\frac{|\tau|}{T}\right)^2\right]\right\}, \tag{6.150}$$

where $\widetilde{G}(\alpha, x)$ is the incomplete gamma function.

Limiting by the first and second terms of Eq. (6.150), which define the variance of fluctuations in the zero detuning domain in frequency and, taking into account the well-known relationships between the incomplete gamma functions[45,46]

$$\widetilde{G}(0.5; x^2) = \sqrt{\pi}\Phi(x); \tag{6.151}$$

$$\widetilde{G}(\alpha+1; x) = \alpha\widetilde{G}(\alpha; x) - x^{\alpha}\cdot e^{-x}; \tag{6.152}$$

$$\widetilde{G}(1+n; x) = n!\left(1 - e^{-x}\sum_{m=0}^{n}\frac{x^m}{m!}\right), \tag{6.153}$$

we can write

$$\begin{aligned}\frac{\sigma_s^2(\tau,\Omega)}{E_{a_1}^2\left(\overline{A^2(t)} - \alpha_0^2\right)} &\simeq \frac{b}{2\xi}\cdot\Phi(b\xi\sqrt{\pi}) - \frac{1}{2\pi\xi^2}\cdot\left(1 - e^{-\pi b^2\xi^2}\right)\\ &-\frac{\pi b}{\xi}\cdot\left(\frac{\Omega}{\Delta\Omega_M}\right)^2\left[0.5\Phi(b\xi\sqrt{\pi}) - b\xi\cdot e^{-\pi b^2\xi}\right]\\ &+\frac{1}{\xi}\cdot\left(\frac{\Omega}{\Delta\Omega_M}\right)^2\left[1 - (1+\pi b^2\xi^2)\cdot e^{-\pi b^2\xi^2}\right] + \cdots,\end{aligned} \tag{6.154}$$

where

$$b = 1 - \frac{|\tau|}{T} \tag{6.155}$$

and

$$\Phi(x) = \frac{2}{\sqrt{\pi}}\int_0^x e^{-t^2}dt \tag{6.156}$$

is the error integral.

The ratio of the power of the signal noise component at the output of the generalized detector at the point

$$\tau = 0 \quad \text{and} \quad \Omega = 0$$

to the total power of the signal noise component at the output of the generalized detector

$$\frac{2\sigma_s^2(0,0)}{E_{a_1}^2\left(\overline{A^2(t)} - \alpha_0^2\right)} \tag{6.157}$$

as a function of the parameter ξ is shown in Fig. 6.9 by the solid line curves. These dependences are determined by Eq. (6.154).

In the case of the slow fluctuating multiplicative noise

$$\xi \ll 1 \tag{6.158}$$

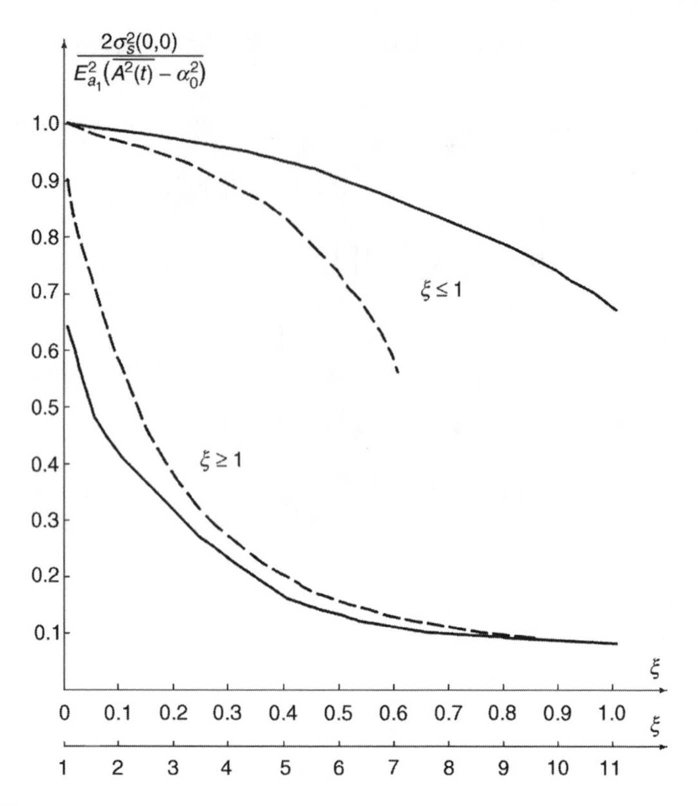

FIGURE 6.9
The ratio $\frac{\sigma_s^2(\tau,\Omega)}{E_{a_1}^2\left(\overline{A^2(t)}-\alpha_0^2\right)}$ as a function of the parameter ξ determined by the exact (——) and approximate (- - - -) formulae. The top x axis is for the case $\xi < 1$; the bottom x axis is for the case $\xi > 1$.

Equation (6.154) can be simplified for the condition

$$\sigma_s^2(\tau, \Omega) = 0,$$

using the series expansion with respect to the functions $\Phi(x)$ and e^{-ax^2} and limiting by the first, second, and third terms.

In the process, we can write

$$\frac{2\sigma_s^2(\tau, 0)}{E_{a_1}^2\left(\overline{A^2(t)} - \alpha_0^2\right)} \simeq b^2\left(1 + \frac{2\pi\xi^2 - \pi b^2\xi^2}{3}\right). \tag{6.159}$$

The dependence defined by Eq. (6.159) is shown in Fig. 6.9 by the dotted line curves. Reference to Fig. 6.9 shows that in the case of the slow fluctuating multiplicative noise we can use the approximate formula in Eq. (6.159) for the condition

$$\xi \text{ is between 0.2 and 0.3} \tag{6.160}$$

The error does not exceed 5 to 7%.

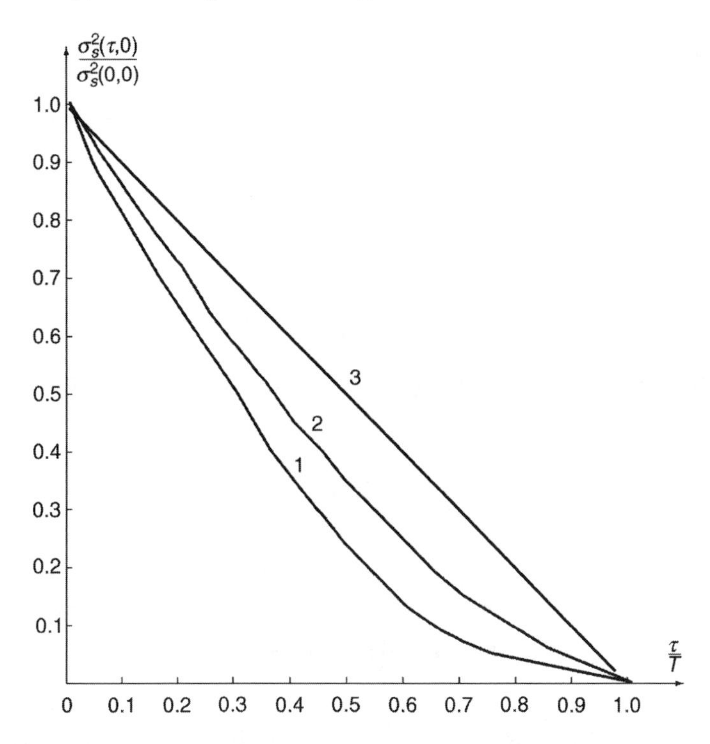

FIGURE 6.10
Normalized power of the signal noise component at the output of the generalized detector at some values of the parameter ξ: for 1, $\xi = 0.1$; for 2, $\xi = 1.0$; for 3, $\xi = 10$.

For the case of the signals with a square wave-form amplitude envelope we can see from Eq. (6.154) that the multiplicative noise gives rise to changing the distribution of power of the signal component at the output of the generalized detector with respect to the delay τ.

The distribution of the normalized power

$$\frac{\sigma_s^2(\tau, 0)}{\sigma_s^2(0, 0)}$$

of the signal noise component at the output of the generalized detector with respect to the delay τ for different values of the parameter ξ is shown in Fig. 6.10.

Reference to Fig. 6.10 shows that with an increase in the parameter ξ the distribution of the normalized power of the signal noise component at the output of the generalized detector is varied from the function

$$\left(1 - \frac{|\tau|}{T}\right)^2 \quad \text{at } \xi \ll 1 \tag{6.161}$$

to the function

$$1 - \frac{|\tau|}{T} \quad \text{at } \xi \gg 1. \tag{6.162}$$

6.3.3 Signals with Phase-Code Modulation

Consider the phase-manipulated signals with binary code when the signal phase takes two values 0 or π within the limits of the interval $\Delta\tau$, depending on the value "0" or "1"—the code of the phase-manipulated signal within the limits of the interval $\Delta\tau$.

Let Δt be the duration of the elementary signal. The total number of code elements is equal to

$$N = \frac{T}{\Delta\tau}. \tag{6.163}$$

The complex amplitude envelope of the signal can be written in the following form:[47,48]

$$\dot{S}(t) = \sum_{n=0}^{N-1} D_n \mathbf{1}\left(\frac{t}{\Delta t} - n - 0.5\right), \tag{6.164}$$

where

$$\mathbf{1}(x) = \begin{cases} 1, & |x| < 0.5; \\ 0.5, & |x| = 0.5; \\ 0, & |x| > 0.5; \end{cases} \tag{6.165}$$

D_n is the coefficient taking the values ± 1 depending on the value of phase of the signal within the limits of the given interval (0 or π).

The autocorrelation function of the signal determined by Eq. (6.164) for the positive and negative shifts τ can be written in the following form:

$$\dot{\rho}(\tau, \Omega) = \begin{cases} \frac{1}{2E_{a_1}} \int\limits_{\tau}^{T} \dot{S}(t) S^*(t-\tau) \cdot e^{j\Omega t}\, dt, & \tau \geq 0; \\ \frac{1}{2E_{a_1}} \int\limits_{0}^{T+\tau} \dot{S}(t) S^*(t-\tau) \cdot e^{j\Omega t}\, dt, & \tau < 0. \end{cases} \tag{6.166}$$

Substituting Eq. (6.164) in Eq. (6.166) for the condition $\tau \geq 0$, we can write

$$\dot{\rho}(\tau, \Omega) = \frac{1}{2E_{a_1}} \int\limits_{\tau}^{T} \sum_{k=0}^{N-1} \sum_{n=0}^{N-1} D_k D_n \mathbf{1}\left(\frac{t}{\Delta t} - k - 0.5\right) \times \mathbf{1}\left(\frac{t-\tau}{\Delta t} - n - 0.5\right) \cdot e^{j\Omega t} dt. \tag{6.167}$$

Assume that

$$\frac{\tau}{\Delta t} = \ell. \tag{6.168}$$

Then we can determine the autocorrelation function at the discretes with the interval Δt. In this case, the autocorrelation function can be written in the following form:

$$\dot{\rho}(\tau, \Omega) = \frac{1}{2E_{a_1}} \int\limits_{\ell\Delta t}^{N\Delta t} \sum_{p=\ell}^{N-1} D_p D_{p-\ell} \mathbf{1}\left(\frac{t}{\Delta t} - p - 0.5\right) \times \mathbf{1}\left(\frac{t}{\Delta t} - p - 0.5\right) \cdot e^{j\Omega t}\, dt. \tag{6.169}$$

During determination of the variance of fluctuations it is required to define the square of the module of the autocorrelation function. Under changing the integral of sums by sum of integrals, the square of the module of the autocorrelation function takes the following form:

$$|\dot{\rho}(\tau, \Omega)|^2 = |\dot{\rho}(\ell\Delta t, \Omega)|^2 = \frac{\sin^2 \frac{\Omega\Delta t}{2}}{N^2\left(\frac{\Omega\Delta t}{2}\right)^2} \cdot \left|\sum_{p=\ell}^{N-1} D_p D_{p-\ell} \cdot e^{jp\Omega}\right|^2, \quad \ell \geq 0, \tag{6.170}$$

where the fact that

$$2E_{a_1} = N\Delta t \tag{6.171}$$

at the unit amplitude of the signal in accordance with the Eq. (6.167) is taken into account. For the case $\tau \geq 0$ we can obtain the analogous formula.

Combining formulae for the autocorrelation function

$$|\dot{\rho}(\ell\Delta t, \Omega)|$$

at the positive and negative values ℓ, we finally can write:

$$|\dot{\rho}(\ell\Delta t, \Omega)|^2 = \frac{\sin^2 \frac{\Omega\Delta t}{2}}{N^2\left(\frac{\Omega\Delta t}{2}\right)^2} \cdot \left|\sum_{p=|\ell|}^{N-1} D_p D_{p-|\ell|} \cdot e^{jp\Omega}\right|^2. \tag{6.172}$$

Consider in more detail the square of the module in Eq. (6.172):

$$\begin{aligned}\left|\sum_{p=|\ell|}^{N-1} D_p D_{p-|\ell|} \cdot e^{jp\Omega}\right|^2 &= \sum_{k=|\ell|}^{N-1} \sum_{m=|\ell|}^{N-1} D_k D_{k-|\ell|} D_m D_{m-|\ell|} \cos k\Omega\Delta t \cos m\Omega\Delta t \\ &\quad + \sum_{k=|\ell|}^{N-1} \sum_{m=|\ell|}^{N-1} D_k D_{k-|\ell|} D_m D_{m-|\ell|} \sin k\Omega\Delta t \sin m\Omega\Delta t \\ &= \sum_{k=|\ell|}^{N-1} \sum_{m=|\ell|}^{N-1} D_k D_{k-|\ell|} D_m D_{m-|\ell|} \cos(k - m)\Omega\Delta t.\end{aligned} \tag{6.173}$$

We proceed to introduce a new variable:

$$r = k - m. \tag{6.174}$$

Then we can write

$$\left|\sum_{p=|\ell|}^{N-1} D_p D_{p-|\ell|} \cdot e^{jp\Omega}\right|^2 = \sum_{k=|\ell|}^{N-1} \sum_{r=k-|\ell|}^{N-1} D_k D_{k-|\ell|} D_{k-r} D_{k-r-|\ell|} \cos r\Omega\Delta t. \quad (6.175)$$

Changing the order of summation in Eq. (6.175), we can write

$$\begin{aligned}\left|\sum_{p=|\ell|}^{N-1} D_p D_{p-|\ell|} \cdot e^{jp\Omega}\right|^2 &= \sum_{r=1}^{N-1-|\ell|} \cos r\Omega\Delta t \sum_{k=r+|\ell|}^{N-1} D_k D_{k-|\ell|} D_{k-r} D_{k-r-|\ell|} \\ &+ \sum_{r=-(N-1-|\ell|)}^{-1} \cos r\Omega\Delta t \\ &\times \sum_{k=|\ell|}^{N-1+r} D_k D_{k-|\ell|} D_{k-r} D_{k-r-|\ell|} + N - |\ell|. \quad (6.176)\end{aligned}$$

By changing variables

$$s = -r \quad (6.177)$$

and

$$q = -r + k \quad (6.178)$$

the second term in Eq. (6.176) can be put in the following form:

$$\sum_{s=1}^{N-1-|\ell|} \cos \Omega\Delta t \sum_{q=s+|\ell|}^{N-1} D_{q-s} D_{q-s-|\ell|} D_q D_{q-|\ell|}. \quad (6.179)$$

Equation (6.179) coincides with the first term in Eq. (6.176). In terms of the statements mentioned above we finally can write:

$$\left|\sum_{p=|\ell|}^{N-1} D_p D_{p-|\ell|} \cdot e^{jp\Omega}\right|^2 = 2 \sum_{r=1}^{N-1-|\ell|} D_{r,\ell} \cos r\Omega\Delta t + N - |\ell|, \quad (6.180)$$

where

$$D_{r,\ell} = \sum_{k=r+|\ell|}^{N-1} D_k D_{k-|\ell|} D_{k-r} D_{k-r-|\ell|}. \quad (6.181)$$

Note that the coefficients $D_{r,\ell}$ are invariant with respect to changing the order of the written form of the indexes.

In terms of Eq. (6.180), the square of the module of the autocorrelation function of the phase-manipulated signal (see Eq. (6.172)) can be written in the following form:

$$|\dot{\rho}(\ell\Delta t, \Omega)|^2 = \frac{\sin^2 \frac{\Omega\Delta t}{2}}{N^2 \left(\frac{\Omega\Delta t}{2}\right)^2} \cdot \left[2 \sum_{r=1}^{N-1-|\ell|} D_{r,\ell} \cos r\Omega\Delta t + N - |\ell|\right]. \quad (6.182)$$

Substituting Eq. (6.182) in Eq. (6.109), we obtain that the variance of fluctuations of the signal distorted by the multiplicative noise at the output of the generalized detector can be written in the following form:

$$\frac{2\sigma_s^2(\ell\Delta t,\Omega)}{E_{a_1}^2\left(\overline{A^2(t)}-\alpha_0^2\right)} = \frac{1}{\pi N^2}\sum_{r=1}^{N-1-|\ell|} D_{r,\ell} \times \int_{-\infty}^{\infty} G_{V_0}(\omega)\cdot\frac{\sin^2\frac{(\omega+\Delta\Omega)\Delta t}{2}}{\frac{(\omega+\Omega)^2\Delta t^2}{4}}\cdot\cos r(\omega+\Omega)\Delta t\, d\omega + \frac{N-|\ell|}{2\pi N^2}\int_{-\infty}^{\infty} G_{V_0}(\omega)\cdot\frac{\sin^2\frac{(\omega+\Omega)\Delta t}{2}}{\frac{(\omega+\Omega)^2\Delta t^2}{4}}\cdot d\omega, \tag{6.183}$$

where

$$G_{V_0}(\omega) = \frac{G_V(\omega)}{R_V(0)} \tag{6.184}$$

is the normalized energy spectrum of fluctuations of the noise modulation function $\dot{M}(t)$ of the multiplicative noise.

Further simplifications can be made if we suppose that the energy spectrum bandwidth of the noise modulation function $\dot{M}(t)$ of the multiplicative noise is much less than the value

$$\frac{2\pi}{\Delta t}$$

that defines the energy spectrum bandwidth of the signal. This fact always occurs in practice, especially at high coefficients of compression.[49–52]

In the process, the function

$$\frac{\sin^2 ax}{x^2}$$

in Eq. (6.183) can be thought of as a constant within the limits of the interval, where $G_{V_0}(\omega)$ does not equal zero. Then for the condition $\Omega = 0$, we obtain the following result:

$$\frac{2\sigma_s^2(\ell\Delta t,0)}{E_{a_1}^2\left(\overline{A^2(t)}-\alpha_0^2\right)} \simeq \frac{1}{\pi N^2}\sum_{r=1}^{N-1-|\ell|} D_{r,\ell}\int_{-\infty}^{\infty} G_{V_0}(\omega)\cos r\omega\Delta t\, d\omega + \frac{N-|\ell|}{N^2} = \frac{2}{N^2}\sum_{r=1}^{N-1-|\ell|} D_{r,\ell}R_{V_0}(r\Delta t) + \frac{N-|\ell|}{N^2}, \quad |\ell| \le N, \tag{6.185}$$

where $R_{V_0}(r\Delta t)$ is the Fourier transform with respect to the value $G_{V_0}(\omega)$.

Equation (6.185) can be simplified by carrying out an additional averaging with respect to an ensemble of possible signals, assuming that the values of the signal phase (0 or π) are equiprobable for each discrete. For this case, the average values of the coefficients $D_{r,\ell}$ do not exceed unity by the absolute value under the condition $\ell \neq 0$ and

$$D_{r,\ell} = N - r. \tag{6.186}$$

Then neglecting the first term in Eq. (6.185) in comparison with the second term for the condition

$$N \gg 1, \tag{6.187}$$

we finally can write

$$\frac{2\sigma_s^2(\ell \Delta t, 0)}{E_{a1}^2\left(\overline{A^2(t)} - \alpha_0^2\right)} = \begin{cases} \frac{2}{N^2} \sum_{r=1}^{N-1} (N-r) R_{V_0}(r\Delta t), & \ell = 0; \\ \frac{N-|\ell|}{N^2}, & 0 < |\ell| \leq N, \end{cases} \tag{6.188}$$

where the condition

$$R_{V_0}(0) = 1$$

is taken into consideration.

We can show that the variance of fluctuations at the point $\tau = 0$ ($\ell = 0$) and $\Omega = 0$ determined by Eq. (6.188) coincides with the value determined by Eq. (6.145) for the variance of fluctuations of the signal with a square wave-form amplitude envelope and constant radio frequency carrier at the corresponding point of the plane (τ, Ω).

For example, we carry out a limiting passage from sum to integral in Eq. (6.188). Then Eq. (6.188) can be written in the simpler form:

$$\frac{2\sigma_s^2(\ell \Delta t, 0)}{E_{a_1}^2\left(\overline{A^2(t)} - \alpha_0^2\right)} \simeq \begin{cases} \frac{2\sigma_s^2(0,0)}{E_{a_1}^2\left(\overline{A^2(t)} - \alpha_0^2\right)}, & \ell = 0; \\ \frac{N-|\ell|}{N^2}, & |\ell| \leq N, \end{cases} \tag{6.189}$$

where $\sigma_s^2(0, 0)$ is defined by Eq. (6.154) under the bell-shaped energy spectrum of the noise modulation function $\dot{M}(t)$ of the multiplicative noise and under the conditions

$$\Omega = 0 \quad \text{and} \quad b = 1. \tag{6.190}$$

As follows from the formulae discussed above, the variance of fluctuations at side lobes of the autocorrelation function is inversely proportional to the coefficient of signal truncation, which is equal to the number N of readings of the signal, and decreases linearly farther and farther away from the main (central) peak of the autocorrelation function with an increase in the value ℓ.

Formally, a decrease in the amplitude of side lobes of the autocorrelation function with an increase in the number of elementary signals in the case of the slow multiplicative noise is a consequence of averaging by an ensemble of the possible signals (codes) that are random under determination of the coefficients $D_{r,\ell}$.

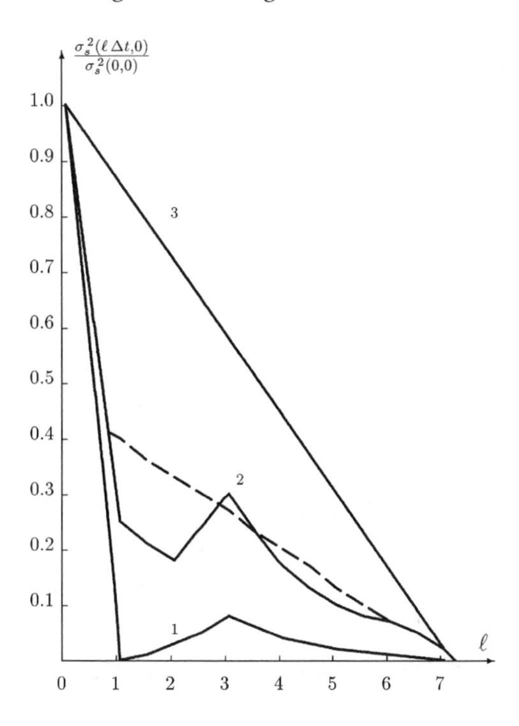

FIGURE 6.11
The ratio $\frac{\sigma_s^2(\ell\Delta t,0)}{\sigma_s^2(0,0)}$ as a function of the variable $\ell = \frac{\tau}{\Delta t}$ at some values of the parameter ξ: for 1, $\xi = 0$; for 2, $\xi = 2.8$; for 3, $\xi \to \infty$; $N = 7$ (the Chaffman code).

Actually, the variance of fluctuations of side lobes of the autocorrelation function

$$\sigma_s^2(\ell\Delta t, 0)$$

as a function of the number N is explained by the fact that slow distortions at the generalized detector are processed in the same way as rapid distortions within the limits of the correlation interval Δt because the generalized detector de-correlates the signal distorted by the multiplicative noise.

The function

$$f(\ell) = \frac{\sigma_s^2(\ell\Delta t, 0)}{\sigma_s^2(0, 0)} \tag{6.191}$$

determined by Eq. (6.183) under the bell-shaped energy spectrum of the noise modulation function $\dot{M}(t)$ of the multiplicative noise for the phase-manipulated signals by 7- and 15-digit Chaffman code is shown in Figs. 6.11 and 6.12 for some values of the parameter ξ. Determined functions

$$\frac{\sigma_s^2(\ell\Delta t, 0)}{\sigma_s^2(0, 0)}$$

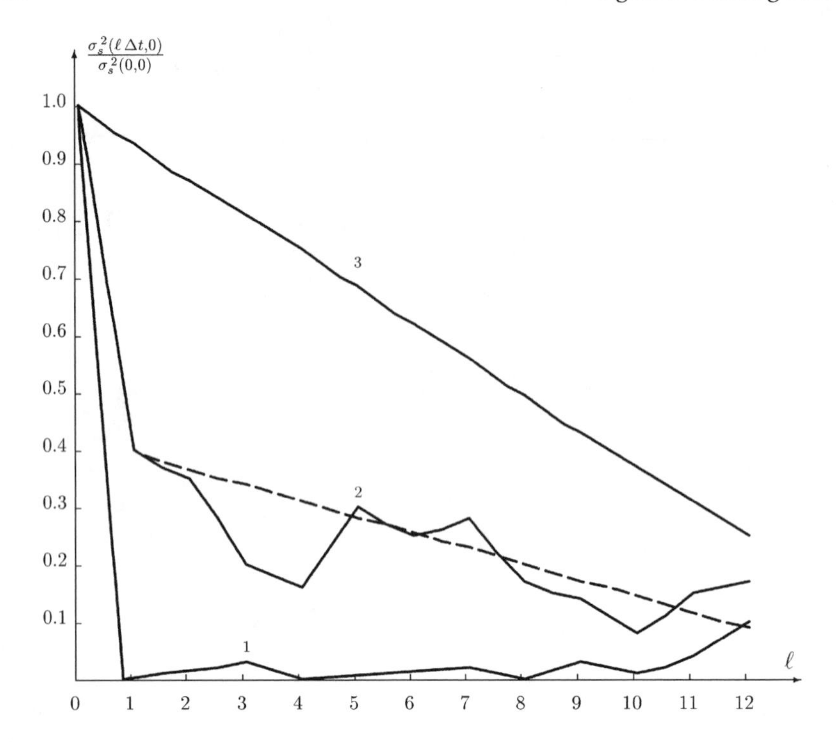

FIGURE 6.12
The ratio $\frac{\sigma_s^2(\ell\Delta t,0)}{\sigma_s^2(0,0)}$ as a function of the variable $\ell = \frac{\tau}{\Delta t}$ at some values of the parameter ξ: for 1, $\xi = 0$; for 2, $\xi = 6$; for 3, $\xi \to \infty$; $N = 15$ (the Chaffman code).

at the points

$$\tau = \ell\Delta t \tag{6.192}$$

are conditionally connected by segments. The same functions determined by Eq. (6.188) are shown by the dotted line.

Reference to Figs. 6.11 and 6.12 shows that the formula in Eq. (6.188) defines an exact dependence in a satisfactory manner. By the way, Eq. (6.188) becomes closer to Eq. (6.183) with an increase in the value N, the number of code elements.

6.4 Signal Noise Component under the Stimulus of Slow and Rapid Multiplicative Noise

General formulae for definition of statistical characteristics of the signal noise component at the output of the generalized detector in Section 6.2 can be simplified for two limiting cases: the first, in which the energy spectrum bandwidth of the noise modulation function $\dot{M}(t)$ of the multiplicative noise is very wide, is the case of rapid multiplicative noise; and the second, in which the energy spectrum bandwidth of the noise modulation function $\dot{M}(t)$ of the multiplicative noise is very narrow, is the case of slow multiplicative noise.

Let us analyze the statistical characteristics of the signal noise component at the output of the generalized detector for these two cases.

6.4.1 Rapid Fluctuating Multiplicative Noise

When the energy spectrum bandwidth $\Delta\Omega_M$ of the noise modulation function $\dot{M}(t)$ of the multiplicative noise is much greater than the energy spectrum bandwidth of the signal $\Delta\Omega_a$, the energy spectrum $G_V(\omega)$ can be approximately considered as the constant value within the limits of the interval of the values ω, where the autocorrelation function $\dot{\rho}(\tau, \Omega + \omega)$ is essential by the magnitude.[53,54] The energy spectrum bandwidth of this interval is equal to $2\Delta\Omega_a$.

In this case, as in Section 6.1, Eq. (6.166) can be transformed in the following approximate form:

$$R_s(\tau_1, \tau_2, \Omega_1, \Omega_2) \simeq \frac{C^2 E_{a_1}^2}{4\pi} \cdot G_V\left(-\frac{\Omega_1 + \Omega_2}{2}\right) \times Re\left\{e^{j\omega_a(\tau_1-\tau_2)} \int_{-\infty}^{\infty} \dot{\rho}(\tau_1, \Omega_1 + \omega)\rho^*(\tau_2, \Omega_2 + \omega)\, d\omega\right\}. \tag{6.193}$$

The integral in Eq. (6.193) is determined:

$$\begin{aligned} J &= \frac{1}{2\pi} \int_{-\infty}^{\infty} \dot{\rho}(\tau_1, \Omega_1 + \omega)\rho^*(\tau_2, \Omega_2 + \omega)\, d\omega \\ &= \frac{1}{8\pi E_{a_1}^2} \int_{-\infty}^{\infty}\int_{-\infty}^{\infty}\int_{-\infty}^{\infty} \dot{S}(t_1)S^*(t_1 - \tau_1)S^*(t_2) \\ &\quad \times \dot{S}(t_2 - \tau_2) \cdot e^{j[(\Omega_1+\omega)t_1 - (\Omega_2+\omega)t_2]}\, dt_1\, dt_2\, d\omega \\ &= \frac{1}{4E_{a_1}^2} \int_{-\infty}^{\infty}\int_{-\infty}^{\infty} \dot{S}(t_1)S^*(t_2)S^*(t_1 - \tau_1) \\ &\quad \times \dot{S}(t_2 - \tau_2) \cdot e^{j(\Omega_1 t_1 - \Omega_2 t_2)}\delta(t_1 - t_2)\, dt_1\, dt_2, \end{aligned} \tag{6.194}$$

where

$$\delta(t_1 - t_2) = \frac{1}{2\pi} \int_{-\infty}^{\infty} e^{j\omega(t_1 - t_2)}\, d\omega \tag{6.195}$$

is the delta function.

Taking into account the main peculiarities of the delta function and using Eq. (6.194), we can write

$$J = \frac{1}{4E_{a_1}^2} \int_{-\infty}^{\infty} |\dot{S}(t_1)|^2 S^*(t_1 - \tau_1)\dot{S}(t_1 - \tau_2) \cdot e^{j(\Omega_1 - \Omega_2)t_1}\, dt_1. \tag{6.196}$$

We introduce the new variables:

$$\tau_2 = \tau; \tag{6.197}$$

$$\tau_1 = \tau + \Delta\tau; \tag{6.198}$$

$$\Omega_1 = \Omega; \tag{6.199}$$

$$\Omega_2 = \Omega - \Delta\Omega. \tag{6.200}$$

In terms of Eqs. (6.196)–(6.200), Eq. (6.193) takes the following form:

$$R_s(\tau + \Delta\tau, \tau, \Omega, \Omega - \Delta\Omega) \simeq \frac{C^2}{8} \cdot G_V\left(\frac{\Delta\Omega}{2} - \Omega\right) Re\left\{ e^{j\omega_a \Delta\tau} \int_{-\infty}^{\infty} |\dot{S}(t)|^2 \times \dot{S}(t - \tau)S^*(t - \tau - \Delta\tau) \cdot e^{j\Delta\Omega t} dt \right\}. \tag{6.201}$$

The integral in Eq. (6.201) can be determined when the signals possess a square wave-form amplitude envelope. Actually, for this case for the condition

$$\Delta\tau \leq -\tau, \tau \geq 0 \tag{6.202}$$

or

$$\Delta\tau \geq -\tau, \tau \leq 0 \tag{6.203}$$

we can write

$$\begin{aligned} J &= \frac{A^2(t)}{4E_{a_1}^2} \int_0^T \dot{S}(t - \tau)S^*(t - \tau - \Delta\tau) \cdot e^{j\Delta\Omega t}\, dt \\ &= \frac{A^2(t)}{4E_{a_1}^2} \cdot e^{j\Delta\Omega\tau} \int_{-\tau}^{T-\tau} \dot{S}(t - \tau)S^*(t - \tau - \Delta\tau) \cdot e^{j\Delta\Omega t}\, dt \\ &= \frac{A^2(t)}{2E_{a_1}^2} \cdot \dot{\rho}(\Delta\tau, \Delta\Omega) \cdot e^{j\Delta\Omega\tau}, \end{aligned} \tag{6.204}$$

where $A(t)$ is the amplitude of the signal and T is the duration of the signal.

Taking into consideration that for signals having the duration T and square wave-form amplitude envelope, the following condition

$$2E_{a_1} = A^2(t)T \tag{6.205}$$

is true, Eq. (6.201) can be written in the following form:

$$R_s(\tau + \Delta\tau, \tau, \Omega, \Omega - \Delta\Omega) \simeq \frac{C^2 E_{a_1}^2}{2T} \cdot G_V\left(\frac{\Delta\Omega}{2} - \Omega\right) \times Re\{\dot{\rho}(\Delta\tau, \Delta\Omega) \cdot e^{j(\omega_a \Delta\tau + \Delta\Omega\tau)}\} \quad (6.206)$$

if the conditions determined by Eqs. (6.202) and (6.203) are true.

With some additional approximations the analogous formulae can be obtained in the case of the wide-band signals for any type of amplitude envelope, but the amplitude envelope must be defined by a function that is slowly varied within the limits of the interval

$$\frac{1}{\Delta\Omega_a},$$

where $\Delta\Omega_a$ is the energy spectrum bandwidth of the signal.[55,56]

Substituting the condition

$$\Delta\tau = \Delta\Omega = 0$$

in Eq. (6.201), we obtain that the variance of fluctuations takes the following form:

$$\sigma_s^2(\tau, \Omega) \simeq \frac{C^2}{8} \cdot G_V(-\Omega) \int_{-\infty}^{\infty} |\dot{S}(t)|^2 |\dot{S}(t-\tau)|^2 \, dt. \quad (6.207)$$

Equation (6.207) shows that when the energy spectrum of fluctuations of the noise modulation function $\dot{M}(t)$ of the multiplicative noise can be considered as uniformly distributed within the limits of the interval $2\Delta\Omega_a$, then the variance of the signal noise component at the output of the generalized detector does not depend on the law of angle (frequency and phase) modulation of the signal. The variance of the signal noise component at the output of the generalized detector is only defined by the amplitude envelope of the signal and the energy spectrum $G_V(\Omega)$.

For the signal with a square wave-form amplitude envelope and duration T, Eq. (6.207) takes the following form:

$$\sigma_s^2(\tau, \Omega) \simeq \begin{cases} \frac{C^2 E_{a_1}^2}{2T} \cdot G_V(-\Omega)\left(1 - \frac{|\tau|}{T}\right), & |\tau| \le T; \\ 0, & |\tau| > T. \end{cases} \quad (6.208)$$

When the amplitude envelope of the signal is bell-shaped, the amplitude envelope of the signal is Gaussian

$$S(t) = A(t) \cdot e^{-\frac{\pi t^2}{T^2}}, \quad (6.209)$$

Eq. (6.207) can be written in the following form:

$$\sigma_s^2(\tau, \Omega) \simeq \frac{C^2 E_{a_1}^2}{2T} \cdot G_V(-\Omega) \cdot e^{-\frac{\pi\tau^2}{T^2}}. \tag{6.210}$$

To estimate errors caused by using the approximate formulae mentioned above with the purpose of defining the limits of use for these formulae, consider a set of examples and compare the results obtained with the results in Section 6.3, where these formulae are defined exactly.

We estimate the limits of use for Eq. (6.208) given the case of a square waveform amplitude envelope of the signal when the energy spectrum of the noise modulation function $\dot{M}(t)$ of the multiplicative noise is bell-shaped (see Eq. (6.121)).

Substituting Eq. (6.121) in Eq. (6.208), we can write

$$\frac{\sigma_s^2(\tau, \Omega)}{E_{a_1}^2\left(\overline{A^2(t)} - \alpha_0^2\right)} \simeq \frac{1 - \frac{|\tau|}{T}}{2} \cdot \xi \cdot e^{-\frac{\Omega^2 T^2}{4\pi\xi^2}}. \tag{6.211}$$

The function

$$f(\xi) = \frac{2\sigma_s^2(0, 0)}{E_{a_1}^2\left(\overline{A^2(t)} - \alpha_0^2\right)} \tag{6.212}$$

determined by Eq. (6.211) is shown in Fig. 6.9 by the dotted line curves.

Comparison of the functions determined by the exact formulae, the solid line curves shown in Fig. 6.9, and by the approximate formulae, the dotted line curves shown in Fig. 6.9, shows that the approximate definition of the variance $\sigma_s^2(0, 0)$ can be used with a low error for the condition $\xi \geq$ a value between 2 and 3, for example, when the energy spectrum bandwidth of the noise modulation function $\dot{M}(t)$ of the multiplicative noise equals or exceeds the energy spectrum bandwidth of the signal multiplied by a factor of value between 2 and 3.

We take the ratio of the variances $\sigma_s^2(\tau, \Omega)$ determined by Eqs. (6.127) and (6.210), with the purpose of comparing the results obtained for the case of a bell-shaped amplitude envelope of the signal defined by the exact formula (see Eq. (6.127)) and the approximate formula (see Eq. (6.210)) when the energy spectrum of the noise modulation function $\dot{M}(t)$ of the multiplicative noise is bell-shaped and determined by Eq. (6.121). This ratio is:

$$\frac{\sigma_{s_{ex}}^2(\tau, \Omega)}{\sigma_{s_{ap}}^2(\tau, \Omega)} = \frac{\xi}{\sqrt{1+\xi^2}} \cdot \exp\left(\frac{\Omega^2}{4\pi\xi^2} \cdot \frac{T^2}{1+\xi^2}\right), \tag{6.213}$$

where $\sigma_{s_{ex}}^2(\tau, \Omega)$ is the exact determination of the variance $\sigma_s^2(\tau, \Omega)$; $\sigma_{s_{ap}}^2(\tau, \Omega)$ is the approximate determination of the variance $\sigma_s^2(\tau, \Omega)$.

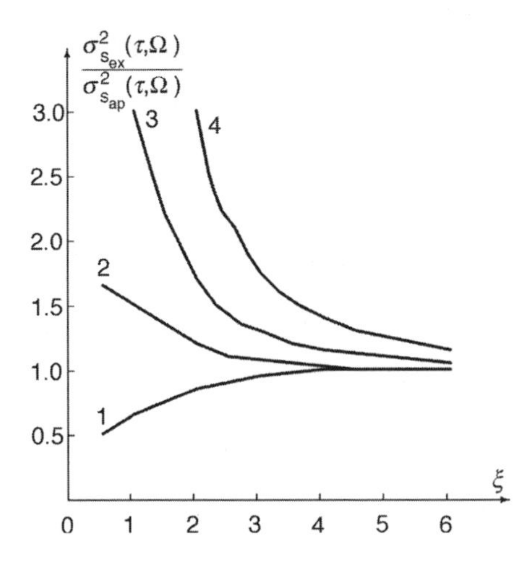

FIGURE 6.13
The ratio $\frac{\sigma^2_{s_{ex}}(\tau,\Omega)}{\sigma^2_{s_{ap}}(\tau,\Omega)}$ as a function of the parameter ξ under fixed values μ: for 1, $\mu = 0$; for 2, $\mu = 0.5$; for 3, $\mu = 1.0$; for 4, $\mu = 2.0$.

The ratio

$$\frac{\sigma^2_{s_{ex}}(\tau, \Omega)}{\sigma^2_{s_{ap}}(\tau, \Omega)}$$

that is determined by Eq. (6.213) is shown in Fig. 6.13 as a function of the parameter ξ at the fixed values of the parameter

$$\mu = \frac{1}{2\pi} \cdot \Omega T. \tag{6.214}$$

The value μ characterizes the ratio between the moving value of the frequency Ω and the energy spectrum bandwidth of amplitude envelope of the signal.

With the purpose of comparing the exact formulae determined by Eqs. (6.132) and (6.133) for the condition $Q_y = 0$, with the approximate formulae determined by Eq. (6.201), we determine the correlation function of the signal noise component at the output of the generalized detector for the case when the signal has a bell-shaped amplitude envelope and constant radio frequency carrier using Eq. (6.201).

Substituting Eq. (6.209) in Eq. (6.201), after uncomplicated transformations, we obtain the following form:

$$R_s(\tau + \Delta\tau, \tau, \Delta\Omega) \simeq \frac{C^2 E^2_{a_1}}{2T} \cdot \exp\left[-\frac{\pi}{T^2}\left(\tau^2 + \tau\Delta\tau + \frac{3}{4}\Delta\tau^2\right) - \frac{\Delta\Omega^2 T^2}{16\pi}\right]$$
$$\times \cos\left[\frac{\Delta\Omega}{4}(2\tau + \Delta\tau) + \omega_a \Delta\tau\right] \cdot G_V\left(\frac{\Delta\Omega}{2} - \Omega\right). \tag{6.215}$$

In accordance with Eq. (6.128), the normalized coefficient of correlation takes the following form:

$$r_s(\Delta\tau, \Delta\Omega) \simeq \frac{G_V\left(\frac{\Delta\Omega}{2} - \Omega\right)}{\sqrt{G_V(-\Omega)G_V(\Delta\Omega - \Omega)}} \cdot e^{-\frac{\pi\Delta\tau^2}{4T^2} - \frac{\Delta\Omega^2 T^2}{16\pi}} \times \cos\left[\frac{\Delta\Omega}{4}(2\tau + \Delta\tau) + \omega_a \Delta\tau\right]. \quad (6.216)$$

In doing so, the correlation intervals in time and frequency are equal to the duration of the autocorrelation function of the signal on the corresponding axis—the time axis or frequency axis.

In many cases, simple approximate relationships for definition of the variance $\sigma_s^2(\tau, \Omega)$ can be used during the softer condition in comparison with the condition of approximate uniform distribution of the energy spectrum $G_V(\Omega)$ within the limits of the interval $2\Delta\Omega_a$, which was just formulated above.

If a cross-section of the autocorrelation function $\dot{\rho}(\tau, \Omega)$ made by the plane $\tau = const$ contains the main peak of the autocorrelation function $\dot{\rho}(\tau, \Omega)$ and the amplitude of the main peak of the autocorrelation function $\dot{\rho}(\tau, \Omega)$ is much greater than the amplitudes of side lobes of the autocorrelation function $\dot{\rho}(\tau, \Omega)$, then the values of the energy spectrum $G_V(\omega)$ lying inside of the frequency region covered by the main peak of the autocorrelation function $\dot{\rho}(\tau, \Omega)$ make the main contribution to the integral in Eq. (6.109).

The bandwidth of this range of the values Ω is defined by the double bandwidth of the energy spectrum of amplitude envelope of the signal. This bandwidth is approximately equal to the value

$$\frac{4\pi}{T},$$

where T is the duration of the signal. This fact takes place, in particular, for all types of pulse signals for the condition

$$|\tau| < \frac{2\pi}{\Delta\Omega_a} \quad (6.217)$$

and for the frequency-modulated signals during all τ, where the autocorrelation function $\dot{\rho}(\tau, \Omega)$ does not equal zero.

For this case, the approximate formulae mentioned above are used when the energy spectrum $G_V(\omega)$ is approximately constant within the limits of the frequency bandwidth covered by the main peak of the autocorrelation function $\dot{\rho}(\tau, \Omega)$. In other words, when the energy spectrum bandwidth is equal to the doubled energy spectrum bandwidth of amplitude envelope of the signal. This is true for the condition $\tau_c \ll T$, where τ_c is the correlation interval in time of the noise modulation function $\dot{M}(t)$ of the multiplicative noise.

Taking into account that[57]

$$\tau_c = \frac{\pi}{\Delta\Omega_M}, \quad (6.218)$$

the given condition also can be written in the following form

$$2\xi \gg 1,$$

where

$$\xi = \frac{\Delta\Omega_M T}{2\pi} = \Delta F_M T. \tag{6.219}$$

For the case of signals without intrapulse modulation, the energy spectrum bandwidth of the signal is equal to the energy spectrum bandwidth of amplitude envelope of the signal and this condition coincides with the condition of use in Eqs. (6.201)–(6.210) that was formulated above.

For the case of signals with intrapulse modulation the condition

$$2\xi \gg 1$$

can be true when the energy spectrum bandwidth of the noise modulation function $\dot{M}(t)$ of the multiplicative noise is much less than the energy spectrum bandwidth of the signal.

Taking into account all statements discussed above for all types of signals for the condition

$$T \gg \tau_c,$$

the variance of the signal noise component at the output of the generalized detector for the condition $\tau = 0$, according to Eqs. (6.208) and (6.210), is determined in the following form:

$$\sigma_s^2(0, \Omega) \simeq \frac{C^2 E_{a_1}^2}{2T} \cdot G_V(-\Omega). \tag{6.220}$$

For signals with linear modulation in frequency at the shift in time τ, the maximum of the main peak of the autocorrelation function $\dot{\rho}(\tau, \Omega)$ occurs for the condition

$$\Omega = \Omega_0 = -\frac{\tau}{T} \cdot \Delta\omega_d, \tag{6.221}$$

where $\Delta\omega_d$ is the frequency deviation (see Fig. 6.14).

In accordance with Eq. (6.221), the module of the autocorrelation function

$$|\dot{\rho}(\tau, \Omega)|^2$$

in Eq. (6.109) takes the maximum when the value of the frequency ω is equal to (see Fig. 6.15)

$$\omega_{max} = \Omega_0 - \Omega = -\left(\frac{\Delta\omega_d}{T}\tau + \Omega\right). \tag{6.222}$$

When the energy spectrum $G_V(\omega)$ is approximately constant within the limits of the frequency bandwidth covered by the main peak of the autocorrelation function $\dot{\rho}(\tau, \Omega + \omega)$, i.e., this frequency bandwidth is equal to

$$\Delta\omega = \frac{4\pi}{T}, \tag{6.223}$$

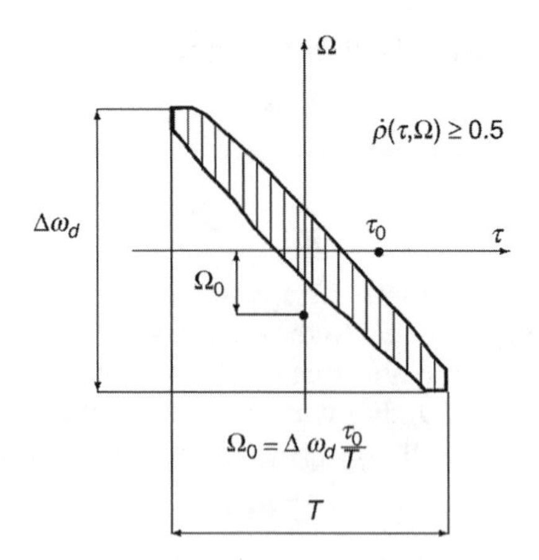

FIGURE 6.14
Definition of maximum of the main peak of the autocorrelation function $\dot{\rho}(\tau, \Omega)$.

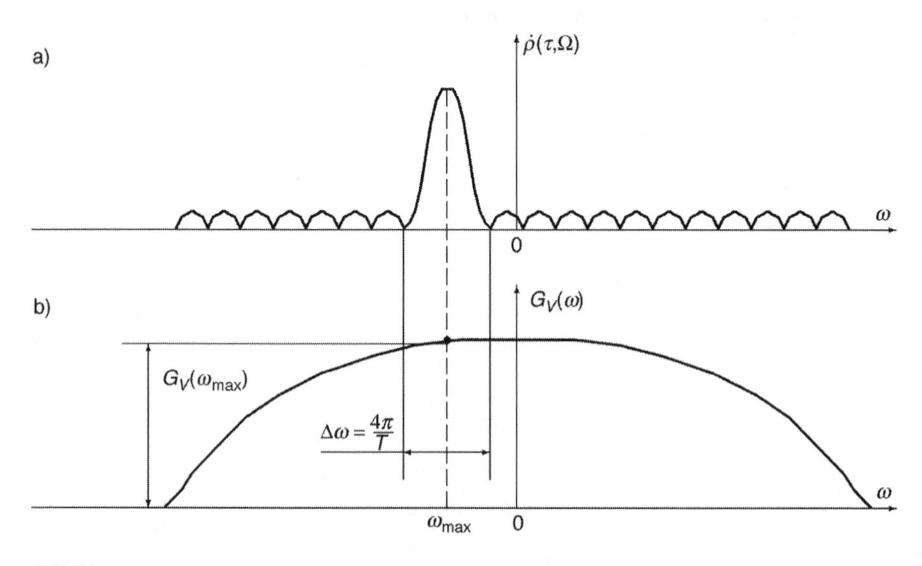

FIGURE 6.15
Definition of maximum of the (a) autocorrelation function $\dot{\rho}(\tau, \Omega)$ and (b) spectral power density $G_V(\omega)$.

the energy spectrum $G_V(\omega)$ in Eq. (6.109) can be changed in the following manner:

$$G_V(\omega_{max}) = G_V\left(-\frac{\Delta\omega_d}{T}\tau - \Omega\right) \qquad \text{(see Fig. 6.15).} \tag{6.224}$$

Then for the case

$$\tau_c \ll T \tag{6.225}$$

we have the following approximate formula with the purpose of defining the variance of signal noise component of the signal with the linear frequency modulation at the output of the generalized detector:

$$\sigma_s^2(\tau, \Omega) \simeq \frac{C^2 E_{a_1}^2}{4\pi} \cdot G_V\left(-\frac{\Delta\omega_d}{T}\tau - \Omega\right) \int\limits_{-\infty}^{\infty} |\dot{\rho}(\tau, \omega)|^2 \, d\omega. \tag{6.226}$$

For the case when the signal has a square wave-form amplitude envelope, taking into account Eq. (6.208), Eq. (6.226) takes the following form:

$$\sigma_s^2(\tau, \Omega) \simeq \begin{cases} \frac{C^2 E_{a_1}^2}{2T} \cdot G_V\left(-\frac{\Delta\omega_d}{T}\tau - \Omega\right)\left(1 - \frac{|\tau|}{T}\right), & |\tau| \leq T; \\ 0, & |\tau| > T. \end{cases} \tag{6.227}$$

When the signal possesses a bell-shaped amplitude envelope Eq. (6.226), in terms of Eq. (6.209), takes the following form:

$$\sigma_s^2(\tau, \Omega) \simeq \frac{C^2 E_{a_1}^2}{2T} \cdot G_V\left(-\frac{\Delta\omega_d}{T}\tau - \Omega\right) \cdot e^{-\frac{\pi\tau^2}{T^2}}. \tag{6.228}$$

Substituting Eq. (6.121) in Eq. (6.228), we obtain that the variance of the signal noise component at the output of the generalized detector, in the case of a bell-shaped energy spectrum of the noise modulation function $\dot{M}(t)$ of the multiplicative noise, takes the following form:

$$\sigma_s^2(\tau, \Omega) \simeq \frac{1}{2\xi} \cdot C^2 E_{a_1}^2 \left(\overline{A^2(t)} - \alpha_0^2\right) \times \exp\left[-\frac{\pi\tau^2}{T^2\xi^2}\left(\xi^2 + Q_y^2 - \frac{\Omega^2 T^2}{4\pi} - Q_y \tau \Omega\right)\right]. \tag{6.229}$$

To estimate the range of the value

$$\xi = \frac{1}{2\pi} \cdot \Delta\Omega_M T, \tag{6.230}$$

under which Eq. (6.228) is true, let us define the ratio of the variances $\sigma_s^2(\tau, \Omega)$ determined by the exact formula in Eq. (6.124) and the approximate formula in Eq. (6.229):

$$\frac{\sigma_{s_{ex}}^2(\tau, \Omega)}{\sigma_{s_{ap}}^2(\tau, \Omega)} = \frac{\xi}{\sqrt{1 + \xi^2}} \cdot \exp\left[\frac{1}{\xi^2(1 + \xi^2)}\left(\pi\tau^2 Q_y^2 + \frac{\Omega^2 T^2}{4\pi} + Q_y \Omega\tau\right)\right]. \tag{6.231}$$

Actually, for the condition $\tau = 0$ Eq. (6.231) coincides with Eq. (6.213). The ratio

$$\frac{\sigma_{s_{ex}}^2(\tau, \Omega)}{\sigma_{s_{ap}}^2(\tau, \Omega)}$$

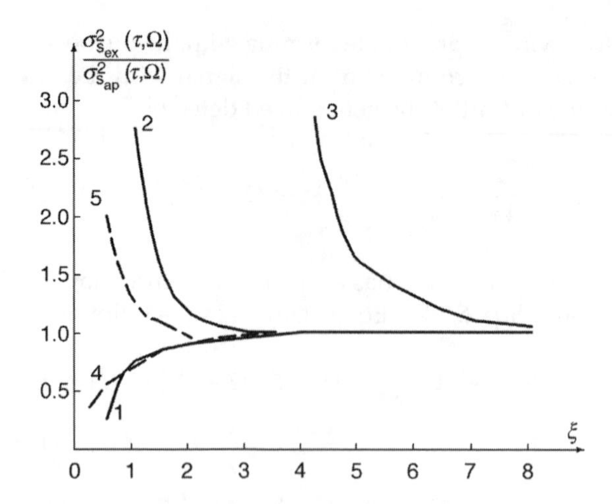

FIGURE 6.16
The ratio $\frac{\sigma^2_{S_{ex}}(\tau,\Omega)}{\sigma^2_{S_{ap}}(\tau,\Omega)}$ as a function of the parameter ξ at some values μ and ν: for 1, $\nu = 0$; for 2, $\nu = 1$; for 3, $\nu = 10$; for 4, $\mu = 0.16$; for 5, $\mu = 0.48$; $Q_y = 100$; (——), $\mu = 0$; (- - - -), $\nu = 0$.

as a function of the argument ξ for some parameters

$$\mu = \frac{\Omega T}{2\pi} \tag{6.232}$$

and

$$\nu = \frac{\tau}{T} \cdot Q_y \tag{6.233}$$

is shown in Fig. 6.16. This ratio allows us to define the area of use for the approximate formula.

For the condition

$$\tau = 0 \quad \text{or} \quad \Omega = 0$$

the simple mathematical formulae are easily obtained using Eq. (6.231) with the purpose of estimating the use of Eq. (6.229).

Let

$$\frac{\sigma^2_{S_{ex}}(\tau,\Omega)}{\sigma^2_{S_{ap}}(\tau,\Omega)} = 1 + \epsilon, \tag{6.234}$$

where ϵ is the relative error.

Then to satisfy the condition

$$\epsilon < \epsilon_{given}$$

there is a need that the conditions

$$\xi \geq \sqrt[4]{\frac{\pi \mu^2}{\epsilon_{given}}} \quad \text{at } \tau = 0 \tag{6.235}$$

and

$$\xi \geq \sqrt[4]{\frac{\pi \nu^2}{\epsilon_{given}}} \quad \text{at } \Omega = 0 \tag{6.236}$$

would be true. For the relative error 20% ($\epsilon = 0.2$) and $\xi \geq 0.75$ as follows from Eqs. (6.235) and (6.236), Eq. (6.229) can be used if

$$\mu = \nu \text{ is between 2.25 and 4.}$$

The equations mentioned above and the dependences presented in Fig. 6.16 show us that the value of the parameter ξ, under which the relative error caused by the use of the approximate formulae defining the variance $\sigma_s^2(\tau, \Omega)$ does not exceed the given magnitude, depends on the parameters ν and μ.

The parameter ν is the ratio between the time τ and the duration of the truncated pulse

$$\frac{T}{Q_y}.$$

The parameter μ is the ratio between the frequency Ω and the energy spectrum bandwidth

$$\frac{2\pi}{T}$$

of amplitude envelope of the signal. If the parameters μ and ν are high, then the value of the parameter ξ is high.

However, there is a need to take into account that the greatest values of the variance $\sigma_s^2(\tau, \Omega)$ take place at low values of the parameters τ and Ω. At high values of the parameters τ and Ω, the relative value of the variance $\sigma_s^2(\tau, \Omega)$ equal to the ratio

$$\frac{\sigma_s^2(\tau, \Omega)}{\sigma_s^2(0, 0)}$$

is very low. Because of this, despite the fact that use of the approximate formulae determining the variance $\sigma_s^2(\tau, \Omega)$ at high values of the parameters τ and Ω leads to the great relative error

$$\frac{\sigma_{S_{ex}}^2(\tau, \Omega)}{\sigma_{S_{ap}}^2(\tau, \Omega)}$$

at not-so-high values of the parameter ξ, the absolute error under determination of the variance $\sigma_s^2(\tau, \Omega)$ is not so high.

The approximate formulae obtained above for determination of the variance $\sigma_s^2(\tau, \Omega)$ of the signal noise component at the output of the generalized detector simplify significantly a definition of statistical characteristics of the signal noise component at the output of the generalized detector for the condition of rapid fluctuating multiplicative noise.

Analysis of these approximated formulae for determination of the variance $\sigma_s^2(\tau, \Omega)$ of the signal noise component at the output of the generalized detector confirms the conclusions made in Section 6.2 regarding the signal noise component at the output of the generalized detector for the condition of fluctuating multiplicative noise.

6.4.2 Slow Fluctuating Multiplicative Noise

The slow fluctuating multiplicative noise

$$\tau_c \gg T \quad \text{and} \quad \Delta\Omega_M \ll \frac{\pi}{T} \tag{6.237}$$

is one of the limiting cases in which general relationships defining the autocorrelation function and the variance of the signal at the output of the generalized detector can be simplified.[58,59]

We use the following formula:

$$R_s(t_1, t_2, \Omega_1, \Omega_2) = \frac{1}{8} \cdot Re\left\{ e^{j\omega_0(t_1 - t_2)} \int\limits_{-\infty}^{\infty}\int\limits_{-\infty}^{\infty} \dot{R}_V(\tau_1 - \tau_2)\dot{S}(\tau_1)S^*(\tau_2) \right.$$
$$\left. \times\, \dot{H}(t_1 - \tau_1)H^*(t_2 - \tau_2) \cdot e^{j(\Omega_1\tau_1 - \Omega_2\tau_2)}d\tau_1\, d\tau_2 \right\}. \tag{6.238}$$

In terms of Eq. (6.76), Eq. (6.238) can be rewritten in the following form:

$$R_s(\tau_1, \tau_2, \Omega_1, \Omega_2) = \frac{C^2}{8} \cdot Re\left\{ e^{j\omega_0(\tau_1 - \tau_2)} \int\limits_{-\infty}^{\infty}\int\limits_{-\infty}^{\infty} \dot{R}_V(\widetilde{\tau}_1 - \widetilde{\tau}_2)\dot{S}(\widetilde{\tau}_1)S^*(\widetilde{\tau}_2) \right.$$
$$\left. \times\, S^*(\widetilde{\tau}_1 - \tau_1)\dot{S}(\widetilde{\tau}_2 - \tau_2) \cdot e^{j(\Omega_1\widetilde{\tau}_1 - \Omega_2\widetilde{\tau}_2)}d\widetilde{\tau}_1\, d\widetilde{\tau}_2 \right\}. \tag{6.239}$$

The integrand in Eq. (6.239) is equal to zero for the condition

$$|\widetilde{\tau}_1 - \widetilde{\tau}_2| > T. \tag{6.240}$$

For this reason, in the case of the slow fluctuating multiplicative noise for the condition

$$\tau_c \gg T$$

the following equality

$$|\widetilde{\tau}_1 - \widetilde{\tau}_2| \ll \tau_c \tag{6.241}$$

is true and we are able to use the approximate formula defining the function

$$\dot{R}_V(\widetilde{\tau}_1 - \widetilde{\tau}_2)$$

containing the first, second, and third terms of the Maclaurin series expansion for the function

$$\dot{R}_V(\widetilde{\tau}_1 - \widetilde{\tau}_2)$$

in the following form:

$$\dot{R}_V(\widetilde{\tau}_1 - \widetilde{\tau}_2) \simeq \dot{R}_V(0) + (\widetilde{\tau}_1 - \widetilde{\tau}_2)\dot{R}'_V(0) + 0.5(\widetilde{\tau}_1 - \widetilde{\tau}_2)^2 \dot{R}''_V(0). \tag{6.242}$$

The values $\dot{R}_V(0)$, $\dot{R}'_V(0)$, and $\dot{R}''_V(0)$ are defined in Appendix II. To define these values it is not required to know a type of the correlation function $\dot{R}_V(\tau)$ or the energy spectrum $G_V(\Omega)$.

Substituting Eq. (6.242) in Eq. (6.239) and carrying out some simple mathematical transformations, we determine that the correlation function of the signal noise component at the output of the generalized detector in the case of the slow multiplicative noise has the following form:

$$\begin{aligned} R_s(\tau_1, \tau_2, \Omega_1, \Omega_2) \simeq \frac{C^2 E_{a_1}^2}{2} \cdot Re\{e^{j\omega_0(\tau_1 - \tau_2)}[\dot{R}(0)\gamma_0(\tau_1, \tau_2, \Omega_1, \Omega_2) \\ + \dot{R}'_V(0)\gamma_1(\tau_1, \tau_2, \Omega_1, \Omega_2) + 0.5\dot{R}''_V(0)\gamma_2(\tau_1, \tau_2, \Omega_1, \Omega_2,)]\}. \end{aligned} \tag{6.243}$$

The following designations are used in Eq. (6.243):

$$\gamma_0(\tau_1, \tau_2, \Omega_1, \Omega_2) = \dot{\rho}(\tau_1, \Omega_1)\rho^*(\tau_2, \Omega_2); \tag{6.244}$$

$$\begin{aligned} \gamma_1(\tau_1, \tau_2, \Omega_1, \Omega_2) = -j\rho^*(\tau_2, \Omega_2) \cdot \frac{\partial \dot{\rho}(\tau_1, \Omega_1)}{\partial \Omega_1} \\ - j\dot{\rho}(\tau_1, \Omega_1) \cdot \frac{\partial \rho^*(\tau_2, \Omega_2)}{\partial \Omega_2}; \end{aligned} \tag{6.245}$$

$$\begin{aligned} \gamma_2(\tau_1, \tau_2, \Omega_1, \Omega_2) = -\dot{\rho}(\tau_1, \Omega_1) \cdot \frac{\partial^2 \rho^*(\tau_2, \Omega_2)}{\partial \Omega_2^2} - \rho^*(\tau_2, \Omega_2) \cdot \frac{\partial^2 \dot{\rho}(\tau_1, \Omega_1)}{\partial \Omega_1^2} \\ - 2 \cdot \frac{\partial \dot{\rho}(\tau_1, \Omega_1)}{\partial \Omega_1} \cdot \frac{\partial \rho^*(\tau_2, \Omega_2)}{\partial \Omega_2}, \end{aligned} \tag{6.246}$$

where

$$\dot{\rho}(\tau, \Omega) = \frac{1}{2E_{a_1}} \int_{-\infty}^{\infty} \dot{S}(\widetilde{\tau}) S^*(\widetilde{\tau} - \tau) \cdot e^{j\Omega\widetilde{\tau}} d\widetilde{\tau} \tag{6.247}$$

is the autocorrelation function of the signal.

Introducing new variables

$$\Omega_1 = \Omega \tag{6.248}$$

and

$$\Omega_2 = \Omega + \Delta\Omega, \tag{6.249}$$

Eqs. (6.243)–(6.246) can be written in the following form:

$$\gamma_0(\tau_1, \tau_2, \Omega_1, \Omega_2) = \dot{\rho}(\tau_1, \Omega)\rho^*(\tau_2, \Omega + \Delta\Omega); \tag{6.250}$$

$$\gamma_1(\tau_1, \tau_2, \Omega_1, \Omega_2) = -j \cdot \frac{\partial}{\partial\Omega}\gamma_0(\tau_1, \tau_2, \Omega, \Omega + \Delta\Omega); \tag{6.251}$$

$$\begin{aligned}\gamma_2(\tau_1, \tau_2, \Omega_1, \Omega_2) &= -j \cdot \frac{\partial}{\partial\Omega}\gamma_1(\tau_1, \tau_2, \Omega, \Omega + \Delta\Omega)\\ &= \frac{\partial^2}{\partial\Omega^2}\gamma_0(\tau_1, \tau_2, \Omega, \Omega + \Delta\Omega).\end{aligned} \tag{6.252}$$

Using Eq. (6.243), we can easily obtain that the approximate formula defining the variance of the signal noise component at the output of the generalized detector in the case of the slow fluctuating multiplicative noise takes the following form:

$$\begin{aligned}\sigma_s^2(\tau, \Omega) = \frac{C^2 E_{a_1}^2}{2} &\cdot Re\{\dot{R}_V(0)\gamma_0(\tau, \Omega)\\ &+ \dot{R}_V'(0)\gamma_1(\tau, \Omega) + 0.5\dot{R}_V''(0)\gamma_2(\tau, \Omega)\},\end{aligned} \tag{6.253}$$

where $\gamma_0(\tau, \Omega)$, $\gamma_1(\tau, \Omega)$, and $\gamma_2(\tau, \Omega)$ are determined by Eqs. (6.250)–(6.252) and for the conditions

$$\tau_1 = \tau_2 = \tau$$

and

$$\Omega_1 = \Omega_2 = \Omega(\Delta\Omega)$$

can be determined in the following manner:

$$\gamma_0(\tau, \Omega) = |\dot{\rho}(\tau, \Omega)|^2; \tag{6.254}$$

$$\gamma_1(\tau, \Omega) = -\, j\dot{\rho}(\tau, \Omega) \cdot \frac{\partial \rho^*(\tau, \Omega)}{\partial\Omega} - j\rho^*(\tau, \Omega) \cdot \frac{\partial \dot{\rho}(\tau, \Omega)}{\partial\Omega}; \tag{6.255}$$

$$\begin{aligned}\gamma_2(\tau, \Omega) = &-\dot{\rho}(\tau, \Omega) \cdot \frac{\partial^2 \rho^*(\tau, \Omega)}{\partial\Omega^2} - \rho^*(\tau, \Omega) \cdot \frac{\partial^2 \dot{\rho}(\tau, \Omega)}{\partial\Omega^2}\\ &-2 \cdot \frac{\partial \dot{\rho}(\tau, \Omega)}{\partial\Omega} \cdot \frac{\partial \rho^*(\tau, \Omega)}{\partial\Omega}.\end{aligned} \tag{6.256}$$

Approximate formulae that are similar to Eqs. (6.243) and (6.253) are true for the arbitrary linear PF at the input linear tract of the generalized detector when the correlation time of the multiplicative noise is much greater in comparison to the duration of the amplitude-frequency response of the PF at the input linear tract of the generalized detector.

For the arbitrary linear PF at the input linear tract of the generalized detector on the basis of Eq. (6.57) we can obtain the formulae that are similar to Eqs. (6.243) and (6.253). However, instead of the autocorrelation function $\dot{\rho}(\tau, \Omega)$ we must use the following autocorrelation function:

$$\dot{\rho}_1(t, \Omega) = \frac{1}{2\pi} \int_{-\infty}^{\infty} \dot{S}(\tilde{\tau}) \dot{H}(t - \tilde{\tau}) \cdot e^{j\Omega\tilde{\tau}}\, d\tilde{\tau}, \tag{6.257}$$

where $\dot{H}(t)$ is the complex amplitude envelope of the amplitude-frequency response of the PF at the input linear tract of the generalized detector.

The autocorrelation function $\dot{\rho}(\tau, \Omega)$ and, consequently, the functions $\gamma_0(\tau, \Omega)$, $\gamma_1(\tau, \Omega)$, and $\gamma_2(\tau, \Omega)$ are defined by the complex amplitude envelope of the signal. Knowledge of these functions for the given type of signals allows us to define easily the main statistical characteristics and parameters of the signal at the output of the generalized detector using the parameters of the fluctuating multiplicative noise (see Eqs. (6.243)–(6.246)) and to estimate distortions of the signal at the output of the generalized detector caused by a stimulus of the fluctuating multiplicative noise.

The autocorrelation functions $\dot{\rho}(\tau, \Omega)$ for the main types of the signals are well known.[60,61] We will proceed to define the functions $\gamma_0(\tau, \Omega)$, $\gamma_1(\tau, \Omega)$, and $\gamma_2(\tau, \Omega)$ for some types of signals.

6.4.2.1 Signal with Constant Radio Frequency Carrier and Square Wave-Form Amplitude Envelope

When the signal possesses the duration T, the complex autocorrelation function of the signal takes the following form:

$$\dot{\rho}(\tau, \Omega) = \begin{cases} \frac{1}{j\Omega T} \cdot \left[e^{j\Omega(T - |\tau|)} - 1\right], & -T \leq \tau \leq 0; \\ \frac{1}{j\Omega T} \cdot \left(e^{j\Omega T} - e^{j\Omega\tau}\right), & 0 \leq \tau \leq T; \\ 0, & |\tau| > T; \end{cases} \tag{6.258}$$

$$\dot{\rho}(\tau, 0) = \begin{cases} \frac{T - |\tau|}{T}, & |\tau| \leq T; \\ 0 & |\tau| > T. \end{cases} \tag{6.259}$$

Substituting Eq. (6.259) in Eqs. (6.254)–(6.256) and carrying out required transformations, we can obtain that the functions $\gamma_0(\tau, \Omega)$, $\gamma_1(\tau, \Omega)$, and $\gamma_2(\tau, \Omega)$

for the signals with a square wave-form amplitude envelope and constant radio frequency carrier have the following form:

$$\gamma_0(\tau, \Omega) = \begin{cases} \frac{4\sin^2 \Omega \frac{T-|\tau|}{2}}{\Omega^2 T^2}, & |\tau| \leq T; \\ 0, & |\tau| > T; \end{cases} \tag{6.260}$$

$$\gamma_0(\tau, 0) = \begin{cases} \frac{(T-|\tau|)^2}{T^2}, & |\tau| \leq T; \\ 0, & |\tau| > T; \end{cases} \tag{6.261}$$

$$\gamma_1(\tau, \Omega) = \begin{cases} \frac{4j\sin^2 \Omega \frac{T-|\tau|}{2}}{\Omega^2 T^2}, & |\tau| \leq T; \\ 0 & |\tau| > T; \end{cases} \tag{6.262}$$

$$\gamma_1(\tau, 0) = 0; \tag{6.263}$$

$$\gamma_2(\tau, \Omega) = \begin{cases} \frac{1}{\Omega^2 T^2} \cdot \left[\frac{8(T-|\tau|)}{\Omega} \cdot \sin \Omega(T - |\tau|) - 2(T - |\tau|)^2 \right. \\ \left. \times \cos \Omega(T - |\tau|) - \frac{24}{\Omega^2} \cdot \sin^2 \Omega \frac{T-|\tau|}{2}\right], & |\tau| \leq T; \\ 0, & |\tau| > T; \end{cases} \tag{6.264}$$

$$\gamma_2(\tau, 0) = \begin{cases} \frac{(T-|\tau|)^4}{6T^2}, & |\tau| \leq T; \\ 0, & |\tau| > T. \end{cases} \tag{6.265}$$

6.4.2.2 Signal with Bell-Shaped Amplitude Envelope and Linear Frequency Modulation

The complex amplitude envelope of the signal with a bell-shaped amplitude envelope and linear law of frequency modulation along with the autocorrelation function of this signal are defined by Eqs. (6.115) and (6.116).

Substituting Eqs. (6.115) and (6.116) in Eqs. (6.254)–(6.256), after differentiation and simple mathematical transformation we can write

$$\gamma_0(\tau_1, \tau_2, \Omega, \Omega + \Delta\Omega) = \dot{\rho}(\tau_1, \Omega) \cdot \rho^*(\tau_2, \Omega + \Delta\Omega); \tag{6.266}$$

$$\gamma_1(\tau_1, \tau_2, \Omega, \Omega + \Delta\Omega) = -0.5\dot{\rho}(\tau_1, \Omega) \cdot \rho^*(\tau_2, \Omega + \Delta\Omega) \times \left[\tau_1 - \tau_2 + jQ_y(\tau_1 + \tau_2) - j \cdot \frac{2\Omega + \Delta\Omega}{2\pi} \cdot T^2\right]; \tag{6.267}$$

$$\gamma_2(\tau_1, \tau_2, \Omega, \Omega + \Delta\Omega) = 0.25\dot{\rho}(\tau_1, \Omega) \cdot \rho^*(\tau_2, \Omega + \Delta\Omega)\left\{\frac{2T^2}{\pi} - \left[(\tau_1 + \tau_2)Q_y - (2\Omega + \Delta\Omega) \cdot \frac{T^2}{2\pi} + j(\tau_1 - \tau_2)\right]^2\right\}, \tag{6.268}$$

where

$$\dot{\rho}(\tau, \Omega) = e^{-\frac{\pi\tau^2}{2T^2}} \cdot e^{-\frac{\Delta\omega_d^2 T^2}{8\pi}\left(\frac{\tau}{T} - \frac{\Omega}{\Delta\omega_d}\right)^2} \cdot e^{j\frac{\Omega\tau}{2}}; \tag{6.269}$$

$$Q_y = \frac{1}{2\pi} \cdot \Delta\omega_d T. \tag{6.270}$$

For the conditions

$$\tau_1 = \tau_2 = \tau$$

and

$$\Delta\Omega = 0$$

the functions $\gamma_0(\tau, \Omega)$, $\gamma_1(\tau, \Omega)$, and $\gamma_2(\tau, \Omega)$ have the following form:

$$\gamma_0(\tau, \Omega) = |\dot{\rho}(\tau, \Omega)|^2; \tag{6.271}$$

$$\gamma_1(\tau, \Omega) = -j|\dot{\rho}(\tau, \Omega)|^2\left(\tau Q_y - \frac{\Omega T^2}{2\pi}\right); \tag{6.272}$$

$$\gamma_2(\tau, \Omega) = |\dot{\rho}(\tau, \Omega)|^2\left[\frac{T^2}{2\pi} - \left(\tau Q_y - \frac{\Omega T^2}{2\pi}\right)^2\right]. \tag{6.273}$$

6.4.2.3 Bell-Shaped Signal with Constant Radio Frequency Carrier

Using Eqs. (6.267) and (6.271)–(6.273) for the condition

$$Q_y = 0$$

or

$$\Delta\omega_d = 0,$$

we can obtain the functions $\gamma_0(\tau, \Omega)$, $\gamma_1(\tau, \Omega)$, and $\gamma_2(\tau, \Omega)$ defining the autocorrelation function and the variance of the signal noise component at the output of the generalized detector, for the case in which the amplitude-frequency response of the PF at the input linear tract of the generalized detector is matched with the bell-shaped amplitude envelope of the input signal with the constant radio frequency carrier:

$$\begin{aligned}\gamma_0(\tau_1, \tau_2, \Omega, \Omega + \Delta\Omega) &= e^{-\frac{\pi(\tau_1^2+\tau_2^2)}{2T^2}} \cdot e^{-\frac{T^2(2\Omega^2+2\Omega\Delta\Omega+\Delta\Omega^2)}{8\pi}} \cdot e^{j\left(\frac{\tau_1-\tau_2}{2}\cdot\Omega - \frac{\tau_2\Delta\Omega}{2}\right)}\\ &= \dot{\rho}(\tau_1, \Omega) \cdot \rho^*(\tau, \Omega + \Delta\Omega);\end{aligned} \tag{6.274}$$

$$\gamma_0(\tau, \Omega) = e^{-\frac{\pi\tau^2}{T^2}} \cdot e^{-\frac{T^2\Omega^2}{4\pi}} = |\dot{\rho}(\tau, \Omega)|^2; \tag{6.275}$$

$$\begin{aligned}\gamma_1(\tau_1, \tau_2, \Omega, \Omega + \Delta\Omega) &= 0.5\dot{\rho}(\tau_1, \Omega) \cdot \rho^*(\tau_2, \Omega + \Delta\Omega)\\ &\quad \times \left[j(\tau_1 - \tau_2) - \frac{T^2}{2\pi}(2\Omega + \Delta\Omega)\right];\end{aligned} \tag{6.276}$$

$$\gamma_1(\tau, \omega) = -j \cdot \frac{\Omega T^2}{2\pi}|\dot{\rho}(\tau, \Omega)|^2; \tag{6.277}$$

$$\gamma_2(\tau_1, \tau_2, \Omega, \Omega + \Delta\Omega) = 0.25\dot{\rho}(\tau, \Omega) \cdot \rho^*(\tau_2, \Omega + \Delta\Omega) \times \left\{ \frac{2T^2}{\pi} - \left[j(\tau_1 - \tau_2) - \frac{T}{2\pi}(2\Omega + \Delta\Omega) \right]^2 \right\}; \quad (6.278)$$

$$\gamma_2(\tau, \Omega) = \frac{T^2}{2\pi}\left(1 - \frac{T^2\Omega^2}{2\pi}\right)|\dot{\rho}(\tau, \Omega)|^2, \quad (6.279)$$

where

$$\dot{\rho}(\tau\Omega) = e^{-\frac{\pi\tau^2}{2T^2} - \frac{\Omega^2T^2}{8\pi}} \cdot e^{-j\cdot\frac{\Omega\tau}{2}}. \quad (6.280)$$

The use of Eqs. (6.243) and (6.253) in combination with the formulae mentioned above, defining the functions $\gamma_0(\tau, \Omega)$, $\gamma_1(\tau, \Omega)$, and $\gamma_2(\tau, \Omega)$, allows us to define the statistical characteristics of the signal at the output of the generalized detector during the slow fluctuating multiplicative noise in a simpler manner.

To define the statistical characteristics of the signal noise component at the output of the generalized detector, less knowledge regarding the characteristics of the multiplicative noise is required than in a general case.

6.5 Signal Distribution Law under the Stimulus of Multiplicative Noise

The most complete statistical characteristic of the signal at the output of the generalized detector is the multivariate probability distribution density during the fluctuating multiplicative noise.

In many cases, for example, in the study of the stimulus of multiplicative noise on the detection performances of the signals, the probability of definition of parameters of the signals, and the probability of resolution of two signals, we can be limited by knowledge of the one-dimensional probability distribution density of the signal at the output of the generalized detector.

Unfortunately, at the present time general methods of transformation analysis of distribution laws of stochastic processes obeying the non-Gaussian probability distribution density during passing through linear systems are not found.[62–64]

If we limit our consideration to linear systems carrying out transformations of the central type, then we can believe—on the basis of the central limiting theorem of the theory of probabilities—that for the cases in which the correlation interval of the input stochastic process τ_c is much less than the time constant τ_s of the linear system, the probability distribution density of the output signal tends to approach the Gaussian probability distribution density as the ratio $\frac{\tau_s}{\tau_c}$ increases.[65,66]

Actually, for the linear systems of the integral type considered in this section—the generalized detector can be considered as this linear system—the input signal distorted by the fluctuating multiplicative noise

$$a_M(t) = Re\{\dot{M}(t)\dot{S}(t) \cdot e^{j\omega_0 t}\} \tag{6.281}$$

and the signal at the output of the generalized detector

$$a_{1_M}(t) = Re\{\dot{S}_{1_M}(t) \cdot e^{j\omega_0 t}\} \tag{6.282}$$

are functionally related in the following form:

$$\begin{aligned} a_{1_M}(t) &= 0.5Re\left\{e^{j\omega_0 t} \int_{-\infty}^{\infty} \dot{H}(t-\tau)\dot{S}_M(\tau)\,d\tau\right\} \\ &= 0.5Re\left\{e^{j\omega_0 t} \int_{0}^{T} \dot{H}(t-\tau)\dot{S}_M(\tau)\,d\tau\right\}, \end{aligned} \tag{6.283}$$

where $\dot{H}(t)$ is the complex amplitude envelope of the pulse transient function of the linear system—the PF at the input linear tract of the generalized detector.

For the condition

$$\tau_s \gg \tau_c$$

the integral in Eq. (6.283) can be represented in the form of the sum

$$a_{1_M}(t) \simeq 0.5Re\left\{\sum_{k=0}^{N-1} \dot{H}(t-k\Delta t)\dot{S}_M(k\Delta t) \cdot e^{j\omega_0 t}\Delta t\right\}, \tag{6.284}$$

where the elementary time interval Δt must satisfy the following conditions:

$$\tau_s \gg \Delta t \quad \text{and} \quad \Delta t \ll \tau_c. \tag{6.285}$$

The number of summing elements is equal to

$$N = \frac{t}{\Delta t}. \tag{6.286}$$

When the condition

$$\Delta t \simeq \tau_c \tag{6.287}$$

is satisfied, the individual terms in Eq. (6.284) can be thought of as uncorrelated and the probability distribution function $F[a_M(t)]$ of the sum in Eq. (6.284) tends to approach the Gaussian probability distribution function for the condition

$$\frac{\tau_s}{\tau_c} \to \infty \tag{6.288}$$

on the basis of the central limiting theorem.

The probability distribution densities of stochastic functions, which are little more than Gaussian, are represented in the form of the Edgeworth series:[67]

$$f[a_M(t)] \simeq f_G[a_M(t)] \sum_{k=0}^{N} \frac{b_k}{k!\sqrt{M_{2_G}^k}} \cdot H_k\left(\frac{a_{1_M}(t) - m_{1_G}}{\sqrt{M_{2_G}}}\right), \tag{6.289}$$

where

$$f_G[a_M(t)] = \frac{1}{\sqrt{2\pi M_{2_G}}} \cdot \exp\left\{-\frac{\left[a_{1_M}(t) - m_{1_G}^2\right]^2}{2M_{2_G}}\right\} \tag{6.290}$$

is the Gaussian probability distribution density; m_{1_G} and M_{2_G} are the initial moment of the first order and the central moment of the second order of the Gaussian probability distribution density $f_G[a_{1_M}(t)]$, respectively; and

$$H_k = (-1)^k \cdot e^{\frac{z^2}{2}} \cdot \frac{d^k}{dz^k} e^{-\frac{z^2}{2}} \tag{6.291}$$

is the one-dimensional Hermite polynomial; and b_k is the quasi-moment[57] determined by

$$b_k = \sqrt{M_{2_G}^k} \int_{-\infty}^{\infty} f[a_M(t)] H_k\left(\frac{a_{1_M}(t) - m_{1_G}}{\sqrt{M_{2_G}}}\right) d\left[a_{1_M}(t)\right]. \tag{6.292}$$

Taking into account Eq. (6.291), Eq. (6.289) can be rewritten in the following form:

$$\begin{aligned} f\left[a_{1_M}(t)\right] &= \sum_{k=0}^{N} \frac{(-1)^k}{k!} \cdot \frac{b_k}{\sqrt{M_{2_G}^k}} \cdot f_G^{(k)}\left[a_{1_M}(t)\right] \\ &= f_G\left[a_{1_M}(t)\right] + \sum_{k=1}^{N} \frac{(-1)^k}{k!} \cdot \frac{b_k}{\sqrt{M_{2_G}^k}} \cdot f_G^{(k)}\left[a_{1_M}(t)\right]. \end{aligned} \tag{6.293}$$

The probability distribution function

$$F\left[a_{1_M}(t)\right] = \int_{-\infty}^{a_{1_M}(t)} f(x)\, dx \tag{6.294}$$

takes the following form:

$$F\left[a_{1_M}(t)\right] = F_G\left[a_{1_M}(t)\right] + \sum_{k=1}^{N} \frac{(-1)^k}{k!} \cdot \frac{b_k}{\sqrt{M_{2_G}^k}} \cdot f_G^{(k-1)}\left[a_{1_M}(t)\right], \tag{6.295}$$

where

$$F_G\left[a_{1_M}(t)\right] = \frac{1}{\sqrt{2\pi M_{2_G}}} \int_{-\infty}^{a_{1_M}(t)} \exp\left\{-\frac{\left[a_{1_M}(t) - m_{1_G}\right]^2}{2M_{2_G}}\right\} d\left[a_{1_M}(t)\right]. \tag{6.296}$$

The number of terms N (in the Edgeworth series) depends on the accuracy required for approximation of the probability distribution function by the series and choice of the normalized initial moment of the first order m_{1_G} and the central moment of the second order M_{2_G}. As a rule, the moments m_{1_G} and M_{2_G} are equal to the mean and variance of the process $a_{1_M}(t)$, respectively. Following from these statements, a convergence of the series in Eqs. (6.289) and (6.295) is the best solution.[68]

Limit the subsequent consideration by a simple case in which all terms in Eq. (6.284) have the same moments. This assumption is equivalent to the case in which the linear system is the ideal integrator for the time period

$$T = N\Delta t \tag{6.297}$$

tuned to the average frequency ω_0 of the signal, and the undistorted input signal is the harmonic oscillation with a constant amplitude within the limits of the interval T, and the noise modulation function $\dot{M}(t)$ of the multiplicative noise is the stationary stochastic process.[69,70]

We can also use this model in the consideration of linear systems matched with the signals possessing a constant amplitude within the limits of the interval T and the arbitrary phase structure of the signal for the condition $\tau = 0$ (see Sections 6.1 and 6.2).

Under the formulated conditions within the limits of the interval $[0, N\Delta t]$, the amplitude envelope of amplitude-frequency response of the PF at the input linear tract of the generalized detector is constant. Henceforth we can believe that

$$|\dot{H}(t)| = 1$$

within the limits of the interval

$$0 \leq t \leq N\Delta t \tag{6.298}$$

for simplicity.

The correlation interval τ_c is equal to the correlation interval of the noise modulation function $\dot{M}(t)$ of the multiplicative noise for the case considered. The number of independent readings is equal to the parameter ξ, determined as before

$$\xi = \frac{1}{2\pi} \cdot \Delta\Omega_M T. \tag{6.299}$$

For the case considered, the mean $m_{1_{a_1}}$ and the variance $M_{2_{a_1}}$ of the signal $a_{1_M}(t)$ at the output of the generalized detector are related to the mean m_1 and the variance M_2 of the harmonic signal

$$a_M(t) = Re\{A(t) \cdot e^{j\varphi(t)} \cdot e^{j\omega_0 t}\} = Re\{\dot{M}(t) \cdot e^{j\omega_0 t}\}, \tag{6.300}$$

which is distorted by the same multiplicative noise as the input signal, by simple functions:

$$m_{1_{a_1}} = Nm_1 \tag{6.301}$$

and

$$M_{2_{a_1}} = NM_2. \tag{6.302}$$

As was noted in Chapter 4, even under the stimulus of stationary multiplicative noise the signal distorted by the multiplicative noise is the non-stationary stochastic process. Moreover, non-stationarability of the signal is caused by both a comparatively slow changing in the amplitude and phase of the undistorted signal—in comparison with the period of the carrier frequency ω_0—and by rapid oscillations with the frequency ω_0.

Under the model of the PF at the input linear tract of the generalized detector and the input stochastic process adopted above, the slow non-stationarability of the input stochastic process is of little interest for the subsequent analysis, since the amplitude of the undistorted signal is constant within the limits of the interval T.

Therefore, the mean and variance of the signal $a_{1_M}(t)$ at the output of the generalized detector can be written in the following form (see Section 4.3):

$$m_{1_{a_1}} = A_1 \cos \omega_0 t; \tag{6.303}$$

$$\begin{aligned} M_{2_{a_1}} &= A_2 + C_2 \cos 2\omega_0 t = M_{2_{a_1}}^{(s)} + M_{2_{a_1}}^{(n)} \\ &= NM_2^{(s)} + NM_2^{(n)}, \end{aligned} \tag{6.304}$$

where the coefficients A_1, A_2, and C_2 do not depend on the time; the index "s" denotes the stationary components of the central moments of the second order M_2 and $M_{2_{a_1}}$; and the index "n" denotes the non-stationary components of the central moments of the second order M_2 and $M_{2_{a_1}}$.

Reference to Eq. (6.304) shows that the mean of the signal $a_{1_M}(t)$ at the output of the generalized detector oscillates with the frequency ω_0, and the variance of the signal $a_{1_M}(t)$ at the output of the generalized detector possesses both the non-oscillating component $M_{2_{a_1}}^{(s)}$ and the oscillating component $M_{2_{a_1}}^{(n)}$ with the frequency $2\omega_0$.

Taking into account the fact that there are no special requirements for the choice of the normalized constants m_1 and M_2 in Eqs. (6.293) and (6.295), in the subsequent discussion we are able to estimate the convergence of the probability distribution density of the signal $a_{1_M}(t)$ at the output of the generalized detector with two types of the Gaussian probability distribution density:[71,72]

- The Gaussian probability distribution density with the non-oscillating central moment of the second order

$$M_{2_G} = M_{2_{a_1}}^{(s)} \tag{6.305}$$

and

$$m_{1_G} = m_{1_{a_1}}. \tag{6.306}$$

- The Gaussian probability distribution density with the oscillating central moment of the second order

$$M_{2_G} = M_{2_{a_1}}^{(n)} \tag{6.307}$$

and

$$m_{1_G} = m_{1_{a_1}}. \tag{6.308}$$

Taking into consideration the Edgeworth polynomials in Eqs. (6.293) and (6.295), and also the function between the central and initial moments of the signal $a_M(t)$ and the signal $a_{1_M}(t)$ at the output of the generalized detector[57,68,73]

$$m_{1_{a_1}} = Nm_1; \tag{6.309}$$

$$M_{2_{a_1}} = NM_2; \tag{6.310}$$

$$M_{3_{a_1}} = NM_3; \tag{6.311}$$

$$M_{4_{a_1}} = NM_4 + 3N(N-1)M_2^2; \tag{6.312}$$

$$M_{5_{a_1}} = NM_5 + 10N(N-1)M_2M_3; \tag{6.313}$$

$$M_{6_{a_1}} = NM_6 + 15N(N-1)M_2M_4 + 10N(N-1)M_3^2 + 15N(N^2-3N+2)M_2^3, \tag{6.314}$$

the quasimoments in Eq. (6.292) can be written in the following form:

$$b_1 = 0; \tag{6.315}$$

$$\frac{b_2}{NM_2^{(s)}} = \frac{M_2^{(n)}}{M_2^{(s)}}; \tag{6.316}$$

$$\frac{b_3}{\sqrt{\left(NM_2^{(s)}\right)^3}} = \frac{M_3}{\sqrt{N\left(M_2^{(s)}\right)^3}}; \tag{6.317}$$

$$\frac{b_4}{\left(NM_2^{(s)}\right)^2} = \frac{M_4 - 3M_2^2}{N\left(M_2^{(s)}\right)^2} + \frac{\left(M_2^{(n)}\right)^2}{\left(M_2^{(s)}\right)^2}; \tag{6.318}$$

$$\frac{b_5}{\sqrt{\left(NM_2^{(s)}\right)^5}} = \frac{M_5 - 10M_2M_3}{\sqrt{N^3\left(M_2^{(s)}\right)^5}} + \frac{10M_2^{(n)}M_3}{\sqrt{N\left(M_2^{(s)}\right)^5}}; \tag{6.319}$$

$$\frac{b_6}{\left(NM_2^{(s)}\right)^3} = \frac{M_6 - 15M_2M_4 - 10M_3^2 + 30M_2^3}{N^2\left(M_2^{(s)}\right)^3} + \frac{10}{N}\left[\frac{M_3^2}{\left(M_2^{(s)}\right)^3} + \frac{3M_2^{(n)}M_4}{2\left(M_2^{(s)}\right)^3} - \frac{9\left(M_2^{(n)}\right)^3}{2\left(M_2^{(s)}\right)^3} - \frac{9M_2^{(n)}}{2M_2^{(s)}}\right] + 15\frac{\left(M_2^{(n)}\right)^3}{\left(M_2^{(s)}\right)^3}. \tag{6.320}$$

Formulae in Eqs. (6.315)–(6.320) are obtained for the case in which the normalization in Eqs. (6.293) and (6.295) has been carried out as applied to the problem of estimation of convergence of the probability distribution density of the signal $a_{1_M}(t)$ (at the output of the generalized detector) and the Gaussian probability distribution density with the non-oscillating central moment of the second order equal to $NM_2^{(s)}$.

If we assume that

$$M_2^{(n)} = 0 \tag{6.321}$$

and

$$M_2^{(s)} = M_2 \tag{6.322}$$

in Eq. (6.315), then we can obtain the quasi-moments for the case in which the normalizing constants in Eqs. (6.293) and (6.295) are chosen in accordance with the relationships

$$M_{2_G} = NM_2 \tag{6.323}$$

and

$$m_{1_G} = Nm_1. \tag{6.324}$$

Substituting Eqs. (6.315)–(6.320) in Eq. (6.293), and ordering the terms of the series by the order

$$\frac{1}{\sqrt{N}},$$

and limiting by the terms with the order equal to

$$\frac{1}{N},$$

and introducing the designation

$$\lambda = \frac{M_2^{(n)}}{M_2^{(s)}}, \tag{6.325}$$

in this case we can write

$$\begin{aligned}
f\left[a_{1_M}(t)\right] &= f_G\left[a_{1_M}(t)\right] - \frac{\lambda}{2}\left\{ f_G^{(1)}\left[a_{1_M}(t)\right]\right. \\
&\quad \left. - \frac{\lambda}{12} f_G^{(4)}\left[a_{1_M}(t)\right] - \frac{\lambda^3}{24} f_G^{(6)}\left[a_{1_M}(t)\right]\right\} \\
&\quad - \frac{M_3}{3!\sqrt{N\left(M_2^{(s)}\right)^3}} \cdot \left\{ f_G^{(3)}\left[a_{1_M}(t)\right] + \frac{\lambda}{2} f_G^{(5)}\left[a_{1_M}(t)\right]\right\} \\
&\quad + \frac{M_4 - 3M_2^2}{4!N\left(M_2^{(s)}\right)^2} \cdot f_G^{(4)}\left[a_{1_M}(t)\right] + \frac{10M_3^2}{6!N\left(M_2^{(s)}\right)^3} \cdot f_G^{(6)}\left[a_{1_M}(t)\right] \\
&\quad + \frac{15\lambda}{6!N} \cdot \left[\frac{M_4 - 3\left(M_2^{(s)}\right)^2}{\left(M_2^{(s)}\right)^2} - 6\lambda - 3\lambda^2\right] \cdot f_g^{(6)}\left[a_{1_M}(t)\right],
\end{aligned} \tag{6.326}$$

where

$$f_G^{(k)}\left[a_{1_M}(t)\right] = \frac{1}{\sqrt{2\pi M_{2_G}}} \cdot \frac{\partial^k}{\partial\left[a_{1_M}(t)\right]^k} \cdot \exp\left\{-\frac{\left[a_{1_M}(t) - m_{1_G}\right]^2}{2M_{2_G}}\right\}. \tag{6.327}$$

If we assume that the moment $M_2^{(s)}$ is determined by Eq. (6.322) and, consequently,

$$\lambda = 0 \tag{6.328}$$

then Eq. (6.326) takes the following well-known form:

$$\begin{aligned} f\left[a_{1_M}(t)\right] &= F_G\left[a_{1_M}(t)\right] - \frac{k}{6\sqrt{N}} \cdot f_G^{(3)}\left[a_{1_M}(t)\right] \\ &\quad + \frac{\gamma}{24N} \cdot f_G^{(4)}\left[a_{1_M}(t)\right] + \frac{k^2}{72N} \cdot f_G^{(6)}\left[a_{1_M}(t)\right], \end{aligned} \tag{6.329}$$

where k and γ are the coefficients of asymmetry and kurtosis, respectively.

Equation (6.329) allows us to estimate a convergence of the probability distribution density of the signal $a_{1_M}(t)$ (at the output of the generalized detector) and the Gaussian probability distribution density with the oscillating central moment of the second order.

The convergence in Eq. (6.329) was investigated by H. Crammer.[68] He proved that the remainder term in Eq. (6.329) has the order of the first rejected term, so that for Eq. (6.329) it is equivalent to

$$\begin{aligned} &-\frac{1}{5!\sqrt{N^3}} \cdot \left(\frac{M_5}{\sqrt{M_2^5}} - k\right) \cdot f_g^{(5)}\left[a_{1_M}(t)\right] \\ &-\frac{35}{7!\sqrt{N^3}} \cdot k\gamma f_g^{(7)}\left[a_{1_M}(t)\right] \\ &-\frac{280}{9!\sqrt{N^3}} \cdot k^3 \cdot f_G^{(9)}\left[a_{1_M}(t)\right]. \end{aligned} \tag{6.330}$$

With an increase in the number N of independent summable terms, Eq. (6.329) converges toward the Gaussian probability distribution density for all finite values of the coefficients practically. In order for Eq. (6.326) to be converged toward the Gaussian probability distribution density, it is required that the condition[74,75]

$$\lambda = \frac{M_2^{(n)}}{M_2^{(s)}} \to 0 \tag{6.331}$$

would be satisfied in addition to the condition

$$N \to \infty. \tag{6.332}$$

Thus, in order that the probability distribution density of the signal $a_{1_M}(t)$ distorted by the multiplicative noise at the output of the generalized detector for the condition

$$N = \frac{\tau_s}{\tau_c} \gg 1 \tag{6.333}$$

could be approximated using the non-oscillating with respect to the central moment of the second order Gaussian probability distribution density, specific limitations would be required on the statistical characteristics of the noise modulation function $\dot{M}(t)$ of the multiplicative noise with the purpose of satisfying the following relationship:

$$|\lambda| = \left| \frac{M_2^{(n)}}{M_2^{(s)}} \right| \ll 1. \tag{6.334}$$

In computer calculations we use the initial moments m_k, which are defined in Section 4.3, of the signal distorted by the stochastic multiplicative noise. These initial moments allow us to determine the coefficients in Eq. (6.326) using the following relationships:

$$k^{(s)} = \frac{M_3}{\sqrt{\left(M_2^{(s)}\right)^3}} = \frac{m_3 - 3m_1m_2 + 2m_1^3}{\sqrt{\left[\left(m_2 - m_1^2\right)^{(s)}\right]^3}}; \tag{6.335}$$

$$\gamma^{(s)} = \frac{M_4 - 3M_2^2}{\left(M_2^{(s)}\right)^2} = \frac{m_4 + 3m_2^2 - 4m_1m_3}{\left[\left(m_2 - m_1^2\right)^{(s)}\right]^2} - 6(\lambda - 1)^2; \tag{6.336}$$

$$\gamma'^{(s)} = \frac{M_4 - 3\left(M_2^{(s)}\right)^2}{\left(M_2^{(s)}\right)^2} = \frac{m_4 + 3m_2^2 - 4m_1m_3}{\left[\left(m_2 - m_1^2\right)^{(s)}\right]^2} - 3(\lambda^2 - 2\lambda + 2). \tag{6.337}$$

The coefficients k and γ are obtained from Eqs. (6.335) and (6.336) by changing the value

$$M_2^{(s)} = \left(m_2 - m_1^2\right)^{(s)} \tag{6.338}$$

by the value

$$M_2 = m_2 - m_1^2. \tag{6.339}$$

In the subsequent analysis, we assume that the functions defining fluctuations in the amplitude and phase of the signal owing to the stimulus of the multiplicative noise are the independent stationary stochastic processes. Then

$$m_k = m_k^A m_k^\varphi, \tag{6.340}$$

where m_k^A is the initial moment of the probability distribution density of distortions in the amplitude $A(t)$ of the signal at the input of the generalized detector and m_k^φ is the initial moment of the probability distribution density of the signal at the input of the generalized detector

$$a_M(t) = Re\{A(t) \cdot \exp j[\omega_0 t + \Psi_a(t) + \varphi_0]\} \tag{6.341}$$

for the condition that distortions in the amplitude of the signal are absent, i.e.,

$$A(t) = 1.$$

Taking into consideration that the mean $m_1^{\varphi}\{\varphi(t)\}$ of distortions in phase of the signal can be taken into account using the initial phase φ_0 of the signal, and assuming that the probability distribution density of centered distortions in phase of the signal is the symmetric function, we obtain that the initial moments of the signal $a_M(t)$ at the input of the generalized detector, in accordance with the main results discussed in Section 4.3, take the following form:

$$m_1 = m_1^A\big|\Theta_1^{\varphi}(1)\big|\cos\alpha(t); \tag{6.342}$$

$$m_2 = 0.5m_2^A\big[1 + \Theta_1^{\varphi}(2)\cos 2\alpha(t)\big]; \tag{6.343}$$

$$m_3 = 0.25m_3^A\big[3\big|\Theta_1^{\varphi}(1)\big|\cos\alpha(t) + \big|\Theta_1^{\varphi}(3)\big|\cos 3\alpha(t)\big]; \tag{6.344}$$

$$m_4 = 0.375m_4^A\left[1 + \frac{4}{3}\big|\Theta_1^{\varphi}(2)\big|\cos 2\alpha(t) + \frac{1}{3}\big|\Theta_1^{\varphi}(4)\big|\cos 4\alpha(t)\right], \tag{6.345}$$

where

$$\alpha(t) = \omega_0 t + \varphi_0 \tag{6.346}$$

for the model of the PF at the input linear tract of the generalized detector and the input signal adopted in this section.

In terms of Eq. (6.342), we can write

$$\begin{aligned} M_2 &= 0.5\big\{\big[m_2^A - \big(m_1^A\big)^2\big|\Theta_1^{\varphi}(1)\big|^2\big] + m_2^A\big|\Theta_1^{\varphi}(2)\big| - \big(m_1^A\big)^2\big|\Theta_1^{\varphi}(1)\big|^2\cos 2\alpha(t)\big\} \\ &= A_2 + C_2\cos 2\alpha(t) = M_2^{(s)} + M_2^{(n)}; \end{aligned} \tag{6.347}$$

$$\lambda = \frac{C_2}{A_2}\cdot\cos 2\alpha(t); \tag{6.348}$$

$$k^{(s)} = \frac{1}{\sqrt{A_2^3}}\cdot[A_3\cos\alpha(t) + C_3\cos 3\alpha(t)]; \tag{6.349}$$

$$k = \frac{A_3\cos\alpha(t) + C_3\cos 3\alpha(t)}{\sqrt{[A_2 + C_2\cos 2\alpha(t)]^3}}; \tag{6.350}$$

$$\begin{aligned} \gamma^{(s)} = &\frac{1}{A_2^2}\cdot\big[A_4 - 3A_2^2C_2 + (C_4 + 12A_2C_2)\cos 2\alpha(t) \\ &+ (D_4 - 3A_2C_2)\cos 4\alpha(t)\big] - 6; \end{aligned} \tag{6.351}$$

$$\begin{aligned} \gamma'^{(s)} = &\frac{1}{A_2^2}\cdot[A_4 - 1.5A_2C_2 + (C_4 + 6A_2C_2)\cos 2\alpha(t) \\ &+ (D_4 - 1.5A_2C_2)\cos 4\alpha(t)] - 6, \end{aligned} \tag{6.352}$$

where

$$A_2 = 0.5\left[m_2^A - \left(m_1^A\right)^2 \left|\Theta_1^\varphi(1)\right|^2\right]; \tag{6.353}$$

$$C_2 = 0.5\left[m_2^A \left|\Theta_1^\varphi(2)\right| - \left(m_1^A\right)^2 \left|\Theta_1^\varphi(1)\right|^2\right]; \tag{6.354}$$

$$\begin{aligned} A_3 = 0.75\big[&m_3^A \left|\Theta_1^\varphi(1)\right| - 2m_1^A m_2^A \left|\Theta_1^\varphi(1)\right| \\ &- m_1^A m_2^A \left|\Theta_1^\varphi(1)\Theta_1^\varphi(2)\right| + 2\left(m_1^A\right)^3 \left|\Theta_1^\varphi(1)\right|^3\big]; \end{aligned} \tag{6.355}$$

$$C_3 = 0.25\left[m_3^A \left|\Theta_1^\varphi(3)\right| - 3m_1^A m_2^A \left|\Theta_1^\varphi(1)\Theta_1^\varphi(2)\right| + 2\left(m_1^A\right)^3 \left|\Theta_1^\varphi(1)\right|^3\right]; \tag{6.356}$$

$$A_4 = 0.375\left[m_4^A + 2\left(m_2^A\right)^2 + \left(m_2^A\right)^2 \left|\Theta_1^\varphi(2)\right|^2 - 4m_1^A m_3^A \left|\Theta_1^\varphi(1)\right|^2\right]; \tag{6.357}$$

$$\begin{aligned} C_4 = 0.5\big[&m_4^A \left|\Theta_1^\varphi(2)\right| + 3\left(m_2^A\right)^2 \left|\Theta_1^\varphi(2)\right| \\ &- 3m_1^A m_3^A \left|\Theta_1^\varphi(1)\right|^2 - m_1^A m_3^A \left|\Theta_1^\varphi(1)\Theta_1^\varphi(3)\right|\big]; \end{aligned} \tag{6.358}$$

$$\begin{aligned} D_4 = 0.125\big[&m_4^A \left|\Theta_1^\varphi(4)\right| + 3\left(m_2^A\right)^2 \left|\Theta_1^\varphi(2)\right|^2 \\ &- 4m_1^A m_3^A \left|\Theta_1^\varphi(1)\Theta_1^\varphi(3)\right|\big]. \end{aligned} \tag{6.359}$$

Equations (6.347) and (6.349)–(6.352) show a dependence of the coefficients in Eqs. (6.326) and (6.329) on time in an explicit form.

In the study of Gaussian stochastic processes in problems of statistical radio engineering, as a rule, those Gaussian stochastic processes with central moments of the second order with a constant or slow changing rate are considered. In principle, this consideration can also be formally propagated on the stochastic processes obeying the Gaussian probability distribution density with the oscillating central moment of the second order (see Eq. (6.303)).

However, for example, the probability of detection of the signals can have a component oscillating with the frequency $2\omega_0$ and higher in such consideration of the problem. To define the mean of the computer-calculated characteristics of the generalized detector in such manner, it is obvious that the characteristics of the generalized detector must be averaged with respect to time within the limits of the period of the carrier frequency ω_0 of the signal.

In experimental investigations of the probability distribution densities of the signals distorted by the multiplicative noise using the standard measurement equipment (the devices), the fast oscillations of the probability distribution density of the signal at the output of the generalized detector cannot be registered. The fact is that the intervals of averaging—the time intervals, within the limits of which the probability distribution density of the signal at the output of the generalized detector is analyzed—are much more than the period of oscillations of parameters of the probability distribution density of the signal at the output of the generalized detector.

For this reason, the probability distribution density of the signal at the output of the generalized detector, averaged with respect to time within the limits of the period of the carrier frequency of the signal, will only be observed as results of measurements. To define the statistical characteristics of non-stationarability, of the probability distribution density of the signal at the output of the generalized detector, there is a need to design specific equipment—apparatus and devices.

The analytical probability distribution density of the signal at the output of the generalized detector, which is averaged within the limits of the period of the carrier frequency ω_0 of the signal, is found by averaging Eq. (6.326) within the limits of the time interval $\pm\frac{\pi}{\omega_0}$:

$$
\begin{aligned}
\langle f\left[a_{1_M}(t)\right]\rangle = {} & f_G\left[a_{1_M}(t)\right] + \frac{1}{48}\cdot\frac{C_2^2}{A_2^2} f_G^{(4)}\left[a_{1_M}(t)\right] \\
& + \frac{1}{128}\cdot\frac{C_2^4}{A_2^4} f_G^{(6)}\left[a_{1_M}(t)\right] \\
& + \frac{1}{4!\,N}\left(\frac{A_4 - 3A_2C_2}{A_2^2} - 6\right) f_G^{(4)}\left[a_{1_M}(t)\right] \\
& + \frac{5}{6!\,N}\cdot\frac{A_3^2 + C_3^2}{A_2^2} f_G^{(6)}\left[a_{1_M}(t)\right] \\
& + \frac{15}{6!\,2N}\cdot\frac{C_2(C_4 + 6A_2C_2)}{A_2^3} f_G^{(6)}\left[a_{1_M}(t)\right]. \qquad (6.360)
\end{aligned}
$$

Reference to Eq. (6.360) shows that the probability distribution density $\langle f[a_{1_M}(t)]\rangle$, averaged within the limits of the period of the carrier frequency ω_0 of the signal, converges faster toward the Gaussian probability distribution density with the non-oscillating central moment of the second order as $N \to \infty$, since the terms of Eq. (6.360) independent of the value N contain the ratio

$$
\frac{C_2}{A_2},
$$

which is always less than unity under the condition of convergence, of a higher order in comparison with the terms in Eq. (6.360).

We carry out a quantitative estimation of accuracy of approximation of the probability distribution function $F[a_{1_M}(t)]$ of the signal distorted by the multiplicative noise at the output of the generalized detector by the Gaussian probability distribution function $F_G[a_{1_M}(t)]$ with the non-oscillating and oscillating central moments of the second order.

The characteristic function $\Theta_1^{\varphi}(x)$ in Eqs. (6.347) and (6.349)–(6.359), required to computer calculate the coefficients in Eqs. (6.326), (6.334), and (6.360), are given in Table II.1 (see Appendix II). The initial moments of the probability distribution density of the amplitude m_k^A of the signal are presented in Table 6.1.

We first estimate a dependence of convergence conditions of the probability distribution density in Eq. (6.326) to the Gaussian probability distribution

TABLE 6.1

Probability Distribution Densities of Amplitude of the Signal

Probability Distribution Density of Amplitude	m_1^A	m_2^A	m_3^A	m_4^A	Comment
$\delta(A - A_0)$	A_0	A_0^2	A_0^3	A_0^4	
$\frac{1}{\sqrt{2\pi\sigma_A^2}} e^{-\frac{(A-A_0)^2}{2\sigma_A^2}}$	A_0	$A_0^2\left(1+\varrho_1^2\right)$	$A_0^3\left(1+3\varrho_1^3\right)$	$A_0^4\left(1+6\varrho_1^2+\varrho_1^4\right)$	$\varrho_1 = \frac{\sigma_A}{A_0}, \varrho_1 \leq 0.3$
$\frac{A}{\nu} e^{-\frac{A^2}{\nu}}$	$\frac{\sqrt{\nu\pi}}{2}$	ν	$\frac{\sqrt{9\nu^3\pi}}{4}$	$2\nu^2$	$A \geq 0$
$\frac{2}{\Gamma(n)} \left(\frac{n}{\nu}\right)^n A^{2n-1} e^{-\frac{A^2 n}{\nu}}$	$\sqrt{\frac{\nu}{n}}\Gamma(n+0.5)$	ν	$3\sqrt{\frac{\nu^3}{n^3} \frac{\Gamma(n+1.5)}{\Gamma(n)}}$	$\frac{n+1}{n}\nu^2$	
$\frac{1}{\pi \Delta A \sqrt{1-\frac{(A-A_0)^2}{\Delta A}}}$	A_0	$A_0^2\left(1+0.5\varrho_2^2\right)$	$A_0^3\left(1+1.5\varrho_2^2\right)$	$A_0^4\left(1+3\varrho_2^2+0.375\varrho_2^4\right)$	$\frac{\lvert A-A_0\rvert}{\Delta A} \leq 1, \varrho_2 = \frac{\Delta A}{A_0}$

density with the non-oscillating central moment of the second order on statistical characteristics of the noise modulation function $\dot{M}(t)$ of the multiplicative noise for the limiting case, when the energy spectrum bandwidth of the noise modulation function $\dot{M}(t)$ of the multiplicative noise is much greater than the energy spectrum bandwidth of the PF at the input linear tract of the generalized detector

$$N \to \infty.$$

As $N \to \infty$ we can see from Eq. (6.326) that

$$F\left[a_{1_M}(t)\right] - F_G\left[a_{1_M}(t)\right] = -\frac{\lambda}{2} \cdot \left\{ f_G\left[a_{1_M}(t)\right] - \frac{\lambda}{12} \cdot f_G^{(3)}\left[a_{1_M}(t)\right] - \frac{\lambda^3}{24} \cdot f_G^{(5)}\left[a_{1_M}(t)\right] \right\}. \tag{6.361}$$

For all probability distribution densities of the amplitude of the signals given in Table 6.1 for the condition

$$\varrho_1 \leq 0.3, \quad \varrho_2 \leq 0.7 \quad \text{and} \quad n \leq 2$$

we obtain with a small error

$$\lambda \simeq \frac{\left|\Theta_1^{\varphi}(2)\right| - \left|\Theta_1^{\varphi}(1)\right|^2}{1 - \left|\Theta_1^{\varphi}(1)\right|^2} \cdot \cos 2\alpha(t) = \Delta \cos 2\alpha(t). \tag{6.362}$$

If the relative accuracy of approximation is given by

$$\frac{\left|F\left[a_{1_M}(t)\right] - F_G\left[a_{1_M}(t)\right]\right|}{F_G\left[a_{1_M}(t)\right]} \leq 10^{-k}, \tag{6.363}$$

then more rigorous requirements to the value Δ will apply for low values of the probability distribution function $F_G[a_{1_M}(t)]$.

When

$$F_G\left[a_{1_M}(t)\right] \leq \text{a value between 0.1 and 0.3} \tag{6.364}$$

one can be limited by the first term in Eq. (6.361).

Then the condition limiting the value Δ takes the following form:

$$|\Delta| \leq 2 \cdot 10^{-k} \frac{F_G\left[a_{1_M}(t)\right]}{f_G\left[a_{1_M}(t)\right]}, \quad F_G\left[a_{1_M}(t)\right] \leq 0.3. \tag{6.365}$$

Under the Gaussian probability distribution function of errors in phase of the signal with the phase variance σ_φ^2, Eq. (6.365) is transformed to the following condition for the mean square value of errors in phase of the signal:

$$\sigma_\varphi \geq 1.58 \cdot k \cdot \frac{F_G\left[a_{1_M}(t)\right]}{f_G\left[a_{1_M}(t)\right]}. \tag{6.366}$$

At $k = 1$ we obtain

$$F_G\left[a_{1_M}(t)\right] = 0.1, \quad \sigma_\varphi \geq 1.28; \tag{6.367}$$

$$F_G\left[a_{1_M}(t)\right] = 0.2, \quad \sigma_\varphi \geq 1.18. \tag{6.368}$$

For other probability distribution functions of random errors in phase of the signal, for example, for the uniformed probability distribution function within the limits of the interval $\pm\varphi_m$ and the probability distribution function corresponding to the sinusoid with the random initial phase and deviation of distortions in phase φ_m (see Table II.1, Appendix II), dependences of the value Δ on the maximal deviation of the phase φ_m of the signal are shown in Fig. 6.17.

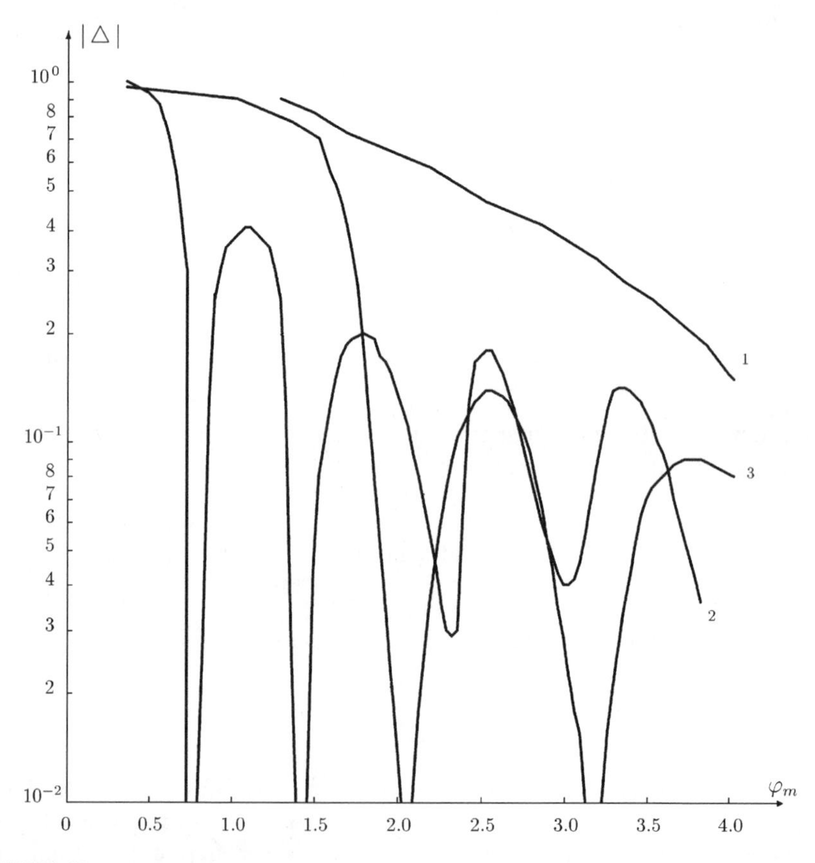

FIGURE 6.17
Dependence of the value $|\Delta|$ on the maximum phase deviation φ_m of the signal: for 1, $\Theta_1^\varphi(1) = e^{-\frac{\sigma_\varphi^2}{2}}$; for 2, $\Theta_1^\varphi(1) = J_0(2\varphi_m)$; for 3, $\Theta_1^\varphi(1) = \frac{\sin\varphi_m}{\varphi_m}$; $\varphi_m = 3\sigma_\varphi$.

The function $|\Delta(\varphi_m)|$, for the case of the Gaussian probability distribution function of distortions in phase of the signal at $\varphi_m = 3\sigma_\varphi$, is shown in Fig. 6.17 by curve 1.

The function $|\Delta(\varphi_m)|$ (see Fig 6.17) indicates that under more general suggestions regarding the probability distribution densities of distortions in amplitude and phase of the signal, the probability distribution function of the signal distorted by the stationary multiplicative noise at the output of the generalized detector (for the given model of the signal) can be believed to have accuracy given as the Gaussian probability distribution function with the non-oscillating central moment of the second order only when two conditions are satisfied:

- The distortions in phase of the signal are sufficiently high, the maximal error in phase of the signal reaches $\pm$ (2 to 3) radian;
- The correlation interval of the signal distorted by the multiplicative noise is much less than the time constant of the PF at the input linear tract of the generalized detector.

The last condition is well known and is the general condition required to normalize the stochastic processes by linear systems. The first condition is caused by specific cases of the signals considered and the requirement of absence in oscillations of the central moment of the second order of the probability distribution density of the signal at the output of the generalized detector.

For the cases when we can be limited by consideration of the probability distribution density, which is averaged within the limits of the period of the carrier frequency of the signal distorted by the multiplicative noise at the output of the generalized detector, the first condition formulated above is not so rigorous and is determined in the following form:

$$|\Delta(\varphi_m)| = \sqrt{24 \cdot 10^{-k} \frac{F_G\left[a_{1_M}(t)\right]}{f_G^{(s)}\left[a_{1_M}(t)\right]}}. \tag{6.369}$$

Equation (6.369) is obtained in the same way as Eq. (6.365), using Eq. (6.360) instead of Eq. (6.326), and ensures the following inequality:

$$\lim_{N\to\infty} \frac{\left|\langle F\left[a_{1_M}(t)\right]\rangle - F_G\left[a_{1_M}(t)\right]\right|}{F_G\left[a_{1_M}(t)\right]} \leq 10^{-k}, \tag{6.370}$$

where $\langle F[a_{1_M}(t)]\rangle$ is the probability distribution function corresponding to the probability distribution density determined by Eq. (6.360) of the signal at the output of the generalized detector (see Eq. (6.360)).

We proceed to estimate the conditions of convergence of the probability distribution density of the signal distorted by the multiplicative noise at the output of the generalized detector toward the Gaussian probability distribution density with the oscillating central moment of the second order.

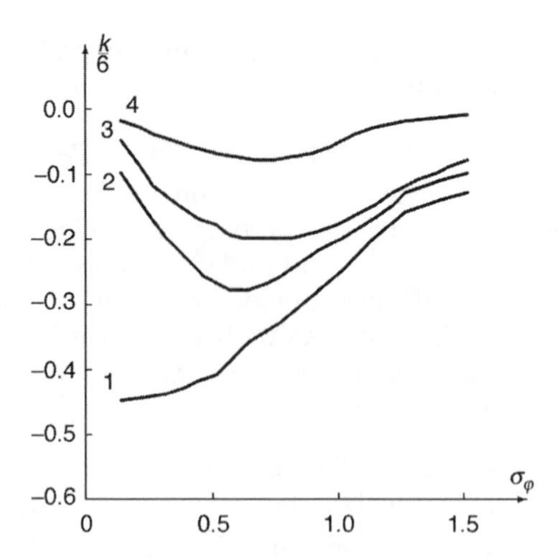

FIGURE 6.18
Coefficient of asymmetry k as a function of phase distortions of the signal at the fixed values of the angle α: for 1, $\alpha = 0$; for 2, $\alpha = \frac{\pi}{6}$; for 3, $\alpha = \frac{\pi}{4}$; for 4, $\alpha = \frac{5\pi}{12}$; $k(\frac{\pi}{2} + \alpha) = -k(\frac{\pi}{2} - \alpha)$.

The series defining the errors of approximation for the corresponding probability distribution density of the signal at the output of the generalized detector is given by Eq. (6.329). As an example consider the case in which the multiplicative noise only gives rise to the distortions in phase of the signal and the distortions in phase obey the Gaussian probability distribution density with the variance σ_φ^2.

Dependences of the coefficients of asymmetry k and kurtosis γ are shown in Figs. 6.18 and 6.19 for the given case for some fixed values of the angle

$$\alpha = \omega_0 + \varphi_0 \tag{6.371}$$

for some time instants.

The analysis shows that at $\sigma_\varphi \leq 1$ the maximal absolute values of the coefficients k and γ are obtained at $\alpha = k\pi$ $(k = 1, 2)$. When $\sigma_\varphi \geq 1.5$, the coefficients of asymmetry k and kurtosis γ weakly depend on the time—the angle α—and are close to the corresponding coefficients of the uniform probability distribution density of errors in phase of the signal as $\sigma_\varphi \to 0$ the coefficients $k \to 0$ and $\gamma \to -1.5$.

Errors of approximation of the probability distribution function of the signal distorted by the stationary multiplicative noise at the output of the generalized detector by the Gaussian probability distribution function with the oscillating central moment of the second order (see Eq. (6.342)–(6.345)) are shown in Figs. 6.20 and 6.21, respectively. Computer calculations take the coefficients of asymmetry k and kurtosis γ to the corresponding maximal errors of approximation, i.e., at $\alpha = k\pi$.

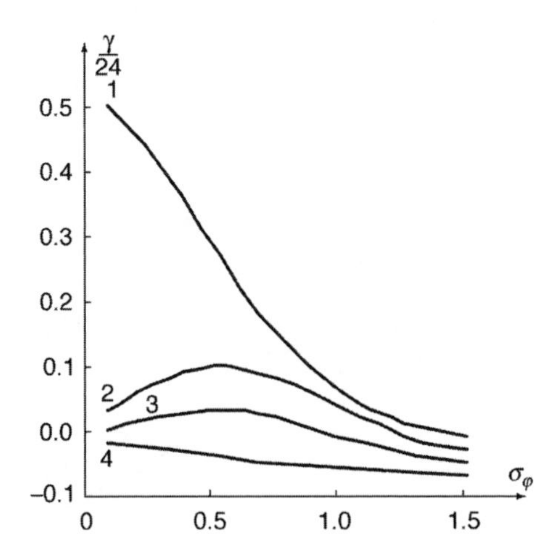

FIGURE 6.19
Coefficient of kurtosis γ as a function of phase distortions of the signal at the fixed values of the angle α: for 1, $\alpha = 0$; for 2, $\alpha = \frac{\pi}{6}$; for 3, $\alpha = \frac{\pi}{4}$; for 4, $\alpha = \frac{\pi}{2}$; $\gamma(\frac{\pi}{2} + \alpha) = \gamma(\frac{\pi}{2} - \alpha)$.

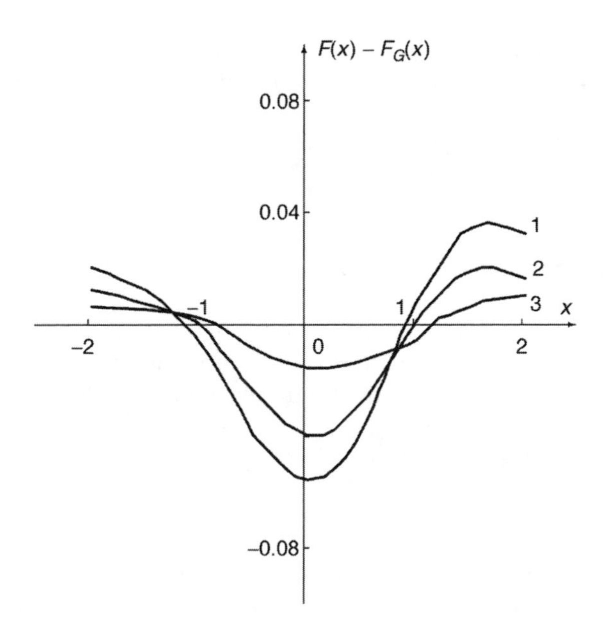

FIGURE 6.20
Errors under approximation of the probability distribution function by Gaussian law: for 1, $\sigma_\varphi = 0.3$; for 2, $\sigma_\varphi = 0.8$; for 3, $\sigma_\varphi = 1.4$: $N = 10$.

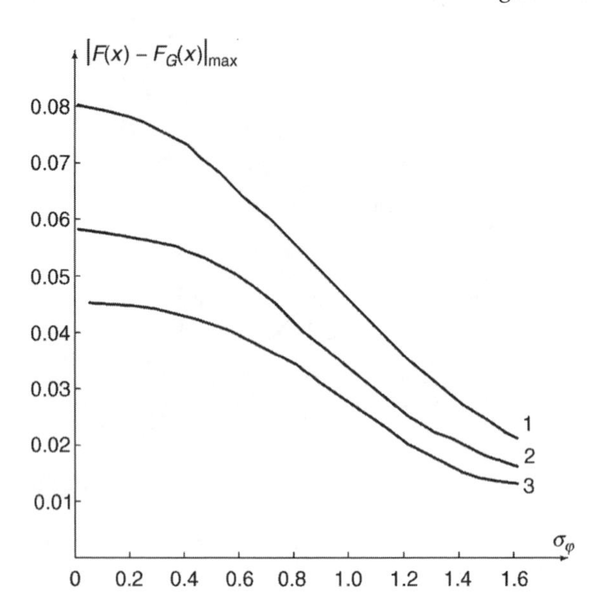

FIGURE 6.21
Maximum errors of approximation as a function of variance of phase distortions of the signal: for 1, $N = 5$; for 2, $N = 10$; for 3, $N = 15$.

Dependences of approximation errors on the argument of the probability distribution density of the signal at the output of the generalized detector

$$x = \frac{a_{1_M}(t) - m_{1_{a_1}}}{\sqrt{2M_{2_{a_1}}}} \tag{6.372}$$

at $N = 10$ and some values of distortions in phase of the signal at the output of the generalized detector—the variance σ_φ^2 of distortions in phase of the signal—are shown in Fig. 6.20.

These dependences have an analogous form for other values of N—at other ratios between the correlation interval of the noise modulation function $\dot{M}(t)$ of the multiplicative noise and the time constant of the PF at the input linear tract of the generalized detector.

Dependence of the maximal errors of approximation

$$|F(x) - F_G(x)|_{max}$$

on distortions in phase σ_φ of the signal are shown in Fig. 6.21.

Reference to Fig. 6.21 shows that even during very slow distortion in phase of the signal, the probability distribution function of the signal distorted by the multiplicative noise at the output of the generalized detector more rapidly tends to approach the Gaussian probability distribution function with the oscillating central moment of the second order with an increase in the value N

of the parameter

$$\xi = \frac{1}{2\pi} \cdot \Delta\Omega_M T. \tag{6.373}$$

6.6 Conclusions

Research on the signals distorted by the multiplicative noise at the output of the PF at the input linear tract of the generalized detector shows that the signal at the output of the PF, for example, is the sum of two components: the undistorted portion of the signal at the output of the PF at the input of the linear tract of the generalized detector, which recurs in the scale α_0 of the signal formed at the output of the PF by an action of the undistorted signal at the input of the generalized detector, and the signal noise component that arises as a consequence of the stimulus of the multiplicative noise.

Statistical characteristics of these two components can be defined using the complex envelope of amplitude of the signal formed at the output of the PF at the input linear tract of the generalized detector by an action of the undistorted signal at the input of the generalized detector and owing to characteristics and parameters of the noise modulation function $\dot{M}(t)$ of the multiplicative noise.

The signal component at the output of the generalized detector has the following peculiarities. Since the autocorrelation function $\dot{\rho}(\tau, \Omega)$, of the signal at the output of the generalized detector, differs from zero only within the limits of the interval

$$-T \leq \tau \leq T,$$

the signal noise component at the output of the generalized detector can exist only within the limits of this interval.

It should be pointed out that the interval

$$-T \leq \tau \leq T,$$

which can be overlapped by the signal noise component at the output of the generalized detector, is much greater than the duration of the main peak of the autocorrelation function of the signal component at the output of the generalized detector for some types of the signals when the multiplicative noise is absent.

With variation from zero to infinity in the energy spectrum bandwidth of the noise modulation function $\dot{M}(t)$ of the multiplicative noise, the correlation interval in time varies from infinity to the value equal to the duration of the undistorted signal component at the output of the generalized detector. The infinite value of the correlation interval corresponds to the infinite correlation interval of the noise modulation function $\dot{M}(t)$ of the multiplicative noise.

Finiteness of the correlation interval with an infinite decrease in the correlation interval of fluctuations of the noise modulation function $\dot{M}(t)$ of the multiplicative noise is caused by the limited bandwidth of the PF at the input linear tract of the generalized detector.

Under the same durations of the signals with a bell-shaped amplitude envelope and constant radio frequency carrier and the frequency-modulated signals with a bell-shaped amplitude envelope, the correlation interval of the second type of signal decreases rapidly in Q_y times with an increase in the parameter ξ under the stimulus of the multiplicative noise in comparison with the first type of signal. This fact is caused by a de-correlation process of the signal distorted by the multiplicative noise in the course of signal processing by the generalized detector and an amplitude-frequency response of the PF at the input linear tract of the generalized detector is matched with the modulated signal using a complex law.

The distribution of the normalized power

$$\frac{\sigma_s^2(\tau, 0)}{\sigma_s^2(0, 0)}$$

of the signal noise component at the output of the generalized detector with an increase in the parameter ξ is varied from the function

$$\left(1 - \frac{|\tau|}{T}\right)^2$$

at $\xi \ll 1$, to the function

$$1 - \frac{|\tau|}{T}$$

at $\xi \gg 1$ for the signals with a square wave-form amplitude envelope and constant radio frequency carrier.

For the case of the phase-manipulated signals with binary code when the signal phase takes two values 0 or π within the limits of the interval $\Delta\tau$, depending on the value "0" or "1"—the code of the phase-manipulated signal within the limits of the interval $\Delta\tau$—the variance of fluctuations at side lobes of the autocorrelation function, of the signal at the output of the generalized detector, is inversely proportional to the coefficient of signal truncation, which is equal to the number N of readings of the signal, and decreases linearly farther and farther away from the main (central) peak of the autocorrelation function of the signal at the output of the generalized detector.

Formally, a decrease in the amplitude of side lobes of the autocorrelation function, of the signal at the output of the generalized detector, with an increase in the number of elementary signals in the case of the slow fluctuating multiplicative noise is a consequence of averaging by an ensemble of the possible signals (codes) that are random under determination of the coefficients $D_{r\ell}$.

The variance of fluctuations of side lobes of the autocorrelation function, of the signal at the output of the generalized detector, as a function of the number N is explained by the fact that slow distortions in the course of the signal processing by the generalized detector are processed in the same way as rapid distortions within the limits of the correlation interval Δt as a result of the generalized detector de-correlating the signal distorted by the multiplicative noise.

There is a need to take into account that the greatest values of the variance $\sigma_s^2(\tau, \Omega)$, of the signal noise component at the output of the generalized detector, occur at low values of the parameters τ and Ω. At high values of the parameters τ and Ω, the relative value of the variance $\sigma_s^2(\tau, \Omega)$ of the signal noise component at the output of the generalized detector equal to the ratio

$$\frac{\sigma_s^2(\tau, \Omega)}{\sigma_s^2(0, 0)}$$

is very low.

Because of this, despite the fact that use of the approximate formulae defining the variance $\sigma_s^2(\tau, \Omega)$ at high values of the parameters τ and Ω leads to the great relative error

$$\frac{\sigma_{s_{ex}}^2(\tau, \Omega)}{\sigma_{s_{ap}}^2(\tau, \Omega)}$$

at not-so-high values of the parameter ξ, the absolute error under determination of the variance $\sigma_s^2(\tau, \Omega)$ is not so high.

The approximate formulae for determination of the variance $\sigma_s^2(\tau, \Omega)$, of the signal noise component at the output of the generalized detector, simplify significantly the process of definition of statistical characteristics of the signal noise component at the output of the generalized detector for the condition of rapid fluctuating multiplicative noise.

For the case of the slow fluctuating multiplicative noise the approximate formulae of definition of the variance, of the signal noise component at the output of the generalized detector, are true for the arbitrary amplitude-frequency response of the PF at the input linear tract of the generalized detector when the correlation time of the multiplicative noise is much greater in comparison to the duration of the amplitude-frequency response of the PF at the input linear tract of the generalized detector.

The autocorrelation function $\dot{\rho}(\tau, \Omega)$, of the signal at the output of the generalized detector, and the functions $\gamma_0(\tau, \Omega)$, $\gamma_1(\tau, \Omega)$, and $\gamma_2(\tau, \Omega)$ are defined by the complex amplitude envelope of the signal. Knowledge of these functions for the given type of signal allows us to easily define the main statistical characteristics and parameters of the signal at the output of the generalized detector using the parameters of the fluctuating multiplicative noise and to estimate distortions of the signal at the output of the generalized detector caused by a stimulus of the fluctuating multiplicative noise.

To define the statistical characteristics of the signal noise component at the output of the generalized detector less knowledge regarding the characteristics of the multiplicative noise is required than in a general case involving the use of the approximate formulae of definition of the variance of the signal noise component at the output of the generalized detector under the stimulus of the slow fluctuating multiplicative noise.

The probability distribution density of the signal $a_{1_M}(t)$, at the output of the generalized detector, can be approximated by the Gaussian probability distribution density with the non-oscillating and oscillating central moment of the second order. In order for the probability distribution density of the signal $a_{1_M}(t)$, at the output of the generalized detector, to be approximated by the Gaussian probability distribution density with the non-oscillating central moment of the second order the condition

$$N = \frac{\tau_s}{\tau_c} \gg 1$$

must be true, and specific limitations would be required on the statistical characteristics of the noise modulation function $\dot{M}(t)$ of the multiplicative noise with the purpose of satisfying the following condition:

$$|\lambda| = \left| \frac{M_2^{(n)}}{M_2^{(s)}} \right| \ll 1.$$

Even during very slow distortions in phase of the signal at the output of the generalized detector the probability distribution density of the signal distorted by the multiplicative noise at the output of the generalized detector more rapidly tends to approach the Gaussian probability distribution density with the oscillating central moment of the second order with an increase in the value N of the parameter ξ.

References

1. Tuzlukov, V., *Signal Processing in Noise: A New Methodology*, IEC, Minsk, 1998.
2. Tuzlukov, V., *Signal Detection Theory*, Springer-Verlag, New York, 2001.
3. Cohen, L., *Time-Frequency Analysis*, Prentice-Hall, Englewood Cliffs, NJ, 1995.
4. Haykin, S., *Adaptive Filter Theory*, 3rd ed. Prentice-Hall, Englewood Cliffs, NJ, 1996.
5. Kailath, T., *Linear Systems*, Prentice-Hall, Englewood Cliffs, NJ, 1980.
6. Heitz, C., Optimum time-frequency representations for the classification and detection of signals, *Appl. Signal Process.*, Vol. 2, No. 3, 1995, pp. 124–143.
7. Brown, W., *Analysis of Linear Time-Invariant Systems*, McGraw-Hill, New York, 1963.

8. Scharf, L., *Statistical Signal Processing: Detection, Estimation, and Time Series Analysis*, Addison-Wesley, Reading, MA, 1991.
9. McFadden, P. and Smith, J., A signal processing technique for detection of local defections in a gear from the signal average of the vibration, *Proc. Inst. Mech. Eng.*, Vol. 199, No. C4, 1985, pp. 287–292.
10. Kelly, E., An adaptive detection algorithm, *IEEE Trans.*, Vol. AES-22, No. 2, 1986, pp. 115–127.
11. Porat, B., *Digital Processing of Random Signals: Theory and Methods*, Prentice-Hall, Englewood Cliffs, NJ, 1994.
12. Skolnik, M., *Radar Handbook*, McGraw-Hill, New York, 1990.
13. Richaczek, A., Signal energy distributions in time and frequency, *IEEE Trans.*, Vol. IT-14, No. 2, 1968, pp. 369–374.
14. Ripley, B., *Stochastic Simulation*, Wiley, New York, 1987.
15. Diniz, P., *Adaptive Filtering: Algorithms and Practical Implementation*, Kluwer Academic Publishers, Boston, 1997.
16. Widrow, B. and Stearns, S., *Adaptive Signal Processing*, Prentice-Hall, Englewood Cliffs, NJ, 1985.
17. Kay, S., *Fundamentals of Statistical Signal Processing: Estimation Theory*, Prentice-Hall, Englewood Cliffs, NJ, 1993.
18. Solo, V. and Kong, X., *Adaptive Signal Processing Algorithms*, Prentice-Hall, Englewood Cliffs, NJ, 1995.
19. Kucera, V., *Analysis and Design of Discrete Linear Control Systems*, Prentice-Hall, Englewood Cliffs, NJ, 1991.
20. Borden, B., *Radar Imaging of Airborne Targets: A Primer for Applied Mathematicians and Physicists*, IOP, Philadelphia, 1999.
21. Benveniste, A., Metivier, M., and Prionret, P., *Adaptive Algorithms and Stochastic Approximation*, Springer-Verlag, New York, 1990.
22. Shirman, Y. and Manjos, V., *Theory and Methods in Radar Signal Processing*, Radio and Svyaz, Moscow, 1981 (in Russian).
23. Fuller, W., *Introduction to Statistical Time Series*, Wiley, New York, 1996.
24. Farhang-Boroujeny, B., *Adaptive Filters: Theory and Applications*, Wiley, New York, 1998.
25. Nahi, N., *Estimation Theory and Applications*, Krieger, Huntington, NJ, 1976.
26. Manikas, A. and Proukakis, C., Modeling and estimation of ambiguities in linear arrays, *IEEE Trans.*, Vol. SP-46, No. 9, 1998, pp. 2166–2179.
27. Wehner, D., *High Resolution Radar*, Artech House, Norwood, MA, 1987.
28. Mensa, D., *High Resolution Radar Cross-Section Imaging*, Artech House, Norwood, MA, 1991.
29. Katkovnik, V., A new concept of additive beam forming for moving sources and impulse noise environment, *Signal Process.*, Vol. 80, No. 9, 2000, pp. 1863–1882.
30. Titterington, D., Smith, A., and Makov, U., *Statistical Analysis of Finite Mixture Distributions*, Wiley, New York, 1985.
31. Besson, O. and Stoica, P., Non-linear least-squares approach to frequency estimation and detection for sinusoidal signals with arbitrary envelope, *Digital Signal Process.: Rev. J.*, Vol. 9, No. 1, 1999, pp. 45–56.
32. Wang, H., Park, H., and Wicks, M., Recent results in space-time processing, *Proc. Nat. Radar Conf.*, Atlanta, GA, 1994, pp. 104–109.
33. Schleher, D., *MTI and Pulsed Doppler Radar*, Artech House, Norwood, MA, 1991.

34. Highes, P., A high resolution range radar detection strategy, *IEEE Trans.*, Vol. AES-19, No. 5, 1989, pp. 663–667.
35. Farina, A., *Antenna-Based Signal Processing Techniques for Radar Systems*, Artech House, Norwood, MA, 1992.
36. Swindlehurst, A. and Stoica, P., Maximum likelihood methods in radar array signal processing, *Proc. IEEE*, Vol. 86, No. 2, 1998, pp. 421–441.
37. Farina, A., Scannapieco, F., and Vinelli, F., Target detection and classification with very high range resolution radar, in *Proc. Int. Conf. Radar*, Versailles, France, April 1989, pp. 20–25.
38. Gradshteyn, I. and Ryzhik, I., *Table of Integrals, Series, and Products*, 5th ed., Academic Press, New York, 1994.
39. Kuo, S. and Morgan, D., *Active Noise Control Systems*, Wiley, New York, 1996.
40. Silverman, B., *Density Estimation for Statistics and Data Analysis*, Chapman & Hall, London, 1986.
41. Poor, H., *Introduction to Signal Detection and Estimation*, 2nd ed., Springer-Verlag, New York, 1995.
42. Stuber, G., *Principles of Mobile Communications*, Kluwer Academic Publishers, Boston, 1996.
43. Helstrom, C., *Elements of Signal Detection and Estimation*, Prentice-Hall, Englewood Cliffs, NJ, 1995.
44. Zwilliger, D., *Standard Mathematical Tables and Formulae*, CRC Press, Boca Raton, FL, 1996.
45. Macchi, O., *Adaptive Processing: The LMS Approach with Applications in Transmission*, Wiley, New York, 1995.
46. Muirhead, R., *Aspects of Multivariate Statistical Theory*, Wiley, New York, 1982.
47. Tufts, D., Kirsteins, I., and Kumaresan, R., Data adaptive detection of a weak signal, *IEEE Trans.*, Vol. AES-19, No. 2, 1983, pp. 313–316.
48. Gerlach, K., Adaptive detection of range distributed targets, *IEEE Trans.*, Vol. SP-47, No. 7, 1999, pp. 1844–1851.
49. Gasser, T., Mocks, J., and Kohler, W., Amplitude probability distribution of noise for flash-evoked potentials and robust response estimates, *IEEE Trans.*, Vol. BME-33, No. 6, 1986, pp. 579–584.
50. Bar-Shalom, Y. and Li, X., *Multitarget-Multisensor Tracking: Principles and Techniques*, YBS, New Orleans, LA, 1995.
51. Li, J., Lin, G., Jiang, N., and Stoica, P., Moving target feature extraction for airborne high-range resolution phased-array radar, *IEEE Trans.*, Vol. SP-49, No. 2, 2001, pp. 277–289.
52. Dickey, F., Labitt, M., and Standaher, F., Development of airborne moving target radar for long range surveillance, *IEEE Trans.*, Vol. AES-27, No. 11, 1991, pp. 959–976.
53. Jacobs, S. and O'Sullivan, J., High resolution radar models for joint tracking and recognition, *Proc. IEEE Nat. Conf. Radar*, Syracuse, NY, May 1997, pp. 99–104.
54. Hudson, S. and Pslatis, D., Correlation filters for aircraft identification from radar range profiles, *IEEE Trans.*, Vol. AES-29, No. 4, 1999, pp. 741–748.
55. Leshem, A. and Van Der Veen, A., Direction-of-arrival estimation for constant modulus signals, *IEEE Trans.*, Vol. SP-47, No. 11, 1999, pp. 3125–3129.
56. Stoica, P., Besson, O., and Gershman, A., Direction-of-arrival estimation of an amplitude-distorted wave-front, *IEEE Trans.*, Vol. SP-49, No. 2, 2001, pp. 269–276.

57. Tichonov, V., *Statistical Radio Engineering*, Radio and Svyaz, Moscow, 1982 (in Russian).
58. Robbins, H. and Monro, S., A stochastic approximation method, *Ann. Math. Statist.*, Vol. 22, 1951, pp. 400–407.
59. Fuller, W., *Measurement Error Models*, Wiley, New York, 1987.
60. Capon, J. and Goodman, N., Probability distributions for estimators of the frequency wave-number spectrum, *Proc. IEEE*, Vol. 58, 1970, pp. 1785–1786.
61. Reed, I., Mallett, J., and Brennan, L., Rapid convergence rate in adaptive arrays, *IEEE Trans.*, Vol. AES-10, No. 6, 1974, pp. 853–863.
62. Middleton, D., Statistical-physical model of man-made radio noise. Part 1: First-order probability models of the instantaneous amplitude, Office of Telecommunications, Report 74-36, April 1974, pp. 1–76.
63. Middleton, D., Statistical-physical model of man-made radio noise. Part 2: First-order probability models of the envelope and phase, Office of Telecommunications, Report 76-86, April 1976, pp. 76–124.
64. Middleton, D., Man-made noise in urban environments and transportation systems: models and measurements, *IEEE Trans.*, Vol. IT-22, No. 4, 1973, pp. 25–57.
65. Weiss, A. and Mitra, D., Digital adaptive filters, conditions for convergence, rates of convergence, effects of noise and errors arising from the implementation, *IEEE Trans.*, Vol. IT-25, No. 3, 1979, pp. 637–652.
66. Bitmead, R., Persistence of excitation conditions and the convergence of adaptive schemes, *IEEE Trans.*, Vol. IT-30, No. 2, 1984, pp. 183–191.
67. Sage, A. and White, C., *Optimum Systems Control*, Prentice-Hall, Englewood Cliffs, NJ, 1977.
68. Crammer, H., *Mathematical Methods of Statistics*, Princeton University Press, Princeton, NJ, 1946.
69. Eweda, E., Convergence analysis of the sign algorithm without the independence and Gaussian assumptions, *IEEE Trans.*, Vol. SP-48, No. 9, 2000, pp. 2535–2544.
70. Tanrikulu, O. and Chambers, J., Convergence and steady-state properties of the least-mean mixed-norm (LMMN) adaptive algorithm, in *Proc. Inst. Elect. Eng., Vis. Image Signal Process.*, Vol. 143, No. 3, 1996, pp. 137–142.
71. Ye, W. and Zhou, X., Criteria of convergence of median filters and perturbation theorem, *IEEE Trans.*, Vol. SP-49, No. 2, 2001, pp. 360–364.
72. Eweda, E., Convergence analysis of adaptive filtering algorithms with singular data covariance matrix, *IEEE Trans.*, Vol. SP-49, No. 2, 2001, pp. 334–343.
73. Levin, B., *Theoretical Foundations of Statistical Radio Engineering*, Parts 1–3, Soviet Radio, Moscow, 1974–1976 (in Russian).
74. Macchi, O. and Eweda, E., Second order convergence analysis of stochastic linear adaptive filtering, *IEEE Trans.*, Vol. AC-28, No. 1, 1983, pp. 76–85.
75. Masry, E. and Bullo, F., Convergence analysis of the signal algorithm for adaptive filtering, *IEEE Trans.*, Vol. IT-41, No. 2, 1995, pp. 489–495.

7

Generalized Approach to Signal Detection in the Presence of Multiplicative and Additive Gaussian Noise

In this chapter we consider the impact of the stimulus of multiplicative noise on the detection of signals. In the process, we take into consideration the additive Gaussian noise as the set noise of the receiver or detector that is always in parallel with the multiplicative noise.

The first part of this chapter focuses on an analysis of the stimulus of multiplicative noise on statistical characteristics of the process at the output of the generalized detector used with the purpose of detecting the signals in the presence of additive Gaussian noise.

This approach to analysis corresponds to the modern design of complex signal processing systems in various areas of application in practice including radar, communications, acoustics, wireless communications, mobile communications, underwater signal processing, sonar, navigation systems, remote sensing, geophysical signal processing, biomedical signal processing, and so on. For this approach the generalized detector is considered as the receiver of the signals in the presence of additive Gaussian noise, and the generalized detector ensures the best detection performances of the signals when the additive Gaussian noise is considered as the set noise of the receiver or detector.

In combination with a corresponding choice of signal parameters, the generalized detector can ensure the detection performances of the signal that are very close to the potential achieved detection performances for some types of natural and man-made interference.[1]

For this reason, an analysis of the stimulus of multiplicative noise on the detection performances of the signals during the use of the generalized detector is of great interest both from the viewpoint of estimation of deterioration of qualitative characteristics of complex signal processing systems caused by the stimulus of the multiplicative noise, and from the viewpoint of definition of conditions, the energy characteristics of the noise and interferences, for which specific methods and receivers used during signal processing in the presence of multiplicative noise are appropriate.

The second part of this chapter is concerned with methods and techniques intended for the generalized approach to signal processing in the presence of additive Gaussian noise and multiplicative noise, and the definition of the detection performances of the signals under the use of these methods and techniques.

7.1 Statistical Characteristics of Signals at the Output of the Generalized Detector

7.1.1 Model of the Generalized Detector

To analyze the impact of the multiplicative noise on detection of the signals in the presence of additive Gaussian noise we use the following model of the generalized detector. Assume that the generalized detector contains N elements of resolution or N channels by instant of the incoming signal, for example, in radar—the range of resolution, or by frequency of the signal—the radial velocity resolution. In accordance with this fact we suppose that the parameter by which a signal resolution is carried out, i.e., the instant of incoming of the signal or frequency of the signal, can take N known discrete values.

Moreover, an interval of sampling is chosen in such a manner that the signals are orthogonal when the multiplicative noise is absent. One of N non-overlapping elements of resolution, or channels of the receiver or detector, corresponds to each value of the signal parameter. Henceforth, we will use the term "channel" for simplicity, also meaning the resolution element. All values of the signal parameter are equiprobable.

From an orthogonality of the signals it follows that the given input stochastic process generates the output signal only using a single channel of the generalized detector when the multiplicative noise is absent. We denote this channel by the index "k." All other channels of the generalized detector will be denoted by the index "m."

Applying this approach to the real signals, the condition of an orthogonality of the signals is approximately satisfied when the sampling interval of the signal parameter is equal to the bandwidth of the main peak of the ambiguity function of the signal on the corresponding axis. In the process, the condition of an orthogonality of the signals is carried out well when we neglect the residuals (side shoots) of the ambiguity function. When the multiplicative noise is present, in some cases we must take into account the residuals of the ambiguity function.

The generalized detector, as is shown in References 1–4, generates the coefficient of the likelihood (or some value dependent on this coefficient) by a monotone function at each channel. In this case signal processing by the generalized detector depends strongly on the signal type.

We will use two types of signal: the signal with the known initial phase and the signal with the unknown (random) initial phase. The amplitude of the signal and the probability distribution density of amplitude of the signal do not act on the signal processing by the generalized detector. These data points are only taken into account during determination of the detection performances of the signals.[5–10]

These types of signals correspond very well to the majority forms of signals used by complex signal processing systems, for example, segments of continuous oscillations, pulses, and coherent pulse burst. The type of signals corresponding to the non-coherent pulse burst will not be analyzed and discussed, since an analysis and discussion of the type of signals with the unknown initial phase allow us to get all of the necessary information to estimate the stimulus of multiplicative noise in the case of the non-coherent pulse burst.

We assume that the additive noise is Gaussian. There is a need to note that the technique used below is appropriate for the case of the correlated Gaussian additive noise under corresponding corrections.

7.1.1.1 Signal with Known Initial Phase

In the case when the multiplicative noise is absent, the signal, the parameters of which correspond to the "k"-th channel of the generalized detector—we will call this signal the signal of the "k"-th channel of the generalized detector, for simplicity—can be written in the following form:

$$a_k(t, A_a) = A_a a_{m_k}(t), \tag{7.1}$$

where

$$a_{m_k}(t) = S_{m_k}(t)\cos[\omega_k t + \Psi(t)] = Re\{\dot{S}_{m_k}(t) \cdot e^{j\varphi_k t}\} \tag{7.2}$$

is the model signal (the reference signal) generated by the MSG (the model signal generator) of the generalized detector and A_a is the factor that is proportional to the amplitude of the signal. This case can be considered as signal processing of the signal with the unknown (stochastic) amplitude by the generalized detector.[1,2,11–21]

Since a scale of the factor A_a can be chosen in an arbitrary way, we assume for simplicity that

$$A_a = \sqrt{E_{a_1}}, \tag{7.3}$$

where E_{a_1} is the energy of the signal $a_k(t)$. In doing so, the energy of the model signal $a_{m_k}(t)$—the reference signal generated by the MSG of the generalized detector—is equal to $\sqrt{E_{a_1}}$, as well.

Then the signal $a_k(t)$ can be written in the following form:

$$a_k(t) = \sqrt{E_{a_1}}\, a_{m_k}(t) = \sqrt{E_{a_1}}\, Re\left\{\dot{S}_{m_k}(t) \cdot e^{j\varphi_k t}\right\}. \tag{7.4}$$

As is well known,[1–4,15–17] for the case considered, when the additive noise is Gaussian, signal processing by each channel of the generalized detector reduces to the following correlation integral related to the likelihood coefficient by the monotone function:

$$z_g^{out}(t) = 2\int_{-\infty}^{\infty} X_{in}(t) a_{m_n}(t)\,dt - \int_{-\infty}^{\infty} X_{in}(t) X_{in}(t-\tau)\,dt$$

$$= \frac{1}{2}\cdot Re\left\{2\int_{-\infty}^{\infty} \dot{S}_{in}(t) S_{m_n}^{*}(t)\,dt - \int_{-\infty}^{\infty} \dot{S}_{in}(t) S_{in}^{*}(t-\tau)\,dt\right\}, \tag{7.5}$$

where $X_{in}(t)$ is the stochastic process at the input of the linear tract (the PF and AF) of the generalized detector—the signal + the additive Gaussian noise; $\dot{S}_{in}(t)$ is the complex envelope of amplitude of the stochastic process at the input of linear tract of the generalized detector;

$$a_{m_n}(t) = Re\{\dot{S}_{m_n}(t)\cdot e^{j\omega_n t}\} \tag{7.6}$$

is the model signal of the given channel of the generalized detector ($n = k, m$), i.e., the reference signal generated by the MSG of the generalized detector.

7.1.1.2 Signal with Unknown Initial Phase

For this case the signal at the "k"-th channel of the generalized detector, by analogy with Eqs. (7.1) and (7.4), can be written in the following form:

$$a_k(t, A_a, \varphi_0) = A_a S_{m_k}(t, \varphi_0) = \sqrt{E_{a_1}}\, Re\left\{\dot{S}_{m_k}(t)\cdot e^{j(\omega_k t + \varphi_0)}\right\}, \tag{7.7}$$

where φ_0 is the unknown (random) initial phase of the signal.

As is shown in References 1, 2, 22–34, during signal processing of the signals with the unknown (random) initial phase (in the presence of additive Gaussian noise) by the generalized detector, the signal at the output of the generalized detector is the complex envelope of amplitude of the correlation integral in Eq. (7.5):

$$z_g^{out}(t) = \left|\dot{z}_g^{out}(t)\right| = \frac{1}{2}\cdot\left|2\int_{-\infty}^{\infty} \dot{S}_{in}(t) S_{m_n}^{*}(t)\,dt - \int_{-\infty}^{\infty} \dot{S}_{in}(t) S_{in}^{*}(t-\tau)\,dt\right|, \tag{7.8}$$

where $\dot{S}_{in}(t)$ is the complex envelope of amplitude of the stochastic process at the input of linear tract of the generalized detector; $\dot{S}_{m_n}(t)$ is the complex envelope of amplitude of the model signal (reference signal) generated by the MSG at the "n"-th channel of the generalized detector ($n = k, m$).

For the signal detection problem, the decision regarding a "yes" or a "no" signal in the stochastic process at the input of linear tract of the generalized detector is taken by comparing the signal at the output of the generalized detector—the correlation integral $z_g^{out}(t)$ or the complex envelope of amplitude $\dot{z}_g^{out}(t)$ of the correlation integral—with the threshold K_g or K_g^*.

As a rule, the Neyman–Pearson criterion is used for solving the signal detection problem. The value of the threshold K_g or K_g^* for each channel of the generalized detector is chosen based on the energy characteristics of the additive Gaussian noise and the *a priori* given probability of false alarm P_{F_k} of the "k"-th channel of the generalized detector related with the total probability of false alarm P_F of the N-channel generalized detector by the function

$$P_{F_k} \simeq \frac{P_F}{N}, \tag{7.9}$$

where N is the total number of channels—resolution elements—of the generalized detector. This relationship is true for low values of the probability of false alarm

$$P_F \ll 1. \tag{7.10}$$

7.1.2 Signal with Known Initial Phase

For the case in which the multiplicative noise is absent the stochastic process at the input of linear tract of the generalized detector is determined in the following form:

$$X_{in}(t) = a_k(t) + n(t), \tag{7.11}$$

where $n(t)$ is the additive Gaussian noise. In the process, the signal at the output of the "n"-th channel of the generalized detector determined by Eq. (7.5) obeys the probability distribution density of the background noise of the generalized detector defined in Chapter 5.

As was shown in Chapter 5, the probability distribution density of the background noise of the generalized detector for definite conditions tends to approach the Gaussian probability distribution density with the mean determined in the following form

$$z_{g_n}^{out}(t) = E_{a_1} \int_{-\infty}^{\infty} a_{m_k}(t) a_{m_n}(t)\, dt = \begin{cases} E_{a_1}, & n = k; \\ 0, & n \neq k \end{cases} \tag{7.12}$$

and with the variance determined by References 1, 2, 23–31, and 34–38

$$\begin{aligned} D\left[z_{g_n}^{out}(t)\right] &= \int_{-\infty}^{\infty}\int_{-\infty}^{\infty} M\left[\eta^2(t_1) - \xi^2(t_2)\right] dt_1\, dt_2 \\ &= \frac{N_0^2 \omega_0^4}{16(\Delta\omega_{PF})^2 T\beta}, \end{aligned} \tag{7.13}$$

where $\frac{N_0}{2}$ is the spectral power density of the additive Gaussian noise; ω_0 is the carrier frequency of the signal, which coincides with the resonant frequency of the PF (preliminary filter) at the input linear tract of the generalized detector (see Chapter 5); ω_{PF} is the bandwidth of the PF at the input linear tract of the generalized detector; $T\beta$ is the signal base; $\xi(t)$ is the additive Gaussian noise formed at the output of the PF at the input linear tract of the generalized detector; $\eta(t)$ is the additive Gaussian noise formed at the output of the AF (additional filter) at the input linear tract of the generalized detector (see Chapter 5). Note we consider the spectral power density $\frac{N_0}{2}$ of the additive Gaussian noise both in the positive and negative frequency regions.

When the multiplicative noise is present the stochastic process at the input of linear tract of the generalized detector can be determined in the following form:

$$X_{in}(t) = a_{M_k}(t) + n(t), \tag{7.14}$$

where

$$a_{M_k}(t) = \sqrt{E_{a_1}}\, Re\left\{\dot{M}(t)\dot{S}_{m_k}(t)\cdot e^{j\omega_k t}\right\} \tag{7.15}$$

is the signal formed as a result of the stimulus of the multiplicative noise on the signal $a_k(t)$.

The resulting signal at the output of the one-channel generalized detector under the stimulus of the multiplicative noise in terms of Eqs. (7.5) and (7.15), and on the basis of the main results discussed in Chapter 5, takes the following form:

$$\begin{aligned}
z_{g_M}^{out}(t) &= 2\int_{-\infty}^{\infty}\left[a_{M_k}(t)+\xi(t)\right]a_{m_n}(t)\,dt \\
&\quad - \int_{-\infty}^{\infty}\left[a_{M_k}(t)+\xi(t)\right]\left[a_{M_k}(t-\tau)+\xi(t-\tau)\right]dt \\
&\quad + \int_{-\infty}^{\infty}\eta(t)\eta(t-\tau)\,dt - \int_{-\infty}^{\infty}\xi(t)\xi(t-\tau)\,dt \\
&= 2\int_{-\infty}^{\infty}a_{M_k}(t)a_{m_n}(t)\,dt - \int_{-\infty}^{\infty}a_{M_k}(t)a_{M_k}(t-\tau)\,dt \\
&\quad +2\int_{-\infty}^{\infty}a_{m_n}(t)\xi(t)\,dt - \int_{-\infty}^{\infty}a_{M_k}(t)\xi(t-\tau)\,dt - \int_{-\infty}^{\infty}a_{M_k}(t-\tau)\xi(t)\,dt \\
&\quad + \int_{-\infty}^{\infty}\eta(t)\eta(t-\tau)\,dt - \int_{-\infty}^{\infty}\xi(t)\xi(t-\tau)\,dt
\end{aligned}$$

$$= \int_{-\infty}^{\infty} \left[2a_{M_k}(t)a_{m_n}(t) - a_{M_k}(t)a_{M_k}(t-\tau)\right] dt$$

$$+ \int_{-\infty}^{\infty} \left[2a_{m_n}(t)\xi(t) - a_{M_k}(t)\xi(t-\tau) - a_{M_k}(t-\tau)\xi(t)\right] dt$$

$$+ \int_{-\infty}^{\infty} [\eta(t)\eta(t-\tau) - \xi(t)\xi(t-\tau)]\, dt, \tag{7.16}$$

where τ is the delay caused by structural elements of the generalized detector.

The delay τ does not act on the resulting signal formed at the output of the generalized detector. For this reason, we can suppose $\tau = 0$ without any loss in accuracy with the purpose of simplifying further consideration and analysis.

In doing so, the last equation can be rewritten in the following form:

$$z_{g_M}^{out}(t) = \int_{-\infty}^{\infty} \left[2a_{M_k}(t)a_{m_n}(t) - a_{M_k}^2(t)\right] dt$$

$$+ \int_{-\infty}^{\infty} \left[2a_{m_n}(t)\xi(t) - 2a_{M_k}(t)\xi(t)\right] dt$$

$$+ \int_{-\infty}^{\infty} [\eta^2(t) - \xi^2(t)]\, dt$$

$$= z'_n(t) + z''(t) + z'''(t). \tag{7.17}$$

Thus, the resulting signal at the output of the "n"-th channel of the generalized detector is the sum of three components. The third term in Eq. (7.17)

$$z'''(t) = \int_{-\infty}^{\infty} [\eta^2(t) - \xi^2(t)]\, dt \tag{7.18}$$

is the background noise of the generalized detector (see Chapter 5).

As is shown in References 1, 2, 23–31, and 34–38, the background noise of the generalized detector is generated by the additive Gaussian noise at the input of linear tract of the generalized detector independently of a "yes" or a "no" multiplicative noise. The background noise obeys the probability distribution density discussed in both Chapter 5 and in References 1, 2, and 35–37 for all channels of the generalized detector. The background noise of the generalized detector possesses the mean equal to zero in the statistical sense and the variance determined by Eq. (7.13).

The second term in Eq. (7.17)

$$z''(t) = \int_{-\infty}^{\infty} \left[2a_{m_n}(t)\xi(t) - 2a_{M_k}(t)\xi(t)\right] dt \tag{7.19}$$

is the total noise component of the generalized detector caused by the interaction between the noise component of the correlation channel

$$2\int_{-\infty}^{\infty} a_{m_n}(t)\xi(t)\,dt$$

and the random component of the autocorrelation channel

$$2\int_{-\infty}^{\infty} a_{M_k}(t)\xi(t)\,dt$$

of the generalized detector.

The noise component of the correlation channel

$$2\int_{-\infty}^{\infty} a_{m_n}(t)\xi(t)\,dt$$

of the generalized detector is caused by the interaction between the model signal $a_{m_n}(t)$ generated by the MSG at each channel of the generalized detector and the noise $\xi(t)$ formed at the output of the PF at the input linear tract of the generalized detector.

The random component of the autocorrelation channel

$$2\int_{-\infty}^{\infty} a_{M_k}(t)\xi(t)\,dt$$

of the generalized detector is caused by the interaction between the signal and noise that are formed at the output of the PF at the input linear tract of the generalized detector.

When the multiplicative noise is present the random component of the autocorrelation channel

$$2\int_{-\infty}^{\infty} a_{M_k}(t)\xi(t)\,dt$$

of the generalized detector exists both at the "k"-th channel of the generalized detector, $n = k$, in which there is the signal component at the output of the

"k"-th channel of the generalized detector for a "no" condition multiplicative noise and at other channels of the generalized detector, $n = m \neq k$, in which a "no" signal is at the output of given channels of the generalized detector when the multiplicative noise is absent.

As was proven rigorously in References 1, 2, and 39–47, and shown experimentally in References 1–4, 21, 30–32, 39–41, and 47–67, the noise component of the correlation channel and the random component of the autocorrelation channel of the generalized detector are completely compensated during the definite conditions in the statistical sense. The definite conditions along with the complete compensation are discussed in more detail in the bibliography mentioned above.

The first term in Eq. (7.17)

$$z'_n(t) = \int_{-\infty}^{\infty} \left[2a_{M_k}(t)a_{m_n}(t) - a^2_{M_k}(t)\right] dt \tag{7.20}$$

is the energy of the signal component at the output of the "k"-th channel of the generalized detector.

When the multiplicative noise is present the signal $a_{M_k}(t)$ generates the first term $z'_n(t)$ (see Eq. (7.20)) both at the "k"-th channel of the generalized detector, $n = k$, in which there is the signal component at the output of the "k"-th channel of the generalized detector for a "no" condition multiplicative noise and at other channels of the generalized detector, $n = m \neq k$, in which a "no" signal component at the output of given channels of the generalized detector when the multiplicative noise is absent. For this reason, there is a need to define the first term $z'_n(t)$ (see Eq. (7.20)) both for the condition $n = k$ and for the condition $n = m \neq k$.

We proceed to consider Eq. (7.20). The model signal at the "n"-th channel of the generalized detector, in accordance with Eq. (7.2), takes the following form:

$$a_{m_n}(t) = Re\left\{\dot{S}_{m_n}(t) \cdot e^{j\omega_n t}\right\}. \tag{7.21}$$

Assume that the "n"-th and "k"-th channels—the resolution elements—of the generalized detector differ in time instant of the incoming signal by the value τ_{nk} and differ in frequency by the value Ω_{nk}. Then the model signal of the "n"-th channel of the generalized detector, using the model signal of the "k"-th channel of the generalized detector, can be written in the following form:

$$a_{m_n}(t) = a_{m_k}(t - \tau_{nk}, \Omega_{nk}) = Re\left\{\dot{S}_{m_k}(t - \tau_{nk}) \cdot e^{j(\omega_k - \Omega_{nk})(t - \tau_{nk})}\right\}. \tag{7.22}$$

In terms of Eqs. (7.15) and (7.22), and taking into account the main condition for the functioning generalized detector, which is mentioned in Chapter 5 and discussed in additional detail in References 1–4, 44, 55, and 64, Eq. (7.20) can be rewritten in the following form:

$$z'_n(\tau_{nk}, \Omega_{nk}) = \frac{E_{a_1}}{2} \cdot Re\left\{e^{j\omega_n\tau_{nk}} \int_{-\infty}^{\infty} \dot{M}(t)\dot{S}_{m_k}(t)S^*_{m_k}(t-\tau_{nk}) \cdot e^{j\Omega_{nk}t}\right\}. \quad (7.23)$$

It is not difficult to see that Eq. (7.23) is analogous to Eq. (6.84) in defining the signal component at the output of the one-channel generalized detector generated by the signal distorted by the multiplicative noise. The value $z'_n(\tau_{nk}, \Omega_{nk})$ is defined by the signal component at the output of the one-channel generalized detector, which is generated by the signal $a_M(t)$ at the definite time instant, with an accuracy of constant factors. Because of this, for definition of the statistical characteristics of the first term $z'_n(\tau_{nk}, \Omega_{nk})$ we can use the results obtained in Chapter 6.

7.1.2.1 Periodic Multiplicative Noise

By analogy with Eqs. (6.84) and (6.87), Eq. (7.23) can be written in the following form:

$$\begin{aligned} z'_n(\tau_{nk}, \Omega_{nk}) &= E_{a_1} Re\left\{e^{j\omega_n\tau_{nk}} \sum_{s=-\infty}^{\infty} \alpha_s \cdot e^{j\beta_s} \dot{\rho}(\tau_{nk}, \Omega_{nk} + s\Omega_M)\right\} \\ &= E_{a_1} r_{nk}, \end{aligned} \quad (7.24)$$

where

$$r_{nk} = \frac{z'_n(\tau_{nk}, \Omega_{nk})}{E_{a_1}}. \quad (7.25)$$

Other designations are the same as in Eq. (6.87), in particular

$$\beta_s = \arg \dot{C}_s - s\vartheta. \quad (7.26)$$

All conclusions made in Section 6.2 for the signal at the output of the one-channel generalized detector, in the case of the periodic multiplicative noise, are true for the first term $z'_n(\tau_{nk}, \Omega_{nk})$ determined by Eq. (7.24).

For the case in which the initial phase ϑ_0 of the noise modulation function $\dot{M}(t)$ of the multiplicative noise is the deterministic variable, then the first term $z'_n(\tau_{nk}, \Omega_{nk})$ determined by Eq. (7.24) is the deterministic variable as well. In doing so, the resulting signal $z^{out}_{g_M}(t)$ at the output of the "n"-th channel of the generalized detector (see Eq. (7.17)) obeys the probability distribution density of the background noise of the generalized detector with the mean

$$\overline{z^{out}_{g_M}(t)} = z'_n(\tau_{nk}, \Omega_{nk}) \quad (7.27)$$

that can be determined by Eq. (7.24) and the variance determined by Eq. (7.13).

For the case when there is the signal component at the output of the "n"-th channel of the generalized detector if the multiplicative noise is absent, $n = k$, the mean of the resulting signal $z_{g_M}^{out}(t)$ at the output of the "n"-th channel of the generalized detector, when the multiplicative noise is present, is determined in the following form:

$$\overline{z_{g_M}^{out}(t)} = z_k'(0,0) = E_{a_1} Re \sum_{s=-\infty}^{\infty} \alpha_s^2 \cdot e^{j(\arg \dot{C}_s - s\vartheta)} \dot{\rho}(0, s\Omega_M). \tag{7.28}$$

If the parameter ϑ is the random variable the conditional probability distribution function of the resulting signal at the output, for example, of the "n"-th channel of the generalized detector is Gaussian at the fixed value of the parameter ϑ.

The probability distribution function of the resulting signal at the output of the "n"-th channel of the generalized detector can be defined by averaging the conditional probability distribution function with respect to the parameter ϑ.

7.1.2.2 Fluctuating Multiplicative Noise

In the case of the fluctuating multiplicative noise, in terms of Eqs. (3.83) and (4.85), Eq. (7.23) gives us the following result:

$$\begin{aligned} z_n'(\tau_{nk}, \Omega_{nk}) = {} & \frac{\alpha_0^2 E_{a_1}}{2} \cdot Re\left\{ e^{j(\omega_n \tau_{nk} + \beta_0)} \int_{-\infty}^{\infty} \dot{S}_{m_k}(t) S_{m_k}^*(t - \tau_{nk}) \cdot e^{j\Omega_{nk}t}\, dt \right\} \\ & + \frac{E_{a_1}}{2} \cdot Re\left\{ e^{j\omega_n \tau_{nk}} \int_{-\infty}^{\infty} \dot{V}_0(t) \dot{S}_{m_k}(t) S_{m_k}^*(t - \tau_{nk}) \cdot e^{j\Omega_{nk}t}\, dt \right\}. \end{aligned} \tag{7.29}$$

The mean of the first term $z_n'(\tau_{nk}, \Omega_{nk})$ at the output of the "n"-th channel of the generalized detector, taking into account the peculiarities of orthogonality of the model signals $a_{m_n}(t)$, is determined in the following form:

$$\overline{z_n'(\tau_{nk}, \Omega_{nk})} = \begin{cases} \alpha_0^2 E_{a_1} \cos \beta_0, & n = k; \\ 0, & n = m \neq k. \end{cases} \tag{7.30}$$

In order to define the variance of the first term $z_n'(\tau_{nk}, \Omega_{nk})$, determined by Eq. (7.29), we can use the formulae mentioned in Section 6.2 defining the variance of the signal noise component $v(t)$ at the output of the one-channel generalized detector. Since the second term in Eq. (7.29) is analogous to the signal noise component $v(t)$ in Eq. (6.105), it is not difficult to obtain on the basis of Eq. (6.107) that

$$D[z_n'(\tau_{nk}, \Omega_{nk})] = E_{a_1}^2 \delta_{n_1}^2(\tau_{nk}, \Omega_{nk}) + E_{a_1}^2 Re\{\dot{\delta}_{n_2}^2(\tau_{nk}, \Omega_{nk}) \cdot e^{2j\omega_n \tau_{nk}}\}, \tag{7.31}$$

where

$$\delta^2_{n_1}(\tau_{nk}, \Omega_{nk}) = \frac{1}{4\pi} \int\limits_{-\infty}^{\infty} G_V(\omega)|\dot{\rho}(\tau_{nk}, \Omega_{nk} + \omega)|^2 \, d\omega; \tag{7.32}$$

$$\dot{\delta}^2_{n_2}(\tau_{nk}, \Omega_{nk}) = \frac{1}{4\pi} \int\limits_{-\infty}^{\infty} \dot{G}_D(\omega)\dot{\rho}(\tau_{nk}, \Omega_{nk} + \omega)\dot{\rho}(\tau_{nk}, \Omega_{nk} - \omega) \, d\omega. \tag{7.33}$$

The functions $\delta^2_{n_1}(\tau_{nk}, \Omega_{nk})$ and $\dot{\delta}^2_{n_2}(\tau_{nk}, \Omega_{nk})$ are completely defined by a shape of the signal and the noise modulation function $\dot{M}(t)$ of the multiplicative noise. For the definite conditions that will be discussed below, the second term in Eq. (7.31) is low in value in comparison to the first term in Eq. (7.31). Thus, we can neglect the second term without any losses in a generality.

We can use the formulae obtained in Sections 6.2–6.4 to define the function $\dot{\delta}^2_{n_2}(\tau_{nk}, \Omega_{nk})$ and use $\sigma^2_s(\tau, \Omega)$ for a definition of the variance for the case when an averaging with respect to the initial phase φ_0 of the signal has been carried out.

To define the function $\delta^2_{n_1}(\tau_{nk}, \Omega_{nk})$ we can suppose that

$$C^2 E^2_{a_1} = 1 \tag{7.34}$$

for these formulae.

Since the signal noise component $v(t)$ at the output of the one-channel generalized detector and the additive Gaussian noise $n(t)$ are statistically independent, the first term $z'_n(\tau_{nk}, \Omega_{nk})$ and the third term $z'''(\tau_{nk}, \Omega_{nk})$ in Eq. (7.17) are statistically independent as well.

The second term $z''(\tau_{nk}, \Omega_{nk})$ is completely compensated and tends to approach zero in the statistical sense. For this reason, this term is not taken into consideration.

Then the variance of the resulting signal at the output of the "n"-th channel of the generalized detector is determined in the following form:

$$\begin{aligned}
\sigma^2_n(\tau_{nk}, \Omega_{nk}) &= D[z'_n(\tau_{nk}, \Omega_{nk})] + \frac{N_0^2\omega_0^4}{16(\Delta\omega_{PF})^2 T\beta} \\
&= E^2_{a_1}\delta^2_{n_1}(\tau_{nk}, \Omega_{nk}) + E^2_{a_1} Re\{\dot{\delta}^2_{n_2}(\tau_{n_k}, \Omega_{nk}) \cdot e^{2j\omega_n\tau_{nk}}\} \\
&\quad + \frac{N_0^2\omega_0^4}{16(\Delta\omega_{PF})^2 T\beta}.
\end{aligned} \tag{7.35}$$

Since the first term $z'_n(\tau_{nk}, \Omega_{nk})$ and the process at the output of the one-channel generalized detector are analogous, all relationships in Section 6.5 are true for the probability distribution function of the first term $z'_n(\tau_{nk}, \Omega_{nk})$. The probability distribution function of the resulting signal at the output of the "n"-th channel of the generalized detector (see Eq. (7.17)) can be defined as a convolution of the probability distribution functions of the first $z'_n(\tau_{nk}, \Omega_{nk})$

and the third $z'''(\tau_{nk}, \Omega_{nk})$ terms in Eq. (7.17). The second term in Eq. (7.17) is completely compensated and tends to approach zero in the statistical sense. Because of this, the second term in Eq. (7.17) is not taken into consideration.

In a general case, the probability distribution function of the first term $z'_n(\tau_{nk}, \Omega_{nk})$ is very complex. For this reason, a definition of the probability distribution function of the resulting signal $z^{out}_{g_{M_n}}(t)$ at the output of the "n"-th channel of the generalized detector is a very difficult process.

However, in some practical cases, we can assume that the probability distribution function of the resulting signal $z^{out}_{g_{M_n}}(t)$ at the output of the "n"-th channel of the generalized detector is very close to the probability distribution function of the background noise of the generalized detector. This occurs during the following conditions:

- For any probability distribution function of the first term $z'_n(\tau_{nk}, \Omega_{nk})$ when the variance $D[z'_n(\tau_{nk}, \Omega_{nk})]$ is much less than the variance of the additive Gaussian noise σ_n^2 at the input of the PF of the linear tract of the generalized detector.
- For any relationships between the variances $D[z'_n(\tau_{nk}, \Omega_{nk})]$ and σ_n^2 when the probability distribution function of the first term $z'_n(\tau_{nk}, \Omega_{nk})$ is close to the Gaussian probability distribution function.

The last condition is true when the probability distribution function of instantaneous variables of the signal $a_M(t)$, at the input of the PF of the linear tract of the generalized detector, is close to the Gaussian probability distribution function or for an arbitrary probability distribution function of the signal $a_M(t)$ at the input of the PF of the linear tract of the generalized detector when the time of correlation of the noise modulation function $\dot{M}(t)$ of the multiplicative noise is much less than a duration of the signal. The latest statement follows from the fact that the first term $z'_n(\tau_{nk}, \Omega_{nk})$ is defined by the process at the output of the one-channel generalized detector for the definite time instant with an accuracy of a constant factor. The conditions of normalization of this process were discussed in Section 6.5.

For the given case, the condition of convergence to the Gaussian probability distribution function with the oscillating moment of the second order is sufficient. These conditions are not rigorous. The extent, to which the probability distribution function of the first term $z'_n(\tau_{nk}, \Omega_{nk})$ is close to the Gaussian probability distribution function, which is necessary that the probability distribution function of the resulting signal $z^{out}_{g_{M_n}}(t)$ at the output of the "n"-th channel of the generalized detector would be close to the Gaussian probability distribution function, depends on the relationship between the variance

$$D[z'_n(\tau_{nk}, \Omega_{nk})]$$

and the variance of the background noise of the generalized detector (see Eq. (7.13)). When the relative value of the variance

$$D[z'_n(\tau_{nk}, \Omega_{nk})]$$

is less, the requirements for the probability distribution function of the variable $z'_n(\tau_{nk}, \Omega_{nk})$ are more rigorous.

In the case when one of the conditions mentioned above is satisfied, the probability distribution function of the resulting signal $z^{out}_{g_{M_n}}(t)$ at the output of the "n"-th channel of the generalized detector can be approximated by the Gaussian probability distribution function with the mean determined by Eq. (7.30) and the variance determined by Eq. (7.35).

7.1.3 Signal with Unknown Initial Phase

When the multiplicative noise is absent the resulting signal $z^{out}_{g_{M_k}}(t)$, at the output of the "k"-th channel of the generalized detector, obeys the exponential probability distribution density (see Eq. (5.181)) for the case in which a duration of the signal is infinitesimal. When a duration of the signal is very large the resulting signal $z^{out}_{g_{M_k}}(t)$, at the output of the "k"-th channel of the generalized detector, obeys the asymptotic Gaussian probability distribution density (see Eq. (5.164)).

The mean of the resulting signal $z^{out}_{g_{M_k}}(t)$, at the output of the "k"-th channel of the generalized detector, is equal to E_{a_1}. There is a need to recall that we previously supposed that the energy of the model signal had been taken equal to energy of the incoming signal. For other channels the mean of the signal at the output of the generalized detector is equal to zero.

The variance at the outputs of all channels of the generalized detector is defined by the variance of the background noise of the generalized detector.

Thus, the probability distribution density of the resulting signal $z^{out}_{g_{M_k}}(t)$, at the output of the "k"-th channel of the generalized detector, is determined in the following form:

- The exponential probability distribution density

$$f^{H_1}_{\xi_2^2-\xi_1^2}\left[\left|z^{out}_{g_{M_k}}(t)\right|\right] = \frac{2(\Delta\omega_{PF})\sqrt{T\beta}}{N_0\omega_0^2} \times \exp\left\{-\frac{4(\Delta\omega_{PF})\sqrt{T\beta}}{N_0\omega_0^2}\left[\left|z^{out}_{g_{M_k}}(t)\right| - E_{a_1}\right]\right\} \tag{7.36}$$

  for the case in which a duration of the signal is infinitesimal.

- The asymptotic Gaussian probability distribution density

$$f^{H_1}_{\eta^2-\xi^2}\left[z^{out}_{g_{M_k}}(t)\right] = \frac{4(\Delta\omega_{PF})T\beta}{\sqrt{2\pi}\,N_0\omega_0^2} \times \exp\left\{-\frac{\left[z^{out}_{g_{M_k}}(t) - E_{a_1}\right]^2 8(\Delta\omega_{PF})^2T^2\beta^2}{N_0^2\omega_0^4}\right\} \tag{7.37}$$

  for the case in which a duration of the signal is very large.

Here $\xi_1(t)$ is the complex envelope of amplitude of the additive Gaussian noise $\xi(t)$ formed at the output of the PF at the input linear tract of the generalized detector; $\xi_2(t)$ is the complex envelope of amplitude of the additive Gaussian noise $\eta(t)$ formed at the output of the AF at the input linear tract of the generalized detector; and $T\beta$ is the signal base. Other designations are the same as before.

For all other channels the probability distribution density at the output of the generalized detector is determined in the following form:

- The exponential probability distribution density

$$f^{H_0}_{\xi_2^2-\xi_1^2}\left[\left|z^{out}_{g_{M_k}}(t)\right|\right] = \frac{2(\Delta\omega_{PF})\sqrt{T\beta}}{N_0\omega_0^2} \times \exp\left\{-\frac{4(\Delta\omega_{PF})\sqrt{T\beta}}{N_0\omega_0^2}\left|z^{out}_{g_{M_k}}(t)\right|\right\} \tag{7.38}$$

  for the case in which a duration of the signal is infinitesimal.
- The asymptotic Gaussian probability distribution density

$$f^{H_0}_{\eta^2-\xi^2}\left[\left|z^{out}_{g_{M_k}}(t)\right|\right] = \frac{4(\Delta\omega_{PF})T\beta}{\sqrt{2\pi}\,N_0\omega_0^2} \times \exp\left\{-\frac{8(\Delta\omega_{PF})^2T^2\beta^2}{N_0^2\omega_0^4}\left[z^{out}_{g_{M_k}}(t)\right]^2\right\} \tag{7.39}$$

  for the case in which a duration of the signal is very large.

When the multiplicative noise is present the signal at the output of the PF of the linear tract of the generalized detector is determined in the following form:

$$a_{M_{in}}(t) = a_{M_k}(t) + \xi(t) = Re\left\{\dot{S}_{M_{in}}(t)\cdot e^{j\omega_k t}\right\}, \tag{7.40}$$

where $\dot{S}_{M_{in}}(t)$ is the complex envelope of amplitude of the signal at the output of the PF of the linear tract of the generalized detector.

Taking into account Eqs. (4.2) and (7.7), the complex envelope of amplitude of the signal at the output of the PF of the input linear tract of the generalized detector can be written in the following form:

$$\dot{S}_{M_{in}}(t) = \sqrt{E_{a_1}}\dot{M}(t)\dot{S}_{m_k}(t)\cdot e^{j\varphi_0} + \dot{N}(t), \tag{7.41}$$

where $\dot{N}(t)$ is the complex envelope of amplitude of the additive Gaussian noise $\xi(t)$ at the output of the PF of the linear tract of the generalized detector;[5,6] and E_{a_1} is the energy of the signal at the output of the PF of the linear tract of the generalized detector.

The resulting signal $z_{g_M}^{out}(t)$ at the output of the generalized detector for the case in which the multiplicative noise is present is determined in terms of Eqs. (7.8), (7.22), and (7.41) in the following form:

$$\dot{z}_{g_M}^{out}(t) = \frac{E_{a_1}}{2} \cdot e^{j\varphi_0} \int_{-\infty}^{\infty} \dot{M}(t)\dot{S}_{m_k}(t)S_{m_k}^*(t - \tau_{nk}) \cdot e^{j\Omega_{nk}t}\, dt + \frac{1}{2} \int_{-\infty}^{\infty} \left[\xi_2^2(t) - \xi_1^2(t)\right] dt. \tag{7.42}$$

7.1.3.1 Periodic Multiplicative Noise

Using Eqs. (4.7) and (7.35), the complex envelope of amplitude of the signal at the output of the PF of the linear tract of the generalized detector can be written in the following form:

$$\dot{S}_{M_{in}}(t) = \sqrt{E_{a_1}}\,\dot{S}_{m_k}(t) \cdot e^{j\varphi_0} \sum_{s=-\infty}^{\infty} \alpha_s \cdot e^{j(\beta_s + s\Omega_M)} + \dot{N}(t), \tag{7.43}$$

where

$$\alpha_s = |\dot{C}_s| \tag{7.44}$$

and

$$\beta_s = \arg \dot{C}_s - s\vartheta; \tag{7.45}$$

$\dot{C}_s$ are the coefficients of the Fourier series expansion of the noise modulation function $\dot{M}(t)$ of the multiplicative noise; ϑ is the initial phase of the noise modulation function $\dot{M}(t)$ of the multiplicative noise with respect to the complex envelope $\dot{S}_{m_k}(t)$ of amplitude of the signal.

Substituting Eq. (7.43) in Eq. (7.42), it is not difficult to obtain that the mean of the resulting signal $\dot{z}_{g_M}^{out}(t)$ at the fixed value of the initial phase ϑ takes the following form:

$$\overline{\dot{z}_{g_M}^{out}(t)} = E_{a_1} \cdot e^{j\varphi_0}\dot{R}_{nk}, \tag{7.46}$$

where

$$\dot{R}_{nk} = \sum_{s=-\infty}^{\infty} \alpha_s^2 \cdot e^{j\beta_s}\dot{\rho}(\tau_{nk}, \Omega_{nk} + s\Omega_M). \tag{7.47}$$

There is also a need to note that the signal at the output of the "k"-th channel of the generalized detector generates the output effect in other channels of the generalized detector. The variance at the output of the "k"-th channel of the generalized detector, as for the case in which the multiplicative noise is absent, is determined by Eq. (7.13) and is the same for all channels of the generalized detector.

Thus, we are able to obtain the conditional probability distribution density of the signal at the output of individual channels of the generalized detector during the fixed value of the initial phase ϑ of the noise modulation function $\dot{M}(t)$ of the multiplicative noise when the multiplicative noise is present and periodic:

$$f_n^{H_1}\left[\dot{z}_{g_M}^{out}(t)\right] = \frac{4(\Delta\omega_{PF})T\beta}{\sqrt{2\pi}\,\mathrm{N}_0\omega_0^2} \times \exp\left\{-\frac{\left\{\left[z_{g_M}^{out}(t)\right]^2 + E_{a_1}^2|\dot{R}_{nk}|^2\right\}8(\Delta\omega_{PF})^2T^2\beta^2}{\mathrm{N}_0^2\omega_0^4}\right\}. \tag{7.48}$$

If the initial phase ϑ of the noise modulation function $\dot{M}(t)$ of the multiplicative noise is the deterministic variable then the probability distribution density of the signal at the output of the generalized detector is equal to the conditional probability distribution density and is determined by Eq. (7.48):

$$f_n^{H_1}\left[z_{g_M}^{out}(t)\right] = f_n^{H_1}\left[z_{g_M}^{out}(t)/\vartheta\right]. \tag{7.49}$$

When the initial phase ϑ, of the noise modulation function $\dot{M}(t)$ of the multiplicative noise, is the random variable the probability distribution density of the signal at the output of the generalized detector is defined by averaging Eq. (7.48) with respect to the random initial phase ϑ. The exception to this rule is the case when the frequency of the multiplicative noise satisfies the condition

$$\Omega_M > \frac{4\pi}{T}. \tag{7.50}$$

In the process, the main peaks of the individual terms in Eq. (7.47) are not overlapped and all terms of the sum, except a single term, are approximately equal to zero. In doing so, there is the only term in Eq. (7.47) characterizing the "k"-th channel of the generalized detector and corresponding to the undistorted part of the signal

$$|\dot{R}_k| = \alpha_0. \tag{7.51}$$

Also, there is the term in Eq. (7.47) characterizing the "m"-th channel of the generalized detector corresponding to the condition

$$s\Omega_M \simeq \Omega_{mk}. \tag{7.52}$$

Because of this, Eqs. (7.48) and (7.49) are true for the condition

$$\Omega_M > \frac{4\pi}{T}. \tag{7.53}$$

7.1.3.2 Fluctuating Multiplicative Noise

Using Eqs. (3.83) and (7.41), the complex envelope of amplitude of the signal at the output of the PF of the linear tract of the generalized detector for the case of the fluctuating multiplicative noise can be written in the following form:

$$\dot{S}_{M_{in}}(t) = \alpha_0\sqrt{E_{a_1}}\,\dot{S}_{m_k}(t)\cdot e^{j(\varphi_0+\beta_0)} + \sqrt{E_{a_1}}\,\dot{V}_0(t)\dot{S}_{m_k}(t)\cdot e^{j\varphi_0} + \dot{N}(t). \quad (7.54)$$

The mean of the resulting signal $z_{g_{M_n}}^{out}(t)$ at the output of the "n"-th channel of the generalized detector is determined by the following form:

$$M\big[z_{g_{M_n}}^{out}(t)\big] = \alpha_0^2 E_{a_1} Re\big\{\dot{\rho}(\tau_{nk}, \Omega_{nk})\cdot e^{j(\varphi_0+\beta_0)}\big\}. \quad (7.55)$$

In the case of the "k"-th channel of the generalized detector for the condition determined by

$$\tau_{nk} = \Omega_{nk} = 0$$

we can write

$$M\big[z_{g_M}^{out}(t)\big] = \alpha_0^2 E_{a_1}. \quad (7.56)$$

For all other channels of the generalized detector the mean of the signal at the output is equal to zero.

We define the variance of the signal at the output of the "n"-th channel of the generalized detector. Taking into account Eqs. (7.54) and (7.55), we can write

$$D\big[z_{g_M}^{out}(t)\big] = m_2\left\{E_{a_1} Re\left[e^{j\varphi_0}\int\limits_{-\infty}^{\infty}\dot{V}_0(t)\dot{S}_{m_k}(t)S_{m_k}^*(t-\tau_{nk})\cdot e^{j\Omega_{nk}t}dt\right]\right\} + \frac{N_0^2\omega_0^4}{16(\Delta\omega_{PF})^2T\beta}. \quad (7.57)$$

Carrying out mathematical transformations in Eq. (7.57) that are analogous to the mathematical transformations made in Sections 6.1 and 6.2 while obtaining the formulae for the variance $\sigma_s^2(\tau, \Omega)$ of the signal noise component at the output of the one-channel generalized detector, and taking into account the following relationships

$$\begin{cases} Im\,\dot{a}\;Im\,\dot{b} = 0.5\cdot[Re\,\dot{a}b^* - Re\,\dot{a}\dot{b}];\\ Re\,\dot{a}\;Im\,\dot{b} = 0.5\cdot[Im\,\dot{a}\dot{b} - Im\,\dot{a}b^*], \end{cases} \quad (7.58)$$

we obtain the variance of the signal at the output of the n-th channel of the generalized detector in the following form:

$$D\big[z_{g_M}^{out}(t)\big] = E_{a_1}^2\delta_{n_1}^2(\tau_{nk}, \Omega_{nk}) + \big|E_{a_1}^2 Re\big\{\dot{\delta}_{n_2}^2(\tau_{nk}, \Omega_{nk})\cdot e^{2j\varphi_0}\big\}\big| + \frac{N_0^2\omega_0^4}{16(\Delta\omega_{PF})^2T\beta}, \quad (7.59)$$

where the functions $\delta_{n_1}^2(\tau_{nk}, \Omega_{nk})$ and $\dot{\delta}_{n_2}^2(\tau_{nk}, \Omega_{nk})$ are determined by Eqs. (7.32) and (7.33), respectively.

We proceed to consider the probability distribution function of the resulting signal $z_{g_M}^{out}(t)$ at the output of the generalized detector. Comparing Eqs. (6.85) and (7.42), it is not difficult to see that the first term of the resulting signal $z_{g_M}^{out}(t)$, at the output of the generalized detector, in Eq. (7.42) can be presented as the process generated by the signal distorted by the multiplicative noise at the output of the generalized detector at some instants with an accuracy of constant factors. This representation allows us to estimate the probability distribution function of this term of the resulting signal $z_{g_M}^{out}(t)$ in Eq. (7.42) using the main results discussed in Section 6.5.

In accordance with this fact, in a general case, the probability distribution function of the resulting signal $z_{g_M}^{out}(t)$ at the output of the generalized detector determined by Eq. (7.42) can be very complex. However, in many cases that are essential in practice, the probability distribution function of the resulting signal $z_{g_M}^{out}(t)$, at the output of the generalized detector (see Eq. (7.42)), can be approximated by the Gaussian probability distribution function.

The conditions, for which this approximation is true, are the same as the conditions formulated for the signal $z_{g_{M_n}}^{out}(t)$ at the output of the "n"-th channel of the generalized detector in Section 7.1.2.2.

We consider a function between the parameters of the multiplicative noise and the functions $\delta_{n_1}^2(\tau_{nk}, \Omega_{nk})$ and $\dot{\delta}_{n_2}^2(\tau_{nk}, \Omega_{nk})$. In Reference 69 it was shown that

$$\left|\dot{\delta}_{n_2}^2(\tau_{nk}, \Omega_{nk})\right| \leq k\left[\delta_{n_1}^2(\tau_{nk}, \Omega_{nk}) + q^{-1}\right], \tag{7.60}$$

where q is the parameter of detection of the generalized detector defined in References 1 and 2 as the ratio

$$q = \frac{E_{a_1}}{N_0}; \tag{7.61}$$

k is the coefficient of proportionality.

Reference to the condition determined by Eq. (7.60) shows that if the permissible value $|\dot{\delta}_{n_2}^2(\tau_{nk}, \Omega_{nk})|$ is high, then the energy characteristics of the additive Gaussian noise are high. During low energy characteristics of the additive Gaussian noise, or during high values of the detection parameter q of the generalized detector, the following condition

$$\left|\dot{\delta}_{n_2}^2(\tau_{nk}, \Omega_{nk}\right| \leq k\delta_{n_1}^2(\tau_{nk}, \Omega_{nk}) \tag{7.62}$$

must be satisfied.

It is easy to show that the condition

$$\left|\dot{\delta}_{n_2}^2(\tau_{nk}, \Omega_{nk})\right| \ll \delta_{n_1}^2(\tau_{nk}, \Omega_{nk}) \tag{7.63}$$

is equivalent to the conditions of convergence of the probability distribution density of the signal at the output of the one-channel generalized detector toward the Gaussian probability distribution density with the non-oscillating variance, which were discussed in Section 6.5.

Determining the values of the functions $\delta^2_{n_1}(\tau_{nk}, \Omega_{nk})$ and $\dot{\delta}^2_{n_2}(\tau_{nk}, \Omega_{nk})$ on the basis of Eqs. (7.33) and (7.34), we can verify how the conditions mentioned above can be satisfied in specific cases of application in practice.

It is also worthwhile to define the general conditions for which the inequalities mentioned above are true. For this purpose, consider the functions $\delta^2_{n_1}(\tau_{nk}, \Omega_{nk})$ and $\dot{\delta}^2_{n_2}(\tau_{nk}, \Omega_{nk})$.

In accordance with Eqs. (7.32) and (7.33) we can write

$$\delta^2_{n_1}(\tau_{nk}, \Omega_{nk}) = \frac{1}{4\pi}\int\limits_{-\infty}^{\infty} G_V(\omega)|\dot{\rho}(\tau_{nk}, \Omega_{nk} + \omega)|^2\, d\omega; \tag{7.64}$$

$$\dot{\delta}^2_{n_2}(\tau_{nk}, \Omega_{nk}) = \frac{1}{4\pi}\int\limits_{-\infty}^{\infty} \dot{G}_D(\omega)\dot{\rho}(\tau_{nk}, \Omega_{nk} + \omega)\dot{\rho}(\tau_{nk}, \Omega_{nk} - \omega)\, d\omega, \tag{7.65}$$

where $G_V(\omega)$ is the Fourier transform of the correlation function $\dot{R}_V(\tau)$ and $G_D(\omega)$ is the Fourier transform of the correlation function $\dot{D}_V(\tau)$.

The correlation function $\dot{R}_V(\tau)$ was defined in Section 3.4. The correlation function $\dot{D}_V(\tau)$ can be defined using the characteristic functions of distortions in amplitude and phase of the signal on the basis of Eqs. (3.63), (4.108), and (4.116) by an analogous method as the correlation function $\dot{R}_V(\tau)$ in Section 3.4.

When there are distortions in phase of the signal we can write

$$\dot{D}_V(\tau) = \Theta_2^{\varphi}(1, 1) - \left[\Theta_1^{\varphi}(1)\right]^2. \tag{7.66}$$

When distortions in amplitude and phase of the signal are independent of each other we can write

$$\dot{D}_V(\tau) = A_0^2(t)\left\{\left[1 + \sigma_\xi^2 r_\xi(\tau)\right]\Theta_2^{\varphi}(1, 1) - \left[\Theta_1^{\varphi}(1)\right]^2\right\}, \tag{7.67}$$

where σ_ξ^2 is the variance of distortions in amplitude of the signal.

When distortions in amplitude and phase of the signal are correlated with each other we can write

$$\dot{D}_V(\tau) = -A_0^2(t)\left\{\frac{\partial^2}{\partial x_1 \partial x_2}\left[\Theta_4^{\xi\varphi}(x_1, x_2, 1, 1)\cdot e^{j(x_1+x_2)}\right]\right\}_{x_1=0, x_2=0} - A_0^2(t)\left\{\Theta_1^{\varphi}(1) - j\left[\frac{\partial}{\partial x_1}\Theta_2^{\xi\varphi}(x_1, 0)\right]_{x_1=0}\right\}^2. \tag{7.68}$$

As was discussed in Section 4.2, when there are distortions only in phase of the signal, or when there are independent distortions in amplitude and phase of the signal, the condition

$$\dot{D}_V(\tau) \equiv 0 \tag{7.69}$$

is satisfied and, consequently, the condition

$$\dot{\delta}^2_{n_2}(\tau_{nk}, \Omega_{nk}) \equiv 0 \tag{7.70}$$

is satisfied, too, if and only if the phase of the signal is uniformly distributed within the limits of the interval $[0, 2\pi]$.

It is not difficult to verify that for the continuous probability distribution functions of distortions in phase of the signal, which are offered in practice, the probability distribution function normalized within the limits of the interval $[0, 2\pi]$ is close to the uniform probability distribution function, the variance σ^2_φ of distortions in phase of the signal is high.

With an increase in the variance σ^2_φ, of distortions in phase of the signal, the symmetric probability distribution functions are converged most rapidly to the uniform probability distribution function within the limits of the interval $[0, 2\pi]$.

For example, when the probability distribution function of distortions in phase of the signal is Gaussian the probability distribution function normalized within the limits of the interval $[0, 2\pi]$ deviates from the uniform probability distribution function by nothing more than 10^{-15} under the condition in[70]

$$\sigma_\varphi > 2\pi. \tag{7.71}$$

Consequently, as the degree of distortions in phase of the signal is increased, the value $|\dot{D}_V(\tau)|$ is decreased and for this reason, the values $|\dot{G}_V(\omega)|$ and $|\dot{\delta}^2_{n_2}(\tau_{nk}, \Omega_{nk})|$ are decreased.

The value of the correlation function $\dot{R}_V(\tau)$ is increased as the degree of distortions in phase of the signal is increased, since the maximum $\dot{R}_V(0)$ of the correlation function $\dot{R}_V(\tau)$ is equal to the power, or energy characteristics, of fluctuations of the noise modulation function $\dot{M}(t)$ of the multiplicative noise.

Because of this, the values of $G_V(\omega)$ and $\delta^2_{n_1}(\tau_{nk}, \Omega_{nk})$ are increased as the degree of distortions in phase of the signal is increased. Thus, the relationship between the values $\delta^2_{n_1}(\tau_{nk}, \Omega_{nk})$ and $|\dot{\delta}^2_{n_2}(\tau_{nk}, \Omega_{nk})|$, and, consequently, the use of Eqs. (7.48) and (7.59) depends on the degree of distortions in phase of the signal.

For the "k"-th channel of the generalized detector the condition

$$\left|\dot{\delta}^2_{k_2}(\tau_{nk}, \Omega_{nk})\right| \ll \delta^2_{k_1}(\tau_{nk}, \Omega_{nk}) \tag{7.72}$$

in terms of Eqs. (7.32) and (7.33) takes the following form:

$$\left| \int_{-\infty}^{\infty} \dot{G}_D(\omega)|\dot{\rho}(0, \omega)|^2 \, d\omega \right| \ll \int_{-\infty}^{\infty} G_V(\omega)|\dot{\rho}(0, \omega)|^2 \, d\omega. \tag{7.73}$$

Since the bandwidths of the energy spectrums $\dot{G}_D(\omega)$ and $G_V(\omega)$ have the same order, the given relationship is true for the condition

$$|\dot{D}_V(0)| \ll \dot{R}_V(0). \tag{7.74}$$

Next we consider the well-known case in which the distortions in amplitude and phase of the signal obey the Gaussian probability distribution density. We proceed to compare the magnitudes of the correlation functions $\dot{R}_V(0)$ and $|\dot{D}_V(0)|$.

The formulae defining the function $\dot{R}_V(0)$ for the given case were obtained in Section 3.4 and Appendix II. On the basis of Eqs. (7.66)–(7.68) we are able to obtain the following relationships for the value $|\dot{D}_V(0)|$:

- When there are only distortions in phase of the signal we can write

$$|\dot{D}_V(0)| = \left|e^{-\sigma_\varphi^2} - e^{-2\sigma_\varphi^2}\right|. \tag{7.75}$$

- When there are distortions in amplitude and phase of the signal, and these distortions are independent, we can write

$$|\dot{D}_V(0)| = A_0^2(t)\left|\left(1+\sigma_\xi^2\right)\cdot e^{-2\sigma_\varphi^2} - e^{-\sigma_\varphi^2}\right|. \tag{7.76}$$

- When there are distortions in amplitude and phase of the signal, and these distortions are correlated, we can write

$$|\dot{D}_V(0)| = A_0^2(t)\Big\{\Big[\left(1+\sigma_\xi^2 - 4r_{\xi\varphi}^2\sigma_\xi^2\sigma_\varphi^2\right)\cdot e^{-2\sigma_\varphi^2} - \left(1 - r_{\xi\varphi}^2\sigma_\xi^2\sigma_\varphi^2\right)\cdot e^{-\sigma_\varphi^2}\Big]^2 + 4r_{\xi\varphi}^2\sigma_\xi^2\sigma_\varphi^2\left(e^{-2\sigma_\varphi^2} - e^{-\sigma_\varphi^2}\right)^2\Big\}^{\frac{1}{2}}, \tag{7.77}$$

where σ_ξ^2 is the variance of distortions in amplitude of the signal; σ_φ^2 is the variance of distortions in phase of the signal; and $r_{\xi\varphi}$ is the coefficient of correlation between the distortions in amplitude and phase of the signal at the coinciding instants.

Dependence of the ratio

$$\frac{|\dot{D}_V(0)|}{\dot{R}_V(0)}$$

on the degree of distortions in phase of the signal determined by Eqs. (7.75)–(7.77) and by formulae in Section 3.4 is shown in Fig. 7.1.

Reference to Fig. 7.1 shows that with an increase in the variance of distortions in phase of the signal, the ratio

$$\frac{|\dot{D}_V(0)|}{\dot{R}_V(0)}$$

and, consequently, the ratio

$$\frac{\left|\dot{\delta}_{n_2}^2(\tau_{nk}, \Omega_{nk})\right|}{\delta_{n_1}^2(\tau_{nk}, \Omega_{nk})}$$

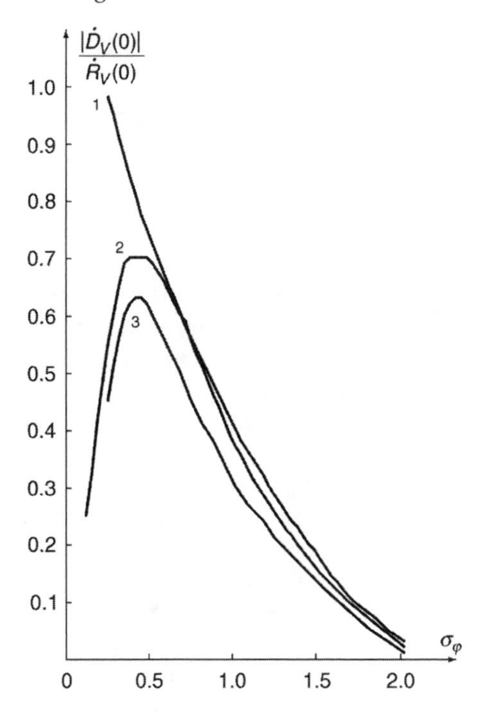

FIGURE 7.1
Ratio $\frac{|\dot{D}_V(0)|}{\dot{R}_V(0)}$ as a function of the degree of distortions in phase σ_φ of the signal: for 1, $\sigma_\xi = 0$; for 2, $\sigma_\xi = 0.15, r_{\xi\varphi} = 0$; for 3, $\sigma_\xi = 0.15, r_{\xi\varphi} = 1.0$.

are sharply decreased. This fact confirms the correctness of the statements mentioned above.

The presence of distortions in amplitude of the signal, which are statistically independent, for example,

$$\sigma_\varphi^2 = 0, \quad r_{\xi\varphi} = 0, \tag{7.78}$$

jointly with the distortions in phase of the signal and the statistical dependence between the distortions in amplitude and phase of the signal

$$\sigma_\xi^2 \neq 0, \quad r_{\xi\varphi} \neq 0, \tag{7.79}$$

are not an essential influence on the ratio

$$\frac{|\dot{D}_V(0)|}{\dot{R}_V(0)}.$$

There is also a need to note that when the condition

$$|\dot{D}_V(0)| \ll \dot{R}_V(0) \tag{7.80}$$

is true, we can neglect the second terms on the right side of Eqs. (4.119), (6.41), (6.42), (6.53), (6.56), and (7.31).

7.2 Detection Performances of the Generalized Detector

We define the probability of detection P_D, of the signal distorted by the multiplicative noise in the presence of additive Gaussian noise, on the basis of the model of the generalized detector discussed in Section 7.1.1.

In accordance with this model, the decision a "yes" signal in the stochastic process at the input of the PF of the linear tract of the generalized detector is taken when the signal at the output of the generalized detector exceeds the threshold K_g. The threshold K_g is chosen on the basis of the probability of false alarm P_F according to the Neyman–Pearson criterion. Because of this, the threshold K_g, when the multiplicative noise is present, is the same value as in the case when the multiplicative noise is absent.

As a rule, real complex signal processing systems with an automatic control of the threshold possess a time constant that is many times greater than a duration of the signal. For this reason, even for those cases in which the signal in the presence of additive Gaussian noise comes in at the input of the threshold control device, the presence of additional side lobe shoots of the signal at the output of the generalized detector (caused by the signal noise component) does not lead to essential changes in the threshold value with respect to the threshold defined by the energy characteristics of the additive Gaussian noise.

As was discussed in previous chapters, particularly in Chapter 6 and Section 7.1, an orthogonality of the signals is broken during the presence of multiplicative noise, and the signal distorted by the multiplicative noise can generate the signal at the output of the generalized detector not only at the channel, in which there is the signal when the multiplicative noise is absent, but in other channels of the generalized detector, in which there is a "no" signal when the multiplicative noise is absent.

In the process, detection of the signal distorted by the multiplicative noise is possible owing to exceeding the threshold by the undistorted component of the signal and the signal noise component at the channel of the generalized detector, which corresponds to true parameters of the undistorted signal—we will call this channel of the generalized detector the "true" channel—and owing to exceeding the threshold by the signal noise component of the signal distorted by the multiplicative noise in other channels of the generalized detector, which corresponds to other values of parameters of the signal.

For the last case the fact a "yes" signal is true, but parameters of the channel of the generalized detector, in which this signal can be detected, do not correspond to the true parameters of the undistorted component of the signal.

For this reason, we can recognize two forms of signal detection:

- Signal detection at the "true" channel of the generalized detector. The probability of detection P_{D_M} of the signal for this case is equal to the probability $P_{D_{M_k}}$ of exceeding the threshold at the "k"-th channel of the generalized detector by the signal at the output of the generalized detector—in other words, at the channel in which there is the signal when the multiplicative noise is absent—during the condition of a "yes" signal at the input of the PF of the linear tract of the generalized detector—the simple binary detection.

- A "yes" signal at the input of the PF of the linear tract of the generalized detector. The probability of detection $P_{D'_M}$ for this case is equal to the probability of exceeding the threshold by the signal at the output of a single channel of the generalized detector during the condition of a "yes" signal at the input of the PF of the linear tract of the generalized detector—the complex binary detection.

It seems likely that the first case—the signal detection at the "true" channel of the generalized detector—is of great interest, since exceeding the threshold in other channels of the generalized detector, in some specific cases, cannot give us useful information or instead gives us false information.

The probabilities of detection of the signals mentioned above can be determined in the following form:

$$P_{D_M} = P_{D_{M_k}}; \tag{7.78}$$

$$P_{D'_M} = 1 - (1 - P_{D_{M_k}}) \prod_{m=1}^{N} (1 - P_{D_{M_m}}), \quad m \neq k, \tag{7.79}$$

where $P_{D_{M_m}}$ is the probability of exceeding the threshold at the "m"-th channel of the generalized detector during the condition that the presence of the signal, at the input of the PF of the linear tract of the generalized detector, generates the signal at the output of the "k"-th channel of the generalized detector. Equation (7.79) is true for the case in which the signals at the outputs of individual channels of the generalized detector are statistically independent.

As was discussed in Section 7.1.2, the output signal generated by the signal $a_M(t)$ at each channel of the generalized detector can be considered as the amplitude of the signal at the output of the one-channel generalized detector or as the complex envelope of amplitude of the signal at the definite instant. Reference to a study of the correlation functions of amplitude of the signal at the output of the one-channel generalized detector, when the fluctuating multiplicative noise is present (see Chapter 6), shows that for the case in which the signal noise component overlaps with a set of channels of the generalized detector the signals at the output of individual channels of the generalized detector are not practically correlated.

We assume that the conditions formulated in Section 7.1.2.2 are satisfied and the signal $z^{out}_{g_{M_n}}(t)$, at the output of the "n"-th channel of the generalized detector, obeys the asymptotic Gaussian probability distribution density (see Eq. (5.164)). In the case of the Gaussian probability distribution density for the output signal $z^{out}_{g_{M_n}}(t)$ we can consider a non-correlatedness is equivalent to independence. Because of this, Eq. (7.79) is true for the conditions mentioned above.

7.2.1 Signal with Known Initial Phase

For this case the threshold K_g, owing to the Neyman–Pearson criterion, is defined by the probability of false alarm P_{F_k} for a single channel of the generalized detector:[68]

$$P_{F_k} = 1 - \Phi\left[\frac{4(\Delta\omega_{PF})\sqrt{T\beta}}{N_0\omega_0^2} \cdot K_g\right] = 1 - \Phi(K_{g_0}), \tag{7.80}$$

where

$$\Phi(x) = \frac{1}{\sqrt{2\pi}} \int\limits_{-\infty}^{x} e^{-\frac{t^2}{2}}\, dt \tag{7.81}$$

is the error integral and

$$K_{g_0} = \frac{4(\Delta\omega_{PF})\sqrt{T\beta}}{N_0\omega_0^2} \cdot K_g; \tag{7.82}$$

$\frac{N_0}{2}$ is the spectral power density of the additive Gaussian noise at the input of the PF of the linear tract of the generalized detector. Other designations are the same as in Section 7.1.

When the multiplicative noise is absent and

$$P_F \ll 1,$$

and the probability of detection of the signals is high in value, the probability of detection P_D at the "true" channel of the generalized detector and the probability of detection $P_{D'}$ are practically equal to each other:[71]

$$P_D \simeq P_{D'}.$$

In doing so,

$$\begin{aligned} P_D = P_{D_k} &= 1 - \Phi\left[\frac{(K_g - E_{a_1})4(\Delta\omega_{PF})\sqrt{T\beta}}{N_0\omega_0^2}\right] \\ &= 1 - \Phi(K_{g_0} - q), \end{aligned} \tag{7.83}$$

where

$$q = \frac{4(\Delta\omega_{PF})\sqrt{T\beta}}{N_0\omega_0^2} \cdot E_{a_1}; \tag{7.84}$$

E_{a_1} is the energy of the undistorted signal at the output of the PF of the linear tract of the generalized detector.

7.2.1.1 Periodic Multiplicative Noise

During periodic multiplicative noise the probability of detection P_{D_M}, of the signal at the "true" channel of the generalized detector, in terms of the results discussed in Section 7.1.2.1 can be determined in the following form:

$$P_{D_M} = P_{D_{M_k}} = \int_{K_g}^{\infty} f_k^{H_1}\left[z_{gM}^{out}(t)\right] d\left[z_{gM}^{out}(t)\right]$$
$$= 1 - \Phi\left\{\frac{4(\Delta\omega_{PF})\sqrt{T\beta}}{N_0\omega_0^2} \cdot \left[K_g - z'_{gM_k}(t)\right]\right\}$$
$$= 1 - \Phi\left(K_{g_0} - qr_k\right), \tag{7.85}$$

where $z'_{gM_k}(t)$ and r_k are determined by Eq. (7.24) for the conditions

$$n = k \qquad \text{and} \qquad \tau_{nk} = \Omega_{nk} = 0. \tag{7.86}$$

The probability of exceeding the threshold at the "m"-th channel of the generalized detector during the condition that there is the signal at the "k"-th channel of the generalized detector, when the multiplicative noise is absent, is determined in the following form:

$$P_{D_{M_m}} = \int_{K_g}^{\infty} f_m^{H_1}\left[z_{gM_m}^{out}(t)\right] d\left[z_{gM_m}^{out}(t)\right]$$
$$= 1 - \Phi\left\{\frac{4(\Delta\omega_{PF})\sqrt{T\beta}}{N_0\omega_0^2} \cdot \left[K_g - z'_{gM_{mk}}(t)\right]\right\}$$
$$= 1 - \Phi\left(K_{g_0} - qr_{mk}\right), \tag{7.87}$$

where $z'_{gM_{mk}}(t)$ and r_{mk} are determined by Eq. (7.24) for the condition

$$n = m.$$

The probability of a "yes" signal at the input of the PF of the linear tract of the generalized detector, in accordance with Eq. (7.79), is determined in the following form:

$$P_{D'_M} = 1 - \prod_{n=1}^{N} \Phi\left(K_{g_0} - qr_{nk}\right). \tag{7.88}$$

Equations (7.85) and (7.88) allow us to define the probability of detection of the signal when the periodic multiplicative noise exists by using the probability of detection and the probability of false alarm that are defined by the parameters K_{g_0} and q when the multiplicative noise is absent and using the parameters of the multiplicative noise and the signal—the parameter r_{nk}.

7.2.1.2 Fluctuating Multiplicative Noise

For the case of the fluctuating multiplicative noise when the conditions of normalization of the signal $z_{g_M}^{out}(t)$ at the output of the generalized detector are satisfied, the probability of detection of the signal at the "true" channel of the generalized detector, in terms of results discussed in Section 7.1.2.2, takes the following form:

$$P_{D_M} = P_{D_{M_k}} = \int_{K_g}^{\infty} f_k^{H_1}\left[z_{g_M}^{out}(t)\right] d\left[z_{g_M}^{out}(t)\right] = 1 - \Phi\left[\frac{K_g - \alpha_0^2 E_{a_1}}{\sigma_k^2}\right], \tag{7.89}$$

where the variance σ_k^2 is determined by Eq. (7.35) in a general case for the conditions

$$n = k, \quad \tau_{nk} = 0, \quad \Omega_{nk} = 0. \tag{7.90}$$

In some cases the conditions, for which

$$\delta_{k_1}^2 \gg \left|\dot{\delta}_{k_2}^2\right|, \tag{7.91}$$

are satisfied (see Section 7.1.3).

In doing so,

$$\sigma_k^2 = E_{a_1}^2 \delta_{k_1}^2 + \frac{N_0^2 \omega_0^4}{16(\Delta\omega_{PF})^2 T\beta}. \tag{7.92}$$

Thus, Eq. (7.89) can be presented in the following form:

$$P_{D_M} = P_{D_{M_k}} = 1 - \Phi\left(\frac{K_{g_0}}{\sqrt{1 + q^2\delta_{k_1}^2}} - \frac{\alpha_0^2 q}{\sqrt{1 + q^2\delta_{k_1}^2}}\right). \tag{7.93}$$

Equation (7.93) allows us to define the probability of detection of the signal at the "true" channel of the generalized detector (when there is fluctuating multiplicative noise) using the probability of detection of the signal and the probability of false alarm that are defined by the parameters K_{g_0} and q when the multiplicative noise is absent and using the parameters of the multiplicative noise and the signal, where α_0 is the relative level of the undistorted component of the signal and $\delta_{k_1}^2$ is proportional to the variance of the signal noise component at the "true" channel of the generalized detector.

If the time of correlation of the noise modulation function $\dot{M}(t)$ of the multiplicative noise is much less than the duration of the signal, then we can use the approximate formula in Eq. (6.220).

In the process, we obtain

$$\delta_{k_1} \simeq \frac{G_V(0)}{2T}. \tag{7.94}$$

We consider the magnitude of the parameter $\delta^2_{k_1}$ for this case. Reference to equalities

$$\dot{R}_V(0) = \overline{A^2(t)} - \alpha_0^2 \tag{7.95}$$

and

$$\dot{R}_V(0) = \frac{1}{2\pi}\int\limits_{-\infty}^{\infty} G_V(\Omega)\, d\Omega = G_V(0)\Delta F_M, \tag{7.96}$$

where ΔF_M is the bandwidth of energy spectrum of the noise modulation function $\dot{M}(t)$ of the multiplicative noise, shows that

$$G_V(0) = \frac{\overline{A^2(t)} - \alpha_0^2}{\Delta F_M}; \tag{7.97}$$

$$\delta_{k_1} = \frac{\overline{A^2(t)} - \alpha_0^2}{2\Delta F_M T} \simeq \left(\overline{A^2(t)} - \alpha_0^2\right) \cdot \frac{\Delta F_{en}}{2\Delta F_M}, \tag{7.98}$$

where

$$\Delta F_{en} \simeq \frac{1}{T} \tag{7.99}$$

is the bandwidth of the energy spectrum of the complex envelope of amplitude of the signal.

Thus, at the given magnitudes of the parameters $\overline{A^2(t)}$ and α_0^2, the parameter δ_{k_1} characterizes the ratio between the bandwidth of the energy spectrum of the noise modulation function $\dot{M}(t)$ of the multiplicative noise and the bandwidth of the energy spectrum of the complex envelope of amplitude of the signal.

The probability of exceeding the threshold at the "m"-th channel of the generalized detector for the condition of the signal at the "k"-th channel of the generalized detector, when the multiplicative noise is absent, is determined in the following form:

$$P_{D_{Mm}} = \int_{K_g}^{\infty} f_m^{H_1}\left[z_{g_{Mm}}^{out}(t)\right] d\left[z_{g_{Mm}}^{out}(t)\right] = 1 - \Phi\left(\frac{K_g}{\sigma_m^2}\right), \tag{7.100}$$

where σ_m^2 is determined by Eq. (7.35) for the condition

$$n = m.$$

If the conditions, in which

$$\delta_{m_1}^2 \gg \left|\dot{\delta}_{m_2}^2\right|, \tag{7.101}$$

are satisfied, Eq. (7.100) can be written in the following form:

$$P_{D_{Mm}} = 1 - \Phi\left(\frac{K_{g_0}}{\sqrt{1 + \delta_{m_1}^2 q^2}}\right), \tag{7.102}$$

where $\delta_{m_1}^2$ is determined by Eq. (7.33) for the condition

$$n = m$$

and by corresponding formulae in Chapter 6. Substitution of Eqs. (7.89) and (7.100) in Eq. (7.79) allows us to define the probability of detection $P_{D'_M}$.

The probability of detection of the signal at the "true" channel of the generalized detector for the case of the fluctuating multiplicative noise determined by Eq. (7.93) as a function of the parameters of the multiplicative noise at some values of the probability of detection P_D of the signal and the probability of false alarm P_F, when the multiplicative noise is absent, is shown in Figs. 7.2–7.5.

The probability of detection $P_{D'_M}$ of the signal for the case in which the spectral power density of the signal noise component is the same for all channels of the generalized detector is shown in Figs. 7.6 and 7.7.

The comparative analysis between the detection performances of the generalized detector and the correlation detector, or the matched filter, for the conditions mentioned above is also presented in Figs. 7.2–7.7.

Figures 7.2–7.7 show the superiority of the generalized detector over the correlation detector (or the matched filter) during use in complex signal processing systems.

7.2.2 Signal with Unknown Initial Phase

The use of the Neyman–Pearson criterion is peculiar to problems of signal detection, since *a priori* probabilities of a "yes" or a "no" signal in the stochastic process at the input of the PF of the linear tract of the generalized detector are unknown by virtue of the initial conditions.

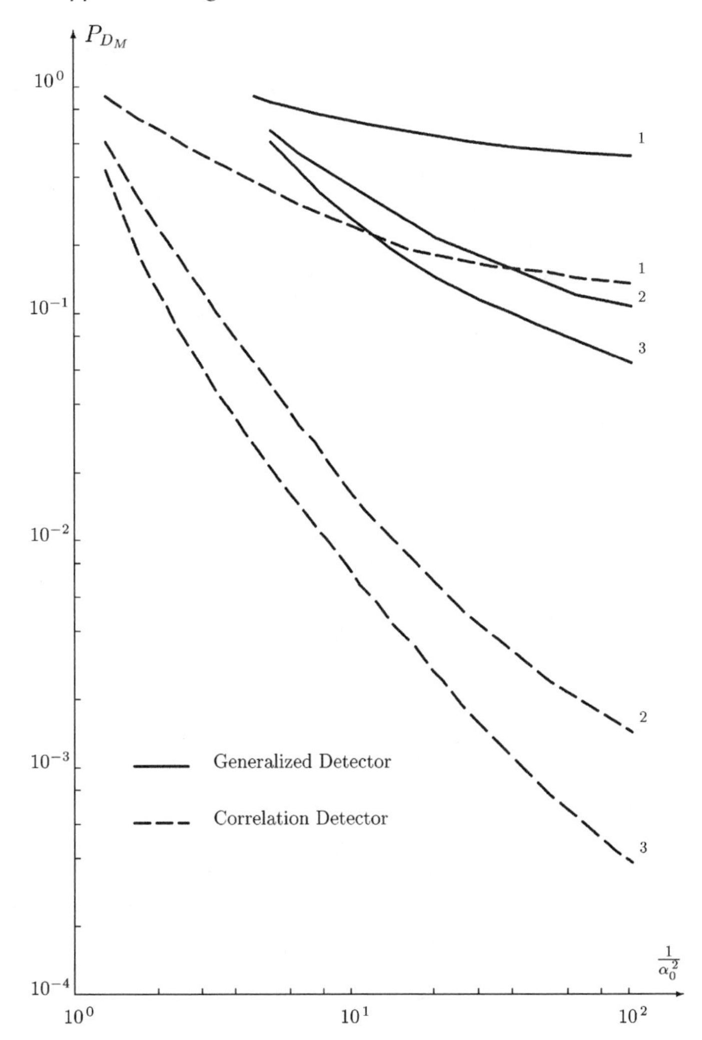

FIGURE 7.2
Detection performances of the signal with known initial phase by "true" channel of the detector: $P_{F_k} = 10^{-4}$; $P_D = 0.9$; for 1, $\delta^2_{k_1} = 0.1$; for 2, $\delta^2_{k_1} = 0.01$; for 3, $\delta^2_{k_1} = 0.001$.

According to the Neyman–Pearson criterion the threshold is given by References 12–14, 72, and 73,

$$K_g = \chi_\alpha \sqrt{D_\Sigma \left[z^{out}_{g_M}(t)\right]}. \tag{7.103}$$

Here χ_α is the α-percentage point of the probability distribution density (see Eqs. (7.36) and (7.37)) of the signal $z^{out}_{g_M}(t)$ at the output of the generalized detector. This α-percentage point is determined by the tabulated distribution law for the fixed and predetermined value of the probability of false alarm P_F.

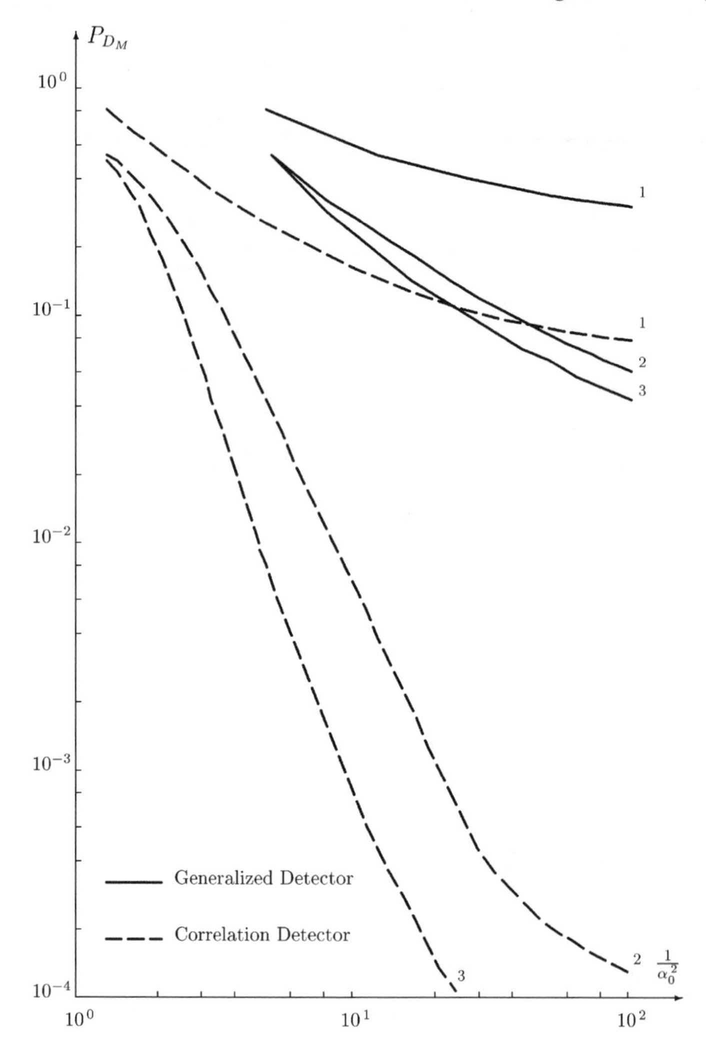

FIGURE 7.3
Detection performances of the signal with known initial phase by "true" channel of the detector: $P_{F_k} = 10^{-6}$; $P_D = 0.9$; for 1, $\delta_{k_1}^2 = 0.1$; for 2, $\delta_{k_1}^2 = 0.01$; for 3, $\delta_{k_1}^2 = 0.001$.

$D[z_{g_M}^{out}(t)]$ is the total variance of the signal $z_{g_M}^{out}(t)$ at the output of the generalized detector.

According to the Neyman–Pearson criterion the probability of false alarm P_F and the probability of detection P_D for the generalized detector are determined in the following form:

$$P_F = \int_{K_g}^{\infty} f_g^{H_0}\left[z_{g_M}^{out}(t)\right] d\left[z_{g_M}^{out}(t)\right]; \tag{7.104}$$

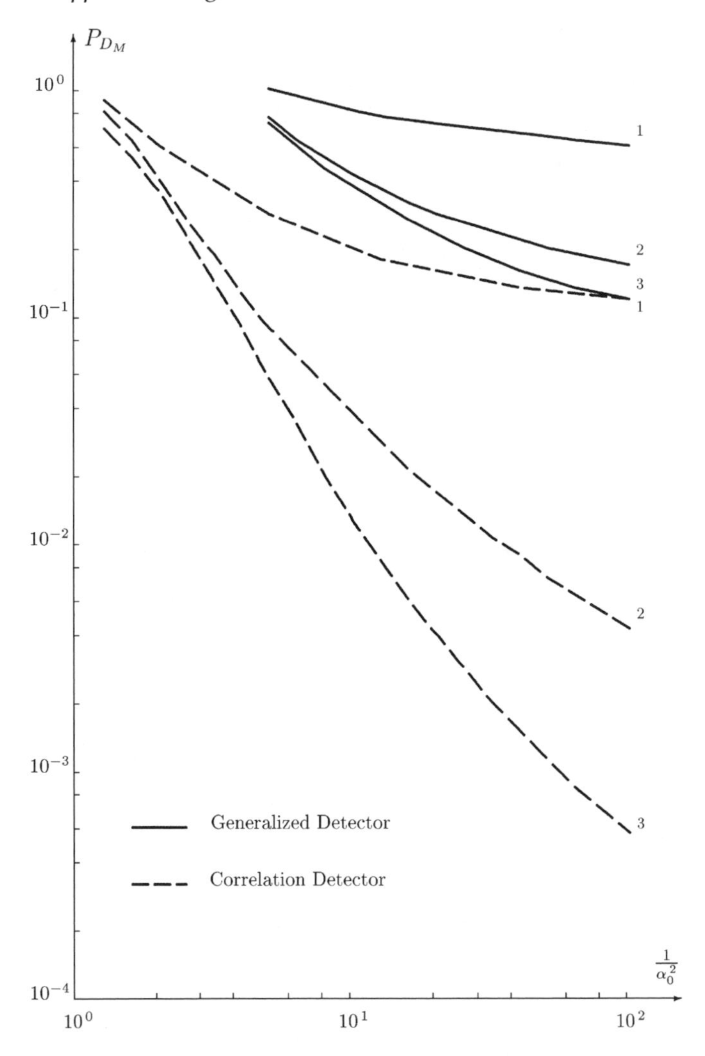

FIGURE 7.4
Detection performances of the signal with known initial phase by "true" channel of the detector: $P_{F_k} = 10^{-4}$; $P_D = 0.99$; for 1, $\delta_{k_1}^2 = 0.1$; for 2, $\delta_{k_1}^2 = 0.01$; for 3, $\delta_{k_1}^2 = 0.001$.

$$P_D = \int_{K_g}^{\infty} f_g^{H_1}\left[z_{g_M}^{out}(t)\right] d\left[z_{g_M}^{out}(t)\right], \tag{7.105}$$

where $f_g^{H_0}[z_{g_M}^{out}(t)]$ is the probability distribution density of the signal $z_{g_M}^{out}(t)$ at the output of the generalized detector defined by Eqs. (7.38) and (7.39); $f_g^{H_1}[z_{g_M}^{out}(t)]$ is the probability distribution density of the signal $z_{g_M}^{out}(t)$ at the output of the generalized detector defined by Eqs. (7.36) and (7.37).

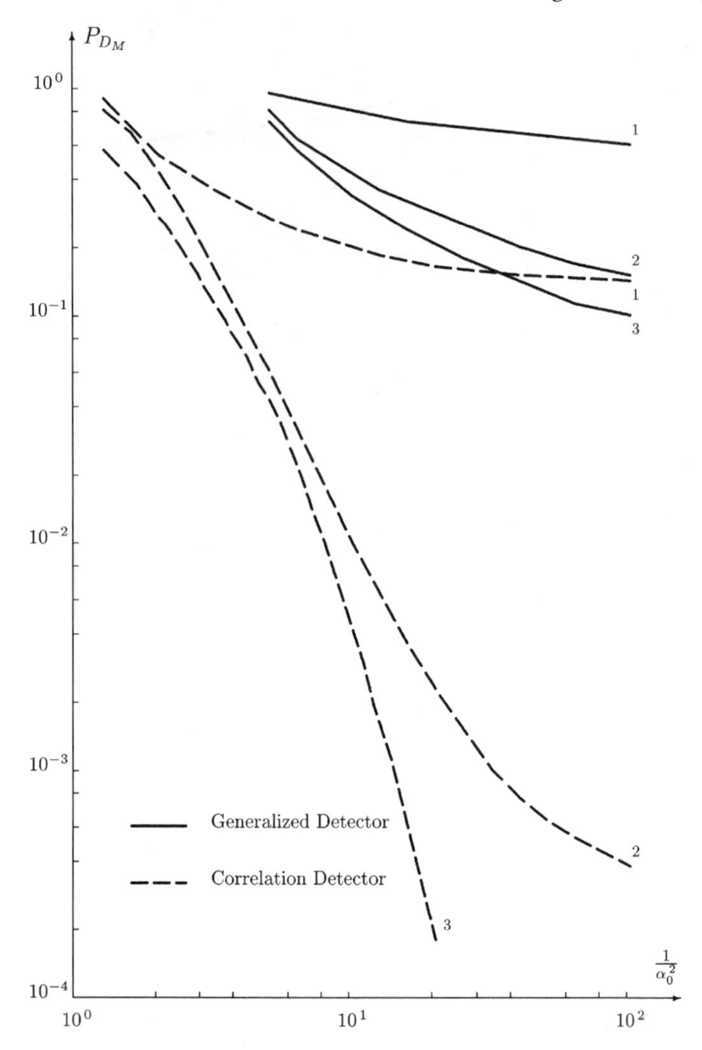

FIGURE 7.5
Detection performances of the signal with known initial phase by "true" channel of the detector: $P_{F_k} = 10^{-6}$; $P_D = 0.99$; for 1, $\delta_{k_1}^2 = 0.1$; for 2, $\delta_{k_1}^2 = 0.01$; for 3, $\delta_{k_1}^2 = 0.001$.

7.2.2.1 Periodic Multiplicative Noise

For the case of the periodic multiplicative noise, when the condition

$$\Omega_M > \frac{4\pi}{T} \tag{7.106}$$

is satisfied, the probability of detection of the signal at the "true" channel of the generalized detector is determined in the following form:

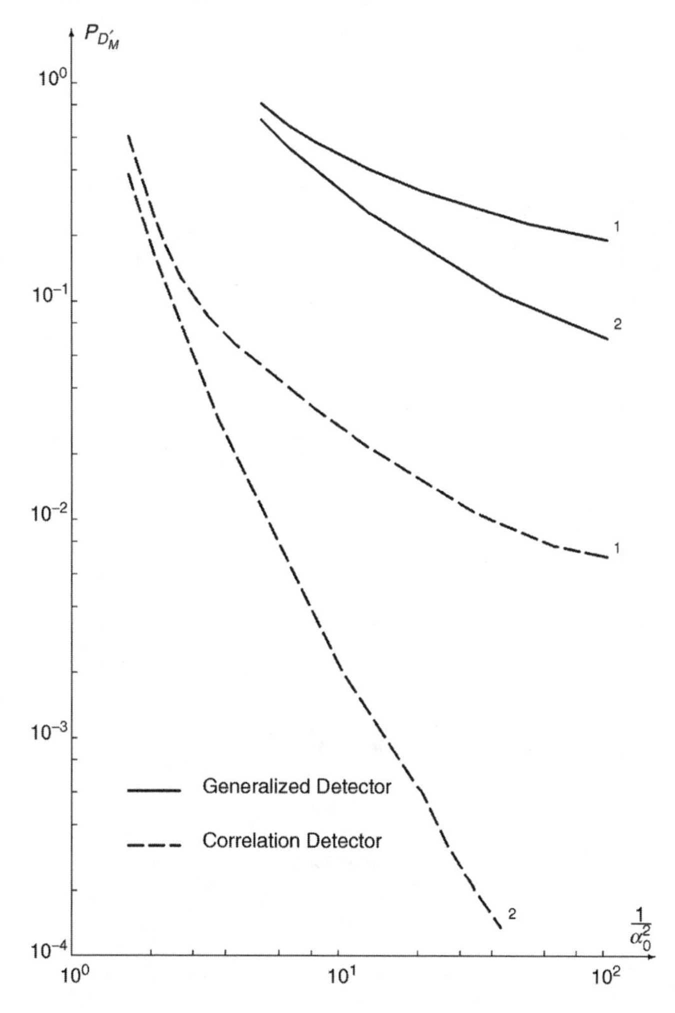

FIGURE 7.6
Detection performances of the signal with known initial phase. Spectral power density of the signal noise component is the same for all channels of detector: $P_D = 0.9$; $N = 100$; $\delta^2_{k_1} \leq 0.001$; for 1, $P_{F_k} = 10^{-4}$; for 2, $P_{F_k} = 10^{-6}$.

- For the case in which a duration of the signal is infinitesimal

$$P_{D_M} = P_{D_{M_k}} = \int_{K_g}^{\infty} f_k^{H_1}\left[\left|z^{out}_{g_{M_k}}(t)\right|\right] d\left[\left|z^{out}_{g_{M_k}}(t)\right|\right]$$

$$= \frac{2(\Delta\omega_{PF})\sqrt{T\beta}}{N_0\omega_0^2} \int_{K_{g0}}^{\infty} \exp\left\{-\frac{\left|z^{out}_{g_{M_k}}(t)\right|^2 + \alpha_0^4 q^2}{2}\right\} d\left[\left|z^{out}_{g_{M_k}}(t)\right|\right]. \tag{7.107}$$

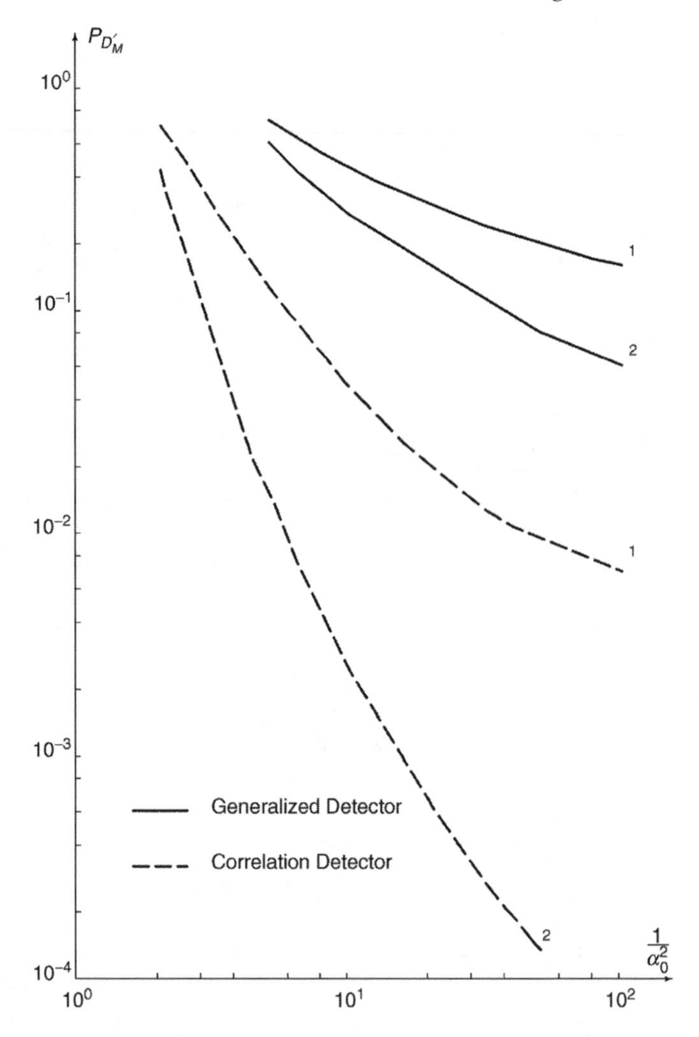

FIGURE 7.7
Detection performances of the signal with known initial phase. Spectral power density of the signal noise component is the same for all channels of detector: $P_D = 0.99$; $N = 100$; $\delta_{k_1}^2 \leq 0.001$; for 1, $P_{F_k} = 10^{-4}$; for 2, $P_{F_k} = 10^{-6}$.

- For the case in which a duration of the signal is very large

$$P_{D_M} = P_{D_{M_k}} = \int_{K_g}^{\infty} f_k^{H_1}\left[z_{g_{M_k}}^{out}(t)\right] d\left[z_{g_{M_k}}^{out}(t)\right]$$

$$= \frac{4(\Delta\omega_{PF})T\beta}{\sqrt{2\pi}\,N_0\omega_0^2} \int_{K_{g0}}^{\infty} \exp\left\{-\frac{\left[z_{g_{M_k}}^{out}(t)\right]^2 + \alpha_0^4 q^2}{2}\right\} d\left[z_{g_{M_k}}^{out}(t)\right]. \qquad (7.108)$$

The probability of exceeding the threshold at the "m"-th channel of the generalized detector for the condition that there is the signal at the "k"-th channel of the generalized detector, when the multiplicative noise is absent, is determined by the following form:

- For the case in which a duration of the signal is infinitesimal

$$P_{D_{Mm}} = \int_{K_g}^{\infty} f_m^{H_1}\left[\left|z_{g_{Mm}}^{out}(t)\right|\right] d\left[\left|z_{g_{Mm}}^{out}(t)\right|\right]$$
$$= \frac{2(\Delta\omega_{PF})\sqrt{T\beta}}{N_0\omega_0^2}\int_{K_{g0}}^{\infty} \exp\left\{-\frac{\left|z_{g_{Mm}}^{out}(t)\right|^2 + q^2|\dot{R}_{mk}|^2}{2}\right\} d\left[\left|z_{g_{Mm}}^{out}(t)\right|\right]. \tag{7.109}$$

- For the case in which a duration of the signal is very large

$$P_{D_{Mm}} = \int_{K_g}^{\infty} f_m^{H_1}\left[z_{g_{Mm}}^{out}(t)\right] d\left[z_{g_{Mm}}^{out}(t)\right]$$
$$= \frac{4(\Delta\omega_{PF})T\beta}{\sqrt{2\pi}\,N_0\omega_0^2}\int_{K_{g0}}^{\infty} \exp\left\{-\frac{\left[z_{g_{Mm}}^{out}(t)\right]^2 + q^2|\dot{R}_{mk}|^2}{2}\right\} d\left[z_{g_{Mm}}^{out}(t)\right], \tag{7.110}$$

where the parameter $|\dot{R}_{mk}|$ is defined by Eq. (7.47) for the condition

$$n = m.$$

Reference to Eqs. (7.107)–(7.110) shows that the probabilities of detection $P_{D_{M_k}}$ and $P_{D_{Mm}}$ can be defined using the detection performances of the signal with the random initial phase by the generalized detector under the probability of false alarm P_{F_k} and the detection parameter equal to $\alpha_0^2 q$ and $q|\dot{R}_{mk}|$, respectively.

7.2.2.2 Fluctuating Multiplicative Noise

In the case of fluctuating multiplicative noise the probability of detection of the signal at the "true" channel of the generalized detector can be determined in the following form:

- For the case when a duration of the signal is infinitesimal

$$P_{D_M} = P_{D_{M_k}} = \int_{K_g}^{\infty} f_k^{H_1}\left[\left|z_{g_{M_k}}^{out}(t)\right|\right] d\left[\left|z_{g_{M_k}}^{out}(t)\right|\right]$$
$$= \frac{2(\Delta\omega_{PF})\sqrt{T\beta}}{N_0\omega_0^2}\int_{K_{g0_M}}^{\infty} \exp\left\{-\frac{\left|z_{g_{M_k}}^{out}(t)\right|^2 + q_M^2}{2}\right\} d\left[\left|z_{g_{Mm}}^{out}(t)\right|\right]. \tag{7.111}$$

- For the case in which a duration of the signal is very large

$$P_{D_M} = P_{D_{M_k}} = \int\limits_{K_g}^{\infty} f_k^{H_1}\left[z_{g_{M_k}}^{out}(t)\right] d\left[z_{g_{M_k}}^{out}(t)\right]$$

$$= \frac{4(\Delta\omega_{PF})T\beta}{\sqrt{2\pi}\,N_0\omega_0^2} \int\limits_{K_{g_{0_M}}}^{\infty} \exp\left\{-\frac{\left[z_{g_{M_k}}^{out}(t)\right]^2 + q_M^2}{2}\right\} d\left[z_{g_{M_k}}^{out}(t)\right], \qquad (7.112)$$

where

$$K_{g_{0_M}} = \frac{K_{g_0}}{\sqrt{1+\delta_{k_1}^2 q^2}}; \qquad (7.113)$$

$$q_M = \frac{\alpha_0^2 q}{\sqrt{1+\delta_{k_1}^2 q^2}}. \qquad (7.114)$$

Reference to Eqs. (7.111)–(7.114) shows that the probability of detection P_{D_M} of the signal for the case, in which the multiplicative noise is present, is defined using the probability of false alarm and the probability of detection of the signal for the case in which the multiplicative noise is absent, which are defined by the parameters K_{g_0} and q using the parameters of the multiplicative noise and the signal—the parameters α_0 and $\delta_{k_1}^2$, respectively.

Also, reference to Eqs. (7.111)–(7.114) shows that for the case considered the probability of detection P_{D_M} of the signal can be defined using the detection performances of the signal with the random initial phase by the generalized detector when the equivalent value of the probability of false alarm is determined by

$$P_{F_{eq}} = P_{F_k}^{\frac{1}{\sqrt{1+\delta_{k_1}^2 q^2}}} \qquad (7.115)$$

and the detection parameter of the generalized detector is equal to q_M (see Eq. (7.114)).

The probability of exceeding the threshold at the "m"-th channel of the generalized detector, during the condition that there is the signal at the "k"-th channel of the generalized detector when the multiplicative noise is absent, takes the following form for the case considered

$$P_{D_{M_m}} = \int\limits_{K_g}^{\infty} f_m^{H_1}\left[z_{g_{M_m}}^{out}(t)\right] = \exp\left[-\frac{K_{g_0}^2}{2\sqrt{1+\delta_{m_1}^2 q^2}}\right]$$

$$= P_{F_k}^{\frac{1}{\sqrt{1+\delta_{m_1}^2 q^2}}}. \qquad (7.116)$$

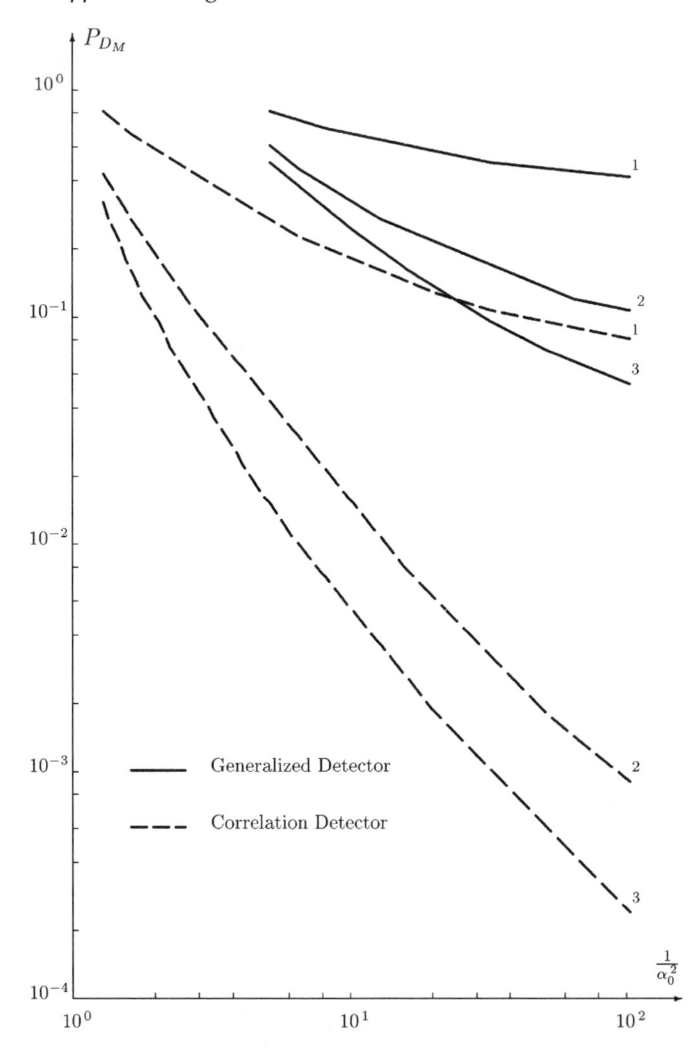

FIGURE 7.8
Detection performances of the signal with unknown initial phase by "true" channel of the detector: $P_{F_k} = 10^{-4}$; $P_D = 0.9$; for 1, $\delta_{k_1}^2 = 0.1$; for 2, $\delta_{k_1}^2 = 0.01$; for 3, $\delta_{k_1}^2 = 0.001$.

The probability of detection of the signal at the "true" channel of the generalized detector, when the fluctuating multiplicative noise is present, as a function of the parameters of the fluctuating multiplicative noise is shown in Figs. 7.8–7.11.

The probability of detection of the signal for the case in which the spectral power density of the signal noise component is the same for all channels of the generalized detector is shown in Figs. 7.12 and 7.13.

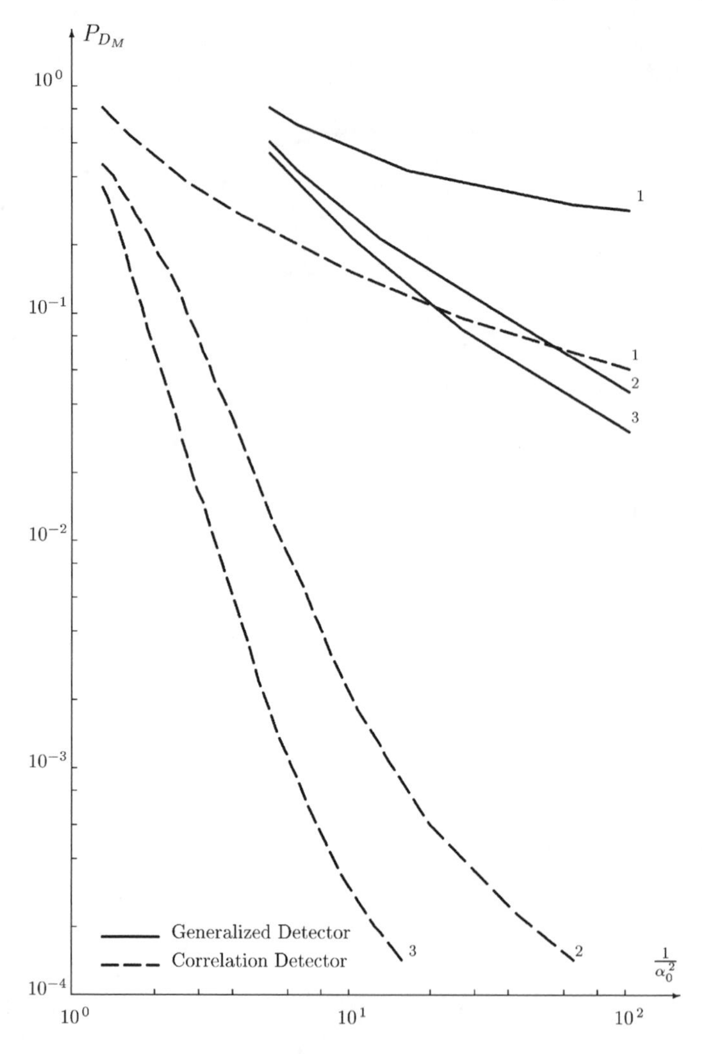

FIGURE 7.9
Detection performances of the signal with unknown initial phase by "true" channel of the detector: $P_{F_k} = 10^{-6}$; $P_D = 0.9$; for 1, $\delta^2_{k_1} = 0.1$; for 2, $\delta^2_{k_1} = 0.01$; for 3, $\delta^2_{k_1} = 0.001$.

Also, the comparative analysis between the detection performances of the generalized detector and the correlation detector (or the matched filter) is presented in Figs. 7.8–7.13 for the conditions mentioned above. Figures 7.8–7.13 show the superiority of the generalized detector over the correlation detector (or the matched filter).

The analysis carried out in this section shows that the impact of the multiplicative noise on detection of the signals by the generalized detector depends both on the parameters of the multiplicative noise and on the parameters of the signal.

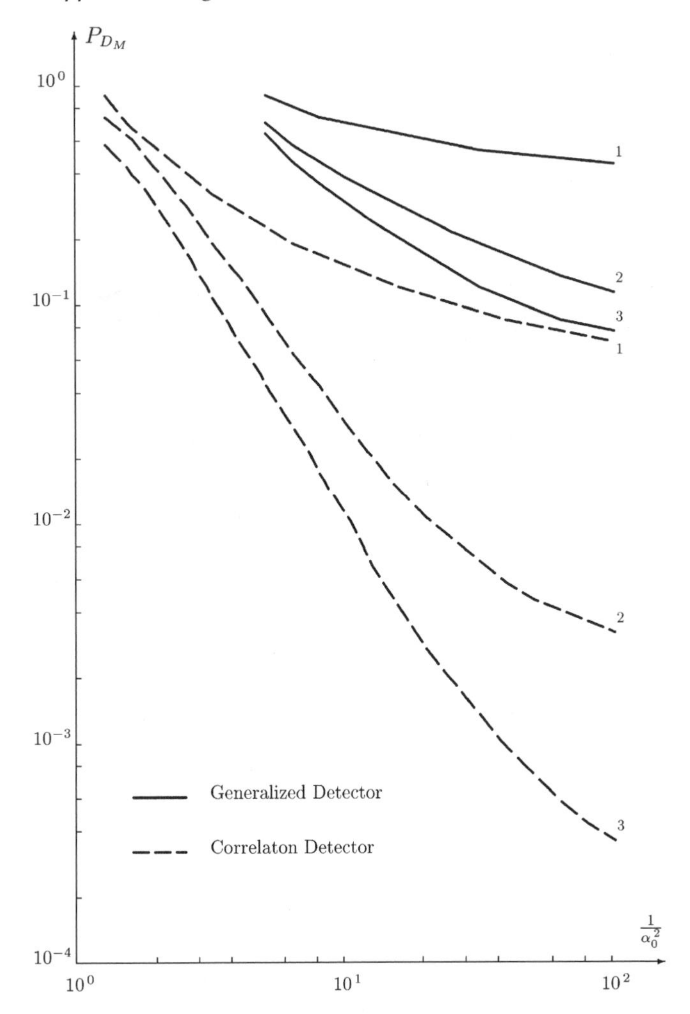

FIGURE 7.10
Detection performances of the signal with unknown initial phase by "true" channel of the detector: $P_{F_k} = 10^{-4}$; $P_D = 0.99$; for 1, $\delta_{k_1}^2 = 0.1$; for 2, $\delta_{k_1}^2 = 0.01$; for 3, $\delta_{k_1}^2 = 0.001$.

The detection performances presented in Figs. 7.2–7.13 confirm the following fact: a decrease in the probability of detection of the signals caused by the multiplicative noise is high, the parameters α_0^2 and $\delta_{k_1}^2$ is low.

The parameter α_0^2, as shown in Section 4.2, is only defined by characteristics of the multiplicative noise, in general, by a degree of distortions in phase of the signal. As to the parameter $\delta_{k_1}^2$, it depends both on the parameters of the multiplicative noise and on the parameters of the signal.

Reference to Eq. (7.94) shows that for the definite value of the energy spectrum $G_V(0)$ of the multiplicative noise in which the value of the parameter $\delta_{k_1}^2$ is low, the duration T of the coherent signal is large. For this reason, complex

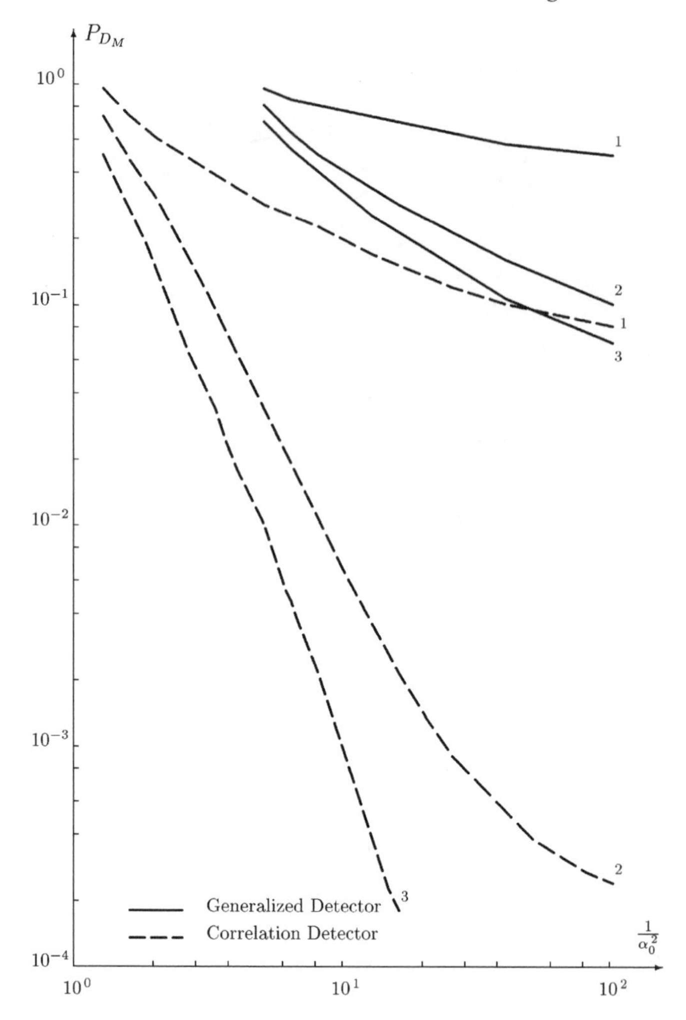

FIGURE 7.11
Detection performances of the signal with unknown initial phase by "true" channel of the detector: $P_{F_k} = 10^{-6}$; $P_D = 0.99$; for 1, $\delta_{k_1}^2 = 0.1$; for 2, $\delta_{k_1}^2 = 0.01$; for 3, $\delta_{k_1}^2 = 0.001$.

signal processing systems using signals with the large signal duration, or the coherent pulse bursts, are more vulnerable with respect to the multiplicative noise. It is characteristic that the duration of the coherent signal plays an essential role. The phase or frequency modulation of the signal is of no concern for the given case.

We compare, for example, two signals that can give the same resolution in time (the range resolution in radar): the non-coherent pulses without an intrapulse modulation—the narrow-band pulses—with the value T_1 of pulse duration and pulses with the frequency or phase intrapulse modulation—the

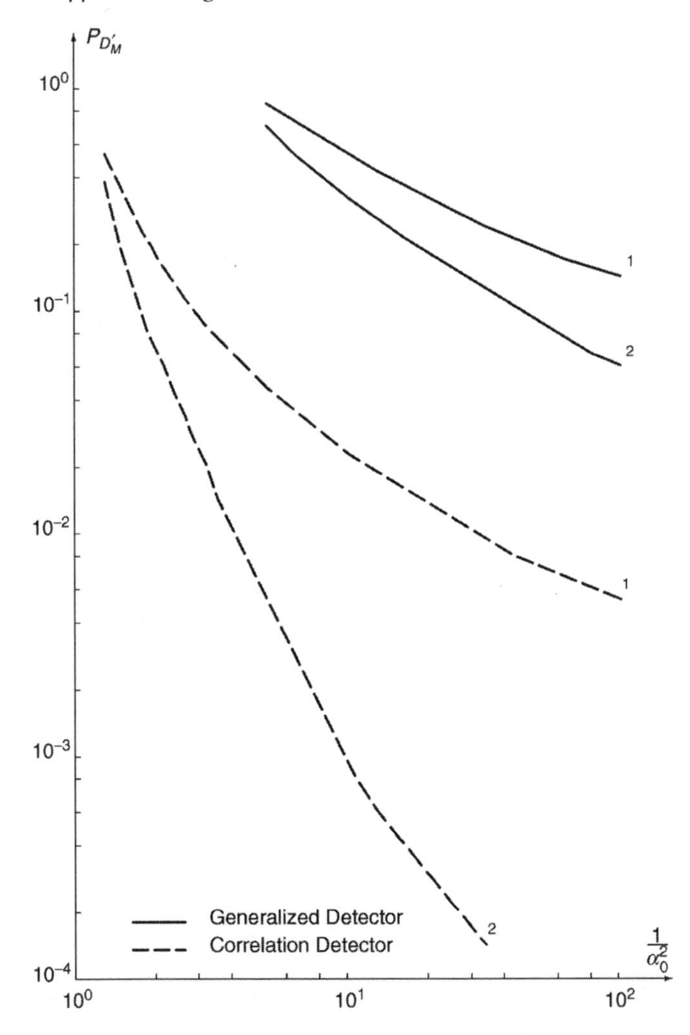

FIGURE 7.12
Detection performances of the signal with unknown initial phase. Spectral power density of the signal noise component is the same for all channels of detector: $P_D = 0.9$; $N = 100$; $\delta^2_{k_1} \leq 0.001$; for 1, $P_{F_k} = 10^{-4}$; for 2, $P_{F_k} = 10^{-6}$.

wide-band pulses—with the value T_2 of pulse duration and with the energy bandwidth equal to

$$\Delta\Omega_p \gg \frac{1}{T_2}. \tag{7.117}$$

For signals that can give the same resolution in time—the range resolution in radar—we can write

$$T_2 = Q_y T_1, \tag{7.118}$$

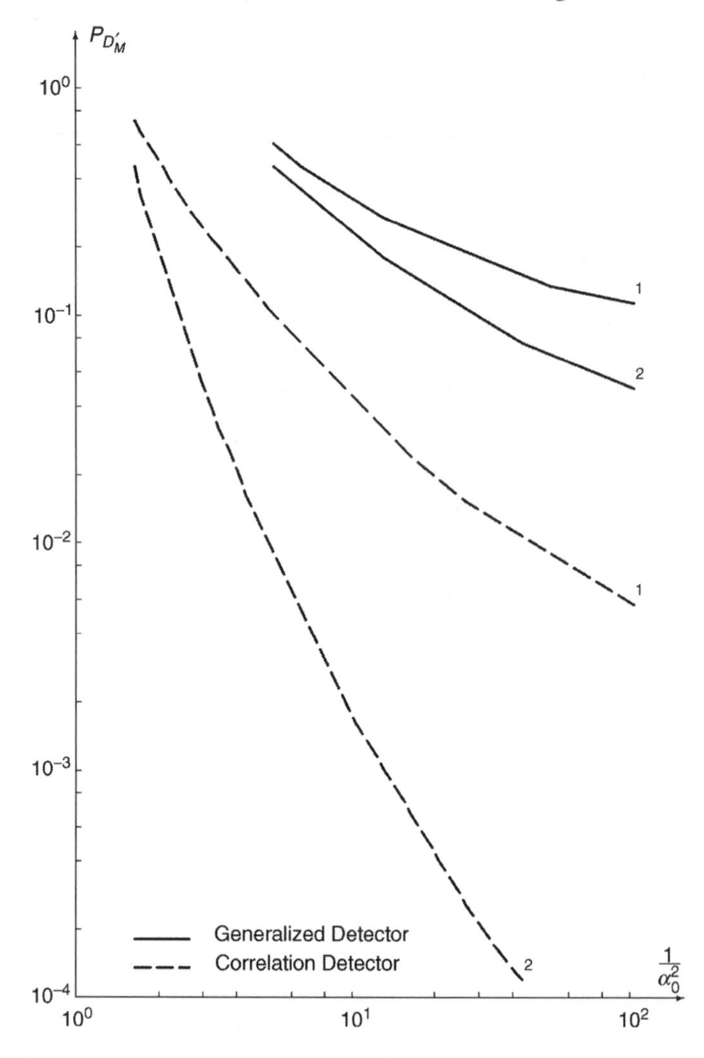

FIGURE 7.13
Detection performances of the signal with unknown initial phase. Spectral power density of the signal noise component is the same for all channels of detector: $P_D = 0.99$; $N = 100$; $\delta^2_{\bar{k}_1} \leq 0.001$; for 1, $P_{F_k} = 10^{-4}$; for 2, $P_{F_k} = 10^{-6}$.

where

$$Q_y = \Delta\Omega_p T_2. \tag{7.119}$$

It is not difficult to see that when there is the same multiplicative noise the parameter $\delta^2_{\bar{k}_1}$ for the case of the wide-band pulses is in Q_y times low and, consequently, a decrease in the probability of detection of the signals is much higher than in the case of the narrow-band pulses.

To illustrate possibilities of the use of the formulae discussed above for solving problems that are more widely used in practice, we consider two examples: radar and communications.

7.2.3 Radar Range

A decrease in the probability of detection of the target, which is caused by the multiplicative noise, can be defined using the formulae mentioned above along with dependences presented in Figs. 7.2–7.13.

To determine a decrease in the radar range there is a need to define what increase in the energy of the target return signal is required, during the presence of the multiplicative noise, to save the same probability of detection of the target for the case in which the multiplicative noise is absent.

Taking into account that the energy of the target return signal is inversely proportional to the fourth order of distance between the radar and the target, the main function defining a decrease in the radar range (that is caused by the multiplicative noise) can be determined using the following form:

$$\frac{R_2}{R_1} = \sqrt[4]{\frac{E'_{a_1}}{E''_{a_1}}} \quad \text{at } P_{D_M} = P_D, \tag{7.120}$$

where P_D is the probability of detection of the signal for the case of absence of the multiplicative noise when the distance to the target is equal to R_1 and the energy of the target return signal is equal to E'_{a_1}; P_{D_M} is the probability of detection of the target for the case of presence of the multiplicative noise when the distance to the target is equal to R_2 and the energy of the target return signal is equal to E''_{a_1}.

To define a decrease in the radar range there is a need to determine the ratio

$$\frac{E'_{a_1}}{E''_{a_1}},$$

using the condition

$$P_D = P_{D_M}$$

and for which this condition is satisfied.

In the case of the periodic multiplicative noise a definition of the ratio

$$\frac{E'_{a_1}}{E''_{a_1}}$$

is not a difficult problem. However, in the case of the fluctuating multiplicative noise a definition of the ratio

$$\frac{E'_{a_1}}{E''_{a_1}}$$

is not so simple.

In radar we must use the formulae that are true for the case of the signal with the unknown (random) initial phase. For the given case based on the probability of detection of the target at the "true" channel of the generalized detector the condition

$$P_D = P_{D_M}$$

leads to the following relationship for the definition of the ratio

$$\frac{E'_{a_1}}{E''_{a_1}}:$$

- For the case in which a duration of the signal is infinitesimal

$$\frac{2(\Delta\omega_{PF})\sqrt{T\beta}}{N_0\omega_0^2} \times \int_{K_g}^{\infty} \exp\left\{-\frac{4(\Delta\omega_{PF})\sqrt{T\beta}\left[\left|z_g^{out}(t)\right| + E'_{a_1}\right]}{N_0\omega_0^2}\right\} d\left[\left|z_g^{out}(t)\right|\right]$$

$$= \frac{(\Delta\omega_{PF})\sqrt{T\beta}}{\sqrt{16(\Delta\omega_{PF})^2 T\beta E''_{a_1}\delta_{k_1}^2 + N_0^2\omega_0^2}}$$

$$\times \int_{K_g}^{\infty} \exp\left\{-\frac{4(\Delta\omega_{PF})\sqrt{T\beta}\left[\left|z_{g_M}^{out}(t)\right| + \alpha_0^2 E''_{a_1}\right]}{N_0\omega_0^2}\right\} d\left[\left|z_{g_M}^{out}(t)\right|\right]. \quad (7.121)$$

- For the case in which a duration of the signal is very large

$$\frac{4(\Delta\omega_{PF})T\beta}{\sqrt{2\pi}\,N_0\omega_0^2} \times \int_{K_g}^{\infty} \exp\left\{-\frac{8(\Delta\omega_{PF})^2 T\beta\left[z_g^{out}(t) + E'_{a_1}\right]}{N_0^2\omega_0^4}\right\} d\left[z_g^{out}(t)\right]$$

$$= \frac{4(\Delta\omega_{PF})T\beta}{\sqrt{2\pi\left[16(\Delta\omega_{PF})^2 T\beta E''_{a_1}\delta_{k_1}^2 + N_0^2\omega_0^4\right]}}$$

$$\times \int_{K_g}^{\infty} \exp\left\{-\frac{8(\Delta\omega_{PF})^2 T\beta\left\{\left[z_{g_M}^{out}(t)\right]^2 + \alpha_0^2 E''_{a_1}\right\}}{16(\Delta\omega_{PF})^2 T\beta E''_{a_1}\delta_{k_1}^2 + N_0^2\omega_0^4}\right\} d\left[z_{g_M}^{out}(t)\right]. \quad (7.122)$$

However, Eqs. (7.121) and (7.122) do not allow us to obtain the ratio

$$\frac{E'_{a_1}}{E''_{a_1}}$$

in the explicit form. To define the ratio

$$\frac{E'_{a_1}}{E''_{a_1}}$$

we use the following technique that is based on Eqs. (7.121) and (7.122).

We denote q_1 as the value of the detection parameter q of the generalized detector when the distance to the target is equal to the value R_1 and, consequently, q_2 is the value of the detection parameter q of the generalized detector when the distance to the target is equal to the value R_2.

Then

$$\frac{E'_{a_1}}{E''_{a_1}} = \frac{q_1}{q_2}. \tag{7.123}$$

Under the given probability of detection P_D, and the probability of false alarm P_{F_k}, the value of the detection parameter q_1 is determined by Eq. (7.105) and can be defined using dependences presented in Figs. 7.8–7.13.

The value of the detection parameter q_2 that ensures the same probability of detection P_D and the probability of false alarm, when the multiplicative noise is present, can be defined using the following algorithm.

- To construct graphically the equivalent probability of false alarm $P_{F_{eq}}$ as a function of the detection parameter q_2 for the given parameters α_0^2 and $\delta_{k_1}^2$ of the multiplicative noise and the given probability of detection P_D. For this purpose:
  - Under the given equivalent probability of false alarm $P_{F_{eq}} = P_F$ there is a need to define the corresponding values of the detection parameter $q_M = q$ of the generalized detector when the probability of detection of the signal is equal to P_D on the basis of the detection performances presented in Figs. 7.8–7.13;
  - For each determined detection parameter q_M of the generalized detector there is a need to define the corresponding detection parameter of the generalized detector for the case in which the multiplicative noise is absent:

$$q_2 = \frac{q_M}{\sqrt{\alpha_0^4 + \delta_{k_1}^2 q_M^2}}; \tag{7.124}$$

  - There is a need to construct graphically the equivalent probability of false alarm $P_{F_{eq}}$ as a function of the detection parameter q_2 of the generalized detector.
- To construct graphically the equivalent probability of false alarm $P_{F_{eq}}$ as a function of the detection parameter q_2 of the generalized detector for the constant value of the probability of false alarm P_{F_k}, which is equal to the predetermined *a priori* value.

The intersection point of the curves plotted in accordance with this algorithm gives us the value of the detection parameter q_2 of the generalized detector. The obtained detection parameter q_2 corresponds to the given probability of detection P_D and the probability of false alarm P_F for the given parameters α_0^2 and $\delta_{k_1}^2$ of the multiplicative noise. Thus, the detection parameter q_2 of the generalized detector is the searched value of the detection parameter of the generalized detector, which is satisfied according to Eqs. (7.121) and (7.122).

Based on the determined detection parameters q_1 and q_2 of the generalized detector, we can define a decrease in the radar range using the following equation

$$\frac{R_2}{R_1} = \sqrt[4]{\frac{q_1}{q_2}}, \tag{7.125}$$

where R_1 is the radar range when the multiplicative noise is absent and R_2 is the radar range when the multiplicative noise is present.

The dependence of a decrease in the radar range caused by the multiplicative noise, and constructed on the basis of the algorithm mentioned above as a function of parameters of the multiplicative noise under the given probability of detection of the target, is shown in Figs. 7.14 and 7.15.

Also, for the purpose of comparison, the analogous dependences for the correlation detector are presented in Figs. 7.14 and 7.15.

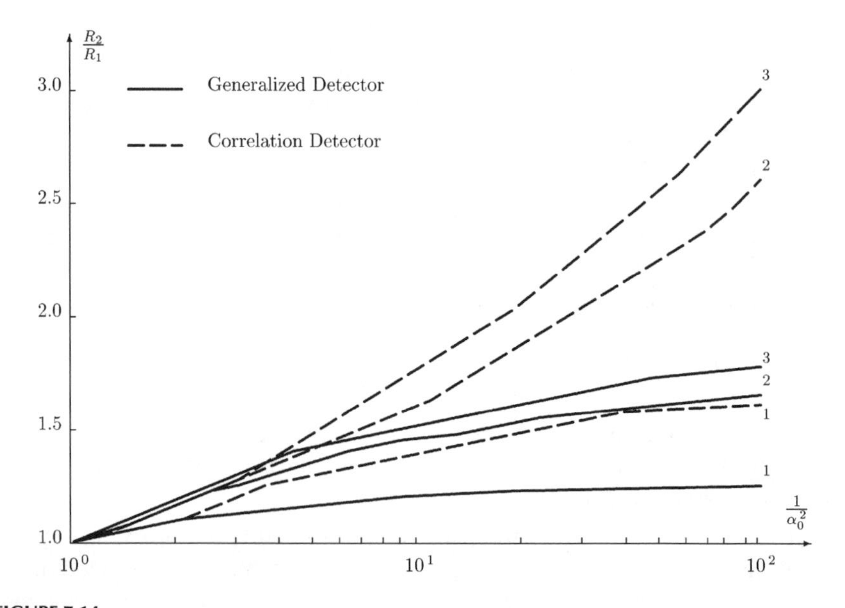

FIGURE 7.14
A decrease in the radar range as a function of parameters of the multiplicative noise: $P_D = 0.5$; $P_{F_k} = 10^{-4}$; for 1, $\delta_{k_1}^2 = 0.1$; for 2, $\delta_{k_1}^2 = 0.01$; for 3, $\delta_{k_1}^2 = 0.001$.

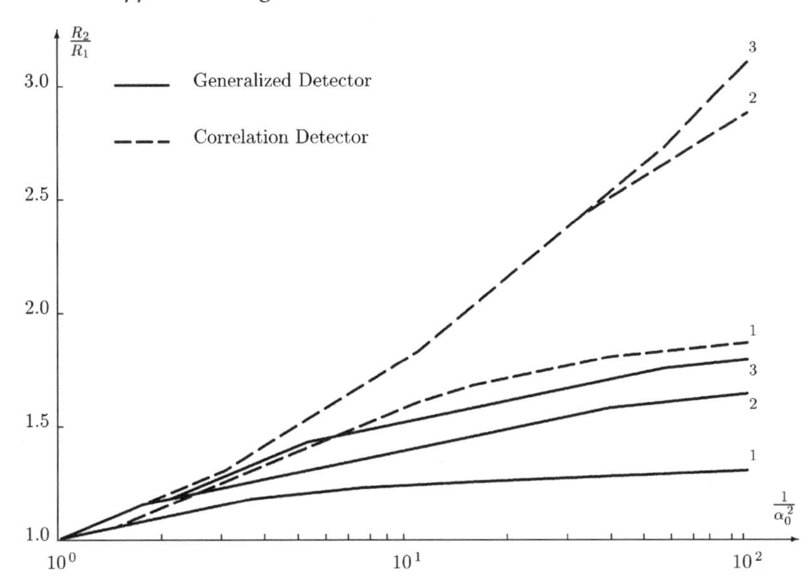

FIGURE 7.15
A decrease in the radar range as a function of parameters of the multiplicative noise: $P_D = 0.7$; $P_{F_k} = 10^{-4}$; for 1, $\delta_{k_1}^2 = 0.1$; for 2, $\delta_{k_1}^2 = 0.01$; for 3, $\delta_{k_1}^2 = 0.001$.

Comparative analysis shows us the superiority of the generalized detector over the correlation detector during use in radar systems.

The estimation of a decrease in the radar range on the basis of the algorithm mentioned above is very cumbersome. Using Eqs. (7.83) and (7.93) for definition of the probability of detection of the signals with the known initial phase by the generalized detector, we are able to obtain the approximate relationships with the purpose of quickly defining the tentative estimation of a decrease in the radar range for some particular cases.

Substituting Eqs. (7.83) and (7.119) in the equality

$$P_{D_M} = P_D,$$

we can write

$$K_{g_0} - q_1 = \frac{K_{g_0}}{\sqrt{1 + \delta_{k_1}^2 q_2^2}} - \frac{\alpha_0^2 q_2}{\sqrt{1 + \delta_{k_1}^2 q_2^2}}. \tag{7.126}$$

We proceed to consider the particular cases.

- The level of energy characteristics of the signal noise component during the presence of the multiplicative noise is much greater than the level of energy characteristics of the additive Gaussian noise:

$$\sigma_k^2 \gg \frac{N_0^2 \omega_0^4}{16(\Delta\omega_{PF})^2 T\beta} \tag{7.127}$$

or

$$\delta_{k_1}^2 q_1^2 \gg 1. \tag{7.128}$$

In this case, taking into account Eq. (7.126), we are able to determine the ratio

$$\frac{R_2}{R_1} = \sqrt[4]{\frac{q_1}{q_2}} \simeq \sqrt[4]{\alpha_0^2 \cdot \frac{q_1}{q_2} - (q_1 - K_{g_0}) \cdot \frac{q_1}{K_{g_0}} \cdot \delta_{k_1}}, \tag{7.129}$$

where q_1 and K_{g_0} are defined by the given probability of detection P_D and the predetermined probability of false alarm P_F. Equation (7.129) can be used for the condition

$$\alpha_0^2 \gg \delta_{k_1}(q_1 - K_{g_0}). \tag{7.130}$$

- The level of energy characteristics of the signal noise component during the presence of the multiplicative noise is much less than the level of energy characteristics of the additive Gaussian noise:

$$\sigma_k^2 \ll \frac{N_0^2 \omega_0^4}{16(\Delta\omega_{PF})^2 T\beta} \tag{7.131}$$

or

$$\delta_{k_1}^2 q_1^2 \ll 1. \tag{7.132}$$

For this case, in terms of Eq. (7.126), we can write

$$\frac{R_2}{R_1} = \sqrt[4]{\frac{q_1}{q_2}} \simeq \sqrt{\alpha_0}. \tag{7.133}$$

The value of the ratio $\frac{R_2}{R_1}$ in Eq. (7.133) is the limiting case, to which the ratio

$$\frac{R_2}{R_1}$$

tends to approach as the energy spectrum of the noise modulation function $\dot{M}(t)$, of the multiplicative noise, is extended for the constant value of relative level of the undistorted component of the signal, in other words, as the parameter $\delta_{k_1}^2$ is decreased. This limiting case is shown by the dotted line in Figs. 7.14 and 7.15.

7.2.4 Wide-Band Noise Signals in Communications

A set of signals is transmitted simultaneously within the limits of the same bandwidth during the use of wide-band noise signals in communications. The spectrum bandwidths of these signals coincide with the bandwidth, within the limits of which the signals are transmitted. The required signal is isolated from the whole totality of the signals, noise, and interferences during the use of the generalized detector. The matched filter and the correlation detector are used for this purpose as well. We will compare the final results for the generalized detector and the correlation detector.

The transmitted information is included in the signal formed at the output of the generalized detector. For some complex signal processing systems the position of the main peak of amplitude of the signal, for example, the narrow-band signal, at the output of the generalized detector on the time axis, or the presence of the signal at the output of the generalized detector, is considered as the information parameter.[74] The speech signals can be transmitted by the pulse-phase modulation.[75]

The main peak of amplitude of the signal at the output of the generalized detector is decreased and the energy level of the side lobe shoots is increased as a result of the stimulus of multiplicative noise. As a result, the probability of detection of the signal containing the transmitted information is decreased; for example, the probability of losses of the transmitted information is increased, and the probability of the fact that false alarm shoots of amplitude of the signal at the output of the generalized detector, which are generated by parasitic signals and interferences, can be thought as the information signal, i.e., the false information would influence the decision-making.

The main interest for us in this section is to estimate a decrease in the probability of detection of the information signal and an increase in the probability of believing that the false side lobe shoots, caused by the multiplicative noise, are the information signal.

The main peculiarity of the case considered is the fact that the additive interference in communication systems arose both from the set noise and from the simultaneous presence of other signals in communication systems within the limits of the energy spectrum bandwidth of the information signal. For this case other signals can be thought of as the interference.

We proceed to define the characteristics of the interference caused by other parasitic signals in communication systems. Assume that $a_1(t)$ is the information signal required to detect; $z_1(t)$ is the amplitude of the information signal at the output of the generalized detector used by communication systems; $a_{M_p}(t)$ is one of the other parasitic signals in communication systems (interference); and $z_{M_p}(t)$ is the amplitude of the interference at the output of the generalized detector used by communication systems, which is generated by parasitic signals.

Since all signals used by communication systems are the quasiorthogonal signals, the following inequality

$$z_1(0) \gg z_{M_p}(0) \tag{7.134}$$

is true when the multiplicative noise is absent in communication systems.

When the multiplicative noise is present in communication systems the amplitude $z_{M_p}(t)$ of the signal at the output of the generalized detector, which is formed by the interference in accordance with Eqs. (6.5) and (6.76), can be written in the following form:

$$z_{M_p}(\tau) = Re\{\dot{S}_{M_p}(\tau) \cdot e^{j(\omega_1 \tau + \varphi_p)}\}, \tag{7.135}$$

where

$$\dot{S}_{M_p}(\tau) = \frac{C}{2} \int_{-\infty}^{\infty} \dot{M}_p(\tau) \dot{S}_{M_p}(t) S_1^*(t-\tau) \cdot e^{j\Omega_{p_1} t}\, dt, \tag{7.136}$$

$$\Omega_{p_1} = \omega_p - \omega_1. \tag{7.137}$$

There is a need to note that only the PF of the input linear tract of the generalized detector is matched with the information signal $a_1(t)$. The signal noise component of the information signal $a_1(t)$, in the case of the periodic multiplicative noise, is defined by Eq. (6.15). For the case of the fluctuating multiplicative noise the variance of the signal noise component is defined by Eq. (6.56) for the condition

$$\Omega = \Omega_{p_1}. \tag{7.138}$$

Comparing these formulae with Eqs. (6.87) and (6.107) that define the signal noise component of the signal $a_{M_1}(t)$ in terms of quasiorthogonality of the signals, one can see that during the same energy characteristics of the signals at the input of the linear tract of the generalized detector the signal noise component of the information signal $a_1(t)$, at the output of the generalized detector, is much greater than the signal noise component of the interference:

$$\dot{S}_1(\tau) \gg \dot{S}_{M_p}(\tau). \tag{7.139}$$

According to the statements mentioned above this relationship is true for the undistorted component of the signal at the output of the generalized detector. For this reason, the multiplicative noise basically acts on detection of the information signal by changing the characteristics of the information signal. The impact of the multiplicative noise on the additive interference generated by the parasitic signals is not so high.

Instants of appearance of individual parasitic signals (interference) are random, independent, and uniformly distributed within the limits of the interval T_r, in which the parasitic signals (interference) appear. Suppose that the number of signals overlapping one another within the limits of the interval, which is equal to the duration of the information signal, is sufficiently high. Then the probability distribution density of amplitude of the signal generated at the output of the generalized detector by all N parasitic signals (interference) is approximately Gaussian. The mean of amplitude of this signal at the output of the generalized detector is equal to zero. The variance of amplitude of this signal is equal to the average total power of the parasitic signals (interference) P_N^{av} at the output of the generalized detector within the limits of the period of replica of the parasitic signals (interference).

Taking into consideration the fact that the amplitudes at the output of the generalized detector generated by the additive set noise and parasitic signals (interference) are independent, the variance of the additive set noise and parasitic signals (interference) at the output of the generalized detector is equal to the sum: $P_N^{av}+$ the variance of the additive set noise.

For this form of the additive set noise and parasitic signals (interference) the optimal signal processing algorithm of the information signal is the generation of choice of the greatest signal at the output of the generalized detector. However, the technique of detection of the information signal by exceeding the threshold (given before) is very often used. This technique allows us to design a simpler apparatus with relatively low losses in the noise immunity.

In this case, taking into account the relationship between the amplitude of the signal at the output of the generalized detector and the value $\overline{z_M(t)}$ (see Section 7.1), we can use the following formulae obtained in Section 7.2.1 with the purpose of estimating an influence of the multiplicative noise:

- Equations (7.77) and (7.89) with the purpose of defining the probability of detection P_{D_M} of the information signal. There is a need to change the value

$$\frac{N_0^2\omega_0^4}{16(\Delta\omega_{PF})^2T\beta} \tag{7.140}$$

  in Eq. (7.35) by the value

$$\left(P_N^{av}\right)^2 + \frac{N_0^2\omega_0^4}{16(\Delta\omega_{PF})^2T\beta}. \tag{7.141}$$

  Moreover, the threshold K_g is defined by the allowable probability of false alarm P_F of exceeding the threshold K_g when the multiplicative noise is absent.

- Equations (7.35) and (7.43) with the purpose of defining the probability of false alarm $P_{F_M}(\tau)$—the probability when the false side lobe shoots are thought of as the information signal; the false side lobe shoots are shifted in time by the value τ with respect to the information signal. There is a need to change the value

$$\frac{N_0^2\omega_0^4}{16(\Delta\omega_{PF})T\beta} \tag{7.142}$$

  in Eq. (7.35) by the value

$$\left(P_N^{av}\right)^2 + \frac{N_0^2\omega_0^4}{16(\Delta\omega_{PF})^2T\beta} \tag{7.143}$$

  and to use the condition

$$\Omega_{mk} = 0 \qquad \text{and} \qquad \tau_{mk} = \tau. \tag{7.144}$$

The value of the probability of detection P_{D_M} can be defined based on Figs. 7.2–7.5. The probability of detection P_D, shown in Figs. 7.2–7.5, must be thought of as the probability of detection of the information signal at the output of the generalized detector used in the communication system when the multiplicative noise is absent and the following condition

$$P_F = P_{F_k}$$

is satisfied.

The dependence $P_{F_M}(\tau)$ as a function of the parameters of the multiplicative noise in the case when the condition

$$\delta^2_{m_1} \gg |\dot{\delta}^2_{m_2}| \tag{7.145}$$

is satisfied and the function

$$\delta^2_{m_1} = \frac{\sigma^2_S}{E^2_{a_1}} \tag{7.146}$$

is defined by Eqs. (6.208) and (7.94) for the conditions

$$\Omega = 0 \tag{7.147}$$

and

$$C = \frac{1}{\sqrt{E_{a_1}}}, \tag{7.148}$$

as shown in Fig. 7.16.

Also, the corresponding curves for the correlation detector, for the same conditions of analysis and the input conditions, are shown in Fig. 7.16.

Comparative analysis of the detection performances of the generalized detector and the correlation detector allows us to draw the conclusion that the generalized detector is clearly superior to the correlation detector during use in communication systems.

The example considered in this section shows us how we can use the main formulae obtained in the course of discussion with the purpose of studying the impact of stimulus of the multiplicative noise on qualitative characteristics of some complex signal processing systems.

The results obtained confirm the fact that for the cases in which a level of energy characteristics of the information signal is much greater than a level of energy characteristics of the additive set noise and parasitic signals (interference), the multiplicative noise can be the only main source of false decision-making in communication systems.

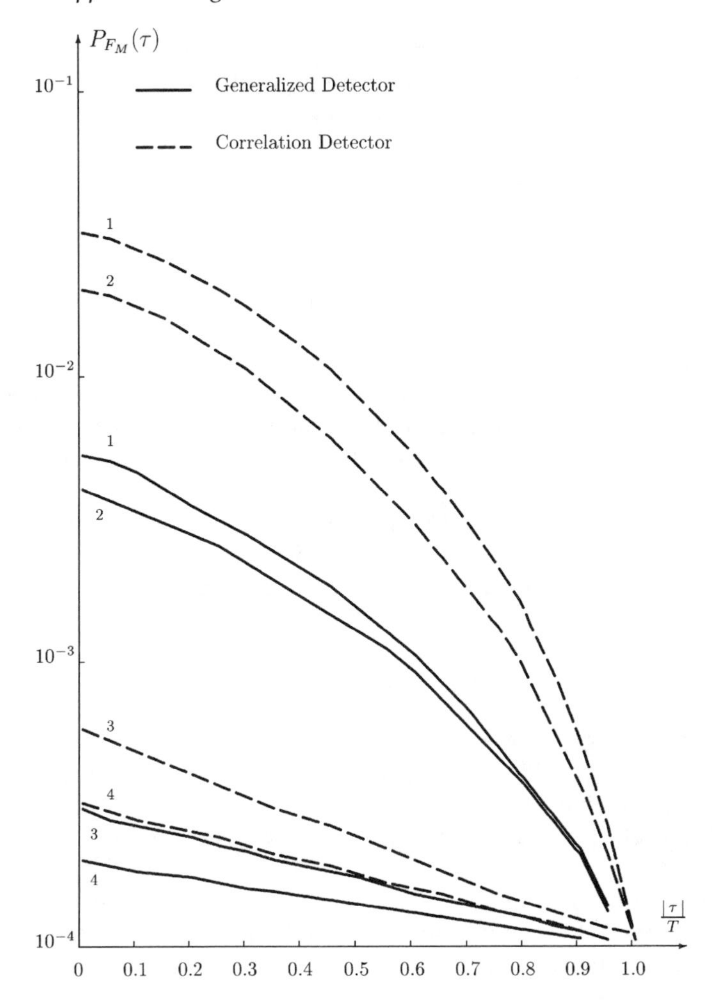

FIGURE 7.16
The probability $P_{F_M}(\tau)$ as a function of parameters of the multiplicative noise: for 1, $P_D = 0.99$, $\delta^2_{k_1} = 0.1$; for 2, $P_D = 0.9$, $\delta^2_{k_1} = 0.1$; for 3, $P_D = 0.99$, $\delta^2_{k_1} = 0.01$; for 4, $P_D = 0.99$, $\delta^2_{k_1} = 0.01$.

7.3 Known Correlation Function of the Multiplicative Noise

In some practical cases, for example, radar, communications, underwater signal processing, sonar, remote sensing, and so on, some characteristics and parameters of the information channel defining characteristics and parameters of the noise modulation function $\dot{M}(t)$ of the multiplicative noise can be *a priori* known.

In this case we can use the generalized approach to signal processing when the fluctuating multiplicative noise is present taking into account *a priori* known information regarding the characteristics and parameters of the noise modulation function $\dot{M}(t)$ of the multiplicative noise.

In this section we consider the case when the correlation function $R_M(\tau)$, or the energy spectrum $G_M(\Omega)$, of the noise modulation function $\dot{M}(t)$ of the multiplicative noise is known. The problem of signal detection under the stimulus of multiplicative noise with the known correlation function applying to the rapid fluctuating multiplicative noise

$$\tau_c \ll T$$

is discussed in References 5, 6, 11, and 68; it is assumed that the distortions in phase of the signal are uniformly distributed within the limits of the interval $[0, 2\pi]$. Consequently, in accordance with the results discussed in Chapter 4, the relative level α_0^2 of the undistorted component of the signal is equal to zero.

Based on results discussed in References 5, 6, 11, and 68 and using the method discussed in Reference 68, we consider the multi-channel generalized detector for the case of the rapid multiplicative noise in a more general case: there is the undistorted component of the signal

$$\alpha_0 \neq 0$$

jointly with the signal noise component–fluctuations of the signal.

We proceed to consider the problem of detection of the signal distorted by the multiplicative noise in the presence of additive Gaussian noise by the generalized detector. Assume that the additive Gaussian noise has the spectral power density $\frac{N_0}{2}$.

As known from References 5, 6, and 11, the optimal procedure of signal detection reduces to a definition of the likelihood ratio and comparison of the likelihood ratio with the threshold that is defined by the predetermined probability of false alarm P_F.

We define the likelihood ratio for the signal distorted by the rapid stationary fluctuating multiplicative noise during the use of the generalized approach to signal detection. We can write the undistorted signal in the following form:

$$a(t) = \sqrt{E_{a_1}}\, a_0(t, \varphi_0) = \sqrt{E_{a_1}}\, Re\left\{\dot{S}_0(t) \cdot e^{j\varphi_0} \cdot e^{j\omega_0 t}\right\}, \tag{7.149}$$

where $a_0(t)$ is the signal with the energy equal to unity and φ_0 is the random initial phase of the signal.

Then the signal distorted by the multiplicative noise has the following form:

$$a_M(t) = \sqrt{E_{a_1}}\, Re\left\{\dot{M}(t)\dot{S}_0(t) \cdot e^{j\varphi_0} \cdot e^{j\omega_0 t}\right\}. \tag{7.150}$$

Using the series expansion of the stationary in the wide sense stochastic function within the limits of the interval $[0, T]$,[5,6,76] the noise modulation function $\dot{M}(t)$ of the multiplicative noise can be written in the following form:

$$\dot{M}(t) = \overline{\dot{M}(t)} + \dot{V}_0(t) = \alpha_0 \cdot e^{j\beta_0} + \sum_{\ell=-\infty}^{\infty} \dot{\upsilon}_\ell \cdot e^{j\ell\Omega_T t}, \quad t \leq T, \tag{7.151}$$

where

$$\Omega_T = \frac{2\pi}{T}. \tag{7.152}$$

The coefficients of series expansion

$$\dot{\upsilon}_\ell = \frac{1}{T}\int_0^T \dot{V}_0(t) \cdot e^{-j\ell\Omega_T t}\, dt \tag{7.153}$$

are the random variables.

As was shown in Reference 76, the mutual correlation of the coefficients $\dot{\upsilon}_\ell$ is decreased as the ratio

$$\frac{T}{\tau_c}$$

is increased, where τ_c is the time of correlation of the noise modulation function $\dot{M}(t)$ of the multiplicative noise. There is a need to note that in the limiting case as

$$\frac{T}{\tau_c} \to \infty \tag{7.154}$$

the coefficients $\dot{\upsilon}_\ell$ are not correlated.

Because of this, in the case of the rapid fluctuating multiplicative noise

$$\tau_c \ll T \tag{7.155}$$

the coefficients $\dot{\upsilon}_\ell$ can be approximately thought of as uncorrelated. For the condition mentioned above, the variance of the coefficients $\dot{\upsilon}_\ell$ is determined in the following form:[76]

$$\begin{aligned}\sigma_\ell^2 = \overline{|\dot{\upsilon}_\ell|^2} &\simeq \frac{1}{T}\int_{-\infty}^{\infty} \dot{R}_V(\tau) \cdot e^{-j\ell\Omega_T \tau}\, d\tau \\ &= \frac{1}{T} G_V(\ell\Omega_T),\end{aligned} \tag{7.156}$$

where $\dot{R}_V(\tau)$ is the correlation function of the fluctuations $\dot{V}_0(t)$ of the noise modulation function $\dot{M}(t)$ of the multiplicative noise and $G_V(\Omega)$ is the energy spectrum of the fluctuations $\dot{V}_0(t)$ of the noise modulation function $\dot{M}(t)$ of the multiplicative noise.

Reference to Eq. (7.156) shows that the series expansion in Eq. (7.151) contains only the terms with the indexes ℓ, for which the function $G_V(\ell\Omega_T)$ differs from zero. The total number of terms of the series expansion in Eq. (7.151) is approximately equal to

$$N \simeq \frac{\Delta\Omega_M}{\Omega_T}, \tag{7.157}$$

where $\Delta\Omega_M$ is the energy spectrum bandwidth of the noise modulation function $\dot{M}(t)$ of the multiplicative noise.

Substituting Eq. (7.151) in Eq. (7.150), the signal distorted by the multiplicative noise can be written as the sum of components

$$\begin{aligned} a_M(t) &= \alpha_0 \cdot \sqrt{E_{a_1}} Re\{\dot{S}_0(t) \cdot e^{j(\varphi_0+\beta_0)} \cdot e^{j\omega_0 t}\} \\ &\quad + \sqrt{E_{a_1}} \sum_{\ell=-\infty}^{\infty} \upsilon_\ell Re\{\dot{S}_0(t) \cdot e^{j\psi_\ell} \cdot e^{j(\omega_0+\ell\Omega_T)t}\} \\ &= \alpha_0 \cdot \sqrt{E_{a_1}} a_0(t, \varphi_0 + \beta_0) + \sqrt{E_{a_1}} \sum_{\ell=-\infty}^{\infty} \upsilon_\ell a_0(t, \psi_\ell, \ell\Omega_T), \end{aligned} \tag{7.158}$$

where

$$\psi_\ell = \varphi_0 + \beta_\ell; \quad \beta_\ell = \arg \dot{\upsilon}_\ell; \quad \upsilon_\ell = |\dot{\upsilon}_\ell|; \tag{7.159}$$

$$a_0(t, \psi, \Omega) = Re\{\dot{S}_0(t) \cdot e^{j\psi} \cdot e^{j(\omega_0+\Omega)t}\}. \tag{7.160}$$

The signal $a_0(t, \psi, \Omega)$ in Eq. (7.160) differs from the signal $a_0(t)$ only by the shift in initial phase by the value ψ and by the shift in carrier frequency ω_0 by the value Ω.

Equations (7.158)–(7.160) allow us to consider the signal distorted by the multiplicative noise as the sum of the undistorted component of the signal and the signal noise component. The signal noise component is the sum of the signals that are similar to the undistorted signal having the stochastic amplitude υ_ℓ and random initial phase ψ_ℓ and are shifted in frequency by the value

$$\ell\Omega_T, \quad \ell = 0, \pm 1, \pm 2, \ldots.$$

The signal $a_M(t)$ in Eq. (7.158) depends on the stochastic parameters φ_0, υ_ℓ, ψ_ℓ. During signal processing of the signal determined by Eq. (7.158) in the presence of additive Gaussian noise $n(t)$ by the generalized detector, the likelihood ratio for the signal at the input of the PF of the linear tract of the generalized detector

$$X(t) = a_M(t) + n(t) \tag{7.161}$$

can be written in the following form:[1,2]

$$\Lambda(X) = \int\limits_{\varphi_0}\int\limits_{\upsilon_\ell}\int\limits_{\psi_\ell} \ldots \Lambda(X/\varphi_0, \dot{\upsilon}_0, \dot{\upsilon}_1, \ldots) \times f(\varphi_0) f(\dot{\upsilon}_0, \dot{\upsilon}_1, \ldots, \psi_0, \psi_1, \ldots)\, d\varphi_0\, d\dot{\upsilon}_0\, d\dot{\upsilon}_1 \ldots d\psi_0\, d\psi_1 \ldots, \quad (7.162)$$

where

$$\Lambda(X/\varphi_0, \dot{\upsilon}_0, \dot{\upsilon}_1, \ldots)$$

is the particular value of the likelihood ratio for the fixed values of the random parameters $\varphi_0, \dot{\upsilon}_0, \dot{\upsilon}_1, \ldots$, which is defined by the following form:

$$\Lambda(X/\varphi_0, \dot{\upsilon}_0, \dot{\upsilon}_1, \ldots) = \exp\left\{-\frac{E_{a_1}(\varphi_0, \dot{\upsilon}_0, \dot{\upsilon}_1, \ldots)}{\sqrt{D_\Sigma}} + \frac{z(X/\varphi_0, \dot{\upsilon}_0, \dot{\upsilon}_1, \ldots)}{\sqrt{D_\Sigma}}\right\}, \quad (7.163)$$

where

$$E_{a_1}(\varphi_0, \dot{\upsilon}_0, \dot{\upsilon}_1, \ldots)$$

is the particular value of the energy of the signal determined by Eq. (7.158);

$$z(X/\varphi_0, \dot{\upsilon}_0, \dot{\upsilon}_1, \ldots)$$

is the particular value of the correlation integral of the generalized detector (see Section 5.4); D_Σ is the variance of the background noise at the output of the generalized detector determined by Eq. (7.13).

We define the values

$$E_{a_1}(\varphi_0, \dot{\upsilon}_0, \dot{\upsilon}_1, \ldots)$$

and

$$z(X/\varphi_0, \dot{\upsilon}_0, \dot{\upsilon}_1, \ldots)$$

for the signal determined by Eq. (7.158). For simplicity assume that the complex envelope of amplitude of the signal takes the square wave-form.

Then, using Eq. (7.158) and taking into account the peculiarity of orthogonality of the functions $e^{j\ell\Omega_T t}$, we can write

$$E_{a_1}(\varphi_0, \dot{\upsilon}_0, \dot{\upsilon}_1, \ldots) = \int\limits_0^T a_M^2(t)\, dt = \alpha_0^2 \cdot E_{a_1} + E_{a_1} \sum_{\ell=-\infty}^{\infty} \upsilon_\ell^2, \quad (7.164)$$

where E_{a_1} is the energy of the undistorted signal $a(t)$;

$$z(X/\varphi_0, \dot{\upsilon}_0, \dot{\upsilon}_1, \ldots) = 2\int_0^T X(t)a_M(t)\,dt - \int_0^T X^2(t)\,dt + \int_0^T \eta^2(t)\,dt$$
$$= \alpha_0 E_{a_1} Z_0 \cos[\varphi_0 + \beta - \vartheta_0]$$
$$+ E_{a_1} \sum_{\ell=-\infty}^{\infty} \upsilon_\ell Z_\ell \cos[\psi_\ell - \vartheta_\ell]; \tag{7.165}$$

where

$$Z_\ell = \frac{1}{2} \cdot \left| 2\int_0^T \dot{X}(t) S_0^*(t) \cdot e^{-j\ell\Omega_T t}\,dt \right|$$
$$- \left| \int_0^T \dot{X}(t) X^*(t) \cdot e^{j\ell\Omega_T t}\,dt \right| + \left| \int_0^T \eta^2(t)\,dt \right|; \tag{7.166}$$

$$\vartheta_\ell = \arg\left\{ \left| 2\int_0^T \dot{X}(t) S_0^*(t) \cdot e^{-j\ell\Omega_T t}\,dt \right| \right.$$
$$\left. - \left| \int_0^T \dot{X}(t) X^*(t) \cdot e^{j\ell\Omega_T t}\,dt \right| + \left| \int_0^T \eta^2(t)\,dt \right| \right\}. \tag{7.167}$$

In terms of Eqs. (7.164) and (7.165) the particular value of the likelihood ratio in Eq. (7.163) can be written in the following form

$$\Lambda(X/\varphi_0, \dot{\upsilon}_0, \dot{\upsilon}_1, \ldots) = \exp\left\{ -\frac{\alpha_0^2 E_{a_1}}{\sqrt{D_\Sigma}} + \frac{\alpha_0 \cdot \sqrt{E_{a_1}}}{\sqrt{D_\Sigma}} \cdot Z_0 \cos[\varphi_0 + \beta_0 - \vartheta_0] \right\}$$
$$\times \prod_{\ell=-\infty}^{\infty} \exp\left\{ -\frac{\upsilon_\ell^2 E_{a_1}}{\sqrt{D_\Sigma}} + \frac{\upsilon_\ell \sqrt{E_{a_1}}}{\sqrt{D_\Sigma}} \cdot Z_\ell \cos[\psi_\ell - \vartheta_\ell] \right\}$$
$$= \Lambda(X_{res}/\varphi_0) \prod_{\ell=-\infty}^{\infty} \Lambda(X_\ell/\upsilon_\ell, \psi_\ell), \tag{7.168}$$

where $\Lambda(X_{res}/\varphi_0)$ is the particular value of the likelihood ratio for the undistorted residual of the signal and $\Lambda(X_\ell/\upsilon_\ell)$ is the particular value of the likelihood ratio for the "ℓ"-th component of the signal noise component.

Reference to Eq. (7.168) shows that the particular value of the likelihood ratio for the signal $a_M(t)$ determined by Eq. (7.158) for the fixed signal parameters is the product of the particular values of the likelihood ratio for individual components of the signal $a_M(t)$.

To determine the value of the likelihood ratio there is a need to average Eq. (7.168) with respect to the random parameters

$$\varphi_0, \upsilon_0, \upsilon_1, \ldots, \upsilon_i, \ldots, \psi_0, \psi_1, \ldots.$$

For this purpose it is required to set the definite statistical characteristics of the multiplicative noise defining the probability distribution densities of the random coefficients $\dot{\upsilon}_0, \dot{\upsilon}_1, \ldots$. Assume that instantaneous values of the fluctuations $\dot{V}_0(t)$ of the noise modulation function $\dot{M}(t)$ of the multiplicative noise obey the Gaussian probability distribution density. Then the random coefficients $\dot{\upsilon}_\ell$ that are a linear transform of the fluctuations $\dot{V}_0(t)$ also obey the Gaussian probability distribution density. The absolute values of the coefficients υ_ℓ obey the Rayleigh probability distribution density.

As was discussed above, the random coefficients $\dot{\upsilon}_\ell$ can be thought of as uncorrelated. As is well known, the uncorrelated random variables obeying the Gaussian probability distribution density are statistically independent. When the random coefficients $\dot{\upsilon}_\ell$ are independent an average of Eq. (7.168) gives us the following result

$$\Lambda(X) = \Lambda(X_{res}) \prod_{\ell=-\infty}^{\infty} \Lambda(X_\ell) \tag{7.169}$$

or

$$\ln \Lambda(X) = \ln \Lambda(X_{res}) + \sum_{\ell=-\infty}^{\infty} \ln \Lambda(X_\ell), \tag{7.170}$$

where

$$\Lambda(X_{res}) = e^{-\frac{\alpha_0^2 E_{a_1}}{\sqrt{D_\Sigma}}} \times \int_{\varphi_0} \exp\left\{\frac{\alpha_0 \cdot \sqrt{E_{a_1}}}{\sqrt{D_\Sigma}} \cdot Z_0 \cos[\varphi_0 + \beta_0 - \vartheta_0]\right\} f(\varphi_0)\, d\varphi_0; \tag{7.171}$$

$$\Lambda(X_\ell) = \int_{\upsilon_\ell}\int_{\psi_\ell} \exp\left\{-\frac{\upsilon_\ell^2 E_{a_1}}{\sqrt{D_\Sigma}} + \frac{\upsilon_\ell \sqrt{E_{a_1}}}{\sqrt{D_\Sigma}} \cdot Z_\ell \cos[\psi_\ell - \vartheta_\ell]\right\} \times f(\upsilon_\ell, \psi_\ell)\, d\upsilon_\ell\, d\psi_\ell. \tag{7.172}$$

Thus, when the coefficients $\dot{v}_\ell$ are independent the likelihood ratio for the signal $a_M(t)$ is the product of values of the likelihood ratio for individual components of the signal $a_M(t)$.

In the case of the Rayleigh probability distribution density of the random variables v_ℓ and the uniform probability distribution density of the phases φ_0 and ψ_ℓ, Eqs. (7.171) and (7.172) can be written in the following form:

$$\Lambda(X_{res}) = \exp\left\{-\frac{\alpha_0^2 E_{a_1}}{\sqrt{D_\Sigma}}\right\} \cdot K_0\left(\frac{\alpha_0\sqrt{E_{a_1}}}{\sqrt{D_\Sigma}} \cdot Z_0\right); \tag{7.173}$$

$$\Lambda(X_\ell) = \frac{\sqrt{D_\Sigma}}{\sigma_\ell^2 E_{a_1} + \sqrt{D_\Sigma}} \cdot \exp\left\{\frac{\sigma_\ell^2 E_{a_1}}{2\sigma_\ell^2 E_{a_1} + \sqrt{D_\Sigma}} \cdot Z_\ell^2\right\}, \tag{7.174}$$

where $K_0(x)$ is the modified Bessel function of the second order of an imaginary argument; σ_ℓ^2 is the variance of the coefficients v_ℓ determined by Eq. (7.156); and D_Σ is the variance of the background noise of the generalized detector determined by Eq. (7.13).

Substituting Eqs. (7.156), (7.173), and (7.174) in Eq. (7.170) and ignoring the constant factor

$$k_1 = \frac{4(\Delta\omega_{PF})\sqrt{T\beta}}{\omega_0^2} \tag{7.175}$$

that is defined only by the parameters of the PF of the input linear tract of the generalized detector, we obtain that the likelihood ratio $\Lambda(X)$ can be determined in the following form:

$$\frac{N_0}{E_{a_1}} \cdot \ln\Lambda(X) = \frac{N_0}{E_{a_1}} \cdot \ln K_0\left(\frac{\alpha_0\sqrt{E_{a_1}}}{N_0} \cdot Z_0\right) + \sum_{\ell=-\infty}^{\infty} \frac{G_V(\ell\Omega_T)}{2qG_V(\ell\Omega_T) + T} \cdot Z_\ell^2 + C, \tag{7.176}$$

where C is the constant value containing the terms that are defined only by *a priori* known parameters of the signal and the multiplicative noise; the sum contains N terms

$$N \simeq \frac{\Delta\Omega_M}{\Omega_T}, \tag{7.177}$$

for which the following condition

$$G_V(\ell\Omega_T) \neq 0 \tag{7.178}$$

is true; and q is the detection parameter of the generalized detector determined by Eq. (7.84).

Thus, the problem of detection of the signal $a_M(t)$ in the presence of additive Gaussian noise, when there is the rapid fluctuating multiplicative noise, is reduced to a definition of the likelihood ratio

$$\ln \Lambda(X)$$

or the proportional value

$$\frac{N_0}{E_{a_1}} \cdot \ln \Lambda(X)$$

for the signal

$$X(t) = a_M(t) + n(t) \tag{7.179}$$

at the input of the PF of the linear tract of the generalized detector in accordance with Eq. (7.176) and comparison of the value of the likelihood ratio with the threshold, where $z_{g_M}^{out}(t)$ is the process formed at the output of the generalized detector.

Each value Z_ℓ in Eq. (7.176) is the complex envelope of amplitude of the signal at the output of the generalized detector generated by the signal $X(t)$ at the input of the PF of the linear tract of the generalized detector at the instant $t = T$ — the PF is matched with the signal $a_0(t, \ell\Omega_T)$. The considered PF of the linear tract of the generalized detector differs from the PF matched with the undistorted signal $a(t)$ only by the shift in the detuning frequency by the value $\ell\Omega_T$.

Thus, reference to Eq. (7.176) shows that for the considered case the generalized detector must contain N frequency channels by the frequencies

$$\omega_0 + \ell\Omega_T, \quad \ell = 0, \pm 1, \pm 2, \ldots; \tag{7.180}$$

see Fig. 7.17.

The shift in frequency between the adjacent channels of the generalized detector is equal to

$$\Omega_T = \frac{2\pi}{T}. \tag{7.181}$$

Each channel of the generalized detector must contain the input linear tract—the PF—matched with the undistorted signal $a(t, \ell\Omega_T)$, the correlation and autocorrelation channels, the compensating channel (see Chapter 5). The AF of the input linear tract of the generalized detector can be common for all channels and must be mismatched with the undistorted signal $a(t, \ell\Omega_T)$ (see Chapter 5).

Amplitudes of the signals at the output of channels of the generalized detector are added with the weight coefficients

$$d_\ell = \frac{G_V(\ell\Omega_T)}{2q G_V(\ell\Omega_T) + T} \tag{7.182}$$

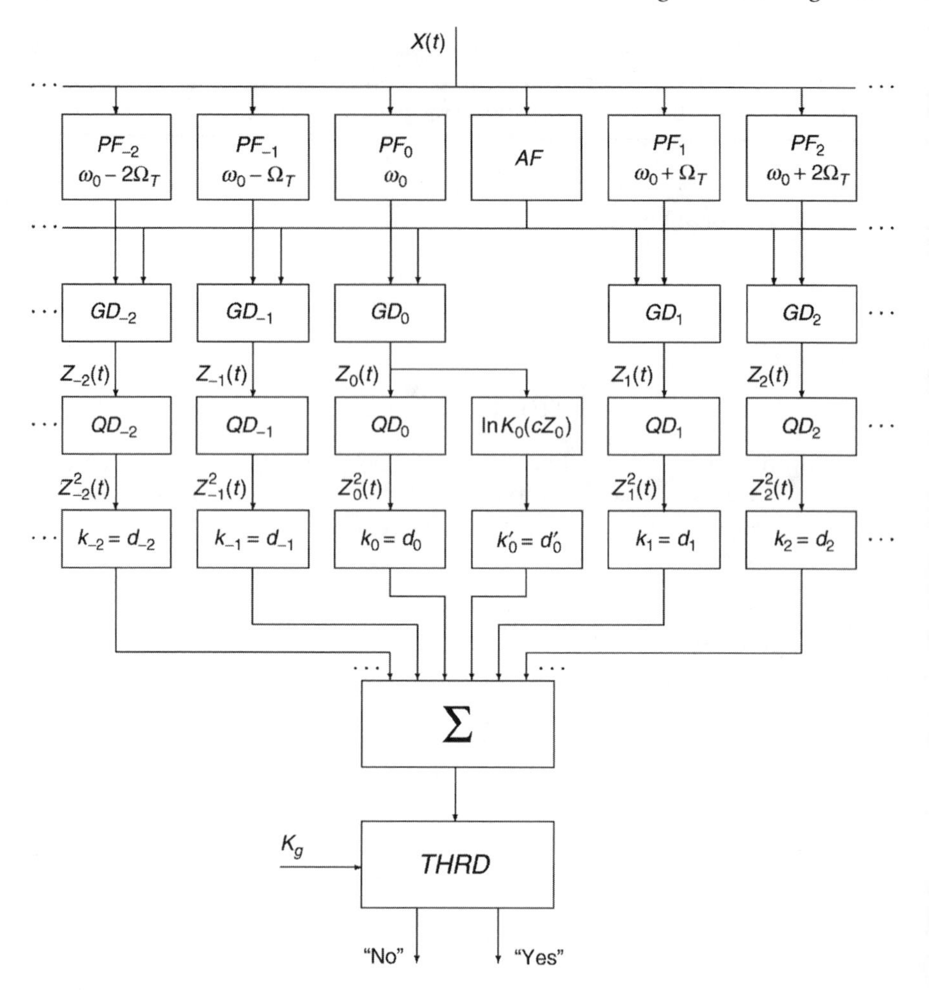

FIGURE 7.17
Multi-channel generalized detector: GD_k is the "k"-th channel of the generalized detector; QD is the quadrature detector; k_k is the amplifier of the "k"-th channel; Σ is the summator; $THRD$ is the threshold device; K_g is the threshold value.

that are defined by characteristics of the multiplicative noise—the energy spectrum of the noise modulation function $\dot{M}(t)$ of the multiplicative noise—and the signal-to-noise ratio at the output of the PF at the input linear tract of the generalized detector. The amplifier coefficients k_ℓ of channels of the generalized detector must be proportional to the corresponding weight coefficients d_ℓ in order to carry out a weight summing.

Dependence of the weight coefficients d_ℓ on the detection parameter q of the generalized detector can be explained by the following fact. To achieve optimal signal processing there is a need to use all frequency components of the signal (see Eq. (7.158)), the level of energy characteristics of which exceeds the level of energy characteristics of the additive Gaussian noise. Because of this, at high values of the signal-to-noise ratio at the input of the linear tract of the generalized detector it is worthwhile to use the large number of frequency channels.

Reference to Eq. (7.182) shows that a difference between the weight coefficients of those channels of the generalized detector, in which the value $G_V(\ell\Omega_T)$ is maximum, and those channels of the generalized detector, in which the value $G_V(\ell\Omega_T)$ is minimum, is decreased with an increase in the signal-to-noise ratio at the input of the PF of the linear tract of the generalized detector that corresponds to an increase in the total numbers of channels of the generalized detector, which are required to be used.

In addition to the functions mentioned above, in order to generate the amplitude of the signal at the output of the generalized detector, which corresponds to the first term on the right side of Eq. (7.176), there is a need to connect the output of the "zero" channel of the linear tract of the generalized detector to the generator forming the transform

$$u = \ln K_0(cZ_0), \tag{7.183}$$

where c is the constant value. The amplitude obtained with the weight coefficient

$$d_0^{'} = \frac{1}{q} \tag{7.184}$$

must be coming in to the input of the summator Σ (see Fig. 7.17).

We consider two particular cases when the ratio between the energy of the undistorted component of the signal and the spectral power density of the additive Gaussian noise

$$\frac{\alpha_0^2 \cdot E_{a_1}}{N_0} = \alpha_0^2 \cdot q \tag{7.185}$$

is very high and very low.

- The case $\alpha_0^2 \cdot q \gg 1$. Then we can write

$$\ln K_0\left(\frac{\alpha_0 \cdot \sqrt{E_{a_1}}}{N_0} \cdot Z_0\right) \simeq \frac{\alpha_0 \cdot \sqrt{E_{a_1}}}{N_0} \cdot Z_0. \tag{7.186}$$

For this case the amplitude of the signal at the output of the "zero" channel of the generalized detector with the weight coefficient

$$d_0^{''} = \frac{\alpha_0 \cdot N_0}{\sqrt{E_{a_1}}} \tag{7.187}$$

is added to amplitudes at the outputs of other channels of the generalized detector.

- The case $\alpha_0^2 \cdot q \ll 1$. For this case

$$\ln K_0\left(\frac{\alpha_0 \cdot \sqrt{E_{a_1}}}{N_0} \cdot Z_0\right) \simeq \frac{\alpha_0^2 \cdot E_{a_1}}{N_0} \cdot Z_0^2. \tag{7.188}$$

Reference to Eq. (7.188) shows that the first term on the right side of Eq. (7.176) can be taken into account during the use of the following weight coefficient of the "zero" channel

$$d_0''' = d_0 + \alpha_0^2. \tag{7.189}$$

The detection performances of the multi-channel generalized detector for the case in which $\alpha_0 = 0$ and the energy spectrum $G_V(\Omega)$ of fluctuations of the noise modulation function $\dot{M}(t)$ of the multiplicative noise has the square wave-form and, consequently, all weight coefficients d_ℓ determined by Eq. (7.182) are the same as will be discussed in Section 7.5.

The formulae mentioned above are true for the case

$$\tau_c \ll T$$

and

$$\tau_c \gg T.$$

For the case

$$\tau_c \simeq T$$

the number of channels and the amplitude-frequency response of the PF and AF can be defined based on results discussed in Reference 77.

Analysis performed in this section allows us to draw some conclusions.

- To detect the signals distorted by the rapid fluctuating multiplicative noise in the presence of additive Gaussian noise it is required to know *a priori* information about the characteristics of the multiplicative noise, the energy of the signal at the input of the PF of the linear tract of the generalized detector, and the spectral power density of the additive Gaussian noise.
- The weight coefficients must be controlled both during variations in characteristics of the multiplicative noise and during variations in the signal-to-noise ratio at the input of the PF of the linear tract of the generalized detector.
- Absence of *a priori* information in some practical cases about the characteristics of the information channel—the noise modulation function $\dot{M}(t)$ of the multiplicative noise—and their variability and *a priori* ignorance of the energy characteristics of the detected signal, very often in practice leads to the employment of complex signal processing systems used for the signals with unknown *a priori* structure.

- The bandwidth of the input linear tract of such complex signal processing systems is wider than a possible energy spectrum bandwidth of fluctuations of the received signals.

The next section is devoted to estimation of the impact of the multiplicative noise on the detection performances of these complex signal processing systems using the generalized detector.

7.4 One-Channel Generalized Detector

As was noted in Chapter 3, the fluctuating multiplicative noise existing at the information channel leads to distortions in amplitude and phase of the signal that can be *a priori* unknown. If a degree of distortions in the amplitude $A(t)$ and the phase $\varphi(t)$ of the signal is very high, then the structure of the received signal at the input of the PF of the linear tract of the generalized detector is unknown.

With the purpose of reducing requirements to the information signals regarding the characteristics of the information channel, which are varied in time, during generation of the model signal—the reference signal—of the generalized detector, only the autocorrelation channel of the generalized detector can be used (see Fig. 7.18). As was shown in References 5, 6, 10–14, 78, and 79, the autocorrelation signal processing method is optimal for the case of the signals with an *a priori* unknown amplitude-phase-frequency structure.

In this section we study and discuss the detection performances of the signals that are processed only by the autocorrelation channel of the generalized detector when the signals are distorted by the multiplicative noise.

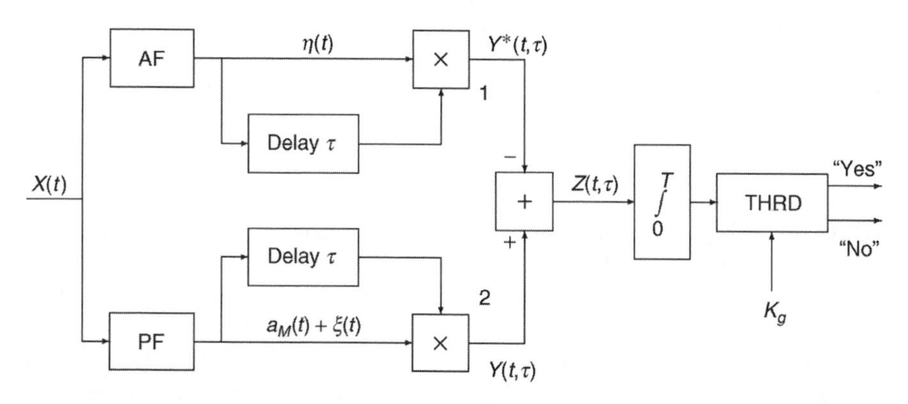

FIGURE 7.18
Autocorrelation channel of the generalized detector.

7.4.1 Detection Performances

The signal at the output of the autocorrelation channel of the generalized detector at the instant $t = T_{int}$, corresponding to the instant of decision-making, is determined by the following form (see Chapter 5):

$$
\begin{aligned}
z_{au}^{out}(T_{int}, t) &= \int_0^{T_{int}} h(T_{int} - t)X(t-\tau)X(t)\,dt \\
&\quad - \int_0^{T_{int}} h(T_{int} - t)X^*(t-\tau)X^*(t)\,dt \\
&= \int_0^{T_{int}} h(t)X(T_{int} - t - \tau)X(T_{int} - t)\,dt \\
&\quad - \int_0^{T_{int}} h(t)X^*(T_{int} - t - \tau)X^*(T_{int} - t)\,dt,
\end{aligned}
\tag{7.190}
$$

where $h(t)$ is the pulse transient response of the integrator (the ripple filter); τ is the delay between the processes $X(t)$ and $X(t-\tau)$ and the processes $X^*(t)$ and $X^*(t-\tau)$;

$$X(t) = a_M(t) + \xi(t) \tag{7.191}$$

is the process at the output of the PF of the linear tract of the generalized detector; $a_M(t)$ is the signal distorted by the multiplicative noise at the output of the PF; $\xi(t)$ is the additive Gaussian noise formed at the output of the PF of the linear tract of the generalized detector under the stimulus of the additive Gaussian noise $n(t)$ at the input of the PF of the linear tract of the generalized detector; and

$$X^*(t) = \eta(t) \tag{7.192}$$

is the process at the output of the AF of the linear tract of the generalized detector, where $\eta(t)$ is the additive Gaussian noise formed at the output of the AF of the linear tract of the generalized detector under the stimulus of the additive Gaussian noise $n(t)$ at the input of the AF of the linear tract of the generalized detector. As was discussed in Chapter 5, the AF does not pass the signal $a(t)$. The relationships between the pulse transient responses of the AF and PF were discussed in additional detail in Chapter 5.

Taking into account Eqs. (7.191) and (7.192), Eq. (7.190) can be written in the following form:

$$z_{au}^{out}(T_{int}, t) = \int_0^{T_{int}} h(t)X(T_{int} - t - \tau)X(T_{int} - t)\,dt - \int_0^{T_{int}} h(t)\eta(T_{int} - t - \tau)\eta(T_{int} - t)\,dt, \quad (7.193)$$

where

$$X(t) = Re\{A_0(t)\dot{M}(t)\dot{S}_0(t) \cdot e^{j(\omega_0 t + \varphi_{0_1})}\} + \xi(t) \quad (7.194)$$

and

$$X(t-\tau) = Re\{A_0(t)\dot{M}(t-\tau)\dot{S}_0(t-\tau) \cdot e^{j(\omega_0 t + \varphi_{0_2})}\} + \xi(t-\tau) \quad (7.195)$$

are the signals at the inputs of the multiplier 1 (see Fig. 7.18) of the autocorrelation channel of the generalized detector taking into consideration the additive set noise $\xi(t)$ and $\xi(t-\tau)$; $A_0(t)$ is the amplitude factor of the received signal; T_{int} is the time of integration (accumulation); and

$$\varphi_{0_2} = \varphi_{0_1} + \omega_0\tau. \quad (7.196)$$

It is taken into consideration in Eq. (7.190) that the PF of the linear tract of the generalized detector has the bandwidth $\Delta\omega'_{PF}$ that is much greater than the equivalent energy spectrum bandwidth $\Delta\Omega_{0_M}$ of the received signal for the condition that the multiplicative noise is present, owing to absence of *a priori* knowledge about the structure of the signal and possible variations in the carrier frequency ω_0 of the signal—an absence of *a priori* knowledge about the carrier frequency.

As a rule, for the autocorrelation channel of the generalized detector the integration time T_{int} is much greater than the expected correlation interval τ_{c_0} of the received signal and the time of correlation τ_c of the noise modulation function $\dot{M}(t)$ of the multiplicative noise:

$$T_{int} \gg \tau_{c_0} \quad (7.197)$$

and

$$T_{int} \gg \tau_c. \quad (7.198)$$

For this case the signal at the output of the integrator of the autocorrelation channel of the generalized detector (see Eq. (7.190)) can be represented as the sum of the large number of uncorrelated samples.

In accordance with the central limit theorem, we can believe that the probability distribution density of the signal at the output of the integrator of the autocorrelation channel (see Eq. (7.193)) is defined by the Gaussian probability distribution density:[5,6,76,80]

$$f_{H_1}\left[z_{au}^{out}(\tau)\right] = \frac{1}{\sqrt{\pi}\sigma_{11}(\tau)} \cdot \exp\left\{-\frac{\left[z_{au}^{out}(\tau) - m_{11}(\tau)\right]^2}{2\sigma_{11}^2(\tau)}\right\} \tag{7.199}$$

with the mean

$$m_{11}(\tau) = \overline{z_{au}^{out}(\tau)} \tag{7.200}$$

and the variance

$$\sigma_{11}^2(\tau) = m_2\left\{z_{au}^{out}(\tau) - \overline{\left[z_{au}^{out}(\tau)\right]^2}\right\}, \tag{7.201}$$

where

$$\begin{aligned} m_2\left\{z_{au}^{out}(\tau)\right\} = & \int_0^{T_{int}}\int_0^{T_{int}} h(T_{int} - t')h(T_{int} - t'') \\ & \times \overline{X(t' - \tau)X(t')X(t'' - \tau)X(t'')}\, dt'\, dt'' \\ & + \int_0^{T_{int}}\int_0^{T_{int}} h(T_{int} - t')h(T_{int} - t'') \\ & \times \overline{\eta(t' - \tau)\eta(t')\eta(t'' - \tau)\eta(t'')}\, dt' d\, t'' \end{aligned} \tag{7.202}$$

and $\sigma_{11}^2(\tau)$ is the variance of the process at the output of the integrator of the autocorrelation channel of the generalized detector under the hypothesis H_1—a "yes" signal at the input of the PF of the linear tract of the generalized detector.

We introduce the coefficient characterizing a quality of filtering by the PF and AF of the linear tract of the generalized detector in the following form:

$$\gamma_1 = \frac{\Delta\omega_{PF}}{\Delta\omega_{in}} = \frac{\Delta\omega_{AF}}{\Delta\omega_{in}}, \tag{7.203}$$

where $\Delta\omega_{PF}$ and $\Delta\omega_{AF}$ are the equivalent bandwidths of the PF and AF of the linear tract of the generalized detector, respectively, and $\Delta\omega_{in}$ is the equivalent bandwidth of the integrator. Henceforth, we will believe that

$$\gamma_1 \gg 1. \tag{7.204}$$

We proceed to define the probability of detection P_{D_M} of the signal at the output of the integrator of the autocorrelation channel of the generalized detector in the presence of additive Gaussian noise when the multiplicative noise is present.

The additive noise $\xi(t)$ and $\eta(t)$ caused by the additive set noise of the generalized detector at the inputs of the multipliers 1 and 2, respectively, are the stationary Gaussian stochastic processes with a zero mean and the correlation function

$$R_n(\tau) = \frac{N_0}{4\pi} \int_{-\infty}^{\infty} |\dot{\mathcal{G}}_{1,2}(\omega)|^2 \cdot e^{j\omega\tau}\, d\omega = Re\{\dot{R}_n(\tau) \cdot e^{j\omega_0\tau}\} \tag{7.205}$$

owing to the conditions discussed in Chapter 5. $\dot{\mathcal{G}}_1(\omega)$ in Eq. (7.205) is the amplitude-frequency response of the PF of the linear tract of the generalized detector and $\dot{\mathcal{G}}_2(\omega)$ is the amplitude-frequency response of the AF of the linear tract of the generalized detector.

Since the correlation functions of the additive Gaussian noise $\xi(t)$ and $\eta(t)$ formed at the outputs of the PF and AF, respectively, are the same, we use the same correlation function $R_n(\tau)$ for both the additive Gaussian noise $\xi(t)$ and for the additive Gaussian noise $\eta(t)$.

For example, during passing the additive Gaussian noise through the PF and AF that are the cascade intermediate-frequency amplifier—the number of cascades is greater than 6—the correlation function determined by Eq. (7.205) has the following form:[14]

$$\begin{aligned} R_n(\tau) &= \sigma_n^2 \cdot \exp\{-\pi(\Delta\omega_{PF})^2\tau^2\} \cos\omega_0\tau \\ &= \sigma_n^2 \cdot \exp\{-\pi(\Delta\omega_{AF})^2\tau^2\} \cos\omega_0'\tau, \end{aligned} \tag{7.206}$$

where $\Delta\omega_{PF}$ and $\Delta\omega_{AF}$ are the bandwidths of the PF and AF of the linear tract of the generalized detector, respectively; ω_0 is the detuning frequency of the PF; ω_0' is the detuning frequency of the AF; and

$$\sigma_n^2 = \frac{N_0}{\Delta\omega_{PF}} = \frac{N_0}{\Delta\omega_{AF}} \tag{7.207}$$

is the variance of the additive Gaussian noise at the outputs of the PF and AF, respectively.

Suppose that the hypothesis H_0 corresponds to a "no" signal at the input of the PF of the linear tract of the generalized detector and the hypothesis H_1 corresponds to a "yes" signal at the input of the PF of the linear tract of the generalized detector.

Then the probability distribution density of the signal, determined by Eq. (7.193) for the hypothesis H_0, can be written in the following form:

$$f_{H_0}\left[z_{au}^{out}(\tau)\right] = \frac{1}{\sqrt{\pi}\sigma_{00}(\tau)} \cdot \exp\left\{-\frac{\left[z_{au}^{out}(\tau) - m_{00}(\tau)\right]^2}{2\sigma_{00}^2(\tau)}\right\}, \tag{7.208}$$

where $m_{00}(\tau)$ is the mean of the signal at the output of the integrator of the autocorrelation channel of the generalized detector under the hypothesis H_0; $\sigma_{00}^2(\tau)$ is the variance of the signal at the output of the integrator of the autocorrelation channel of the generalized detector under the hypothesis H_0.

The probability distribution density of the signal at the output of the integrator of the autocorrelation channel of the generalized detector under the hypothesis H_1 is determined by Eq. (7.199) with the mean $m_{11}(\tau)$ and the variance $\sigma_{11}^2(\tau)$ determined by Eqs. (7.200)–(7.202), respectively.

The probability of false alarm P_F defined by the probability of choosing the hypothesis H_1, when the hypothesis H_0 is true, is determined by the following form:

$$P_F = \int_{K_g}^{\infty} f_{H_0}\left[z_{au}^{out}(\tau)\right] d\left[z_{au}^{out}(\tau)\right] = 1 - \Phi\left[\frac{K_g - m_{00}(\tau)}{\sigma_{00}(\tau)}\right], \tag{7.209}$$

where

$$\Phi(x) = \frac{1}{\sqrt{2\pi}} \int_{-\infty}^{x} e^{-\frac{t^2}{2}}\, dt \tag{7.210}$$

is the error integral.

As given by the probability of false alarm P_F, we can determine the threshold value:

$$K_g = \sigma_{00}(\tau)\Phi^{-1}(1 - P_F) + m_{00}(\tau), \tag{7.211}$$

where $\Phi^{-1}(x)$ is the reciprocal function with respect to the error integral $\Phi(x)$.

The probability of detection P_{D_M} of the signal at the output of the integrator of the autocorrelation channel of the generalized detector—the probability that the hypothesis H_1 is true—is determined by the following form:

$$P_{D_M} = \int_{K_g}^{\infty} f_{H_1}\left[z_{au}^{out}(\tau)\right] d\left[z_{au}^{out}(\tau)\right] = 1 - \Phi\left[\frac{K_g - m_{11}(\tau)}{\sigma_{11}(\tau)}\right]. \tag{7.212}$$

Substituting Eq. (7.211) that defines the threshold value K_g in Eq. (7.212), the probability of detection P_{D_M} of the signal at the output of the integrator of the autocorrelation channel of the generalized detector is determined in the following form:

$$P_{D_M} = 1 - \Phi\left\{\frac{\sigma_{00}(\tau)}{\sigma_{11}(\tau)}\left[\Phi^{-1}(1 - P_F) - \frac{m_{11}(\tau) - m_{00}(\tau)}{\sigma_{00}(\tau)}\right]\right\}. \tag{7.213}$$

Analogous to References 11–13 the signal-to-noise ratio is the ratio between the absolute value of the mean differential of the signal at the output of the integrator of the autocorrelation channel of the generalized detector determined by Eq. (7.193) under the hypothesis H_1 with respect to the mean of the signal at the output of the integrator of the autocorrelation channel of the generalized detector determined by Eq. (7.193) under the hypothesis H_0 and the variance of fluctuations of the signal at the output of the integrator of the autocorrelation channel of the generalized detector under the hypothesis H_0.

In line with this statement we can rewrite Eq. (7.213) in the following form:

$$P_{D_M} = 1 - \Phi\left\{\frac{\sigma_{00}(\tau)}{\sigma_{11}(\tau)}\left[\Phi^{-1}(1 - P_F) - q_{au}^{out}(\tau)\right]\right\}, \tag{7.214}$$

where

$$q_{au}^{out}(\tau) = \frac{|m_{11}(\tau) - m_{00}(\tau)|}{\sigma_{00}(\tau)} \tag{7.215}$$

is the signal-to-noise ratio at the output of the integrator of the autocorrelation channel of the generalized detector.

Equations (7.214) and (7.215) allow us to define a set of the detection performances of the signals by the autocorrelation channel of the generalized detector for the known values of the statistical characteristics $m_{11}(\tau)$, $m_{00}(\tau)$, $\sigma_{11}(\tau)$, and $\sigma_{00}(\tau)$ of the signal determined by Eq. (7.193) and the predetermined probability of false alarm P_F.

7.4.2 Statistical Characteristics

We define the mean and variance of fluctuations of the signal determined by Eq. (7.193), at the output of the integrator of the autocorrelation channel of the generalized detector, when the integration time is equal to T_{int} and the signal duration is equal to T:

$$T_{int} \geq T.$$

Taking into consideration the results discussed in Section 4.2, we can define the mean of the process at the output of the multiplier 1 (see Fig. 7.18)

$$\overline{Y(t,\tau)} = \overline{X(t)X(t-\tau)}, \tag{7.216}$$

which is the initial moment of the second order and can be written in the following form:

$$\begin{aligned}\overline{Y(t,\tau)} &= \overline{X(t)X(t-\tau)} = R_Y(t, t-\tau)\\ &= Re\{\dot{R}_\xi(\tau)\cdot e^{j\omega_0\tau}\} + \frac{A_0^2(t)}{2}\cdot Re\{\dot{R}_M(\tau)\dot{S}_0(t)S_0^*(t-\tau)\cdot e^{j\omega_0\tau}\}\\ &\quad + \frac{A_0^2(t)}{2}\cdot Re\{\dot{D}_M(t,\tau)\dot{S}_0(t)\dot{S}_0(t-\tau)\cdot e^{j(2\omega_0 t-\omega_0\tau+2\varphi_{0_1})}\},\end{aligned} \tag{7.217}$$

where

$$\dot{D}_M(t,\tau) = \overline{\dot{M}(t)\dot{M}(t-\tau)}. \tag{7.218}$$

The mean of the process at the output of the multiplier 2 (see Fig. 7.18)

$$\overline{Y^*(t,\tau)} = \overline{X^*(t)X^*(t-\tau)} \tag{7.219}$$

has the following form:

$$\overline{Y^*(t,\tau)} = \overline{X^*(t)X^*(t-\tau)} = Re\{\dot{R}_\eta(\tau)\cdot e^{j\omega_0\tau}\}. \tag{7.220}$$

The mean of the process at the output of the summator "+" (see Fig. 7.18) can be determined in the following form:

$$\begin{aligned}\overline{Z(t,\tau)} &= \overline{Y(t,\tau)} - \overline{Y^*(t,\tau)}\\ &= Re\{\dot{R}_M(\tau)\dot{S}_0(t)S_0^*(t-\tau)\cdot e^{j\omega_0\tau}\}\\ &\quad + \frac{A_0^2(t)}{2}\cdot Re\{\dot{D}_M(t,\tau)\dot{S}_0(t)\dot{S}_0(t-\tau)\cdot e^{j(2\omega_0 t-\omega_0\tau+2\varphi_{0_1})}\}\\ &\quad + Re\{[\dot{R}_\xi(\tau)-\dot{R}_\eta(\tau)]\cdot e^{j\omega_0\tau}\}.\end{aligned} \tag{7.221}$$

The mean of the signal formed at the output of the integrator of the autocorrelation channel of the generalized detector, in terms of Eqs. (7.217)–(7.221), can be written in the following form:

$$\begin{aligned}\overline{z_{au}^{out}(\tau)} = m_{11}(\tau) &= \frac{A_0^2(t)}{2}\cdot Re\{\dot{R}_M(\tau)\dot{\rho}_a(\tau,0)\cdot e^{j\omega_0\tau}\}\\ &\quad + g_0(t,\tau) + K_T(0)Re\{[\dot{R}_\xi(\tau)-\dot{R}_\eta(\tau)]\cdot e^{j\omega_0\tau}\},\end{aligned} \tag{7.222}$$

where

$$\dot{\rho}_a(\tau,0) = \int_0^{T_{int}} h(T_{int}-t)\dot{S}_0(t)S_0^*(t-\tau)\,dt; \tag{7.223}$$

$$\begin{aligned}g_0(t,\tau) = \frac{A_0^2(t)}{2}Re\Bigg\{\int_0^{T_{int}} & h(T_{int}-t)\dot{D}_M(t,\tau)\\ & \times \dot{S}_0(t)\dot{S}_0(t-\tau)\cdot e^{j(2\omega_0 t-\omega_0\tau+2\varphi_{0_1})}\,dt\Bigg\};\end{aligned} \tag{7.224}$$

$$K_T(0) = \int_0^{T_{int}} h(T_{int}-t)\,dt \tag{7.225}$$

is the value of the frequency response $K_T(\Omega)$ of the integrator (the ripple filter) at the frequency $\Omega = 0$.

The term $g_0(t, \tau)$, determined by Eq. (7.224), is infinitesimal in comparison with other terms in Eq. (7.222). For this reason, we can neglect the term $g_0(t, \tau)$ in Eq. (7.222) by the analogous method to Section 7.1.

Thus, the mean of the signal at the output of the integrator of the autocorrelation channel of the generalized detector in the case of the hypothesis H_1 is determined in the following form:

$$\overline{z_{H_1}^{out}(\tau)} = m_{11}(\tau) = \frac{A_0^2(t)}{2} \cdot Re\{\dot{R}_M(\tau)\dot{\rho}_a(\tau, 0) \cdot e^{j\omega_0\tau}\} + K_T(0)Re\{[\dot{R}_\xi(\tau) - \dot{R}_\eta(\tau)] \cdot e^{j\omega_0\tau}\}. \quad (7.226)$$

The mean of the signal at the output of the integrator of the autocorrelation channel of the generalized detector in the case of the hypothesis H_0 is determined in the following form:

$$\overline{z_{H_0}^{out}(\tau)} = m_{00}(\tau)K_T(0)Re\{[\dot{R}_\xi(\tau) - \dot{R}_\eta(\tau)] \cdot e^{j\omega_0\tau}\}. \quad (7.227)$$

Taking into consideration that the additive Gaussian noise $\xi(t)$ and $\eta(t)$, formed at the outputs of the PF and AF of linear tract of the generalized detector, are uncorrelated and independent of each other and have the same correlation function

$$R_\xi(\tau) = R_\eta(\tau) \quad (7.228)$$

owing to the conditions of choice of the amplitude-frequency responses of the PF and AF, therefore Eqs. (7.226) and (7.227) can be written in the following form in the statistical sense:

$$\overline{z_{H_1}^{out}(\tau)} = m_{11}(\tau) = \frac{A_0^2(t)}{2} \cdot Re\{\dot{R}_M(\tau)\dot{\rho}_a(\tau, 0) \cdot e^{j\omega_0\tau}\}; \quad (7.229)$$

$$\overline{z_{H_0}^{out}(\tau)} = m_{00}(\tau) = 0. \quad (7.230)$$

We proceed to define the variance of fluctuations of the signal $z_{au}^{out}(\tau)$, at the output of the integrator of the autocorrelation channel of the generalized detector, considering that the probability distribution density of instantaneous values of the signal $X(t)$ determined by Eq. (7.191) at the input of the PF of the linear tract of the generalized detector is the non-stationary Gaussian probability distribution density with a nonzero mean (see Section 4.4).

The variance of fluctuations of the signal $z_{au}^{out}(\tau)$, at the output of the integrator of the autocorrelation channel of the generalized detector, can be written in a general form:

$$\sigma_z^2(\tau) = \int_0^{T_{int}}\int_0^{T_{int}} h(T_{int} - t')h(T_{int} - t'')\mu_1[Y(t', t' - \tau), Y(t'', t'' - \tau)]\, dt'\, dt''$$

$$+ \int_0^{T_{int}}\int_0^{T_{int}} h(T_{int} - t')h(T_{int} - t'')$$

$$\times\, \mu_1[Y^*(t', t' - \tau), Y^*(t'', t'' - \tau)]\, dt' d\, t'', \tag{7.231}$$

where

$$Y(t, t - \tau) = X(t)X(t - \tau); \tag{7.232}$$

$$Y^*(t, t - \tau) = X^*(t)X^*(t - \tau); \tag{7.233}$$

the term

$$\mu_1[Y(t', t' - \tau), Y(t'', t'' - \tau)]$$

is the central moment of the second order of the signal at the output of the multiplier 1 and the term

$$\mu_1[Y^*(t', t' - \tau), Y^*(t'', t'' - \tau)]$$

is the central moment of the second order of the signal at the output of the multiplier 2.

To simplify the following mathematics we will define the variance of fluctuations of the signal $z_{au}^{out}(t)$, at the output of the integrator of the autocorrelation channel (see Eq. (7.231)), only near the maximum of the mean of the signal $z_{au}^{out}(\tau)$ at the output of the integrator of the autocorrelation channel of the generalized detector (see Eqs. (7.226)–(7.230)).

In other words, we will define the variance of fluctuations of the signal $z_{au}^{out}(\tau)$, at the output of the integrator of the autocorrelation channel of the generalized detector, for the condition

$$\tau \ll \tau_{c_0}, \tag{7.234}$$

where τ_{c_0} is the time of correlation of the received signal for the case in which there is multiplicative noise.

In this case, to solve the problems in practice with an allowable accuracy, the variance of fluctuations of the signal $z_{au}^{out}(\tau)$, at the output of the integrator of the autocorrelation channel of the generalized detector, in the region near the maximum of the mean determined by Eq. (7.229) is determined by the variance for the condition $\tau = 0$:

$$\sigma_z^2(0) = \int_0^{T_{int}}\int_0^{T_{int}} h(T_{int} - t')h(T_{int} - t'')\mu_1[Y(t'), Y(t'')]\, dt'\, dt''$$

$$+ \int_0^{T_{int}}\int_0^{T_{int}} h(T_{int} - t')h(T_{int} - t'')\mu_1[Y^*(t'), Y^*(t'')]\, dt'\, dt'', \qquad (7.235)$$

where the central moment of the second order under the hypothesis H_1—a "yes" signal at the input of the PF of the linear tract of the generalized detector—in terms of the well known formula for the initial moment of the fourth order of the stationary Gaussian process[81,82] is determined by the following form:

$$\begin{aligned}\mu_1[Y(t'), Y(t'')] + \mu_1[Y^*(t'), Y^*(t'')] &= \mu_1[X^2(t'), X^2(t'')] + \mu_1[X^{*^2}(t'), X^{*^2}(t'')]\\ &= \frac{1}{4}\cdot Re\big\{A_0^4(t)|\dot{R}_V(t''-t')|^2|\dot{S}_0(t')|^2|S_0^*(t'')|^2\\ &\quad + 2\alpha_0^2 A_0^4(t)\dot{R}_V(t''-t')|\dot{S}_0(t')|^2|S_0^*(t'')|^2\\ &\quad + 2A_0^2(t)\dot{R}_\xi(t''-t')\dot{R}_V(t''-t')\dot{S}_0(t')S_0^*(t'')\\ &\quad + 2\alpha_0^2 A_0^2(t)\dot{R}_\xi(t''-t')\dot{S}_0(t')S_0^*(t'')\\ &\quad + |\dot{R}_\xi(t''-t')|^2 + |\dot{R}_\eta(t''-t')|^2\big\}. \qquad (7.236)\end{aligned}$$

Based on Eq. (7.236), and the well-known transform with respect to the double integral during changing the integration variable $\theta = t'' - t'$,

$$\int_0^{T_{int}}\int_0^{T_{int}} h(T_{int} - t')h(T_{int} - t'')\dot{R}(t'' - t')\, dt'\, dt''$$

$$= 2\int_0^{T_{int}} Re\{\dot{R}(\theta)\}\left[\int_0^{T_{int}-\theta} h(T_{int} - x)h(T_{int} - \theta - x)\right] d\theta. \qquad (7.237)$$

After calculation we obtain that the variance of fluctuations determined by Eq. (7.235) of the signal at the output of the integrator of the generalized detector takes the following form:

$$\sigma_{11}^2(0) = \frac{A_0^4(t)}{2}\int_0^{T_{int}} \rho_{T_s}(\theta)|\dot{R}_V(\theta)|^2\, d\theta + A_0^2(t)\int_0^{T_{int}} \rho_{T_c}(\theta)\dot{R}_V(\theta)R_\xi^*(\theta)\, d\theta$$

$$+ \alpha_0^2 A_0^2(t) Re\left\{\int_0^{T_{int}} \rho_{T_s}(\theta)\left[A_0^2(t)\dot{R}_V(\theta) + \dot{R}_\xi(\theta)\right] d\theta\right\}$$

$$+ \frac{1}{2}\int_0^{T_{int}} \rho_T(\theta)|\dot{R}_\xi(\theta)|^2\, d\theta + \frac{1}{2}\int_0^{T_{int}} \rho_T(\theta)|\dot{R}_\eta(\theta)|^2\, d\theta; \qquad (7.238)$$

$$\sigma_{00}^2(0) = \frac{1}{2}\int_0^{T_{int}} \rho_T(\theta)|\dot{R}_\xi(\theta)|^2\,d\theta + \frac{1}{2}\int_0^{T_{int}} \rho_T|\dot{R}_\eta(\theta)|^2\,d\theta$$

$$= \int_0^{T_{int}} \rho_T|\dot{R}_\xi(\theta)|^2\,d\theta = \int_0^{T_{int}} \rho_T|\dot{R}_\eta(\theta)|^2\,d\theta, \tag{7.239}$$

since

$$\dot{R}_\xi(0) = \dot{R}_\eta(0) \tag{7.240}$$

owing to the choice of the amplitude-frequency responses of the PF and AF of the linear tract of the generalized detector.

There are the following designations in Eqs. (7.238) and (7.239):

$$\rho_{T_s}(\theta) = \int_0^{T_{int}-\theta} h(T_{int} - x)h(T_{int} - \theta - x)\,|S_0^*(x)|^2\,|S_0^*(\theta + x)|^2\,dx; \tag{7.241}$$

$$\rho_{T_c}(\theta) = \int_0^{T_{int}-\theta} h(T_{int} - x)h(T_{int} - \theta - x)\,S_0^*(x)\dot{S}_0(\theta + x)\,dx; \tag{7.242}$$

$$\rho_T(\theta) = \int_0^{T_{int}-\theta} h(T_{int} - x)h(T_{int} - \theta - x)\,dx. \tag{7.243}$$

The function $\rho_T(\theta)$, in Eq. (7.243), is the truncated autocorrelation function of the integrator (the ripple filter).[11] In the case of the ideal integrator[11] we can write

$$h(t) = \frac{1}{T_{int}}, \quad 0 \le t \le T_{int}; \tag{7.244}$$

$$\rho_T(\theta) = \frac{1}{T_{int}}\left(1 - \frac{|\theta|}{T_{int}}\right), \quad |\theta| \le T_{int}. \tag{7.245}$$

As was noted above, to decrease fluctuations of the signal determined by Eq. (7.193) the ratio between the bandwidths of the PF and AF and the bandwidth of the integrator (the ripple filter) of the autocorrelation channel of the generalized detector chosen must be much greater than unity[83–85]

$$\gamma_1 = \frac{\Delta\omega_{PF}}{\Delta\omega_{int}} = \frac{\Delta\omega_{AF}}{\Delta\omega_{int}} \gg 1. \tag{7.246}$$

In some practical examples the coefficient γ_1 can be equal to a value in the hundreds, for example, during detection of the wide-band signals.[79]

When the conditions

$$\tau_{c_n} \ll T_{int} \tag{7.247}$$

and

$$\tau_c \ll T_{int} \tag{7.248}$$

are satisfied, where τ_{c_n} is the correlation interval of the multiplicative noise, the integrands in Eqs. (7.238) and (7.239) are not equal to zero if and only if the following condition

$$|\theta| \ll T_{int} \tag{7.249}$$

is satisfied.

For these values of the parameter θ the functions $\rho_{T_s}(\theta)$, $\rho_{T_c}(\theta)$, and $\rho_T(\theta)$, in Eqs. (7.238) and (7.239), can be changed by the approximate formulae during the condition

$$\theta = 0.$$

Then for the signals with the square wave-form envelope of amplitude

$$S_0(t) = 1, \quad 0 \leq t \leq T \tag{7.250}$$

and with the duration satisfied to the condition $T \leq T_{int}$, in the case of the ideal integrator, we can write

$$\rho_T(\theta) \simeq \frac{1}{T_{int}}; \tag{7.251}$$

$$\rho_{T_s}(\theta) \simeq \frac{T}{T_{int}^2}; \tag{7.252}$$

$$\rho_{T_c}(\theta) \simeq \frac{T}{T_{int}^2}; \tag{7.253}$$

$$|\theta| < \tau_{c_0}. \tag{7.254}$$

Reference to Eqs. (7.238) and (7.239) shows that when taking into consideration the condition

$$\tau_c \ll T_{int},$$

we can write

$$\frac{1}{R_V(0)}\int_0^{T_{int}} |\dot{R}_V(\theta)|\,d\theta = \tau_c; \tag{7.255}$$

$$\frac{1}{R_V^2(0)}\int_0^{T_{int}} |\dot{R}_V(\theta)|^2\,d\theta \simeq \tau_c; \tag{7.256}$$

$$\frac{1}{R_\xi^2(0)}\int_0^{T_{int}} |\dot{R}_\xi(\theta)|^2\,d\theta = \frac{1}{R_\eta^2(0)}\int_0^{T_{int}} |\dot{R}_\eta(\theta)|^2\,d\theta \simeq \tau_{c_n}. \tag{7.257}$$

However, when the condition

$$\tau_c \geq \tau_{c_n}$$

is satisfied we can write that

$$\frac{1}{R_\xi(0)R_\eta(0)}\int_0^{T_{int}} |R_\xi(\theta)||R_V(\theta)|\,d\theta \simeq \tau_{c_n}. \tag{7.258}$$

After straightforward mathematical transforms we can define the variance of fluctuations of the signal determined by Eq. (7.193), at the output of the integrator of the autocorrelation channel of the generalized detector, for the case of the ideal integrator in the region of maximum of the mean determined by Eq. (7.226):

- A "yes" signal at the input of the PF of the linear tract of the generalized detector—the hypothesis H_1:

$$\sigma_{11}^2(0) = \frac{\left[A_0^4(t)R_V(0)R_M(0) + \alpha_0^2 A_0^4(t)R_V(0)\right]T\tau_c}{2T_{int}^2} + \frac{A_0^2(t)R_M(0)\sigma_n^2 T\tau_{c_n}}{T_{int}^2} + \frac{2\sigma_n^4\tau_{c_n}}{T_{int}}; \tag{7.259}$$

- A "no" signal at the input of the PF of the linear tract of the generalized detector—the hypothesis H_0:

$$\sigma_{00}^2 = \frac{2\sigma_n^4\tau_{c_n}}{T_{int}}, \tag{7.260}$$

where σ_n^2 is the variance of the additive Gaussian noise $\xi(t)$ and $\eta(t)$ formed at the outputs of the PF and AF of the linear tract of the generalized detector, respectively.

After defining the means (see Eqs. (7.226) and (7.227)) and the variances of fluctuations (see Eqs. (7.259) and (7.260)) of the signal at the output of the integrator of the autocorrelation channel of the generalized detector under the hypotheses H_1 and H_0, respectively, we can define the signal-to-noise ratio $q_{au}^{out}(0)$ (see Eq. (7.215)) and the ratio between the variances of fluctuations

$$\mathcal{Q} = \frac{\sigma_{00}^2(0)}{\sigma_{11}^2(0)} \tag{7.261}$$

of the signal at the output of the integrator, of the autocorrelation channel of the generalized detector, for the case of the ideal integrator under the hypotheses H_0 and H_1, respectively:

$$q_{au}^{out}(0) = \sqrt{\frac{\gamma_1}{2}} \cdot \frac{E_{a_{1M}} R_M(0)}{\sigma_n^4 T_{int}}; \tag{7.262}$$

$$\mathcal{Q} = \left\{ \frac{T}{T_{int}} \left[\frac{E_{a_{1M}}^2 R_V(0) R_M(0) + \alpha_0^2 E_{a_{1M}}^2 R_V(0)}{4\sigma_n^4} \cdot \frac{\tau_c}{\tau_{c_n}} + \frac{E_{a_{1M}} R_M(0)}{2\sigma_n^2} \cdot \frac{\tau_{c_0}}{\tau_{c_N}} \right] + 1 \right\}^{-1}; \tag{7.263}$$

$$\gamma_1 = \frac{\Delta\omega_{PF}}{\Delta\omega_{int}} = \frac{\Delta\omega_{AF}}{\Delta\omega_{int}} = \frac{T_{int}}{\tau_{c_n}}, \tag{7.264}$$

where $E_{a_{1M}}$ is the energy of the signal distorted by the multiplicative noise.

We rewrite Eqs. (7.262) and (7.263), using the signal-to-noise ratio at the input of linear tract of the generalized detector, as the ratio between the energy of the signal E_{a_1}, when the multiplicative noise is absent, and the spectral power density $\frac{N_0}{2}$ of the additive Gaussian noise.

Taking into account that

$$N_0 = \sigma_n^2 \tau_{c_n} \tag{7.265}$$

and

$$E_{a_1} = E_{a_{1M}} \tau_{c_0}, \tag{7.266}$$

Eqs. (7.262) and (7.263) can be rewritten in the following form:

$$q_{au}^{out}(0) = \frac{R_M(0)}{\sqrt{\gamma_1}} \cdot q; \tag{7.267}$$

$$\mathcal{Q} = \left\{ \frac{R_M(0) R_V(0) + \alpha_0^2 R_V(0)}{\gamma_1 \gamma_2} \cdot q^2 + \frac{R_M(0)}{\gamma_1} \cdot q + 1 \right\}^{-1}, \tag{7.268}$$

where

$$q = \frac{E_{a_1}}{N_0}; \tag{7.269}$$

$$R_M(0) = \overline{A_0^2(t)}; \tag{7.270}$$

$$R_V(0) = R_M(0) - \alpha_0^2 = \overline{A_0^2(t)} - \alpha_0^2; \tag{7.271}$$

$A_0(t)$ is the amplitude factor defining distortions in amplitude of the signal—the absolute value of the noise modulation function $\dot{M}(t)$ of the multiplicative noise; $R_V(0)$ is the variance of fluctuations of the noise modulation function $\dot{M}(t)$ of the multiplicative noise (see Section 4.2); and

$$\gamma_2 = \frac{T}{\tau_c} \tag{7.272}$$

is the coefficient defining the ratio between the signal duration and the correlation interval of the noise modulation function $\dot{M}(t)$ of the multiplicative noise.

Reference to Eqs. (7.267) and (7.268) shows that for the condition

$$q = const,$$

and with an increase in the coefficient γ_1 that is equivalent to an extension of the bandwidth of the PF of the input linear tract of the generalized detector for the condition

$$T_{int} = const,$$

the signal-to-noise ratio at the output of the integrator of the autocorrelation channel of the generalized detector is decreased for the condition

$$\tau = 0.$$

In the process, the coefficient $\mathcal{Q}$ that defines the ratio between the variance of fluctuations of the signal determined by Eq. (7.193) under the hypothesis H_1 and the variance of fluctuations of the same signal under the hypothesis H_0 is increased and tends to approach unity with an increase in the coefficient γ_1.

The dependence of the ratio $\mathcal{Q}$ (of the variances), which is determined by Eq. (7.263), as a function of the parameter q in the case of the uniform probability distribution density of fluctuations of the signal phase within the limits of the interval $[0, 2\pi]$ is shown in Fig. 7.19.

Also, the same function for the universally adopted autocorrelation detector is shown in Fig. 7.19 as well.

Reference to Fig. 7.19 shows us superiority of the use of the autocorrelation channel of the generalized detector in problems of detecting the signals with an *a priori* unknown structure compared with employment of the universally adopted autocorrelation detector. Superiority will be shown

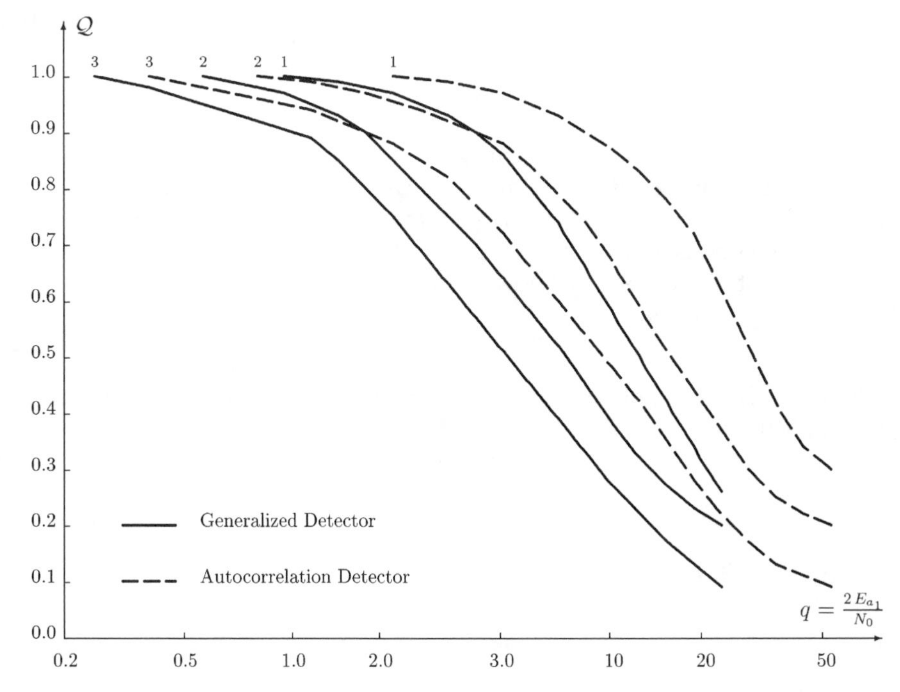

FIGURE 7.19
The ratio $\mathcal{Q}$ as a function of the signal-to-noise ratio at the output of the detector: $d^2 = \frac{R_M(0)R_V(0)}{\gamma_1\gamma_2}$; $\gamma_2 = 10$; for 1, $d^2 = 0.02$; for 2, $d^2 = 0.03$; for 3, $d^2 = 0.04$.

more clearly in the next section in a comparative analysis of the detection performances.

Taking into consideration the obtained formulae defining the signal-to-noise ratio at the output of the autocorrelation channel of the generalized detector (see Eq. (7.267)), the variance in fluctuations (see Eqs. (7.259) and (7.260)) of the signal at the output of the integrator of the autocorrelation channel of the generalized detector, and dependences shown in Fig. 7.19, we are able to carry out the comparative analysis of the detection performances of the autocorrelation channel of the generalized detector and the detection performances of the universally adopted autocorrelation detector.

7.4.3 Comparative Analysis of Detection Performances

Reference to Eq. (7.268) and Fig. 7.19 shows that at the values of the parameter q satisfying the following conditions

$$q \ll \sqrt{\frac{\gamma_1\gamma_2}{R_M^2(0) - \alpha_0^4}} \tag{7.273}$$

and

$$q \ll \gamma_1 \tag{7.274}$$

the parameter Q is very close to unity.

In doing so, in accordance with Eq. (7.214) the detection performances of the autocorrelation channel of the generalized detector are defined by the method analogous to the detection performances of the generalized detector in the case of the completely known signal:

$$P_{D_M} = 1 - \Phi\left[\Phi^{-1}(1 - P_{F_0}) - q_{au}^{out}(0)\right], \tag{7.275}$$

where the signal-to-noise ratio $q_{au}^{out}(0)$, determined by Eq. (7.158) for the case of the ideal integrator, satisfies a role of the detection parameter.

Reference to Eq. (7.267) shows that during the condition

$$\sqrt{\gamma_1} R_M(0) < 1 \tag{7.276}$$

the following inequality

$$q_{au}^{out}(0) < q \tag{7.277}$$

takes place.

In the process, in accordance with Eq. (7.275) the detection performances of the autocorrelation channel of the generalized detector at the same value of the signal-to-noise ratio q are worse than the detection performances of the generalized detector for the case of the completely known signal.

However, the detection performances of the autocorrelation channel of the generalized detector are better than the detection performances of the universally adopted autocorrelation detector, considered as the optimal detector for the signals with an *a priori* unknown amplitude-phase-frequency structure, and are even better than the detection performances of the correlation detector for the case of the completely known signal, which is thought of as the optimal detector.

If the degradations in detection performances of the autocorrelation channel of the generalized detector in comparison to the detection performances of the generalized detector for the case of the completely known signal are high, then the ratio

$$\gamma_1 = \frac{T_{int}}{\tau_{c_n}} \tag{7.278}$$

is high, since the degradations are proportional to the parameter γ_1 in the case of non-coherent signal processing.[9]

A decrease in the degradations in the detection performances can be reached owing to a decrease in the bandwidth of the PF of the linear tract of the generalized detector, bearing the autocorrelation channel of the generalized detector, that leads to a decrease in the parameter γ_1. The limit of reduction in the bandwidth of the PF of the linear tract of the generalized detector is defined by the bandwidth of the energy spectrum of the signal when multiplicative noise is present.

In the case in which the bandwidth of energy spectrum of the detected signal is known, the minimum degradations for the case of the wide-band signals during a low value of the signal-to-noise ratio q is determined by the following form:

$$\gamma_1 = \frac{\Delta\Omega_{0_M} T_{int}}{2\pi}, \tag{7.279}$$

where $\Delta\Omega_{0_M}$ is the bandwidth of the energy spectrum of the received signal.

As may be seen from Eqs. (7.215) and (7.226), the rapid fluctuations in phase of the signal during signal processing by the autocorrelation channel of the generalized detector deteriorate the probability of detection P_{D_M} of the signal if $\tau \neq 0$ owing to the decreasing region of correlation on the axis τ of the processed signal, and do not influence the probability of detection P_{D_M} of the signal at $\tau = 0$ using the condition determined by Eq. (7.273).

If the condition

$$q \gg \sqrt{\frac{\gamma_1\gamma_2}{R_M^2(0) - \alpha_0^4}} \tag{7.280}$$

is satisfied, as may be seen from Eqs. (7.214) and (7.268), the probability of detection P_{D_M} of the signal can be determined by the following form:[86]

$$P_{D_M} = 1 - \frac{1}{\sqrt{2\pi\gamma_2} - \frac{1}{q}\cdot\sqrt{\frac{\gamma_1\gamma_2}{R_M^2(0)-\alpha_0^4}}\cdot\Phi^{-1}(1-P_{F_0})} \times \exp\left\{-\sqrt{\frac{\gamma_2}{2}} + \frac{1}{q}\cdot\sqrt{\frac{\gamma_1\gamma_2}{R_M^2(0)-\alpha_0^4}}\cdot\Phi^{-1}(1-P_{F_0})\right\}. \tag{7.281}$$

Since during the condition

$$q \gg \frac{1}{2}\cdot\sqrt{\frac{\gamma_1\gamma_2}{R_M^2(0)-\alpha_0^4}} \tag{7.282}$$

and if the probability of false alarm $P_{F_0} > 10^{-5}$, the condition

$$\frac{1}{2}\cdot\sqrt{\frac{\gamma_1\gamma_2}{R_M^2(0)-\alpha_0^4}}\cdot\Phi^{-1}(1-P_{F_0}) < 1 \tag{7.283}$$

is true in essence, Eq. (7.281) as $q \to \infty$ can be rewritten in the following form:

$$P_{D_M} \simeq 1 - \frac{1}{\sqrt{2\pi\gamma_2}}\cdot\exp\left\{-\sqrt{\frac{\gamma_2}{2}}\right\}, \tag{7.284}$$

where the coefficients γ_1 and γ_2 are determined by Eqs. (7.264) and (7.272), respectively. Moreover

$$\gamma_1 \gg 1.$$

Reference to Eq. (7.281) shows that when the condition in Eq. (7.282) is true, i.e., the signal-to-noise ratio at the input of the PF of the linear tract of the generalized detector $q \gg 1$, the probability of detection P_{D_M} depends very weakly on the energy of the received signal, and when there is multiplicative noise the probability of detection P_{D_M} does not equal unity even as $q \to \infty$.

As may be seen from Eq. (7.282), if a difference by absolute value of the probability of detection P_{D_M} as $q \to \infty$ from unity depends on the ratio between a duration of the detected signal and the time of correlation of the noise modulation function $\dot{M}(t)$ of the multiplicative noise is less, then the probability of detection P_{D_M} of the signal under the condition $\tau = 0$ is high.

This phenomenon can be explained in the following way. The variance of fluctuations of the signal at the output of the integrator (the ripple filter) of the autocorrelation channel of the generalized detector, as may be seen from Eq. (7.259), depends not only on the energy of the signal and the power of the additive Gaussian noise, but also on the time of correlation τ_c—the bandwidth of the energy spectrum of fluctuations $\Delta\Omega_V$ of the noise modulation function $\dot{M}(t)$ of the multiplicative noise—and on the time of integration—the bandwidth of the integrator.

At the same time, as may be seen from Eqs. (7.211), (7.227), and (7.260), the value of the threshold K_g depends on the energy characteristics of the additive Gaussian noise, on the time of integration, and does not depend on the bandwidth of the energy spectrum of the signal when there is multiplicative noise. Consequently, during a low signal-to-noise ratio (see Eq. (7.273)) a variation in the bandwidth of the energy spectrum of the signal noise component leads to a variation in the variance of fluctuations of the signal determined by Eq. (7.193), but an influence of this effect on the probability of detection P_{D_M} of the signal during the condition $\tau = 0$ is very low. The probability distribution density is the distribution law of the additive Gaussian noise.

When the signal-to-noise ratio is high (see Eq. (7.282)), an increase in the bandwidth of the energy spectrum of the signal noise component for the condition

$$\tau_c \to \tau_{c_n}$$

leads to a decrease in the variance σ_{11}^2 (see Eq. (7.259)) of fluctuations of the signal determined by Eq. (7.193), at the output of the integrator of the autocorrelation channel of the generalized detector. As

$$q \to \infty$$

the probability distribution density of fluctuations is essentially defined by the signal noise component caused by the multiplicative noise.

In this conjunction, when the threshold K_g, chosen in accordance with the energy characteristics of the additive Gaussian noise at the output of the integrator of the autocorrelation channel of the generalized detector (see Eq. (7.211)), is not variable and when the signal noise component caused by the stimulus of the multiplicative noise exists, then there is always a definite value of the probability of non-detection of the signal.

Because of this, the probability of detection P_{D_M} of the signal, which is not equal to unity (see Eq. (7.284)), is the constant value that is independent of the energy of the signal at the input of the PF of the linear tract of the generalized detector and tends to approach unity when the bandwidth of energy spectrum of fluctuations of the signal is extended

$$\tau_c \to \tau_{c_n}.$$

It is not difficult to see that an influence of increasing the signal-to-noise ratio

$$q \to \infty$$

at the input of the PF of the linear tract of the generalized detector for the condition $\tau = 0$, is the same both on increasing the mean (see Eq. (7.226)) of the signal at the output of the integrator of the autocorrelation channel of the generalized detector and on increasing the variance of fluctuations (see Eq. (7.259)):

$$\frac{m_{11}^2(0)}{\sigma_{11}^2(0)} \simeq \left[1 + \frac{\alpha_0^2}{R_V(0)}\right] \cdot \frac{\gamma_2}{2} \quad \text{as} \quad q \to \infty \qquad \text{and} \qquad \gamma_1 \gg 1. \tag{7.285}$$

Dependences of the probability of detection P_{D_M}, of the signal at the output of the integrator of the autocorrelation channel of the generalized detector (see Eq. (7.214)) for the condition $\tau = 0$, as a function of the signal-to-noise ratio q at the input of the linear tract of the generalized detector at various values of the ratio

$$\frac{\gamma_1}{\gamma_2}$$

during the conditions

$$\gamma_2 = 10 \qquad \text{and} \qquad R_M(0) = R_V(0) = 1, \tag{7.286}$$

the case of distortions only in phase of the signal that is distributed uniformly within the limits of the interval $[0, 2\pi]$, are shown in Fig. 7.20.

Dependences of the probability of detection P_{D_M}, of the signal at the output of the integrator of the autocorrelation channel of the generalized detector for the case in which the bandwidth of the PF of the linear tract of the generalized

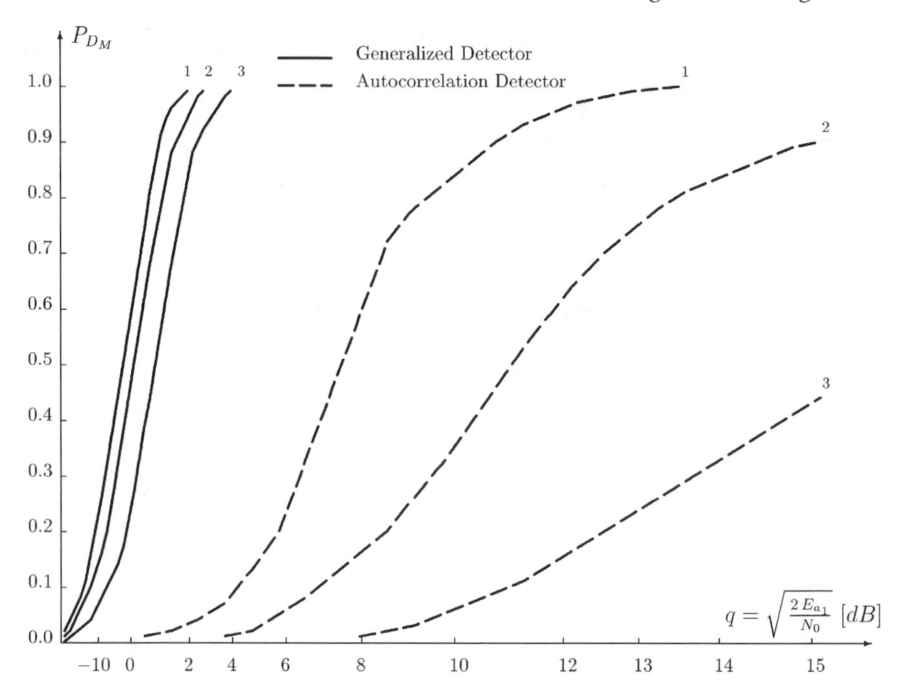

FIGURE 7.20
Detection performances at various values of the ratio $\frac{\gamma_1}{\gamma_2}$: $R_M(0) = R_V(0) = 1$; $P_{F_0} = 10^{-4}$; $\gamma_2 = 10$; for 1, $\frac{\gamma_1}{\gamma_2} = 10$; for 2, $\frac{\gamma_1}{\gamma_2} = 50$; for 3, $\frac{\gamma_1}{\gamma_2} = 100$.

detector is equal to the bandwidth of the energy spectrum of the wide-band signal—the potentially achieved case—during the conditions

$$\gamma_1 = \gamma_2 = 100 \quad \text{and} \quad R_V(0) = 1, \tag{7.287}$$

as a function of the signal-to-noise ratio at the input of the PF of the linear tract of the generalized detector are shown in Fig. 7.21.

The detection performances of the universally adopted autocorrelation detector, which are defined at the same conditions, are also shown in Figs. 7.20 and 7.21.

Comparative analysis of the detection performances of the autocorrelation channel of the generalized detector and the universally adopted autocorrelation detector demonstrates a superiority of the first detector over the second detector.

The detection performances of the correlation detector for the cases of the completely known signal, and the signal with the random initial phase that is distributed uniformly within the limits of the interval $[0, 2\pi]$ under the probability of false alarm P_{F_0} equal to 10^{-5}, are shown in Fig. 7.22.

With the purpose of comparing the detection performances of the autocorrelation channel of the generalized detector under the same probability

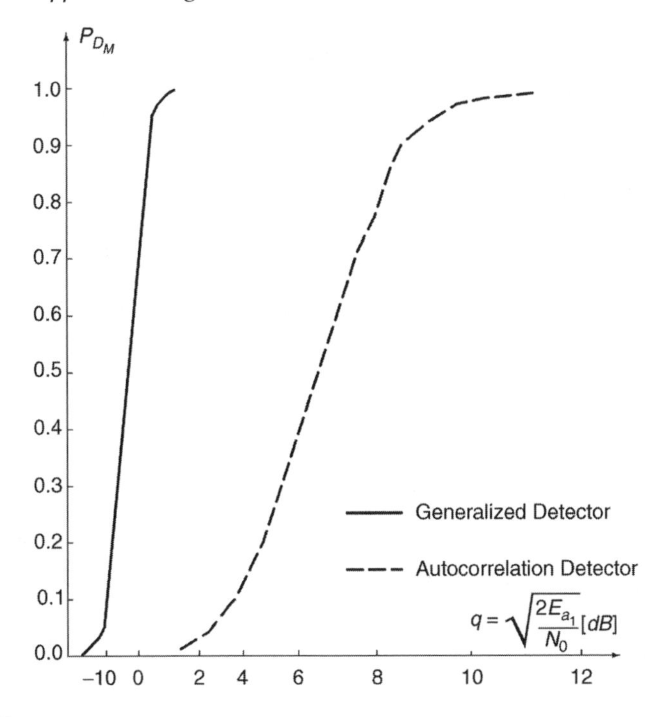

FIGURE 7.21
Detection performances. The bandwidth of linear tract of the generalized detector is equal to the bandwidth of the wide-band signal (potentially achieved case): $R_V(0) = 1$; $\gamma_1 = \gamma_2 = 100$; $P_{F_0} = 10^{-3}$.

of false alarm P_{F_0} and during distortions in phase of the signal that is distributed uniformly within the limits of the interval $[0, 2\pi]$, and the variance of fluctuations

$$R_M(0) = R_V(0) = 1 \tag{7.288}$$

of the noise modulation function $\dot{M}(t)$ of the multiplicative noise for the case of *a priori* unknown information about the amplitude-phase-frequency structure of the signal, are also shown in Fig. 7.22.

The most favorable condition

$$\gamma_1 = \gamma_2$$

when the bandwidth of the PF of the linear tract of the generalized detector is equal to the bandwidth of the energy spectrum of the detected signal when the multiplicative noise is present is taken for comparison.

As may be seen in Fig. 7.22, the autocorrelation channel of the generalized detector is superior to the correlation detector that is thought of as the

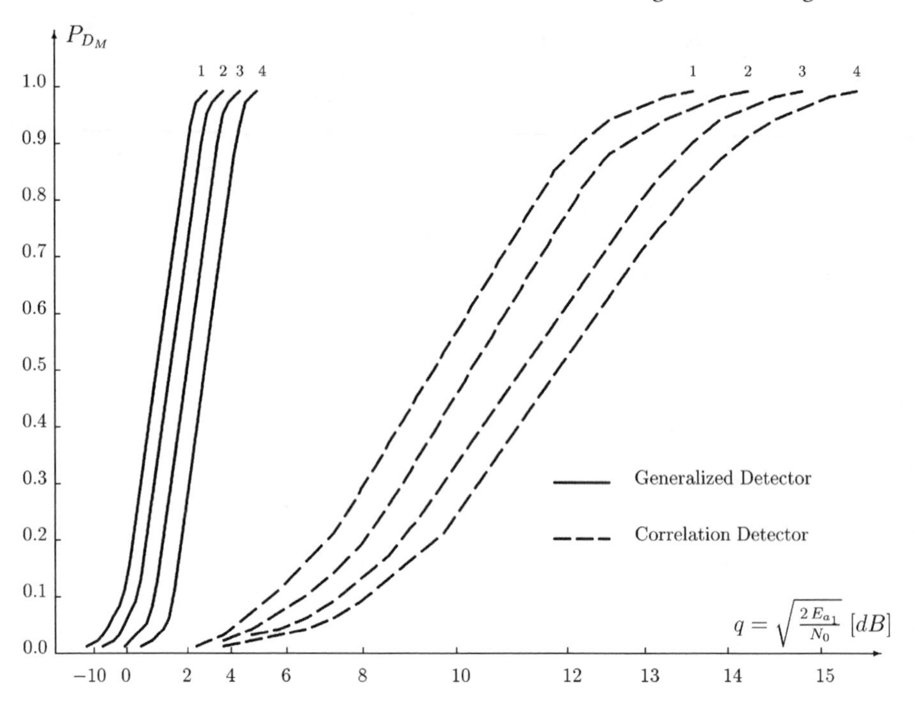

FIGURE 7.22
Detection performances, $P_{F_0} = 10^{-5}$: (a) completely known signal: for 1, $\gamma_1 = \gamma_2 = 50$; for 2, $\gamma_1 = \gamma_2 = 100$; (b) signal with random initial phase: for 3, $\gamma_1 = \gamma_2 = 50$; for 4, $\gamma_1 = \gamma_2 = 100$.

optimal detector. When the superiority is high, the ratio

$$\gamma_2 = \frac{T}{\tau_c}$$

is low and the signal-to-noise ratio

$$q = \frac{E_{a_1}}{N_0}$$

is high.

7.5 Diversity Signal Detection

Diverse complex multi-channel signal processing systems, constructed on the basis of the generalized detector, are used in the case of transmission of the information signals with the purpose of reducing to a minimum the errors caused by fading of the complex envelope of amplitude of the received signal.

There are some versions of the diverse multi-channel signal processing on the basis of the generalized approach to signal processing in the presence of noise[1,2] that exist for the purpose of reducing to a minimum an influence of fading action on an accuracy of transmission of the information signals:[87,88]

- Signal processing using the diversity antennas and the distance between antennas must be greater than the three-dimensional correlation interval—the received signals are considered as the independent signals.
- Separation of the input channels of the diverse complex multi-channel signal processing systems, constructed on the basis of the generalized detector, by frequency when the same signal is transmitted using two or more carrier frequencies.
- Signal processing of the multiple recurring signals with the period that is greater than an interval of the expected fading.

The additive set noise of the input channels of the diverse complex multi-channel signal processing system, constructed on the basis of the generalized detector, is mutually uncorrelated (independent), as a rule.

In a general case, signal processing by the diverse complex multi-channel signal processing system constructed on the basis of the generalized detector can be presented as the sum with the weight coefficients:

$$\mathcal{R}^{out}(\tau, \Omega) = \sum_{i=1}^{N} d_i \mathcal{Y}_{M_i}^{out}(\tau, \Omega), \tag{7.289}$$

where d_i is the weight coefficient; N is the number of the input channels of the diverse complex multi-channel signal processing system constructed on the basis of the generalized detector; and $\mathcal{Y}_{M_i}^{out}(\tau, \Omega)$ is the signal at the output of the "i"-th channel of the diverse complex multi-channel signal processing system constructed on the basis of the generalized detector.

We have the following diverse complex multi-channel signal processing systems, constructed on the basis of the generalized detector, by the technique of signal generation determined by Eq. (7.289):

- The diverse complex multi-channel signal processing systems, constructed on the basis of the generalized detector, with an automatic control of choice of the greatest signal and autoranging the input channels by the considered signal processing system. For this case the output signals $\mathcal{Y}_{M_i}^{out}(\tau, \Omega)$ are compared with the predetermined threshold value until any signal $\mathcal{Y}_{M_i}^{out}(\tau, \Omega)$ can be detected, i.e., the energy of the signal $\mathcal{Y}_{M_i}^{out}(\tau, \Omega)$ exceeds the threshold. After that, the signal $\mathcal{Y}_{M_i}^{out}(\tau, \Omega)$ is switched to the output of the diverse complex multi-channel signal processing system constructed on the basis of the generalized detector. When the energy of the signal $\mathcal{Y}_{M_i}^{out}(\tau, \Omega)$ is low in comparison with the predetermined threshold, the search of the signal is repeated. For this case the weight coefficient is determined in the following form:

$$d_i = 1 \quad \text{at } i = j \qquad \text{and} \qquad d_j = 0 \quad \text{at } i \neq j. \tag{7.290}$$

- The diverse complex multi-channel signal processing system, constructed on the basis of the generalized detector, with a simultaneous comparison of the signals at the outputs of N channels and choice of the greatest signal by power—the diverse signal processing with the optimal autochoice system. For this case the channel of the greatest signal $\mathcal{Y}_{M_i}^{out}(\tau, \Omega)$ is switched to the output of the diverse complex multi-channel signal processing system constructed on the basis of the generalized detector, i.e., the condition in Eq. (7.290) is true.
- The diverse complex multi-channel signal processing system, constructed on the basis of the generalized detector, with simultaneous summing of all received signals at the outputs of N channels.

Among all techniques of construction of the diverse complex multi-channel signal processing systems, on the basis of the generalized detector, only two methods can be widely used.

The first method—the method of linear summing—is to sum all signals with some weight coefficients:

$$d_1 = d_2 = \cdots = d_N. \tag{7.291}$$

During the use of the second method, the method of optimal summing,[14,88] the weight coefficients are not equal to each other.

For this reason, we must take into consideration characteristics of the signal. For this case the weight coefficients d_i are automatically controlled in such manner that the maximum signal-to-noise ratio can be reached if the output signal is determined by Eq. (7.289).

An impact of the multiplicative noise on the detection performances of the diverse complex multi-channel signal processing system, constructed on the basis of the generalized detector, with the optimal autochoice system and with the simultaneous comparison and choice of the greatest signal are not considered, since for this case this problem reduces to an estimation of an impact of the multiplicative noise on the detection performances of the generalized detector discussed in Section 7.2.

Using the main results discussed in Section 7.2, and taking into account the number of the input channels of the diverse complex multi-channel signal processing system constructed on the basis of the generalized detector, we are able to define the probability of detection P_{D_M} of the signal by the optimal autochoice system of the diverse complex multi-channel signal processing system constructed on the basis of the generalized detector.

As was shown in References 1 and 2, the maximum value of the signal-to-noise ratio for linear summing of the signals is 1 dB less than for the optimal summing of the signals when the number of the input channels of the diverse complex multi-channel signal processing system constructed on the basis of the generalized detector is satisfied for the condition

$$N \leq 24.$$

For this reason, we consider the diverse complex multi-channel signal processing system, constructed on the basis of the generalized detector with the linear summing of the signals determined by Eq. (7.289), at the outputs of N channels for the condition

$$d_1 = d_2 = \cdots = d_N = 1. \tag{7.292}$$

Based on the results discussed in References 87 and 88 and in Section 7.3, each input channel of the diverse complex multi-channel signal processing system, constructed on the basis of the generalized detector, must contain the generalized detector that can detect the signals with fluctuations of the complex envelope of amplitude of the signal obeying the Rayleigh probability distribution density and with the signal phase distributed uniformly within the limits of the interval $[0, 2\pi]$ and the quadratic detector.

Moreover, the PF of the input linear tract of the generalized detector used by the diverse complex multi-channel signal processing system must be matched with the incoming signal without fading. The signals at the outputs of the diverse complex multi-channel signal processing system, constructed on the basis of the generalized detector, are summarized and come in at the input of the decision-making device (the threshold device—THRD, see Fig. 7.17).

Let us consider a function of the diverse complex multi-channel signal processing system, constructed on the basis of the generalized detector, during detection of the signals distorted by the multiplicative noise when the existing multiplicative noise distorts the structure of the signal.

The detection performances of the considered diverse complex multi-channel signal processing system, constructed on the basis of the generalized detector, depend on the probability distribution density of the resulting output signal determined by Eq. (7.289). Since the signals at the outputs of individual channels, of the considered diverse complex multi-channel signal processing system constructed on the basis of the generalized detector, are independent, then the probability distribution density of the resulting output signal $\mathcal{R}_M^{out}(\tau, \Omega)$ is defined by the probability distribution densities of the signals $\mathcal{Y}_{M_i}^{out}(\tau, \Omega)$ at the outputs of each channel.

The signal at the input of the PF of the linear tract of the "i"-th channel of the diverse complex multi-channel signal processing system, constructed on the basis of the generalized detector, can be written in the following form:

$$X_i(t + \tau_i) = Re\{\dot{M}(t + \tau_i)\dot{S}(t + \tau_i) \cdot e^{j(\omega_i t + \varphi_i)}\} + n(t + \tau_i), \tag{7.293}$$

where

$$\tau_i = \frac{l_i \sin \beta_i}{c} \tag{7.294}$$

is the delay caused by the distance l_i between the input channels of the diverse complex multi-channel signal processing system constructed on the basis of the generalized detector—in the case of frequency diversity $\tau_i = 0$; β_i is the angle between the perpendicular to the distance line between two adjacent inputs and a direction to a source of electromagnetic waves; ω_i is the frequency of the "i"-th input channel of the diverse complex multi-channel signal processing system constructed on the basis of the generalized detector—in the case of spacing $\omega_i = \omega_0$; and φ_i is the random initial phase.

In terms of filtering of the input signal determined by Eq. (7.293), by the PF of the linear tract of each input channel of the diverse complex multi-channel signal processing system constructed on the basis of the generalized detector, and signal processing by the generalized detector the signal at the input of the quadratic detector at the instant τ takes the following form:

$$z_{M_i}^{out}(\tau, \Omega) = Z_{M_i}^{out}(\tau, \Omega)\cos[\omega_i \tau + \varphi_i(t)], \tag{7.295}$$

where $Z_{M_i}^{out}(\tau, \Omega)$ is the complex envelope of amplitude of the signal at the output of the generalized detector of the "i"-th channel of the diverse complex multi-channel signal processing system constructed on the basis of the generalized approach to signal processing in the presence of noise in terms of the additive set noise (see Fig. 7.17), which is determined by Eq. (7.42);

$$\Omega = \omega_i - \omega_0 \tag{7.296}$$

is the detuning in frequency of the signal with respect to the frequency of filter tuning.

The probability distribution density of the complex envelope $Z_{M_i}^{out}(\tau, \Omega)$ of amplitude of the signal at the output of the generalized detector (see Fig. 7.17), when the probability distribution density of instantaneous values of the input signal $X_i(t + \tau_i)$ determined by Eq. (7.293) is Gaussian, takes the following form

$$f_{H_1}\left[Z_{M_i}^{out}(\tau, \Omega)\right] = \frac{Z_{M_i}^{out}(\tau, \Omega)}{\sqrt{\sigma_i^2}} \times \exp\left\{-\frac{\left[Z_{M_i}^{out}(\tau, \Omega)\right]^2 + \alpha_0^2 \cdot E_{a_1}|\dot{\rho}(\tau, \Omega)|^2}{2\sigma_i^2}\right\}$$
$$\times K_0\left[\frac{\alpha_0 \cdot \sqrt{E_{a_1}} Z_{M_i}^{out}(\tau, \Omega)|\dot{\rho}(\tau, \Omega)|}{\sigma_i^2}\right], \tag{7.297}$$

where

$$Z_{M_i}^{out}(\tau, \Omega) \geq 0 \tag{7.298}$$

and

$$\sigma_i^2 = \sigma_s^2 + 4\sigma_n^4 \tag{7.299}$$

is the variance of the signal at the output of the generalized detector of the "i"-th channel of the diversity complex multi-channel signal processing system constructed on the basis of the generalized approach to signal processing in the presence of noise, which is generated by the input signal determined by Eq. (7.293); σ_s^2 is the variance of the signal noise component at the output of the generalized detector (see Eq. (6.109));

$$\sigma_n^2 = \frac{N_0 \omega_0^2}{8(\Delta\omega_{PF})\sqrt{T\beta}} \tag{7.300}$$

is the variance of the additive set noise at the output of the PF of the linear tract of the input channel of the diverse complex multi-channel signal processing system constructed on the basis of the generalized detector.

The probability distribution density of complex envelope of the signal at the output of the quadratic detector of the "i"-th channel

$$\mathcal{Y}_{M_i}^{out}(\tau, \Omega) = \left[Z_{M_i}^{out}(\tau, \Omega)\right]^2 \tag{7.301}$$

can be written in the following form based on the results discussed in Reference 14:

$$f_{H_1}\left[\mathcal{Y}_{M_i}^{out}(\tau, \Omega)\right] = \frac{1}{2\sqrt{\mathcal{Y}_{M_i}^{out}(\tau, \Omega)}} \times \left\{ f_{H_1}\left[\sqrt{\mathcal{Y}_{M_i}^{out}(\tau, \Omega)}\right] + f_{H_1}\left[-\sqrt{\mathcal{Y}_{M_i}^{out}(\tau, \Omega)}\right]\right\}, \tag{7.302}$$

where

$$\mathcal{Y}_{M_i}^{out}(\tau, \Omega) \geq 0. \tag{7.303}$$

In terms of Eq. (7.297) the probability distribution density of the signal $\mathcal{Y}_{M_i}^{out}(\tau, \Omega)$ during the conditions

$$\tau = 0 \quad \text{and} \quad \Omega = 0$$

at the output of the "i"-th channel of the diverse complex multi-channel signal processing system constructed on the basis of the generalized detector if a "yes" signal at the input of the linear tract of the generalized detector has the following form:

$$f_{H_1}\left[\mathcal{Y}_{M_i}^{out}(\tau, \Omega)\right] = \frac{1}{2\sigma_i^2} \cdot \exp\left\{-\frac{\left[\mathcal{Y}_{M_i}^{out}(\tau, \Omega)\right]^2 + m_1^2}{2\sigma_i^2}\right\} \times K_0\left[\frac{m_1 \mathcal{Y}_{M_i}^{out}(\tau, \Omega)}{\sigma_i^2}\right], \tag{7.304}$$

where

$$\mathcal{Y}_{M_i}^{out}(\tau, \Omega) \geq 0 \tag{7.305}$$

and

$$m_1 = \alpha_0^2 \cdot E_{a_1} \tag{7.306}$$

is the mean of the signal at the output of the quadratic detector.

For the case of a "no" signal at the input of the PF of linear tract of the generalized detector of the diverse complex multi-channel signal processing system

$$E_{a_1} = 0$$

and when there is only the additive set noise $n(t)$ at the input channels of the diverse complex multi-channel signal processing system, constructed on the basis of the generalized detector, the probability distribution density of the signal at the output of the quadratic detector is defined by the well-known formula

$$f_{H_0}\left[\mathcal{Y}_{M_i}^{out}(\tau, \Omega)\right] = \frac{1}{8\sigma_n^4} \cdot \exp\left[-\frac{\mathcal{Y}_{M_i}^{out}(\tau, \Omega)}{8\sigma_n^4}\right], \quad \mathcal{Y}_{M_i}^{out}(\tau, \Omega) \geq 0, \tag{7.307}$$

where σ_n^2 is the variance of the additive set noise at the input of the PF of the linear tract of the input channels of the diverse complex multi-channel signal processing system constructed on the basis of the generalized detector.

The mean

$$m_1 = \alpha_0 \cdot \sqrt{E_{a_1}} \tag{7.308}$$

in Eq. (7.304) depends on the form of the probability distribution density and the degree of distortions in amplitude $A(t)$ and phase $\varphi(t)$ of the signal (see Sections 4.2 and 4.4).

Knowing the probability distribution density of the signal at the output of the "i"-th input channel of the diverse complex multi-channel signal processing system, constructed on the basis of the generalized detector, for the cases of a "yes" signal—the hypothesis H_1—and a "no" signal—the hypothesis H_0—at the input of the PF of the linear tract of the generalized detector used by the diversity complex multi-channel signal processing system, we can define the probability distribution density of the resulting summing signal $\mathcal{R}^{out}(\tau, \Omega)$ determined by Eq. (7.289) at the input of the threshold device (THRD, see Fig. 7.17) of the diverse complex multi-channel signal processing system constructed on the basis of the generalized detector under the hypotheses H_1 and H_0, respectively, when the weight coefficients are determined by Eq. (7.292).

It is worthwhile to define the probability distribution density of the sum of independent signals for the condition

$$N \gg 1, \tag{7.309}$$

by first defining the characteristic function of the resulting summing signal determined by Eq. (7.289).

The characteristic function of the signal $\mathcal{Y}_{M_i}^{out}(\tau, \Omega)$ at the output of the "i"-th channel of the diverse complex multi-channel signal processing system, constructed on the basis of the generalized detector, under the hypothesis H_1 in terms of Eq. (7.304) after integration can be written in the following form:[89]

$$\Theta_1^i(\upsilon) = \int_0^\infty f_{H_1}\left[\mathcal{Y}_{M_i}^{out}(\tau, \Omega)\right] \cdot e^{j\upsilon \cdot \mathcal{Y}_{M_i}^{out}(\tau,\Omega)} d\left[\mathcal{Y}_{M_i}^{out}(\tau, \Omega)\right]$$

$$= \frac{1}{\sqrt{1 - 2j\upsilon\sigma_i^2}} \cdot \exp\left\{-q_p\left(1 - \frac{1}{\sqrt{1 - 2j\upsilon\sigma_i^2}}\right)\right\}, \tag{7.310}$$

where

$$q_p = \frac{\alpha_0^2 \cdot E_{a_1}}{2\sqrt{\sigma_i^2}} = \frac{\alpha_0^2 \cdot q}{2\sqrt{(1 + \delta_1^2 q^2)}} \tag{7.311}$$

is the ratio between the undistorted component of the signal and the sum of the signal noise component and the additive set noise at the input of the PF of the linear tract of the generalized detector, used by the diverse complex multi-channel signal processing system when there is the multiplicative noise; q is the signal-to-noise ratio

$$q = \frac{E_{a_1}}{N_0} \tag{7.312}$$

at the output of the generalized detector used by the diverse complex multi-channel signal processing system when the multiplicative noise is absent; and

$$\delta_1^2 = \frac{\sigma_s^2}{E_{a_1}^2} \tag{7.313}$$

is the normalized variance of the signal noise component at the output of the generalized detector used by the diverse complex multi-channel signal processing system determined by Eq. (6.110) when there is the multiplicative noise.

We assume that the energy of the signals, spectral power density of the additive set noise, and statistical characteristics of the multiplicative noise for all channels of the diverse complex multi-channel signal processing system, constructed on the basis of the generalized detector, are the same.

Thus, the variances σ_i^2 are the same for all channels of the diverse complex multi-channel signal processing system, constructed on the basis of the generalized detector, and are equal to σ^2.

The characteristic function of the sum of the independent signals determined by Eq. (7.289) at the output of the diverse complex multi-channel signal processing system constructed on the basis of the generalized detector for the conditions mentioned above, taking into account Eq. (7.310) and using the series expansion of the exponential function, can be written in the following form:

$$\Theta_1^N(\upsilon) = \prod_{i=1}^{N} \Theta_1^i(\upsilon) = e^{-q_p N} \sum_{m=0}^{\infty} \frac{q_p^m N^m}{m!} \cdot \frac{1}{\sqrt{(1 - 2j\upsilon\sigma^2)^{N+m}}}. \tag{7.314}$$

Equation (7.314) is true for the diverse complex multi-channel signal processing system, constructed on the basis of the generalized detector, when the input channels are spaced apart or separated by frequency. In the case of separation by frequency of the input channels of the diverse complex multi-channel signal processing system, constructed on the basis of the generalized detector, it is suggested that the transmitter of the information signals generates N signals using corresponding frequencies.

When the transmitter generates the information signals using a single carrier frequency and the input channels, of the diverse complex multi-channel signal processing system constructed on the basis of the generalized detector, are separated by frequency (see Section 7.3), taking into account Eq. (7.310) and denoting the input frequency channels corresponding to the carrier frequency of the information signal by the index k, we can write

$$\Theta_1^N(\upsilon) = \exp\left\{-q_p\left(1 - \frac{1}{\sqrt{1 - 2j\upsilon\sigma_k^2}}\right)\right\} \prod_{i=1}^{N} \frac{1}{\sqrt{1 - 2j\upsilon\sigma_i^2}}, \quad i \neq k, \tag{7.315}$$

where contrary to Eq. (7.314)

$$\sigma_i^2 = 4\sigma_n^4 + \sigma_s^2(\tau, \Omega_i); \tag{7.316}$$

$$\sigma_k^2 = 4\sigma_n^4 + \sigma_s^2(\tau, 0); \tag{7.317}$$

$$\Omega_i = (i - k)\Omega_T \tag{7.318}$$

is the detuning by frequency between the input channels of the diverse complex multi-channel signal processing system constructed on the basis of the generalized detector;

$$\Omega_T = \frac{\pi}{T} \tag{7.319}$$

is the shift in frequency between the adjacent channels of the diversity complex multi-channel signal processing system constructed on the basis of the generalized detector.

The value of the variance $\sigma_s^2(\tau, \Omega)$, determined by Eq. (6.109), of the signal at the output of the generalized detector used by the diverse complex multi-channel signal processing system depends on the form of the energy spectrum of fluctuations $G_V(\Omega)$ of the noise modulation function $\dot{M}(t)$ of the multiplicative noise (see Sections 3.4 and 6.2).

Taking into consideration, for simplicity, that the energy spectrum $G_V(\Omega)$ is uniform within the limits of the frequency range for all N input channels of the diverse complex multi-channel signal processing system constructed on the basis of the generalized detector, i.e.,

$$G_V(\Omega) = G_V(0) \tag{7.320}$$

and

$$|\Omega| \leq N\Omega_T \tag{7.321}$$

and taking into account the main results discussed in Section 6.4 for the variance of the signal noise component at the output of the generalized detector, used by the diverse complex multi-channel signal processing system for the condition

$$\tau = 0,$$

Eq. (7.315) can be written in the following form:

$$\Theta_1^N(\upsilon) = e^{-q_p} \sum_{m=0}^{\infty} \frac{q_p^m}{m!} \cdot \frac{1}{\sqrt{(1 - 2j\upsilon\sigma^2)^{N+m}}}, \tag{7.322}$$

where

$$\sigma_s^2(0, \Omega_T) = \sigma_s^2(0, 2\Omega_T) = \cdots = \sigma_s^2(0, N\Omega_T) = \frac{E_{a_1}^2}{2T} \cdot G_V(0)\,; \tag{7.323}$$

T is the duration of the signal.

Since the characteristics function and the probability distribution density are the joint pair of the Fourier transform,[14] taking into account Eqs. (7.314) and (7.322), we can write after integration

$$f_{H_1}[\mathcal{R}^{out}(\tau, \Omega)] = \frac{1}{2\pi} \int_{-\infty}^{\infty} \Theta_1^N(\upsilon) \cdot e^{-j\upsilon\mathcal{R}^{out}(\tau,\Omega)}\, d\upsilon = \frac{1}{2\sigma^2} \cdot e^{-q_{eq}} \sum_{m=0}^{\infty} \frac{q_{eq}^m}{m!} \left[\frac{\mathcal{R}^{out}(\tau, \Omega)}{2\sigma^2}\right]^{N+m-1} \times \frac{1}{\Gamma(N+m)} \cdot e^{-\frac{\mathcal{R}^{out}(\tau,\Omega)}{2\sigma^2}}, \tag{7.324}$$

where

$$q_{eq} = Nq_p \tag{7.325}$$

is the case of spacing or separation by frequency of the input channels of the diverse complex multi-channel signal processing system constructed on the basis of the generalized detector.

In the last case the signals are generated using N various carrier frequencies

$$q_{eq} = q_p \tag{7.326}$$

for the case when only the carrier frequency of the signal is used for N various frequencies (see Section 7.3).

When a "no" signal at the input of the PF of the linear tract of the generalized detector used by the diverse complex multi-channel signal processing system and there is only the additive set noise at the input channels of the diverse complex multi-channel signal processing system constructed on the basis of the generalized detector, as can be seen from Eqs. (7.307) and (7.324), the probability distribution density of the signal at the output of the quadratic detector (see Fig. 7.17) has the following form:

$$f_{H_0}[\mathcal{R}^{out}(\tau,\Omega)] = \frac{1}{8\sigma_n^4} \cdot e^{-\frac{\mathcal{R}^{out}(\tau,\Omega)}{8\sigma_n^4}} \cdot \frac{[\mathcal{R}^{out}(\tau,\Omega)]^{N-1}}{\left(8\sigma_n^4\right)^{N-1}(N-1)!}. \tag{7.327}$$

The probability distribution density in Eq. (7.327) is called the χ^2-distribution that defines the probability distribution density of the sum of N independent random variables obeying the exponential probability distribution density (see Eq. (7.304)).

Comparing Eqs. (7.324) and (7.327), we are able to obtain that the probability distribution density of the resulting summing signal, determined by Eq. (7.289), at the output of the quadratic detector of the diverse complex multi-channel signal processing system, constructed on the basis of the generalized detector, is defined by a formula that is similar to Eq. (7.327) when distortions in phase of the signal are distributed uniformly within the limits of the interval $[0, 2\pi]$ for the conditions

$$\alpha_0 = 0 \qquad \text{and} \qquad q_{eq} = 0.$$

However, for this case we must use the variance σ^2 that is determined by Eq. (7.299).

Taking into account Eq. (7.327), we can define the probability of false alarm P_{F_N} equal to the probability of exceeding the threshold K_g if a "no" signal at the input of the PF of the linear tract of the generalized detector used by the diverse complex multi-channel signal processing system—the hypothesis H_0:

$$P_{F_N} = \int_{K_g}^{\infty} f_{H_0}[\mathcal{R}^{out}(\tau,\Omega)]\, d[\mathcal{R}^{out}(\tau,\Omega)] = 1 - \gamma[K_g', N], \tag{7.328}$$

where

$$\gamma[K'_g, N] = \frac{1}{\Gamma(N)} \int\limits_0^{K'_g} [\mathcal{R}^{out}(\tau, \Omega)]^{N-1} \cdot e^{-\mathcal{R}^{out}(\tau,\Omega)} d[\mathcal{R}^{out}(\tau, \Omega)] \tag{7.329}$$

is the incomplete gamma function;[89] $\Gamma(N)$ is the gamma function; and

$$K'_g = \frac{K_g}{2\sigma^2} \tag{7.330}$$

is the normalized threshold.

Using the power series expansion of the incomplete gamma function[89]

$$\gamma(N, x) = 1 - e^{-x} \sum_{k=0}^{N-1} \frac{x^k}{k!}, \tag{7.331}$$

Eq. (7.328) can be written in the following form

$$P_{F_N} \simeq \frac{(K'_g)^N}{N!} \cdot e^{-K'_g}, \quad N < K'_g, \tag{7.332}$$

$$P_{F_N} = e^{-K'_g} \sum_{k=0}^{N-1} \frac{(K'_g)^k}{k!} = P_N(N-1, K'_g), \tag{7.333}$$

where $P_N(N-1, K'_g)$ is the Poisson probability distribution function.[89]

It is very difficult in the mathematical sense to solve Eq. (7.333). As a rule, the empirical formula allowing us to define the threshold for the predetermined probability of false alarm P_{F_N} is used in practice:

$$K'_g = \left|\ln P_{F_N}\right| + 0.55(N-1)\sqrt{1 + 1.3\left|\ln P_{F_N}\right|}, \quad N \leq 10. \tag{7.334}$$

Dependences of the threshold K'_g as a function of the number N of the input channels, of the diverse complex multi-channel signal processing system constructed on the basis of the generalized detector, determined by exact (see Eq. (7.328)) and approximated (see Eq. (7.334)) formulae, are shown in Fig. 7.23.

The same dependences for the diverse complex multi-channel signal processing system constructed on the basis of the correlation detector are also shown in Fig. 7.23 for the purpose of comparison.

Comparative analysis shows us a superiority of the generalized detector over the correlation detector during the use by the diverse complex multi-channel signal processing systems.

When the number of the input channels of the diverse complex multi-channel signal processing system constructed on the basis of the generalized detector is much greater than unity

$$N \gg 1$$

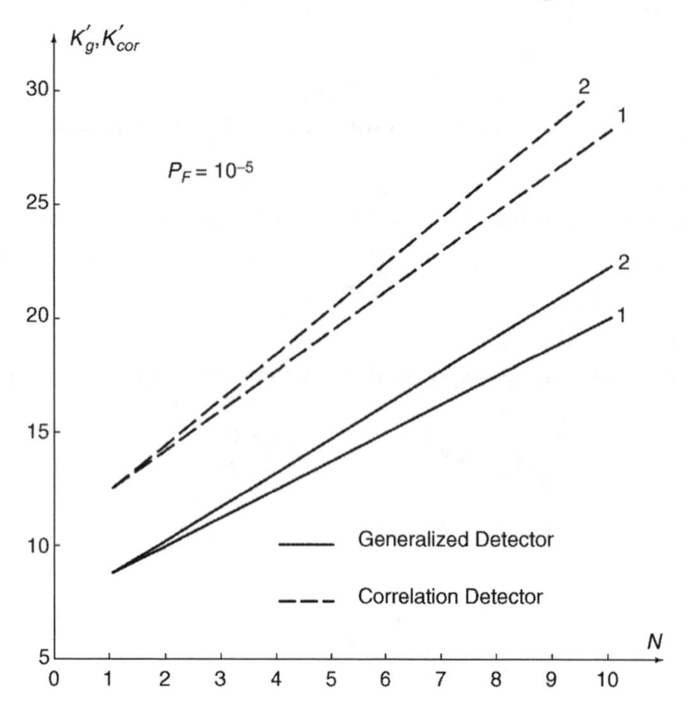

FIGURE 7.23
Threshold value as a function of the number N of the input channels of a diverse signal processing system: for 1, Eq. (7.258) is used; for 2, Eq. (7.264) is used.

Eqs. (7.328) and (7.332) are approximated with sufficient precision in the following form:

$$P_{F_N} \simeq 1 - \Phi\left[2\left(\sqrt{K'_g} - \sqrt{N}\right)\right], \tag{7.335}$$

where $\Phi(x)$ is the error integral determined by Eq. (7.210).

We define the probability of exceeding the threshold K_g by the resulting summing signal determined by Eq. (7.289), at the output of the diverse complex multi-channel signal processing system constructed on the basis of the generalized detector, when a "yes" signal is at the input of the PF of the linear tract of the generalized detector—the hypothesis H_1—used by the diverse complex multi-channel signal processing system.

The probability of detection $P_{D_{M_N}}$ of the signal can be written in terms of Eq. (7.324) in the following form

$$P_{D_{M_N}} = \int\limits_{K_g}^{\infty} f_{H_1}[\mathcal{R}^{out}(\tau, \Omega)]\, d[\mathcal{R}^{out}(\tau, \Omega)]$$

$$= 1 - e^{-q_{eq}} \sum_{m=0}^{\infty} \frac{q_{eq}^m}{m!} \cdot \gamma\left(N + m, K_{g_{eq}}\right), \tag{7.336}$$

where

$$K_{g_{eq}} = \frac{K'_g}{\sqrt{1 + \delta_1^2(\tau, \Omega)q^2}} \tag{7.337}$$

is the detection parameter defining the probability of detection $P_{D_{M_N}}$ of the signal, and

$$\delta_1^2(\tau, \Omega) = \frac{\sigma_s^2(\tau, \Omega)}{E_{a_1}^2} \tag{7.338}$$

is the normalized variance of the signal noise component determined by Eq. (6.110).

Taking into consideration a convergence of the incomplete gamma function for the condition

$$N \gg 1$$

to the error integral determined by Eq. (7.210),[14] Eq. (7.336) can be written in the following form:

$$P_{D_{M_N}} \simeq 1 - e^{-q_{eq}} \sum_{m=0}^{\infty} \frac{q_{eq}^m}{m!} \cdot \Phi\left[2\left(\sqrt{K_{g_{eq}}} - \sqrt{N+m}\right)\right]. \tag{7.339}$$

The parameter

$$q_p = \frac{\alpha_0^2 q}{2\sqrt{1 + \delta_1^2(\tau, \Omega)q^2}} \tag{7.340}$$

tends to approach the constant equal to the ratio between the energy of the undistorted component of the signal and the energy of the signal noise component at the input of the quadratic detector of the diverse complex multichannel signal processing system constructed on the basis of the generalized detector, when the energy of the signal at the input of the considered signal processing system is increased

$$q = \frac{E_{a_1}}{N_0} \to \infty. \tag{7.341}$$

For this case we can write

$$q_p \simeq \frac{\alpha_0^2}{2\sqrt{\delta_1^2(\tau, \Omega)}} \tag{7.342}$$

when

$$q \gg \frac{1}{\sqrt{\delta_1^2(\tau, \Omega)}}. \tag{7.343}$$

The detection parameter $K_{g_{eq}}$ is varied within the limits of the interval

$$0 \leq K_{g_{eq}} \leq K'_g \tag{7.344}$$

in the course of variation of the ratio between the variance of the signal noise component and the variance of the additive Gaussian noise, at the output of the quadratic detector of the diverse complex multi-channel signal processing system constructed on the basis of the generalized detector

$$\frac{\sigma_s^2}{4\sigma_n^4} = \delta_1^2(\tau, \Omega)q^2 \tag{7.345}$$

within the limits of the interval $[0, \infty)$.

Reference to Fig. 7.23 shows that during the condition $N \leq 10$ the detection parameter $K_{g_{eq}}$ is less than 21 if the probability of false alarm P_{F_N} is varied within the limits of the interval

$$10^{-5} \leq P_{F_N} \leq 10^{-1}. \tag{7.346}$$

For the same interval of variation of the probability of false alarm P_{F_N} (see Eq. (7.275)) the value of the detection parameter during the use of the correlation detector by the diverse complex multi-channel signal processing system is less than 30.

Thus, the generalized detector has a great superiority over the correlation detector that is thought of as the optimal detector during use by the diverse complex multi-channel signal processing system for various and specific applications.

In that case when distortions in phase of the signal are distributed uniformly within the limits of the interval $[0, 2\pi]$, the undistorted component of the signal is equal to zero

$$\alpha_0 = 0$$

and the probability of detection $P_{D_{M_N}^{(un)}}$, of the resulting summing signal at the output of the diverse complex multi-channel signal processing system constructed on the basis of the generalized detector (see Eq. (7.336)), has the following form:

$$P_{D_{M_N}^{(un)}} = 1 - \gamma\left(N, K_{g_{eq}}\right). \tag{7.347}$$

The detection performances based on Eq. (7.347) as a function of the number N of the input channels of the diverse complex multi-channel signal processing system constructed on the basis of the generalized detector for the case of distortions in phase of the signal, which are distributed uniformly within the limits of the interval $[0, 2\pi]$ for various values of the ratio determined by Eq. (7.345), are shown in Fig. 7.24.

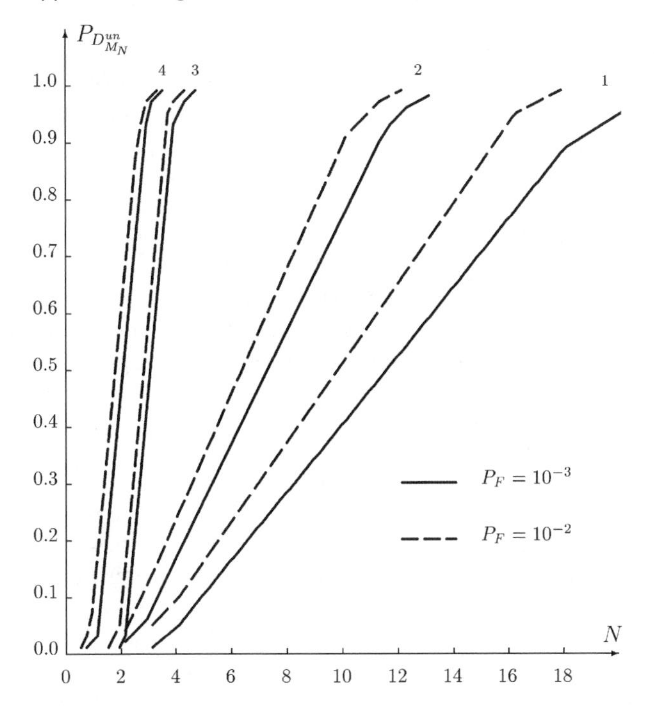

FIGURE 7.24
Detection performances as a function of the number N of the input channels: for 1, $\frac{\sigma_s^2}{4\sigma_n^4} = 0.5$; for 2, $\frac{\sigma_s^2}{4\sigma_n^4} = 1.0$; for 3, $\frac{\sigma_s^2}{4\sigma_n^4} = 5.0$; for 4, $\frac{\sigma_s^2}{4\sigma_n^4} = 10$.

The detection performances are plotted for the probability of false alarm P_{F_N} equal to 10^{-2} and 10^{-3}.

As may be seen from Fig. 7.24, an increase in the number N of the input channels of the diverse complex multi-channel signal processing system constructed on the basis of the generalized detector for

$$P_{F_N} = const \qquad \text{and} \qquad \frac{\sigma_s^2}{4\sigma_n^4} = const \tag{7.348}$$

leads to an increase in the probability of detection of the signal.

Also, it may be seen that a character of the detection performances for the given case during the condition

$$\alpha_0 = 0$$

depends significantly on the product $\delta_1^2(\tau, \Omega)q^2$.

If

$$\delta_1^2(\tau, \Omega)q^2 < 1 \tag{7.349}$$

the probability of detection of the signal is increased smoothly with an increase in the number of the input channels of the diverse complex multi-channel signal processing system constructed on the basis of the generalized detector. But if the condition

$$\delta_1^2(\tau, \Omega)q^2 > 1 \tag{7.350}$$

is satisfied this dependence has a threshold character.

For the last case an increase in the probability of detection with an increase in the number of the input channels, of the diverse complex multi-channel signal processing system constructed on the basis of the generalized detector, is insignificant starting with some number N_{ef}.

For example, for the condition

$$\delta_1^2(\tau, \Omega)q^2 = 10 \tag{7.351}$$

the effective number of the input channels of the diverse complex multi-channel signal processing system constructed on the basis of the generalized detector is equal to $N_{ef} = 3$.

An increase in the probability of detection of the signal with an increase in the number of the input channels, of the diverse complex multi-channel signal processing system constructed on the basis of the generalized detector, for relatively low values of the product $\delta_1^2(\tau, \Omega)q^2$, for example,

$$\delta_1^2(\tau, \Omega)q^2 < 10, \tag{7.352}$$

and under the constant probability of false alarm P_{F_N} can be explained in the following way.

Taking into account Eqs. (7.307) and (7.324), the normalized mean m_Σ and the variance σ_Σ^2 of fluctuations of the signal at the output of the diverse complex N-channel signal processing system constructed on the basis of the generalized detector for the condition

$$\alpha_0 = 0$$

can be written in the following form:[10]

$$m_\Sigma = 2N\sigma_n^2\sqrt{1 + \delta_1^2(\tau, \Omega)q^2}; \tag{7.353}$$

$$\sigma_\Sigma^2 = 4N\sigma_n^4\left[1 + \delta_1^2(\tau, \Omega)q^2\right]. \tag{7.354}$$

The ratio between the square of the mean m_Σ^2 and the variance σ_Σ^2 is determined in terms of Eqs. (7.353) and (7.354) by the following form

$$\frac{m_\Sigma^2}{\sigma_\Sigma^2} = \frac{N}{4}. \tag{7.355}$$

Thus, reference to Eqs. (7.347) and (7.355) shows that with an increase in the number of the input channels of the diverse complex multi-channel signal processing system constructed on the basis of the generalized detector—the number N can be considered the number of degrees of freedom of the probability distribution density—the role of fluctuations of the signal is decreased.

In other words, the variance of fluctuations of the resulting summing signal is increased with an increase in the number of input channels of the diverse complex multi-channel signal processing system constructed on the basis of the generalized detector in N times slower than the square of the mean of the resulting summing signal.

When a "yes" signal at the input of the PF of the linear tract of the generalized detector used by the diverse complex multi-channel signal processing system—the hypothesis H_1—during the condition

$$\delta_1^2(\tau, \Omega)q^2 > 0 \tag{7.356}$$

the mean of the resulting summing signal at the output of the quadratic detector of the diverse complex multi-channel signal processing system constructed on the basis of the generalized detector differs from the mean of the output signal under the hypothesis H_0. If the difference is high, then the number of the input channels of the diverse complex multi-channel signal processing system constructed on the basis of the generalized detector is high.

As a consequence, with an increase in the number N of the input channels of the diverse complex multi-channel signal processing system constructed on the basis of the generalized detector, the probability of detection of the signal is significantly increased.

Reference to Fig. 7.24 shows that this case corresponds to values

$$N < N_{ef}$$

for the condition

$$\frac{\sigma_s^2}{4\sigma_n^4} = \delta_1^2(\tau, \Omega)q^2 < 10. \tag{7.357}$$

If the number N of the input channels, of the diverse complex multi-channel signal processing system constructed on the basis of the generalized detector, is much greater in value than the detection parameter $K_{g_{eq}}$ then, as is well known,[14] the probability distribution density function defined by the incomplete gamma function tends asymptotically toward the Gaussian probability distribution function.

It is well known[1,2] that when the instantaneous values of the signal and the additive Gaussian noise at the input of the PF of the linear tract of the generalized detector, used by the diverse complex multi-channel signal processing system constructed on the basis of the generalized detector, obey the Gaussian probability distribution density a proportional increase in the power of the signal and the additive Gaussian noise does not lead to an increase in the probability of detection of the signal when the probability of false alarm is constant

$$P_{F_N} = const.$$

This phenomenon corresponds to the values

$$N > N_{ef},$$

shown in Fig. 7.24.

We define the value N_{ef} during the following condition: when the mean m_Σ exceeds the threshold

$$K_g = 4\sigma_n^4 k_g \tag{7.358}$$

determined in accordance with the predetermined probability of false alarm P_{F_N}, a further increase in the number N of the input channels of the diverse complex multi-channel signal processing system constructed on the basis of the generalized detector acts insignificantly on the probability of detection of the signal.

Taking into account Eq. (7.353), we are able to define the estimation

$$N_{ef} > \frac{k_g}{2\sqrt{1 + \delta_1^2(\tau, \Omega)q^2}}, \tag{7.359}$$

where

$$k_g = \sqrt{2}k'_g \tag{7.360}$$

is the coefficient defined by the probability of false alarm P_{F_N}.

Reference to Eq. (7.359) shows that the energy of the signal at the input of the PF of the linear tract of the generalized detector used by the diversity complex multi-channel signal processing system is high, the number N of the input channels required that the probability of detection $P_{D_{M_N}^{(un)}}$ of the signal could be close to unity is low.

We estimate how the undistorted component of the signal acts on the probability of detection $P_{D_{M_N}}$ of the signal determined by Eq. (7.336) that is characteristic of the case when distortions in phase of the signal within the limits of the interval $[0, 2\pi]$ obey the probability distribution density differed from the uniform distribution, as follows from Sections 4.2 and 4.4.

During estimation we compare the detection performances of the signal when the undistorted component of the signal is not equal to zero

$$\alpha_0 \neq 0$$

with the detection performances of the signal if the undistorted component of the signal is absent (see Eq. (7.347)).

Equations (7.336) and (7.339) are not convenient under analysis of the stimulus of the multiplicative noise on the detection performances of the diverse complex multi-channel signal processing system constructed on the basis of the generalized detector when the undistorted component of the signal is not equal to zero. For simplicity, we consider two cases.

Case 1. The variance of the signal noise component $\sigma_s^2(\tau, \Omega)$ determined by Eq. (6.107) is much greater than the predetermined threshold K_g:

$$K'_g \ll \delta_1^2(\tau, \Omega) q^2. \tag{7.361}$$

As was noted above,

$$K'_g \gg 1 \quad \text{at} \quad P_{F_N} < 10^{-1}. \tag{7.362}$$

Then the detection parameter can be determined in the following form:

$$K_{g_{eq}} = \frac{K'_g}{\sqrt{1 + \delta_1^2(\tau, \Omega) q^2}} \ll 1 \tag{7.363}$$

and the parameter q_{eq} determined by Eq. (7.325) takes the following form:

$$q_{eq} \simeq \frac{N\alpha_0^2}{2\sqrt{\delta_1^2(\tau, \Omega)}}. \tag{7.364}$$

Imaging the incomplete gamma function as the power series with respect to the detection parameter $K_{g_{eq}}$[90]

$$\gamma(N + m, K_{g_{eq}}) = \sum_{l=0}^{\infty} \frac{(-1)^l K_{g_{eq}}^{N+m+l}}{(N + m - 1)!\, l!\, (N + Ml)} \tag{7.365}$$

and limiting for the conditions

$$K_{g_{eq}} \ll 1 \tag{7.366}$$

and

$$N + m \geq 1 \tag{7.367}$$

by a single term ($l = 0$) of alternating series, the probability of detection $P_{D_{M_N}}$ of the resulting summing signal determining by Eq. (7.336) at the output of the diverse complex multi-channel signal processing system constructed on the basis of the generalized detector can be written in the following form:

$$P_{D_{M_N}} = 1 - e^{-q_{eq}} \sum_{m=0}^{\infty} \frac{q_{eq}^m K_{g_{eq}}^{N+M}}{m!\, (N + m)!}. \tag{7.368}$$

Taking into account a representation of the Bessel function in the form of the power series

$$J_n(x) = \sum_{m=0}^{\infty} \frac{(-1)^m \left(\frac{x}{2}\right)^{2m+n}}{m!(m+n)!}, \tag{7.369}$$

we can finally obtain that the probability of detection $P_{D_{M_N}}$ of the signal determined by Eq. (7.368) can be written in the following form:

$$P_{D_{M_N}} = 1 - \left(\frac{K_{g_{eq}}}{q_{eq}}\right)^{\frac{N}{2}} \cdot e^{-q_{eq}} J_N\left(2q_{eq}\sqrt{K_{g_{eq}}}\right), \tag{7.370}$$

where

$$q \gg \frac{K'_g}{\sqrt{\delta_1^2(\tau, \Omega)}}. \tag{7.371}$$

Taking into account Eqs. (7.347) and (7.371), the relative differential of the probability of detection of the signal caused by the nonzero undistorted component of the signal can be written in the following form:

$$\mathcal{D} = \frac{P_{D_{M_N}} - P_{D_{M_N}^{(un)}}}{P_{D_{M_N}^{(un)}}} = \frac{\gamma(N, K_{g_{eq}})}{1 - \gamma(N, K_{g_{eq}})} \times \left\{1 - \left(\frac{K_{g_{eq}}}{q_{eq}}\right)^{\frac{N}{2}} \cdot \frac{J_N(2q_{eq}\sqrt{K_{g_{eq}}})}{\gamma(N, K_{g_{eq}})} \cdot e^{-q_{eq}}\right\}, \tag{7.372}$$

where

$$q_{eq} = \frac{N\alpha_0^2 q}{\sqrt{1 + \delta_1^2(\tau, \Omega)q^2}}; \tag{7.373}$$

$$K_{g_{eq}} = \frac{k'_g}{\sqrt{1 + \delta_1^2(\tau, \Omega)q^2}}. \tag{7.374}$$

Case 2. The variance of the signal noise component $\sigma_s^2(\tau, \Omega)$ is much less than the variance of the additive set noise at the output of the quadratic detector of the diverse complex multi-channel signal processing system constructed on the basis of the generalized detector

$$\delta_1^2(\tau, \Omega)q^2 \ll 1. \tag{7.375}$$

In this case when the condition

$$q_{eq} < 1 \qquad \text{or} \qquad N\alpha_0^2 < 1 \tag{7.376}$$

is satisfied we can be limited by two terms of the power series determined by Eq. (7.336).

Thus, we can write

$$P_{D_{MN}} = 1 - \left[\gamma\left(N, K_{g_{eq}}\right) + q_{eq}\gamma\left(N+1, K_{g_{eq}}\right)\right] \cdot e^{-q_{eq}}, \quad q_{eq} < 1. \tag{7.377}$$

Taking into consideration the recurrence relation[91]

$$\gamma\left(N+1, K_{g_{eq}}\right) = \gamma\left(N, K_{g_{eq}}\right) - \frac{K_{g_{eq}}^{N} e^{-K_{g_{eq}}}}{N!}, \tag{7.378}$$

the relative differential of the probability of detection of the signal caused by the nonzero undistorted component of the signal for the condition $q_{eq} < 1$ can be written in the following form:

$$\begin{aligned} \mathcal{D} &= \frac{\gamma\left(N, K_{g_{eq}}\right)}{1-\gamma\left(N, K_{g_{eq}}\right)}\left\{1 - \left[1 - q_{eq}\left(1 - \frac{K_{g_{eq}}^{N} e^{-K_{g_{eq}}}}{\gamma\left(N, K_{g_{eq}}\right) N!}\right)\right] \cdot e^{-q_{eq}}\right\} \\ &\simeq q_{eq} \cdot \frac{\gamma\left(N, K_{g_{eq}}\right)}{1-\gamma\left(N, K_{g_{eq}}\right)}, \quad q_{eq} < 1, \end{aligned} \tag{7.379}$$

where

$$q_{eq} \simeq N\alpha_0^2 q. \tag{7.380}$$

Since during the condition determined by Eq. (7.375) we can believe that

$$K_{g_{eq}} \simeq K'_g \tag{7.381}$$

and Eq. (7.379) can be written in terms of Eq. (7.328) in the following form:

$$\mathcal{D} \simeq q_{eq} \cdot \frac{1 - P_{F_N}}{P_{F_N}}, \quad q_{eq} < 1, \tag{7.382}$$

where P_{F_N} is the predetermined probability of false alarm in the case of the diverse complex N-channel signal processing system constructed on the basis of the generalized detector.

Dependences of the relative differential $\mathcal{D}$, of the probability of detection of the signal determined by Eqs. (7.336) and (7.347) as a function of the ratio between the energy of the undistorted component of the signal and the energy of the additive set noise of the diverse complex multi-channel signal processing system, constructed on the basis of the generalized detector with two input channels during the probability of false alarm $P_F = 10^{-2}$, are shown in Fig. 7.25.

The ratio between the normalized variance of the signal noise component, at the output of the generalized detector used by the diverse complex multi-channel signal processing system caused by the stimulus of the multiplicative

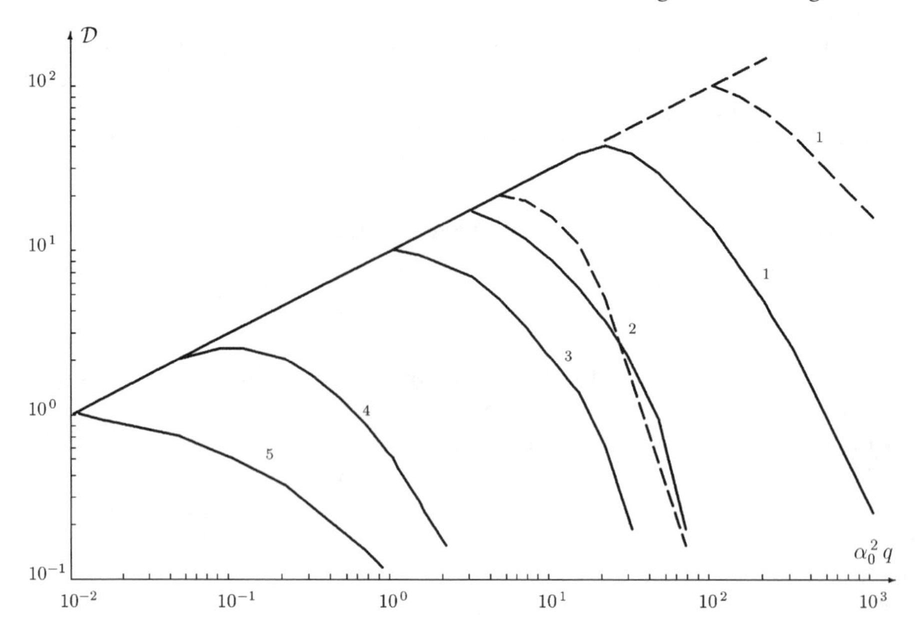

FIGURE 7.25
Relative differential $\mathcal{D}$ of the probability of detection of the signal as a function of the ratio between the energy of the undistorted component of the signal and the energy of the additive set noise $\alpha_0^2 q$; $N = 2$; $P_F = 10^{-2}$: for 1, $\chi^2 = 0.1$; for 2, $\chi^2 = 0.5$; for 3, $\chi^2 = 1.0$; for 4, $\chi^2 = 5.0$; for 5, $\chi^2 = 10$.

noise and the square of the relative energy level of the undistorted component of the signal is taken as the parameter

$$\chi^2 = \frac{2\delta_1^2(\tau, \Omega)}{\alpha_0^2(\tau, \Omega)} \tag{7.383}$$

in Fig. 7.25.

The dotted lines shown in Fig. 7.25 are determined by Eqs. (7.372) and (7.382) and allow us to make comparison with dependences defined on the basis of Eqs. (7.336) and (7.347), and to define a range of values $\alpha_0^2(\tau, \Omega)$ on the basis of comparative analysis for the given precision. Using the defined range of the values $\alpha_0^2(\tau, \Omega)$, we are able to apply Eqs. (7.372) and (7.382).

Reference to Fig. 7.25 shows that with an increase in the ratio between the energy of the undistorted component of the signal and the power of the additive set noise, of the diverse complex multi-channel signal processing system constructed on the basis of the generalized detector, the relative differential of the probability of detection of the resulting summing signal at the output of the diverse complex multi-channel signal processing system constructed on the basis of the generalized detector is increased at first, and after that, starting with some ratio

$$q_\alpha = \alpha_0^2 \cdot q, \tag{7.384}$$

is decreased, tending to approach zero.

In conclusion, we note the following.

- The detection performances of the diverse complex multi-channel signal processing system, constructed on the basis of the generalized detector, depend significantly on the ratio between the variance of the signal noise component and the power of the undistorted component of the signal when the parameter χ^2 is satisfied to the following condition

$$\chi^2 = \frac{2\delta_1^2(\tau, \Omega)}{\alpha_0^2} < 10. \qquad (7.385)$$

- An increase in the number of the input channels, of the diverse complex multi-channel signal processing system constructed on the basis of the generalized detector, allows us to obtain the higher probability of detection of the signals when there is multiplicative noise. However, if

$$\delta_1^2(\tau, \Omega)q^2 = \chi^2 q_\alpha^2 \geq 1 \qquad (7.386)$$

  a great increase in the number of the input channels, of the diverse complex multi-channel signal processing system constructed on the basis of the generalized detector, is not worthwhile, since starting with some number of input channels the probability of detection of the signal is increased insignificantly with an increase in the number of the input channels (see Fig. 7.24).

7.6 Conclusions

We summarize the main results discussed in this chapter:

- To analyze the impact of the multiplicative noise on detection of the signals in the presence of additive Gaussian noise the following model of the generalized detector is used. The generalized detector contains N channels—N elements of resolution by instant of the incoming signal or by frequency of the signal. Accordingly, the parameter by which a signal resolution is carried out can take N known discrete values. The model of the generalized detector is considered for two types of signal: the signal with the known initial phase and the signal with the unknown (random) initial phase.
- For the case of the signal with known initial phase the resulting signal at the output of the "n"-th channel of the generalized detector is the sum of three components: the background noise of the generalized detector; the total noise component of the generalized detector caused by the interaction between the noise component of the

correlation channel

$$2\int_{-\infty}^{\infty} a_{m_n}(t)\xi(t)\,dt$$

and the random component of the autocorrelation channel

$$2\int_{-\infty}^{\infty} a_{M_k}(t)\xi(t)\,dt$$

of the generalized detector; and the signal component at the output of the generalized detector.

— The background noise of the generalized detector is caused by the additive set noise. This was discussed in more detail in Chapter 5.

— The noise component of the correlation channel of the generalized detector is caused by the interaction between the model signal generated by the MSG at each channel of the generalized detector and the additive set noise. The random component of the autocorrelation channel of the generalized detector is caused by the interaction between the incoming signal and the additive set noise of the generalized detector. Under the stimulus of the multiplicative noise the random component of the autocorrelation channel of the generalized detector exists both at the "k"-th channel of the generalized detector

$$n = k,$$

in which there is the signal component at the output of the "k"-th channel of the generalized detector for a "no" multiplicative noise condition, and at other channels of the generalized detector

$$n = m \neq k,$$

in which a "no" signal is at the output of the given channels of the generalized detector when the multiplicative noise is absent. For the definite conditions the noise component of the correlation channel and the random component of the autocorrelation channel of the generalized detector are completely compensated in the statistical sense.

— Under the stimulus of the multiplicative noise the input signal generates the signal component both at the output of the "k"-th channel of the generalized detector

$$n = k,$$

in which there is the signal component at the output of the "k"-th channel of the generalized detector for a "no" multiplicative noise condition, and at the outputs of other channels of the generalized detector

$$n = m \neq k,$$

in which a "no" signal component at the output of the given channels of the generalized detector when the multiplicative noise is absent.

— In the case of periodic multiplicative noise if the initial phase of the noise modulation function $\dot{M}(t)$ of the multiplicative noise is the random variable the conditional probability distribution function of the resulting signal at the output, for example, of the "n"-th channel of the generalized detector, is Gaussian at the fixed value of the initial phase of the noise modulation function $\dot{M}(t)$ of the multiplicative noise. The probability distribution function of the resulting signal at the output of the "n"-th channel of the generalized detector can be defined by averaging the conditional probability distribution function with respect to the initial phase of the noise modulation function $\dot{M}(t)$ of the multiplicative noise.

— For the case of fluctuating multiplicative noise there is a need to note that the variance of the signal component at the output of the "n"-th channel of the generalized detector is defined by a shape of the signal and the noise modulation function $\dot{M}(t)$ of the multiplicative noise. The variance of the resulting signal at the output of the "n"-th channel of the generalized detector is defined by the variance of the signal component and the variance of the background noise of the generalized detector. The probability distribution function of the resulting signal at the output of the "n"-th channel of the generalized detector can be defined as a convolution of the probability distribution functions of the signal component and the background noise. However, for some definite conditions the probability distribution function of the resulting signal at the output of the "n"-th channel of the generalized detector can be approximated by the Gaussian probability distribution function with the mean determined by Eq. (7.30) and the variance determined by Eq. (7.35).

- For the case of the signal with unknown (random) initial phase the conditional probability distribution density of the signal at the output of individual channels of the generalized detector during the fixed value of the initial phase ϑ of the noise modulation function $\dot{M}(t)$ of the multiplicative noise, when the multiplicative noise is present and periodic, is determined by Eq. (7.48) for the case in which the initial phase ϑ of the noise modulation function $\dot{M}(t)$ of the multiplicative noise is the deterministic variable. When the initial phase, of the noise modulation function $\dot{M}(t)$ of the multiplicative noise, is the random variable, the probability distribution density of the signal at the output of the generalized detector is defined by averaging Eq. (7.48) with respect to the random initial phase. For the case of the fluctuating multiplicative noise the probability distribution density of the resulting signal at the output of the generalized detector that is determined by Eq. (7.42) can be approximated by the Gaussian probability distribution density.

- There are two forms of signal detection for the chosen model of the generalized detector:
  - Signal detection at the "true" channel of the generalized detector. The probability of detection P_{D_M} of the signal for this case is equal to the probability $P_{D_{M_k}}$ of exceeding the threshold at the "k"-th channel of the generalized detector by the signal at the output of the generalized detector, in other words, at the channel, in which there is the signal when the multiplicative noise is absent, during the condition of a "yes" signal at the input of the PF of the linear tract of the generalized detector—the simple binary detection.
  - A "yes" signal at the input of the PF of the linear tract of the generalized detector. The probability of detection $P_{D'_M}$ for this case is equal to the probability of exceeding the threshold by the signal at the output of a single channel of the generalized detector during the condition of a "yes" signal at the input of the PF of the linear tract of the generalized detector—the complex binary detection.
  - In the case of the signal with the known initial phase the probability of detection of the signals by the generalized detector is defined using the probability of detection and the probability of false alarm that are defined by the parameters K_{g_0} and q when the multiplicative noise is absent and using the parameters of the multiplicative noise and the signal—the parameter r_{nk}—when the multiplicative noise is present and periodic (see Eqs. (7.85) and (7.88)). When there is the fluctuating multiplicative noise the probability of detection of the signal at the "true" channel of the generalized detector is defined using the probability of detection of the signal and the probability of false alarm that are defined by the parameters K_{g_0} and q when the multiplicative noise is absent and using the parameters of the multiplicative noise and the signal—α_0^2 is the relative level of the undistorted component of the signal and $\delta_{k_1}^2$ is proportional to the variance of the signal noise component at the "true" channel of the generalized detector.
  - In the case of the signal with the unknown initial phase, the probability of detection of the signal is defined using the detection performances of the signal with the random initial phase by the generalized detector under the predetermined probability of false alarm P_{F_k}, and the detection parameter $\alpha_0^2 q$ under definition of the probability of detection $P_{D_{M_k}}$, and the detection parameter $q\,|\dot{R}_{mk}|$ under definition of the probability of detection $P_{D_{M_m}}$ when the multiplicative noise is periodic. In the case of the fluctuating multiplicative noise, the

probability of detection P_{D_M} of the signal by the generalized detector can be defined using the detection performances of the signal with the random initial phase by the generalized detector when the equivalent magnitude of the probability of false alarm is determined by Eq. (7.115) and the detection parameter of the generalized detector is equal to q_M (see Eq. (7.114)).

- Analysis of the detection performances of the signals by the generalized detector shows us the superiority of the generalized detector over the correlation detector or the matched filter during use in complex signal processing systems in various areas of application. For example, employment of the generalized detector in radar systems ensures greater radar range compared with the correlation detector or the matched filter under the same probabilities of detection P_D and false alarm P_F both when the multiplicative noise is present and when the multiplicative noise is absent. As is well known, in practice the multiplicative noise is always present.
- The multiplicative noise can be the only main source of false decision-making in communication systems when a level of energy characteristics of the information signal is much greater than a level of energy characteristics of the additive set noise and parasitic signals (interference). Comparative analysis between the detection performances of the generalized detector and the correlation detector allows us to draw the conclusion that the superiority of the generalized detector over the correlation detector during use in complex communication systems is clear.
- In some practical cases, for example, radar, communication, underwater signal processing, sonar, remote sensing, and so on, some characteristics and parameters of the information channel defining characteristics and parameters of the noise modulation function $\dot{M}(t)$ of the multiplicative noise can be *a priori* known. To detect the signals distorted, for example, by the rapid fluctuating multiplicative noise in the presence of additive Gaussian noise it is required to know *a priori* information about the characteristics of the multiplicative noise, for example, the correlation function or the energy spectrum of the noise modulation function $\dot{M}(t)$ of the multiplicative noise, the energy of the signal at the input of the PF of linear tract of the generalized detector, and the spectral power density of the additive noise (see Eq. (7.176)). The weight coefficients must be controlled both during variations in characteristics of the multiplicative noise and during variations in the signal-to-noise ratio at the input of the PF of the linear tract of the generalized detector. Absence of *a priori* information in some practical cases about the characteristics of the information channel—the noise modulation function $\dot{M}(t)$ of the multiplicative noise—and their variability and *a priori* ignorance of the energy characteristics of the detected signal, very often in practice leads to

employment of complex signal processing systems used for the signals with unknown *a priori* structure.

- Analysis of the detection performances shows us a superiority during the use of the generalized detector in problems of detection of the signals with an *a priori* unknown amplitude-phase-frequency structure in comparison with an employment of the universally adopted autocorrelation detector. The detection performances of the autocorrelation channel of the generalized detector at the same value of the signal-to-noise ratio are worse than the detection performances of the generalized detector for the case of the completely known signal. However, the detection performances of the autocorrelation channel of the generalized detector are better than the detection performances of the universally adopted autocorrelation detector, that is considered the optimal detector for the signals with an *a priori* unknown amplitude-phase-frequency structure, and are even better than the detection performances of the correlation detector for the case of the completely known signal, which is thought of as the optimal detector. If the losses in the detection performances of the autocorrelation channel of the generalized detector in comparison to the detection performances of the generalized detector for the case of the completely known signal are high, then the ratio between the time of integration T_{int} of the autocorrelation channel of the generalized detector and the correlation interval of the multiplicative noise τ_{c_n} is high, since the losses are proportional to this ratio. A decrease in the losses in the detection performances can be reached owing to a decrease in the bandwidth of the PF of the linear tract of the generalized detector. The limit of reduction in the bandwidth of the PF of the linear tract of the generalized detector is defined by the bandwidth of the energy spectrum of the signal when there is multiplicative noise present.
- The detection performances of the diverse complex multi-channel signal processing system, constructed on the basis of the generalized detector, depend significantly on the ratio between the variance of the signal noise component and the power of the undistorted component of the signal, when the ratio between the normalized variance of the signal noise component at the output of the generalized detector used by the diverse complex multi-channel signal processing system caused by the stimulus of the multiplicative noise and the square of the relative energy level of the undistorted component of the signal is not more than 10.
- An increase in the number of the input channels of the diverse complex multi-channel signal processing system constructed on the basis of the generalized detector allows us to obtain the higher probability of detection of the signals when there is multiplicative noise. However, if the condition determined in Eq. (7.386) is satisfied a great increase in the number of the input channels, of the diverse complex multi-channel signal processing system constructed on the basis of the generalized

detector, is not worthwhile, since starting with some number of the input channels the probability of detection of the signals is insignificantly increased with an increase in the number of the input channels (see Fig. 7.24).

- Comparative analysis of the detection performances of the diverse complex multi-channel signal processing system, constructed both on the basis of the generalized detector and on the basis of the correlation detector, shows a great superiority during the use of the generalized detector over the correlation detector (see Figs. 7.23–7.25).

References

1. Tuzlukov, V., *Signal Detection Theory,* Springer-Verlag, New York, 2001.
2. Tuzlukov, V., *Signal Processing in Noise: A New Methodology,* IEC, Minsk, 1998.
3. Tuzlukov, V., *Experimental Study of Correlation and FACP Signal Detection Algorithms,* IEC, Minsk, 1990.
4. Tuzlukov, V., A new approach to signal detection theory, *Digital Signal Processing: Review Journal,* 1998, Vol. 8., No. 3, pp. 166–184.
5. Helstrom, C., *Statistical Theory of Signal Detection,* 2nd ed., Pergamon Press, Oxford, 1968.
6. Helstrom, C., *Elements of Signal Detection and Estimation,* Prentice-Hall, Englewood Cliffs, NJ, 1995.
7. Scharf, L., *Statistical Signal Processing: Detection, Estimation and Time Series Analysis,* Addison-Wesley, New York, 1991.
8. Varshney, P., *Distributed Detection and Data Fusion,* Springer-Verlag, New York, 1996.
9. Shirman, Y. and Manjos, V., *Theory and Methods in Radar Signal Processing,* Radio and Svyaz, Moscow, 1981 (in Russian).
10. Tichonov, V., *Statistical Radio Engineering,* Radio and Svyaz, Moscow, 1982 (in Russian).
11. Middleton, D., *An Introduction to Statistical Communication Theory,* McGraw-Hill, New York, 1960.
12. Van Trees, H., *Detection, Estimation and Modulation Theory. Part I: Detection, Estimation and Linear Modulation Theory,* Wiley, New York, 1968.
13. Van Trees, H., *Detection, Estimation and Modulation Theory. Part III: Radar–Sonar Signal Processing and Gaussian Signals in Noise,* Wiley, New York, 1972.
14. Levin, B., *Theoretical Foundations of Statistical Radio Engineering,* Parts I–III, Soviet Radio, Moscow, 1974–1976 (in Russian).
15. Tuzlukov, V., Detection of deterministic signal in noise, *Radio Eng.,* 1986, No. 4, pp. 57–60.
16. Tuzlukov, V., Signal detection in noise in communications, *Radio Eng. Electron. Phys.,* 1986, Vol. 15, pp. 6–12.
17. Tuzlukov, V., Detection of deterministic signal in noise, *Telecomm. and Radio Eng.,* 1988, Vol. 41, No. 10.

18. Tuzlukov, V., Generalized signal detection algorithm in additive noise, *News of Academy of Belarus. Ser.: Phys.-Tech. Sci.*, 1991, No. 3, pp. 101–109.
19. Tuzlukov, V., Signal detection algorithm based on jointly sufficient statistics, *Problems of Increase in Efficiency in Military*, 1992, Vol. 3, No. 2, pp. 48–55.
20. Tuzlukov, V., Generalized methodology of signal detection in noise, in *Proc. 1992 Korean Int. Conf. on Automatic Control*, Seoul, 1992, pp. 255–260.
21. Tuzlukov, V., Signal-to-noise ratio improvement by employment of generalized signal detection algorithm, in *Proc. SPIE's 1995 Int. Symposium on OE/Aerospace Sensing and Dual Use Photonics*, Orlando, FL, April 17–21, 1995, Vol. 2496, pp. 811–822.
22. Tuzlukov, V., Increase in efficiency under signal processing of quasideterministic signals in additive noise, *News of Academy of Belarus. Ser.: Phys.-Tech. Sci.*, 1985, No. 4, pp. 98–104.
23. Tuzlukov, V., Detection of signals with stochastic amplitude and random initial phase, *Radio Eng.*, 1988, No. 9, pp. 59–61.
24. Tuzlukov, V., Interference compensation in signal detection for a signal of arbitrary amplitude and initial phase, *Telecomm. and Radio Eng.*, 1989, Vol. 44, No. 10, pp. 131–133.
25. Tuzlukov, V., The generalized algorithm of signal detection in statistical pattern recognition, *Pattern Recognition and Image Analysis*, 1993, Vol. 3, No. 4, pp. 474–485.
26. Tuzlukov, V., Detection of signals with stochastic parameters by employment of generalized algorithm, in *Proc. SPIE's 1997 Int. Symposium on AeroSense: Aerospace/Defense Sensing, Simulation, and Control*, Orlando, FL, April 21–25, 1997, Vol. 3079, pp. 302–313.
27. Tuzlukov, V., Detection of signals with random initial phase by employment of generalized algorithm, in *Proc. SPIE's 1997 Int. Symposium on Optical Science, Engineering, and Instrumentation*, San Diego, CA, July 27–August 1, USA, 1997, Vol. 3163, pp. 61–72.
28. Tuzlukov, V., Generalized detection algorithm for signals with stochastic parameters, in *Proc. IEEE (IGARSS'97)*, Singapore, August 4–8, 1997, Vol. 2, pp. 139–141.
29. Tuzlukov, V., Noise reduction by employment of generalized algorithm, in *Proc. IEEE (DSP'97)*, Santorini, Greece, July 2–4, 1997, pp. 617–620.
30. Tuzlukov, V., Tracking systems for statistical signal processing by employment of generalized algorithm, in *Proc. IEEE (ICICSP'97)*, Singapore, September 9–12, 1997, pp. 311–315.
31. Tuzlukov, V., Signal-to-noise ratio improvement under detection of stochastic signals using generalized detector, in *Proc. 1998 Int. Conf. on Applications of Photonics Technology (ICAPT '98)*, July 27–30, Ottawa, Canada, 1998.
32. Tuzlukov, V., Detection of signals using the generalized detector, in *Proc. IASTED Int. Conf. on Signal and Image Processing (SIP'98)*, Las Vegas, NV, October 28–31, 1998.
33. Tuzlukov, V., Detection of signals using digital generalized detector, in *Proc. 5th Int. Symposium on Signal Processing and Its Applications (ISSPA '99)*, Brisbane, Australia, August 23–25, 1999, pp. 171–174.
34. Tuzlukov, V., Detection of stochastic signals using generalized detector, in *Proc. 1999 IASTED Int. Conf. on Signal and Image Processing (SIP'99)*, Nassau, Bahamas, October 18–21, 1999, pp. 95–99.
35. Tuzlukov, V., Background noise distribution law of generalized detector, *News of Academy of Belarus. Ser.: Phys.-Tech. Sci.*, 1993, No. 4, pp. 63–70.

36. Tuzlukov, V., Distribution law of process at the output of the generalized detector, in *Proc. PRIA'95,* Minsk, Belarus, September 19–21, 1995, pp. 145–150.
37. Tuzlukov, V., Statistical characteristics of processes at the output of the generalized detector, in *Proc. PRIA'95,* Minsk, Belarus, September 19–21, 1995, pp. 151–156.
38. Tuzlukov, V., Method of correlation function estimation of process at the output of the generalized detector, in *Proc. 1998 SPIE's Int. Symposium on AeroSense: Aerospace/Defense Sensing, Simulations, and Controls,* Orlando, FL, April 13–17, 1998, Vol. 3373, pp. 54–65.
39. Tuzlukov, V., Signal processing in noise in communications: a new approach. Tutorial No. 3, in *Proc. 6th IEEE Int. Conf. on Electron. Circuit and Syst. (ICECS'99),* Paphos, Cyprus, September 5–8, 1999.
40. Tuzlukov, V., New remote sensing algorithms on the basis of generalized approach to signal processing in noise. Tutorial No. 2, in *Proc. 2nd Int. ICSC Symposium on Engineering of Intelligent Systems (EIS'2000),* University of Paisley, Paisley, Scotland, June 27–30, 2000.
41. Tuzlukov, V., Signal processing in noise. Tutorial No. 3, in *Proc. 2000 IASTED Int. Conf. on Signal Processing and Communications (SPC'2000),* Marbella, Spain, September 27–30, 2000.
42. Tuzlukov, V., Variance of noise components of optimal and generalized signal detection algorithms, *News of Academy of Belarus. Ser.: Phys.-Tech. Sci.,* 1992, No. 3, pp. 72–79.
43. Tuzlukov, V., Estimation of background noise correlation function at the output of the generalized detector, *News of Academy of Belarus. Ser.: Phys.-Tech. Sci.,* 1993, No. 1, pp. 97–107.
44. Tuzlukov, V., Functioning principles of the generalized detector, *News of Academy of Belarus. Ser.: Phys.-Tech. Sci.,* 1994, No. 1, pp. 51–58.
45. Tuzlukov, V., Comparative analysis of correlation and FACP detectors, in *Proc. Int. Conf. on Theory and Techniques under Space-Time Signal Processing,* Swerdlovsk, U.S.S.R., June 7–9, 1989, pp. 165–176.
46. Tuzlukov, V., Signal processing by employment of generalized algorithm, in *Proc. 7th IEEE Digital Signal Processing Workshop,* Loen, Norway, September 1–4, 1996, pp. 478–481.
47. Tuzlukov, V., New remote sensing algorithms under detection of minefields in littoral waters, in *Proc. Int. Conf. on Remote Sensing Technologies for Minefield Detection and Monitoring in Humanitarian Operations,* Easton, MA, May 17–20, 1999, pp. 237–314.
48. Tuzlukov, V., Experimental investigations of correlation and FACP detectors for weak signals, *News of Academy of Belarus. Ser.: Phys.-Tech. Sci.,* 1990, No. 4, pp. 102–107.
49. Tuzlukov, V., Experimental investigations of correlation and FACP algorithms for powerful signals, *News of Academy of Belarus. Ser.: Phys.-Tech. Sci.,* 1992, No. 1, pp. 108–118.
50. Tuzlukov, V., Weak signal detection under the use of correlation and FACP algorithms, *News of Academy of Belarus. Ser.: Phys.-Math. Sci.,* 1992, No. 2, pp. 99–106.
51. Tuzlukov, V., Investigations of correlation and FACP detectors for powerful signals, *Radio Eng.,* 1995, No. 1–2, pp. 33–37.
52. Tuzlukov, V., Detection of powerful signals using optimal and FACP detectors: comparative analysis, *Telecomm. Radio Eng.,* 1996, Vol. 51, No. 1, pp. 53–62.

53. Tuzlukov, V., Signal-to-noise ratio improvement in video signal processing, in *Proc. 1993 SPIE's Int. Conf on High-Definition Video*, Berlin, Germany, April 5–9, 1993, Vol. 1976, pp. 346–358.
54. Tuzlukov, V., Digital signal processing by employment of generalized algorithm in nondestructive testing systems, in *Proc. CMNDT'95*, Minsk, Belarus, November 21–24, 1995, pp. 314–318.
55. Tuzlukov, V., Signal-to-noise improvement by employment of generalized signal detection algorithm, in *Proc. SPIE's 1995 Int. Symposium on Optical Science, Engineering, and Instrumentation*, San Diego, CA, July 9–14, 1995, Vol. 2561, pp. 555–566.
56. Tuzlukov, V., Digital signal processing by employment of generalized algorithm in detection systems for mine and mine-like targets, in *Proc. SPIE's 1996 Int. Symposium on AeroSense: Aerospace/Defense Sensing, Simulations, and Controls*, Orlando, FL, April 8–12, 1996, Vol. 2765, pp. 287–298.
57. Tuzlukov, V., New features of signal detection by employment of generalized algorithm, in *Proc. 1st Int. Conf. on Information Fusion (FUSION'98)*, Las Vegas, NV, July 6–9, 1998, pp. 121–129.
58. Tuzlukov, V., Detection of powerful signals by generalized detector, in *Proc. IEEE (IGARSS'98)*, Seattle, WA, July 6–10, 1998, pp. 157–160.
59. Tuzlukov, V., Generalized detector with digital threshold device under high-speed data transmission, in *Proc. Int. Conf. on Performance Evaluation: Theory, Techniques, and Applications*, The University of Aizu, Aizu-Wakamatsu City, Fukushima, Japan, September 20–22, 2000, pp. 44–50.
60. Tuzlukov, V., Generalized detector for signals with random amplitude and phase, *Proc. SPIE's Int. Symposium on Optical Science, Engineering, and Instrumentation*, San Diego, CA, July 19–24, 1998, Vol. 3462, pp. 72–81.
61. Tuzlukov, V., Low-power signal detection using the generalized detector in communications, in *Proc. 1998 MidWest Symposium on Circuits and Systems*, Notre Dame, IN, August 9–12, 1998.
62. Tuzlukov, V., Detection of powerful signals using generalized detector, in *Proc. Int. Symposium on Information, Decision, and Control (IDC'99)*, Adelaide, Australia, February 8–10, 1999, pp. 539–544.
63. Tuzlukov, V., Detection of weak signals by generalized detector, in *Proc. IEEE (IGARSS'99)*, Hamburg, Germany, June 28–July 2, 1999, Vol. 2, pp. 1384–1386.
64. Tuzlukov, V., Experimental study and functioning principles of generalized detector, in *Proc. 1999 SPIE's Int. Symposium on Optical Sciences, Engineering, and Instrumentation*, Denver, CO, July 18–23, 1999, Vol. 3809, pp. 470–481.
65. Tuzlukov, V., Detection of low-power signals using generalized detector, in *Proc. Int. Conf. on Signal Processing Applications and Technology (ICSPAT'99)*, Orlando, FL, November 1–4, 1999.
66. Tuzlukov, V., New features of signal detection by employment of generalized detector, in *Proc. 1999 IEEE Int. Symposium on Intelligent Signal Processing and Communications Systems (ISPACS'99)*, Phurket, Thailand, December 8–10, 1999.
67. Tuzlukov, V., New features of low-power signal detection using generalized detector, in *Proc. 2nd IEEE Int. Conf. on Information, Communications and Signal Processing (ICICS'99)*, Singapore, December 7–10, 1999.
68. Shirman, Y., Golikov, V., and Busygin, I., *Theoretical Basis of Radar*, Soviet Radio, Moscow, 1970.

69. Kremer, I., Vladimirov, V., and Karpuhin, V., *Multiplicative Noise and Radio Signal Processing*, Soviet Radio, Moscow, 1972 (in Russian).
70. Levin, B. and Serov, V., About distribution of the periodic function of random variable, *Radio Eng. Electron. Phys.*, 1964, Vol. 9, No. 6.
71. Falkovich, S., *Radar Signal Processing under Fluctuating Additive Noise*, Soviet Radio, Moscow, 1961 (in Russian).
72. Lehman, E., *Testing Statistical Hypotheses*, 2nd ed., Wiley, New York, 1986.
73. Zakai, M. and Ziv, J., On the threshold effect in radar range estimation, *IEEE Trans.*, 1969, Vol. IT-15, No. 1, pp. 167–178.
74. Proakis, J., *Digital Communications*, 3rd ed., McGraw-Hill, New York, 1995.
75. Stuber, J., *Principles of Mobile Communications*, Kluwer Academic Publishers, Boston, 1996.
76. Davenport, V. and Root, V., *Introduction to Theory of Stochastic Signals and Noise*, McGraw-Hill, New York, 1959.
77. Shirman, Y., Some problems of detection of Gaussian signals in noise, *Radio Eng. Electron, Phys.*, 1971, Vol. 16, No. 2.
78. Okunev, Yu. and Yakovlev, L., To justification of autocorrelation signal processing method, *Radio Eng.*, 1969, Vol. 24, No. 7.
79. Petrovich, N. and Razmahnin, M., *Communication Systems with Noise Signals*, Soviet Radio, Moscow, 1969 (in Russian).
80. Skolnik, M., *Radar Applications*, IEEE Press, New York, 1988.
81. Ventzel, E. and Ovcharov, L., *Probability Theory*, Nauka, Moscow, 1973 (in Russian).
82. Pugachev, V., *Theory of Probabilities and Mathematical Statistics*, Nauka, Moscow, 1979 (in Russian).
83. Middleton, D., Statistical–physical models of man-made radio noise. Part 1: First order probability models of the instantaneous amplitude, Office of Telecommunications, Report 76-86, 1976, April, pp. 1–76.
84. Middleton, D., Statistical–physical models of man–made radio noise. Part 1: First order probability models of the envelope and phase, Office of Telecommunications, Report 76-86, 1976, April, pp. 76–124.
85. Middleton, D., Man–made noise in urban environments and transportation systems: models and measurements, *IEEE Trans.*, 1973, Vol. IT-22, No. 4, pp. 25–57.
86. Lebedev, N., *Specific Functions and Their Applications*, Fizmathgis, Moscow, 1963 (in Russian).
87. Zuko, A., *Noise Immunity and Efficiency in Communications*, Svyaz, Moscow, 1963 (in Russian).
88. Phink, L., *Theory of Discrete Signal Transmission*, Soviet Radio, Moscow, 1963 (in Russian).
89. Gradshtein, I. and Ryzhik, I., *Table of Integral, Series, and Products*, 5th ed., Academic Press, San Diego, CA, 1994.
90. Crammer, H., *Mathematical Methods of Statistics*, Princeton University Press, Princeton, NJ, 1946.
91. Pagurova, V., *Table of Incomplete Gamma Function*, Academy of the USSR, Moscow, 1963 (in Russian).

8

Signal Parameter Measurement Precision

This chapter examines the stimulus of the fluctuating stationary multiplicative noise jointly with the additive Gaussian noise and the resulting impact on measurement precision of non-energy parameters of the signals.

In the course of analysis we consider the generalized detector for signal processing in the presence of the additive Gaussian noise when the signals are distorted by the multiplicative noise and it is known *a priori* a "yes" signal exists in the stochastic process at the input of the generalized detector.

Methodologically, it is worthwhile to consider two cases of measurement of parameters of the signals:

- The measurement of parameters of the signal during the weak multiplicative and additive Gaussian noise when a value of estimation of the measured signal parameters is close to the true value of the signal parameters with the probability equal approximately to unity—the foolproof measurement.
- The measurement of parameters of the signal during the high multiplicative and additive Gaussian noise—the non-foolproof measurement.

The terms "weak" and "high" noise are widely used and applied to the additive Gaussian noise, for example, and are defined by the ratio between the energy of the signal and the spectral power density of the additive Gaussian noise

$$q = \frac{E_a}{N_0} \tag{8.1}$$

in the case of the generalized detector. If the parameter

$$q \gg 1,$$

the additive Gaussian noise is considered weak additive Gaussian noise. Otherwise, the additive Gaussian noise is considered high additive Gaussian noise.

Under a combined stimulus of the multiplicative noise and the additive Gaussian noise the sense of these terms must be clarified further. The fact is that in the case of the weak multiplicative and additive Gaussian noise measurement errors of the signal parameters are low because the multiplicative noise and the additive Gaussian noise distort only a shape of the signal at the output of the generalized detector shifting the maximum of the amplitude of the output signal on the time axis. When multiplicative noise is present, the low distortions in shape of the signal at the output of the generalized detector can occur in two cases:

- For the slow fluctuating multiplicative noise when the correlation interval of the noise modulation function $\dot{M}(t)$ of the multiplicative noise is comparable with the duration of the received signal.
- An arbitrary correlation interval of the noise modulation function $\dot{M}(t)$ of the multiplicative noise when the energy of the undistorted component of the signal is much greater in value than the energy of fluctuations caused by the stimulus of the multiplicative noise, i.e., the power or energy of the signal noise component at the output of the generalized detector for the condition $l = l_0$, where l_0 is the true value of parameter of the received signal.

Taking into account Eqs. (6.104) and (6.109), the last condition can be written in the following form:

$$\frac{\alpha_0^2(\tau, \Omega) E_{a_1} C^2}{\sigma_s^2(l_0)} \gg 1, \tag{8.2}$$

where $\sigma_s^2(l_0)$ is the variance of the signal noise component at the output of the generalized detector and C is the constant factor that is proportional to an amplification coefficient of the generalized detector, which is equal to unity, for simplicity.

As indicated by the condition formulated above, the weak multiplicative noise is very close in sense to the condition of the weak additive Gaussian noise. This condition is used in this chapter for analysis of the stimulus of the multiplicative noise and the resulting impact on measurement precision of the signal parameters using a non-tracking analyzer.

In the case of the tracking analyzer investigations are carried out for the condition that measurement errors are within the limits of linear part of discrimination characteristic.[1]

The condition of the weak additive Gaussian noise, when there is multiplicative noise, takes the following form:

$$\alpha_0^2(\tau, \Omega) q \gg 1. \tag{8.3}$$

Equation (8.3) takes into consideration an increase in the relative level $\alpha_0^2(\tau, \Omega)$ of the undistorted component of the signal resulting from the stimulus of the multiplicative noise.

The mean of the measurement error $\overline{\Delta l}$ (shift) and the variance σ_l^2 are considered the main characteristics of precision of the signal parameter measurement in the case of the weak multiplicative and additive Gaussian noise.

An impact of the multiplicative noise on measurement precision of the signal parameter is defined by the ratio between the variance σ_l^2 of measurement error when the multiplicative noise exists and the variance $\sigma_{l_0}^2$ of measurement errors when the multiplicative noise is absent.

During the high multiplicative and additive Gaussian noise a character of possible measurement errors is changed essentially. Great values of amplitude shoots of the signal at the output of the generalized detector, owing to the signal noise component caused by the stimulus of the multiplicative noise and owing to the stimulus of the additive Gaussian noise, can be observed when values of the measured signal parameter differ essentially from the true value of the signal parameter.

In this case, the main problem during the process of signal parameter measurement is to distinguish an amplitude shoot of the signal at the output of the generalized detector corresponding to the true value of the signal parameter from false amplitude shoots of the output signal.

The probability of amplitude shoot definition of the signal at the output of the generalized detector, corresponding to the true value of the measured signal parameter during the condition of a "yes" signal in the stochastic process at the input of the generalized detector, defines for this case the probability P_{0_k} of true signal parameter measurement. The probability of the error measurement of parameters of the signal can be characterized by totality of the conditional probabilities P_{0_m} of false signal parameter measurement corresponding to the "m"-th value of the signal parameter, $m \neq k$.

Further, we will use the probabilities P_{0_k} and P_{0_m} with the purpose of estimating the stimulus of the multiplicative noise and the additive Gaussian noise and the resulting impact on measurement precision of the signal parameter during the condition of the high multiplicative and additive Gaussian noise.

8.1 A Single Signal Parameter Measurement under a Combined Stimulus of Weak Multiplicative and Additive Gaussian Noise

We consider the generalized detector constructed with the purpose of estimation of a single unknown non-energy parameter l_0 of the signal $a(t, l_0)$ with the random initial phase φ_0 distributed uniformly within the limits of the interval $[0, 2\pi]$ in the presence of additive Gaussian noise $n(t)$ possessing the complex amplitude envelope $\dot{N}(t)$. This generalized detector is discussed in more detail in References 2 and 3.

If the stochastic process at the input of the generalized detector is determined in the following form

$$X_{in}(t) = a(t, l_0) + n(t) \tag{8.4}$$

the likelihood function

$$\Lambda(l) = \ln K_0\left[Z_g^{out}(l)\right] \tag{8.5}$$

is formed at the output of the generalized detector, where $K_0(x)$ is the modified Bessel function of the second order of an imaginary argument and $Z_g^{out}(l)$ is the envelope of amplitude of the signal at the output of the generalized detector, and can be written in a similar way as in Eq. (7.8):

$$Z_g^{out}(l) = \frac{1}{2}\left|e^{j\varphi_0}\int\limits_0^T \dot{S}(t, l_0)\dot{S}_m(t, l)\cdot e^{j[\Psi(t,l_0)-\Psi(t,l)]}\,dt\right| - \left|\int\limits_0^T [\dot{N}(t)N^*(t-\tau) - \dot{N}(t-\tau)N^*(t)]\,dt\right|, \tag{8.6}$$

where $[0, T]$ is the time interval of observation of the signal.

Equations (8.5) and (8.6) describe the structure of the generalized detector for definition and measurement of a single parameter of the signal with the random initial phase in the presence of additive Gaussian noise (see Fig. 8.1).

Reference to Fig. 8.1 shows that the generalized detector in this case is the multi-channel detector. Moreover, the model signal or the reference signal at each channel of the generalized detector possesses one of possible values of the parameter l.

There is the envelope detector with the characteristic $\ln K_0[Z_g^{out}(l)]$ at the output of each channel of the generalized detector. The decision device (DD) selects the channel with the greatest value of the parameter l, for which the function $\Lambda(l)$ can reach the maximum. Since the function $\Lambda(l)$ can possess

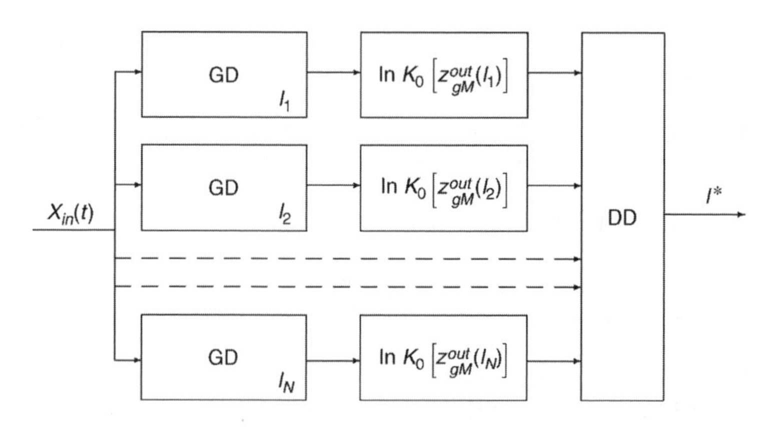

FIGURE 8.1
The structure of the generalized detector.

some peaks (maximums) at the output of each channel of the generalized detector that is a consequence of the stimulus of the additive Gaussian noise, the greatest value of the function $\Lambda(l)$ is taken as an estimation of the signal parameter l.

When there is the weak additive Gaussian noise, i.e., the foolproof measurement, the peaks (maximums) of the function $\Lambda(l)$ caused by the stimulus of the additive Gaussian noise with the high probability are lower in comparison to the main peak (maximum) of the function $\Lambda(l)$—the signal at the input of the DD—and for this reason these peaks of the function $\Lambda(l)$ cannot be considered the true signal.

In doing so, the additive Gaussian noise distorts only a shape of the main peak (maximum) of the function $\Lambda(l)$. For the given case, the estimation l^* of the signal parameter l is defined as a solution of the likelihood equation with respect to the parameter l:[3]

$$\frac{d}{dl}\Lambda(l) = \frac{d\,Z_g^{out}(l)}{dl} = 0. \tag{8.7}$$

The solution of Eq. (8.7), when there is the additive Gaussian noise and for the statistical characteristics of errors of the signal parameter measurement—the mean and variance (mentioned above)—is discussed in Reference 4 in more detail.

In the present section the procedure and main results discussed in Reference 4 are generalized for the case in which the signal parameter measurement is carried out by the generalized detector during the simultaneous stimulus of the multiplicative and additive Gaussian noise on the signal.

We proceed to use the presentation of the signal introduced in Chapter 4. Thus the signal distorted by the multiplicative noise can be presented by two additive components:

$$\begin{aligned} a_M(t) &= Re\{\dot{M}(t)\dot{S}(t)\cdot e^{j(\omega_0 t+\varphi_0)}\} \\ &= Re\{\alpha_0(\tau,\Omega)\sqrt{E_{a_1}}\,\dot{S}_m(t)\cdot e^{j(\omega_0 t+\beta_0+\varphi_0)}\} \\ &\quad + Re\{\dot{V}(t)\sqrt{E_{a_1}}\,\dot{S}_m(t)\cdot e^{j(\omega_0 t+\varphi_0)}\}, \end{aligned} \tag{8.8}$$

where the amplitude of the received signal relative to the model signal—the reference signal—when the multiplicative noise is absent, is equal to $\sqrt{E_{a_1}}$, analogously as in Section 7.1.

In terms of this representation the envelope of amplitude of the sum of the signal, which is distorted by the multiplicative noise, and the additive Gaussian noise at the output of the generalized detector takes the following form:

$$Z_g^{out}(l) = \left|\dot{S}_{1_1}(l-l_0)\cdot e^{j(\beta_0+\varphi_0)} + \dot{S}_{1_2}(l-l_0)\cdot e^{j\varphi_0} + \dot{n}(l)\right|, \tag{8.9}$$

where

$$\dot{S}_{1_1}(l - l_0) = \frac{\alpha_0(\tau, \Omega)}{2} \int_0^T \dot{S}(t, l_0) S_m^*(t, l)\, dt$$
$$= \alpha_0(\tau, \Omega)\dot{S}_1(l - l_0)$$
$$= \alpha_0(\tau, \Omega) S_1(l - l_0) \cdot e^{j\beta(l - l_0)}$$
$$= \alpha_0(\tau, \Omega) E_{a_1} \dot{\rho}(l - l_0); \tag{8.10}$$

$$\dot{S}_{1_2}(l - l_0) = \frac{1}{2} \int_0^T \dot{V}(t)\dot{S}(t, l_0) S_m^*(t, l)\, dt; \tag{8.11}$$

$$\dot{n}(l) = \frac{1}{2} \int_0^T [\dot{N}(t) N^*(t - \tau) - \dot{N}(t - \tau) N^*(t)]\, dt. \tag{8.12}$$

Reference to Eqs. (8.9)–(8.12) shows that the function

$$\dot{S}_{1_1}(l - l_0)$$

defines the complex amplitude envelope of the undistorted component of the signal transformed by the generalized detector. The function

$$\dot{n}(l)$$

defines the complex amplitude envelope of the additive Gaussian noise transformed by the generalized detector. The function

$$\dot{S}_{1_2}(l - l_0)$$

defines the complex amplitude envelope of the signal noise component at the output of the generalized detector, which is caused by the stimulus of the multiplicative noise.

The function

$$\dot{S}_{1_1}(l - l_0)$$

is proportional to the autocorrelation function of the signal with respect to the parameter

$$l\dot{\rho}(l - l_0)$$

and possesses the following features that will henceforth be used:

$$\dot{S}_{1_1}(l - l_0) = S_{1_1}^*(l - l_0); \tag{8.13}$$

$$\dot{S}_{1_1}(0) = \alpha_0(\tau, \Omega) S_1(0) \geq \left|\dot{S}_{1_1}(l - l_0)\right|. \tag{8.14}$$

In other words, the real component of the function

$$\dot{S}_{1_1}(l - l_0)$$

has an even symmetry with respect to the point $l = l_0$ and the imaginary component of the function

$$\dot{S}_{1_1}(l - l_0)$$

has an odd symmetry with respect to the point $l = l_0$. At the point $l = l_0$ the function

$$\dot{S}_{1_1}(l - l_0)$$

becomes real and the module of the function

$$\dot{S}_{1_1}(l - l_0)$$

is maximum.

For the weak multiplicative and additive Gaussian noise considered in this section, the ratio between the energy of the undistorted component of the signal

$$\dot{S}_{1_1}(l - l_0)$$

at the output of the generalized detector at the point $l = l_0$ and the energy of sum of the additive fluctuations $\dot{n}(l)$ and fluctuations of the signal $\dot{S}_{1_2}(l - l_0)$ caused by the stimulus of the multiplicative noise at the same point $l = l_0$ is much greater than unity.

In the process, the amplitude of the signal determined by Eq. (8.9) can be written in the following form:

$$Z_g^{out}(l) \simeq \alpha_0(\tau, \Omega) S_1(l - l_0) \times \sqrt{1 + 2\,Re\left\{\frac{\dot{S}_{1_1}(l - l_0) S_{1_2}^*(l - l_0) \cdot e^{j\beta_0} + \dot{S}_{1_1}(l - l_0)\, n^*(l) \cdot e^{j(\varphi_0 + \beta_0)}}{\alpha_0^2(\tau, \Omega) S_1^2(l - l_0)}\right\}}. \tag{8.15}$$

In Eq. (8.15) we can neglect the term

$$\frac{\left|\dot{S}_{1_1}(l - l_0) S_{1_2}^*(l - l_0) \cdot e^{j\beta_0} + \dot{S}_{1_1}(l - l_0)\, n^*(l) \cdot e^{j(\varphi_0 + \beta_0)}\right|^2}{\alpha_0^4(\tau, \Omega) S_1^4(l - l_0)} \tag{8.16}$$

because it is infinitesimal during the weak multiplicative and additive Gaussian noise.

Next we carry out the series expansion for the complex amplitude envelope $Z_g^{out}(l)$ at the output of the generalized detector by powers of the function

$$Re\left\{\frac{\dot{S}_{1_1}(l - l_0) S_{1_2}^*(l - l_0) \cdot e^{j\beta_0} + \dot{S}_{1_1}(l - l_0)\, n^*(l) \cdot e^{j(\varphi_0 + \beta_0)}}{\alpha_0^2(\tau, \Omega) S_1^2(l - l_0)}\right\} \tag{8.17}$$

that is much less than unity in the statistical sense during the weak multiplicative and additive Gaussian noise.

Limiting by two terms of the power series expansion mentioned above, we can write

$$\begin{aligned} Z_g^{out}(l) &\simeq \alpha_0(\tau, \Omega) S_1(l - l_0) + \frac{Re\{\dot{S}_{1_1}(l - l_0) S_{1_2}^*(l - l_0) \cdot e^{j\beta_0}\}}{\alpha_0(\tau, \Omega) S_1(l - l_0)} \\ &\quad + \frac{Re\{\dot{S}_{1_1}(l - l_0) n^*(l) \cdot e^{j(\varphi_0 + \beta_0)}\}}{\alpha_0(\tau, \Omega) S_1(l - l_0)} \\ &= \alpha_0(\tau, \Omega) S_1(l - l_0) + n_M(l - l_0) + n_a(l). \end{aligned} \tag{8.18}$$

The function in Eq. (8.17), the root mean square deviation of which is equal to the ratio between the sum of energies of the additive Gaussian noise and fluctuations caused by the multiplicative noise—the signal noise component—and the energy of the undistorted component of the signal at the output of the PF of the linear input tract of the generalized detector is less. When this ratio is low, Eq. (8.18) more exactly describes the complex amplitude envelope $Z_g^{out}(l)$.

The first and second terms in Eq. (8.18) define the complex amplitude envelope of the signal at the output of the generalized detector distorted by the multiplicative noise. The third term in Eq. (8.18) defines fluctuations of the complex amplitude envelope of the signal at the output of the generalized detector caused by the additive Gaussian noise.

Reference to Eq. (8.18) shows that the use of presentation of the signal, at the output of the generalized detector distorted by the multiplicative noise, in the form of two additive components allows us to reduce an effect of the stimulus of the multiplicative noise on signal parameter measurement to an appearance of additional additive fluctuations in the first approximation.

In the presence of the weak multiplicative and additive Gaussian noise we can use the well-known procedure,[5,6] based on the Taylor series expansion by powers of the difference $l - l_0$ for the complex amplitude envelope determined by Eq. (8.18), for the purpose of solving the likelihood equation determined by Eq. (8.10). We can limit the series expansion of the functions $S_1(l - l_0)$ and $n_M(l - l_0)$ by the first, second, and third terms because the measurement errors are small.

Substituting the Taylor series expansion for the functions $S_1(l - l_0)$ and $n_M(l - l_0)$ in Eq. (8.7) and taking into account that

$$\left. \frac{d S_1(l - l_0)}{dl} \right|_{l=l_0} = 0, \tag{8.19}$$

we obtain that the errors of the signal parameter measurement, when the multiplicative and additive Gaussian noise are present, can be determined in the following form:

$$\begin{aligned} \Delta l &= l^* - l_0 \\ &= -\frac{\frac{d}{dl}[n_M(l - l_0) + n_a(l)]_{l=l_0}}{\frac{d^2}{dl^2}[\alpha_0(\tau, \Omega) S_1(l - l_0) + n_M(l - l_0)]_{l=l_0}}. \end{aligned} \tag{8.20}$$

As would be expected, Eq. (8.20) coincides with the first approximation for errors of the signal parameter measurement (defined in Reference 4) during the use of the low signal parameter method when the multiplicative noise is absent. During the weak multiplicative and additive Gaussian noise the second term of the denominator in Eq. (8.20) is much less than the first term of the denominator in the statistical sense. For this reason, we can neglect the second term of the denominator in Eq. (8.20) compared to the function

$$\alpha_0(\tau, \Omega) S_1(l - l_0).$$

Taking into consideration that the functions

$$n_M(l - l_0)$$

and

$$n_a(l)$$

in Eq. (8.18) are not correlated and possess zero means, and neglecting the second term of the denominator in Eq. (8.20) in comparison to the first term of the denominator, we obtain that the mean of errors of the signal parameter measurement is equal to zero

$$\overline{\Delta l} = 0$$

and the variance of errors of the signal parameter measurement is determined in the following form:

$$\sigma_l^2 = \overline{\Delta l^2} = \sigma_{l_a}^2 + \sigma_{l_M}^2$$
$$= \frac{\left\{ \frac{\partial^2 [n_a(l_1) n_a(l_2)]}{\partial l_1 \partial l_2} + \frac{\partial^2 [n_M(l_1 - l_0) n_M(l_2 - l_0)]}{\partial l_1 \partial l_2} \right\}_{l_1 = l_2 = l = l_0}}{\left\{ \alpha_0^2(\tau, \Omega) \frac{d^2 S_M(l - l_0)}{d^2 l} \Big|_{l = l_0} \right\}^2}. \tag{8.21}$$

Thus, in the presence of the weak multiplicative and additive Gaussian noise the estimation of a single signal parameter is unbiased and the variance of estimation of a single signal parameter is the sum of two terms. The first term $\sigma_{l_a}^2$ is the variance of estimation of the signal parameter when there is the additive Gaussian noise only with an accuracy of the constant factor $\alpha_0^2(\tau, \Omega)$, the multiplicative noise is absent. The second term $\sigma_{l_M}^2$ takes into account the stimulus of the multiplicative noise.

In terms of results discussed in Reference 4, the variance $\sigma_{l_a}^2$ can be written in the following form:

$$\sigma_{l_a}^2 = \frac{\sigma_{l_0}^2}{\alpha_0^2(\tau, \Omega)} = -\frac{N_0}{E_{a_1} \alpha_0^2(\tau, \Omega) \cdot \frac{d^2}{dl^2} |\dot{\rho}(l - l_0)|_{l = l_0}}, \tag{8.22}$$

where the parameter q determined by Eq. (8.1) is the detection parameter of the generalized detector, $\sigma_{l_0}^2$ is the variance of errors of the signal parameter measurement when the multiplicative noise is absent.

To determine the value of the variance $\sigma_{l_M}^2$ there is a need to define the correlation function of the fluctuation component $n_M(l - l_0)$ with respect to the parameter l:

$$\overline{n_M(l_1 - l_0)n_M(l_2 - l_0)} = 0.5\,Re\{\overline{S^*_{1_2}(l_1 - l_0)S^*_{1_2}(l_2 - l_0)} \times e^{2j\beta_0} \cdot e^{j[\beta(l_1 - l_0) + \beta(l_2 - l_0)]} + \dot{S}_{1_2}(l_2 - l_0)S^*_{1_2}(l_1 - l_0) \cdot e^{j[\beta(l_1 - l_0) - \beta(l_2 - l_0)]}\}. \quad (8.23)$$

Reference to Eq. (8.23) shows that the correlation function

$$\overline{n_M(l_1 - l_0)n_M(l_2 - l_0)}$$

depends not only on the parameters l_1 and l_2, but also on the parameter β_0, where β_0 is the argument of the mean of the noise modulation function $\dot{M}(t)$ of the multiplicative noise.

For the cases when the probability distribution density of distortions in phase of the signal is the symmetry function, and distortions in amplitude of the signal are functionally related with distortions in phase of the signal and possess the symmetric probability distribution density or are independent of distortions in phase of the signal, the argument of the constant component β_0 of the noise modulation function $\dot{M}(t)$ of the multiplicative noise is equal to zero (see Section 4.4). In the process, Eq. (8.23) does not depend on the parameter β_0. Henceforth, we will consider only the multiplicative noise in this form.

Consider the second term in Eq. (8.23), denoting it by $K_{M_1}(l_1, l_2)$:

$$K_{M_1}(l_1, l_2) = 0.5\,Re\{\dot{S}_{1_2}(l_2 - l_0)S^*_{1_2}(l_1 - l_0) \cdot e^{j[\beta(l_1 - l_0) - \beta(l_2 - l_0)]}\}. \quad (8.24)$$

The moment of the second order of the function $\dot{S}_{1_2}(l - l_0)$ coincides with the correlation function of the signal distorted by the multiplicative noise at the output of the generalized detector by the parameter l and, in terms of Eq. (6.166), can be determined in the following form:

$$\overline{S^*_{1_2}(l_1 - l_0)\dot{S}_{1_2}(l_2 - l_0)} = \frac{E_{a_1}}{2\pi}\int_{-\infty}^{\infty} G_V(\Omega)\rho^*(l_1 - l_0, \Omega)\dot{\rho}(l_2 - l_0, \Omega)\,d\Omega = \frac{1}{2\pi}\int_{-\infty}^{\infty} G_V(\Omega)S_1(l_1 - l_0, \Omega)S_1(l_2 - l_0, \Omega) \times e^{j[\beta(l_2 - l_0, \Omega) - \beta(l_1 - l_0, \Omega)]}\,d\Omega, \quad (8.25)$$

where $G_V(\Omega)$ is the energy spectrum of fluctuations of the noise modulation function $\dot{M}(t)$ of the multiplicative noise, and

$$E_{a_1}\dot{\rho}(l - l_0, \Omega) = \dot{S}_1(l - l_0, \Omega) \quad (8.26)$$

is the response of the generalized detector on the undistorted component of the signal shifted in frequency with respect to the expected signal on the value Ω.

In terms of Eq. (8.25) we can write

$$K_{M_1}(l_1, l_2) = \frac{1}{4\pi}\int\limits_{-\infty}^{\infty} G_V(\Omega) S_1(l_1 - l_0, \Omega) S_1(l_2 - l_0, \Omega) \times \cos[\beta(l_1 - l_0) - \beta(l_2 - l_0) - \beta(l_1 - l_0, \Omega) + \beta(l_2 - l_0, \Omega)]\, d\Omega. \tag{8.27}$$

After analogous transformations, regarding the first term in Eq. (8.23), we can write

$$K_{M_2}(l_1, l_2) = \frac{1}{4\pi}\int\limits_{-\infty}^{\infty} G_D(\Omega) S_1(l_1 - l_0, \Omega) S_1(l_2 - l_0, \Omega) \times \cos[\beta(l_1 - l_0) + \beta(l_2 - l_0) - \beta(l_1 - l_0, \Omega) - \beta(l_2 - l_0, \Omega)]\, d\Omega, \tag{8.28}$$

where $G_D(\Omega)$ is the Fourier transform with respect to the correlation function of fluctuations of the noise modulation function $\dot{M}(t)$ of the multiplicative noise

$$\dot{D}_V(t_1 - t_2) = [\dot{M}(t_1) - \alpha_0(\tau, \Omega) \cdot e^{j\beta_0}][\dot{M}(t_2) - \alpha_0(\tau, \Omega) \cdot e^{j\beta_0}]. \tag{8.29}$$

To determine the variance $\sigma^2_{l_M}$ in accordance with Eq. (8.21) there is a need to define the mixed second derivative with respect to the correlation function

$$\overline{n_M(l_1 - l_0) n_M(l_2 - l_0)}$$

at the point

$$l_1 = l_2 = l = l_0.$$

After fulfilling a differentiation in Eqs. (8.27) and (8.28) that is carried out independently of integration, and simple transformations, we can write:

$$\left.\frac{\partial^2 K_{M_1}(l_1, l_2)}{\partial l_1 \partial l_2}\right|_{l_1=l_2=l_0} = K''_{M_1}(0, 0) = \frac{1}{4\pi}\int\limits_{-\infty}^{\infty} G_V(\Omega)\{[S'_1(0, \Omega)]^2 + S_1^2(0, \Omega)[\beta'(0) - \beta'(0, \Omega)]^2\}\, d\Omega, \tag{8.30}$$

$$\left.\frac{\partial^2 K_{M_2}(l_1, l_2)}{\partial l_1 \partial l_2}\right|_{l_1=l_2=l_0} = K''_{M_2}(0, 0) = \frac{1}{4\pi}\int\limits_{-\infty}^{\infty} G_D(\Omega)\Big\{\big[[S'_1(0, \Omega)]^2 - S_1^2(0, \Omega)[\beta'(0) - \beta'(0, \Omega)]^2\big]\cos 2[\beta(0) - \beta(0, \Omega)] - 2S'_1(0, \Omega) S_1(0, \Omega)[\beta'(0) - \beta'(0, \Omega)] \times \sin 2[\beta(0) - \beta(0, \Omega)]\Big\}^2 d\Omega. \tag{8.31}$$

Substituting Eqs. (8.30) and (8.31) in Eq. (8.21), we can write the variance of estimation of the signal parameter measurement in terms of a simultaneous stimulus of the multiplicative and additive Gaussian noise.

Equations (8.30) and (8.31) are very sophisticated. However, in some cases these equations can be simplified. Consider these cases:

- Let

$$\beta(0) = \beta(0, \Omega) = 0. \tag{8.32}$$

This condition occurs, for example, during measurement of frequency of the non-modulated signal. Taking into account the given condition, we can write

$$\sigma^2_{l_M} = \frac{1}{4\pi\alpha_0^2(\tau, \Omega)[S_1''(0)]^2} \int\limits_{-\infty}^{\infty} [S_1'(0, \Omega)]^2 [G_V(\Omega) + G_D(\Omega)]\, d\Omega. \tag{8.33}$$

- Let

$$S_1(l, \Omega) = g(l) r(\Omega) \tag{8.34}$$

and

$$\left.\frac{dg(l)}{dl}\right|_{l=0} = 0. \tag{8.35}$$

The presentation of the function

$$S_1(l, \Omega)$$

in the form given by Eqs. (8.34) and (8.35) takes place, for example, when measuring appearance time of the signals with the constant radio frequency carrier and when measuring the phase-code-modulated signals and the signals with the noise modulation in the presence of additive Gaussian noise.[7–9]

For the case considered we can write

$$\begin{aligned}\sigma^2_{l_M} = {} & \frac{1}{4\pi\alpha_0^2(\tau, \Omega)[S_1''(0)]^2} \\ & \times \int\limits_{-\infty}^{\infty} \{G_V(\Omega) S_1^2(0, \Omega)[\beta'(0) - \beta'(0, \Omega)]^2 \\ & - G_D(\Omega) S_1^2(0, \Omega)[\beta'(0) - \beta'(0, \Omega)]^2 \\ & \times \cos 2[\beta(0) - \beta(0, \Omega)]\}\, d\Omega.\end{aligned} \tag{8.36}$$

- Let

$$\beta(l, \Omega) = \frac{\mu}{\Omega}, \tag{8.37}$$

where the argument $\beta(l, \Omega)$ is proportional to the measured signal parameter and frequency for the condition

$$\beta(0) = \beta(0, \Omega) = 0.$$

Then

$$\sigma^2_{l_M} = \frac{1}{4\pi\alpha_0^2(\tau, \Omega)[S_1''(0)]^2} \times \int\limits_{-\infty}^{\infty} G_V(\Omega)\{[S_1'(0, \Omega)]^2 + S_1^2(0, \Omega)[\beta'(0, \Omega)]^2\}\, d\Omega + \frac{1}{4\pi\alpha_0^2(\tau, \Omega)[S_1''(0)]^2} \times \int\limits_{-\infty}^{\infty} G_D(\Omega)\{[S_1'(0, \Omega)]^2 - S_1^2(0, \Omega)[\beta'(0, \Omega)]^2\}\, d\Omega. \quad (8.38)$$

Reference to Eqs. (8.33), (8.36), and (8.38) shows that the variance of errors of the signal parameter measurement caused by the multiplicative noise depends essentially on spectral characteristics of the noise modulation function $\dot{M}(t)$ of the multiplicative noise.

As we can see from Eq. (8.33), the greatest errors in the signal parameter measurement caused by the weak multiplicative noise occur for the case when the energy spectrum bandwidth of fluctuations of the noise modulation function $\dot{M}(t)$ of the multiplicative noise is commensurate with the energy spectrum bandwidth of complex amplitude envelope of the signal at the output of the generalized detector by frequency axis, in other words, with the bandwidth of the autocorrelation function

$$|\dot{\rho}(0, \Omega)|.$$

Thus, the slow fluctuating multiplicative noise is the most dangerous. The analogous conclusion can be made for the cases determined by Eqs. (8.36) and (8.38) for the condition that the functions

$$S_1^2(0, \Omega)$$

and

$$[S_1'(0, \Omega)]^2$$

are decreased with an increase in frequency more rapid than the function

$$[\beta'(0, \Omega)]^2.$$

The general formulae for the variance of errors in the signal parameter measurement, when there is the multiplicative noise, are true for the condition in which the function determined by Eq. (8.15) is infinitesimal. In the statistical sense, this condition can be written in the following form:

$$\frac{0.5\overline{|\dot{n}(l)|^2} + \overline{[n_M(l - l_0)]^2}}{\alpha_0^2(\tau, \Omega) S_m^2(l - l_0)} \ll 1. \tag{8.39}$$

Taking into account that

$$\overline{|\dot{n}(l)|^2} = N_0; \tag{8.40}$$

$$\overline{[n_M(l - l_0)]^2} = K_{M_1}(l_1, l_2) + K_{M_2}(l_1, l_2); \tag{8.41}$$

$$l_1 = l_2 = l, \tag{8.42}$$

instead of Eq. (8.39) we can write

$$\left.\frac{0.5N_0 + K_{M_1}(l_1, l_2) + K_{M_2}(l_1, l_2)}{\alpha_0^2(\tau, \Omega) S_1^2(l - l_0)}\right|_{l_1 = l_2 = 0} \ll 1. \tag{8.43}$$

Since the solution of the likelihood equation in Eq. (8.10), during the low errors in the signal parameter measurement, can be found in the neighborhood of the point

$$l = l_0,$$

Eq. (8.43) can take this condition with the purpose of determining the approximate estimation.

Then the condition of the use of the formulae obtained above can be written in the following form:

$$\alpha_0^2(\tau, \Omega) \gg \frac{N_0}{E_{a_1}} + \frac{1}{4\pi} \int_{-\infty}^{\infty} |\dot{\rho}(0, \Omega)|^2 \{G_V(\Omega) + G_D(\Omega) \cos 2[\beta(0, \Omega)]\}\, d\Omega, \tag{8.44}$$

where the condition

$$\beta(0) = 0$$

is satisfied.

We proceed to apply the generalized formulae obtained above to some examples of measurement of the frequency and the time of appearance of the signal on the time axis.

8.1.1 Signal Frequency Measurement Precision

When measuring the frequency ω we can write

$$S_1(\omega, \Omega) \cdot e^{j\beta(\omega, \Omega)} = \frac{E_{a_1}}{2} \int_0^T S_m^2(t) \cdot e^{j(\omega + \Omega)t}\, dt = E_{a_1} \dot{\rho}(\omega + \Omega), \tag{8.46}$$

where

$$\omega = l - l_0. \tag{8.47}$$

In the majority of cases the envelope $S_0(t)$ of amplitude of the signal is the even function. Then, for corresponding choice of the origin

$$\beta(\omega) = \beta(\omega, \Omega) = 0 \tag{8.48}$$

and for the condition that the signal is within the limits of the interval $[0, T]$ we can write

$$\left.\frac{d^2 S_1(\omega)}{d\omega^2}\right|_{\omega=0} = -\frac{E_{a_1}}{2} \int\limits_{-\infty}^{\infty} S_m^2(t) t^2\, dt = -E_{a_1}\overline{t^2}, \tag{8.49}$$

where $\overline{t^2}$ is the mean square duration of the signal.[10,11]

We define the function

$$\left.\frac{d S_1(\omega, \Omega)}{d\omega}\right|_{\omega=0} = \frac{E_{a_1}}{2} \cdot \frac{d}{d\omega}\left\{\int\limits_{-\infty}^{\infty} S_m^2(t)\cos(\omega+\Omega)t\, dt\right\}_{\omega=0}$$
$$= E_{a_1} G'_{en}(\Omega), \tag{8.50}$$

which allows us to determine the variance of estimation of the signal parameter measurement.

In Eq. (8.50) the function

$$G'_{en}(\omega) = \frac{1}{2\pi}\int\limits_{-\infty}^{\infty} S^2(t)\cos\omega t\, dt \tag{8.51}$$

is the normalized spectrum of the quadratic amplitude envelope of the undistorted signal

$$G_{en}(\omega) = |\dot{\rho}(0, \omega)|, \tag{8.52}$$

where

$$G'_{en} = \frac{dG_{en}(\omega)}{d\omega}. \tag{8.53}$$

Since for the considered case the condition

$$\beta(\omega) = \beta(\omega, \Omega) = 0 \tag{8.54}$$

is satisfied, the variance of measurement errors of the signal frequency caused by the multiplicative noise is determined by Eq. (8.33).

Taking into account Eqs. (8.22), (8.49), and (8.50), the total variance of measurement errors of the signal frequency can be written in the following form:

$$\sigma_\omega^2 = \frac{N_0}{\alpha_0^2(\tau, \Omega) E_{a_1}\overline{t^2}}\left\{1 + \frac{E_{a_1}}{4\pi N_0 \overline{t^2}}\int\limits_{-\infty}^{\infty}[G'_{en}(\Omega)]^2[G_V(\Omega) + G_D(\Omega)]\, d\Omega\right\}. \tag{8.55}$$

Reference to Eq. (8.55) shows that an impact of the multiplicative noise on the precision of the signal frequency measurement is completely defined by the spectrum of the quadratic amplitude envelope of the signal and spectral characteristics of the noise modulation function $\dot{M}(t)$ of the multiplicative noise. If the power of the additive Gaussian noise is low, i.e., the value q is high, then the impact of the multiplicative noise is high.

Equation (8.55) can be essentially simplified for two limiting cases: the slow and rapid fluctuating multiplicative noise. In the case of the slow fluctuating multiplicative noise the function $G'_{en}(\Omega)$ in Eq. (8.55) is varied more slowly in comparison with the functions $G_V(\Omega)$ and $G_D(\Omega)$. Because of this, the function $G'_{en}(\Omega)$ can be presented by the first, second, and third terms of the Taylor series expansion in the region, where the functions $G_V(\Omega)$ and $G_D(\Omega)$ are essentially differed from zero.[12]

Taking into account that the condition

$$G'_{en}(\Omega) = 0 \tag{8.56}$$

is satisfied for the signals with the even amplitude envelopes, we can write that

$$[G'_{en}(\Omega)] \simeq \Omega^2[G''_{en}(0)]^2. \tag{8.57}$$

In terms of Eqs. (8.49) and (8.50) we can write

$$[G''_{en}(0)]^2 = (\overline{t^2})^2. \tag{8.58}$$

Substituting Eq. (8.57) in Eq. (8.55), we can write

$$\begin{aligned} \sigma_\omega^2 &= \frac{N_0}{\alpha_0^2(\tau, \Omega) E_{a_1} \overline{t^2}} + \frac{1}{4\pi\alpha_0^2(\tau, \Omega)} \int\limits_{-\infty}^{\infty} \Omega^2[G_V(\Omega) + G_D(\Omega)]\, d\Omega \\ &= \frac{N_0}{\alpha_0^2(\tau, \Omega) E_{a_1} \overline{t^2}} - \frac{1}{2\alpha_0^2(\tau, \Omega)}[R''_V(0) + D''_V(0)], \end{aligned} \tag{8.59}$$

where

$$R''_V(0) = -\frac{1}{2\pi} \int\limits_{-\infty}^{\infty} \Omega^2 G_V(\Omega)\, d\Omega \tag{8.60}$$

is the second derivative of the correlation function of fluctuations of the noise modulation function $\dot{M}(t)$ of the multiplicative noise. The function $G''_V(0)$ is defined in an analogous way.

In the case of the rapid fluctuating multiplicative noise the spectral power densities $G_V(\Omega)$ and $G_D(\Omega)$ in the region, where the function

$$[G'_{en}(\Omega)]^2$$

is essentially differed from zero, can be thought constant and equal to $G_V(0)$ and $G_D(0)$, respectively.[13,14]

Then

$$\sigma_\omega^2 = \frac{N_0}{\alpha_0^2(\tau,\Omega)E_{a_1}\overline{t^2}}\left\{1 + \frac{E_{a_1}}{4\pi N_0\overline{t^2}}[G_V(0) + G_D(0)]\int_{-\infty}^{\infty}[G'_{en}(\Omega)]^2\,d\Omega\right\}. \quad (8.61)$$

We consider in more detail the integral in Eq. (8.61). In terms of Eq. (8.50) we can write

$$\int_{-\infty}^{\infty}[G'_{en}(\Omega)]^2 d\Omega = \frac{1}{4}\cdot\frac{\partial^2}{\partial\omega_1\partial\omega_2}\int_{-\infty}^{\infty}\int_{-\infty}^{\infty}S_m^2(t_1)S_m^2(t_2) \times \int_{-\infty}^{\infty}Re\{e^{j[(\omega_1-\Omega)t_1+(\omega_2-\Omega)t_2]} + e^{j[(\omega_1-\Omega)t_1-(\omega_2-\Omega)t_2]}\}dt_1\,dt_2\,d\Omega. \quad (8.62)$$

Carrying out an integration with respect to the variable Ω and taking into account the well-known formula for the delta function

$$\delta(\tau) = \frac{1}{2\pi}\int_{-\infty}^{\infty}e^{j\omega\tau}d\omega, \quad (8.63)$$

after integration with respect to the variable t_2 we can write

$$\int_{-\infty}^{\infty}[G'_{en}(\Omega)]^2\,d\Omega = \frac{\pi}{2}\cdot\frac{\partial^2}{\partial\omega_1\partial\omega_2} \times Re\left\{\int_{-\infty}^{\infty}\left[S_m^4(t) + S_m^2(t)S_m^2(-t)\right]\cdot e^{j(\omega_1-\omega_2)t}dt\right\}. \quad (8.64)$$

For corresponding choice of the origin for the signals with the even amplitude envelopes the following condition

$$S_m(t) = S_m(-t) \quad (8.65)$$

is satisfied.

Then

$$\int_{-\infty}^{\infty}[G'_{en}(\Omega)]^2\,d\Omega = \pi\int_{-\infty}^{\infty}S_m^4(t)t^2\,dt. \quad (8.66)$$

In the case of the square wave-form amplitude envelope of the signal with the duration T, the amplitude of the model signal—the reference signal—of the generalized detector $S_m(t)$ is equal to $\sqrt{\frac{2}{T}}$. The amplitude of the model signal of the generalized detector is determined by the following formula

$$0.5 \int_{-\frac{T}{2}}^{\frac{T}{2}} S_m^2(t)\, dt = 1. \tag{8.67}$$

For this case we can write

$$\int_{-\infty}^{\infty} S_m^4(t) t^2\, dt = \frac{2}{T} \int_{-\frac{T}{2}}^{\frac{T}{2}} S_m^2(t) t^2\, dt = \frac{4}{T} \cdot \overline{t^2}. \tag{8.68}$$

Then the variance of measurement errors of the signal frequency has the following form:

$$\sigma_\omega^2 \simeq \frac{N_0}{\alpha_0^2(\tau, \Omega) E_{a_1} \overline{t^2}} \cdot \left\{1 + \frac{E_{a_1}}{N_0 T}[G_V(0) + G_D(0)]\right\}. \tag{8.69}$$

8.1.2 Signal Appearance Time Measurement Precision

During measurement of the signal appearance time τ, determined in the form

$$\tau = l - l_0$$

on the time axis, we can write

$$S_1(\tau, \Omega) \cdot e^{j\beta(\tau,\Omega)} = \frac{E_{a_1}}{2} \int_{-\infty}^{\infty} \dot{S}_m(t) S_m^*(t - \tau) \cdot e^{j\Omega t} dt, \tag{8.70}$$

where

$$\dot{S}_m(t) = S_m(t) \cdot e^{j\Psi(t)} \tag{8.71}$$

is the complex amplitude envelope of the model signal—the reference signal—of the generalized detector.

We proceed to define the statistical characteristics of the functions $S_1(\tau, \Omega)$ and $\beta(\tau, \Omega)$ that are used by general formulae for definition of the variance of measurement errors of the signal appearance time

$$\left.\frac{d^2 S_1(\tau)}{d\tau^2}\right|_{\tau=0} = -E_{a_1} \overline{\Omega^2}, \tag{8.72}$$

and

$$S_1(0, \Omega) = \frac{E_{a_1}}{2} \int_{-\infty}^{\infty} S_m^2(t) \cos \Omega t\, dt = E_{a_1} G_{en}(\Omega), \tag{8.73}$$

where

$$\Omega^2 = \frac{1}{2E_{a_1}} \int_{-\infty}^{\infty} \Omega^2 |F_a(\Omega)|^2 \, d\Omega \tag{8.74}$$

is the square mean bandwidth of the energy spectrum of the searching signal $F_a(\Omega)$.[10,11]

Next we define the derivative of the first order of the function $S_1(\tau, \Omega)$ with respect to the parameter τ for the condition $\tau = 0$, i.e.:

$$\left. \frac{dS_1(\tau, \Omega)}{d\tau} \right|_{\tau=0}.$$

Then we can write

$$S_1(\tau, \Omega) = \frac{E_{a_1}}{2} \times \sqrt{\left\{Re\left\{\int_{-\infty}^{\infty} \dot{S}_m(t) S_m^*(t-\tau) \cdot e^{j\Omega t} dt\right\}\right\}^2 + \left\{Im\left\{\int_{-\infty}^{\infty} \dot{S}_m(t) S_m^*(t-\tau) \cdot e^{j\Omega t} dt\right\}\right\}^2}; \tag{8.75}$$

$$\begin{aligned} \left. \frac{dS_1(\tau, \Omega)}{d\tau} \right|_{\tau=0} = & -E_{a_1} \left\{ Re\left\{ \int_{-\infty}^{\infty} \dot{S}_m(t) S_m^*(t-\tau) \cdot e^{j\Omega t} dt \right\} \right. \\ & \times \frac{d}{d\tau} Re\left\{ \int_{-\infty}^{\infty} \dot{S}_m(t) S_m^*(t-\tau) \cdot e^{j\Omega t} dt \right\} \\ & + Im\left\{ \int_{-\infty}^{\infty} \dot{S}_m(t) S_m^*(t-\tau) \cdot e^{j\Omega t} dt \right\} \\ & \left. \times \frac{d}{d\tau} Im\left\{ \int_{-\infty}^{\infty} \dot{S}_m(t) S_m^*(t-\tau) \cdot e^{j\Omega t} dt \right\} \right\}_{\tau=0} \\ & \times \left\{ 2\left\{ \left\{ Re\left\{ \int_{-\infty}^{\infty} \dot{S}_m(t) S_m^*(t-\tau) \cdot e^{j\Omega t} dt \right\} \right\}^2 \right. \right. \\ & \left. \left. + \left\{ Im\left\{ \int_{-\infty}^{\infty} \dot{S}_m(t) S_m^*(t-\tau) e^{j\Omega t} dt \right\} \right\}^2 \right\}^{\frac{1}{2}} \right\}_{\tau=0}^{-1}. \end{aligned} \tag{8.76}$$

For corresponding choice of the origin and the condition that the signal has the even amplitude envelope and is also within the limits of the time interval of observation $[0, T]$ we can write

$$\frac{d}{d\tau} Re\left\{\int_{-\infty}^{\infty} \dot{S}_m(t) S_m^*(t-\tau) \cdot e^{j\Omega t} dt\right\}_{\tau=0} = -\frac{E_{a_1}}{2} \int_{-\infty}^{\infty} S_m^2(t) \Psi(t) \sin \Omega t \, dt; \quad (8.77)$$

$$Im\left\{\int_{-\infty}^{\infty} \dot{S}_m(t) S_m^*(t-\tau) \cdot e^{j\Omega t} dt\right\}_{\tau=0} = 0. \quad (8.78)$$

Taking into account Eqs. (8.77) and (8.78), we can write

$$\left.\frac{d S_1(\tau, \Omega)}{d\tau}\right|_{\tau=0} = \frac{E_{a_1}}{2} \int_{-\infty}^{\infty} S_m^2(t) \Psi'(t) \sin \Omega t \, dt = E_{a_1} \gamma(\Omega). \quad (8.79)$$

As was shown in Reference 9, for the signals possessing the even function $\Psi(t)$ the following equality

$$\beta(\tau, \Omega) = \frac{\tau \Omega}{2} \quad (8.80)$$

is true. For this case the variance of measurement errors of the signal appearance time caused by the fluctuating multiplicative noise is determined by Eq. (8.38).

The total variance of measurement errors of the signal appearance time can be determined in the following form:

$$\sigma_\tau^2 = \frac{N_0}{\alpha_0^2(\tau, \Omega) E_{a_1} \overline{\Omega^2}} \times \left\{1 + \frac{E_{a_1}}{4\pi N_0 \overline{\Omega^2}} \int_{-\infty}^{\infty} G_V(\Omega) \left[\frac{\Omega^2}{4} \cdot G_{en}^2(\Omega) + \gamma^2(\Omega)\right] d\Omega - \frac{E_{a_1}}{4\pi N_0 \overline{\Omega^2}} \int_{-\infty}^{\infty} G_D(\Omega) \left[\frac{\Omega^2}{4} \cdot G_{en}^2(\Omega) - \gamma^2(\Omega)\right] d\Omega\right\}. \quad (8.81)$$

Reference to Eq. (8.81) shows that the component of the variance of measurement errors of the signal appearance time, caused by the fluctuating multiplicative noise, depends both on the amplitude envelope of the signal using the function $G_{en}(\Omega)$ and on the phase structure of the signal using the function $\gamma(\Omega)$.

In the case of the non-modulated signal, for which the following conditions

$$\Psi(t) = 0 \quad (8.82)$$

and

$$\gamma(\Omega) = 0 \quad (8.83)$$

are true, taking into account Eq. (8.81), we can write

$$\sigma_\tau^2 = \frac{N_0}{\alpha_0^2(\tau, \Omega)\overline{\overline{\Omega^2}}}\left\{1 + \frac{E_{a_1}}{16\pi\overline{\overline{\Omega^2}}}\int\limits_{-\infty}^{\infty} G_{en}^2(\Omega)[G_V(\Omega) - G_D(\Omega)]\Omega^2\, d\Omega\right\}. \tag{8.84}$$

Comparing Eq. (8.84) with Eq. (8.55), we can see that the spectrum energy characteristics of the multiplicative noise $G_V(\Omega)$ and $G_D(\Omega)$ are used by Eqs. (8.55) and (8.84) in the form of the sum and difference, respectively.

The main results of analysis carried out on the basis of the formulae in Section 3.4 allows us to be sure that the function $G_D(\Omega)$ is less than zero when distortions in amplitude and phase of the signal, and distortions only in phase of the signal, obey the Gaussian probability distribution density.

Taking into consideration the statement mentioned above and comparing Eqs. (8.55) and (8.84), we can conclude that an impact of the weak fluctuating multiplicative noise on precision of measurement of the signal appearance time is stronger than an impact of the weak fluctuating multiplicative noise on precision of measurement of the signal frequency. Supposedly, this phenomenon can be explained by the fact that the signal distorted by the weak fluctuating multiplicative noise is the non-stationary process, as was shown in Sections 4.4, 6.5, and 7.1.

As the degree of distortions of the signal is increased, the ratio

$$\frac{G_D(\Omega)}{G_V(\Omega)}$$

is decreased and tends to approach zero in the limiting case. In the process, the signal distorted by the stationary fluctuating multiplicative noise becomes closer to the stationary signal.

The formulae in Eqs. (8.55) and (8.84) become closer to each other owing to the condition

$$G_V(\Omega) \gg G_D(\Omega), \tag{8.85}$$

and the stimulus of the fluctuating multiplicative noise on precision of measurement of the signal frequency and signal appearance time is the same in practice.

We carry out a transformation in Eq. (8.74) for the cases of the slow and rapid fluctuating multiplicative noise.

For the case of the slow multiplicative noise the term in the square brackets in Eq. (8.81) can be presented in the form of the power series expansion in the neighborhood of the point

$$\Omega = 0.$$

We can limit the power series expansion by two terms.

After transformations that are analogous to the transformations carried out in accordance with definition of the variance of measurement errors of the signal frequency for the slow fluctuating multiplicative noise we can write

$$\sigma_\tau^2 = \frac{N_0}{\alpha_0^2(\tau,\Omega)E_{a_1}\overline{\Omega^2}}\left\{1 - \frac{E_{a_1}R_V''(0)}{8N_0\overline{\Omega^2}}[1+4[\gamma'(0)]^2] + \frac{E_{a_1}D_V''(0)}{8N_0\overline{\Omega^2}}[1-4[\gamma'(0)]^2]\right\}. \tag{8.86}$$

In the case of the rapid fluctuating multiplicative noise we assume that the spectral power density of fluctuations of the noise modulation function $\dot{M}(t)$ of the multiplicative noise is not varied in frequency within the limits if the interval, where the functions

$$G_{en}^2(\Omega)\Omega^2$$

and

$$\gamma^2(\Omega)$$

are essentially differed from zero.

For this case there are the following integrals in Eq. (8.81) that must be determined:

$$I_1 = \frac{1}{16\pi}\int_{-\infty}^{\infty} G_{en}^2(\Omega)\Omega^2\,d\Omega \tag{8.87}$$

and

$$I_2 = \frac{1}{4\pi}\int_{-\infty}^{\infty} \gamma^2(\Omega)\,d\Omega. \tag{8.88}$$

In terms of Eq. (8.50) that defines the function $G_{en}(\Omega)$ we can write

$$I_1 = \frac{1}{128\pi} \times Re\left\{\int_{-\infty}^{\infty}\int_{-\infty}^{\infty}\int_{-\infty}^{\infty} S_m^2(t_1)S_m^2(t_2)\Omega^2[e^{j\Omega(t_1+t_2)} + e^{j\Omega(t_1+t_2)}]\,dt_1\,dt_2\,d\Omega\right\}. \tag{8.89}$$

Taking into consideration the following relationships

$$\Omega^2 \cdot e^{j\Omega(t_1-t_2)} = -\frac{d^2 e^{j\Omega(t_1-t_2)}}{dt_1^2}; \tag{8.90}$$

$$\frac{1}{2\pi}\cdot\frac{d^2}{dt_1^2}\int_{-\infty}^{\infty} e^{j\Omega(t_1-t_2)}d\Omega = \frac{d^2}{dt_1^2}\delta(t_1-t_2); \tag{8.91}$$

$$\int_{-\infty}^{\infty} g(t)\cdot\frac{d^2\delta(t)}{dt^2}\,dt = \left.\frac{d^2 g(t)}{dt^2}\right|_{t=0} \tag{8.92}$$

for the signals, the amplitude envelope of which is the even function, we can write

$$I_1 = -\frac{1}{32}\int_{-\infty}^{\infty} S_m^2(t)[S_M''(t)]^2\,dt. \tag{8.93}$$

Taking into account Eq. (8.79), we can write

$$I_2 = \frac{1}{32\pi}\cdot Re\left\{\int_{-\infty}^{\infty}\int_{-\infty}^{\infty}\int_{-\infty}^{\infty} S_m^2(t_1)S_m^2(t_2)\Psi'(t_1)\,\Psi'(t_2)\right.$$

$$\left.\times\,[e^{j\Omega(t_1-t_2)} - e^{j\Omega(t_1+t_2)}]\,dt_1\,dt_2\,d\Omega\right\}. \tag{8.94}$$

Reference to Eq. (8.94) shows that for the signals, the amplitude envelope of which is the even function and for which the following conditions

$$\Psi(t) = \Psi(-t) \tag{8.95}$$

and

$$\Psi'(t) = -\Psi'(-t) \tag{8.96}$$

are true, we can write

$$I_2 = \frac{1}{8}\int_{-\infty}^{\infty} S_m^4(t)[\Psi'(t)]^2 dt. \tag{8.97}$$

Substituting Eqs. (8.93) and (8.97) in Eq. (8.81), we obtain that for the case of the rapid fluctuating multiplicative noise the variance of measurement errors of the signal appearance time can be determined in the following form:

$$\sigma_\tau^2 = \frac{N_0}{\alpha_0^2(\tau,\Omega)\,E_{a_1}\overline{\Omega^2}}$$

$$\times\left\{1 + \frac{E_{a_1}G_V(0)}{8N_0\overline{\Omega^2}}\int_{-\infty}^{\infty} S_m^2(t)\left[S_m^2(t)[\Psi'(t)]^2 - \frac{1}{4}[S_m''(t)]^2\right]dt\right.$$

$$\left.+\,\frac{E_{a_1}G_D(0)}{8N_0\overline{\Omega^2}}\int_{-\infty}^{\infty} S_m^2(t)\left[S_m^2(t)[\Psi'(t)]^2 + \frac{1}{4}[S_m''(t)]^2\right]dt\right\}. \tag{8.98}$$

The approximate formulae determined by Eqs. (8.86) and (8.98), defining the variance of measurement errors of the signal appearance time, contain the simple characteristics of the multiplicative noise—the derivatives of the second order of the correlation functions $R_V(\tau)$ and $D_V(\tau)$ for the condition

$\tau = 0$, which are proportional to the energy spectrum bandwidth of the noise modulation function $\dot{M}(t)$ of the multiplicative noise, and the spectral power densities of the noise modulation function $\dot{M}(t)$ of the multiplicative noise for the condition $\Omega = 0$ ($G_V(0)$ and $G_D(0)$).

In many practical cases these formulae allow us to estimate the stimulus of the multiplicative noise on precision of measurement of the signal appearance time or signal delay for each form of the signal.

8.1.3 Estimation of Precision of the Signal Frequency and Signal Appearance Time Measurement

We determine the estimation of precision of measurement of the signal frequency and signal appearance time during low distortions only in phase of the received signal. Assume that distortions in phase of the received signal obey the Gaussian probability distribution density with a zero mean and the correlation function

$$R_\varphi(\tau) = \sigma_\varphi^2 r_\varphi(\tau). \tag{8.99}$$

The spectral characteristics $G_V(\Omega)$ and $G_D(\Omega)$ of the noise modulation function $\dot{M}(t)$ of the multiplicative noise are the Fourier transform of the corresponding correlation functions $R_V(\tau)$ and $D_V(\tau)$ determined for the case considered in Section 3.4 and Chapter 4:

$$\begin{aligned} R_V(\tau) &= \Theta_2^\varphi(1, -1) - \left|\Theta_1^\varphi(1)\right| \\ &\simeq \sigma_\varphi^2\left(1 - \sigma_\varphi^2\right) r_\varphi(\tau) + 0.5\sigma_\varphi^4 r_\varphi^2(\tau); \end{aligned} \tag{8.100}$$

$$D_V(\tau) \simeq -\sigma_\varphi^2\left(1 - \sigma_\varphi^2\right) r_\varphi(\tau) + 0.5\sigma_\varphi^4 r_\varphi^2(\tau); \tag{8.101}$$

$$\alpha_0^2(\tau, \Omega) \simeq 1 - \sigma_\varphi^2 + 0.5\sigma_\varphi^4. \tag{8.102}$$

8.1.3.1 Signal with Bell-Shaped Amplitude Envelope and Constant Radio Frequency Carrier

Assume that the received signal possessing the complex amplitude envelope

$$\dot{S}(t) = e^{-\frac{\pi}{T^2}\cdot t^2 + j\omega t} \tag{8.103}$$

is within the limits of the time interval of observation $[0, T]$.

Then for the corresponding choice of the origin we can write

$$S_1(\omega, \Omega) = E_{a_1} \cdot e^{-\frac{(\omega-\Omega)^2 T^2}{8\pi}}; \tag{8.104}$$

$$\beta(\omega, \Omega) = 0; \tag{8.105}$$

$$E_{a_1} = \frac{E_{a_1}^{el} T}{\sqrt{2}}, \tag{8.106}$$

where $E_{a_1}^{el}$ is the energy of the elementary signal;

$$S_1(\tau, \Omega) = E_{a_1} \cdot e^{-\frac{\pi\tau^2}{2T^2} - \frac{\Omega^2 T^2}{8\pi}}; \tag{8.107}$$

$$\beta(\tau, \Omega) = -\frac{\Omega T}{2}. \tag{8.108}$$

Reference to Eqs. (8.104)–(8.108) shows that for estimation of the appearance time of the signal with a bell-shaped amplitude envelope

$$S(\tau, \Omega) = g(\tau) r(\Omega), \tag{8.109}$$

or for determination of the variance of measurement errors of the signal appearance time, we can use Eq. (8.36) for this case.

Substituting Eqs. (8.104)–(8.108) in Eq. (8.84), we can write

$$\sigma_{\omega_a}^2 = \frac{4\pi N_0}{\alpha_0^2(\tau, \Omega) E_{a_1} T^2}; \tag{8.110}$$

$$\sigma_{\omega_M}^2 = \frac{1}{4\pi \alpha_0^2(\tau, \Omega)} \int_{-\infty}^{\infty} G_\Sigma(\Omega)\Omega^2 \cdot e^{-\frac{\Omega^2 T^2}{4\pi}} d\Omega; \tag{8.111}$$

$$\sigma_{\tau_a}^2 = \frac{N_0 T^2}{\pi \alpha_0^2(\tau, \Omega) E_{a_1}}; \tag{8.112}$$

$$\sigma_{\tau_M}^2 = \frac{T^4}{16\pi^3 \alpha_0^2(\tau, \Omega)} \int_{-\infty}^{\infty} G_d(\Omega)\Omega^2 \cdot e^{-\frac{\Omega^2 T^2}{4\pi}} d\Omega, \tag{8.113}$$

where

$$G_\Sigma(\Omega) = G_V(\Omega) + G_D(\Omega); \tag{8.114}$$

$$G_d(\Omega) = G_V(\Omega) - G_D(\Omega); \tag{8.115}$$

$\sigma_{\omega_a}^2$ is the variance of measurement errors of the signal frequency caused by the additive Gaussian noise; and $\sigma_{\omega_M}^2$ is the variance of measurement errors of the signal frequency caused by the fluctuating multiplicative noise; $\sigma_{\tau_a}^2$ is the variance of measurement errors of the signal appearance time caused by the additive Gaussian noise; and $\sigma_{\tau_M}^2$ is the variance of measurement errors of the signal appearance time caused by the multiplicative noise.

In terms of Eqs. (8.100)–(8.102) we can write

$$G_\Sigma(\Omega) \simeq \sigma_\varphi^4 G_{\varphi^2}(\Omega) \tag{8.116}$$

and

$$G_d(\Omega) = 2\sigma_\varphi^2 \left(1 - \sigma_\varphi^2\right) G_\varphi(\Omega), \tag{8.117}$$

where the functions $G_\varphi(\Omega)$ and $G_{\varphi^2}(\Omega)$ are the Fourier transform of the functions $r_\varphi(\tau)$ and $r_\varphi^2(\tau)$, respectively.

Substituting Eqs. (8.116) and (8.117) in Eqs. (8.110)–(8.113), we finally obtain that the variances of measurement errors of the signal frequency and signal appearance time for the case of low distortions only in phase of the signal obeying the Gaussian probability distribution density can be determined in the following form:

$$\sigma_\omega^2 = \sigma_{\omega_a}^2 + \sigma_{\omega M}^2$$
$$= \frac{\sigma_{\omega_0}^2}{\alpha_0^2(\tau, \Omega)} \left\{ 1 + \frac{E_{a_1} T^2 \sigma_\varphi^2}{16\pi^2 N_0} \int_{-\infty}^{\infty} G_{\varphi^2}(\Omega)\Omega^2 \cdot e^{-\frac{\Omega^2 T^2}{4\pi}} d\Omega \right\}$$
$$= \sigma_{\omega_0}^2 \eta_{\omega M} \tag{8.118}$$

and

$$\sigma_\tau^2 = \sigma_{\tau_a}^2 + \sigma_{\tau M}^2$$
$$= \frac{\sigma_{\tau_0}^2}{\alpha_0^2(\tau, \Omega)} \left\{ 1 + \frac{E_{a_1} T^2}{8\pi^2 N_0} \sigma_\varphi^2 (1 - \sigma_\varphi^2) \int_{-\infty}^{\infty} G_\varphi(\Omega)\Omega^2 \cdot e^{-\frac{\Omega^2 T^2}{4\pi}} d\omega \right\}$$
$$= \sigma_{\tau_0}^2 \eta_{\tau M}, \tag{8.119}$$

where $\sigma_{\omega_0}^2$ and $\sigma_{\tau_0}^2$ are the variances of measurement errors of the signal frequency and signal appearance time when the fluctuating multiplicative noise is absent.

The coefficients $\eta_{\omega M}$ and $\eta_{\tau M}$ are always greater than unity

$$\eta_{\omega M} > 1 \tag{8.120}$$

and

$$\eta_{\tau M} > 1. \tag{8.121}$$

The coefficients $\eta_{\omega M}$ and $\eta_{\tau M}$ define a degree of increasing the variance of measurement errors of the signal frequency and signal appearance time owing to the stimulus of the multiplicative noise.

Let us assume that the coefficient of correlation $r_\varphi(\tau)$ between distortions in phase of the signal possesses a bell-shaped amplitude envelope

$$r_\varphi(\tau) = e^{-\mu^2\tau^2}. \tag{8.122}$$

Then

$$G_\varphi(\Omega) = \frac{2\pi}{\Delta\Omega_\varphi} \cdot e^{-\frac{\pi\Omega^2}{\Delta\Omega_\varphi^2}} \tag{8.123}$$

and

$$G_{\varphi^2}(\Omega) = \frac{\sqrt{2}\pi}{\Delta\Omega_\varphi} \cdot e^{-\frac{\pi\Omega^2}{2\Delta\Omega_\varphi^2}}, \tag{8.124}$$

where $\Delta\Omega_\varphi$ is the equivalent bandwidth of the energy spectrum of distortions in phase $\varphi(t)$ of the signal.

In terms of Eqs. (8.104)–(8.108), (8.118), and (8.119) the coefficients η_{ω_M} and η_{τ_M} can be determined in the following form:

$$\eta_{\omega M} = \frac{1 + \frac{E_{a_1}\sigma_\varphi^4\xi_\varphi^2}{N_0\sqrt{(2\xi_\varphi^2+1)^3}}}{1 - \sigma_\varphi^2 + 0.5\sigma_\varphi^4} \tag{8.125}$$

and

$$\eta_{\tau_M} = \frac{1 + 0.5\sigma_\varphi^2(1-\sigma_\varphi^2)\cdot\frac{E_{a_1}\xi_\varphi^2}{N_0\sqrt{(1+\xi_\varphi^2)^3}}}{1 - \sigma_\varphi^2 + 0.5\sigma_\varphi^4}, \tag{8.126}$$

where

$$\xi_\varphi = \frac{T\Delta\Omega_\varphi}{2\pi}. \tag{8.127}$$

The coefficient η_{ω_M} is maximum for the following condition

$$\xi_\varphi = 1. \tag{8.128}$$

The coefficient η_{τ_M} is maximum for the following condition

$$\xi_\varphi = \sqrt{2}. \tag{8.129}$$

Substituting Eqs. (8.128) and (8.129) in Eqs. (8.125) and (8.126), respectively, we can write

$$\eta_{\omega M}^{max} = \frac{N_0 + 0.096E_{a_1}\sigma_\varphi^4}{N_0(1 - \sigma_\varphi^2 + 0.5\sigma_\varphi^4)} \tag{8.130}$$

and

$$\eta_{\tau_M}^{max} = \frac{N_0 + 0.19\sigma_\varphi^2(1-\sigma_\varphi^2)E_{a_1}}{N_0(1 - \sigma_\varphi^2 + 0.5\sigma_\varphi^4)}. \tag{8.131}$$

The application of the results obtained (see Eq. (8.44)) for the conditions introduced above in the case of the considered fluctuating multiplicative noise, if the condition

$$\frac{\alpha_0^2(\tau,\Omega)E_{a_1}}{N_0} \gg 1 \tag{8.132}$$

is true, has the following form:

$$\alpha_0^2(\tau,\Omega) \gg \frac{\sigma_\varphi^4}{4\pi}\int_{-\infty}^{\infty} G_{\varphi^2}(\Omega)\cdot e^{-\frac{\Omega^2T^2}{4\pi}}\,d\Omega \tag{8.133}$$

or

$$2\sqrt{2\xi_\varphi^2+1} \gg \frac{\sigma_\varphi^4}{1 - \sigma_\varphi^2 + 0.5\sigma_\varphi^4}. \tag{8.134}$$

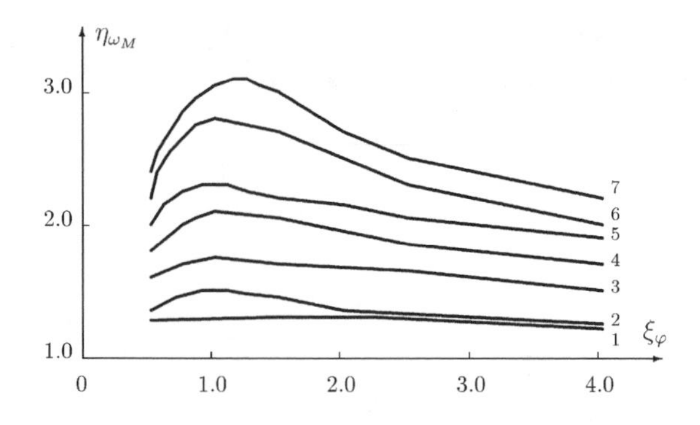

FIGURE 8.2
The coefficient η_{ω_M} as a function of the parameter ξ_φ: for 1, $\sigma_\varphi = 0.4$; $q = 20$; for 2, $\sigma_\varphi = 0.4$; $q = 40$; for 3, $\sigma_\varphi = 0.6$; $q = 10$; for 4, $\sigma_\varphi = 0.6$; $q = 20$; for 5, $\sigma_\varphi = 0.7$; $q = 10$; for 6, $\sigma_\varphi = 0.6$; $q = 40$; for 7, $\sigma_\varphi = 0.7$; $q = 20$.

The functions $\eta_{\omega_M}(\xi_\varphi)$ and $\eta_{\tau_M}(\xi_\varphi)$ under the condition

$$\sigma_\varphi = const$$

are shown in Figs. 8.2 and 8.3, respectively, taking into consideration Eqs. (8.133) and (8.134).

Two interesting features of the results obtained are noteworthy. Low distortions in phase of the signal act on measurement precision of the signal appearance time more than on measurement precision of the signal frequency. The maximum of the variances $\sigma^2_{\omega_M}$ and $\sigma^2_{\tau_M}$ is reached for various values of the correlation interval of distortions in phase of the received signal.

As was noted above, during discussion of the general formulae for the variances σ^2_ω and σ^2_τ (see Eqs. (8.61) and (8.81)), at the high and rapid fluctuating distortions in phase of the received signal the inequality

$$G_V(\Omega) \gg G_D(\Omega) \tag{8.135}$$

is true in accordance with the results discussed in Sections 3.4 and 7.1.

In doing so, for the considered case the functions $\eta_{\omega_M}(\xi_\varphi)$ and $\eta_{\tau_M}(\xi_\varphi)$ coincide with each other.

In other words, the signal distorted by the fluctuating multiplicative noise becomes closer to the stationary stochastic process.

8.1.3.2 Frequency-Modulated Signal

Consider the frequency-modulated signal with a bell-shaped complex amplitude envelope

$$\dot{S}(t) = e^{-\frac{\pi t^2}{T^2} - j\cdot\frac{\Delta\Omega_d}{2T}\cdot t^2}, \tag{8.136}$$

where

$$T = \int\limits_{-\infty}^{\infty} |\dot{S}(t)|\, dt \tag{8.137}$$

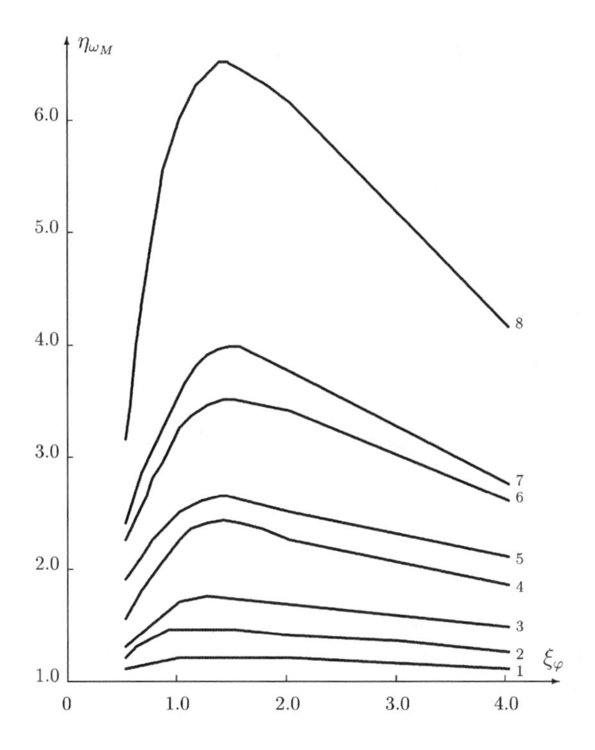

FIGURE 8.3
The coefficient η_{τ_M} as a function of the parameter ξ_φ: for 1, $\sigma_\varphi = 0.2$; $q = 20$; for 2, $\sigma_\varphi = 0.2$; $q = 40$; for 3, $\sigma_\varphi = 0.4$; $q = 10$; for 4, $\sigma_\varphi = 0.4$; $q = 20$; for 5, $\sigma_\varphi = 0.6$; $q = 10$; for 6, $\sigma_\varphi = 0.4$; $q =$ 40; for 7, $\sigma_\varphi = 0.6$; $q = 20$; for 8, $\sigma_\varphi = 0.6$; $q = 40$.

is the equivalent duration of the received signal; $\Delta\Omega_d$ is the deviation in frequency within the limits of the time interval $[0, T]$.

During measurement of the signal appearance time the functions in Eq. (8.81) take the following form:

$$q = \frac{E_{a_1}}{N_0} = \frac{E_{a_1}^{el} T}{2\sqrt{2} N_0}; \tag{8.138}$$

$$\overline{\Omega^2} = \frac{\pi\left(1 + Q_y^2\right)}{T^2}; \tag{8.139}$$

$$G_{en}(\Omega) = e^{-\frac{\Omega^2 T^2}{8\pi}}; \tag{8.140}$$

$$\gamma(\Omega) = \frac{Q_y \Omega}{2} \cdot e^{-\frac{\Omega^2 T^2}{8\pi}}, \tag{8.141}$$

where

$$Q_y = \frac{T \Delta\Omega_d}{2\pi} \tag{8.142}$$

is the coefficient of cutting the received signal by the generalized detector.

In terms of Eqs. (8.138)–(8.142) the variance of measurement errors of the signal appearance time or the signal delay determined in Eq. (8.81) can be written in the following form:

$$\sigma^2_{\tau_{F-M}} = \frac{\sigma_{\tau_{0_{F-M}}}}{\alpha_0^2(\tau, \Omega)} \times \left\{1 + \frac{E_{a_1} T^2}{16\pi^2 N_0} \int\limits_{-\infty}^{\infty} G_V(\Omega)\Omega^2 \cdot e^{-\frac{\Omega^2 T^2}{4\pi}} d\Omega - \frac{E_{a_1} T^2 (1 - Q_y^2)}{16\pi^2 N_0 (1 + Q_y^2)} \int\limits_{-\infty}^{\infty} G_D(\Omega)\Omega^2 \cdot e^{-\frac{\Omega^2 T^2}{4\pi}} d\Omega \right\} = \sigma^2_{\tau_{0_{F-M}}} \cdot \eta_{\tau_{M_{F-M}}}, \tag{8.143}$$

where

$$\sigma^2_{\tau_{0_{F-M}}} = \frac{2\sqrt{2} N_0 T}{\pi (1 + Q_y^2)} \tag{8.144}$$

is the variance of measurement errors of the signal appearance time when the multiplicative noise is absent.

Reference to Eqs. (8.110)–(8.113) and (8.119) shows that for the condition

$$Q_y = 0$$

the variance of measurement errors of the signal appearance time of the frequency-modulated signal is equal to the variance of measurement errors of the signal appearance time of the non-modulated signal with a bell-shaped complex amplitude envelope. If

$$Q_y \gg 1,$$

this condition is usual in practice for real complex signal processing systems that are constructed on the basis of the generalized detector and use the frequency-modulated signals, Eq. (8.143) coincides with an accuracy of a constant factor with the variance of measurement errors of frequency of the signal with a bell-shaped complex amplitude envelope (see Eqs. (8.110)–(8.113)).

Consequently, in the case when the correlation function of distortions in phase of the signal has a bell-shaped form we can write that

$$\eta_{\tau_{M_{F-M}}} = \eta_{\omega M} \tag{8.145}$$

and

$$\eta^{max}_{\tau_{M_{F-M}}} = \eta^{max}_{\omega M}; \tag{8.146}$$

see Eqs. (8.125), (8.126), (8.130), (8.131), and Fig. 8.3.

The coincidence of Eqs. (8.86) and (8.143) with an accuracy of constant factor during the condition

$$Q_y \gg 1$$

can be explained from the physical viewpoint in the following manner: the autocorrelation function of the frequency-modulated signal has the linear functional relationship between the shift in delay τ and frequency Ω, which causes a proportional dependence between the measurement errors of the signal appearance time when the signal frequency is known and the measurement errors when the signal appearance time is known.

8.2 Simultaneous Measurement of Two Signal Parameters under a Combined Stimulus of Weak Multiplicative and Additive Gaussian Noise

In this section we carry out an estimation of the stimulus of the fluctuating multiplicative noise on precision of the joint measurement of two signal parameters. We consider the generalized detector during the joint measurement of two signal parameters in the presence of additive Gaussian noise.

The stochastic process at the input of the generalized detector has the following form:

$$\dot{S}_{in}(t) = \dot{S}(t, l_0, p_0, \varphi_0) + \dot{N}(t), \tag{8.147}$$

where $\dot{S}_{in}(t)$ is the complex amplitude envelope of the input stochastic process; $\dot{S}(t, l_0, p_0, \varphi_0)$ is the complex amplitude envelope of the received signal with two non-energy unknown parameters l_0 and p_0; φ_0 is the random initial phase of the received signal distributed uniformly within the limits of the interval $[0, 2\pi]$; and $\dot{N}(t)$ is the complex amplitude envelope of the additive Gaussian noise.

The signal at the output of the generalized detector has the following form:

$$\Lambda(l, p) = \ln K_0\left[Z_g^{out}(l, p)\right], \tag{8.148}$$

where

$$Z_g^{out}(l, p) = 0.5 \cdot \left| 2\int_0^T \dot{S}_{in}(t) S_m^*(t, l, p)\, dt - \int_0^T \dot{S}_{in}(t) S_{in}^*(t - \tau)\, dt \right| \tag{8.149}$$

is the complex amplitude envelope (the module) of two-dimensional stochastic function with respect to parameters l and p formed at the output of the generalized detector during an action of the stochastic process determined by Eq. (8.147) at the input of the generalized detector.

As we can see from Eq. (8.149) the complex amplitude envelope

$$Z_g^{out}(l, p)$$

of the signal at the output of the generalized detector is determined by the formula that is analogous to Eq. (8.6).

Estimations of the signal parameters are defined using the maximum of the function $\Lambda(l, p)$. For the case of the weak multiplicative and additive Gaussian noise these estimations can be defined on the basis of the following likelihood equations:[4,15]

$$\frac{\partial \Lambda(l, p)}{\partial l} = \frac{\partial Z_g^{out}(l, p)}{\partial l} = 0 \tag{8.150}$$

and

$$\frac{\partial \Lambda(l, p)}{\partial p} = \frac{\partial Z_g^{out}(l, p)}{\partial p} = 0. \tag{8.151}$$

The block diagram of the generalized detector is analogous to the block diagram shown in Fig. 8.1, but the model signal, i.e., the reference signal

$$S_m(t, l, p) \cdot e^{j\Psi(t,l,p)}$$

is used instead of the model signal

$$S_m(t, l) \cdot e^{j\Psi(t,l)}.$$

Under a combined stimulus of the weak fluctuating multiplicative and additive Gaussian noise the complex amplitude envelope of the signal at the output of the generalized detector can be written in the following form:

$$\begin{aligned} Z_g^{out}(l, p) &\simeq \alpha_0(\tau, \Omega) S_1(l - l_0, p - p_0) \\ &\quad + \frac{Re\{\dot{S}_{1_1}(l - l_0, p - p_0) S_{1_2}^*(l - l_0, p - p_0) \cdot e^{j\beta_0}\}}{\alpha_0(\tau, \Omega) S_1(l - l_0, p - p_0)} \\ &\quad + \frac{Re\{\dot{S}_{1_1}(l - l_0, p - p_0) n^*(l - l_0, p - p_0) \cdot e^{j(\varphi_0 + \beta_0)}\}}{\alpha_0(\tau, \Omega) S_1(l - l_0, p - p_0)} \\ &= \alpha_0(\tau, \Omega) S_1(l - l_0, p - p_0) + \dot{n}(l - l_0, p - p_0), \end{aligned} \tag{8.152}$$

where

$$\dot{n}(l - l_0, p - p_0) = n_M(l - l_0, p - p_0) + n_a(l, p). \tag{8.153}$$

The function

$$n_M(l - l_0, p - p_0)$$

is caused by the fluctuating multiplicative noise and is defined by the second term in Eq. (8.152). The function

$$n_a(l, p)$$

is caused by the additive Gaussian noise and is defined by the third term in Eq. (8.152). The function

$$\dot{S}_{1_1}(l - l_0, p - p_0) = \alpha_0(\tau, \Omega) S_1(l - l_0, p - p_0) \cdot e^{j\beta(l - l_0, p - p_0)} \tag{8.154}$$

and the functions

$$\dot{S}_{1_2}(l - l_0, p - p_0)$$

and

$$\dot{n}(l - l_0, p - p_0)$$

are determined by formulae that are analogous to Eqs. (8.9)–(8.12).

The condition defining that there is the weak multiplicative and additive Gaussian noise, for which Eq. (8.152) is true, has the form that is analogous to Eq. (8.17) and can be written in the following manner:

$$\begin{aligned} 1 \gg & \frac{1}{\alpha_0^2(\tau, \Omega)\, S_1^2(l - l_0, p - p_0)} \\ & \times Re\{\dot{S}_{1_1}(l - l_0, p - p_0) S_{1_2}^*(l - l_0, p - p_0) \cdot e^{j\beta_0} \\ & + \dot{S}_{1_1}(l - l_0, p - p_0) n^*(l - l_0, p - p_0) \cdot e^{j(\varphi_0 + \beta_0)}\}. \end{aligned} \tag{8.155}$$

The condition in Eq. (8.155) must be satisfied in the statistical sense. Using the procedure discussed in References 4, 16, and 17, we can verify that estimations of the signal parameters l and p are unbiased in the first approximation if the condition in Eq. (8.155) is satisfied.

The variances of estimations can be determined in the following form:

$$\sigma_l^2 = \frac{\left\{\overline{\left(\frac{\partial n(l,p)}{\partial l}\right)^2}\left(\frac{\partial^2 S_1(l,p)}{\partial p^2}\right)^2 - 2\overline{\frac{\partial n(l,p)}{\partial l}\frac{\partial n(l,p)}{\partial p}} \cdot \frac{\partial^2 S_1(l,p)}{\partial l \partial p}\frac{\partial^2 S_1(l,p)}{\partial p^2} + \left(\frac{\partial^2 S_1(l,p)}{\partial l \partial p}\right)^2 \overline{\left(\frac{\partial n(l,p)}{\partial p}\right)^2}\right\}}{\alpha_0^2(\tau, \Omega)\left[\frac{\partial^2 S_1(l,p)}{\partial l^2} \cdot \frac{\partial^2 S_1(l,p)}{\partial p^2} - \left(\frac{\partial^2 S_1(l,p)}{\partial l \partial p}\right)^2\right]^2} \tag{8.156}$$

and

$$\sigma_p^2 = \frac{\left\{\overline{\left(\frac{\partial n(l,p)}{\partial p}\right)^2}\left(\frac{\partial^2 S_1(l,p)}{\partial l^2}\right)^2 - 2\overline{\frac{\partial n(l,p)}{\partial l}\frac{\partial n(l,p)}{\partial p}} \cdot \frac{\partial^2 S_1(l,p)}{\partial l^2}\frac{\partial^2 S_1(l,p)}{\partial l \partial p} + \overline{\left(\frac{\partial n(l,p)}{\partial l}\right)^2}\left(\frac{\partial^2 S_1(l,p)}{\partial l \partial p}\right)^2\right\}}{\alpha_0^2(\tau, \Omega)\left[\frac{\partial^2 S_1(l,p)}{\partial l^2}\frac{\partial^2 S_1(l,p)}{\partial p^2} - \left(\frac{\partial^2 S_1(l,p)}{\partial l \partial p}\right)^2\right]^2}, \tag{8.157}$$

for the condition

$$l = l_0 \quad \text{and} \quad p = p_0. \tag{8.158}$$

Taking into consideration that the fluctuating multiplicative noise and the additive Gaussian noise are not correlated with each other, i.e.,

$$\overline{n_a(l, p) n_M(l - l_0, p - p_0)} = 0, \tag{8.159}$$

Eqs. (8.156) and (8.157) can be written in the following form:

$$\sigma_l^2 = \sigma_{l_a}^2 + \sigma_{l_M}^2 \tag{8.160}$$

and

$$\sigma_p^2 = \sigma_{p_a}^2 + \sigma_{p_M}^2, \tag{8.161}$$

where $\sigma_{l_a}^2$ and $\sigma_{p_a}^2$ are the variances caused by the additive Gaussian noise; $\sigma_{l_M}^2$ and $\sigma_{p_M}^2$ are the variances caused by the fluctuating multiplicative noise.

The variances of measurement errors of the signal parameters, caused by the additive Gaussian noise $\sigma^2_{l_a}$ and $\sigma^2_{p_a}$ and the fluctuating multiplicative noise $\sigma^2_{l_M}$ and $\sigma^2_{p_M}$, are determined by Eqs. (8.156) and (8.157) during substitution of the functions

$$n_a(l, p)$$

or

$$n_M(l - l_0, p - p_0)$$

instead of the function

$$n(l, p),$$

respectively.

The variances $\sigma^2_{l_a}$ and $\sigma^2_{p_a}$ coincide with an accuracy of constant factor with the variances $\sigma^2_{l_0}$ and $\sigma^2_{p_0}$ of errors of simultaneous measurement of two signal parameters l and p when the multiplicative noise is absent and, taking into account the results discussed in Reference 4, are determined for the condition in Eq. (8.158) in the following form:

$$\sigma^2_{l_a} = \frac{\sigma^2_{l_0}}{\alpha_0^2(\tau, \Omega)} = -\frac{N_0 \cdot \frac{\partial^2 S_1(l,p)}{\partial p^2}}{\alpha_0^2(\tau, \Omega) E_{a_1} \left\{ \frac{\partial^2 S_1(l,p)}{\partial p^2} \cdot \frac{\partial^2 S_1(l,p)}{\partial l^2} - \left[\frac{\partial^2 S_1(l,p)}{\partial l \partial p} \right]^2 \right\}}, \tag{8.162}$$

and

$$\sigma^2_{p_a} = \frac{\sigma^2_{p_0}}{\alpha_0^2(\tau, \Omega)} = -\frac{N_0 \cdot \frac{\partial^2 S_1(l,p)}{\partial l^2}}{\alpha_0^2(\tau, \Omega) E_{a_1} \left\{ \frac{\partial^2 S_1(l,p)}{\partial p^2} \cdot \frac{\partial^2 S_1(l,p)}{\partial l^2} - \left[\frac{\partial^2 S_1(l,p)}{\partial l \partial p} \right]^2 \right\}}. \tag{8.163}$$

To define the variances $\sigma^2_{l_M}$ and $\sigma^2_{p_M}$ of errors of simultaneous measurement of two signal parameters l and p caused by the stimulus of the fluctuating multiplicative noise, we must to determine the following functions for the condition in Eq. (8.158):

$$\overline{\left[\frac{\partial n_M(l - l_0, p - p_0)}{\partial l} \right]^2} \tag{8.164}$$

and

$$\overline{\frac{\partial n_M(l - l_0, p - p_0)}{\partial l} \cdot \frac{\partial n_M(l - l_0, p - p_0)}{\partial p}} \tag{8.165}$$

that are defined by the correlation function of the noise component of the signal distorted by the fluctuating multiplicative noise at the output of the generalized detector:

$$\overline{\left[\frac{\partial n_M(l - l_0, p - p_0)}{\partial l}\right]^2} = \frac{\partial^2 \overline{n_M(l_1 - l_0, p - p_0) n_M(l_2 - l_0, p - p_0)}}{\partial l_1 \partial l_2}$$
$$= \frac{\partial^2 K_M(l_1, p_0, l_2, p_0)}{\partial l_1 \partial l_2} \tag{8.166}$$

if the condition

$$l_1 = l_2 = l_0 \tag{8.167}$$

is satisfied and

$$\overline{\frac{\partial n_M(l - l_0, p - p_0)}{\partial l} \cdot \frac{\partial n_M(l - l_0, p - p_0)}{\partial p}} = \frac{\partial^2 K_M(l, p)}{\partial l \partial p} \tag{8.168}$$

for the condition determined by Eq. (8.158)

The two-dimensional correlation function

$$K_M(l_1, p_1, l_2, p_2)$$

of noise component of the signal at the output of the generalized detector is defined using the same procedure considered in Section 8.1 in an analogous way on the basis of Eqs. (8.25), (8.27), (8.28), (8.30), and (8.31) and can be determined in the following form:

$$K_M(l_1, p_1, l_2, p_2) = K_{M_1}(l_1, p_1, l_2, p_2) + K_{M_2}(l_1, p_1, l_2, p_2)$$
$$= \frac{1}{4\pi} \int_{-\infty}^{\infty} G_V(\Omega) S_1(l_1 - l_0, p_1 - p_0, \Omega) S_1(l_2 - l_0, p_2 - p_0, \Omega)$$
$$\times \cos[\beta(l_1 - l_0, p_1 - p_0) - \beta(l_2 - l_0, p_2 - p_0)$$
$$- \beta(l_1 - l_0, p_1 - p_0, \Omega) + \beta(l_2 - l_0, p_2 - p_0, \Omega)]\, d\Omega$$
$$+ \frac{1}{4\pi} \int_{-\infty}^{\infty} G_D(\Omega) S_1(l_1 - l_0, p_1 - p_0, \Omega) S_1(l_2 - l_0, p_2 - p_0, \Omega)$$
$$\times \cos[\beta(l_1 - l_0, p_1 - p_0) + \beta(l_2 - l_0, p_2 - p_0)$$
$$- \beta(l_1 - l_0, p_1 - p_0, \Omega) - \beta(l_2 - l_0, p_2 - p_0, \Omega)]\, d\Omega, \tag{8.169}$$

where

$$S_1(l_i - l_0, p_i - p_0, \Omega)$$

is the module and

$$\beta(l_i - l_0, p_i - p_0, \Omega)$$

is the argument of the complex amplitude envelope of the signal at the output of the generalized detector if there is the signal with the parameters l_i

and p_i and the shift in frequency Ω, which is undistorted by the fluctuating multiplicative noise at the input of the generalized detector.

In doing so,

$$\beta(l_i - l_0, p_i - p_0) = \beta(l_i - l_0, p_i - p_0, \Omega) = 0\,. \tag{8.170}$$

We define the derivatives of the function

$$K_M(l_1, l_2, p_1, p_2)$$

in Eqs. (8.156) and (8.157) that can be determined after simple mathematical transformations:

$$\begin{aligned}
&\overline{\left[\frac{\partial n_M(l - l_0, p - p_0)}{\partial l}\right]^2}\Bigg|_{l=l_0,\,p=p_0} = \frac{1}{4\pi}\int_{-\infty}^{\infty} G_V(\Omega)\Bigg\{\left[\frac{\partial S_1(l - l_0, 0, \Omega)}{\partial l}\right]^2 \\
&+ S_1^2(0, 0, \Omega)\left[\frac{\partial \beta(l - l_0, 0)}{\partial l} - \frac{\partial \beta(l - l_0, 0, \Omega)}{\partial l}\right]^2\Bigg\}_{l=l_0} d\Omega \\
&+ \frac{1}{4\pi}\int_{-\infty}^{\infty} G_D(\Omega)\Bigg\{\Bigg[\left[\frac{\partial S_1(l - l_0, 0, \Omega)}{\partial l}\right]^2 - S_1^2(0, 0, \Omega)\Bigg[\frac{\partial \beta(l - l_0, 0)}{\partial l} \\
&- \frac{\partial \beta(l - l_0, 0, \Omega)}{\partial l}\Bigg]^2\Bigg]\cos 2[\beta(0, 0) - \beta(0, 0, \Omega)] - 2\frac{\partial S_1(l - l_0, 0, \Omega)}{\partial l}\cdot S_1(0, 0, \Omega) \\
&\times\left[\frac{\partial \beta(l - l_0, 0)}{\partial l} - \frac{\partial \beta(l - l_0, 0, \Omega)}{\partial l}\right]\sin 2[\beta(0, 0) - \beta(0, 0, \Omega)]\Bigg\}_{l=l_0} d\Omega
\end{aligned} \tag{8.171}$$

and

$$\begin{aligned}
&\overline{\left[\frac{\partial n_M(l - l_0, p - p_0)}{\partial p}\right]^2}\Bigg|_{l=l_0,\,p=p_0} = \frac{1}{4\pi}\int_{-\infty}^{\infty} G_V(\Omega)\Bigg\{\left[\frac{\partial S_1(0, p - p_0, \Omega)}{\partial p}\right]^2 \\
&+ S_1^2(0, 0, \Omega)\left[\frac{\partial \beta(0, p - p_0)}{\partial p} - \frac{\partial \beta(0, p - p_0, \Omega)}{\partial p}\right]^2\Bigg\}_{p=p_0} d\Omega \\
&+ \frac{1}{4\pi}\int_{-\infty}^{\infty} G_D(\Omega)\Bigg\{\Bigg[\left[\frac{\partial S_1(0, p - p_0, \Omega)}{\partial p}\right]^2 - S_1^2(0, 0, \Omega)\Bigg[\frac{\partial \beta(0, p - p_0, 0)}{\partial p} \\
&- \frac{\partial \beta(0, p - p_0, 0, \Omega)}{\partial p}\Bigg]^2\Bigg]\cos 2[\beta(0, 0) - \beta(0, 0, \Omega)] \\
&- 2\frac{\partial S_1(0, p - p_0, 0, \Omega)}{\partial p}\cdot S_1(0, 0, \Omega) \\
&\times\left[\frac{\partial \beta(0, p - p_0, 0)}{\partial p} - \frac{\partial \beta(0, p - p_0, 0, \Omega)}{\partial p}\right]\sin 2[\beta(0, 0) - \beta(0, 0, \Omega)]\Bigg\}_{p=p_0} d\Omega.
\end{aligned} \tag{8.172}$$

Furthermore, we can write

$$\overline{\frac{\partial n_M(l-l_0, p-p_0)}{\partial l}\cdot\frac{\partial n_M(l-l_0, p-p_0)}{\partial p}}\Bigg|_{l=l_0, p=p_0} = \frac{1}{2\pi}$$

$$\times \int_{-\infty}^{\infty} G_V(\Omega)\left[\frac{\partial S_1(l, 0, \Omega)}{\partial l}\cdot\frac{\partial S_1(0, p, \Omega)}{\partial p} + S_1(0, 0, \Omega)\cdot\frac{\partial^2 S_1(l, p, \Omega)}{\partial l\partial p}\right]_{l=p=0} d\Omega$$

$$+\frac{1}{2\pi}\int_{-\infty}^{\infty} G_D(\Omega)\left\{\frac{\partial S_1(l, 0, \Omega)}{\partial l}\cdot\frac{\partial S_1(0, p, \Omega)}{\partial p}\cos 2[\beta(0, 0) - \beta(0, 0, \Omega)]\right.$$

$$+ S_1(0, 0, \Omega)\frac{\partial^2 S_1(l, p, \Omega)}{\partial l\partial p}\cos 2[\beta(0, 0) - \beta(0, 0, \Omega)]$$

$$- 2S_1(0, 0, \Omega)\frac{\partial S_1(l, 0, \Omega)}{\partial l}\cdot\frac{\partial[\beta(0, p) - \beta(0, p, \Omega)]}{\partial p}\sin 2[\beta(0, 0) - \beta(0, 0, \Omega)]$$

$$- 2S_1(0, 0, \Omega)\frac{\partial S_1(0, p, \Omega)}{\partial p}\cdot\frac{\partial[\beta(l, 0) - \beta(l, 0, \Omega)]}{\partial l}\sin 2[\beta(0, 0) - \beta(0, 0, \Omega)]$$

$$- 2S_1^2(0, 0, \Omega)\frac{\partial[\beta(l, 0) - \beta(l, 0, \Omega)]}{\partial l}\cdot\frac{\partial[\beta(0, p) - \beta(0, p, \Omega)]}{\partial p}$$

$$\times \cos 2[\beta(0, 0) - \beta(0, 0, \Omega)] - S_1^2(0, 0, \Omega)\frac{\partial^2[\beta(l, p) - \beta(l, p, \Omega)]}{\partial l\partial p}$$

$$\left.\times \sin 2[\beta(0, 0) - \beta(0, 0, \Omega)]\right\}_{l=0, p=0} d\Omega. \tag{8.173}$$

The variances $\sigma_{l_M}^2$ and $\sigma_{p_M}^2$ of errors of simultaneous measurement of two signal parameters caused by the fluctuating multiplicative noise are defined by substitution of Eqs. (8.171)–(8.173) in Eqs. (8.156) and (8.157), respectively.

During simultaneous measurement of two signal parameters the errors can be correlated. The coefficient of correlation is determined in the following form:

$$r_{l,p}(l, p) = \frac{\overline{(l^* - l_0)(p^* - p_0)}}{\sigma_l \sigma_p} = \frac{\overline{\Delta l \Delta p}}{\sigma_l \sigma_p}, \tag{8.174}$$

where

$$\overline{\Delta l \Delta p} = \frac{1}{\alpha_0^2(\tau, \Omega)\left[\frac{\partial^2 S_1(l,p)}{\partial l^2}\cdot\frac{\partial^2 S_1(l,p)}{\partial p^2} - \left(\frac{\partial^2 S_1(l,p)}{\partial l\partial p}\right)^2\right]^2}\left\{\overline{\frac{\partial n(l, p)}{\partial l}\cdot\frac{\partial n(l, p)}{\partial p}}\right.$$

$$\times \frac{\partial^2 S_1(l, p)}{\partial l^2}\frac{\partial^2 S_1(l, p)}{\partial p^2} - \overline{\left(\frac{\partial n(l, p)}{\partial p}\right)^2}\cdot\frac{\partial^2 S_1(l, p)}{\partial l\partial p}\cdot\frac{\partial^2 S_1(l, p)}{\partial l^2}$$

$$- \overline{\left(\frac{\partial n(l, p)}{\partial l}\right)^2}\cdot\frac{\partial^2 S_1(l, p)}{\partial l\partial p}\frac{\partial^2 S_1}{\partial p^2}$$

$$\left.+ \overline{\frac{\partial n(l, p)}{\partial l}\cdot\frac{\partial n(l, p)}{\partial p}}\cdot\left(\frac{\partial^2 S_1(l, p)}{\partial l\partial p}\right)^2\right\}_{l=l_0, p=p_0}. \tag{8.175}$$

Taking into account that

$$n(l, p) = n_a(l, p) + n_M(l, p) \tag{8.176}$$

and

$$\overline{n_a(l, p) n_M(l - l_0, p - p_0)} = 0, \tag{8.177}$$

we can write

$$\overline{\Delta l \Delta p} = \overline{\Delta l_a \Delta p_a} + \overline{\Delta l_M \Delta p_M}. \tag{8.178}$$

Moreover, in accordance with the main results discussed in Reference 3 we can write

$$\overline{\Delta l_a \Delta p_a} = \frac{N_0 \frac{\partial^2 S_1(l,p)}{\partial l \partial p}}{\alpha_0^2(\tau, \Omega) E_{a_1} \left[\frac{\partial^2 S_1(l,p)}{\partial l^2} \cdot \frac{\partial^2 S_1(l,p)}{\partial p^2} - \left[\frac{\partial^2 S_1(l,p)}{\partial l \partial p}\right]^2\right]} \tag{8.179}$$

if the condition determined by Eq. (8.158) is satisfied.

Substituting Eqs. (8.171)–(8.173) in Eq. (8.175), we can define the value

$$\overline{\Delta l_M \Delta p_M}.$$

As can be seen from the formulae obtained above, the general formulae for the variances of errors and the coefficient of correlation of errors during simultaneous measurement of two parameters of the signal distorted by the fluctuating multiplicative noise are very cumbersome.

To carry out analysis of these formulae for some arbitrarily measured parameters of the signal is not easy. We proceed to carry out this analysis for a particular case. Assume that the measured parameter of the signal is frequency—the parameter l. Then the function $S_1(l, p, \Omega)$ can be determined in the following form

$$S_1(l, p, \Omega) = \mathcal{G}_1(l + \Omega)\mathcal{G}_2(p). \tag{8.180}$$

Moreover

$$\left.\frac{d\mathcal{G}_2(p)}{dp}\right|_{p=0} = 0 \tag{8.181}$$

and

$$\left.\frac{d\,\mathcal{G}_1(l)}{d\,l}\right|_{l=0} = 0. \tag{8.182}$$

In doing so, the errors of measurement caused by the additive Gaussian noise are not correlated

$$\overline{\Delta l_a \Delta p_a} = 0. \tag{8.183}$$

Substituting Eqs. (8.180)–(8.182) in Eq. (8.162), we obtain that the variance of measurement errors caused by the additive Gaussian noise during simultaneous measurement of two signal parameters can be determined in the following form:

$$\sigma_{l_a}^2 = -\frac{N_0}{\alpha_0^2(\tau, \Omega) E_{a_1} \cdot \frac{d^2 \dot{\rho}(l,0)}{dl^2}} \tag{8.184}$$

and

$$\sigma_{p_a}^2 = -\frac{N_0}{\alpha_0^2(\tau, \Omega)E_{a_1} \cdot \frac{d^2\dot{\rho}(0,p)}{dp^2}}, \qquad p = 0. \tag{8.185}$$

Comparing Eqs. (8.184) and (8.185) with Eq. (8.173), one can see that the components $\sigma_{l_a}^2$ and $\sigma_{p_a}^2$ of the variance of measurement errors caused by the additive Gaussian noise during simultaneous measurement of two signal parameters are equal to the corresponding component of the variance of measurement errors during a single signal parameter measurement when the conditions in Eqs. (8.180)–(8.182) are satisfied.

However, the conditions in Eqs. (8.180)–(8.182) do not ensure a non-correlatedness of measurement errors caused by the fluctuating multiplicative noise. In order that the variances of measurement errors $\sigma_{l_M}^2$ and $\sigma_{p_M}^2$, caused by the fluctuating multiplicative noise during simultaneous measurement of two signal parameters, can be coincided with the variances of measurement errors caused by the multiplicative noise during individual measurements of two signal parameters (see Section 8.1) there is a need for the condition

$$\beta(l, p, \Omega) = kp(l + \Omega) \tag{8.186}$$

to be satisfied in addition to the conditions determined by Eqs. (8.180)–(8.182), where k is the constant factor.

Substituting Eqs. (8.180)–(8.182), and (8.186) in Eqs. (8.171), (8.173), and (8.174), respectively, it is not difficult to verify that, during simultaneous satisfaction of the conditions mentioned above, the coefficient of correlation of measurement errors caused by the fluctuating multiplicative noise is equal to zero. The variances of measurement errors during simultaneous and individual measurements of the signal parameters are the same.

Thus, when the signal frequency is one of simultaneous measured signal parameters the conditions in Eqs. (8.180)–(8.182) and (8.186) are the sufficient conditions to ensure the condition that the measurement errors, when the fluctuating multiplicative noise and the additive Gaussian noise act simultaneously, would be uncorrelated and the variances of measurement errors would be equal to the variances of measurement errors during the individual measurements of each signal parameter.

These conditions can be satisfied, for example, during simultaneous measurements of frequency and appearance time of the signal with a bell-shaped amplitude envelope and constant radio frequency carrier. Also, these conditions can be satisfied in the case of the phase-modulated signals for some approximation.[8,18]

The general formulae of the variances and the coefficient of correlation of measurement errors caused by the multiplicative noise during simultaneous measurements of two signal parameters are very difficult, as can be seen from the statements mentioned above.

However, in some cases, for example, during simultaneous measurement of the signal frequency and the signal appearance time, the mathematics can be simplified.

8.2.1 Simultaneous Measurement of Frequency and Appearance Time of the Signal

For simultaneous measurement of the signal frequency ω that is determined by

$$\omega = l - l_0$$

and the signal appearance time τ that is determined by

$$\tau = p - p_0,$$

using the generalized detector (see Eq. (8.70)), we can write

$$S_1(\tau, \omega, \Omega) \cdot e^{j\beta(\tau,\omega,\Omega)} = \frac{E_{a_1}}{2} \int\limits_{-\infty}^{\infty} \dot{S}_m(t) S_m^*(t-\tau) \cdot e^{j(\omega+\Omega)t} dt. \tag{8.187}$$

If the signal has the even function $\Psi(t)$ describing the intrapulse angle modulation law we can write

$$\beta(\tau, \omega, \Omega) = 0.5\tau(\omega + \Omega). \tag{8.188}$$

For the condition that the signal is within the limits of the observation time interval $[0, T]$, and for corresponding choice of the origin, the functions

$$S_1(\tau, \omega, \Omega)$$

and

$$\beta(\tau, \omega, \Omega)$$

and their derivatives in Eqs. (8.11)–(8.18), (8.21), (8.22), (8.25), and (8.28) in terms of the results discussed in Section 8.1 can be determined in the following form:

$$S_1(0, 0, \Omega) = E_{a_1} G_{en}(\Omega); \tag{8.189}$$

$$\beta(0, 0, \Omega) = \beta(0, 0) = 0; \tag{8.190}$$

$$\left.\frac{\partial S_1(0, 0, \Omega)}{\partial \omega}\right|_{\omega=0} = E_{a_1} G'_{en}(\Omega); \tag{8.191}$$

$$\left.\frac{\partial \beta(\tau, \omega)}{\partial \tau}\right|_{\omega=0,\tau=0} = \left.\frac{\partial \beta(\tau, \omega)}{\partial \omega}\right|_{\omega=0,\tau=0} = 0; \tag{8.192}$$

$$\left.\frac{\partial \beta(\tau, 0, \Omega)}{\partial \tau}\right|_{\tau=0} = \frac{\Omega}{2}; \tag{8.193}$$

$$\left.\frac{\partial \beta(0, \omega, \Omega)}{\partial \omega}\right|_{\omega=0} = 0; \tag{8.194}$$

$$\left.\frac{\partial^2 S_1(\tau, 0)}{\partial \tau^2}\right|_{\tau=0} = -E_{a_1}\overline{\Omega^2}; \tag{8.195}$$

$$\left.\frac{\partial^2 S_1(0,\Omega)}{\partial \omega^2}\right|_{\omega=0} = -E_{a_1}\overline{t^2}; \tag{8.196}$$

$$\left.\frac{\partial S_1(\tau,\Omega)}{\partial \tau}\right|_{\tau=0} = E_{a_1}\gamma(\Omega). \tag{8.197}$$

We determine the function

$$\frac{\partial^2 S_1(\tau,\omega,\Omega)}{\partial \tau \partial \omega}$$

that was not considered in Section 8.1.

Carrying out a serial differentiation in Eq. (8.75) with respect to the parameters ω and τ, we can write

$$\left.\frac{\partial^2 S_1(\tau,\omega,\Omega)}{\partial \tau \partial \omega}\right|_{\tau=0,\omega=0} =$$

$$\frac{E_{a_1}}{2\sqrt{\left\{Re\left\{\int\limits_{-\infty}^{\infty}\dot{S}_m(t)S_m^*(t-\tau)\cdot e^{j\Omega t}dt\right\}\right\}^2+\left\{Im\left\{\int\limits_{-\infty}^{\infty}\dot{S}_m(t)S_m^*(t-\tau)\cdot e^{j\Omega t}dt\right\}\right\}^2}}$$

$$\times\left\{\frac{\partial}{\partial\omega}Re\left\{\int\limits_{-\infty}^{\infty}\dot{S}_m(t)S_m^*(t-\tau)\cdot e^{j\Omega t}dt\right\}\frac{\partial}{\partial\tau}Re\left\{\int\limits_{-\infty}^{\infty}\dot{S}_m(t)S_m^*(t-\tau)\cdot e^{j\Omega t}dt\right\}\right.$$

$$+Re\left\{\int\limits_{-\infty}^{\infty}\dot{S}_m(t)S_m^*(t-\tau)\cdot e^{j\Omega t}dt\right\}\frac{\partial^2}{\partial\tau\,\partial\omega}Re\left\{\int\limits_{-\infty}^{\infty}\dot{S}_m(t)S_m^*(t-\tau)\cdot e^{j\Omega t}dt\right\}$$

$$+\frac{\partial}{\partial\omega}Im\left\{\int\limits_{-\infty}^{\infty}\dot{S}_m(t)S_m^*(t-\tau)\cdot e^{j\Omega t}dt\right\}\frac{\partial}{\partial\tau}Im\left\{\int\limits_{-\infty}^{\infty}\dot{S}_m(t)S_m^*(t-\tau)\cdot e^{j\Omega t}dt\right\}$$

$$+Im\left\{\int\limits_{-\infty}^{\infty}\dot{S}_m(t)S_m^*(t-\tau)\cdot e^{j\Omega t}dt\right\}\frac{\partial^2}{\partial\omega\,\partial\tau}Im\left\{\int\limits_{-\infty}^{\infty}\dot{S}_m(t)S_m^*(t-\tau)\cdot e^{j\Omega t}dt\right\}$$

$$-\left\{Re\left\{\int\limits_{-\infty}^{\infty}\dot{S}_m(t)S_m^*(t-\tau)\cdot e^{j\Omega t}dt\right\}\frac{\partial}{\partial\tau}Re\left\{\int\limits_{-\infty}^{\infty}\dot{S}_m(t)S_m^*(t-\tau)\cdot e^{j\Omega t}dt\right\}\right.$$

$$\left.+Im\left\{\int\limits_{-\infty}^{\infty}\dot{S}_m(t)S_m^*(t-\tau)\cdot e^{j\Omega t}dt\right\}\frac{\partial}{\partial\omega}Im\left\{\int\limits_{-\infty}^{\infty}\dot{S}_m(t)S_m^*(t-\tau)\cdot e^{j\Omega t}dt\right\}\right\}$$

$$\times\left\{Re\left\{\int\limits_{-\infty}^{\infty}\dot{S}_m(t)S_m^*(t-\tau)\cdot e^{j\Omega t}dt\right\}\frac{\partial}{\partial\tau}Re\left\{\int\limits_{-\infty}^{\infty}\dot{S}_m(t)S_m^*(t-\tau)\cdot e^{j\Omega t}dt\right\}\right.$$

$$\left.\left.+Im\left\{\int\limits_{-\infty}^{\infty}\dot{S}_m(t)S_m^*(t-\tau)\cdot e^{j\Omega t}dt\right\}\frac{\partial}{\partial\tau}Im\left\{\int\limits_{-\infty}^{\infty}\dot{S}_m(t)S_m^*(t-\tau)\cdot e^{j\Omega t}dt\right\}\right\}\right\}. \tag{8.198}$$

After straightforward mathematical transformations we can write

$$Re\left\{\int_{-\infty}^{\infty} \dot{S}_m(t)S_m^*(t-\tau)\cdot e^{j\Omega t}dt\right\}_{\tau=0,\omega=0} = 2E_{a_1}G_{en}(\Omega); \qquad (8.199)$$

$$Im\left\{\int_{-\infty}^{\infty} \dot{S}_m(t)S_m^*(t-\tau)\cdot e^{j\Omega t}dt\right\}_{\tau=0,\omega=0} = 0; \qquad (8.200)$$

$$\frac{\partial}{\partial\omega}Re\left\{\int_{-\infty}^{\infty} \dot{S}_m(t)S_m^*(t-\tau)\cdot e^{j\Omega t}dt\right\}_{t=0,\omega=0} = 2E_{a_1}G'_{en}(\Omega); \qquad (8.201)$$

$$\frac{\partial}{\partial\omega}Im\left\{\int_{-\infty}^{\infty} \dot{S}_m(t)S_m^*(t-\tau)\cdot e^{j\Omega t}dt\right\}_{\tau=0,\omega=0} = 0; \qquad (8.202)$$

$$\frac{\partial}{\partial\tau}Re\left\{\int_{-\infty}^{\infty} \dot{S}_m(t)S_m^*(t-\tau)\cdot e^{j\Omega t}dt\right\}_{\tau=0,\omega=0} = -2E_{a_1}\gamma(\Omega); \qquad (8.203)$$

$$\frac{\partial^2}{\partial\omega\,\partial\tau}Re\left\{\int_{-\infty}^{\infty} \dot{S}_m(t)S_m^*(t-\tau)\cdot e^{j\Omega t}dt\right\}_{\tau=0,\omega=0} = -2E_{a_1}\gamma'(\Omega). \qquad (8.204)$$

Substituting Eqs. (8.199)–(8.204) in Eq. (8.198), we can write

$$\left.\frac{\partial^2 S_1(\tau,\omega,\Omega)}{\partial\omega\,\partial\tau}\right|_{\tau=0,\omega=0} = E_{a_1}\gamma'(\Omega) + 2E_{a_1}\cdot\frac{G'_{en}(\Omega)}{G_{en}(\Omega)}\cdot\gamma(\Omega). \qquad (8.205)$$

We proceed to determine the function

$$\left.\frac{\partial^2 S_1(\tau,\omega)}{\partial\tau\,\partial\omega}\right|_{\tau=0,\omega=0}.$$

Taking into account that

$$G'_{en}(0) = 0$$

and using Eq. (8.205), we can write

$$\left.\frac{\partial^2 S_1(\tau,\omega)}{\partial\tau\partial\omega}\right|_{\tau=0,\omega=0} = E_{a_1}\gamma'(0). \qquad (8.206)$$

Note that the parameter

$$\gamma(0) = \frac{1}{2}\int_{-\infty}^{\infty} S_m^2(t)\Psi'(t)t\,dt \qquad (8.207)$$

characterizes the product between the energy spectrum bandwidth of the signal and the duration of the signal, and is much greater than unity for the case of the wide-band signal.[10,11,19]

In terms of Eqs. (8.187)–(8.207) after straightforward mathematical transformations, we obtain that the variances of measurement errors of the signal frequency and signal appearance time during simultaneous measurement of these two signal parameters can be determined in the following form:

- The variances of measurement errors caused by the additive Gaussian noise:

$$\sigma^2_{\omega_a} = \frac{N_0 \overline{\Omega^2}}{\alpha_0^2(\tau,\Omega) E_{a_1} \left\{ \overline{\Omega^2}\, \overline{t^2} - [\gamma'(0)]^2 \right\}}; \tag{8.208}$$

$$\sigma^2_{\tau_a} = \frac{N_0 \overline{t^2}}{\alpha_0^2(\tau,\Omega) E_{a_1} \left\{ \overline{\Omega^2}\, \overline{t^2} - [\gamma'(0)]^2 \right\}}. \tag{8.209}$$

- The variances of measurement errors caused by the fluctuating multiplicative noise:

$$\begin{aligned}
\sigma^2_{\omega_M} &= \frac{1}{2\alpha_0^2(\tau,\Omega)\left\{ \overline{\Omega^2}\, \overline{t^2} - [\gamma'(0)]^2 \right\}^2} \\
&\times \left\{ \frac{1}{2\pi} \int_{-\infty}^{\infty} [G_V(\Omega) + G_D(\Omega)] \{ (\overline{\Omega^2})^2 [G'_{en}(\Omega)]^2 \right. \\
&+ 4\overline{\Omega^2}\gamma'(0)[G_{en}(\Omega)\gamma'(\Omega) + G'_{en}(\Omega)\gamma(\Omega)]\} \, d\Omega \\
&+ [\gamma'(0)]^2 \left\{ \frac{1}{2\pi} \int_{-\infty}^{\infty} G_V(\Omega)\left[\gamma^2(\Omega) + 0.25 G^2_{en}(\Omega)\Omega^2\right] d\Omega \right. \\
&+ \left. \left. \frac{1}{2\pi} \int_{-\infty}^{\infty} G_D(\Omega)\left[\gamma^2(\Omega) - 0.25 G^2_{en}(\Omega)\Omega^2\right] d\Omega \right\} \right\};
\end{aligned} \tag{8.210}$$

$$\begin{aligned}
\sigma^2_{\tau_M} &= \frac{1}{2\alpha_0^2(\tau,\Omega)\left\{ \overline{\Omega^2}\, \overline{t^2} - [\gamma'(0)]^2 \right\}^2} \\
&\times \left\{ (\overline{t^2})^2 \left\{ \frac{1}{2\pi} \int_{-\infty}^{\infty} G_V(\Omega)\left[\gamma^2(\Omega) + 0.25 G^2_{en}(\Omega)\Omega^2\right] d\Omega \right. \right. \\
&+ \left. \frac{1}{2\pi} \int_{-\infty}^{\infty} G_D(\Omega)\left[\gamma^2(\Omega) - 0.25 G^2_{en}(\Omega)\Omega^2\right] d\Omega \right\} \\
&+ \frac{1}{2\pi} \int_{-\infty}^{\infty} [G_V(\Omega) + G_D(\Omega)] \{ [\gamma'(0)]^2 [G'_{en}(\Omega)]^2 \\
&+ \left. 4\overline{t^2}\gamma'(0)[G_{en}(\Omega)\gamma'(\Omega) + G'_{en}(\Omega)\gamma(\Omega)]\} \, d\Omega \right\}.
\end{aligned} \tag{8.211}$$

Equations (8.210) and (8.211) can be essentially simplified during measurement of the signal frequency and signal appearance time when the angle modulation of the signal phase is absent, i.e.,

$$\Psi(t) = 0.$$

Really, for this case we can write

$$\gamma(\Omega) = \gamma(0) = 0 \tag{8.212}$$

and Eqs. (8.208)–(8.211) are transformed to Eqs. (8.59) and (8.81), respectively.

Thus, the variances of measurement errors of the signal frequency and signal appearance time are the same both during individual measurements of the signal parameters and during simultaneous measurements of the signal parameters for the signals without any angle modulation of the signal phase.

Determination of the coefficient of correlation of measurement errors for the given case using Eq. (8.174) allows us to prove that the measurement errors are not correlated.

The given example confirms the general conclusion (formulated above) regarding the characteristics of simultaneous and individual measurements of the signal parameters when the correlation between the measurement errors is absent.

Formulae in Eqs. (8.210) and (8.211), defining the variances of measurement errors during simultaneous measurements of the signal frequency and signal appearance time, can also be simplified for the cases of the slow and rapid fluctuating multiplicative noise.

During the slow fluctuating multiplicative noise the functions $G_V(\Omega)$ and $G_D(\Omega)$ are varied more rapidly in comparison with other integrand functions in Eqs. (8.210) and (8.211). Then the functions $G_{en}(\Omega)$ and $\gamma(\Omega)$, and also their derivatives in the integrand, can be used for the Taylor series expansion with respect to the variable Ω. Limiting in the given Taylor series expansion by the terms with the order $\overline{\Omega^2}$ and taking into account the following formulae

$$\frac{1}{2\pi}\int_{-\infty}^{\infty} G_V(\Omega)\Omega^2\, d\Omega = -R_V''(0); \tag{8.213}$$

$$\frac{1}{2\pi}\int_{-\infty}^{\infty} G_D(\Omega)\Omega^2\, d\Omega = -D_V''(0); \tag{8.214}$$

$$\frac{1}{2\pi}\int_{-\infty}^{\infty} G_V(\Omega)\, d\Omega = R_V(0); \tag{8.215}$$

$$\frac{1}{2\pi}\int_{-\infty}^{\infty} G_D(\Omega)\, d\Omega = D_V(0), \tag{8.216}$$

we can write

$$\sigma^2_{\omega_M} = \frac{1}{2\alpha_0^2(\tau,\Omega)\{\overline{\Omega^2}\,\overline{t^2} - [\gamma'(0)]^2\}^2}$$
$$\times \left\{-R''_{V_\Sigma}(0)\left[(\overline{\Omega^2}\,\overline{t^2})^2 + 2\overline{\Omega^2}\gamma'(0)\gamma'''(0) + 6\overline{\Omega^2}\,\overline{t^2}[\gamma'(0)]^2\right.\right.$$
$$\left.+[\gamma'(0)]^4 + 4R_{V_\Sigma}(0)\overline{\Omega^2}[\gamma'(0)]^2 - 0.25R''_{V_d}(0)[\gamma'(0)]^2\right\} \quad (8.217)$$

and

$$\sigma^2_{\tau_M} = \frac{1}{2\alpha_0^2(\tau,\Omega)\{\overline{\Omega^2}\,\overline{t^2} - [\gamma'(0)]^2\}^2}$$
$$\times \left\{-2R''_{V_\Sigma}(0)[4(\overline{t^2})^2[\gamma'(0)]^2 + \overline{t^2}\gamma'(0)\gamma'''(0)]\right.$$
$$\left.-0.25R''_{V_d}(0)(\overline{t^2})^2 + 4R_{V_\Sigma}(0)\overline{t^2}[\gamma'(0)]^2\right\}, \quad (8.218)$$

where

$$R_{V_\Sigma}(0) = R_V(0) + D_V(0) \quad (8.219)$$

and

$$R_{V_d}(0) = R_V(0) - D_V(0). \quad (8.220)$$

For the case of the rapid multiplicative fluctuating noise the functions $G_V(\Omega)$ and $G_D(\Omega)$ in Eqs. (8.210) and (8.211) are varied slowly in comparison with other integrand functions.

With the purpose of obtaining the approximate formulae, we can use the power series expansion of the functions $G_V(\Omega)$ and $G_D(\Omega)$ with respect to the parameter Ω, and carry out transformations that are analogous to that discussed in Section 8.1; however, these formulae are very cumbersome.

8.2.2 Simultaneous Measurement of Frequency and Appearance Time of the Frequency-Modulated Signal

In the case of simultaneous measurement of the signal frequency ω and signal appearance time τ of the frequency-modulated signal with the complex amplitude envelope

$$\dot{S}(t,\tau,\omega) = e^{-\frac{\pi(t-\tau)^2}{T^2} - j\cdot\frac{\Delta\Omega_d}{2T}(t-\tau)^2 + j\omega t} \quad (8.221)$$

the majority of the functions in Eqs. (8.208)–(8.211) were defined in Section 8.1.

In addition, there is a need to define the functions $\gamma'(\Omega)$ and $\gamma'(0)$. In terms of Eqs. (8.138)–(8.142) we can write

$$\gamma'(\Omega) = \frac{d}{d\Omega}\left[\frac{Q_y\Omega}{2}\cdot e^{-\frac{\Omega^2T^2}{8\pi}}\right]$$
$$= \frac{Q_y}{2}\left(1 - \frac{\Omega^2T^2}{4\pi}\right)\cdot e^{-\frac{\Omega^2T^2}{8\pi}}; \quad (8.222)$$

$$\gamma'(0) = \frac{Q_y}{2}. \quad (8.223)$$

Substituting Eqs. (8.104)–(8.108), (8.138)–(8.142), (8.222) and (8.223) in Eq. (8.208), we obtain that the variances of measurement errors caused by the additive Gaussian noise can be determined in the following form:

$$\sigma^2_{\omega_a} = \frac{4\pi N_0(1+Q_y^2)}{\alpha_0^2(\tau,\Omega)E_{a_1}T^2} = \frac{\sigma^2_{\omega_0}}{\alpha_0^2(\tau,\Omega)}; \tag{8.224}$$

$$\sigma^2_{\tau_a} = \frac{N_0 T^2}{\pi\alpha_0^2(\tau,\Omega)E_{a_1}} = \frac{\sigma^2_{\tau_0}}{\alpha_0^2(\tau,\Omega)}. \tag{8.225}$$

Comparing Eqs. (8.224) and (8.225) with the formulae defining the variances of measurement errors caused by the additive Gaussian noise during the individual measurements of the signal parameters (see Eqs. (8.110)–(8.113), and (8.143)), we can see that during simultaneous measurement of the signal frequency and signal appearance time of the frequency-modulated signal the variances of measurement errors caused by the additive Gaussian noise are increased $(1+Q_y^2)$ times in comparison to the case of individual measurements of the signal parameters.

In the process, the variance of measurement errors of the signal appearance time is the same as for the case of measurement of the signal appearance time of the non-modulated signal with a bell-shaped complex amplitude envelope and the duration T (see Eqs. (8.110)–(8.113)). This phenomenon is caused by a linear function between shifts in frequency and in delay that is characteristic of the autocorrelation function of the signals with the linear frequency modulation law.

The variances of measurement errors of the signal frequency and signal appearance time caused by the fluctuating multiplicative noise in the case of the considered example can be determined in the following form:

$$\begin{aligned}\sigma^2_{\omega_M} &= \frac{E_{a_1}(1+Q_y^2)}{N_0}\cdot\frac{\sigma^2_{\omega_0}}{\alpha_0^2(\tau,\Omega)}\\ &\times\left\{\frac{T^2}{16\pi^2}\int\limits_{-\infty}^{\infty} G_\Sigma(\Omega)\Omega^2\cdot e^{-\frac{\Omega^2T^2}{4\pi}}\,d\Omega\right.\\ &+\frac{2Q_y^2}{1+Q_y^2}\cdot\frac{1}{2\pi}\int\limits_{-\infty}^{\infty} G_\Sigma(\Omega)\cdot e^{-\frac{\Omega^2T^2}{4\pi}}\,d\Omega\\ &+\frac{Q_y^2}{1+Q_Y^2}\cdot\frac{T^2}{16\pi^2}\int\limits_{-\infty}^{\infty} G_V(\Omega)\Omega^2\cdot e^{-\frac{\Omega^2T^2}{4\pi}}\,d\Omega\\ &\left.+\frac{O_y^2(Q_y^2-1)}{(1+Q_y^2)^2}\cdot\frac{T^2}{16\pi^2}\int\limits_{-\infty}^{\infty} G_D(\Omega)\Omega^2\cdot e^{-\frac{\Omega^2T^2}{4\pi}}\,d\Omega\right\};\end{aligned} \tag{8.226}$$

$$\sigma_{\tau_M}^2 = \frac{E_{a_1}(1+Q_y^2)}{N_0} \cdot \frac{\sigma_{\tau_0}^2}{2\pi\alpha_0^2(\tau,\Omega)}$$
$$\times \left[\frac{Q_y^2}{1+Q_y^2} \cdot \frac{T^2}{8\pi^2} \int\limits_{-\infty}^{\infty} G_\Sigma(\Omega)\Omega^2 \cdot e^{-\frac{\Omega^2 T^2}{4\pi}} d\Omega \right.$$
$$+ \frac{2Q_y^2}{1+Q_y^2} \int\limits_{-\infty}^{\infty} G_\Sigma(\Omega) \cdot e^{-\frac{\Omega^2 T^2}{4\pi}} d\Omega$$
$$+ \frac{T^2}{8\pi^2} \int\limits_{-\infty}^{\infty} G_V(\Omega)\Omega^2 \cdot e^{-\frac{\Omega^2 T^2}{4\pi}} d\Omega$$
$$\left. + \frac{Q_y^2 - 1}{1+Q_y^2} \cdot \frac{T^2}{8\pi^2} \int\limits_{-\infty}^{\infty} G_D(\Omega)\Omega^2 \cdot e^{-\frac{\Omega^2 T^2}{4\pi}} d\Omega \right], \tag{8.227}$$

where

$$G_\Sigma(\Omega) = G_V(\Omega) + G_D(\Omega) \tag{8.228}$$

and the functions $G_V(\Omega)$ and $G_D(\Omega)$ are defined in analogous way as in Section 8.1 (see Eqs. (8.25), (8.27), and (8.28)).

Reference to Eqs. (8.226) and (8.227) shows that as

$$Q_y^2 \to 0$$

the variance $\sigma_{\omega_M}^2$ of measurement errors of the signal frequency and the variance $\sigma_{\tau_M}^2$ of measurement errors of the signal appearance time, caused by the fluctuating multiplicative noise during simultaneous measurements of the signal parameters, coincide with the variances of measurement errors of the signal frequency and signal appearance time for individual measurements of the signal parameters, as would be expected.

For the condition

$$Q_y^2 \gg 1, \tag{8.229}$$

that is characteristic of the use in practice of the frequency modulated signals, for the case of simultaneous measurements of the signal frequency and signal appearance time we can write

$$\sigma_\omega^2 = \sigma_{\omega_a}^2 + \sigma_{\omega_M}^2 = \sigma_{\omega_0}^2 \eta_{\omega M} \tag{8.230}$$

and

$$\sigma_\tau^2 = \sigma_{\tau_a}^2 + \sigma_{\tau_M}^2 = \sigma_{\tau_0}^2 \eta_{\tau M}, \tag{8.231}$$

where the coefficients η_{ω_M} and η_{τ_M}, indicating a degree of increasing the variances of measurement errors owing to the stimulus of the fluctuating multiplicative noise, can be determined in the following form:

$$\eta_{\tau_M} = \eta_{\omega_M}$$
$$= \frac{1}{\alpha_0^2(\tau, \Omega)}\left\{1 + \frac{E_{a_1} Q_y^2}{N_0}\left[\frac{T^2}{8\pi^2}\int_{-\infty}^{\infty} G_\Sigma(\Omega)\Omega^2 \cdot e^{-\frac{\Omega^2 T^2}{4\pi}}\, d\Omega\right.\right.$$
$$\left.\left. + \frac{1}{\pi}\int_{-\infty}^{\infty} G_\Sigma(\Omega)\cdot e^{-\frac{\Omega^2 T^2}{4\pi}}\, d\Omega\right]\right\}. \quad (8.232)$$

Reference to Eq. (8.231) shows that during simultaneous measurements of the signal frequency and signal appearance time of the wide-band frequency-modulated signal the stimulus of the fluctuating multiplicative noise is much more than during individual measurements of these signal parameters. The variances $\sigma^2_{\omega_M}$ and $\sigma^2_{\tau_M}$ of measurement errors are increased by more than $2Q_y^2$ times.

We define the coefficient of correlation of measurement errors. For this purpose we define the value $\overline{\Delta l \Delta p}$ in accordance with Eq. (8.174).

For the frequency-modulated signal considered we can write

$$\overline{\Delta l \Delta p} = \overline{\Delta \tau \Delta \omega} = \frac{2Q_y N_0}{\alpha_0^2(\tau, \Omega) E_{a_1}}\left\{1 + \frac{E_{a_1}(1 + Q_y^2)}{N_0}\right.$$
$$\times\left[\frac{1}{2\pi}\int_{-\infty}^{\infty} G_\Sigma(\Omega)\cdot e^{-\frac{\Omega^2 T^2}{4\pi}}\, d\Omega\right.$$
$$+ \frac{T^2}{16\pi^2}\int_{-\infty}^{\infty} G_V(\Omega)\Omega^2 \cdot e^{-\frac{\Omega^2 T^2}{4\pi}}\, d\Omega$$
$$+ \frac{Q_y^2 - 1}{1 + Q_y^2}\cdot\frac{T^2}{16\pi^2}\int_{-\infty}^{\infty} G_D(\Omega)\,\Omega^2 \cdot e^{-\frac{\Omega^2 T^2}{4\pi}}\, d\Omega$$
$$+ \frac{Q_y^2}{1 + Q_y^2}\cdot\frac{1}{2\pi}\int_{-\infty}^{\infty} G_\Sigma(\Omega)\cdot e^{-\frac{\Omega^2 T^2}{4\pi}}\, d\Omega$$
$$\left.\left. + \frac{T^2}{16\pi^2}\int_{-\infty}^{\infty} G_\Sigma(\Omega)\Omega^2 \cdot e^{-\frac{\Omega^2 T^2}{4\pi}}\, d\Omega\right]\right\}. \quad (8.233)$$

For the condition determined by Eq. (8.229) we can write that

$$\overline{\Delta l \Delta p} = \frac{2Q_y N_0}{\alpha_0^2(\tau, \Omega) E_{a_1}}\cdot \eta_{\omega_M} \quad (8.234)$$

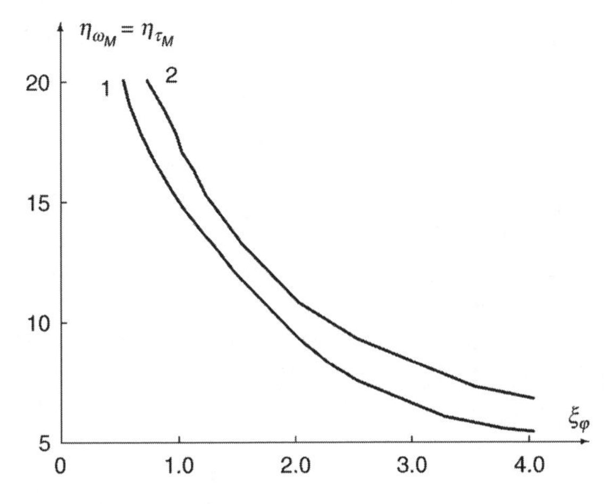

FIGURE 8.4
The coefficient η_{ω_M} as a function of the parameter ξ_φ: for 1, $\sigma_\varphi^2 = 0.10$; $qQ_y = 500$; for 2, $\sigma_\varphi^2 = 0.31$; $qQ_y = 50$.

and the coefficient of correlation of measurement errors, as follows from Eqs. (8.174), (8.232), and (8.233), tends to approach unity with an increase in the cutting coefficient Q_y of the generalized detector.

Thus, the example considered of simultaneous measurements of frequency and appearance time of the frequency-modulated signal is the limiting case in the sense that the coefficient of correlation of measurement errors is very close to unity. The other opposite case—the coefficient of correlation of measurement errors is equal to zero—was discussed during analysis of the signals without any angle modulation of the signal phase.

We assume that the condition determined by Eq. (8.229) is satisfied and the fluctuating multiplicative noise generates only weak distortions in phase of the signal with a bell-shaped correlation function (see Eqs. (8.123) and (8.124)). Then for the case of the frequency-modulated signal with a bell-shaped complex amplitude envelope we can write (see Eq. (8.231))

$$\eta_{\tau_M} = \eta_{\omega_M} = \frac{1 + \frac{E_{a_1} Q_y^2 \sigma_\varphi^4}{N_0} \cdot \frac{2+5\xi_\varphi^2}{\sqrt{(1+2\xi_\varphi^2)^3}}}{1 - \sigma_\varphi^2 + 0.5\sigma_\varphi^4}, \tag{8.235}$$

where σ_φ^2 is the variance of distortions in phase of the signal, and

$$\xi_\varphi = \frac{\Delta\Omega_\varphi T}{2\pi}. \tag{8.236}$$

The function $\eta_{\omega_M}(\xi_\varphi)$ is shown in Fig. 8.4 for the following conditions

$$\sigma_\varphi^2 = const \tag{8.237}$$

and

$$\frac{E_{a_1} Q_y^2}{N_0} = const. \tag{8.238}$$

8.3 A Single Parameter Measurement under a Combined Stimulus of High Multiplicative and Additive Gaussian Noise

The problem of estimating the stimulus of the fluctuating multiplicative noise on measurement precision of non-energy signal parameters under the weak fluctuating multiplicative and additive Gaussian noise was just considered in the previous sections.

During the high fluctuating multiplicative and additive Gaussian noise the procedure used previously cannot be applied because of drastic changes in character of possible errors. Really, the probability of appearance of high shoots of complex amplitude envelope of the signal at the output of the generalized detector at the values of the measured signal parameter l, which differ essentially from the true value of the signal parameter, increases as the fluctuating multiplicative and additive Gaussian noise become increasingly high. The shoots of complex envelope amplitude of the signal at the output of the generalized detector (mentioned above) can be caused both by the stimulus of the additive Gaussian noise and the signal noise component that arise from the stimulus of the fluctuating multiplicative noise.

Thus, during the high fluctuating multiplicative and additive Gaussian noise the problem of distinguishing the true signal parameters from false shoots of complex amplitude envelope of the signal at the output of the generalized detector is the primary problem. For the given formalization, this problem is very close to the problem of estimating the stimulus of the fluctuating multiplicative and additive Gaussian noise on precision of measurement of the discrete signal parameter that can take any value from the number of channels N of the generalized detector with the same probability.[20]

For these conditions we consider the generalized detector for the signals with the random initial phase φ_0 distributed uniformly within the limits of the interval $[0, 2\pi]$ and the unknown amplitude in the presence of additive Gaussian noise (see Section 7.1).

The problem of measurement of the discrete signal parameter is simpler and, for this reason, allows us to make a decision under arbitrary level of energy characteristics of the fluctuating multiplicative and additive Gaussian noise if the signals differ from each other by the measured parameter on the value δl of signal parameter discreteness are orthogonal:[21]

$$\left| \int_{-\infty}^{\infty} \dot{S}(l) S^*(l + m\delta l)\, dl \right|_{m \neq 0} = 2E_{a_1} |\dot{\rho}(m\delta l)|_{m \neq 0} = 0, \tag{8.239}$$

where $\dot{\rho}(l)$ is the autocorrelation function of the received signal with respect to the signal parameter l.

When the fluctuating multiplicative noise is absent, the interval of discreteness of the signal parameter δl, for which Eq. (8.220) is approximately true, is equal to the bandwidth of the autocorrelation function of the signal with respect to the signal parameter l in practice. This phenomenon is caused by the fact that the real signals, as a rule, do not ensure the condition in Eq. (8.220) rigorously if the value δl is less than the value L, where the value L is the full interval of the signal parameter l.

For example, if the signal parameter l is the time, then the value L is the full duration of the signal; if the signal parameter l is the frequency, then the value L is the bandwidth of the energy spectrum of the signal.

Accordingly, in practice the value δl is defined on the basis of the condition of approximate orthogonality:

$$|\dot{\rho}(\delta l)| \leq \epsilon, \tag{8.240}$$

where $\epsilon < 1$ is some origin. By far the most frequently used type of the interval δl is the value equal to the equivalent bandwidth of the function

$$|\dot{\rho}(l)|^2$$

with respect to the signal parameter l:

$$\delta l = \int_{-\infty}^{\infty} |\dot{\rho}(l)|^2\, dl. \tag{8.241}$$

When there is fluctuating multiplicative noise jointly with the additive Gaussian noise, the correlation interval of fluctuations of the signal at the output of the generalized detector is thought of as the interval of orthogonality δl that is equal to the bandwidth of the autocorrelation function of the undistorted signal during the rapid fluctuating multiplicative noise (see Sections 6.3 and 6.4)

$$\xi = \frac{1}{2\pi} \cdot \Delta\Omega_M T \geq 3. \tag{8.242}$$

Henceforth, we will consider only this case.

Thus, the model of the generalized detector considered in this section coincides with the model of the N-channel generalized detector discussed in

Sections 7.1 and 7.2, and the signals at the output of adjacent channels of the generalized detector, both when the fluctuating multiplicative noise is present and when the fluctuating multiplicative noise is absent, are statistically independent. The value of the measured signal parameter corresponding to that channel of the generalized detector, in which the signal is greatest at the instant of measurement, is taken as the value of the measured signal parameter for the model considered of the generalized detector.

Assume that it is known *a priori* there is a "yes" signal in the stochastic process at the input of the generalized detector and the true value of the signal parameter corresponds to the "k"-th channel of the generalized detector. The probability distribution density of the signal at the output of the "k"-th channel of the generalized detector is denoted by

$$f_k^{H_1}\left[z_{gM}^{out}(t,l)\right].$$

The probability distribution density of the signal at the output of the "m"-th channel of the generalized detector is denoted by

$$f_m^{H_1}\left[z_{gM}^{out}(t,l)\right], \quad m \neq k.$$

Then the conditional probability of that the signal at the output of the "k"-th channel of the generalized detector is greatest among the signals at the outputs of other channels of the generalized detector and is determined in the following form:

$$\begin{aligned} P_{0_k}\left[z_{gM}^{out}(t,l)\right] &= \prod_{i=1,i\neq k}\left\{1-\int_{K_g}^{\infty} f_i^{H_1}\left[z_{gM}^{out}(t,l)\right] d\left[z_{gM}^{out}(t,l)\right]\right\} \\ &= \prod_{i=1,i\neq k}\left\{1-P_{D_{ik}}\left[z_{gM}^{out}(t,l)\right]\right\}. \end{aligned} \tag{8.243}$$

The probability of measurement of the true signal parameter l for the condition a "yes" signal in the stochastic process at the input of the generalized detector is determined in terms of Eq. (8.242) in the following form:

$$P_{0_k}\left[z_{gM}^{out}(t,l)\right] = \int_0^{\infty} f_k^{H_1}\left[z_k^{out}(t,l)\right] \prod_{i=1,i\neq k}\left\{1-P_{D_{ik}}\left[z_{gM}^{out}\right]\right\} d\left[z_{gM}^{out}(t,l)\right]. \tag{8.244}$$

The probability of the measured parameter of the signal at the output of the "m"-th channel of the generalized detector, i.e., the probability of false reading, is thought of as the true signal parameter and is determined in the following form:

$$P_{0_m}\left[z_{gM}^{out}(t,l)\right] = \int_0^{\infty} f_m^{H_1}\left[z_{gM}^{out}(t,l)\right] \prod_{i=1,i\neq k}\left\{1-P_{D_{ik}}\left[z_{gM}^{out}(t,l)\right]\right\} d\left[z_{gM}^{out}(t,l)\right]. \tag{8.245}$$

The probability of measurement of the true signal parameter in Eq. (8.244) and the probability of false reading in Eq. (8.245) are true for the condition that a decision of a "yes" signal in the stochastic process at the input of the generalized detector has been made, i.e., the signal has been detected.

In other words, the probabilities

$$P_{0_k}\left[z_{g_M}^{out}(t,l)\right]$$

and

$$P_{0_m}\left[z_{g_M}^{out}(t,l)\right]$$

are the conditional probabilities. The totality of values of the probability of measurement of the true signal parameter

$$P_{0_k}\left[z_{g_M}^{out}(t,l)\right]$$

and the probability of false reading

$$P_{0_m}\left[z_{g_M}^{out}(t,l)\right]$$

allows us to define the probability distribution density of errors of measurement of the discrete signal parameter

$$\Delta l = l^* - l_0,$$

i.e.,

$$f(\Delta l) = P_{0_k}\left[z_{g_M}^{out}(t,l)\right]\delta(0) + \sum_{m=1, m\neq k}^{N} P_{0_m}\left[z_{g_M}^{out}(t,l)\right]\delta(k-m). \tag{8.246}$$

Biased estimation of measurement of the signal parameter will be absent when the following conditions are satisfied:

- The probability distribution density of errors $f(\Delta l)$ of measurement of the discrete signal parameter is within the limits of the interval

$$1 \leq \frac{|\Delta l|}{\delta l} \leq N. \tag{8.247}$$

- The probability distribution density of errors $f(\Delta l)$ of measurement of the discrete signal parameter is the symmetric function with respect to the signal parameter

$$l = k\delta l,$$

i.e.,

$$f(\Delta l) = f(-\Delta l). \tag{8.248}$$

When the biased estimation of measurement of the signal parameter is absent the variance of measurement errors is determined in terms of Eq. (8.227) in the following form:

$$\sigma_l^2 = \delta l^2 \sum_{m=1}^{N} (m-k)^2 P_{0_m}\left[z_{gM}^{out}(t,l)\right]. \tag{8.249}$$

As was mentioned above, the model of the generalized detector considered in this section coincides with the model of the N-channel generalized detector discussed in Sections 7.1 and 7.2 during analysis of the stimulus of the fluctuating multiplicative noise on the detection performances.

In terms of the main results discussed in Section 7.1, the probability distribution densities of the signals at the outputs of the "k"-th and "m"-th channels of the generalized detector, under the stimulus of the high fluctuating multiplicative and additive Gaussian noise, are determined in the following forms:

$$f_k^{H_1}\left[z_{gM}^{out}(t)\right] = \frac{k_1}{\sqrt{D_{\Sigma_k}}} \cdot \exp\left\{-\frac{\left[kz_{gM}^{out}(t)\right]^2 + \alpha_0^2 E_{a_1}}{D_{\Sigma_k}}\right\}; \tag{8.250}$$

$$f_m^{H_1}\left[z_{gM}^{out}(t)\right] = \frac{k_1}{\sqrt{D_{\Sigma_k}}} \exp\left\{-\frac{\left[kz_{gM}^{out}(t)\right]^2}{D_{\Sigma_k}}\right\}, \tag{8.251}$$

where the coefficients k_1 and k were defined in Section 7.1; D_{Σ_k} and D_{Σ_m} are the variances defined in Section 7.1.

Substituting Eqs. (8.250) and (8.251) in Eqs. (8.245), (8.246), and (8.249), we can define all values characterizing a precision of measurement of the discrete signal parameter during the high fluctuating multiplicative and additive Gaussian noise. However, when the number of channels N of the generalized detector is high there are great mathematical difficulties when using Eqs. (8.250) and (8.251) at once.

For this reason, consider the simplest case of the two-channel generalized detector when the measured parameter has two discrete values before an analysis of the general case when the number of channels N of the generalized detector is high. The given case is of interest when, for example, the fluctuating multiplicative noise acts on the communication system with "active pause."

At $N = 2$, taking into consideration Eq. (8.244), we can write

$$P_{0_1}\left[z_{gM}^{out}(t)\right] = \int_0^\infty f_1^{H_1}\left[z_{gM}^{out}(t)\right]\left\{1 - P_{D_{21}}\left[z_{gM}^{out}(t)\right]\right\} d\left[z_{gM}^{out}(t)\right], \tag{8.252}$$

where the true value of the signal parameter corresponds to the first channel of the generalized detector.

When the number of channels N of the generalized detector is high we can make approximation of the function

$$P_{0_{k,z}} = \prod_{i=1, i\neq k} \left\{1 - P_{D_{ik}}\left[z_{gM}^{out}(t)\right]\right\} \tag{8.253}$$

by the step-function

$$P_{0_{k,z}} \simeq \begin{cases} 0, & z_{gM}^{out}(t) < K_g; \\ 1, & z_{gM}^{out}(t) > K_g. \end{cases} \tag{8.254}$$

This approximation is used in Reference 9 for the case of measurement of the signal parameters in the presence of additive Gaussian noise when parameters of the probability distribution densities of the signals at the outputs of individual channels of the generalized detector are the same for all $m \neq k$.

For the given case, the approximation in Eq. (8.253) allows us to ensure the high precision of definition of the probability

$$P_{0_k}\left[z_{gM}^{out}(t)\right]$$

at $N \simeq 16$. For the considered case of a combined stimulus of the fluctuating multiplicative and additive Gaussian noise, the parameters of the probability distribution density

$$f_m^{H_1}\left[z_{gM}^{out}(t)\right]$$

are not, as a rule, the same as a consequence of difference between the energies of noise components of the signals distorted by the fluctuating multiplicative noise at various channels of the generalized detector.

However, as is shown in Reference 22, during the high fluctuating multiplicative and additive Gaussian noise and the high number of channels N of the generalized detector, the changes in behavior of the function $P_{0_{k,z}}$ caused by the parameters of the probability distribution densities

$$f_m^{H_1}\left[z_{gM}^{out}(t)\right]$$

are not the same and are small; consequently, the approximation in Eq. (8.254) can be applied for estimation of the probability P_{0_k}.

Substituting Eq. (8.254) in Eq. (8.243), we can write

$$P_{0_k} \simeq \int_{K_g}^{\infty} f_k^{H_1}\left[z_{gM}^{out}(t)\right] d\left[z_{gM}^{out}(t)\right] = P_{D_{M_k}}(K_g), \tag{8.255}$$

where $P_{D_{M_k}}(K_g)$ is the probability of detection of the signal at the "k"-th channel of the generalized detector under the threshold K_g.

Thus, the probability of measurement of the true signal parameter during the high fluctuating multiplicative and additive Gaussian noise is approximately equal to the probability of detection of the signal at the channel of the generalized detector under the threshold K_g, where there is the undistorted component of the signal.

8.4 Conclusions

We summarize briefly the main results discussed in this chapter:

- In the presence of the weak fluctuating multiplicative and additive Gaussian noise the estimation of a single signal parameter is unbiased and the variance of measurement errors of a single signal parameter is the sum of two terms. The first term $\sigma_{l_a}^2$ is the variance of measurement errors of the signal parameter when there is additive Gaussian noise only with an accuracy of the constant factor $\alpha_0^2(\tau, \Omega)$; the multiplicative noise is absent. The second term $\sigma_{l_M}^2$ takes into account the stimulus of the fluctuating multiplicative noise.
- The variance of measurement errors of the signal parameter caused by the fluctuating multiplicative noise depends essentially on spectral performances of the noise modulation function $\dot{M}(t)$ of the multiplicative noise. The greatest errors in the signal parameter measurement caused by the weak fluctuating multiplicative noise occur for the case when the energy spectrum bandwidth of fluctuations of the noise modulation function $\dot{M}(t)$ of the fluctuating multiplicative noise is commensurate with the energy spectrum bandwidth of complex amplitude envelope of the signal at the output of the generalized detector.
- An impact of the weak fluctuating multiplicative noise on precision of measurement of the signal appearance time is stronger compared to the impact of the weak fluctuating multiplicative noise on precision of measurement of the signal frequency.
- Low distortions in phase of the received signal caused by the weak fluctuating multiplicative noise act on precision of measurement of the signal appearance time more than on precision of measurement of the signal frequency. The maximum of the variances $\sigma_{\omega M}^2$ and $\sigma_{\tau M}^2$ is reached for various values of the correlation interval of distortions in phase of the received signal.
- The variances of measurement errors of the signal frequency and signal appearance time are the same both during individual measurements of the signal parameters and during simultaneous measurements of the signal parameters for the signals without any angle modulation of the signal phase. Determination of the coefficient of correlation of measurement errors for the given case allows us to prove that the measurement errors are not correlated.
- The probability of measurement of the true signal parameter during the high fluctuating multiplicative and additive Gaussian noise is approximately equal to the probability of detection of the signal at the channel of the generalized detector under the given threshold, where there is the undistorted component of the signal.

References

1. Blum, R., Kozick, R., and Salder, B., An adaptive spatial diversity receiver for non-Gaussian interference and noise, *IEEE Trans.*, 1999, Vol. SP-47, No. 8, pp. 2100–2111.
2. Tuzlukov, V., *Signal Detection Theory*, Springer-Verlag , New York, 2000.
3. Tuzlukov, V., *Signal Processing in Noise: A New Methodology*, IEC, Minsk, 1998.
4. Kulikov, E., *Estimation of Signal Parameters in Noise*, Soviet Radio, Moscow, 1969 (in Russian).
5. Papoulis, A., *Signal Analysis*, McGraw-Hill, New York, 1977.
6. Oppenheim, A. and Willsky, A., *Signals and Systems*, Prentice-Hall, Englewood Cliffs, NJ, 1983.
7. Bacut, P. et al. *Problems in Statistical Radar Theory, Parts 1 and 2*, Soviet Radio, Moscow, 1963–1964 (in Russian).
8. Doktorov, A., Spectrum and mutual correlation function of modulation of signals with phase modulation, *Radio Eng.*, 1966, No. 7, pp. 69–74 (in Russian).
9. Falkovich, S., *Signal Processing of Radar Signals in Fluctuating Noise*, Soviet Radio, Moscow, 1961 (in Russian).
10. Helstrom, C., *Statistical Theory of Signal Detection*, 2nd ed., Pergamon Press, Oxford, U.K., 1968.
11. Helstrom, C., *Elements of Signal Detection and Estimation*, Prentice-Hall, Englewood Cliffs, NJ, 1995.
12. Stoica, P. and Moses, R., *Introduction to Spectral Analysis*, Prentice-Hall, Englewood Cliffs, NJ, 1997.
13. Haykin, S. and Stenhardt, A., *Adaptive Radar Detection and Estimation*, Wiley, New York, 1992.
14. Poor, H., *Introduction to Signal Detection and Estimation*, Springer-Verlag, New York, 1988.
15. Rappaport, T., *Wireless Communications: Principles and Practice*, Prentice-Hall, Englewood Cliffs, NJ, 1996.
16. Parsons, J., *The Mobile Radio Propagation Channels*, Wiley, New York, 1996.
17. Proakis, J., *Digital Communications*, 3rd ed., McGraw-Hill, New York, 1995.
18. Verdu, S., *Multiuser Detection*, Cambridge University Press, Cambridge, U.K., 1998.
19. Therrien, C., *Discrete Random Signals and Statistical Signal Processing*, Prentice-Hall, Englewood Cliffs, NJ, 1992.
20. Vaynshtein, L. and Zubakov, V., *Signal Processing in Noise*, Soviet Radio, Moscow, 1960 (in Russian).
21. Porat, B., *Digital Processing of Random Signals*, Prentice-Hall, Englewood Cliffs, NJ, 1994.
22. Kremer, I., Vladimirov, V., and Karpuhin, V., *Multiplicative Noise and Radio Signal Processing*, Soviet Radio, Moscow, 1972 (in Russian).

9

Signal Resolution under the Generalized Approach to Signal Processing in the Presence of Noise

9.1 Estimation Criteria of Signal Resolution

In parallel with the primary characteristics of complex signal processing systems defining a quality of functioning, such as the probability of detection of the signals and precision of measurement of the signal parameters, some indicators defining possibilities of complex signal processing systems to detect the signals individually or to carry out measurements of parameters of the signals, responses of which at the output of the generalized detector are very close on the time or frequency axis, are very important for various types of complex signal processing systems.

The problem of signal resolution—individual definition and measurement of the signal parameters—occurs, for example, in radar while searching two or more targets that are very close to each other.[1–4] For this case, all target return signals are the useful or information signals.

The problem of signal resolution can arise in other complex signal processing systems, for example, when there are some signals, but only a single signal of this totality of the signals is the useful signal or information signal. The other signals are the nuisance signals. The nuisance signals exist as a result of the stimulus, for example, of other complex signal processing systems located near the system considered.[5,6]

It should be pointed out that for many cases, when the multiplicative noise is absent, there is no problem of signal resolution because a mutual interaction between the signals and between the complex signal processing systems for each signal is not essential. When the multiplicative noise exists this mutual interaction can be high.

The signal noise component caused by the stimulus of the multiplicative noise generates shoots of amplitude of the signal at the output of the generalized detector during the same signal parameters, for example, the signal appearance time and shift in frequency, which are absent or very low when there is no multiplicative noise.

The impact of the multiplicative noise on the signal resolution by complex signal processing systems constructed on the basis of the generalized detector, the input linear tract of which contains a filter matched with the undistorted signal, deteriorates the power of the signal at the output of the generalized detector with respect to the power of the additive Gaussian noise.

The simplest criterion of qualitative estimation of the signal resolution was introduced by Rayleigh and applied to problems in the theory of optical devices. In accordance with this criterion, two identical point sources are thought of as resolved if the summing signal at the output of optical device with respect to the corresponding coordinate, for example, the parameter l has two maximums.

Obviously, an interval between two maximums, i.e., the resolution interval, within the limits of which the condition mentioned above is satisfied, coincides with the bandwidth of the energy spectrum of the summing signal at the output of optical device, which is defined in a corresponding manner. Woodward was the first to apply the Rayleigh resolution criterion to radio signals.[7]

The shape of the signal at the output of the generalized detector or receiver with respect to the parameter l, when the signal at the input of complex signal processing system is undistorted, is defined by the autocorrelation function of this signal $\dot{\rho}(l)$.

For this reason, the Rayleigh resolution criterion coincides with the bandwidth of the main peak (maximum) of the autocorrelation function of the signal. The resolution interval is very often defined using the bandwidth of the function $|\dot{\rho}(l)|^2$. Evidently, both procedures of definition of the resolution interval are equivalent in practice.[8–11]

We use the bandwidth of the function $|\dot{\rho}(l)|^2$ as a measure of the resolution interval. The bandwidth of the main peak (maximum) of the function $|\dot{\rho}(l)|^2$ or the resolution interval l_r can be estimated quantitatively in various ways, for example, using the bandwidth of rectangle that is equivalent in area or a level provided previously. The second method of estimation is widely used in practice owing to simplicity of definition and uniqueness of obtained numerical results.[12] During the use of the first method of estimation there is some indefiniteness caused by an arbitrary chosen origin.

When there is fluctuating multiplicative noise, as was discussed in Chapter 6, the signal at the output of the generalized detector is the nonstationary stochastic process. Consequently, the resolution interval can be defined only in the statistical sense, for example, as the equivalent bandwidth of the function defining relationships between the average power of the signal at the output of the generalized detector and the signal parameter l.[13,14]

The Woodward criterion is thought of as the conditional one under definition and estimation of the signal resolution. This criterion is meaningful only under definition of resolution of the signals with the same energy characteristics.

Under resolution of the weak signal against the background of the powerful signal there is a need to take into consideration a character of the autocorrelation function $\dot{\rho}(l)$ of the signal for all magnitudes of the interval Δl between the signals, and not just in the neighborhood of the main peak (maximum) of the autocorrelation function $\dot{\rho}(l)$ of the signal.[15,16]

In other words,

$$\Delta l \leq L, \tag{9.1}$$

where L is the full extension of the signal with respect to the signal parameter l.

This feature is essentially important when there is multiplicative noise. The signal at the output of the generalized detector is distorted as a result of the stimulus of the multiplicative noise, as was noted above. In the process, the main peak (maximum) of the signal at the output of the generalized detector is decreased and the relative level of the power of the signal beyond the main maximum is increased.

As a result of this, if the multiplicative noise is present, an effect of mutual interaction between the signals is increased beyond the main maximum of the autocorrelation function of the undistorted signal. Beyond that point the Woodward criterion has one more weakness: this criterion does not take into consideration the stimulus of the additive Gaussian noise on the resolution performances.

Applying the Woodward criterion, with the purpose of quantitative estimation of the signal resolution if there is multiplicative noise, additional limitations arise. These additional limitations are caused by a feature of the given problem—there are two components of the signal distorted by the multiplicative noise, which are discussed in Chapter 4.

Shoots of amplitude of the signal at the output of the generalized detector caused by the undistorted component of the signal have the same bandwidths with respect to the resolution parameter as shoots of amplitude of the signal at the output of the generalized detector if there is only the undistorted signal at the input of the generalized detector. The bandwidth of amplitude shoots of the signal at the output of the generalized detector, which is caused by the signal noise component, can essentially be greater.

For the cases when a distribution of the total power of the signal distorted by the multiplicative noise at the output of the generalized detector with respect to the parameter l, i.e., the sum of powers of the undistorted signal and the signal noise component, does not have the sharp shoot in the neighborhood of the points, where the undistorted signal exists, the Woodward criterion is applied to estimate the impact of the multiplicative noise on the signal resolution performances.[17,18]

This fact takes place during the following suggestions regarding the energy characteristics of the signal at the output of the generalized detector:

- The power of the undistorted component of the signal at the output of the generalized detector is low in comparison to the variance $\sigma_s^2(l)$ of the signal noise component at the point, where the undistorted component of the signal is maximum.
- The function $\sigma_s^2(l)$ is sufficiently smooth and convex within the limits of the interval that is not less than the equivalent bandwidth of this function.

For these cases, when one or both conditions formulated above are not satisfied, the use of the Woodward criterion can generate errors. So, for example, under the confident distinguishing of the undistorted component of the signal against the background of the signal noise component, the resolution is defined by the bandwidth of the amplitude peak at the output of the generalized detector caused by the undistorted component of the signal, and the signal noise component cannot take it into consideration.

The second condition mentioned above is not satisfied, for example, in the case of the phase-modulated signals and also with other noise-similar signals, if the resolution parameter l is the delay of the signal. This case is discussed in the next section.

Despite the fact that the conditions mentioned above essentially limit some opportunities to use the Woodward criterion during analysis of an impact of the multiplicative noise on the conditions of the signal resolution, this criterion is very convenient because of simplicity. For the cases when the conditions mentioned above are satisfied, this criterion allows us to obtain simple formulae that are applicable in practice.

Analysis of the stimulus of the multiplicative noise on the signal resolution of complex signal processing systems constructed on the basis of the generalized detector using the Woodward criterion is carried out in the next section.

The limitations mentioned above, which are peculiarities of the Woodward criterion, cannot be taken into consideration during estimation of the stimulus of the multiplicative noise on the signal resolution performances for the use of statistical criteria of the signal resolution.

Since the main problem of any complex signal processing system is to extract information from each incoming signal, it is worthwhile to define the resolution based on the quality of functioning for the complex signal processing system when there are many information signals.[19–22]

For example, in radar during the resolution of two signals in the presence of additive Gaussian noise the approach mentioned above can be formulated in the following manner.[22]

We assume that the input stochastic process $X_{in}(t)$ can contain two overlapping signals in the presence of additive Gaussian noise

$$X_{in}(t) = \zeta_1 a_1(t, l_{11}, l_{12}, \ldots, p_{11}, p_{12}, \ldots) + \zeta_2 a_2(t, l_{21}, l_{22}, \ldots, p_{21}, p_{22}, \ldots) + n(t), \tag{9.2}$$

where ζ_1 and ζ_2 are the discrete random parameter taking the values 1 and 0 in accordance with presence or absence of the signals

$$a_1(t, l_{1i}, p_{1i}, \ldots)$$

and

$$a_2(t, l_{2i}, p_{2i}, \ldots)$$

in the input stochastic process $X_{in}(t)$; and l_{1i}, l_{2i}, p_{1i}, and p_{2i} are the measured and unmeasured signal parameters, respectively.

If the presence of the first signal

$$a_1(t, l_{1i}, p_{1i}, \ldots)$$

does not impair the detection performances or measurement of parameters of the second signal

$$a_2(t, l_{2i}, p_{2i}, \ldots)$$

so much that the detection performances or measurement of parameters of the second signal

$$a_2(t, l_{2i}, p_{2i}, \ldots)$$

are not lower than the permissible level, we can believe that the signal

$$a_2(t, l_{2i}, p_{2i}, \ldots)$$

is resolved with respect to the signal

$$a_1(t, l_{1i}, p_{1i}, \ldots)$$

in the detection sense or in the measurement sense.

If the same statement is true for the signal

$$a_1(t, l_{1i}, p_{1i}, \ldots)$$

when there are the second signal

$$a_2(t, l_{2i}, p_{2i}, \ldots)$$

and the multiplicative and additive Gaussian noise, we can believe that both signals

$$a_1(t, l_{1i}, p_{1i}, \ldots)$$

and

$$a_2(t, l_{2i}, p_{2i}, \ldots)$$

are mutually resolved.

In the case when some permissible levels of the detection performances or measurement of the signal parameters were provided before based on the mutual interaction of the signals

$$a_1(t, l_{1i}, p_{1i}, \ldots)$$

and

$$a_2(t, l_{2i}, p_{2i}, \ldots),$$

we can consider the resolution interval l_r, i.e., the distance between the signals

$$a_1(t, l_{1i}, p_{1i}, \ldots)$$

and

$$a_2(t, l_{2i}, p_{2i}, \ldots)$$

with respect to the parameter l, in which the characteristics of the complex signal processing system are within permissible limits.

Naturally, if the resolution interval l_r is high, then the difference in the energy characteristics of the received signals is high and the energy characteristics of the multiplicative noise and the additive Gaussian noise are high as well.

If the statistical characteristics of the input signals, the multiplicative noise, and the additive Gaussian noise are known, then on the basis of the statistical decision-making theory we can construct the generalized detector with the purpose of carrying out the signal resolution in the sense mentioned.[23,24] However, analysis of the generalized detector for this case is very difficult and complex even when only taking into consideration the additive Gaussian noise.

Accordingly, we consider the simple example with the purpose of estimating the stimulus of the multiplicative noise on the resolution performances. Consider the generalized detector applied to the resolution problem of two signals with unknown amplitudes and random initial phases, distributed uniformly within the limits of the interval $[0, 2\pi]$ in the presence of additive Gaussian noise. In doing so, we assume that both signals

$$a_1(t, l_{1i}, p_{1i}, \ldots)$$

and

$$a_2(t, l_{2i}, p_{2i}, \ldots)$$

do not contain some measured or unmeasured parameters except for the random initial phase, and $\zeta_1 = 1$, i.e., it is known *a priori* that the first signal

$$a_1(t, l_{1i}, p_{1i}, \ldots)$$

is present in the input stochastic process.

Then the problem is reduced to the detection problem of the second signal

$$a_2(t, l_{2i}, p_{2i}, \ldots)$$

against the background of the first signal

$$a_1(t, l_{1i}, p_{1i}, \ldots),$$

the multiplicative noise, and the additive Gaussian noise.

In doing so, the input stochastic process can be determined in the following form:

$$X_{in}(t) = a_1(t, \varphi_{01}) + \zeta_2 a_2(t, \varphi_{02}) + n(t). \tag{9.3}$$

The choice of this example is caused by the fact that an analysis of an impact of the multiplicative noise on the resolution performances of complex signal processing systems constructed on the basis of the generalized approach to signal processing in the presence of noise and the generalized detector, in particular, can be carried out on the basis of the technique and results discussed in Chapters 6 and 7. The results obtained on the basis of consideration of the example given allow us to define an impact of the multiplicative noise on the resolution performances of two signals and to define practical situations, in which a consideration of the stimulus of the multiplicative noise is necessary.

Moreover, a comparison of the resolution intervals obtained under the use of the statistical criterion and the Woodward criterion allows us to estimate the use of the Woodward criterion and to explain some equivalent or conditional statistical sense of the Woodward criterion. This problem is considered in Section 9.3.

In principle, we can consider such complex signal processing systems constructed on the basis of the generalized detector, in which the multiplicative noise does not act on the signal resolution performances. We will not discuss the problem of constructing these complex signal processing systems, but rather only mention that the signals that differ by the parameter l can save differences under the stimulus of the multiplicative noise, and these differences can be used during the resolution of the signals mentioned above. Potential possibilities of the resolution of two signals are, evidently, the functions of degree of differences between these signals in the parameter l.

We estimate an impact of the multiplicative noise on the degree of difference Δ of two signals. This difference Δ can be defined as the integral difference in the statistical sense of complex amplitude envelopes of two signals:

$$\Delta = \int_{-\infty}^{\infty} \overline{\left|\dot{S}_{1_M}(l) - \dot{S}_{2_M}(l - \Delta l)\right|^2}\, dl$$

$$= 2E_{a_{(1)}} \overline{|\dot{M}_1(t)|^2} + 2E_{a_{(2)}} \overline{|\dot{M}_2(t)|^2} - 2\,Re\left\{ \int_{-\infty}^{\infty} \dot{S}_{1_M}(l) S^*_{2_M}(l - \Delta l)\, dl \right\}, \tag{9.4}$$

where $E_{a_{(1)}}$ is the energy of the signal $a_1(t, l)$; $E_{a_{(2)}}$ is the energy of the signal $a_2(t, l - \Delta l)$; Δl is the difference in parameters of the signals $a_1(t, l)$ and $a_2(t, l - \Delta l)$; $\dot{S}_{1_M}(l)$ is the complex amplitude envelope of the signal $a_1(t, l)$; $\dot{S}_{2_M}(l - \Delta l)$ is the complex amplitude envelope of the signal $a_2(t, l - \Delta l)$; $\dot{M}_1(t)$ is the noise modulation function of the multiplicative noise for the signal $a_1(t, l)$; and $\dot{M}_2(t)$ is the noise modulation function of the multiplicative noise for the signal $a_2(t, l - \Delta l)$. We assume that the noise modulation functions $\dot{M}_1(t)$ and $\dot{M}_2(t)$ are the stationary and stationary related functions.

It follows from Eq. (9.4) that for the known energies of the signals

$$a_1(t, l)$$

and

$$a_2(t, l - \Delta l)$$

and the known statistical characteristics of the multiplicative noise, only the last term of Eq. (9.4) depends on the difference Δl in the signal parameters. The last term in Eq. (9.4) defines a degree of difference between the signals. We denote the last term in Eq. (9.4) by Δ_l. If the value Δ_l is low, then the difference between two signals is high.

Consider the value Δ_l for two cases: the parameter l of the signal resolution is the frequency Ω and the parameter l of the signal resolution is the delay τ.

For the first case we can write

$$\begin{aligned}\Delta_l = \Delta_\Omega &= 4\sqrt{E_{a_{(1)}} E_{a_{(2)}}}\, Re\left\{\dot{\rho}(\Delta\Omega)\overline{\dot{M}_1(t)\dot{M}_2^*(t)}\right\} \\ &= 4\sqrt{E_{a_{(1)}} E_{a_{(2)}}}\, Re\left\{\dot{\rho}(\Delta\Omega)\dot{R}_{12}^M(0)\right\},\end{aligned} \tag{9.5}$$

where

$$\dot{R}_{12}^M(\Delta\tau) = \overline{\dot{M}_1(t)\dot{M}_2^*(t - \Delta\tau)} \tag{9.6}$$

is the function of mutual correlation between the noise modulation functions $\dot{M}_1(t)$ and $\dot{M}_2(t)$; $\Delta\Omega$ is the detuning of the signals in frequency.

When the following conditions

$$E_{a_{(1)}} = E_{a_{(2)}} = E_a; \tag{9.7}$$

$$\dot{M}_1(t) = \dot{M}_2(t); \tag{9.8}$$

$$\overline{|\dot{M}_1(t)|^2} = \overline{A^2(t)} = const \tag{9.9}$$

are satisfied we can write

$$\Delta_\Omega = 4E_a\overline{A^2(t)}\, Re\{\dot{\rho}(\Delta\Omega)\}. \tag{9.10}$$

In other words, under the same distortions of the signals by the multiplicative noise there is no impact of multiplicative noise on the integral mean square difference between the signals that differ in frequency.

For the case when

$$\dot{M}_1(t) \neq \dot{M}_2(t) \tag{9.11}$$

the dependence of the mean square difference of the signals on the value $\Delta\Omega$ is decreased as the coefficient of mutual correlation of the noise modulation functions $\dot{M}_1(t)$ and $\dot{M}_2(t)$ of the multiplicative noise for the signals considered is decreased. In this case the value $\Delta\Omega$ is less in comparison to the case when the condition

$$\dot{M}_1(t) = \dot{M}_2(t) \tag{9.12}$$

is satisfied for all values $\Delta\Omega$.

For the second case $l = \tau$, we can write

$$\Delta_l = \Delta_\tau = 4\sqrt{E_{a_{(1)}} E_{a_{(2)}}}\, Re\left\{\dot{\rho}(\Delta\tau)\dot{R}_{12}^{M}(\Delta\tau)\right\}. \tag{9.13}$$

When the conditions determined by Eqs. (9.8) and (9.9) are satisfied and there are no statistical or functional relationships between distortions in the amplitude $A(t)$ and the phase $\varphi(t)$ of the signal, i.e.,

$$\dot{M}(t) = A(t) \cdot e^{j\varphi(t)}, \tag{9.14}$$

as was shown in Chapter 4, the correlation function $\dot{R}_{11}^{M}(\Delta\tau)$ is the real function and has the maximum for the condition

$$\Delta\tau = 0$$

and is decreased when the value $\Delta\tau$ is increased.

Because of this, for the given case the multiplicative noise causes a decrease in the value $\Delta\tau$. In other words, the integral mean square difference is increased if even distortions in amplitude and phase of the signals are the same.

Thus, the multiplicative noise does not lead to a decrease in the difference between the signals. On the contrary, the integral mean square difference between the signals is increased in the majority of cases as a result of disruption in correlation between the signals as a consequence of the stimulus of the multiplicative noise.

9.2 Signal Resolution by Woodward Criterion

In this section we consider the stimulus of the quasideterministic and fluctuating multiplicative noise on the resolution performances of complex signal processing systems constructed on the basis of the generalized detector with respect to the signal appearance time τ and signal frequency ω on the basis of the Woodward criterion.

In terms of definitions given in Section 9.1, the resolution intervals with respect to the signal appearance time τ and signal frequency ω, when the multiplicative noise is absent, can be determined in the following form:

$$\tau_{r_0} = \int_{-\infty}^{\infty} |\dot{\rho}(\tau, 0)|^2 \, d\tau; \tag{9.15}$$

$$\omega_{r_0} = \int_{-\infty}^{\infty} |\dot{\rho}(0, \omega)|^2 \, d\omega, \tag{9.16}$$

where $\dot{\rho}(\tau, \omega)$ is the autocorrelation function of the signal.

Equations (9.12) and (9.13) define the resolution intervals in the form of the bandwidth of a rectangle with unit height; the rectangle area is equivalent to the area of the function $|\dot{\rho}(\tau, 0)|^2$ or of the function $|\dot{\rho}(0, \omega)|^2$.

As was discussed in Section 9.1, if the quasideterministic or fluctuating multiplicative noise is present only the mean with respect to ensemble of values in Eqs. (9.15) and (9.16) is discussed.

Recall that if multiplicative noise is present the use of the Woodward criterion is only possible for the cases in which the energy level of the undistorted component of the signal at the output of the generalized detector is less in comparison to the energy characteristics of the signal noise component at the output of the generalized detector at the same value of the parameter τ or ω, for which the undistorted component of the signal is maximum and the function $\sigma_s^2(\tau, \omega)$ defining the distribution of the energy characteristics of the signal noise component in the coordinate system (τ, ω) is smooth.

In doing so, we can neglect the undistorted component of the signal and define the resolution intervals as the equivalent bandwidth of domain occupied by the signal noise component at the output of the generalized detector. Then, taking into account results discussed in Chapter 6, the resolution intervals with respect to the signal appearance time τ and signal frequency ω if multiplicative noise is present can be determined in the following form:

$$\tau_{r_M} = \frac{1}{\sigma_s^2(0, 0)} \int_{-\infty}^{\infty} \sigma_s^2(\tau, 0) \, d\tau; \tag{9.17}$$

$$\omega_{r_M} = \frac{1}{\sigma_s^2(0, 0)} \int_{-\infty}^{\infty} \sigma_s^2(0, \omega) \, d\omega, \tag{9.18}$$

where $\sigma_s^2(\tau, \omega)$ is the variance of the signal noise component at the output of the generalized detector (see Eq. (6.109)).

We suppose in Eqs. (9.17) and (9.18) that the function $\sigma_s^2(\tau, \omega)$ has only the maximum coinciding with respect to the coordinates τ and ω with the maximum of the function $|\dot{\rho}(\tau, \omega)|^2$. Obviously, the last condition is satisfied if the energy spectrum of the noise modulation function $\dot{M}(t)$ of the multiplicative noise is symmetric with respect to zero, i.e., the correlation function $\dot{R}_V(\tau)$ of the noise modulation function $\dot{M}(t)$ of the multiplicative noise is the real function.

In the case of the narrow-band multiplicative noise, as was discussed in Chapter 6, the following relationship

$$\sigma_s^2(\tau, \Omega) = \frac{C^2 E_{a_1}\left[\overline{A^2(t)} - \alpha_0^2(\tau, \Omega)\right]}{2} \cdot |\dot{\rho}(\tau, \Omega)|^2 \tag{9.19}$$

takes place. As we would expect, Eqs. (9.17) and (9.18) are transformed to Eqs. (9.12) and (9.13), respectively.

We proceed to define the resolution intervals of the signals distorted by the multiplicative noise with respect to the signal appearance time and signal frequency (see Eqs. (9.17) and (9.18)) using the statistical characteristics of the input stochastic process and the noise modulation function $\dot{M}(t)$ of the multiplicative noise.

9.2.1 Signal Frequency Resolution Interval

For definition of the frequency resolution interval in accordance with Eqs. (9.17) and (9.18) there is a need to consider the following intergral:

$$I_1 = \int_{-\infty}^{\infty} \sigma_s^2(0, \omega)\, d\omega. \tag{9.20}$$

Taking into consideration the general formula for definition of the variance of fluctuations of the signal distorted by the fluctuating multiplicative noise at the output of the generalized detector when the preliminary filter (PF) of input linear tract of the generalized detector is matched with the undistorted signal (see Section 6.2), the integral I_1 can be determined in the following form:

$$I_1 = \frac{C^2 E_{a_1}^2}{4\pi} \int_{-\infty}^{\infty}\int_{-\infty}^{\infty} G_V(\Omega) |\dot{\rho}(0, \omega + \Omega)|^2\, d\Omega\, d\omega. \tag{9.21}$$

Since by definition the following equality is satisfied

$$\int_{-\infty}^{\infty} |\dot{\rho}(0, \omega + \Omega)|^2\, d\omega = \int_{-\infty}^{\infty} |\dot{\rho}(0, x)|^2\, dx = \omega_{r_0}, \tag{9.22}$$

then during the condition that the multiplicative noise does not change the average power of the signal, i.e.,

$$\overline{A^2(t)} = 1, \tag{9.23}$$

we can write

$$I_1 = \frac{C^2 E_{a_1}^2}{2} \left[1 - \alpha_0^2(0, \Omega)\right] \omega_{r_0}, \tag{9.24}$$

in terms of that

$$G_V(0) = \frac{2\pi \left[1 - \alpha_0^2(0, \Omega)\right]}{\Delta\Omega_M}. \tag{9.25}$$

In the process, if multiplicative noise is present, then frequency resolution interval ω_{r_M} can be written in the following form:

$$\omega_{r_M} = \frac{\omega_{r_0} \left[1 - \alpha_0^2(0, \Omega)\right]}{2\delta_1^2(0, 0)}, \tag{9.26}$$

where

$$\delta_1^2(\tau, \Omega) = \frac{\sigma_s^2(\tau, \Omega)}{C^2 E_{a_1}^2} \tag{9.27}$$

is the normalized variance of the signal noise component determined by Eq. (6.110).

The value $\delta_1^2(0, 0)$ can be defined using the statistical characteristics of the noise modulation function $\dot{M}(t)$ of the multiplicative noise and the complex amplitude envelope of the signal during the slow and rapid fluctuating multiplicative noise. For the case of the slow fluctuating multiplicative noise in accordance with the results discussed in Section 6.4 we can write

$$2\delta_1^2(0, 0) \simeq Re\,\{U_0(0, 0)\dot{R}_V(0) + U_1(0, 0)\dot{R}'_V(0) + 0.5U_2(0, 0)\dot{R}''_V(0)\}, \tag{9.28}$$

where

$$U_0(0, 0) = |\dot{\rho}(0, 0)|^2; \tag{9.29}$$

$$U_1(0, 0) = -2j\,Re \left\{ \dot{\rho}(0, 0) \cdot \frac{\partial \rho^*(0, \Omega)}{\partial \Omega} \right\}_{\Omega=0}; \tag{9.30}$$

$$U_2(0, 0) = -2\,Re \left\{ \dot{\rho}(0, 0) \cdot \frac{\partial^2 \rho^*(0, \Omega)}{\partial \Omega^2} \right\}_{\Omega=0} - 2\left| \frac{\partial \dot{\rho}(0, \Omega)}{\partial \Omega} \right|^2_{\Omega=0}. \tag{9.31}$$

Taking into account that

$$|\dot{\rho}(0, 0)| = 1 \tag{9.32}$$

and if the signals possess the even complex amplitude envelope for the corresponding choice of the origin the following conditions

$$\left. \frac{\partial \dot{\rho}(0, \Omega)}{\partial \Omega} \right|_{\Omega=0} = 0 \tag{9.33}$$

and

$$\left.\frac{\partial^2 \dot{\rho}(0, \Omega)}{\partial \Omega^2}\right|_{\Omega=0} = -\overline{t^2}, \tag{9.34}$$

are true, where $\overline{t^2}$ is the mean square duration of the signal, we can write instead of Eq. (9.28) the following formula

$$2\delta_1^2(0, 0) \simeq \dot{R}_V(0) + \overline{t^2}\dot{R}_V''(0). \tag{9.35}$$

Next we assume that the correlation function $\dot{R}_V(\tau)$ of the noise modulation function $\dot{M}(t)$ of the multiplicative noise is the real function for the condition that distortions in amplitude and phase of the incoming signal are independent and the total power of the signal distorted by the multiplicative noise is normalized

$$\overline{|\dot{M}(t)|^2} = 1. \tag{9.36}$$

Then we can write

$$\dot{R}_V(0) = 1 - \alpha_0^2(0, \Omega); \tag{9.37}$$

$$\dot{R}_V''(0) = -\left[1 - \alpha_0^2(0, \Omega)\right]\overline{\Delta\Omega_M^2}; \tag{9.38}$$

$$\omega_{r_M} \simeq \frac{\omega_{r_0}}{1 - \overline{\Delta\Omega_M^2} \cdot \overline{t^2}}, \tag{9.39}$$

where

$$\Delta\Omega_M^2 = \frac{\int\limits_{-\infty}^{\infty} G_V(\Omega)\Omega^2 \, d\Omega}{\int\limits_{-\infty}^{\infty} G_V(\Omega) \, d\Omega} \tag{9.40}$$

is the mean square bandwidth of the energy spectrum of the noise modulation function $\dot{M}(t)$ of the multiplicative noise.

In the case of the rapid fluctuating noise we can write

$$\delta_1^2(0, 0) = \frac{1}{2\pi} \int\limits_{-\infty}^{\infty} G_V(\Omega)|\dot{\rho}(0, \Omega)|^2 \, d\Omega. \tag{9.41}$$

In Eq. (9.41) the function $G_V(\Omega)$ is varied more slowly in comparison to the function $|\dot{\rho}(0, \Omega)|^2$.

Using the Taylor series expansion for the function $G_V(\Omega)$ in the neighborhood of the point $\Omega = 0$ and limiting by the first, second, and third terms of this Taylor series expansion, during the following conditions

$$G_V(\Omega) = G_V(-\Omega); \tag{9.42}$$

$$\left.\frac{dG_V(\Omega)}{d\Omega}\right|_{\Omega=0} = 0 \tag{9.43}$$

we can write

$$\omega_{r_M} \simeq \frac{\Delta\Omega_M}{1 + 0.5\overline{\omega_{r_0}^2} \cdot \left.\frac{d^2 G_{V_0}(\Omega)}{d\Omega^2}\right|_{\Omega=0}}, \tag{9.44}$$

where

$$\overline{\omega_{r_0}^2} = \frac{\int\limits_{-\infty}^{\infty} \omega^2 |\dot{\rho}(0,\omega)|^2\, d\omega}{\int\limits_{-\infty}^{\infty} |\dot{\rho}(0,\omega)|^2\, d\omega} \tag{9.45}$$

and

$$G_{V_0}(\Omega) = \frac{G_V(\Omega)}{G_V(0)}. \tag{9.46}$$

As the bandwidth $\Delta\Omega_M$ of the energy spectrum of the noise modulation function $\dot{M}(t)$ of the multiplicative noise is increased, the function $G''_V(0)$ is monotonically decreased during the condition that the function $G_V(\Omega)$ is the smooth convex function. In the process, as we can see from Eq. (9.44) the frequency resolution interval ω_{r_M} tends to approach the value $\Delta\Omega_M$.

In other words, the frequency resolution of the signal is defined by the bandwidth of the energy spectrum of the noise modulation function $\dot{M}(t)$ of the multiplicative noise. The last conclusion can be made on the basis of the results discussed in Section 6.4. However, the formula in Eq. (9.44) allows us to estimate in a general form the limits of correctness for this conclusion—to determine the value $\Delta\Omega_M$, for which

$$\omega_{r_M} \simeq \Delta\Omega_M. \tag{9.47}$$

9.2.2 Signal Appearance Time Resolution Interval

Under definition of the signal appearance time resolution interval in accordance with Eqs. (6.109), (9.17), and (9.18) there is a need to consider the following integral

$$I_2 = \frac{1}{2\pi} \int\limits_{-\infty}^{\infty}\int\limits_{-\infty}^{\infty} G_V(\Omega)|\dot{\rho}(\tau,\Omega)|^2\, d\Omega\, d\tau. \tag{9.48}$$

In terms of that the function $|\dot{\rho}(\tau, \Omega)|^2$ can be written as the double integral of complex amplitude envelope of the signal with respect to variables t_1 and t_2 we can carry out a serial integration with respect to the variables τ and Ω. After that we can change variables

$$t_1 = t \tag{9.49}$$

and

$$t_1 - t_2 = x. \tag{9.50}$$

After integration with respect to the variable t we can write

$$I_2 = \int_{-\infty}^{\infty} \dot{R}_V(x)|\dot{\rho}(x, 0)|^2\, dx. \tag{9.51}$$

Substituting Eq. (9.51) in Eqs. (9.17) and (9.18), we can determine the signal appearance time resolution interval if the multiplicative noise is present in the following form

$$\tau_{r_M} = \frac{\int_{-\infty}^{\infty} \dot{R}_V(\tau)|\dot{\rho}(\tau, 0)|^2\, d\tau}{2\delta_1^2(0, 0)}, \tag{9.52}$$

where $\delta_1^2(0, 0)$ is determined by Eq. (9.27).

Before consideration of the cases of the slow and rapid fluctuating multiplicative noise we estimate an impact of the multiplicative noise on the signal appearance time resolution interval for the signals, for which the following condition

$$|\dot{\rho}(\tau, \Omega)| = f(\tau)r(\Omega) \tag{9.53}$$

is true.

For the signals satisfying the condition in Eq. (9.53) we can write

$$I_2 = 2\tau_{r_0}\delta_1^2(0, 0). \tag{9.54}$$

Substituting Eq. (9.54) in Eqs. (9.17) and (9.18), it is not difficult to prove that

$$\tau_{r_M} = \tau_{r_0}. \tag{9.55}$$

So, the multiplicative noise does not act on the signal appearance time resolution interval during the use of the signals, the autocorrelation function of which can be determined in the form in Eq. (9.53). In particular, the signals with a bell-shaped complex amplitude envelope and constant radio frequency carrier, the signals with a square wave-form complex amplitude envelope and constant radio frequency carrier, the signals with the noise modulation, and the signals with the phase-code modulation possessing the high signal base have the autocorrelation function mentioned above.[19,25,26] The signals with the phase-code modulation will be considered below.

In the case of the slow multiplicative noise the correlation function $\dot{R}_V(\tau)$ in Eq. (9.52) can be used during the Taylor series expansion in the neighborhood of the point $\tau = 0$. We can be limited by the first, second, and third terms of the Taylor series expansion since the function $\dot{R}_V(\tau)$ is varied more slowly in comparison to the function $|\dot{\rho}(\tau, 0)|^2$.

In terms of Eqs. (9.37)–(9.39), and (9.41) we can write

$$\tau_{r_M} \simeq \tau_{r_0} \cdot \frac{1 - 0.5\overline{\Delta\Omega_M^2} \cdot \overline{\tau_{r_0}^2}}{1 - \overline{\Delta\Omega_M^2} \cdot \overline{t^2}}, \tag{9.56}$$

where

$$\overline{\tau_{r_0}^2} = \frac{\int\limits_{-\infty}^{\infty} \tau^2 |\dot{\rho}(\tau, 0)|^2 \, d\tau}{\int\limits_{-\infty}^{\infty} |\dot{\rho}(\tau, 0)|^2 \, d\tau} = \frac{1}{\tau_{r_0}} \int\limits_{-\infty}^{\infty} \tau^2 |\dot{\rho}(\tau, 0)|^2 \, d\tau. \tag{9.57}$$

In the case of the rapid fluctuating multiplicative noise using the Taylor series expansion for the autocorrelation function $\dot{\rho}(\tau, 0)$ of the signal, which is varied more slowly in comparison to the correlation function $\dot{R}_V(\tau)$ of fluctuations in the noise modulation function $\dot{M}(t)$ of the multiplicative noise, the approximate formula of the signal appearance time resolution interval can be determined in terms of Eqs. (9.44) and (9.52) in the following form:

$$\tau_{r_M} \simeq \frac{2\pi}{\omega_{r_0}} \cdot \frac{1 - \overline{\tau_{c_V}^2} \cdot \overline{\Omega^2}}{1 - 0.5\,\overline{\omega_{r_0}^2} \cdot \overline{\tau_{c_V}^2}}, \tag{9.58}$$

where

$$\overline{\tau_{c_V}^2} = \frac{\int\limits_{-\infty}^{\infty} \dot{R}_V(\tau)\tau^2 \, d\tau}{\int\limits_{-\infty}^{\infty} \dot{R}_V(\tau) \, d\tau} \tag{9.59}$$

is the mean square correlation interval of the noise modulation function $\dot{M}(t)$ of the multiplicative noise and $\overline{\Omega^2}$ is the mean square bandwidth of the energy spectrum of the signal.

With a decrease in the correlation interval of the noise modulation function $\dot{M}(t)$ of the multiplicative noise, i.e., with an increase in the bandwidth of the energy spectrum of the noise modulation function $\dot{M}(t)$ of the multiplicative noise, the value $\overline{\tau_{c_V}^2}$ tends monotonically toward zero. For this case the limiting value of the signal appearance time resolution interval can be determined in the following form at $\overline{\tau^2_{c_V}} \to 0$

$$\tau_{r_M} \simeq \frac{2\pi}{\omega_{r_0}}. \tag{9.60}$$

In accordance with Eq. (9.16) we can write

$$\begin{aligned} \omega_{r_0} &= \int\limits_{-\infty}^{\infty} |\dot{\rho}(0, \omega)|^2 \, d\omega \\ &= \frac{1}{4E^2_{a_1}} \int\limits_{-\infty}^{\infty}\int\limits_{-\infty}^{\infty}\int\limits_{-\infty}^{\infty} \dot{S}^2(t_1)\dot{S}^2(t_2) \cdot e^{j\omega(t_1-t_2)} \, dt_1 \, dt_2 \, d\omega, \end{aligned} \tag{9.61}$$

where $S(t)$ is the complex amplitude envelope of the signal.

Taking into account that

$$\int\limits_{-\infty}^{\infty} e^{j\omega(t_1-t_2)} \, d\omega = 2\pi\delta(t_1 - t_2) \tag{9.62}$$

and in terms of Eqs. (9.60) and (9.61), we can write

$$\tau_{r_M} \simeq \frac{4E^2_{a_1}}{\int\limits_{-\infty}^{\infty} \dot{S}^4(t) \, dt}. \tag{9.63}$$

If the signal has a square wave-form complex amplitude envelope with the duration T we can write

$$2E_{a_1} = \dot{S}^2(t)T \tag{9.64}$$

and

$$\tau_{r_M} = T. \tag{9.65}$$

If the signal possesses a bell-shaped complex amplitude envelope

$$\dot{S}(t) = e^{-\frac{\pi t^2}{T^2}} \tag{9.66}$$

the limiting value of the signal appearance time resolution interval is defined by the duration T of the signal.

Thus, in the case of the wide-band multiplicative noise the signal appearance time resolution interval is defined only by the complex amplitude envelope of the signal and does not depend on the phase structure of the signal. For the signals with a square wave-form and bell-shaped complex amplitude envelope the signal appearance time resolution interval is equal to the duration of the signal.

9.2.3 Examples of Resolution Interval Definition

The immediate determination of the resolution intervals will be carried out in the following sequence. First, we will define the resolution intervals during the slow and rapid fluctuating multiplicative noise applied to the signals determined only by forms of the complex amplitude envelope and energy spectrum. Then we will determine the frequency and signal appearance time resolution intervals for the specific searching signals, using the general formulae in Eqs. (9.28) and (9.52).

We consider the signal, the complex amplitude envelope of which can be determined in the following form:

$$\dot{S}(t) = \begin{cases} e^{-\frac{\pi}{\Delta_T^2}\cdot\left(t-\frac{T-\Delta_T}{2}\right)^2}, & 0.5(T-\Delta_T) < t < \infty; \\ e^{-\frac{\pi}{\Delta_T^2}\cdot\left(t+\frac{T-\Delta_T}{2}\right)^2}, & -\infty < t < -0.5(T-\Delta_T). \end{cases} \tag{9.67}$$

The energy spectrum of the considered signal can be determined in the following form:

$$G_0(\omega) = \begin{cases} e^{-\frac{\pi}{\Delta_\omega^2}\cdot\left(\omega-\frac{\Delta\omega_0-\Delta_\omega}{2}\right)^2}, & 0.5(\Delta\omega_0-\Delta_\omega) < \omega < \infty; \\ e^{-\frac{\pi}{\Delta_\omega^2}\cdot\left(\omega+\frac{\Delta\omega_0-\Delta_\omega}{2}\right)^2}, & -\infty < \omega < -0.5(\Delta\omega_0-\Delta_\omega); \\ 1, & |\omega| \le 0.5(\Delta\omega_0-\Delta_\omega), \end{cases} \tag{9.68}$$

where T is the equivalent duration of the signal and $\Delta\omega_0$ is the equivalent bandwidth of the energy spectrum of the signal.

Introduced approximations of the complex amplitude envelope and energy spectrum of the signal are very convenient because they allow us to define a very broad class of the signals close to the real signals by changing the parameters T, Δ_T, $\Delta\omega_0$, and Δ_ω.

In particular, for the condition

$$\Delta_T = 0$$

the formula in Eq. (9.67) defines the signal with a square wave-form amplitude envelope.

For the condition

$$\Delta_T = T$$

we can define the signals with a bell-shaped complex amplitude envelope using Eq. (9.67). When the value Δ_T is within the limits of the interval $[0, T]$, i.e.,

$$0 < \Delta_T < T, \tag{9.69}$$

we can define the signals, the complex amplitude envelopes of which have flat peaks and leading and trailing edges with the finite duration.

Under definition of the signals, using Eq. (9.67) and (9.68), there is a need to take into account the following fact: the complex amplitude envelope and energy spectrum of the signals with the constant radio frequency carrier are concerned with each other uniquely by the Fourier transform. This function is absent in the case of the wide-band signals, for example, for the signals with the intrapulse angle modulation. Accordingly, formulae determined by Eqs. (9.67) and (9.68) are very convenient for definition of an impact of the multiplicative noise on the conditions of resolution of the signals with the intrapulse angle modulation.

We proceed to define the parameters ω_{r_0}, $\overline{\omega^2_{r_0}}$, τ_{r_0}, and $\overline{\tau^2_{r_0}}$, characterizing the peculiarities of the undistorted signals in Eqs. (9.15), (9.44), (9.56), and (9.61) using the complex amplitude envelope and energy spectrum of the signal determined by Eqs. (9.67) and (9.68), respectively.

Taking into account the well-known formulae for the autocorrelation function of the signal

$$\dot{\rho}(0,\omega) = \frac{\int\limits_{-\infty}^{\infty} S^2(t)\cdot e^{j\omega t}\,dt}{\int\limits_{-\infty}^{\infty} S^2(t)\,dt} \tag{9.70}$$

and

$$\dot{\rho}(\tau,0) = \frac{\int\limits_{-\infty}^{\infty} G_0(\omega)\cdot e^{-j\omega\tau}\,d\omega}{\int\limits_{-\infty}^{\infty} G_0(\omega)\,d\omega}, \tag{9.71}$$

we can write

$$\omega_{r_0} = \frac{2\pi\int\limits_{-\infty}^{\infty} S^4(t)\,dt}{\left\{\int\limits_{-\infty}^{\infty} S^2(t)\,dt\right\}^2}; \tag{9.72}$$

$$\overline{\omega^2_{r_0}} = \frac{2\pi\int\limits_{-\infty}^{\infty} S^2(t)[S^2(t)]''\,dt}{\left\{\int\limits_{-\infty}^{\infty} S^2(t)\,dt\right\}^2}; \tag{9.73}$$

$$\tau_{r_0} = \frac{2\pi\int\limits_{-\infty}^{\infty} G_0^2(\omega)\,d\omega}{\left\{\int\limits_{-\infty}^{\infty} G_0(\omega)\,d\omega\right\}^2}; \tag{9.74}$$

and

$$\overline{\tau_{r_0}^2} = -\frac{2\pi \int\limits_{-\infty}^{\infty} G_0(\omega) G_0''(\omega)\, d\omega}{\left\{\int\limits_{-\infty}^{\infty} G_0(\omega)\, d\omega\right\}^2}. \tag{9.75}$$

Substituting Eqs. (9.67) and (9.68) in Eqs. (9.72)–(9.75), respectively, and using the formulae defining the parameters ω_{r_M} and τ_{r_M}, we can write:

- For the case of the slow fluctuating multiplicative noise

$$\frac{\omega_{r_M}}{\omega_{r_0}} \simeq \frac{1}{1 - \overline{\Omega_M^2} \cdot \frac{T^2}{4\pi\sqrt{2}} \cdot \frac{\delta T^3 + 2\sqrt{2}\delta T^2(1-\delta T) + \pi\delta T(1-\delta T)^2 + \frac{\pi\sqrt{2}}{3}(1-\delta T)^3}{1 - \delta T[1-(\sqrt{2})^{-1}]}} \tag{9.76}$$

and

$$\frac{\tau_{r_M}}{\tau_{r_0}} \simeq \frac{1 - 0.5\overline{\Delta\Omega_M^2} \cdot \frac{\pi}{\Delta\omega_0^2 \delta\omega\sqrt{2}\{1-\delta\omega[1-(\sqrt{2})^{-1}]\}}}{1 - \overline{\Delta\Omega_M^2} \cdot \frac{T^2}{4\pi\sqrt{2}} \cdot \frac{\delta T^3 + 2\sqrt{2}\delta T^2(1-\delta T) + \pi\delta(1-\delta T)^2 + \frac{\pi\sqrt{2}}{3}(1-\delta T)^3}{1-\delta T[1-(\sqrt{2})^{-1}]}}. \tag{9.77}$$

- For the case of the rapid fluctuating multiplicative noise

$$\frac{\Delta\omega_{r_M}}{\Delta\Omega_M} \simeq \frac{1}{1 + 0.5 G_{V_0}''(0) \cdot \frac{\pi}{T^2 \delta T(1-0.5\delta T)}} \tag{9.78}$$

and

$$\tau_{r_M}\omega_{r_0} \simeq \frac{1 - 0.5\overline{\tau_{c_V}^2} \cdot \frac{\Delta\omega_0^2}{2\pi} \cdot \left[\delta\omega^3 + 2\delta\omega^2(1-\delta\omega) + \frac{\pi}{2}\delta\omega(1-\delta\omega)^2 + \frac{\pi}{6}(1-\delta\omega)^3\right]}{\frac{1}{2\pi} \cdot \left\{1 - 0.5\overline{\tau_{c_V}^2} \cdot \frac{\pi}{T^2\delta T(1-0.5\delta T)}\right\}}, \tag{9.79}$$

where

$$\delta T = \frac{\Delta T}{T}, \tag{9.80}$$

$$\delta\omega = \frac{\Delta\omega}{\Delta\omega_0}. \tag{9.81}$$

Table 9.1 defines the resolution intervals ω_{r_M} and τ_{r_M}, determined by Eqs. (9.76)–(9.79), for various forms of the complex amplitude envelope and energy spectrum of the signal during the condition that the energy spectrum of fluctuations of the noise modulation function $\dot{M}(t)$ of the multiplicative noise has a bell-shaped form.

Consider the stimulus of the multiplicative noise on the frequency and signal appearance time resolution intervals for specific narrow-band and wide-band signals using Eqs. (9.17), (9.18), (9.26), and (9.52). We assume that the

TABLE 9.1

Signal Appearance Time Resolution Interval τ_{r_M} and Signal Frequency Resolution Interval ω_{r_M}

Envelope	Spectrum	Slow Fluctuating Multiplicative Noise	Rapid Fluctuating Multiplicative Noise
	Signal Frequency Resolution Interval		
Square Wave-Form ($\delta T = 0$)	Any	$\frac{\omega_{r_M}}{\omega_{r_0}} = \left(1 - \frac{\pi \xi^2}{6}\right)^{-1}$	
Bell-Shaped ($\delta T = 1$)	Any	$\frac{\omega_{r_M}}{\omega_{r_0}} = \left(1 - \frac{\xi^2}{2}\right)^{-1}$	$\omega_{r_M} = \Delta\omega_M \left(1 - \frac{1}{2\xi^2}\right)^{-1}$
	Signal Appearance Time Resolution Interval		
Square Wave-Form $\delta T = 0$	Bell-Shaped $\delta\omega = 1$	$\frac{\tau_{r_M}}{\tau_{r_0}} = \frac{1 - \frac{\xi^2}{4Q_y^2}}{1 - \frac{\pi\xi^2}{6}}$	
Bell-Shaped $\delta T = 1$	Bell-Shaped $\delta\omega = 1$	$\frac{\tau_{r_M}}{\tau_{r_0}} = \frac{1 - \frac{\xi^2}{2Q_y^2}}{1 - 0.5\xi^2}$	$\tau_{r_M} = \frac{2\pi}{\omega_{r_0}} \cdot \frac{2\xi^2 - Q_y^2}{2\xi^2 - 1}$, $\xi^2 \gg Q_y^2$
Bell-Shaped	Square Wave-Form		$\tau_{r_M} = \frac{2\pi}{\omega_{r_0}} \cdot \frac{12\xi^2 - 2\pi Q_y^2}{12\xi^2 - 6}$, $\xi^2 \gg \frac{\pi}{6} Q_y^2$

energy spectrum of fluctuations of the noise modulation function $\dot{M}(t)$ of the multiplicative noise takes a bell-shaped form.

9.2.3.1 *Signal with a Bell-Shaped Amplitude Envelope and Constant Radio Frequency Carrier*

Substituting the formula defining the variance $\sigma_s^2(0, \Omega)$ of the signal noise component for the signal with a bell-shaped complex amplitude envelope (see Section 6.4) in Eq. (9.26), we can write the following relationship for the frequency resolution interval:

$$\frac{\omega_{r_M}}{\omega_{r_0}} = \sqrt{1 + \xi^2}. \tag{9.82}$$

It is not difficult to see that during the slow and rapid fluctuating multiplicative noise the formula of Eq. (9.82) coincides with the corresponding formulae shown in Table 9.1. For example, for the signals with a bell-shaped complex amplitude envelope and constant radio frequency carrier the fluctuating multiplicative noise does not act on the signal appearance time resolution interval.

As we can see, some formulae are missing from Table 9.1 since in the case of a square wave-form complex amplitude envelope of the signal

$$\overline{\omega_{r_0}^2} \to \infty,$$

and in the case of a square wave-form energy spectrum of the signal

$$\overline{\tau_{r_0}^2} \to \infty.$$

For these cases the total formulae for definition τ_{r_0} and ω_{r_0} must be used.

In Table 9.1 we use the following designations:

$$Q_y = \frac{\Delta\omega_0 T}{2\pi} \tag{9.83}$$

and

$$\xi = \frac{\Delta\Omega_M T}{2\pi}. \tag{9.84}$$

9.2.3.2 Linear Frequency-Modulated Signal

In the case of a bell-shaped complex amplitude envelope of the signal during substitution of Eq. (6.127) in Eq. (9.17) we can write

$$\tau_{r_M} = \tau_{r_0} \cdot \frac{\sqrt{1+Q_y^2} \cdot \sqrt{1+\xi^2}}{\sqrt{1+\xi^2+Q_y^2}}. \tag{9.85}$$

It is not difficult to see that for the condition

$$Q_y^2 \gg 1 + \xi^2 \tag{9.86}$$

Eq. (9.85) with an accuracy of a constant factor can be transformed in Eq. (9.82), since there is a linear function between the shift in time and frequency in the autocorrelation function of the linear frequency-modulated signal.

When the linear frequency-modulated signal has a square wave-form complex amplitude envelope we can use the approximate formula for definition of the variance $\sigma_s^2(0, \tau)$ obtained in Section 6.4, which is true for the condition

$$\xi = \frac{1}{2\pi} \cdot \Delta\Omega_M T > 3, \tag{9.87}$$

with the purpose of determining the signal appearance time resolution interval.

Then we can write

$$\tau_{r_M} = \frac{1}{G_V(0)} \int\limits_{-T}^{T} G_V\left(\frac{\tau Q_y}{T^2}\right)\left(1 - \frac{|\tau|}{T}\right) d\tau. \tag{9.88}$$

When the energy spectrum of fluctuations of the noise modulation function $\dot{M}(t)$ of the multiplicative noise takes a bell-shaped form

$$G_V(\Omega) = G_V(0) \cdot e^{-\frac{\pi\Omega^2}{\Delta\Omega_M^2}} \tag{9.89}$$

in terms of Eq. (9.88) we can write

$$\tau_{r_M} = 2\tau_{r_0}\xi\left\{\Phi\left(\frac{\sqrt{2}Q_y}{\xi}\right) - \frac{\xi}{2\pi Q_y}\left(1 - e^{-\frac{\pi Q_y^2}{\xi^2}}\right)\right\}, \tag{9.90}$$

where $\Phi(x)$ is the error integral determined in the previous chapters.

For the conditions

$$Q_y^2 \gg 1 \tag{9.91}$$

and

$$\xi \ll Q_y \tag{9.92}$$

the following approximate formula

$$\tau_{r_M} \simeq \tau_{r_0}\xi \tag{9.93}$$

is true. For this case Eq. (9.99) coincides with Eq. (9.85).

Thus, for the condition

$$3 \leq \xi \ll Q_y \tag{9.94}$$

the stimulus of the fluctuating multiplicative noise on the signal appearance time resolution interval in the case of the linear frequency-modulated signals does not depend on the complex amplitude envelope form of the signal and is only defined by the bandwidth of the energy spectrum of the noise modulation function $\dot{M}(t)$ of the multiplicative noise.

9.2.3.3 Phase-Modulated Signal with Square Wave-Form Amplitude Envelope

After substitution of Eq. (6.185) in Eq. (9.17) and carrying out all necessary mathematics, we can write

$$\tau_{r_M} = \tau_{r_0}\left[1 + \frac{1 - \alpha_0^2(\tau, \Omega)}{2\delta_1^2(0, 0)}\right], \tag{9.95}$$

where $2\delta_1^2(0, 0)$ is determined by Eq. (6.154) for the condition

$$\Omega = \tau = 0. \tag{9.96}$$

Note that Eq. (9.95), as well as Eq. (6.185), is true during the condition

$$\xi \ll N, \tag{9.97}$$

where N is the number of code elements of the phase-modulated signal.

Moreover, we assume that

$$N \gg 1. \tag{9.98}$$

Taking into account the following condition

$$\xi > 3\delta_1^2(0, 0) \simeq \frac{1 - \alpha_0^2(\tau, \Omega)}{\xi}, \tag{9.99}$$

it is not difficult to note that Eq. (9.95) coincides with Eq. (9.93).

In other words, the Woodward criterion indicates that an impact of the fluctuating multiplicative noise on the signal appearance time resolution interval both of the frequency-modulated signal and the phase-modulated signal is the same.

However, the use of the Woodward criterion during analysis of the stimulus of the fluctuating multiplicative noise on the conditions of resolution of the phase-modulated signals for the condition

$$\xi \ll N$$

leads to errors even for the case in which the undistorted component of the signal is equal to zero.

The fact is that for the given case the function $\delta_1^2(\tau, 0)$, as was shown in Section 6.3, has the clearly defined narrow shoot at the point

$$\tau = 0.$$

Moreover, the bandwidth of this shoot is equal to τ_{r_0}.

Thus, for the given case, even if the equality

$$\alpha_0^2(\tau, \Omega) = 0, \tag{9.100}$$

is true, one of the conditions of the use of the Woodward criterion, which were formulated in Section 9.1 for the purpose of estimating an impact of the fluctuating multiplicative noise on the conditions of resolution, is not satisfied.

The ratio between the power of the signal noise component at the output of the generalized detector at the point

$$\tau = 0$$

and the power of the signal noise component at the output of the generalized detector at the point

$$\tau = \tau_{r_0}$$

for the considered case is equal to the ratio

$$\frac{N}{\xi}$$

for the conditions

$$\xi > 3 \quad \text{and} \quad N \gg 1,$$

as follows from Section 6.3.

If the ratio mentioned is high, then two signals can be distinguished when a difference in the signal appearance time of each signal is close to the resolution interval τ_{r_0}. This statement is true if N is high, i.e., the number of code elements of the phase-modulated signal is high.

Thus, we can conclude that the Woodward criterion cannot be used for the purpose of estimating the signal appearance time resolution interval of the phase-modulated signal in a general case when there is the fluctuating multiplicative noise. The Woodward criterion can be used for the limiting case only if

$$\xi \simeq N$$

and we can neglect the shoot of the function $\delta_1^2(0, \tau)$ at the point

$$\tau = 0.$$

9.3 Statistical Criterion of Signal Resolution

In this section we estimate the stimulus of the multiplicative noise on the resolution characteristics of the signals under signal processing by the generalized detector.[7,8] We assume that there are two signals with the random initial phase and unknown amplitude that differ from each other by the signal appearance time τ and shift in frequency on the value Ω at the input of the generalized detector.

In doing so, it is known *a priori* that the first signal with the complex amplitude envelope

$$\dot{S}_{(1)}(t)$$

exists and the parameters τ and Ω of this signal are equal to τ_0 and Ω_0, respectively.

Without disturbing a general sense, we assume that

$$\tau_0 = 0 \qquad \text{and} \qquad \Omega_0 = 0. \tag{9.101}$$

The second signal at the input of the generalized detector with the complex amplitude envelope

$$\dot{S}_{(2)}(t - \tau, -\Omega)$$

differs from the first signal by the amplitude, the shift in the signal appearance time on the value τ, and the shift in the signal frequency on the value Ω. All of these parameters of the second signal are considered as known.

There is the additive Gaussian noise with the complex amplitude envelope $\dot{N}(t)$ simultaneously with the signals at the input of the generalized detector. Despite the fact that the probability of detection of the second signal does not depend on a "yes" or a "no" first signal at the input of the generalized detector, we believe that the second signal is resolved relative to the first signal in the detection sense.

For the given case the probability of detection of the second signal can be considered as a quantitative measure of resolution of two signals, which depends on the parameters τ and Ω, energy characteristics of the signals and noise. We will call this probability the probability of resolution P_{D_r}.

The signal at the output of the considered model of the generalized detector, under resolution of the second signal in the background of both the first signal and the additive Gaussian noise, takes the following form:[23,24]

$$z_g^{out}(t) = 0.5 \cdot \left| 2 \int_{-\infty}^{\infty} \dot{S}_{in}(t) Q^*(t)\, dt - \int_{-\infty}^{\infty} \dot{S}_{in}(t) S_{in}^*(t)\, dt \right|, \tag{9.102}$$

where

$$\begin{aligned} \dot{S}_{in}(t) &= \dot{S}_{(1)}(t) + \dot{S}_{(2)}(t-\tau, -\Omega) + \dot{N}(t) \\ &= \sqrt{E_{a_{(1)}}}\, \dot{S}_m(t) + \sqrt{E_{a_{(2)}}}\, \dot{S}_m(t-\tau) \cdot e^{-j\Omega(t-\tau)} + \dot{N}(t) \end{aligned} \tag{9.103}$$

is the complex amplitude envelope of the additive mixture of two signals and the Gaussian noise;

$$\dot{Q}(t) = \dot{S}_m(t-\tau) \cdot e^{-j\Omega(t-\tau)} - \rho^*(\tau, \Omega) \dot{S}_m(t) \cdot e^{j\Omega\tau} \tag{9.104}$$

is the complex amplitude envelope of the model signal, i.e., the reference signal, of the considered model of the generalized detector; and $\dot{S}_m(t)$ is the complex amplitude envelope of the model signal with the energy equal to the energy of the received signal.

As was noted above, the probability of resolution of the second signal relative to the first signal, for the considered model of the generalized detector, is defined by the probability of detection of the second signal in the background of the first signal and the additive Gaussian noise, in other words, by the probability of exceeding the process at the output of the considered model of the generalized detector over the given threshold K_g that is defined by the predetermined probability of false alarm of the second signal when the second signal is absent. This probability is defined in the same manner as the probability of false alarm in the signal detection theory. We denote this probability of false alarm as P_{F_r}.

The multiplicative noise distorts both received signals. In doing so, in a general case the noise modulation functions of the multiplicative noise for the first and second signals can differ. These noise modulation functions of the multiplicative noise are denoted as $\dot{M}_1(t)$ and $\dot{M}_2(t)$, respectively.

We assume that the noise modulation functions $\dot{M}_1(t)$ and $\dot{M}_2(t)$ of the multiplicative noise are the stationary and stationary related stochastic processes. When the multiplicative noise is present the complex amplitude envelope of the signal at the input of the considered model of the generalized detector has the following form:

$$\dot{S}_{M_{in}}(t) = \dot{M}_1(t)\sqrt{E_{a_{(1)}}}\,\dot{S}_m(t) + \dot{M}_2(t)\sqrt{E_{a_{(2)}}}\,\dot{S}_m(t-\tau)\cdot e^{-j\Omega(t-\tau)} + \dot{N}(t). \tag{9.105}$$

Henceforth, we consider the case when the multiplicative noise acts essentially on the probability of resolution P_{D_r} and if the multiplicative noise is absent the resolution of the first and second signals is high.

It is not difficult to see that for this case the following condition

$$|\dot{\rho}(\tau,\Omega)| \ll 1 \tag{9.106}$$

is true.

Taking into account the conditions formulated above, the model signal, i.e., the reference signal, $\dot{Q}(t)$, during estimation of the stimulus of the multiplicative noise can be approximately determined in the following form:

$$\dot{Q}(t) \simeq \dot{S}_m(t-\tau)\cdot e^{-j\Omega(t-\tau)}. \tag{9.107}$$

Substituting Eq. (9.105) in Eq. (9.102) and taking into account Eq. (9.107), the complex amplitude envelope of the signal at the output of the considered model of the generalized detector, if both additive Gaussian noise and multiplicative noise are present, can be determined in the following form:

$$\begin{aligned}\dot{z}^{out}_{gM}(t) = {} & E_{a_{(2)}}\int_{-\infty}^{\infty}\dot{M}_2(t)|\dot{S}_m(t-\tau)|^2\,dt \\ & + E_{a_{(1)}}\int_{-\infty}^{\infty}\dot{M}_1(t)\dot{S}_m(t)S^*_m(t-\tau)\cdot e^{j\Omega(t-\tau)}\,dt \\ & + 0.5\cdot\left\{\int_{-\infty}^{\infty}\dot{N}(t)N^*(t-\tau)\,dt - \int_{-\infty}^{\infty}N^*(t)\dot{N}(t-\tau)\,dt\right\}.\end{aligned} \tag{9.108}$$

To determine the probability of resolution P_{D_r} there is a need to define statistical characteristics of the complex amplitude envelope of the signal at the output of the considered model of the generalized detector.

Taking into account that

$$\overline{\dot{M}_1(t)} = \alpha_{0_1}(\tau,\Omega)\cdot e^{j\beta_{0_1}} \tag{9.109}$$

and

$$\overline{\dot{M}_2(t)} = \alpha_{0_2}(\tau, \Omega) \cdot e^{j\beta_{0_2}} \tag{9.110}$$

and assuming that

$$\beta_{0_1} = \beta_{0_2} = 0, \tag{9.111}$$

we can write

$$\overline{\dot{z}^{out}_{g_M}(t)} = \alpha_{0_2}(\tau, \Omega) E_{a_{(2)}} + \alpha_{0_1}(\tau, \Omega) E_{a_{(1)}} \dot{\rho}(\tau, \Omega) \cdot e^{-j\Omega\tau}. \tag{9.112}$$

Recall that, as was shown in Section 7.1, the variances of quadrature components and their mutual correlation function can be determined in the following form:

$$\sigma^2_X = 0.5\{\overline{Re\, \dot{z}'_0(t)\, z^{*''}_0(t)} + \overline{Re\, \dot{z}'_0(t)\, \dot{z}''_0(t)}\}; \tag{9.113}$$

$$\sigma^2_Y = 0.5\{\overline{Re\, \dot{z}'_0(t)\, z^{*''}_0(t)} - \overline{Re\, \dot{z}'_0(t)\, \dot{z}''_0(t)}\}; \tag{9.114}$$

and

$$R_{XY}(\tau) = 0.5\{\overline{Im\, \dot{z}'_0(t)\, z^*_0(t-\tau)} + \overline{Im\, \dot{z}'_0(t)\, \dot{z}''_0(t-\tau)}\}, \tag{9.115}$$

where

$$\dot{z}'_0(t) = \left[\dot{z}^{out}_{g_M}(t)\right]' - \overline{\left[\dot{z}^{out}_{g_M}(t)\right]'} \tag{9.116}$$

and

$$\dot{z}''_0(t) = \left[\dot{z}^{out}_{g_M}(t)\right]'' - \overline{\left[\dot{z}^{out}_{g_M}(t)\right]''}. \tag{9.117}$$

The top indexes of the designations $\dot{z}^{out}_{g_M}(t)$ and $\dot{z}_0(t)$ indicate a difference in variables of integration in Eq. (9.108).

In other words, the process $[\dot{z}^{out}_{g_M}(t)]'$ is determined by Eq. (9.108) when the integration variable is t'; $[\dot{z}^{out}_{g_M}(t)]''$ is determined by Eq. (9.108) when the integration variable is t''.

We introduce the designation

$$\left[\dot{z}^{out}_{g_M}(t)\right]' = \dot{b}'_1 + \dot{b}'_2 + \dot{b}'_3, \tag{9.118}$$

where $\dot{b}'_1$ corresponds to the first term in Eq. (9.108), $\dot{b}'_2$ corresponds to the second term in Eq. (9.108), and $\dot{b}'_3$ corresponds to the third term in Eq. (9.108).

Then we can write

$$\dot{z}'_0(t) z^{*''}_0(t) = \sum_{i=1}^{3}\sum_{k=1}^{3} \dot{b}'_i b^{*''}_k \tag{9.119}$$

and

$$\dot{z}_0'(t)\dot{z}_0''(t) = \sum_{i=1}^{3}\sum_{k=1}^{3} \dot{b}_i'\dot{b}_k''. \tag{9.120}$$

Carrying out mathematics in Eqs. (9.119) and (9.120), we can write

$$\overline{\dot{z}_0'(t)\, z_0^{*\prime\prime}(t)} = \delta_{12}^2(0,0)E_{a_{(2)}}^2 + \delta_{11}^2(\tau,\Omega)E_{a_{(1)}}^2 + E_{a_{(1)}}E_{a_{(2)}}\, Re\left\{\dot{R}_{S_{21}}(0,\tau,0,\Omega)\cdot e^{j\Omega\tau}\right\} + N_0^2; \tag{9.121}$$

$$\overline{\dot{z}_0'(t)\, \dot{z}_0''(t)} = \delta_{22}^2(0,0)E_{a_{(2)}}^2 + \delta_{21}^2(\tau,\Omega)E_{a_{(1)}}^2 + E_{a_{(1)}}E_{a_{(2)}}\dot{D}_{S_{21}}(0,\tau,0,\Omega)\cdot e^{-j\Omega\tau}, \tag{9.122}$$

where the following designations are used:

$$\delta_{1i}^2(\tau,\Omega) = \frac{1}{4\pi}\int_{-\infty}^{\infty} G_{V_i}(\omega)|\dot{\rho}(\tau,\Omega+\omega)|^2\, d\omega; \tag{9.123}$$

$$\delta_{2i}^2(\tau,\Omega) = \frac{1}{4\pi}\int_{-\infty}^{\infty} \dot{G}_{D_i}(\omega)\dot{\rho}(\tau,\Omega+\omega)\dot{\rho}(\tau,\Omega-\omega)\, d\omega; \tag{9.124}$$

$$\dot{R}_{S_{21}}(0,\tau,0,\Omega) = \frac{1}{4\pi}\int_{-\infty}^{\infty} \dot{G}_{V_{21}}(\omega)\dot{\rho}(0,\omega)\rho^*(\tau,\Omega+\omega)\, d\omega; \tag{9.125}$$

$$\dot{D}_{S_{21}}(0,\tau,0,\Omega) = \frac{1}{4\pi}\int_{-\infty}^{\infty} \dot{G}_{D_{21}}(\omega)\dot{\rho}(0,\omega)\dot{\rho}(\tau,\Omega-\omega)\, d\omega, \tag{9.126}$$

where $G_{V_1}(\omega)$ is the energy spectrum of fluctuations of the noise modulation function $\dot{M}_1(t)$ of the multiplicative noise; $G_{V_{21}}(\omega)$ is the mutual energy spectrum of the differences

$$\dot{M}_1(t) - \alpha_{0_1}(\tau,\Omega)\cdot e^{j\beta_{0_1}} \tag{9.127}$$

and

$$\dot{M}_2(t) - \alpha_{0_2}(\tau,\Omega)\cdot e^{j\beta_{0_2}}; \tag{9.128}$$

$\dot{G}_{D_{21}}(\omega)$ is the Fourier transform of the function

$$\dot{D}_{V_{21}}(t_1 - t_2) = \overline{\left[\dot{M}_2(t_1) - \alpha_{0_2}(\tau,\Omega)\cdot e^{j\beta_{0_2}}\right]\left[\dot{M}_1(t_2) - \alpha_{0_1}(\tau,\Omega)\cdot e^{j\beta_{0_1}}\right]}. \tag{9.129}$$

The formulae for $\delta_{12}^2(\tau,\Omega)$ and $\delta_{22}^2(\tau,\Omega)$ characterizing the second signal have the form that is analogous to $\delta_{11}^2(\tau,\Omega)$ and $\delta_{21}^2(\tau,\Omega)$.

Comparing Eqs. (9.113)–(9.126) with the results discussed in Section 7.1, we can see that the complex amplitude envelope of the signal at the output of the considered model of the generalized detector can be thought of as the flat vector with correlated components possessing the various variances. However, when the multiplicative noise essentially impairs the conditions of the resolution, in other words, if there is the rapid and high fluctuating multiplicative noise, then the quadrature components become the non-correlated stochastic processes, which are very close to the Gaussian processes, with the same variances in accordance with the results discussed in Section 7.1.

Analogous results can be obtained in the cases when the noise modulation functions $\dot{M}_1(t)$ and $\dot{M}_2(t)$ of the multiplicative noise are the stationary normal stochastic process. For both cases the probability distribution density of the process at the output of the considered model of the generalized detector can be determined in the following form:

$$f^{H_1}\left[z_{g_M}^{out}(t)\right] = \frac{z_{g_M}^{out}(t)}{\sigma^2} \cdot e^{-\frac{\left[z_{g_M}^{out}(t)\right]^2 + \widetilde{E_a}}{2\sigma^2}} K_0\left[\frac{\widetilde{E_a} z_{g_M}^{out}(t)}{\sigma^2}\right], \tag{9.130}$$

where $K_0(x)$ is the modified Bessel function of the second order of imaginary argument; $\widetilde{E_a}$ is the mean of the process at the output of the considered model of the generalized detector determined by the following form:

$$\widetilde{E_a} = Re\left\{\overline{\dot{z}_{g_M}^{out}(t)}\right\}; \tag{9.131}$$

$\overline{\dot{z}_{g_M}^{out}(t)}$ is determined by Eq. (9.112); and σ^2 is the variance of the process at the output of the considered model of the generalized detector determined by the following form:

$$\begin{aligned}\sigma^2 = \delta_{12}^2(0,0)E_{a_{(2)}}^2 + \delta_{11}^2(\tau,\Omega)E_{a_{(1)}}^2 \\ +0.5E_{a_{(1)}}E_{a_{(2)}} Re\left\{R_{S_{21}}(0,\tau,0,\Omega)\cdot e^{j\Omega\tau}\right\} + N_0^2.\end{aligned} \tag{9.132}$$

It should be pointed out that we can neglect the third term in Eq. (9.132) because during the rapid fluctuating multiplicative noise the coefficient of correlation defined by the mutual correlation function $R_{S_{21}}(0,\tau,0,\Omega)$ at the values τ and Ω, for which the probability of resolution P_{D_r} is high if the fluctuating multiplicative noise is absent, is low.

Then we can write

$$\sigma^2 = \delta_{12}^2(0,0)E_{a_{(2)}}^2 + \delta_{11}^2(\tau,\Omega)E_{a_{(1)}}^2 + N_0^2. \tag{9.133}$$

The probability of resolution of the second signal relative to the first signal, if the additive Gaussian noise and fluctuating multiplicative noise exist, in terms of Eq. (9.130) takes the following form:

$$P_{D_r} = \int_{K_g}^{\infty} \frac{z_{g_M}^{out}(t)}{\sigma^2} \cdot e^{-\frac{\left[z_{g_M}^{out}(t)\right]^2 + \widetilde{E_a}^2}{2\sigma^2}} K_0\left[\frac{\widetilde{E_a} z_{g_M}^{out}(t)}{\sigma^2}\right] d\left[z_{g_M}^{out}(t)\right], \tag{9.134}$$

where K_g is the threshold defined by the previously given probability of false alarm P_{F_r} of the second signal in the background of the first signal and the additive Gaussian noise and the fluctuating multiplicative noise.

The threshold K_g can be determined in the following form ($\widetilde{E}_a = 0$):

$$K_g = \sigma_1 \sqrt{2 \ln \frac{1}{P_{F_r}}}, \tag{9.135}$$

where

$$\sigma_1^2 = \delta_{11}^2(\tau, \Omega) E_{a_{(1)}}^2 + N_0^2. \tag{9.136}$$

Equations (9.134) and (9.135) can be used for the purpose of estimating an impact of the fluctuating multiplicative noise on resolution of two signals.

Reference to Eqs. (9.134) and (9.135) shows that the probability of resolution of the second signal is defined as the probability of detection of the sum of the undistorted component and the signal noise component of the second signal in the background of the additive Gaussian noise and the signal noise component of the first signal at the point of the undistorted component of the second signal.

When the ratio between the energy of the undistorted component of the signal at the output of the considered model of the generalized detector $\widetilde{E}_a$ and the power of fluctuations caused by the fluctuating multiplicative noise and the additive Gaussian noise (the variance σ^2, see Eq. (9.133)) at the point of the second signal is much more than unity, the probability of resolution can be defined using the Laplace function, i.e., the error integral.

Taking into account the results discussed in Reference 27, we can define the probability of resolution in the following form:

$$P_{D_r} = 1 - \Phi(\gamma), \tag{9.137}$$

where

$$\gamma = U - \frac{\sigma}{2\widetilde{E}_a} + \frac{U}{4} \cdot \frac{\sigma^2}{\widetilde{E}_a^2} - \frac{U^2 + 0.5}{6} \cdot \left(\frac{\sigma}{\widetilde{E}_a}\right)^3; \tag{9.138}$$

$$U = \frac{K_g - \widetilde{E}_a}{\sigma} = \frac{\sigma_1}{\sigma} \cdot \sqrt{2 \ln \frac{1}{P_{F_r}}} - \frac{\widetilde{E}_a}{\sigma}; \tag{9.139}$$

and

$$\Phi(\gamma) = \frac{1}{\sqrt{2\pi}} \int_{-\infty}^{\gamma} e^{-\frac{x^2}{2}} \, dx \tag{9.140}$$

is the error integral.

As we can see from Eqs. (9.137)–(9.140), the approximate formula in Eq. (9.137) is only true when the condition

$$\frac{\widetilde{E}_a}{\sigma} > 1 \tag{9.141}$$

is satisfied.

In other words, Eq. (9.137) is true if the peak power of the undistorted component of the signal is more than the power of fluctuations caused by the fluctuating multiplicative noise and the additive Gaussian noise at the point of the second signal.

If the probability of resolution P_{D_r} and the probability of false alarm P_{F_r} are given before the signal appearance time resolution interval τ_r and signal frequency resolution interval Ω_r corresponding to the given probability of resolution P_{D_r} and the probability of false alarm P_{F_r} can be defined on the basis of Eqs. (9.137)–(9.139). In fact, the powers of fluctuations σ^2 and σ_1^2 (see Eqs. (9.133) and (9.135), respectively) depend on the shift in the signal appearance time and signal frequency of the signals. Moreover,

$$\sigma^2(\tau, \Omega) = \sigma_1^2(\tau, \Omega) + \delta_{12}^2(0, 0) E_{a_{(2)}}^2. \tag{9.142}$$

Thus, to define the resolution intervals using the given values of the probability of resolution P_{D_r} and the probability of false alarm P_{F_r} there is a need to define the function

$$\sigma_1^2(\tau, \Omega) = \delta_{11}^2 E_{a_{(1)}}^2(\tau, \Omega) + N_0^2 \tag{9.143}$$

on the basis of Eqs. (9.137)–(9.140) in terms of Eq. (9.135) and after that to define the resolution intervals when the function $\delta_{11}^2(\tau, \Omega)$ has been determined.

Determination of the variance $\sigma_1^2(\tau, \Omega)$ is very difficult in the mathematical sense during substitution of the value $\gamma_0(P_{D_r})$ instead of the value γ on the left side of Eqs. (9.138) and (9.139). For this reason, the value $\sigma_1^2(\tau, \Omega)$ can be determined using numerical procedures only.

We limit by determination of the first approximation for the variance $\sigma_1^2(\tau, \Omega)$, meaning that this approximation is true only if the ratio

$$\frac{\widetilde{E}_a}{\sigma}$$

is high. Saving the first and second terms of the function $\gamma(U)$ on the right side of Eqs. (9.138) and (9.139), we can write

$$\gamma_0 = \frac{K_g - \widetilde{E}_a}{\sigma}\left(1 + \frac{\sigma^2}{2\widetilde{E}_a^2}\right). \tag{9.144}$$

Assume that the condition

$$\frac{2\widetilde{E}_a^2}{\sigma^2} \gg 1 \tag{9.145}$$

is satisfied.

We proceed to introduce the following designations

$$\sigma^2 = \sigma_1^2 + a^2 \tag{9.146}$$

and

$$K_g = \sigma_1 b. \tag{9.147}$$

Then we can write the following equation instead of Eq. (9.144) with respect to the variable σ_1:

$$\sigma_1^2(\gamma_0^2 - b^2) + 2\widetilde{E_a}\sigma_1 b + \gamma_0^2 a^2 - \widetilde{E_a} = 0, \tag{9.148}$$

where

$$a^2 = \delta_{12}^2(0, 0)E_{a_{(2)}}^2 \tag{9.149}$$

and

$$b^2 = 2\ln\frac{1}{P_{F_r}}. \tag{9.150}$$

Note that the parameters a^2 and b^2 are defined by the statistical characteristics of the fluctuating multiplicative noise and the previously given probability of false alarm P_{F_r} and do not depend on the signal appearance time resolution interval τ_r and the signal frequency resolution interval Ω_r between the signals.

The solution of Eq. (9.148) can be written in the following form:

$$\sigma_1^2(\tau, \Omega) = \left\{\frac{b\widetilde{E_a} - \sqrt{b^2\widetilde{E_a}^2 + (\gamma_0^2 a^2 - \widetilde{E_a}^2)(b^2 - \gamma_0^2)}}{b^2 - \gamma_0^2}\right\}^2. \tag{9.151}$$

Substituting the determined value $\sigma_1^2(\tau, \Omega)$ (see Eq. (9.151)) in Eq. (9.135) and using the known function $\delta_{11}^2(\tau, \Omega)$, we can define the signal appearance time resolution interval τ_r and the signal frequency resolution interval Ω_r.

As was noted in Sections 6.3 and 6.4, in many cases there are simple functions that are reciprocal to the function $\delta_{11}^2(\tau, \Omega)$. Denoting the function that is reciprocal to the function $\delta_{11}^2(\tau, \Omega)$ as arc $\delta_{11}^2(\tau, \Omega)$, we can write

$$\tau_r, \Omega_r = \text{arc}\,\delta_{11}^2\left(\frac{\sigma_1^2}{E_{a_{(1)}}^2} - \frac{N_0^2}{E_{a_{(1)}}^2}\right). \tag{9.152}$$

For example, if

$$\delta_{11}^2(\tau, 0) = e^{-a^2\tau^2}, \tag{9.153}$$

then

$$\text{arc } \delta_{11}^2(\tau, 0) = \frac{1}{a} \cdot \sqrt{-\ln \delta_{11}^2(\tau, 0)} \tag{9.154}$$

and the resolution intervals can be determined in the following form:

$$\tau_r, \Omega_r = \frac{1}{a} \sqrt{-\ln\left(\frac{\sigma_1^2}{E_{a(1)}^2} - \frac{N_0^2}{E_{a(1)}^2}\right)}. \tag{9.155}$$

Consider another method of definition of the resolution intervals. This method is based on the linear approximation of the functional relationship between arguments of the incomplete Toronto function.[28] The Q-function determined by

$$Q(x\sqrt{2}, y\sqrt{2}) = 1 - T_y(1, 0, x) \tag{9.156}$$

is the particular case of the Toronto function, where $T_y(1, 0, x)$ is the incomplete Toronto function. Note that for our problem

$$x^2 = \frac{\widetilde{E}_a^2}{2\sigma^2} \tag{9.157}$$

and

$$y^2 = \frac{K_g^2}{2\sigma^2}. \tag{9.158}$$

The function

$$y^2 = f(x^2) \tag{9.159}$$

is shown in Fig. 9.1 at some fixed values of the probability of resolution P_{D_r} (the solid lines). Also, the linear approximation of this function determined in the following form

$$y^2 = \mu x^2 - \nu \tag{9.160}$$

is shown in Fig. 9.1 by the dotted line.

As we can see from Fig. 9.1, the function determined by Eq. (9.160) approximates very well the real functions within the wide intervals of variation of the values x^2 and y^2.

The values of the coefficients μ and ν for some probabilities of resolution P_{D_r} and the boundary values x and y, for which the introduced approximation is true, are shown in Table 9.2.

Two values of the coefficients μ and ν are given for each value of the probability of resolution P_{D_r}. Also, two values of limits of the validity intervals are given for each value of the probability of resolution P_{D_r}.

TABLE 9.2

Coefficients μ and ν at Various Values of the Probability of Resolution P_{D_r}

P_{D_r}	μ	ν	x_{min}	x_{max}	y_{min}	y_{max}
0.5	1.125	0.0	1.41	3.74	1.58	3.74
0.5	1.0	−0.7	0.0	2.45	0.89	2.55
0.7	0.925	0.6	1.58	3.6	1.41	3.4
0.7	0.69	−0.2	0.7	2.0	0.7	1.7
0.9	0.75	1.5	2.0	4.5	1.2	3.6
0.9	0.33	0.0	0.7	1.73	0.39	1.0
0.99	0.55	5.0	3.6	5.0	1.41	3.0

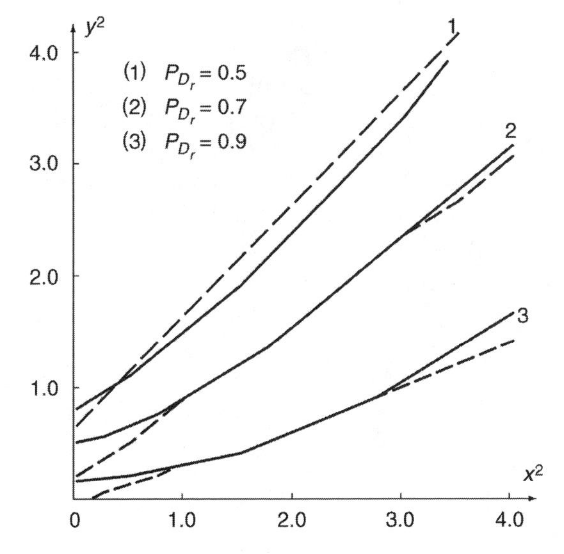

FIGURE 9.1
Dependences $y^2 = f(x^2)$ at some fixed values of the probability of resolution P_{D_r}.

Dependences shown in Fig. 9.1 correspond to the values of the coefficients μ and ν shown in the top string of Table 9.2 for each value of the probability of resolution P_{D_r}.

During variation of values x and y within the limits of intervals shown in Table 9.2, the relative errors if the condition

$$P_{D_r} = const \tag{9.161}$$

is true are greater than (a value between) 3% and 5%. The function determined by Eq. (9.159) is plotted on the basis of the condition

$$P_{D_r} = const. \tag{9.162}$$

In terms of Eq. (9.160), the following equality for definition of the resolution intervals

$$\frac{\sigma_1^2 b^2}{\sigma_1^2 + a^2} = \mu \cdot \frac{\widetilde{E_a}^2}{\sigma_1^2 + a^2} - 2\nu \tag{9.163}$$

can be used.

Using Eqs. (9.135) and (9.160) and in terms of Eq. (9.149), we can write

$$\tau_r, \Omega_r \simeq \text{arc}\, \delta_{11}^2 \left\{ \frac{\mu \cdot \frac{\widetilde{E_a}^2}{E_{a(1)}^2} - 2\nu \cdot \frac{E_{a(2)}^2}{E_{a(1)}^2} \cdot \delta_{12}^2(0,0)}{2\ln \frac{1}{P_{F_r}} + 2\nu} - \frac{N_0^2}{E_{a(1)}^2} \right\}. \tag{9.164}$$

The formula in Eq. (9.163) is simpler in comparison to the formula in Eq. (9.144). However, Eq. (9.163) can be used only under limitations with respect to the range of variation of the parameters x and y indicated in Table 9.2.

Since the parameters x and y are defined by the statistical characteristics of the additive Gaussian noise, fluctuating multiplicative noise, and signals, there are formulae defining limitations with respect to the statistical characteristics of the signals and the fluctuating multiplicative noise and the additive Gaussian noise, in which Eq. (9.164) is true:

$$x_{min}^2 \le \frac{\widetilde{E_a}^2}{2\sigma^2} = \frac{\widetilde{E_a}^2}{N_0^2} \cdot \frac{1}{\frac{E_{a(2)}^2}{N_0^2} \cdot \delta_{12}^2(0,0) + \frac{E_{a(1)}^2}{N_0^2} \cdot \delta_{11}^2(\tau_r, \Omega_r) + 1} \le x_{max}^2; \tag{9.165}$$

$$y_{min}^2 \le \frac{K_g^2}{2\sigma^2} = 0.5 \ln \frac{1}{P_{F_r}} \cdot \frac{\frac{E_{a(1)}^2}{N_0^2} \cdot \delta_{11}^2(\tau_r, \Omega_r) + 1}{\frac{E_{a(2)}^2}{N_0^2} \cdot \delta_{12}^2(0,0) + \frac{E_{a(1)}^2}{N_0^2} \cdot \delta_{11}^2(\tau_r, \Omega_r) + 1} \le y_{max}^2. \tag{9.166}$$

Reference to Eqs. (9.151) and (9.164) shows that the signal appearance time resolution interval τ_r and the signal frequency resolution interval Ω_r are uniquely defined by the probability of resolution P_{D_r} and the probability of false alarm P_{F_r} under the known statistical characteristics of the signals, the additive Gaussian noise, and the fluctuating multiplicative noise.

The probability of resolution P_{F_r} and the probability of false alarm P_{F_r} are taken into account in Eqs. (9.151) and (9.164) using the parameters γ_0 and b in Eq. (9.151) and the parameters μ and ν in Eq. (9.164). The approximate formulae determined by Eqs. (9.151) and (9.164) allow us to define quantitative values of the resolution intervals within the limits of the sufficiently wide interval, in which the statistical parameters of the additive Gaussian noise and the fluctuating multiplicative noise are varied.

At the same time, as noted in Section 9.1, the simplest criterion of quantitative estimation of the resolution intervals for two signals with the same energy characteristics, if there is fluctuating multiplicative noise, based on the Woodward criterion can only be used under very hard limitations applied to the statistical characteristics of the additive Gaussian noise and the fluctuating multiplicative noise.

In particular, as noted in Section 7.1, the Woodward criterion does not allow us to take into consideration an impact of the additive Gaussian noise on conditions of the signal resolution. If fluctuating multiplicative noise is present, then the use of the Woodward criterion can bring us the true results only if the power or energy of the undistorted component of the signal at the output of the considered model of the generalized detector is much greater in comparison to the power or energy of the signal noise component.

Of particular interest is to determine limitations on the energy characteristics of the additive Gaussian noise and the fluctuating multiplicative noise, in which the resolution intervals determined both by the Woodward criterion and by the statistical criterion are very close. Since we are interested in the case for which the relative level of the undistorted component of the signal at the output of the considered model of the generalized detector is low, we can use Eq. (9.164) for determination of the resolution intervals of the signal.

Further analysis is carried out under the following assumptions:

- The energy of the resolved signals are the same:

$$E_{a_{(1)}} = E_{a_{(2)}} = E_a. \tag{9.167}$$

  There is a need to satisfy this condition for the purpose of comparing the statistical criterion with the Woodward criterion.

- Both signals are distorted by the same fluctuating multiplicative noise or different fluctuating multiplicative noise with the same statistical characteristics. So, the following conditions

$$\alpha_{0_1}^2(\tau, \Omega) = \alpha_{0_2}^2(\tau, \Omega) \tag{9.168}$$

  and

$$\delta_{11}^2(\tau, \Omega) = \delta_{12}^2(\tau, \Omega) = \delta_1^2(\tau, \Omega) \tag{9.169}$$

  are satisfied.

Under these assumptions, when the resolution of the signals is high and if the fluctuating multiplicative noise is absent, the parameter $\widetilde{E_a}$ (see Eq. (9.164)) is determined in the following form:

$$\widetilde{E_a} = \alpha_0^2(\tau, \Omega) E_{a_1}. \tag{9.170}$$

Then the formula in Eq. (9.164) can be transformed in the following form:

$$\tau_r, \Omega_r = \text{arc}\, \delta_1^2 \left[\frac{\mu \alpha_0^2(\tau, \Omega) - 2\nu \delta_1^2(0, 0)}{2 \ln \frac{1}{P_{F_r}} + 2\nu} - \frac{N_0^2}{E_{a_1}^2} \right]. \tag{9.171}$$

We define an action of the parameters

$$\alpha_0^2(\tau, \Omega)$$

and

$$\frac{E_a^2}{N_0^2}$$

on the resolution intervals under the fixed values of the probability of resolution P_{D_r} and the probability of false alarm P_{F_r}.

Next we assume that the resolution intervals defined on the basis of the Woodward criterion and the statistical criterion coincide with each other during the following conditions

$$\alpha_0^2(\tau, \Omega) = 0 \tag{9.172}$$

and

$$\frac{E_{a_1}^2}{N_0^2} \to \infty \tag{9.173}$$

and are equal to τ_r and Ω_r, respectively.

If the conditions in Eq. (9.148) are satisfied, then the probability of resolution defined by the statistical criterion is determined in the following form:

$$P_{D_r} = \exp\left(-\frac{K_g^2}{2\sigma^2}\right)\exp\left\{-\frac{\ln\frac{1}{P_{F_r}}}{1+\frac{\delta_1^2(0,0)}{\delta_1^2(\tau,\Omega)}}\right\} \tag{9.174}$$

(compare with Eq. (9.134)).

If the signal appearance time resolution interval τ_r and the signal frequency resolution interval Ω_r are defined based on the Woodward criterion, for example, using the bandwidth of a rectangle with the equivalent area, as was discussed in Section 9.2, then during comparison of the results obtained by the Woodward criterion and the statistical criterion there is a need to take into account the fact that the only and definite value of the probability of false alarm P_{F_r} corresponds to the probability of resolution P_{D_r} in accordance with Eq. (9.174):

$$P_{F_r} = \exp\left\{-\left[1+\frac{\delta_1^2(0,0)}{\delta_1^2(\tau_r,\Omega_r)}\right]\ln\frac{1}{P_{D_r}}\right\}. \tag{9.175}$$

If the probability of resolution P_{D_r} and the probability of false alarm P_{F_r} are previously given then to define the area of applicability of the Woodward criterion, when there is the fluctuating multiplicative noise, there is a need to search a procedure of definition of the signal appearance time resolution interval τ_r and the signal frequency resolution interval Ω_r on the basis of the Woodward criterion.

In other words, we must define the origin of the bandwidth of the interval, within the limits of which there is the signal noise component at the output of the considered model of the generalized detector, in which the signal appearance time resolution interval τ_r and the signal frequency resolution interval Ω_r determined by the Woodward criterion coincide with the resolution intervals determined by the following formula that follows from Eq. (9.174):

$$\tau_r, \Omega_r = \text{arc}\, \delta_1^2 \left[\delta_1^2(0, 0) \cdot \frac{\ln P_{D_r}}{\ln \frac{P_{F_r}}{P_{D_r}}} \right]. \tag{9.176}$$

Henceforth, we assume that the probability of resolution P_{D_r} and the origin of the bandwidth of the interval, within the limits of which there is the signal noise component at the output of the considered model of the generalized detector, defining the signal appearance time resolution interval τ_{r_W} and the signal frequency resolution interval Ω_{r_W} on the basis of the Woodward criterion, were given previously.

For the conditions mentioned above we compare the Woodward criterion and the statistical criterion using two examples for the case in which the energy spectrum of the noise modulation function $\dot{M}(t)$ of the fluctuating multiplicative noise has a bell-shaped form and there is the rapid fluctuating multiplicative noise:

$$\xi = \frac{\Delta\Omega_M T}{2\pi} \geq 3. \tag{9.177}$$

- *Example 1.* During resolution of two signals that differ from each other by the shift in frequency, the conditions mentioned above and Eqs. (6.121) and (6.220) allow us to write

$$\delta_1^2(0, 0) = \frac{\overline{A^2(t)} - \alpha_0^2(\tau, \Omega)}{2\xi} \cdot e^{-\frac{\pi\Omega^2}{\Delta\Omega_M}} = \frac{\overline{A^2(t)} - \alpha_0^2(\tau, \Omega)}{2\xi} \cdot e^{-\frac{\Omega^2 T^2}{4\pi\xi^2}}. \tag{9.178}$$

- *Example 2.* During resolution of two linearly modulated signals that differ from each other by the signal appearance time resolution interval if the condition

$$3 \leq \xi \leq Q_y^2 \tag{9.179}$$

is satisfied, taking into consideration Eqs. (6.227) and (6.228), we can write

$$\delta_1^2(\tau, \Omega) = \frac{\overline{A^2(t)} - \alpha_0^2(\tau, \Omega)}{2\xi} \cdot e^{-\frac{\pi\tau^2}{T^2} \cdot \frac{Q_y^2}{\xi^2}}. \tag{9.180}$$

The formulae determined by Eqs. (9.178) and (9.180) can be replaced by a single formula for simplicity

$$\delta_1^2(l) = \frac{1 - \alpha_0^2(\tau, \Omega)}{2\xi} \cdot e^{-\lambda^2 \frac{\pi^2 l^2}{\xi^2}}, \tag{9.181}$$

where for example 1 we can write

$$\lambda = \frac{T}{2\pi} \tag{9.182}$$

and

$$l = \Omega \tag{9.183}$$

and for example 2 we can write

$$\lambda = \frac{Q_y}{T} \tag{9.184}$$

and

$$l = \tau. \tag{9.185}$$

We assume that

$$\overline{A^2(t)} = 1$$

in Eq. (9.181). In other words, the fluctuating multiplicative noise does not act on the power of the signal.

We consider a relative difference l_{rW} between the resolution intervals determined on the basis of the Woodward criterion and on the basis of the formula in Eq. (9.171), which takes into consideration the stimulus of the undistorted component of the signal at the output of the considered model of the generalized detector and the stimulus of the additive Gaussian noise (l_r), as a measure defining the area of applicability of the Woodward criterion:

$$\Delta = 1 - \frac{l_r}{l_{rW}}. \tag{9.186}$$

Substituting Eq. (9.181) in Eq. (9.164) and taking into account that

$$b^2 = 2 \ln \frac{1}{P_{F_r}}, \tag{9.187}$$

where the probability of false alarm P_{F_r} is determined by Eq. (9.175), we can write

$$l_r = \frac{\xi}{\sqrt{\pi}} \sqrt{-\ln \left\{ \frac{\mu \cdot \frac{\alpha_0^2(\tau,\Omega)}{1-\alpha_0^2(\tau,\Omega)} - \nu}{\left[1 + e^{\frac{\pi l_{rW}^2 \lambda^2}{\xi^2}}\right] \ln \frac{1}{P_{D_r}} + \nu} - \frac{2\xi}{1 - \alpha_0^2(\tau, \Omega)} \cdot \frac{N_0^2}{E_{a_1}^2} \right\}}. \tag{9.188}$$

Taking into account that[29]

$$e^{\frac{\pi l_{rW}^2 \lambda^2}{\xi^2}} \gg 1 \tag{9.189}$$

and

$$\ln \frac{1}{P_{D_r}} \cdot e^{\frac{\pi l_{rW}^2 \lambda^2}{\xi^2}} \gg \nu, \tag{9.190}$$

we can write

$$l_r = \frac{\xi}{\lambda\sqrt{\pi}} \sqrt{-\ln\left\{ \frac{\mu \cdot \frac{\alpha_0^2(\tau,\Omega)\xi}{1-\alpha_0^2(\tau,\Omega)} - \nu}{\ln \frac{1}{P_{D_r}}} \cdot e^{-\frac{\pi l_{rW}^2 \lambda^2}{\xi^2}} - \frac{2\xi}{1-\alpha_0^2(\tau,\Omega)} \cdot \frac{N_0^2}{E_{a_1}^2} \right\}}. \tag{9.191}$$

We estimate an impact of the relative level

$$\alpha_0^2(\tau, \Omega)$$

of the undistorted component of the signal at the output of the considered model of the generalized detector on the signal appearance time resolution interval τ_r and the signal frequency resolution interval Ω_r if the additive Gaussian noise is low and the following condition

$$\frac{2\xi}{1-\alpha_0^2(\tau,\Omega)} \cdot \frac{N_0^2}{E_{a_1}^2} \to 0 \tag{9.192}$$

is satisfied.

Substituting Eq. (9.191) in Eq. (9.186), we can write the following formula

$$\Delta \simeq \frac{\xi^2}{2\pi\lambda^2 l_{rW}^2} \ln\left\{ \frac{\mu \cdot \frac{\alpha_0^2(\tau,\Omega)\xi}{1-\alpha_0^2(\tau,\Omega)} - \nu}{\ln \frac{1}{P_{D_r}}} \right\} \tag{9.193}$$

for low magnitudes of the value Δ.

The origin of the interval, within the limits of which there is a need to define the resolution interval during the use of the Woodward criterion, is absent in an explicit form in Eq. (9.193). For the example considered, the following formulae allowing us to define a function between the resolution interval l_r and the origin

$$\zeta = \frac{\delta_1^2(l_{rW})}{\delta_1^2(0)} \tag{9.194}$$

follow from Eq. (9.181):

$$\frac{\pi\lambda^2 l_{rW}^2}{\xi^2} = \ln \frac{\delta_1^2(0)}{\delta_1^2(l_{rW})} = \ln \frac{1}{\zeta}. \tag{9.195}$$

In terms of Eq. (9.195), Eq. (9.193) can be rewritten in the following form:

$$\Delta \simeq \frac{1}{\ln\frac{1}{\zeta}} \cdot \ln\left\{\frac{\mu \cdot \frac{\alpha_0^2(\tau,\Omega)\xi}{1-\alpha_0^2(\tau,\Omega)} - \nu}{\ln\frac{1}{P_{D_r}}}\right\}. \tag{9.196}$$

If the resolution interval is defined as the bandwidth of rectangle that is equal by area to the function $\delta_1^2(l)$ for the example considered we can write

$$\ln\frac{\delta_1^2(0)}{\delta_1^2(l_{rW})} = \ln\frac{1}{\zeta} = \pi. \tag{9.197}$$

Reference to Eq. (9.193) shows that the relative error Δ of definition of the resolution interval by the Woodward criterion that arises as a result of a presence of the undistorted component of the signal at the output of the considered model of the generalized detector depends on the following parameter

$$\frac{\alpha_0^2(\tau, \Omega)\xi}{1 - \alpha_0^2(\tau, \Omega)}, \tag{9.198}$$

which defines the ratio between the power of the undistorted component of the signal at the output of the considered model of the generalized detector and the power of fluctuations caused by the fluctuating multiplicative noise at the point, where the undistorted component of the signal at the output of the considered model of the generalized detector has the maximum.

Reference to Eq. (9.193) shows that if this ratio is high, then the origin ζ used for a definition of the resolution interval l_{rW} is low. This circumstance is clear in the physical sense.

Really, a decrease in the level of the origin of the resolution interval l_{rW}, as can be seen from Eq. (9.175), corresponds to a decrease in the probability of false alarm P_{F_r}. In other words, the resolution or detection of the second signal is carried out more carefully under the less probability of false alarm P_{F_r}.

In this case, a high relative level of the undistorted component of the signal at the output of the considered model of the generalized detector is required to carry out a high resolution of the signals, if there is the fluctuating multiplicative noise in comparison with the case of high values of the probability of false alarm P_{F_r}—during the higher level of the origin l_{rW}.

Dependence of the relative error Δ as a function of the parameter $\alpha_0^2(\tau, \Omega)\xi$ is shown in Fig. 9.2 for the following values in Eq. (9.193):

$$\overline{1 - \alpha_0^2(\tau, \Omega)}.$$

The origin ζ is considered as the parameter. The curves shown in Fig. 9.2 are plotted taking into consideration limitations following from Eqs. (9.165) and (9.166).

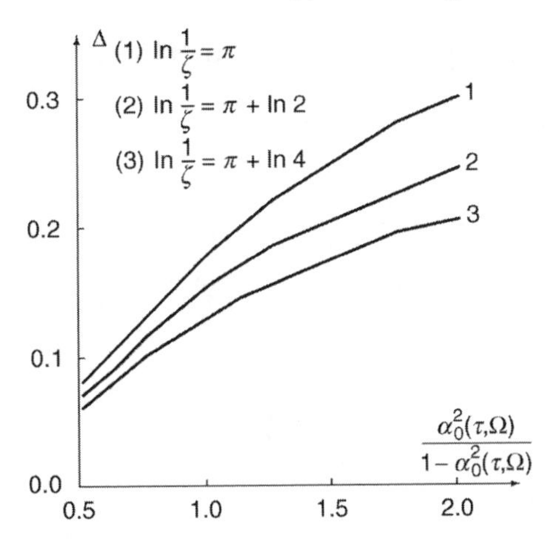

FIGURE 9.2
Dependence of the relative error Δ as a function of the parameter $\frac{\alpha_0^2(\tau,\Omega)\xi}{1-\alpha_0^2(\tau,\Omega)}$ at $P_{D_r} = 0.9$; $\mu = 0$; $\nu = 0$.

Reference to Fig. 9.2 shows that the Woodward criterion allows us to obtain satisfactory results if the power of the undistorted component of the signal at the output of the considered model of the generalized detector is less than or moderately greater than the power of the signal noise component at the output of the considered model of the generalized detector at the same point, where the undistorted component of the signal at the output of the considered model of the generalized detector is maximum.

Note that for the following magnitudes of the origin ζ and the probability of resolution P_{D_r} equal to 0.9 in accordance with Eq. (9.150) we can write

$$P_{F_r} = \begin{cases} 10^{-1} & \text{if } \ln\zeta = -\pi; \\ 10^{-2} & \text{if } \ln\zeta = -\pi - \ln 2; \\ 7\cdot 10^{-5} & \text{if } \ln\zeta = -\pi - \ln 4. \end{cases} \tag{9.199}$$

Now we estimate the stimulus of the additive Gaussian noise within the limits of the resolution interval assuming that

$$\alpha_0^2(\tau, \Omega) = 0. \tag{9.200}$$

For this purpose there is a need to use Eq. (9.174) as an initial formula, since Eq. (9.191) during the condition

$$\alpha_0^2(\tau, \Omega) = 0$$

is not applied because the condition in Eq. (9.165) is not satisfied.

Taking into consideration Eqs. (9.132) and (9.135) and using Eq. (9.174), we can write

$$l_r = \frac{\xi}{\lambda\sqrt{\pi}}\sqrt{-\ln\left\{e^{-\frac{\pi l_{r_W}^2\lambda^2}{\xi^2}} - 2\xi \cdot \frac{N_0^2}{E_{a_1}^2}\right\}}. \tag{9.201}$$

Taking into consideration that the resolution interval l_r is a function of the origin ζ, we can write

$$l_r = \frac{\xi}{\lambda\sqrt{\pi}}\sqrt{-\ln\left(\zeta - 2\xi \cdot \frac{N_0^2}{E_{a_1}^2}\right)}. \tag{9.202}$$

The clearly defined dependence of the resolution interval on the probability of resolution P_{D_r} and the probability of false alarm P_{F_r} is absent in an explicit form in Eq. (9.202).

Evidently, this phenomenon can be explained in the following way: at the previously given resolution interval l_r on the basis of the Woodward criterion the stimulus of the additive Gaussian noise on the resolution interval l_r is the same for any values of the probability of resolution P_{D_r} and the probability of false alarm P_{F_r} satisfying Eq. (9.174). Relative difference between the resolution intervals defined on the basis both of the Woodward criterion and of the statistical criterion (see Eq. (9.186)) in terms of Eq. (9.202) can be determined in the following form:

$$\Delta = 1 - \sqrt{1 + \frac{1}{\ln\zeta} \cdot \ln\left(1 - 2 \cdot \frac{\xi}{\zeta} \cdot \frac{N_0^2}{E_{a_1}^2}\right)}. \tag{9.203}$$

During the relatively low error Δ that is of interest to us, the second term in Eq. (9.203) can be used for the power series expansion and can be limited by the first and the second terms of the power series expansion.

Also, the function *ln* can be used for the Taylor series expansion and we can use the first and the second terms of the Taylor series expansion in the neighborhood of the point

$$2 \cdot \frac{\xi}{\zeta} \cdot \frac{N_0^2}{E_{a_1}^2} = 1. \tag{9.204}$$

Finally, we can write

$$\Delta \simeq \frac{\xi}{\zeta\ln\zeta} \cdot \frac{N_0^2}{E_{a_1}^2}. \tag{9.205}$$

Equation (9.205) allows us to define the permissible energy level of the additive Gaussian noise using the previously given magnitude of the error Δ, in which the use of the Woodward criterion is still possible.

Reference to Eq. (9.205) shows that when the permissible ratio between the signal energy E_{a_1} and the spectral power density N_0 of the additive Gaussian noise is less, then the energy spectrum of the noise modulation function $\dot{M}(t)$ of the fluctuating multiplicative noise is narrow (the parameter ξ is less).

If the resolution interval l_{rw} is defined as the equivalent bandwidth of the function defining a distribution of the power of fluctuations of the signal at the output of the considered model of the generalized detector, as stated in Section 9.2, then we can believe that

$$\zeta = e^{-\pi} \tag{9.206}$$

and the following condition

$$\frac{\zeta}{\xi} \cdot \frac{E_{a_1}^2}{N_0^2} \geq \frac{1}{\pi|\Delta|} \tag{9.207}$$

must be satisfied.

The value

$$\frac{\zeta}{\xi} \cdot \frac{E_{a_1}^2}{N_0^2}$$

can be thought of as the ratio between the power of fluctuations of the signal distorted by the fluctuating multiplicative noise at the point

$$l = l_{rw}$$

and the power of the additive Gaussian noise at the output of the considered model of the generalized detector.

Thus, the use of the Woodward criterion for the purpose of definition of the resolution intervals during a presence of the additive Gaussian noise does not lead to high errors only if the power of the signal noise component at the output of the considered model of the generalized detector at the point

$$l = l_{rw}$$

is higher in comparison to the power of the additive Gaussian noise.

9.4 Conclusions

In this chapter we briefly considered the problem of signal resolution or individual definition and measurement of the signal parameters under the use of the generalized approach to signal processing in the presence of noise and, in particular, the signal resolution by complex signal processing systems constructed on the basis of the generalized detector. We discussed two approaches to the signal resolution problem: the Woodward criterion and the statistical criterion.

When a distribution of the total power of the signal distorted by the fluctuating multiplicative noise at the output of the considered model of the generalized detector with respect to the parameters l—the sum of powers of the undistorted component of the signal and the signal noise component—does not have a sharp shoot in the neighborhood of the points, where the undistorted component of the signal exists, the Woodward criterion is not difficult to use to estimate an impact of the fluctuating multiplicative noise on the signal resolution.

This fact takes place during the following assumptions with respect to the energy characteristics of the signal at the output of the considered model of the generalized detector: the power of the undistorted component of the signal at the output of the considered model of the generalized detector is low in comparison to the variance of the signal noise component at the point, where the undistorted component of the signal is maximum, and the variance of the signal noise component is the sufficiently smooth and convex function within the limits of the interval that is not less than the equivalent bandwidth of this variance. If one or both conditions are not satisfied, then the use of the Woodward criterion generates additional errors.

The limitations mentioned above that are peculiarities of the Woodward criterion cannot be taken into consideration during estimation of the stimulus of the fluctuating multiplicative noise on the signal resolution for the use of the statistical criteria.

Comparison of the resolution intervals obtained under the use of the statistical criteria and the Woodward criterion allows us to estimate the use of the Woodward criterion and to explain some equivalent (or conditional) statistical sense of the Woodward criterion.

The main results discussed in this chapter are the following:

- During the use of the Woodward criterion if there is the fluctuating multiplicative noise the frequency resolution interval of the signal can be defined by the bandwidth of the energy spectrum of the noise modulation function $\dot{M}(t)$ of the fluctuating multiplicative noise. The fluctuating multiplicative noise does not act on the signal appearance time resolution interval during the use of the signals, the autocorrelation function of which is determined by Eq. (9.53). In particular, the signals with a bell-shaped complex amplitude envelope and constant radio frequency carrier, the signals with a square wave-form complex amplitude envelope and constant radio frequency carrier, the signals with the noise modulation, and the signals with the phase-code modulation possessing the high signal base have the autocorrelation function mentioned above.

- In the case of the wide-band fluctuating multiplicative noise the signal appearance time resolution interval is defined only by the complex amplitude envelope of the signal and does not depend on the phase structure of the signal. For the signals with a square wave-form and bell-shaped complex amplitude envelope the signal appearance time resolution interval is equal to the duration of the signal. The Woodward criterion cannot be used for the purpose of estimating the signal appearance time resolution interval of the phase-modulated signal in a general case when there is fluctuating multiplicative noise.
- Under the use of the statistical criteria the probability of resolution of the second signal relative to the first signal, for the considered model of the generalized detector, is defined by the probability of detection of the second signal in the background of the first signal and the additive Gaussian noise and the fluctuating multiplicative noise, in other words, by the probability of exceeding the process at the output of the considered model of the generalized detector over the given threshold K_g that is defined by the predetermined probability of false alarm of the second signal when the second signal is absent from the input stochastic process.
- The signal appearance time resolution interval and the signal frequency resolution interval are uniquely defined by the probability of resolution and the probability of false alarm under the known statistical characteristics of the signals, additive Gaussian noise, and fluctuating multiplicative noise.
- Comparative analysis of the use of the Woodward criterion and the statistical criteria with the purpose of definition of the signal resolution shows that the use of the Woodward criterion when there is the additive Gaussian noise does not lead to high errors only if the power of the signal noise component at the output of the considered model of the generalized detector is greater in comparison to the power of the additive Gaussian noise at the point

$$l = l_{r_W}.$$

References

1. Di Franco and Rubin, L., *Radar Detection*, Artech House, Norwood, MA, 1980.
2. Wehner, D., *High-Resolution Radar*, 2nd ed., Artech House, Norwood, MA, 1994.
3. Rihaczek, A., *Principles of High-Resolution Radar*, Artech House, Norwood, MA, 1996.
4. Benjamin, R., *Modulation Resolution and Signal Processing Radar, Sonar, and Related Systems*, Franklin Book Company, New York, 1966.

5. Bouvet, M. and Bienvenu, G., *High-Resolution Methods in Underwater Acoustics*, Springer-Verlag, New York, 1991.
6. Mensa, D., *High Resolution Radar Cross-Section Imaging*, Artech House, Norwood, MA, 1991.
7. Woodward, F., *Theory of Probabilities and Information Theory Applied to Radar*, Wiley, New York, 1953.
8. Ingle, N., Manolakis, D., and Kogon, S., *Statistical and Adaptive Signal Processing: Spectral Estimation, Signal Modeling, Adaptive Filtering, and Array Processing*, McGraw-Hill, New York, 1999.
9. Hansen, R., *Phased Array Antennas*, Wiley, New York, 1997.
10. Haykin, S., Litva, J., and Shepherd, T., *Radar Array Processing*, Springer-Verlag, New York, 1993.
11. Shin, D., *Wireless Communication Using Dual Antenna Arrays*, Kluwer Academic Publishers, New York, 1999.
12. Brown, W. and Palermo, C., Effect of phase errors on resolution, *IEEE Trans.*, 1965, Vol. MJL-9, No. 1, pp. 77–89.
13. Sabatini, S. and Tarantion, M., *Multi-Function Array Radar System Design and Analysis*, Artech House, Norwood, MA, 1994.
14. Fourikis, N., *Phased Array-Based Systems and Applications*, Wiley, New York, 1996.
15. Unnikrishna Pillai, S. and Burrus, C., *Array Signal Processing*, Springer-Verlag, New York, 1989.
16. Fourikis, N., *Advanced Array Systems, Applications and RF Technologies*, Academic Press, New York, 2000.
17. Naidu, P., *Sensor Array Signal Processing*, CRC Press, Boca Raton, FL, 2000.
18. Steinberg, B., *Microwave Imaging with Large Antenna Arrays: Radio Camera Principles and Techniques*, Wiley, New York, 1983.
19. Bacut, P. et al. *Problems in Statistical Radar Theory. Parts 1 and 2*, Soviet Radio, Moscow, 1963–1964 (in Russian).
20. Helstrom, C., *Statistical Theory of Signal Detection*, 2nd ed., Pergamon Press, Oxford, 1968.
21. Helstrom, C., *Elements of Signal Detection and Estimation*, Prentice-Hall, Englewood Cliffs, NJ, 1965.
22. Shirman, Y. and Manjos, V., *Theory and Methods in Radar Signal Processing*, Radio and Svyaz, Moscow, 1981 (in Russian).
23. Tuzlukov, V., *Signal Detection Theory*, Springer-Verlag, New York, 2001.
24. Tuzlukov, V., *Signal Processing in Noise: A New Methodology*, IEC, Minsk, 1998.
25. Doktorov, A., Spectrum and mutual correlation function of modulation of signals with phase modulation, *Radio Eng.*, 1966, No. 7, pp. 53–59 (in Russian).
26. Falkovich, S., *Signal Processing of Radar Signals in Fluctuating Noise*, Soviet Radio, Moscow, 1961 (in Russian).
27. Bunimovich, V., Approximate determination of the probability of detection of the signals with unknown phase, *Radio Eng. Electron. Phys.*, 1958, Vol. 3, No. 4, pp. 445–467.
28. Marcum, J., A statistical theory of target detection by pulsed radar, *IEEE Trans.*, 1960, Vol. IT-6, No. 2, pp. 135–158.
29. Kremer, I., Vladimirov, V., and Karpuhin, V., *Multiplicative Noise and Radio Signal Processing*, Soviet Radio, Moscow, 1972 (in Russian).

Appendix I

Delta Function

The delta function

$$\delta(x - x_0)$$

is called the function tending to approach ∞ if an argument of this function tends to approach zero, and equal to zero for all other values of argument:

$$\delta(x - x_0) = \begin{cases} \infty & \text{at } x = x_0; \\ 0 & \text{at } x \neq x_0 \end{cases} \tag{I.1a}$$

and

$$\int\limits_{x_0-\varepsilon}^{x_0+\varepsilon} \delta(x - x_0)\, dx = 1, \quad \forall\varepsilon > 0. \tag{I.1b}$$

Occasionally the delta function is defined as the even function:

$$\delta(x - x_0) = \delta(x_0 - x). \tag{I.2}$$

In this case we can write

$$\int\limits_{x_0-\varepsilon}^{x_0} \delta(x - x_0)\, dx = \int\limits_{x_0}^{x_0+\varepsilon} \delta(x - x_0)\, dx = 0.5, \quad \varepsilon > 0. \tag{I.3}$$

The delta function can be considered as a limit of the unlimited sequence of usual functions. We assume that there is the function $f(x)$ that is the continuous function at the point x_0 and there is a set of usual functions $\varphi_\alpha(x)$ satisfying the condition

$$\lim_{\alpha\to\alpha_0} \int\limits_a^b f(x)\,\varphi_\alpha(x - x_0)\, dx = f(x_0), \quad a < x_0 < b. \tag{I.4}$$

Then the delta function $\delta(x - x_0)$ can be written in the following form:

$$\delta(x - x_0) = \lim_{\alpha\to\alpha_0} \varphi_\alpha(x - x_0), \tag{I.5}$$

where the function $f(x_0)$ determined by Eq. (I.5) follows from the formula

$$\int_a^b f(x)\,\delta(x-x_0)\,dx = \begin{cases} f(x_0) & \text{at } a < x_0 < b; \\ 0 & \text{at } x_0 < a \quad \text{or} \quad x_0 > b. \end{cases} \tag{I.6}$$

Consider the function in a form of pulse

$$\varphi_\alpha(x-x_0) = \begin{cases} \frac{1}{\alpha} & \text{at } x_0 - \frac{\alpha}{2} < x_0 + \frac{\alpha}{2}; \\ 0 & \text{otherwise}. \end{cases} \tag{I.7}$$

This function satisfies the equality

$$\int_{-\infty}^{\infty} \varphi_\alpha(x-x_0)\,dx = 1, \quad \forall\alpha > 0. \tag{I.8}$$

For the condition

$$\alpha \to 0$$

a duration of pulse tends to approach zero and a height of pulse tends to approach ∞, and an area of pulse is equal to unity. Because of this, we can believe that

$$\delta(x-x_0) = \lim_{\alpha\to 0} \varphi_\alpha(x-x_0). \tag{I.9}$$

Although the function in the form of a pulse is a simple precursor of the delta function it is the discontinuous function.

In many practical problems it is convenient to use an initial set of functions possessing derivatives. We indicate some classes of this function:

$$\begin{aligned} \delta(x-x_0) &= \frac{1}{\pi} \cdot \lim_{\alpha\to\infty} \left[\frac{\sin\alpha(x-x_0)}{x-x_0}\right] \\ &= \lim_{\alpha\to 0} \left\{\frac{1}{\alpha\sqrt{2\pi}} \cdot \exp\left[-\frac{(x-x_0)^2}{2\alpha^2}\right]\right\} \\ &= \frac{1}{\pi} \cdot \lim_{\alpha\to\infty} \left[\frac{1-\cos\alpha(x-x_0)}{\alpha(x-x_0)^2}\right] \\ &= \frac{1}{\pi} \cdot \lim_{\alpha\to\infty} \left[\frac{\sin^2\alpha(x-x_0)}{\alpha(x-x_0)^2}\right] \\ &= \frac{1}{\pi} \cdot \lim_{\alpha\to\infty} \frac{\alpha}{\alpha^2(x-x_0)^2+1} \\ &= \frac{1}{\pi^2(x-x_0)} \cdot \lim_{\alpha\to 0} \int_{(x-x_0)-\alpha}^{(x-x_0)+\alpha} \frac{dy}{y} \\ &= \lim_{N\to\infty} \sum_{n=0}^{N} \varphi_n(x)\,\varphi_n(x_0), \end{aligned} \tag{I.10}$$

where $\varphi_n(x)$ is the total orthonormalized system of functions.

The following relationships are also true:

$$\delta(x - x_0) = 0.5 \cdot \frac{\partial^2}{\partial x^2}|x - x_0|; \tag{I.11a}$$

$$\delta(x - x_0) = \frac{1}{h} \cdot \frac{\partial}{\partial x}\varphi(x); \tag{I.11b}$$

$$\varphi(x) = \begin{cases} c & \text{at } x < x_0; \\ c + h & \text{at } x > x_0. \end{cases} \tag{I.11c}$$

In many cases the use of the delta function allows us to simplify calculations significantly. This is possible as a result of some peculiarities of the delta function. Equation (I.6) is one of the main peculiarities of the delta function and is called the filtering feature of the delta function.

Equation (I.6) can be obtained in the following manner, without using Eq. (I.4). Since

$$\delta(x - x_0)$$

is equal to zero except for the point x_0, and the continuous function $f(x)$ is equal to the function $f(x_0)$ in the neighborhood of the point x_0 we can use Eqs. (I.1a) and (I.1b) to obtain Eq. (I.6). For this purpose there is a need to factor the function $f(x)$ outside the integral sign.

If the function $f(x)$ is considered as the input signal—the input stochastic process—acting on a linear filter that has the pulse response $\delta(x)$, then according to Eq. (I.6) only one value of the input signal corresponding to the zero argument of the delta function is formed at the output of the linear filter.

Hence it follows that the function

$$\delta(x - x_0)$$

has a dimensional representation that is inverse to the value x. Note if the condition

$$x_0 = a$$

or

$$x_0 = b$$

is satisfied the integral in Eq. (I.6) is indefinite.

Sometimes we can suppose that

$$\int_a^b f(x)\,\delta(x - x_0)\,dx = \begin{cases} \frac{f(a)}{2} & \text{at } x_0 = a; \\ \frac{f(b)}{2} & \text{at } x_0 = 0. \end{cases} \tag{I.12}$$

If x_0 is the break point of the first order of the function $f(x)$ then

$$\int_a^b f(x)\,\delta(x-x_0)\,dx = 0.5\cdot\left[f(x_0^+)+f(x_0^-)\right], \qquad a < x_0 < b, \tag{I.13}$$

where $f(x_0^+)$ or $f(x_0^-)$ is the function $f(x)$ to the right or to the left of the break point.

If the function $\varphi(x)$ is the continuous function at the point x_0 then

$$\varphi(x)\,\delta(x-x_0) = \varphi(x_0)\,\delta(x-x_0), \tag{I.14}$$

since

$$\int_a^b f(x)\,\varphi(x)\,\delta(x-x_0)\,dx = f(x_0)\,\varphi(x_0)$$

$$= \int_a^b f(x)\,\varphi(x_0)\,\delta(x-x_0)\,dx,$$

$$a < x_0 < b. \tag{I.15}$$

In particular

$$\int_a^b \delta(x-u)\,\delta(x-v)\,dx = \delta(u-v) = \delta(v-u),$$

$$a < u, \quad v < b. \tag{I.16}$$

Introducing a change of variable

$$y = cx$$

and using Eq. (I.6), we can write

$$\int_a^b f(x)\,\delta(cx-x_0)\,dx = \frac{1}{c}\int_{ac}^{bc} f\left(\frac{y}{c}\right)\delta(y-x_0)\,dy$$

$$= \frac{1}{|c|}\cdot f\left(\frac{x_0}{c}\right), \qquad a < \frac{x_0}{|c|} < b. \tag{I.17}$$

Because of this, we can write

$$\delta(cx-x_0) = \frac{1}{|c|}\cdot\delta\left(x-\frac{x_0}{c}\right). \tag{I.18}$$

Applying Eq. (I.6) separately to the functions

$$\alpha\delta(x - x_0)$$

and

$$\delta\left(\frac{x - x_0}{\alpha}\right),$$

we can believe that the equality

$$|\alpha| \cdot \delta(x - x_0) = \delta\left(\frac{x - x_0}{\alpha}\right) \tag{I.19}$$

is true.

We consider a more general case. The function $\alpha(x)$ is the monotone increasing function within the limits of the interval $[a, b]$ and the following condition

$$\alpha(a) < \alpha(x_0) = 0 < \alpha(b) \tag{I.20}$$

is satisfied.

For the monotone increasing function the equation

$$\alpha(x) = t$$

has an unambiguous inverse function $x(t)$; in doing so,

$$x(0) = x_0.$$

Introducing a change of variable, we can write in terms of Eq. (I.6)

$$\begin{aligned}\int_a^b f(x)\,\delta[\alpha(x)]\,dx &= \int_{\alpha(a)}^{\alpha(b)} \frac{f[x(t)]}{\alpha'[x(t)]} \cdot \delta(t)\,dt \\ &= \frac{f[x(0)]}{\alpha'[x(0)]} = \frac{f(x_0)}{\alpha'(x_0)}.\end{aligned} \tag{I.21}$$

If $\alpha(x)$ is the monotone decreasing function, then the sign of obtained function is inverse as a result of interchanging the limits of integration.

Thus the integral mentioned above in both cases is equal to

$$\frac{f(x_0)}{|\alpha'(x_0)|}.$$

Because of this, we can assume in Eq. (I.6) that

$$\delta[\alpha(x)] = \frac{\delta(x - x_0)}{|\alpha'(x_0)|}. \tag{I.22}$$

If the condition determined by Eq. (I.20) is not satisfied, then the integral is equal to zero.

Suppose that the equality

$$\alpha(x) = 0$$

is carried out for the condition

$$\alpha(x_n) = 0$$

and at this point the function $\alpha(x)$ possesses the continuous derivative

$$\alpha'(x_n) \neq 0.$$

Then it is easy to show that

$$\delta[\alpha(x)] = \sum_n \frac{\delta(x - x_n)}{|\alpha'(x_n)|}. \tag{I.23}$$

In other words, the function

$$\delta[\alpha(x)]$$

is equal to the sequence of delta pulses for the condition

$$x = x_n$$

with the area equal to

$$|\alpha'(x_n)|^{-1}.$$

In order to prove Eq. (I.23), we divide the axis x on the intervals $[c_i, c_{i+1}]$ so that the function $\alpha(x)$ would be changed monotonically within the limits of the interval $[c_i, c_{i+1}]$.

If the integral with unlimited limits from the function

$$f(x)\,\delta[\alpha(x)]$$

can be presented as a sum of integrals within the limits of the intervals $[c_i, c_{i+1}]$ then, in terms of Eq. (I.21), we can obtain Eq. (I.23).

Now consider again Eq. (I.10):

$$\delta(x - x_0) = \lim_{\alpha\to\infty} \frac{\sin\alpha(x - x_0)}{\pi(x - x_0)}. \tag{I.24}$$

Using Eq. (I.24), we can write

$$\lim_{\alpha\to\infty} \int_{-\infty}^{\infty} f(x) \cdot \frac{\sin\alpha(x - x_0)}{\pi(x - x_0)}\,dx = f(x_0). \tag{I.25}$$

A very important identity

$$\frac{1}{2\pi}\int_{-\infty}^{\infty} e^{\pm j(x-x_0)u}\,du = \int_{-\infty}^{\infty} e^{\pm 2\pi j(x-x_0)\,v}\,dv = \delta(x-x_0) \tag{I.26}$$

is a corollary of Eq. (I.24).

In fact

$$\frac{1}{2\pi}\int_{-\infty}^{\infty} e^{\pm j(x-x_0)u}\,du = \lim_{\alpha\to\infty}\frac{1}{2\pi}\cdot\int_{-\alpha}^{\alpha} e^{\pm j(x-x_0)u}\,du$$
$$= \lim_{\alpha\to\infty}\frac{\sin\alpha(x-x_0)}{\pi(x-x_0)}. \tag{I.27}$$

Considering in Eq. (I.26) the variable x as t (the time) and the variable u as ω (the cyclic frequency), we obtain the delta function

$$\delta(t-t_0)$$

in the form of Fourier integral

$$\delta(t-t_0) = \frac{1}{2\pi}\int_{-\infty}^{\infty} e^{j\omega(t-t_0)}\,d\omega = \frac{1}{\pi}\int_{0}^{\infty}\cos\omega(t-t_0)\,d\omega. \tag{I.28}$$

Using the inverse Fourier transform in terms of Eq. (I.6), we can define the spectral density for the delta function

$$\delta(t-t_0)$$

in the following form:

$$\int_{-\infty}^{\infty}\delta(t-t_0)\cdot e^{-j\omega t}dt = e^{-j\omega t_0}. \tag{I.29}$$

Reference to Eq. (I.29) shows that for the condition

$$t_0 = 0$$

a spectrum of the function $\delta(t)$ is uniform for all frequencies and a spectrum intensity (spectral density) is equal to unity:

$$\int_{-\infty}^{\infty}\delta(t)\cdot e^{-j\omega t}dt = 1. \tag{I.30}$$

If the spectral density of the delta function is not varied in time during the condition

$$t_0 = 0$$

then the spectral density for half-sum of two delta functions

$$\delta(t + t_0)$$

and

$$\delta(t - t_0),$$

which are symmetric with respect to the origin, is a cosine curve:

$$\frac{1}{2}\int_{-\infty}^{\infty} [\delta(t + t_0) + \delta(t - t_0)] \cdot e^{-j\omega t}\, dt = 0.5 \cdot \left(e^{j\omega t_0} + e^{-j\omega t_0}\right) = \cos \omega t_0. \quad (I.31)$$

Using the inverse Fourier transform, we can write

$$\frac{1}{2}[\delta(t + t_0) + \delta(t - t_0)] = \frac{1}{2\pi}\int_{-\infty}^{\infty} \cos \omega t_0 \cdot e^{j\omega t}\, d\omega$$

$$= \frac{1}{\pi}\int_{0}^{\infty} \cos \omega t_0 \, \cos \omega t \, d\omega. \quad (I.32)$$

Equation (I.26) allows us to establish a relationship between the delta function for the universally adopted frequency f and the delta function for the cyclic frequency ω.

Considering in Eq. (I.26) the variable x as f—universally adopted frequency—and introducing a change of variable

$$s = \frac{t}{2\pi},$$

we can write

$$\delta(t - t_0) = \int_{-\infty}^{\infty} e^{\pm 2\pi j(f - f_0)t}\, dt = \int_{-\infty}^{\infty} e^{\pm j(\omega - \omega_0)t}\, dt$$

$$= 2\pi \int_{-\infty}^{\infty} e^{\pm 2\pi j(\omega - \omega_0)s}\, ds = 2\pi\, \delta(\omega - \omega_0). \quad (I.33)$$

Consequently,

$$\delta(t - t_0) = 2\pi\, \delta(\omega - \omega_0). \quad (I.34)$$

We consider the very useful equality that follows from Eqs. (I.19) and (I.26):

$$\int_0^\infty \cos\left(\frac{\omega \pm \omega_0}{\alpha}\right) x\,dx = \alpha\pi\,\delta(\omega \pm \omega_0) = \pi\,\delta\left(\frac{\omega \pm \omega_0}{\alpha}\right). \tag{I.35}$$

Using Eq. (I.26), we can define the Fourier transform for cosine, sine, and unit jump:

$$\int_{-\infty}^{\infty} e^{-j\omega t} \cos\omega_0 t\,dt = \frac{1}{2}\int_{-\infty}^{\infty} e^{-j\omega t}\left(e^{j\omega_0 t} + e^{-j\omega_0 t}\right)dt$$
$$= \pi[\delta(\omega - \omega_0) + \delta(\omega + \omega_0)]; \tag{I.36}$$

$$\int_{-\infty}^{\infty} e^{-j\omega t} \sin\omega_0 t\,dt = \frac{1}{2j}\int_{-\infty}^{\infty} e^{-j\omega t}\left(e^{j\omega_0 t} + e^{-j\omega_0 t}\right)dt$$
$$= j\pi[\delta(\omega + \omega_0) - \delta(\omega - \omega_0)]; \tag{I.37}$$

$$\frac{1}{2}\int_{-\infty}^{\infty} (1 + \operatorname{sgn} t)\cdot e^{-j\omega t}\,dt = \int_0^\infty e^{-j\omega t}\,dt = \pi\,\delta(\omega) + \frac{1}{j\omega}. \tag{I.38}$$

Equation (I.38) follows from the fact that the Fourier transform of the function

$$\operatorname{sgn} t = \begin{cases} 1 & \text{at } t > 0; \\ -1 & \text{at } t < 0 \end{cases} \tag{I.39}$$

is equal to

$$\frac{2}{j\omega}.$$

In fact, the inverse transform for the function

$$\frac{2}{j\omega}$$

is given by

$$\frac{1}{2\pi}\int_{-\infty}^{\infty} \frac{2}{j\omega}\cdot e^{j\omega t}\,d\omega = \int_{-\infty}^{\infty} \frac{\sin\omega t}{\pi\omega}\,d\omega = 2\int_0^\infty \frac{\sin\omega t}{\pi\omega}\,d\omega = \operatorname{sgn} t. \tag{I.40}$$

The periodical sequence of the delta functions

$$\delta\left(x - x_0 - \frac{m}{T_0}\right), \quad m = 0, \pm 1, \pm 2, \ldots$$

can be presented by the Fourier series

$$\sum_{n=-\infty}^{\infty} e^{2\pi jn(x-x_0)T_0} = \sum_{n=0}^{\infty} \varepsilon_n \cos 2\pi n(x-x_0)T_0$$
$$= \frac{1}{T}\sum_{m=-\infty}^{\infty} \delta\left(x - x_0 - \frac{m}{T_0}\right), \tag{I.41}$$

where

$$\varepsilon_0 = 1, \quad \varepsilon_n = 2 \quad \text{at } n \neq 0.$$

The following equations are also true:

$$\sum_{n=-\infty}^{\infty} (-1)^n \varepsilon_n \cos 2\pi n(x-x_0)T_0 = \sum_{m=-\infty}^{\infty} \delta\left(x - x_0 - \frac{2m+1}{2T_0}\right); \tag{I.42}$$

$$\sum_{n=-\infty}^{\infty} \varepsilon_n \cos 2\pi(x-x_0)T_0 \cos 2\pi ny$$
$$= \frac{1}{2T_0}\sum_{m=-\infty}^{\infty}\left[\delta\left(x - x_0 + y + \frac{m}{T_0}\right) + \delta\left(x - x_0 - y + \frac{m}{T_0}\right)\right]; \tag{I.43}$$

$$\sum_{n=-\infty}^{\infty} \varepsilon_n \cos 4\pi n(x-x_0)T_0 = \frac{1}{2T_0}\sum_{m=-\infty}^{\infty} \delta\left(x - x_0 - \frac{m}{T_0}\right). \tag{I.44}$$

Integrating by parts, we can write

$$\int_{x_0-\varepsilon}^{x_0+\varepsilon} f(x)\,\delta^{(n)}(x-x_0)\,dx = (-1)^n f^{(n)}(x_0), \quad \varepsilon > 0, \tag{I.45}$$

where

$$\delta^{(n)}(x-x_0)$$

is the derivative of the n-th order of the delta function.

If the derivative $f^{(n)}(x)$ possesses the jump of the first order at the point x_0, then

$$\int_{x_0-\varepsilon}^{x_0+\varepsilon} f(x)\,\delta^{(n)}(x-x_0)\,dx = 0.5\cdot(-1)^n\left[f^{(n)}(x_0^+) + f^{(n)}(x_0^-)\right], \quad \varepsilon > 0. \tag{I.46}$$

Using Eq. (I.45) with respect to the integral

$$J = \int_a^b \int_a^b f(x, y)\delta''(x - y)\, dx\, dy, \tag{I.47}$$

we obtain that for the condition

$$f_x''(y, y) = \left. \frac{\partial^2 f(x, y)}{\partial x^2} \right|_{x=y} = f_y''(x, x) = \left. \frac{\partial^2 f(x, y)}{\partial y^2} \right|_{y=x} \tag{I.48}$$

the following equality

$$J = \int_a^b f_x''(y, y)\, dy = \int_a^b f_y''(x, x)\, dx \tag{I.49}$$

is true.

We can introduce the two-dimensional delta function

$$\delta_2(x - x_0, y - y_0) = \delta(x - x_0, y - y_0) \tag{I.50}$$

in an analogous way as the one-dimensional delta function.

The following equalities are true for the two-dimensional delta function

$$\int_{-\infty}^{\infty} \int_{-\infty}^{\infty} \delta_2(x - x_0, y - y_0)\, dx\, dy = 1; \tag{I.51}$$

$$\int_{-\infty}^{\infty} \int_{-\infty}^{\infty} f(x, y)\, \delta_2(x - x_0, y - y_0)\, dx\, dy = f(x_0, y_0); \tag{I.52}$$

$$\int_{-\infty}^{\infty} f(x, y)\, \delta_2(x - x_0, y - y_0)\, dy = f(x, y_0)\, \delta(x - x_0). \tag{I.53}$$

Appendix II

Correlation Function and Energy Spectrum of Noise Modulation Function

II.1 Correlation Functions $\dot{R}_M(\tau)$ and $\dot{R}_V(\tau)$ under the Stimulus of Slow Fluctuating Multiplicative Noise

Let the stochastic functions $\varphi(t)$, $A(t)$, and $\xi(t)$ be differentiable in the mean square sense. Then the Maclaurin series expansion can be used for the correlation functions $\dot{R}_M(\tau)$ and $\dot{R}_V(\tau)$ of the noise modulation function $\dot{M}(t)$ of the multiplicative noise.

Based on the physical grounds it follows that during consideration of the stimulus of the multiplicative noise on the signal with the duration T we are interested in values of the correlation functions $\dot{R}_M(\tau)$ and $\dot{R}_V(\tau)$ for the condition

$$|\tau| \leq T.$$

This statement is supported by Eqs. (3.118) and (3.119).

If the time of correlation of these functions equal to τ_c is much greater than the signal duration T, then for the values satisfying the condition

$$|\tau| \leq T \ll \tau_c$$

we can be limited by the first, second, and third terms of Maclaurin's series:

$$\dot{R}_M(\tau) \simeq \dot{R}_M(0) + \tau \dot{R}'_M(0) + 0.5\tau^2 \dot{R}''_M(0) \tag{II.1}$$

and

$$\dot{R}_V(\tau) \simeq \dot{R}_V(0) + \tau \dot{R}'_V(0) + 0.5\tau^2 \dot{R}''_V(0), \tag{II.2}$$

where

$$\dot{R}'_M(\tau) = \frac{d}{d\tau}\dot{R}_M(\tau); \tag{II.3}$$

$$\dot{R}''_M(0) = \frac{d^2}{d\tau^2}\dot{R}_M(\tau); \tag{II.4}$$

$$\dot{R}'_V(\tau) = \frac{d}{d\tau}\dot{R}_V(\tau); \tag{II.5}$$

$$\dot{R}''_V(0) = \frac{d^2}{d\tau^2}\dot{R}_V(\tau). \tag{II.6}$$

We define the values $\dot{R}_M(0)$, $\dot{R}'_M(0)$, $\dot{R}''_M(0)$, $\dot{R}_V(0)$, $\dot{R}'_V(0)$, and $\dot{R}''_V(0)$ when distortions in amplitude $\xi(t)$ and phase $\varphi(t)$ of the signal caused by the multiplicative noise are independent.

By definition, the correlation functions $\dot{R}_M(\tau)$ and $\dot{R}_V(\tau)$ can be written in the following form:

$$\dot{R}_M(\tau) = m_1\left[A(t)A(t-\tau)\cdot e^{j[\varphi(t)-\varphi(t-\tau)]}\right] \tag{II.7}$$

and

$$\dot{R}_V(\tau) = \dot{R}_M(\tau) - \left|\overline{\dot{M}}\right|^2. \tag{II.8}$$

Then we can write

$$\dot{R}_M(0) = \overline{A^2} = A_0^2\left(1+\sigma_\xi^2\right) \tag{II.9}$$

and

$$\dot{R}_V(0) = \overline{A^2} - \left|\overline{\dot{M}}\right|^2 = A_0^2\left(1+\sigma_\xi^2\right) - \left|\overline{\dot{M}}\right|^2, \tag{II.10}$$

where

$$A^2 = m_1[A^2(t)] \tag{II.11}$$

and σ_ξ^2 is the variance of the stochastic process $\xi(t)$.

Taking into account that operations of differentiating and statistical averaging can be interchanged,[1] the derivative of the first order of the correlation functions $\dot{R}_M(\tau)$ and $\dot{R}_V(\tau)$ is determined in the following form:

$$\dot{R}'_M(\tau) = \dot{R}'_V(\tau) = m_1\left[A(t)A'(t-\tau)\cdot e^{j[\varphi(t)-\varphi(t-\tau)]}\right] - jm_1\left[A(t)A(t-\tau)\varphi'(t-\tau)\cdot e^{j[\varphi(t)-\varphi(t-\tau)]}\right], \tag{II.12}$$

where

$$A'(t-\tau) = \frac{d}{d\tau}A(t-\tau) \tag{II.13}$$

and

$$\varphi'(t-\tau) = \frac{d}{d\tau}\varphi(t-\tau). \tag{II.14}$$

Then we can write

$$\begin{aligned}\dot{R}'_M(0) = \dot{R}'_V(0) &= m_1[A(t)A'(t)] - jm_1[A^2(t)\varphi'(t-\tau)] \\ &= -jm_1[A^2(t)\varphi'(t)],\end{aligned} \tag{II.15}$$

where

$$\varphi'(t) = \varphi'(t-\tau)|_{\tau=0} \tag{II.16}$$

and

$$A'(t) = A'(t-\tau)|_{\tau=0}. \tag{II.17}$$

Equation (II.15) takes into consideration that

$$m_1[A(t)A'(t)] = 0$$

when $A(t)$ is the stationary stochastic function [2].

When the stochastic processes $A(t)$ and $\varphi(t)$ are independent then

$$m_1[A^2(t)\varphi'(t)] = 0, \tag{II.18}$$

since in the case of the stationary stochastic process the value

$$m_1[\varphi'(t)] = 0 \tag{II.19}$$

and, consequently,

$$\dot{R}'_M(0) = \dot{R}'_V(0) = 0. \tag{II.20}$$

Differentiating Eq. (II.12) with respect to τ, we can write

$$\begin{aligned}\dot{R}''_M(0) = \dot{R}''_V(0) &= m_1[A(t)A''(t)] - 2jm_1[A(t)A'(t)\varphi'(t)] \\ &- jm_1[A(t)A'(t)\varphi''(t)] - m_1\left[A^2(t)\varphi'^2(t)\right].\end{aligned} \tag{II.21}$$

Since the stochastic processes $A(t)$ and $\varphi(t)$ are independent, we can write

$$\begin{aligned}\dot{R}''_M(0) = \dot{R}''_V(0) &= m_1[A(t)A''(t)] - m_1[A^2(t)]m_1\left[\varphi'^2(t)\right] \\ &= m_1\left[A(t)A''(t)\right] - \overline{A^2}\sigma^2_\omega,\end{aligned} \tag{II.22}$$

where σ^2_ω is the variance of the derivative of phase changes

$$\frac{d\varphi(t)}{dt}.$$

It is easy to verify that σ_ω^2 is the variance of changes of the instantaneous frequency of the signal caused by distortions in phase of the signal. The first term on the right side of Eq. (II.22) can be represented in the following form:[2]

$$m_1[A(t)A''(t)] = -\sigma_{A'}^2, \tag{II.23}$$

where $\sigma_{A'}^2$ is the variance of the derivative $A'(t)$.

Then we can write

$$\dot{R}''_M(0) = \dot{R}''_V(0) = -\sigma_{A'}^2 - \overline{A^2}\sigma_\omega^2 \tag{II.24}$$

or when the condition

$$A(t) = A_0[1 + \xi(t)] \tag{II.25}$$

is satisfied we can write

$$\dot{R}''_M(0) = \dot{R}''_V(0) = -A_0^2\left[\sigma_{\xi'} + \left(1 + \sigma_\xi^2\right)\sigma_\omega^2\right]. \tag{II.26}$$

In terms of the derivatives obtained, determined by Eqs. (II.24) and (II.26), the correlation functions $\dot{R}_M(\tau)$ and $\dot{R}_V(\tau)$ of the noise modulation function $\dot{M}(t)$ of the multiplicative noise during independent distortions in amplitude and phase of the signal at the condition

$$\tau \ll \tau_c$$

can be written in the following form:

$$\dot{R}_M(\tau) \simeq \overline{A^2} - 0.5\tau^2\left(\sigma_{A'}^2 + \overline{A^2}\sigma_\omega^2\right) \tag{II.27}$$

and

$$\dot{R}_V(\tau) \simeq \overline{A^2} - \left|\overline{\dot{M}}\right|^2 - 0.5\tau^2\left(\sigma_{A'}^2 + \overline{A^2}\sigma_\omega^2\right) \tag{II.28}$$

or for the condition determined by Eq. (II.25) we can write

$$\dot{R}_M(\tau) \simeq A_0^2\left(1 + \sigma_\xi^2\right) - 0.5\tau^2 A_0^2\left[\sigma_{\xi'}^2 + \left(1 + \sigma_\xi^2\right)\sigma_\omega^2\right] \tag{II.29}$$

and

$$\dot{R}_V(\tau) \simeq A_0^2\left(1 + \sigma_\xi^2\right) - \left|\overline{\dot{M}}\right|^2 - 0.5\tau^2 A_0^2\left[\sigma_{\xi'}^2 + \left(1 + \sigma_\xi^2\right)\sigma_\omega^2\right]. \tag{II.30}$$

Thus, for the condition

$$\tau \ll \tau_c$$

the correlation function $\dot{R}_V(\tau)$ can be expressed using the variance of the stochastic function $\xi(t)$ defining distortions in amplitude of the signal, the variance of derivative of the stochastic function $\xi(t)$, and the variance of derivative of changes in phase of the signal σ_ω^2.

When there are only distortions in phase of the signal we can write:

$$\dot{R}_M(\tau) \simeq 1 - 0.5\tau^2\sigma_\omega^2 \tag{II.31}$$

and

$$\dot{R}_V(\tau) \simeq 1 - \left|\overline{\dot{M}}\right|^2 - 0.5\tau^2\sigma_\omega^2. \tag{II.32}$$

When distortions in amplitude and phase of the signal obey the Gaussian probability distribution density, Eqs. (II.20) and (II.26) can be obtained by differentiating Eq. (3.114) with respect to the correlation function $\dot{R}_V(\tau)$.

When distortions in amplitude and phase of the signal obey the Gaussian probability distribution density and are correlated with each other, we can obtain the following coefficients of the Taylor series expansion of the correlation functions $\dot{R}_M(\tau)$ and $\dot{R}_V(\tau)$ of the noise modulation function $\dot{M}(t)$ of the multiplicative noise using Eqs. (3.111) and (3.112):

$$\dot{R}_M(0) = A_0^2(1 + \sigma_\xi^2); \tag{II.33}$$

$$\dot{R}_V(0) = A_0^2\{1 + \sigma_\xi^2 - [1 + \sigma_\xi^2\sigma_\varphi^2 r_{\xi\varphi}^2(0)] \cdot e^{-\sigma_\varphi^2}\}; \tag{II.34}$$

$$\dot{R}'_M(0) = \dot{R}'_V(0) = 2jA_0^2\sigma_\xi\sigma_\varphi r_{\xi\varphi}(0); \tag{II.35}$$

$$\dot{R}''_M(0) = \dot{R}''_V(0) = -A_0^2[\sigma_{\xi'}^2 + 2\sigma_\xi^2\sigma_\omega^2 r_{\xi\omega}(0) + (1 + \sigma_\xi^2)\sigma_\omega^2], \tag{II.36}$$

where $r_{\xi\varphi}(0)$ is the coefficient of mutual correlation between the stochastic function $\xi(t)$ and the derivative of changes in phase $\varphi'(t)$ of the signal. Other designations are the same as before.

In this case the correlation functions $\dot{R}_M(\tau)$ and $\dot{R}_V(\tau)$ for the condition

$$\tau \ll \tau_c$$

have the following form:

$$\begin{aligned}\dot{R}_M(\tau) \simeq A_0^2\{&1 + \sigma_\xi^2 + 2j\tau\sigma_\xi\sigma_\omega r_{\xi\omega}(0)\\ &- 0.5\tau^2[\sigma_{\xi'}^2 + 2\sigma_\xi^2\sigma_\omega^2 r_{\xi\omega}(0) + (1 + \sigma_\xi^2)\sigma_\omega^2]\};\end{aligned} \tag{II.37}$$

$$\begin{aligned}\dot{R}_V(\tau) \simeq A_0^2\{&1 + \sigma_\xi^2 - [1 + \sigma_\xi^2\sigma_\varphi^2 r_{\xi\varphi}(0)] \cdot e^{-\sigma_\varphi^2} + 2j\tau\sigma_\xi\sigma_\omega r_{\xi\omega}(0)\\ &-0.5\tau^2[\sigma_{\xi'}^2 + 2\sigma_\xi^2\sigma_\omega^2 r_{\xi\omega}(0) + (1 + \sigma_\xi^2)\sigma_\omega^2]\}.\end{aligned} \tag{II.38}$$

Approximate formulae of the correlation functions $\dot{R}_M(\tau)$ and $\dot{R}_V(\tau)$, of the noise modulation function $\dot{M}(t)$ of the multiplicative noise for the condition

$$\tau \ll \tau_c$$

and when distortions in amplitude and phase of the signal caused by the multiplicative noise obey the Gaussian probability distribution density, can be obtained using Eqs. (3.111) and (3.112) during the use of the Taylor series expansion of correlation functions of distortions in amplitude and phase of the signal and their mutual correlation function.

In terms of the equality

$$R'_{\varphi}(0) = R'_{\xi}(0) = 0 \tag{II.39}$$

the Taylor series expansion for these functions has the following form:

$$R_{\varphi}(\tau) = \sigma_{\varphi}^2 r_{\varphi}(\tau) = R_{\varphi}(0) + 0.5R''_{\varphi}(0) + \cdots = \sigma_{\varphi}^2 - 0.5\tau^2\sigma_{\omega}^2 + \cdots; \tag{II.40}$$

$$R_{\xi}(\tau) = \sigma_{\xi}^2 r_{\xi}(\tau) = R_{\varphi}(0) + 0.5R''_{\xi}(0) + \cdots = \sigma_{\xi}^2 - 0.5\tau^2\sigma_{\xi'}^2 + \cdots; \tag{II.41}$$

$$\begin{aligned} R_{\xi\varphi}(\tau) &= \sigma_{\xi}\sigma_{\varphi} r_{\xi\varphi}(\tau) = R_{\xi\varphi}(0) + \tau R'_{\xi\varphi}(0) + 0.5\tau^2 R''_{\xi\varphi}(0) + \cdots \\ &= \sigma_{\xi}\sigma_{\varphi} r_{\xi\varphi}(0) - \tau\sigma_{\xi}\sigma_{\omega} r_{\xi\omega}(0) - 0.5\tau^2\sigma_{\xi'}^2\sigma_{\omega} r_{\xi'\omega}(0) + \cdots. \end{aligned} \tag{II.42}$$

Substituting these values in Eqs. (3.111) and (3.112) and limiting by terms that contain the value τ, the order of which is not greater than 2, we can write

$$\begin{aligned} \dot{R}_M(\tau) \simeq A_0^2\{1 + \sigma_{\xi}^2 + 2j\tau\sigma_{\xi}\sigma_{\omega} r_{\xi\omega}(0) \\ - 0.5\tau^2\left[\sigma_{\xi'}^2 + 2\sigma_{\xi}^2\sigma_{\omega}^2 r_{\xi\omega}(0)\right]\} \cdot e^{-0.5\tau^2\sigma_{\omega}^2}; \end{aligned} \tag{II.43}$$

$$\begin{aligned} \dot{R}_V(\tau) \simeq A_0^2\{1 + \sigma_{\xi}^2 + 2j\tau\sigma_{\xi}\sigma_{\omega} r_{\xi\omega}(0) \\ - 0.5\tau^2\left[\sigma_{\xi'}^2 + 2\sigma_{\xi}^2\sigma_{\omega}^2 r_{\xi\omega}(0)\right]\} \cdot e^{-0.5\tau^2\sigma_{\omega}^2} \\ - A_0^2\left[1 + \sigma_{\xi}^2\sigma_{\varphi}^2 r_{\xi\varphi}(0)\right] \cdot e^{-\sigma_{\varphi}^2}. \end{aligned} \tag{II.44}$$

It is easy to verify that if we change the factor

$$e^{-0.5\tau^2\sigma_{\omega}^2}$$

by its Taylor series expansion and limit by terms containing the value τ, the order of which is not greater than 2, Eqs. (II.43) and (II.44) convert to Eqs. (II.37) and (II.38), respectively.

II.2 Correlation Function of the Noise Modulation Function When Signal Distortions Are the Narrow-Band Stationary Gaussian Processes

The energy spectrum $G_M(\Omega)$ of the noise modulation function $\dot{M}(t)$ of the multiplicative noise is defined by the well-known Fourier transform of the correlation function $\dot{R}_M(\tau)$. If there is a functional relationship between the amplitude and phase distortions of the signal given by Eq. (3.65) the correlation function $\dot{R}_M(\tau)$ coincides formally with the two-dimensional complex characteristic function of distortions in phase of the signal:

$$\dot{R}_M(\tau) = m_1\left[e^{j[\varphi(t)\dot{\psi} - \varphi(t-\tau)\dot{\psi}^*]}\right] = \dot{\Theta}_2^{\varphi}(\dot{\psi}, -\dot{\psi}^*, \tau), \tag{II.45}$$

where

$$\dot{\psi} = 1 + j\nu \tag{II.46}$$

is the complex coefficient.

The complex characteristic function considered in Eq. (II.45) is understood in the sense that there is the regular function of the complex argument $j\dot{\psi}$ coinciding with the characteristic function of the real values

$$Re\{j\dot{\psi}\}$$

and analytical function within the limits of the range

$$|\dot{\psi}| < R,$$

where R is the radius.

When there are distortions only in phase of the signal, i.e.,

$$\nu = 0$$

we can write

$$\dot{R}_M(\tau) = \Theta_2^{\varphi}(1, -1, \tau). \tag{II.47}$$

As can be seen from Eq. (II.47), the energy spectrum $G_M(\Omega)$ depends on the form of the two-dimensional characteristic function of fluctuations in phase of the signal and, consequently, the coefficient of correlation between fluctuations in phase of the signal

$$r_{\varphi}(\tau) = R_{\varphi}(\tau)\cos\Omega_M\tau, \tag{II.48}$$

where Ω_M is the central frequency of the energy spectrum of narrow-band fluctuations.

We use the Taylor series expansion with respect to the variable $r_{\varphi}(\tau)$ at the point

$$r_{\varphi} = 0$$

for the two-dimensional characteristic function

$$\dot{\Theta}_2^{\varphi}[\dot{x}_1, \dot{x}_2, r_{\varphi}(\tau)].$$

Then we can write

$$\dot{\Theta}_2^{\varphi}[\dot{x}_1, \dot{x}_2, r_{\varphi}(\tau)] = \sum_{n=0}^{\infty} \frac{r_{\varphi}^n(\tau)}{n!} \cdot C_n, \tag{II.50}$$

where

$$C_n = \frac{d^n}{dr_\varphi^n(\tau)} \dot{\Theta}_2^\varphi[\dot{x}_1, \dot{x}_2, r_\varphi(\tau)]\big|_{r_\varphi=0}; \tag{II.51}$$

$$\dot{x}_1 = -j\dot{\psi}; \tag{II.52}$$

and

$$\dot{x}_2 = -j\dot{\psi}^*. \tag{II.53}$$

When the stochastic process $\varphi(t)$ obey the Gaussian probability distribution density the following equalities

$$\begin{aligned} C_n &= \sigma_\varphi^{2n}|\dot{x}_1|^n|\dot{x}_2|^n \cdot \exp\left\{-\frac{\sigma_\varphi^2}{2}\left[|\dot{x}_1|^2 + |\dot{x}_2|^2\right]\right\} \\ &= \sigma_\varphi^{2n}|\dot{x}_1|^n|\dot{x}_2|^n C_0 \end{aligned} \tag{II.54}$$

and

$$\begin{aligned} C_0 &= \exp\left\{-\frac{\sigma_\varphi^2}{2}\left[|\dot{x}_1|^2 + |\dot{x}_2|^2\right]\right\} \\ &= \Theta_1^\varphi(\dot{x}_1) \cdot \Theta_1^\varphi(\dot{x}_2) \end{aligned} \tag{II.55}$$

are true.

Taking into consideration Eqs. (II.50) and (II.54), we can write the energy spectrum of the noise modulation function $\dot{M}(t)$ of the multiplicative noise in the following form:

$$G_M(\Omega) = C_0 \sum_{n=0}^{\infty} \frac{\sigma_\varphi^{2n}|\dot{x}_1\,|^n\,|\dot{x}_2|^n}{n!} \int_{-\infty}^{\infty} R_\varphi^n(\tau) \cdot e^{j\Omega\tau} \cos^n \Omega_M\tau \, d\tau, \tag{II.56}$$

where

$$\dot{x}_1 = -j\dot{\psi} \tag{II.57}$$

and

$$\dot{x}_2 = -j\dot{\psi}^* \tag{II.58}$$

if the amplitude-phase distortions of the signal are determined by Eq. (3.65), and

$$\dot{x}_1 = 1$$

and

$$\dot{x}_2 = -1$$

if there are distortions only in phase of the signal.

TABLE II.1

One-Dimensional Characteristic Functions

No.	Distribution Law $f(\varphi)$	Characteristic Function $\Theta_1^{\varphi}(x)$	Variance σ_φ^2
1	Gaussian		
	$\frac{1}{\sqrt{2\pi}\sigma_\varphi}\exp\left[-\frac{(\varphi-\varphi_0)^2}{2\sigma_\varphi^2}\right]$	$\exp\left[-\frac{\sigma_\varphi^2 x^2}{2}+j\varphi_0 x\right]$	σ_φ^2
2	Uniform		
	$\frac{1}{2\varphi_m}, \quad \lvert\varphi\rvert \leq \varphi_m$	$\frac{\sin\varphi_m x}{\varphi_m x}$	$\sigma_\varphi^2 = \frac{\varphi_m^3}{3}$
3	Laplace		
	$\frac{1}{2b}\exp\left[-\frac{\lvert\varphi\rvert}{b}\right], \quad b \geq 0$	$\frac{1}{1+b^2x^2}$	$\sigma_\varphi^2 = 2b^2$
4	$\frac{1}{\pi\sqrt{\varphi_m^2-\varphi^2}}, \quad \lvert\varphi\rvert \leq \varphi_m$	$J_0(x\varphi_m)$	$\sigma_\varphi^2 = \frac{\varphi_m^2}{2}$

When

$$n = 0$$

the integral in Eq. (II.56) is transformed to the delta function $\delta(\Omega)$ independently of character of the envelope $R_\varphi(\tau)$ of the coefficient of correlation. The presence of the delta function $\delta(\Omega)$ in the energy spectrum of the noise modulation function $\dot{M}(t)$ of the multiplicative noise indicates the presence of the discrete component of the energy spectrum at the frequency

$$\Omega = 0.$$

The relative value of the energy spectrum discrete component does not depend on the form of the correlation function $R_\varphi(\tau)$ and is defined only by the form of the one-dimensional characteristic function—the one-dimensional probability distribution density—since

$$\begin{aligned} C_0 &= \Theta_2^{\varphi}(\dot{x}_1, \dot{x}_2, 0) \\ &= \lim_{\tau\to\infty} \int_{-\infty}^{\infty}\int_{-\infty}^{\infty} f(\varphi_1, \varphi_2, \tau) \cdot e^{j(\dot{x}_1\varphi_1 + \dot{x}_2\varphi_2)}\, d\varphi_1\, d\varphi_2 \\ &= \Theta_1^{\varphi}(\dot{x}_1) \cdot \Theta_1^{\varphi}(\dot{x}_2). \end{aligned} \tag{II.59}$$

Some one-dimensional characteristic functions of changes in phase of the signal, which define the discrete component of the energy spectrum of the noise modulation function $\dot{M}(t)$ of the multiplicative noise, are presented in Table II.1.

We change the function

$$\cos^{2n} y$$

in Eq. (II.56) by the sum of cosines of multiple arcs:

$$\cos^{2n} y = \frac{1}{2^n}\left\{\frac{2n!}{(n!)^2} + \sum_{k=0}^{n-1} 2 \cdot \frac{2n!}{k!(2n-k)!} \cdot \cos\left[2(n-k)y\right]\right\}; \tag{II.60}$$

$$\cos^{2n} y = \frac{1}{2^{2n-1}} \sum_{k=0}^{n-1} \frac{(2n-1)!}{k!(2n-k-1)!} \cdot \cos\left[2(n-k)-1\right]y. \tag{II.61}$$

We proceed to introduce the designation

$$m = n - k.$$

Then Eq. (II.56) takes the following form:

$$\begin{aligned} G_M(\Omega) = {} & C_0 \sum_{n=0}^{\infty} \frac{\sigma_{\varphi_{eq}}^{4n}}{2^{2n}(n!)^2} \int_{-\infty}^{\infty} R_{\varphi}^{2n}(\tau) \cdot e^{j\Omega\tau} d\tau \\ & + C_0 \sum_{m=1}^{\infty} \sum_{n=m}^{\infty} \frac{\sigma_{\varphi_{eq}}^{4n}}{2^{2n}(n-m)!(n+m)!} \\ & \times \int_{-\infty}^{\infty} R_{\varphi}^{2n}(\tau) \cdot e^{j(\Omega \pm 2m\Omega_M)\tau} d\tau \\ & + C_0 \sum_{m=1}^{\infty} \sum_{n=m}^{\infty} \frac{\left(\sigma_{\varphi_{eq}}^2\right)^{2n-1}}{2^{2n-1}(n+m-1)!(n-m)!} \\ & \times \int_{-\infty}^{\infty} R_{\varphi}^{2n-1}(\tau) \cdot e^{j[\Omega \pm (2m-1)\Omega_M]\tau} d\tau, \end{aligned} \tag{II.62}$$

where

$$\sigma_{\varphi_{eq}}^2 = \begin{cases} +|\dot{\psi}|^2 \sigma_{\varphi}^2 & \text{at } \nu \neq 0; \\ \sigma_{\varphi}^2 & \text{at } \nu = 0. \end{cases} \tag{II.63}$$

For degree of distortions in phase

$$\sigma_{\varphi_{eq}}^2 < 1,$$

as can be seen from Eq. (II.62), we can be limited by the term with the index

$$n = 0$$

for the first component, with the indexes

$$n = 1$$

and

$$m = 1$$

for the second and third components, and neglect other terms by virtue of their small values.

In this case we can write

$$G_M(\Omega) = C_0\left\{\delta(\Omega) + \frac{\sigma^4_{\varphi_{eq}}}{4} \cdot G_{2\varphi}(\Omega) + \frac{\sigma^4_{\varphi_{eq}}}{8} \cdot G_{2\varphi}(\Omega \pm 2\Omega_M) + \frac{\sigma^2_{\varphi_{eq}}}{4} \cdot G_{\varphi}(\Omega \pm \Omega_M) + \cdots\right\}, \tag{II.64}$$

where

$$G_{n\varphi}(\Omega \pm \Omega_M) = \int_{-\infty}^{\infty} R^n_{\varphi}(\tau) \cdot e^{j(\Omega \pm m\Omega_M)\tau}\, d\tau. \tag{II.65}$$

As is well known,[2] if there are distortions only in phase of the signal and when $\varphi(t)$ is the wide-band stochastic function, the energy spectrum $G_M(\Omega)$ of the noise modulation function $\dot{M}(t)$ of the multiplicative noise contains the discrete component at the frequency

$$\Omega = 0$$

and continuous component in the neighborhood at the frequency

$$\Omega = 0.$$

As can be seen from Eq. (II.64), the continuous component of the energy spectrum is concentrated in the neighborhood at the points

$$\Omega = m\Omega_M, \quad m = 0, \pm 1, \pm 2, \ldots \tag{II.66}$$

during the narrow-band distortions in phase of the signal.

Consider the case of high distortions in phase of the signal

$$\sigma^2_{\varphi_{eq}} \gg 1.$$

Reference to Eq. (II.62) shows that the power of continuous components of the energy spectrum of the noise modulation function $\dot{M}(t)$ of the multiplicative noise at the frequencies

$$\Omega = \pm m\Omega_M \tag{II.67}$$

is defined by sum of the coefficients in the following form:

$$\mathcal{P}_{con}(\Omega = \pm 2m\Omega_M) = \sum_{n=m}^{\infty} \frac{a(m,n)}{2\pi}$$
$$= \sum_{n=m}^{\infty} \frac{\sigma_{\varphi_{eq}}^{4n}}{2^{2n}2\pi(n-m)!(n+m)!}. \tag{II.68}$$

The terms $a(m, n)$ in Eq. (II.68) increase toward the terms with the index n_0 and after that decrease sharply tending to approach zero with an increase in the index n and for the condition

$$\sigma_{\varphi_{eq}}^2 \gg 1. \tag{II.69}$$

The rate of convergence can be estimated in the following form

$$q(\ell) = \frac{a(m, n_0 + \ell)}{a(m, n_0)} = \prod_{\ell=-n_0}^{\infty} \frac{\sigma_{\varphi_{eq}}^2}{(n_0 + \ell)(\ell + m + n_0)}, \tag{II.70}$$

where

$$a(m, n_0)$$

is the term defining the maximal power;

$$a(m, \ell + n_0)$$

is the term with the index

$$n_0 + m$$

at

$$m = 0, \pm 1, \pm 2, \ldots.$$

Dependence given by Eq. (II.70) is shown in Figs. II.1 and II.2 for the conditions

$$\sigma_{\varphi_{eq}} = 4$$

and

$$\sigma_{\varphi_{eq}} = 8$$

at

$$m = 0.$$

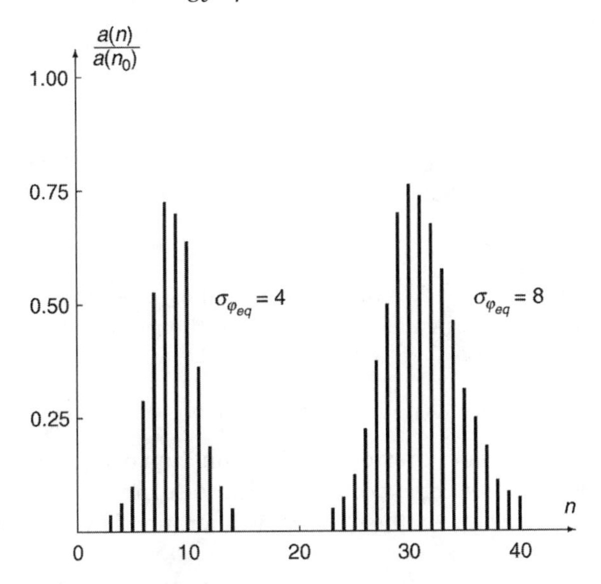

FIGURE II.1
Rate of convergence as a function of n.

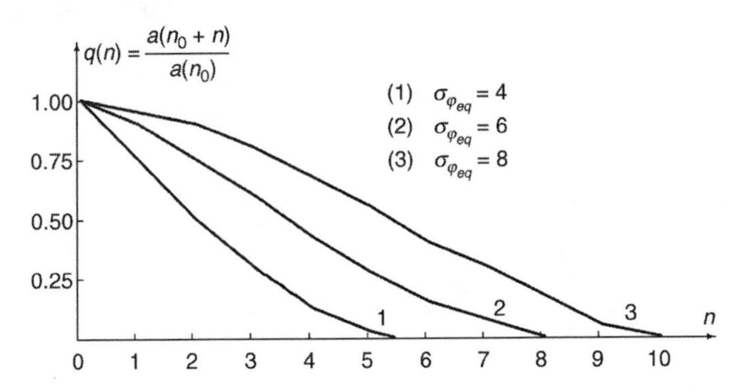

FIGURE II.2
Definition of the value n_0.

The index n_0 of the term with the maximal amplitude depends essentially on the value $\sigma^2_{\varphi_{eq}}$.

We define the index n_0. For this purpose divide the term

$$a(m, n)$$

into the next term

$$a(m, n+1)$$

and suppose that this ratio is equal to unity, believing that their ratio is equal approximately to unity at the maximal point.

So, we can write

$$2n_0 \simeq \sqrt{4m^2 + \sigma^4_{\varphi_{eq}}} - 1. \tag{II.71}$$

The number of terms (see Eq. (II.71)), each determined by the power that is greater, for example, than

$$0.1a\,(m, n_0),$$

as can be seen from Fig. II.2, is not greater than $\sigma_{\varphi_{eq}}$, i.e., to a first approximation, we can sum the terms in Eq. (II.71) with the numbers within the limits of the interval

$$n_0 - \sigma_{\varphi_{eq}} \leq n \leq n_0 + \sigma_{\varphi_{eq}}. \tag{II.72}$$

In the process, the relative error for determination of the power of continuous component of the energy spectrum of the noise modulation function $\dot{M}(t)$ of the multiplicative noise at the frequency determined by Eq. (II.67) does not exceed 10%.

Taking into consideration the statements mentioned above, the energy spectrum $G_M(\Omega)$ determined by Eq. (II.62) of the noise modulation function $\dot{M}(t)$ of the multiplicative noise can be written in the following form:

$$\begin{aligned}
G_M(\Omega) &= C_0 \sum_{n=0}^{n_0+\sigma_{\varphi_{eq}}} \frac{\sigma^4_{\varphi_{eq}}}{2^{2n}(n!)^2} \cdot G_{2n\varphi}(\Omega) \\
&+ C_0 \sum_{m=1}^{\infty} \sum_{n=m}^{n_0+\sigma_{\varphi_{eq}}} \frac{\sigma^4_{\varphi_{eq}}}{(n-m)!(n+m)!} \cdot G_{2n\varphi}(\Omega \pm 2m\Omega_M) \\
&+ C_0 \sum_{m=1}^{\infty} \sum_{n=m}^{n_0+\sigma_{\varphi_{eq}}} \frac{\left(\sigma^2_{\varphi_{eq}}\right)^{2n-1}}{2^{2n-1}(n-m)!\,(n+m-1)!} \cdot G_{(2n-1)\varphi}[\Omega \pm (2m-1)\Omega_M].
\end{aligned} \tag{II.73}$$

The index n in Eq. (II.73), in terms of Eqs. (II.71) and (II.72), is limited within the boundaries of the interval

$$\frac{\sqrt{4m^2 + \sigma^4_{\varphi_{eq}}} - 1}{2} - \sigma_{\varphi_{eq}} \leq n \leq \frac{\sqrt{4m^2 + \sigma^4_{\varphi_{eq}}} - 1}{2} + \sigma_{\varphi_{eq}}. \tag{II.74}$$

During the condition

$$\sigma^2_{\varphi_{eq}} \gg 1,$$

as can be seen from Eqs. (II.72) and (II.74), changes in the index n do not practically impact on the function

$$G_{n\varphi}(\Omega \pm m\Omega_M)$$

determined by Eq. (II.65).

For this reason, we can believe that the energy spectrum components in Eq. (II.73) are related

$$G_{n\varphi}(\Omega \pm m\Omega_M) \simeq G_{n_0\varphi}(\Omega \pm M\Omega_M) \tag{II.75}$$

and, consequently, after transformation for the condition

$$\ell = n - m, \tag{II.76}$$

taking into account that

$$\sum_{\ell=0}^{\infty} \frac{(0.5x)^{m+2\ell}}{\ell!(m+\ell)!} = I_m(x), \tag{II.77}$$

we finally obtain that the energy spectrum of the noise modulation function $\dot{M}(t)$ of the multiplicative noise can be determined in the following form:

$$G_M(\Omega) = C_0 \sum_{m=0}^{\infty} I_m\left(\frac{\sigma_{\varphi_{eq}}^2}{2}\right) \cdot G_{n_0\varphi}(\Omega \pm \Omega_M), \tag{II.78}$$

where

$$n_0 = \frac{\sqrt{4m^2 + \sigma_{\varphi_{eq}}^4} - 1}{2}. \tag{II.79}$$

II.3 Energy Spectrum of the Noise Modulation Function of the Fluctuating Multiplicative Noise

Let distortions in phase of the signal be the uncorrelated stochastic sequence of pulses with the random amplitude φ_n and constant function $y(t)$,

$$0 < t \leq T_M.$$

In terms of that, the function

$$y(t_n - nT_M)$$

is equal to zero outside the interval

$$nT_M < t \leq (n+1)T_M$$

at the fixed n the i-th realization of the noise modulation function $\dot{M}(t)$ of the pulse-fluctuating multiplicative noise can be written in the following form:

$$\dot{M}(t) = \sum_{n=-\infty}^{\infty} \dot{M}_n(t), \tag{II.80}$$

where

$$\dot{M}_n(t) = \begin{cases} e^{j\varphi_n^{(i)} y(t-nT_M)} & \text{at } nT_M < t \leq (n+1)T_M; \\ 0 & \text{otherwise;} \end{cases} \tag{II.81}$$

$\varphi_n^{(i)}$ is the i-th realization of φ_n.

The energy spectrum of the non-stationary stochastic process in Eq. (II.80) can be defined by averaging the instantaneous spectral density with respect to the time and a set[2]

$$G_M(\Omega) = m_1\left[\lim_{N\to\infty} \frac{2}{(2N+1)T_M} \cdot |\dot{S}_{iN}(\Omega)|^2\right], \tag{II.82}$$

where

$$\dot{S}_{iN}(\Omega, t) = \int_{-\infty}^{t} \dot{M}(t) \cdot e^{-j\Omega t} dt \tag{II.83}$$

is the instantaneous spectral density of the i-th realization of the noise modulation function $\dot{M}(t)$ of the pulse-fluctuating multiplicative noise; N is the number of pulses into the sequence.

The spectral density of the noise modulation function $\dot{M}(t)$ of the pulse-fluctuating multiplicative noise in Eq. (II.80) can be written in the following form:

$$\dot{S}_{iN}(\Omega, N) = \sum_{n=-N}^{N} \dot{g}(\Omega, n, i) \cdot e^{-jn\Omega T_M}, \tag{II.84}$$

where

$$\dot{g}(\Omega, n, i) = \int_{0}^{T_M} e^{j\varphi_n^{(i)} y(t')} \cdot e^{-j\Omega t'} dt' \tag{II.85}$$

is the spectral density of the noise modulation function $\dot{M}(t)$ of the pulse-fluctuating multiplicative noise with the duration T_M.

The mean of spectral density with respect to a set takes the following form:

$$m_1\left[|\dot{S}_{iN}(\Omega)|^2\right] = \sum_{n=-N}^{N} K_0(\Omega, n) + \sum_{n=-N}^{N} \sum_{\ell=-N}^{N} h_{n-\ell}(\Omega), \tag{II.86}$$

where

$$K_0(\Omega, n) = m_1\left[|\dot{g}(\Omega, n, i)|^2\right]; \tag{II.87}$$

$$h_{n-\ell}(\Omega) = m_1\left[\dot{g}(\Omega, n, i) g^*(\Omega, \ell, i) \cdot e^{-j(n-\ell)\Omega T_M}\right]; \tag{II.88}$$

and

$$\sum_{n=-N}^{N} K_0(\Omega, n) = (2N+1)K_0(\Omega). \tag{II.89}$$

Taking into account Eqs. (II.82), (II.84), and (II.86), we can write

$$G_M(\Omega) = \frac{2}{T_M} \cdot \left\{ K_0(\Omega) + \lim_{N\to\infty} \sum_{m=1}^{\infty} \left(1 - \frac{m}{2N+1}\right)[h_m(\Omega) + h_m(-\Omega)] \right\}. \tag{II.90}$$

Equation (II.90) defines the total energy spectrum of the noise modulation function $\dot{M}(t)$ of the pulse-fluctuating multiplicative noise, taking into account the mutual correlation of parameters, and allows us to define the energy spectrum both as

$$N \to \infty$$

and during the finite number of modulated pulses.

The functions

$$g(\Omega, i)$$

and

$$e^{-j(n-\ell)\Omega T_M}$$

in Eq. (II.88) are independent as the amplitude and clock time interval are independent parameters.

For this reason, Eq. (II.88) can be rewritten in the following form:

$$h_m(\Omega) = K_m(\Omega) H_m(\Omega), \tag{II.91}$$

where

$$K_m(\Omega) = \int_{-\infty}^{\infty}\int_{-\infty}^{\infty} \dot{g}(\Omega, n, \varphi_n) g^*(\Omega, \ell, \varphi_\ell) f(\varphi_n, \varphi_\ell, m) d\varphi_n \, d\varphi_\ell; \tag{II.92}$$

$$H_m(\Omega) = m_1[e^{-jm\Omega T_M}] = e^{-jm\Omega T_M}; \tag{II.93}$$

$f(\varphi_n, \varphi_\ell, m)$ is the probability distribution density of degree of distortions in phase of the signal.

If changes in the amplitude φ_n from pulse to pulse are independent, then Eq. (II.92) can be written in the following form:

$$K_m(\Omega) = |m_1[\dot{g}(\Omega, \varphi)]|^2 = K_\infty(\Omega). \tag{II.94}$$

After transformation, taking into account Eqs. (II.90)–(II.93) and in terms of Reference 2 we can write

$$\lim_{N\to\infty}\left[1+2\sum_{m=1}^{2N}\left(1+\frac{m}{2N+1}\right)\cos m\Omega T_M\right]=\frac{2\pi}{T_M}\sum_{r=-\infty}^{\infty}\delta\left(\Omega-\frac{2\pi}{T_M}r\right). \tag{II.95}$$

Finally, we can write

$$G_M(\Omega)=\frac{2}{T_M}\cdot\left\{K_0(\Omega)-K_\infty(\Omega)+\frac{2\pi}{T_M}\cdot K_\infty(\Omega)\sum_{r=-\infty}^{\infty}\delta(\Omega-r\Omega_M)\right\}, \tag{II.96}$$

where

$$K_0(\Omega)=m_1\left[\left|\int_0^{T_M}e^{j[\varphi_n y(t)-\Omega t]}\,dt\right|^2\right]; \tag{II.97}$$

$$K_\infty(\Omega)=m_1\left|\left[\int_0^{T_M}e^{j[\varphi_n y(t)-\Omega t]}\,dt\right]\right|^2, \tag{II.98}$$

where

$$\Omega_M=\frac{2\pi}{T_M}. \tag{II.99}$$

Reference to Eq. (II.96) shows that the energy spectrum of the noise modulation function $\dot{M}(t)$ of the pulse-fluctuating multiplicative noise with deterministic clock time interval is the sum of continuous and discrete components.

References

1. Tikhonov, V., *Statistical Radio Engineering*, 2nd ed., Radio and Svyaz, Moscow, 1982 (in Russian).
2. Levin, B., *Theoretical Foundations of Statistical Radio Engineering*, Parts 1–3, Soviet Radio, Moscow, 1974–1976 (in Russian).

Notation Index

AF	additional filter
MSG	model signal generator
PF	preliminary filter
A	random event
$\overline{A}$	random opposite event
$A(t)$	factor of amplitude envelope of the signal $a(t)$
$A_\varphi(t)$	amplitude envelope of the coefficient of correlation of distortions in phase $\varphi(t)$ of the signal $a(t)$
$a(t)$	signal
$\dot{a}(t)$	received signal
$\dot{a}^*(t)$	complex conjugate of the received signal $\dot{a}(t)$
$\widehat{a}(t)$	normalized signal
$\vec{a}(t)$	vector signal
$a_0(t)$	transmitted signal
$a_M(t)$	signal distorted by the multiplicative noise
$a_m(t)$	model signal
$\dot{C}_n$	coefficient under the Fourier series expansion of the noise modulation function $\dot{M}(t)$ of the multiplicative noise
$\dot{C}_0^T$	coefficient defining the undistorted component of the signal $a_M(t)$ for individual realizations
$\widehat{D}$	estimation of the variance of stochastic process
D_D	variance of the statistical variance D_{XX}^{**}
D_{m_X}	variance of the statistical mean m_X^*
D_{XX}	variance of the random variable X
D_{XX}^{**}	statistical variance of the random variable X
D_ξ	variance of the stochastic process $\xi(t)$
$D[X]$	variance of the random variable X
$D_k^*[X]$	statistical central moment of the k-th order of the random variable X
$D[K_{M_i}(\tau)]$	variance of the covariance function $K_{M_i}(\tau)$
$\dot{D}_V(\tau)$	correlation function of complex amplitude envelope of noise component of the signal $a_M(t)$
$D[v(t)]$	variance of noise component of the signal $a_M(t)$
d_ℓ	weight coefficient

E_a	energy of the signal
E_a^*	energy of the model signal
E_{a_1}	energy of the signal at the output of the PF
$F(x)$	probability distribution function
$F(x_1; t_1)$	one-dimensional probability distribution function
$F(x_1, x_2; t_1, t_2)$	two-dimensional probability distribution function
f_M	frequency of the periodic multiplicative noise
$f(x)$	probability distribution density
$f(x_1; t_1)$	one-dimensional probability distribution density
$f(x_1, x_2; t_1, t_2)$	two-dimensional probability distribution density
$f(\vec{X} \mid H_0)$	likelihood function at the hypothesis H_0
$f(\vec{X} \mid H_1)$	likelihood function at the hypothesis H_1
$f[t, \xi(t)]$	non-linear transformation with respect to the stochastic process $\xi(t)$
$G(x)$	gamma function
$\widetilde{G}(\alpha, x)$	incomplete gamma function
$G_M(\Omega)$	energy spectrum of the noise modulation function $\dot{M}(t)$ of the multiplicative noise
$G_V(\Omega)$	energy spectrum of fluctuations of the noise modulation function $\dot{M}(t)$ of the multiplicative noise
$G_{V_0}(\omega)$	normalized energy spectrum of fluctuations of the noise modulation function $\dot{M}(t)$ of the multiplicative noise
$G_c(\Omega)$	continuous component of the energy spectrum $G_M(\Omega)$
$G_d(\omega)$	discrete component of the energy spectrum $G_M(\Omega)$
$G'_{en}(\omega)$	normalized spectrum of quadratic amplitude envelope of the signal $a(t)$
$G_\xi(\Omega)$	energy spectrum of distortions in amplitude $\xi(t)$ of the signal $a_M(t)$
$G_\varphi(\Omega)$	energy spectrum of distortions in phase $\varphi(t)$ of the signal $a_M(t)$
$G_\upsilon(\omega)$	energy spectrum of noise component of the signal $a_M(t)$
H_0	hypothesis a "no" signal in the input stochastic process
H_1	hypothesis a "yes" signal in the input stochastic process
$\dot{H}(\tau)$	complex amplitude envelope of pulse transient response
$h(t)$	pulse transient response of linear system
I_n	modified Bessel function
$J_n(x)$	Bessel function of the n-th order of real argument
$\mathbf{K}$	covariance matrix
K	threshold
K_g	threshold for the generalized detector
K_{op}	threshold for the optimal detector
$K(j\omega)$	frequency response of linear system
$K(P_F)$	threshold function

$K_{M_i}(\tau)$	covariance function of the i-th individual realization of the signal $a_M(t)$
$K_0(x)$	modified second-order Bessel function of an imaginary argument or McDonald's function
$K_\xi(t_1, t_2)$	covariance function of the stochastic process $\xi(t)$
$K_{\xi\eta}(\tau)$	mutual covariance function of stochastic processes $\xi(t)$ and $\eta(t)$
$\dot{K}(\omega, t)$	transfer characteristic
$K_\upsilon(\tau)$	covariance function of noise component of the signal $a_M(t)$
k	coefficient of asymmetry of the probability distribution density
$\ell(X)$	likelihood ratio
l_0	unknown non-energy parameter of the signal $a(t, l_0)$
l	measured parameter of the signal $a(t, l_0)$
$\dot{M}(t)$	noise modulation function of the multiplicative noise
$\overline{\dot{M}(t)}$	mean of the noise modulation function $\dot{M}(t)$ of the multiplicative noise
$M[X]$	mean of the random variable X
$M[X^k]$	initial moment of the k-th order of the random variable X
$M_k[X]$	central moment of the k-th order of the random variable X
$M^*[X^k]$	statistical initial moment of the k-th order of the random variable X
$M[\xi^{\nu_1}(t)]$	initial moment of the ν_1-th order of the stochastic process $\xi(t)$
$M_{\nu_1}(t)$	central moment of the ν_1-th order of the stochastic process $\xi(t)$
$\widehat{m}$	estimation of the mean of the stochastic process
m_X	mean of the random variable X
m_X^*	statistical mean of the random variable X
$\overline{m_X^*}$	statistical mean of the ergodic stationary stochastic process $X(t)$
$m_k[X]$	initial moment of the k-th order of the random variable X
$m_k^*[X]$	statistical initial moment of the k-th order of the random variable X
$m_{11}(t_1, t_2)$	covariance function of stochastic process
$m_\xi(t)$	mean of the stochastic process $\xi(t)$
$m_{\nu_1}(t)$	initial moment of the ν_1-th order of the stochastic process $\xi(t)$
N	sample size
$\frac{N_0}{2}$	spectral power density of the additive noise
$n(t), \vec{n}(t)$	additive noise
P_1	*a priori* probability of a "yes" signal

P_0	*a priori* probability of a "no" signal
P_D	probability of detection
P_F	probability of false alarm
P_M	probability of signal omission
P_{er}	probability of error
$P(A)$	probability of the random event A
$P(\bar{A})$	probability of the random opposite event $\bar{A}$
$P(A \mid S)$	probability that the random event A occurs during the condition S
$P_N(x)$	Poisson probability distribution function
$P\{\sum_{i=1}^{n} A_i\}$	probability of sum of random events $A_1, \ldots, A_n$
$P\{\prod_{i=1}^{n} A_i\}$	probability of product of random events $A_1, \ldots, A_n$
Q_y	coefficient of signal truncation
$\dot{Q}(t)$	complex amplitude envelope of the model signal of the generalized detector
q	signal-to-noise ratio at the output of the generalized detector
q_T^2	signal-to-noise ratio at the output of optimal detector
$\mathbf{R}$	correlation matrix
R_1, R_2	radar range
$R_n(t_1, \ldots, t_n)$	correlation function of the n-th order
$R_a(\tau)$	correlation function of the signal $a(t)$
$R_n(\tau)$	correlation function of the additive noise $n(t)$
$\dot{R}_M(\tau)$	correlation function of the noise modulation function $\dot{M}(t)$ of the multiplicative noise
$\dot{R}_V(\tau)$	correlation function of fluctuations of the noise modulation function $\dot{M}(t)$ of the multiplicative noise
$R_s(\tau)$	correlation function of the noise component $s(\tau, \Omega)$ at the output of the PF
$R_{XX}(\tau)$	correlation function of the ergodic stationary stochastic process $X(t)$
$R^*_{XX}(\tau)$	statistical correlation function of the ergodic stationary stochastic process $X(t)$
$\overline{R^*_{XX}(\tau)}$	statistical correlation function of the ergodic stationary stochastic process $X(t)$
$R_\xi(\tau)$	correlation function of the stochastic process $\xi(t)$
$\widehat{R}_\xi(\tau)$	estimation of the correlation function $R_\xi(\tau)$ of the stochastic process $\xi(t)$
$R_\upsilon(\tau)$	correlation function of the noise component $\upsilon(t)$ of the signal $a_M(t)$
$r_\xi(t_1, t_2)$	normalized correlation function of the stochastic process $\xi(t)$
$r_{\xi\eta}(t_1, t_2)$	normalized mutual correlation function of stochastic processes $\xi(t)$ and $\eta(t)$

$r_{XX}(\tau)$	normalized correlation function of the stationary stochastic process $X(t)$
r_{ξ}	coefficient of correlation during distortions in amplitude of the signal $a_M(t)$
r_{φ}	coefficient of correlation during distortions in phase of the signal $a_M(t)$
$r_{\xi\varphi}$	coefficient of mutual correlation between distortions in amplitude $\xi(t)$ and phase $\varphi(t)$ of the signal $a_M(t)$
$S(t)$	amplitude envelope of the signal
$S_m(t+\tau)$	amplitude envelope of the model signal
$S(f)$	spectral density
$S_0(f)$	spectral density of stationary in a broad sense centered stochastic process
$S(\omega)$	spectral density
$\dot{S}(t)$	complex amplitude envelope of the signal
$S_M(t)$	amplitude envelope of the signal $a_M(t)$
$\dot{S}_M(t)$	complex amplitude envelope of the signal $a_M(t)$
$\dot{S}_M(\Omega)$	spectrum of the noise modulation function $\dot{M}(t)$ of the multiplicative noise
$\dot{S}_S(\Omega)$	spectrum of complex amplitude envelope of the signal $a(t)$
$\dot{S}_a(\omega)$	Fourier transform of the signal $a(t)$
$S_m(t)$	amplitude envelope of the model signal $a_m(t)$
$S_{\xi}(f)$	spectral density of the stochastic process $\xi(t)$
$S_{\eta}(f)$	spectral density of the stochastic process $\eta(t)$
$s(t, \Omega)$	noise component of the signal at the output of the PF
T	equivalent duration of the signal
$T\Delta F, T\beta$	signal base
T_M	period of the multiplicative noise
$T_r(x)$	Chebyshev polynomial
$U_r(x)$	Chebyshev polynomial
$\dot{V}_0(t)$	fluctuations of the noise modulation function $\dot{M}(t)$ of the multiplicative noise
X	random variable
$X(t), \vec{X}(t)$	input stochastic process
x_{med}	median of the random variable X
x_{mod}	mode of the random variable X
Z_N	decision statistic
$\vec{Z}(t)$	vector sufficient decision statistic
$Z(\vec{R})$	rank statistic
$Z^{out}(t)$	decision statistic at the output of detector
α	true variable of probability characteristic
α^*	statistical variable of probability characteristic

$\alpha_0(\tau, \Omega)$	relative level of undistorted component of the signal $a_M(t)$
β	shift in initial phase of the signal $a_M(t)$
$\beta(\varphi \mid \vec{\vartheta})$	power function
γ	coefficient of kurtosis of the probability distribution density
$\tilde{\gamma}$	useful parameter
Δ	deviation of the mean
Δ_X	deviation of the random variable X
Δ_φ	phase of the received signal $\dot{a}(t)$
$\Delta\Omega$	detuning in frequency
$\Delta\tau$	detuning in signal appearance time
$\Delta\ell$	error of signal parameter measurement
$\Delta\Omega_a$	bandwidth of the signal $a(t)$
$\Delta\Omega_\varphi$	equivalent bandwidth of energy spectrum of distortions in phase $\varphi(t)$ of the signal $a_M(t)$
ΔF_{ef}	effective spectrum bandwidth
ΔF_a	spectrum bandwidth of the signal $a(t)$
$\Delta\omega_{AF}$	bandwidth of the AF
$\Delta\omega_{PF}$	bandwidth of the PF
δ_{ij}	Kronecker symbol
$\delta(x - x_0)$	delta function
$\delta_1^2(\tau, \Omega)$	normalized variance of signal noise component
$\zeta(t)$	stochastic process
$\eta(t)$	stochastic process
$\Theta(j\vartheta)$,	one-dimensional characteristic function
$\Theta_n(j\vartheta_1, \ldots, j\vartheta_n)$	n-dimensional characteristic function
$\Theta^{A(t)\varphi(t)}(x)$	characteristic function of distortions in amplitude and phase of the signal
ϑ	initial phase of the noise modulation function $\dot{M}(t)$ of the multiplicative noise
κ_ν	semi-invariant or cumulant of the ν-th order
$\Lambda(X_1, \ldots, X_n)$	likelihood ratio
λ_j	eigenvalue
$\xi(t), \vec{\xi}(t)$	stochastic process
$\tilde{\xi}(t)$	normalized stochastic process
Π_{ij}	loss matrix
$\tilde{\Pi}$	range of the nuisance parameter
$\tilde{\pi}$	nuisance parameter
$\dot{\rho}(\tau, \Omega)$	normalized complex autocorrelation function
σ_D	root mean square deviation of the statistical variance D_{XX}^{**}
σ_D^2	variance of the statistical variance D_{XX}^{**}
$\sigma_R^2(\tau)$	variance of the statistical correlation function $R_{XX}^*(\tau)$

σ_X	root mean square or standardized deviation of the random variable X
σ_X^2	variance of the random variable
σ_X^*	statistical mean square deviation of the random variable X
σ_{m_X}	root mean square deviation of the statistical mean m_X^*
$\sigma_{m_X}^2$	variance of the statistical mean m_X^*
σ_ℓ^2	variance of errors of signal parameter measurement
σ_ξ^2	variance of distortions in amplitude
σ_φ^2	variance of distortions in phase
σ_υ^2	variance of noise component of the signal $a_M(t)$
$\sigma_{C_0}^2$	variance of the value $\dot{C}_0^T$
$\sigma_s^2(\tau, \Omega)$	variance of the noise component $s(\tau, \Omega)$
σ_τ^2	total variance of measurement errors of signal appearance time
$\sigma_{\tau_0}^2$	total variance of measurement errors of signal appearance time when the multiplicative noise is absent
$\sigma_{\tau_M}^2$	variance of measurement errors of signal appearance time caused by the multiplicative noise
$\sigma_{\tau_a}^2$	variance of measurement errors of signal appearance time caused by the additive noise
σ_ω^2	total variance of measurement errors of signal frequency
$\sigma_{\omega_0}^2$	total variance of measurement errors of signal frequency when the multiplicative noise is absent
$\sigma_{\omega_M}^2$	variance of measurement errors of signal frequency caused by the multiplicative noise
$\sigma_{\omega_a}^2$	variance of measurement errors of signal frequency caused by the additive noise
τ_M	shift in time of the main peak of the autocorrelation function
τ_c	correlation interval or correlation time of the noise modulation function $\dot{M}(t)$ of the multiplicative noise
τ_{c_0}	expected correlation interval
τ_{r_0}	resolution interval with respect to signal appearance time
τ_φ	time of correlation of distortions in phase
τ_{r_M}	resolution interval with respect to signal appearance time if there is the multiplicative noise
$\upsilon(t)$	noise component of the signal $a_M(t)$
$\Phi(x)$	error integral
φ_0	random initial phase of the signal
φ_m	maximal deviation of phase of the signal $a_M(t)$

$\varphi(\vec{X})$	optimal decision-making function
χ^2	relative energy level of undistorted component of the signal $a_M(t)$
$\Psi_a(t)$	signal phase modulation law
$\psi_{\hat{a}}(t)$	total instantaneous phase of the normalized signal $\widehat{a}(t)$
Ω_M	average frequency of the multiplicative noise
$\Omega(t)$	frequency modulation law of the signal
ω_0	carrier frequency of the signal
ω_{r_0}	resolution interval with respect to signal frequency
ω_{r_M}	resolution interval with respect to signal frequency if there is the multiplicative noise
$\Re$	average risk
$\Re(P_0^*, P_1^*)$	average Baye's risk
$\mathcal{I}(x)$	identity function
$\mathcal{V}(A)$	frequency of the random event A
$\mathcal{D}$	relative differential of the probability of detection
$\lVert \ldots \rVert$	normalized determinant
$\langle .; . \rangle$	scalar product
$\mathrm{sgn}(x)$	sign function

Index

T

U

V

W